Bionanomaterials for Industrial Applications

Bionanomaterials for Industrial Applications is a comprehensive guide to the current state of bionanomaterials research and their prospective applications in a variety of industrial sectors. The book discusses the properties of bionanomaterials, types, and their potential applications in various disciplines, such as biomedicine, food industry, environment, and so forth. It provides a comprehensive overview of the current state of bionanomaterials research and their potential applications, making it an indispensable resource for anyone interested in learning more about this dynamic and rapidly developing field.

Features:

- Discusses properties, classifications, and synthesis of bionanomaterials in addition to industrial applications.
- Covers circular economy and life cycle assessment of bionanomaterials.
- Explores impact of bionanomaterials on environment and human health.
- Includes individual chapters specifically focusing on a particular application of bionanomaterials.
- Reviews detailed industrial applications in particular field, as follows: environmental, food sciences, biomedical, and so forth.

This book is designed for researchers, scientists, engineers, and graduate students working in the field of bionanomaterials, as well as industrial professionals who could benefit from the use of bionanomaterials.

Bionanomaterials for Industrial Applications

Edited by
Shakeel Ahmed

CRC Press is an imprint of the
Taylor & Francis Group, an informa business

First edition published 2025
by CRC Press
2385 NW Executive Center Drive, Suite 320, Boca Raton FL 33431

and by CRC Press
4 Park Square, Milton Park, Abingdon, Oxon, OX14 4RN

CRC Press is an imprint of Taylor & Francis Group, LLC

ISBN: 9781032458779 (hbk)
ISBN: 9781032558943 (pbk)
ISBN: 9781003432791 (ebk)

DOI: 10.1201/9781003432791

Typeset in Times
by Newgen Publishing UK

Contents

Preface

In recent years, the field of bionanomaterials has experienced remarkable growth, with researchers and scientists from a variety of disciplines working to develop new industrial materials and technologies. Bionanomaterials are materials derived from renewable natural resources, such as proteins, carbohydrates, and lipids, and possessing unique properties that make them desirable for a wide range of industrial applications.

The present book, *Bionanomaterials for Industrial Applications*, provides a comprehensive overview of the current state of bionanomaterials research and their prospective applications across a variety of industrial sectors. Each chapter focuses on a particular aspect of bionanomaterials research and application. Part 1 provides an introduction to bionanomaterials and their properties, as well as an overview of their many varieties and potential applications. Part 2 concentrates on synthesis and characterization of bionanomaterials, whereas Part 3 discusses applications of bionanomaterials in food industry, where specific chapters focus on food packaging, shelf life of food, nutraceuticals, improving food quality and safety and their risk assessment. Part 4 discusses the use of bionanomaterials in biomedical applications, such as drug delivery, dentistry, wound dressings, skin rejuvenation, tissue engineering, and diagnostics and cytotoxicity and biocompatibility. Part 5 examines the application of bionanomaterials in the environment.

This book is designed for researchers, scientists, engineers, and graduate students working in the field of bionanomaterials, as well as industrial professionals who could benefit from the use of bionanomaterials. It provides a comprehensive overview of the current state of bionanomaterials research and their potential applications and is a valuable resource for anyone interested in learning more about this dynamic and rapidly evolving field.

I hope that this book will inspire readers to investigate the potential of bionanomaterials and to consider how these materials can be used to create new and innovative products and technologies for industrial applications. I would like to thank all the authors in this book for their valuable contributions, and CRC Press for its publication.

Shakeel Ahmed, PhD
2023

About the Editor

Shakeel Ahmed is working as Assistant Professor of Chemistry at the Higher Education Department, Government of Jammu and Kashmir and Assistant Professor at the Department of Chemistry, Government Degree College Mendhar, Jammu and Kashmir, India. He obtained a first degree in general science from Government Postgraduate College Rajouri (University of Jammu) followed by a master's degree and a doctoral degree in chemistry from Jamia Millia Islamia, a central university, New Delhi. He gained post-doctoral experience in biocomposite materials at the Indian Institute of Technology, Delhi. He has published several research works in the areas of green nanomaterials and biopolymers for various applications, including biomedical, packaging, and water treatment. He is an active reviewer and member of editorial boards for many reputed journals. He has published more than 30 books in the area of nanomaterials and green materials, with publishers of international repute. His named has been listed among the top 2 per cent of the scientists of the world published by Stanford University for last three years. He is a fellow of the Linnean Society of London and the International Society for Development and Sustainability (ISDS), Japan, and the recipient of many awards from various agencies.

Contributors

Shakeel Ahmed
Department of Chemistry, Government Degree College Mendhar, Jammu and Kashmir, India

Vidushi Ahuja
University Institute of Biotechnology, Chandigarh University, Mohali, Punjab, India

Cansu Aydın
Department of Molecular Biology and Genetics, Faculty of Science and Art, Kırsehir Ahi Evran University, Kırsehir, Turkey

Zahra Vahedi Azad
School of Dentistry, Islamic Azad University Tehran Medical Sciences, Tehran, Iran
Department of Biomedical Engineering, University of Connecticut, Mansfield, USA

Ashkan Badkoobeh
Department of Oral and Maxillofacial Surgery, School of Dentistry, Shahid Beheshti University of Medical Sciences, Tehran, Iran

Morteza Banakar
Dental Research Center, Dentistry Research Institute, Tehran University of Medical Sciences, Tehran, Iran
Health Policy Research Center, Institute of Health, Shiraz University of Medical Sciences, Shiraz, Iran

Amritha Bemplassery
Department of Chemistry, National Institute of Technology, Calicut, Kerala, India

Shailendra Bhatt
Department of Pharmacy, School of Medical and Allied Sciences, GD Goenka University, Sohna, Haryana, India

Saptak Bhattacharjee
Department of Home Science, University of Calcutta, Kolkata, India

Tanima Bhattacharya
Nondestructive Bio-Sensing Laboratory, Department of Biosystems Machinery Engineering College of Agriculture and Life Science, Chungnam National University, Yuseong-GuDaejeon, Republic of Korea

Medini Bheemappa
Department of Biotechnology, School of Applied Sciences, REVA University, Bangalore, Karnataka, India

Hemavathi Brijesh
Department of Biotechnology, School of Applied Sciences, REVA University, Bangalore, Karnataka, India

Tanmoy Das
School of System Semiconductor Engineering, Yonsei University, Seoul, Republic of Korea

Durmus Burak Demirkaya
Department of Molecular Biology and Genetics, Faculty of Science and Art, Kırsehir Ahi Evran University, Kırsehir Turkey

H. Dheeraj
Department of Microbiology, School of Basic and Applied Sciences (SBAS), Dayananda Sagar University, Bengaluru, Karnataka

Prishila Dutta
Department of Home Science, University of Calcutta, Kolkata, India

Duraisamy Elango
Department of Environmental Science, Periyar University, Salem, Tamil Nadu, India

Kirti Garg
Department of Biotechnology, Himachal Pradesh University, Summerhill, Shimla, INDIA

Rupesh K. Gautam
Department of Pharmacology, Indore Institute of Pharmacy, IIST Campus, Rau, Indore, India

Abhishek Gowda
Department of Microbiology, School of Basic and Applied Sciences (SBAS), Dayananda Sagar University, Bengaluru, Karnataka, India

B.K. Nithin Gowda
Department of Microbiology, School of Basic and Applied Sciences, Dayananda Sagar University, Bengaluru, Karnataka, India

Chestha Goyal
Clarivate India, Pvt. Ltd.

Sonam Grewal
M.M. College of Pharmacy, Maharishi Markendeshwar (Deemed to be University), Mullana, Ambala, Haryana, India

Reena Gupta
Department of Biotechnology, Himachal Pradesh University, Summerhill, Shimla, India

Ramakrishna V. Hosur
Department of Biosciences and Bioengineering, Indian Institute of Technology Bombay, Mumbai, India

C.G. Iwuela
Federal University of Technology Owerri, Imo State, Nigeria

Aswathy Jayakumar
Department of Food and Nutrition, BioNanocomposite Research Center, Kyung Hee University, Seoul 02447, Republic of Korea

Ramit Kapoor
Clarivate India, Pvt. Ltd.

Sanjukta Kar
Department of Dietetics and Applied Nutrition, Amity University, Kolkata, West Bengal, India

Jasila Karayil
Department of Applied Science, Government Engineering College West Hill, Kozhikode, Kerala, India

Sukhminderjit Kaur
Department of Biotechnology, University Institute of Biotechnology, Chandigarh University, Mohali, Punjab, India

Jun Tae Kim
Department of Food and Nutrition, Kyung Hee University, Seoul, Republic of Korea
BioNanocomposite Research Center, Kyung Hee University, Seoul, Republic of Korea

Ashutosh Kumar
Department of Biosciences & Bioengineering, Indian Institute of Technology Bombay, Mumbai, India

Manish Kumar
School of Pharmaceutical Sciences, CT University, Ludhiana, Punjab, India

Chin Wei Lai
Nanotechnology and Catalysis Research Centre (NANOCAT), Level 3, Block A, Institute for Advanced Studies (IAS), Universiti Malaya (UM), Kuala Lumpur, Malaysia

Senzekile Majola
Department of Chemistry, Durban University of Technology, Durban, South Africa

M.U. Makgobole
Chiropractic and Somatology, Durban University of Technology (DUT), Durban, South Africa

N.G. Manjula
Department of Microbiology, School of Basic and Applied Sciences (SBAS), Dayananda Sagar University, Bengaluru, Karnataka, India

Ramya Manjunath
Department of Biotechnology, School of Applied Sciences, REVA University, Bangalore, Karnataka, India

Shilpa Borehalli Mayegowda
Department of Psychology, CHRIST (Deemed to be University), Bangalore Kengeri Campus, Kanmanike, Kumbalgodu, Mysore Road, Bangalore, Karnataka, India

P.S. Mdluli
Department of Chemistry, Durban University of Technology (DUT), Durban, South Africa

Brahmeshwar Mishra
Department of Pharmaceutical Engineering & Technology, Indian Institute of Technology (BHU), Varanasi, India

A.K. Misra
Department of Chemistry, Durban University of Technology (DUT), Durban, South Africa

Pooja Mittal
Chitkara College of Pharmacy, Chitkara University, Rajpura, Punjab, India

S.C. Mkhize
Department of Chemistry, Durban University of Technology (DUT), Durban, South Africa

B.N. Mkhwanazi
Dietetic and Human Nutrition, University of KwaZulu-Natal, Durban, South Africa

T.H. Mokhothu
Department of Chemistry, Durban University of Technology (DUT), Durban, South Africa

Seyyed Mojtaba Mousavi
Department of Chemical Engineering, National Taiwan University of Science and Technology, Taiwan

N. Mpofana
Chiropractic and Somatology, Durban University of Technology (DUT), Durban, South Africa

Tulika Mukhopadhyay
Department of Home Science, University of Calcutta, Kolkata, India

Neeraj Narwat
Department of Biotechnology, University Institute of Biotechnology, Chandigarh University, Mohali, Punjab, India

Z. Obiechefu
Department of Chemistry, Durban University of Technology (DUT), Durban, South Africa

C.S. Okonkwo
Federal Collage of Health Science and Technology, Makarfi, Kaduna state, Nigeria

S.C. Onwubu
Department of Chemistry, Durban University of Technology (DUT), Durban, South Africa

Jeyakumar Saranya Packialakshmi
Department of Food and Nutrition, Kyung Hee University, Seoul, Republic of Korea

Pankaj Popli
M.M. College of Pharmacy, Maharishi Markendeshwar (Deemed to be University), Mullana, Ambala, Haryana, India

Arumugam Priyadharsan
Department of Conservative Dentistry and Endodontics, Saveetha Dental College and Hospitals, Saveetha Institute of Medical and Technical Sciences (SIMATS), Chennai, India

Sabarish Radoor
Department of Polymer-Nano Science and Technology, Jeonbuk National University, Deokjin-gu, Jeonju-si, Korea

Parteek Rana
Chitkara College of Pharmacy, Chitkara University, Rajpura, Punjab, India

Nitu Rani
Department of Biotechnology, University Institute of Biotechnology, Chandigarh University, Mohali, Punjab, India

Bhagya Venkanna Rao
Department of Pharmacology, KLE College of Pharmacy, KLE Academy of Higher Education and Research (KAHER), Bengaluru, Karnataka, India

Anindita Ray
Department of Food and Nutrition, Maharani Kasiswari College, University of Calcutta, Kolkata, India

Y. Harinath Reddy
Department of Microbiology, School of Basic and Applied Sciences (SBAS), Dayananda Sagar University, Bengaluru, Karnataka, India

Myalowenkosi Sabela
Department of Chemistry, Durban University of Technology, Durban, South Africa

Rutika Sehgal
Department of Biotechnology, Himachal Pradesh University, Summerhill, Shimla, India

H.V. Shambhavi
Department of Psychology, CHRIST (Deemed to be University), Bangalore Kengeri Campus, Kanmanike, Kumbalgodu, Mysore Road, Bangalore, Karnataka

Divya Sharma
Department of Biotechnology, Himachal Pradesh University, Summerhill, Shimla, India

Kamini Sharma
M.M. College of Pharmacy, Maharishi Markendeshwar (Deemed to be University), Mullana, Ambala, Haryana, India

Suchart Siengchin
Materials and Production Engineering, The Sirindhorn International Thai-German Graduate School of Engineering (TGGS), King Mongkut's University of Technology North Bangkok, Bangkok, Thailand

Madhav Singla
Chitkara College of Pharmacy, Chitkara University, Rajpura, Punjab, India

Murari Lal Soni
Quality Assurance Laboratory, M.P. Council of Science & Technology, Bhopal, Madhya Pradesh, India

Tajunnisa
Department of Microbiology, School of Basic and Applied Sciences (SBAS), Dayananda Sagar University, Bengaluru, Karnataka, India

Babita Thakur
Department of Biotechnology, University Institute of Biotechnology, Chandigarh University, Mohali, Punjab 140413, India

Duygu Aslan Türker
Erciyes University, Engineering Faculty, Department of Food Engineering, Kayseri, Türkiye
Department of Microbiology, School of Basic and Applied Sciences (SBAS), Dayananda Sagar University, Bengaluru, Karnataka-560100

Mansi Upadhyay
Department of Biosciences & Bioengineering, Indian Institute of Technology Bombay, Mumbai, India

Serap Yalcın
Department of Medical Pharmacology, Faculty of Medicine, Kırsehir Ahi Evran University, Kırsehir Turkey

Mehmethan Yıldırım
Department of Molecular Biology and Genetics, Faculty of Science and Art, Kırsehir Ahi Evran University, Kırsehir Turkey

Part 1

Introduction

Synthesis and Characterization of Bionanomaterials

1 Introduction to Bionanomaterials

Pankaj Popli, Sonam Grewal, Kamini Sharma, Manish Kumar, Shailendra Bhatt, Shakeel Ahmed, and Rupesh K. Gautam

1.1 INTRODUCTION

Biomolecules are compounds that are found in living things on a regular basis. Macromolecules—including lipids, proteins, nucleic acids, and carbohydrates—are examples of biomolecules. Small molecules like primary, secondary metabolites, and organic compounds are also included. Nitrogen, sulphur, oxygen, and phosphorus are also present in small volumes of biomolecules. Covalently bonded large molecules with many atoms made the biomolecules (Mondal, 2018). Because of their distinct structure and role in reproduction, sustainability, and death, all kinds of life, including bacteria, animals, and plants, depend on biomolecules. The primary components of life are oligomers, monomers along with macromolecules, which includes amino acids, lipids, peptides, proteins, oligonucleotides, monosaccharides, oligosaccharides, nucleotides, nucleobases, and nucleic acids (DNA/RNA) (Malafaya et al., 2007). One fascinating property of biomolecules is their ability to organise themselves hierarchically to generate rigid and flexible biological materials and structures (Ariga et al., 2016). Examples include biomacromolecules like keratin, collagen, and elastin's capacity to come together to create useful assemblies, the formation of robust and palatable gels by gelatin, and the production of high-strength fibers (functional amyloid), gels by silk, and the production of toxic amyloid structures by some peptides and proteins. The organization of biomolecules and the incredible inter- and intramolecular associations generated by molecular recognition are essential for the development of biological materials (Datta et al., 2020). The biomaterials formed from biomolecules can be designed for use on an array of shapes that, either by themselves or in combination with other components of are part of an intricate structure that interacts with aspects of living things and is expected to have therapeutic or diagnosis impacts within human or animal healthcare. Natural or synthetic materials, as well as hybrids in two material types, utilized for creating biomaterials needed for a number of biological uses, which have immense special capacity for increasing biological, chemical, and mechanical capabilities. The molecular, nano, micro, and macro scales are all included in the hierarchical organization of biological materials (Stevens and George, 2005). The design and manufacture of biomaterials often include choosing an appropriate natural or synthetic material and processing it into the necessary format with the necessary mechanical properties using biocompatible chemical and mechanical transformations. In this sense, the engineering and creation of biomolecules into useful biomaterials must take into account the appropriate ratio of conformation rigidity resulting from covalent alterations and multicomponent assembly techniques' flexibility and functional significance. To get beyond the

DOI: 10.1201/9781003432791-2

translational limitations, improved and high throughput technologies have been created, such as stimulus-responsive or shape-memory materials, lithographic techniques, microarrays, and micro- and nano-fabrication, to create smart biomaterials with well-defined topographies and functions (Datta et al., 2020). Agriculture, transportation, electronics, communication, the food industry, and medical are just a few areas where nanotechnology is being used in daily life (Roco et al., 2011). The use of matter between 1 and 100 nm in a nanoscale to make new particles and gadgets is known as nanotechnology. Nanomaterials have improved during the previous improved during the previous decade, leading for creating numerous molecular imaging devices based on nanoparticles (NPs). Nearly all of the most used imaging methods have been improved by the introduction of NPs, such as optical imaging, positron emission tomography (PET), and magnetic resonance imaging (MRI) (Rosado-de-Castro et al., 2018). Nanotechnology that enhances material quality by rearranging or improving its nanostructures is known as incremental nanotechnology. Applications are regarded to be examples of what has been referred to as evolutionary nanotechnology when materials' physico-chemical properties alter as their size is reduced, which offer the possibility for issue solving and new business prospects (Oluwasanu et al., 2019). Nanomaterials with distinctive physical characteristics are currently attracting an abundance of interest because they are quite versatile, especially in the area of biological science, which includes gene and delivery for medication applications, tissue engineering, cancer treatment, monitoring systems, and so forth (Díez-Pascual, 2022). A large variety of nanomaterials, including nanoparticles, nanowires, nanotubes, and nanoplates offer a great deal of potential for use in biomedicine. They must, however, meet particular requirements for employing in biomedical applications. In this sense, it is necessary to evaluate both their biocompatibility and possible cytotoxicity, which might be brought on by, for instance, their structure, chemical composition, or characteristics. Additionally, under physiological settings, their colloidal stability should be preserved, ideally across a wide pH range (Kobayashi et al., 2014). Nanomaterials' constituents exist as anything between metals or their oxides to carbon or polymers, depending on the desired ultimate qualities. It is useful to employ metallic nanoparticles (such as gold or silver) for imaging and medication delivery systems, however additional investigation of their safety is necessary in order to avoid unfavourable side effects in humans (Guadagnini et al., 2015). Iron oxide is the most often used substance for the core of functionalized nanoparticles because of its exceptional magnetic characteristics. Drug delivery applications commonly use silica NPs. Amino acids, nucleic acids, medicines, and other tiny compounds are frequently loaded onto tunable-pore-size mesoporous silica nanoparticles (MSN) (Hirayama et al., 2021). However, there are problems with biocompatibility when using silica nanoparticles for nanomedicine because of reactive surface silanol groups. There have also been reports of concerns about the dangers of nanoparticles made of carbon which includes graphene, quantum dots, and carbon nanotubes. Because of their highly customizable physicochemical features, good biocompatibility, and capacity to continuously release chemicals, polymeric nanoparticles are a popular choice for biomedical applications (Abd Ellah and Abouelmagd, 2017). When seeking a less dangerous substitute for common dangerous nanomaterials for medical purposes, these bionanomaterials is the best option for ideal solution. Bionanomaterials described as nanomaterials that are produced using biomolecules or that use biomolecules to enclose or immobilise another form of nanomaterial, as shown in Figure 1.1 (Jeevanandam et al., 2022). Bio-reduction procedures are used to create bionanomaterials; the reduction is thought to be mediated by reducing sugar, proteins, enzymes, and phenolic chemicals. The pH, concentrations (reaction mixture), type of source materials, and other factors have an impact on the synthesis process. After synthesis, the materials are characterized using numerous approaches, including atomic force microscopy (AFM), energy-dispersive X-ray spectroscopy (EDS), scanning electron microscopy (SEM), transmission electron microscopy (TEM) and X-ray photoelectron spectroscopy (XPS). Even though there are various uses for metal nanoparticles in the fields of visually active thin-film coatings, catalysis, optoelectronics, sensing, display devices, diagnostic biological probes, and pharmaceuticals, bionanoparticles are crucial to agriculture (Sahayaraj, 2012).

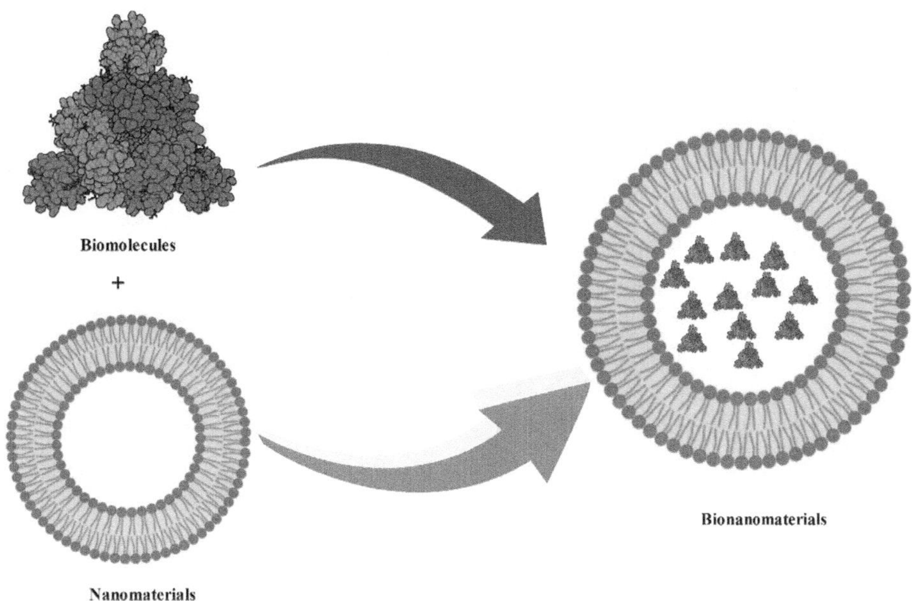

FIGURE 1.1 Nanomaterial Encloses the Biomolecules and Formulates the Bionanomaterial. (Compiled by the Author.)

1.2 SOURCES OF BIOMOLECULES

The majority of biomolecules will be present inside the ecosystem. They can be easily obtained from plants, aquatic life, and microalgae, as shown in Figure 1.2. For industrial purposes, bacteria, yeast, and microalgae produce biomolecules including pigments, enzymes, proteins, lipids, and organic acids. These valuable macromolecules are usually bioaccumulated inside of cells, and in order to retrieve them, numerous additional processes must be taken, the most crucial of which is cell wall breakdown (Gomes et al., 2020).

1.2.1 Plants as Sources of Biomolecules

Health care has benefited greatly from plants. Plants are a valuable source of physiologically active substances, including essential phytochemicals such as aromatic compounds, phytochemicals, organic acids and vitamins (A, B group, C, and E), used to make food additives, natural medicines, and other ingredients (Donno et al., 2016). Many different plant parts are utilised to cure a variety of diseases because they contain a variety of biomolecules as shown in Table 1.1, or biologically active components, including alkaloids, terpenoids, phenolic compounds, glycosides, flavonoids, vitamins, tannins, carbohydrates, coumarins, and organic acids.

1.2.2 Aquatic Life as a Source of Biomolecules

Fishery combustion products can be a source of high-quality bioactive compounds like amino acids, polyunsaturated fatty acids, collagen, carbohydrates, chitosan, polyphenolic components, alkaloids, terpenoids, tocopherols, and vitamins, although their biotechnological prospective is mainly unrealized at this time (Müller and Laibach, 2022). Marine-derived biomolecules may be used in a variety of ways in the food business, agriculture, and biotechnological (chemical, industrial, or environmental) domains, based on the functional and structural features (Ghosh et al., 2022b). The fish gut

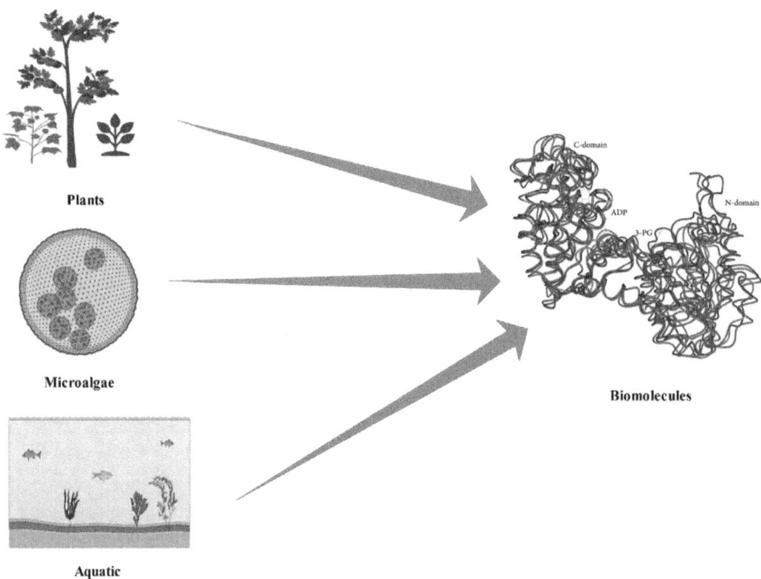

FIGURE 1.2 Different Sources of Biomolecules. (Compiled by the Author.)

TABLE 1.1
Plants as a Source of Biomolecules

Name	Biomolecule	Reference
Nicotiana Tabacum	Nicotine	(Yuan et al., 2019)
Catharanthus Roseus	Vincristine and Vinblastine	(Khan et al., 2021)
Hyoscyamus Niger	Atropine	(Otimenyin, 2022)
Digitalis Purpurea	Digitoxin	(Chaki et al., 2022)
Panax Ginseng	Ginsenoside	(Im, 2020)
Ammi Visnaga	Khelin, khellol	(Abu-Serie et al., 2019)
Maclura Tinctoria (Hedge apple)	Macluraxanthone	(Salehi et al., 2019)
Artemisia Annua	Artemisinin	(Abate et al., 2021)

microbiota is an excellent source of probiotic prospects and biosynthesizes vital or short-chain fatty acids, minerals, enzymes, or vitamins as described in Table 1.2. Additionally, it makes bioactive substances with antimicrobial and biosurfactant/emulsifier activity. Moreover, underutilized supply for bioactive chemicals is found in the internal organs of fish. Viscera, heads, scales, bones, and the creation of additional fish processing by products are produced in enormous quantities, between more than 25 and 70 per cent (Caruso et al., 2020).

1.2.3 Microalgae as Sources of Biomolecules

According to well-established research, algae mostly contain proteins, carbs, lipids, and trace nutrients like vitamins, antioxidants, and trace minerals. These all have the qualities to function as natural supplements in human and animal feed to replace synthetic components or satisfy the rising demand for such substances (Kovač et al., 2013). Different kinds of organisms include eukaryotes as well as prokaryotes, referred to as "algae". Cyanobacteria (blue-green algae), that live in watery

TABLE 1.2
Aquatic as a Source of Biomolecules

Name	Biomolecule	Reference
Skipjack tuna (Katsuwonus pelamis)	Alanine, Glycine, Valine, Leucine, cystine, Tryptophan, Tyrosine (amino acids)	(Chanmangkang et al., 2022)
Monkfish	Eicosapentaenoic acid (omega-3 fatty acids), collagen	(Caruso et al., 2020)
Spanish mackerel (*Scomberomorus niphonius*)	Collagen, trypsin, pepsin	(Zhao et al., 2019)
yellowfin tuna (*Thunnus albacares*)	Collagen	(Nurilmala et al., 2020)

TABLE 1.3
Microalgae as a Source of Biomolecules

Name	Biomolecule	Reference
Haematococcus microalgae	Astaxanthin	(Gherabli et al., 2023)
Muriellopsis species	Lutein	(Yaakob et al., 2014)
Dunaliella species	Beta-carotene	(Goswami et al., 2022)
Aphanizomenon flos-aquae	Phycobiliproteins	(Nuzzo et al., 2019)
Crypthecodinium cohnii	Arachidonic acid, docosahexaenoic acid, and -linoleic acid are examples of polyunsaturated fatty acids.	(Hu et al., 2007)

settings and range in size from microscopic cells to enormous kelp. Algae can be classified as either microalga (whose algal bodies must be viewed under a microscope), or macroalgae based on their body size and structure (sufficiently big for the unaided eye to see) (Barsanti and Gualtieri, 2005). Microalgae appear to have the potential to help resolve the global energy, environmental, and food crises that are currently in place. Many biomolecules such as beta-carotene, lutein, chlorophyll, astaxanthin, phycobiliprotein, and beta-1, 3-glucan, polyunsaturated fatty acids, and medicinal and nutraceutical substances, can be produced by algae as shown in Table 1.3 (Odjadjare et al., 2017).

1.3 TARGETING OF BIONANOMATERIALS

The idea of targeting the disease is not new. Ehrlich initially proposed the "magic bullet" in 1906. Though its popularity is evident from the idea's endurance, it is still challenging to implement the "magic bullet" in a clinical setting. Finding the right drug, identifying the right target, it has been difficult to figure out how to efficiently remove foreign material from the body while preventing immunological and unspecific interactions when delivering medication in a consistent form to specified areas (Zhou and Rossi, 2014; Fahmy et al., 2005). When combined with targeted ligands, bionanomaterials have the potential to be beneficial as carriers of active pharmaceutical ingredients and may possess many characteristics of a "magic bullet" (Maiti and Sen, 2017; Rashid et al., 2019). The crucial part of the delivery of bionanomaterials is the selective and site-specific delivery to target the disease (D. Friedman et al., 2013). Bionanomaterials' structure and size are useful for utilising the disease, such as in the case of cancer, the higher uptake of drugs via bionanomaterials will act as distinct changes towards the pathology of cancer (Clemons et al., 2018). Targeting is done in two ways: active or passive target systems.

1.3.1 Passive Targeting

Diffusion of bionanomaterials that circumvent the body's defence mechanism allows for passive targeting. The bionanomaterials enter the bloodstream or nervous system and, once absorbed by the target receptor, their hydrophobic or hydrophilic nature, surface charge, and molar weight are the multi-properties that enable passive targeting (Chaturvedi et al., 2019). Similar to how the tumour microenvironment causes leaky vasculature in cancer, a hierarchical structure is created, which lowers lymph clearance and facilitates the entry of bionanomaterials using the concept of enhanced permeability and retention impact (Gmeiner and Ghosh, 2014). As a result, bionanomaterials build up the tumour. In this case, passive targeting focusses on the mechanism of the bionanomaterials' size and surface properties, which need to be controlled to prevent absorption by the reticuloendothelial system or to lengthen circulation times and the targeting ability. The ideal particle size diameter should not exceed 100 nm, and the hydrophilic surface facilitates macrophage clearance (Wang et al., 2007; Kim and Nie, 2005).

1.3.2 Active Targeting

The ligand and monoclonal antibody homing moiety of the particular problematic sites linked to the bionanomaterials enable active targeting, which takes advantage of the particular receptor that is overexpressed at the site of action (Alavi and Hamidi, 2019; Kumari et al., 2016). Bionanomaterials are frequently used to target the blood-brain barrier through a variety of receptors, including the folic acid receptor, low-density lipoprotein, insulin, and transferrin (Leonor Pinzon-Daza et al., 2013). In cancer, the receptors are frequently different from those on normal tissue; they are overexpressed to provide sustenance for their hypermetabolism for cancerous cells—thus, these overexpressed receptors are the key to targeting cancer via choosing different types of ligands that can be utilized on bionanomaterials such as small targeting molecules, vitamins, proteins, and oligopeptides (Salahpour Anarjan, 2019; Kaur et al., 2023).

1.4 TYPES OF BIONANOMATERIALS

Organic bionanomaterials have received a lot of attention since entering the biomedical field. Thus, an unexpected rise is visible in the field of bone and tissue regeneration, scaffolds, and so forth. For utilization of bionanomaterials, it must meet specific requirements which includes biodegradability, biocompatibility, and anti-toxin. Chitosan, silk fibroin along with polymer (biodegradable), and like poly(lactic-co-glycolic acid) are some examples of organic bionanomaterials (Singh et al., 2021). Silk fibroin is a high molar weighted natural polymer and is derived from several types of spiders and silkworms (Kundu et al., 2013). It has various medical uses such as regenerative treatments, and efficiently manage drug distribution (Wani et al., 2020). Furthermore, adding the functional group in bionanomaterials can be helpful to attain more advantages such as functionalized silk fibroin aids cancer cells by delivering genes and drugs and is also helpful in managing the kinetics of drug release (Webster et al., 2013). It efficiently distributed the proteins and drugs and even maintained their efficacy (Wenk et al., 2011). Chitosan, a straight-chain polysaccharide consisting of linked de-acetylated units of D-glucosamine and acetylated units of N-acetyl-D-glucosamine, is another type of organic bionanomaterial (Prasanthi et al., 2016). One of the most common cationic polymers, chitosan is obtained through the process of Chitin (after cellulose, the second-most frequent carbohydrate in plants is alkaline de-acetylated), moreover amine and hydroxyl groups are present. It creates inter- as well as intramolecular hydrogen bonds, giving it a hard crystalline structure (Choi et al., 2016). Chitosan demonstrated the possibilities for chitosan nanomaterials-based scaffolds development in bone formation using collagen to support tissue regeneration (McMahon et al., 2013). PLGA is another organic bionanomaterial composed of lactic and glycolic acid monomers; it is known as poly(lactic-co-glycolic acid, PLGA) (Gentile et al., 2014). In terms of design and functionality, PLGA is the most well-defined biomaterial currently used for drug delivery. A drug moiety of any size may be enclosed by PLGA, which is formulated into any form and size. It is soluble in a variety of widely used solvents, such as acetone, tetrahydrofuran, and chlorinated solvents (Makadia and Siegel, 2011). Given that PLGA nanomaterials are composed of a variety of natural and synthetic polymers, they may be easily utilised in tissue engineering and medication delivery technologies to create nanofibers and nanoscaffolds. Several anti-cancer medications, including doxorubicin, paclitaxel, and cisplatin, are effectively incorporated into PLGA nanoparticles (Suri et al., 2007).

Polymeric bionanomaterials have also emerged as a novel method for the controlled release of drugs, likewise silica is also an organic bionanomaterial and is considered as a most versatile component due to its abundance in nature and ease of co-existence with other minerals. Silica is a less reactive material, moreover, various plants, algae, and animals can acquire it in the form of nanostructures in their tissues. This silica is well recognized as a biogenic silica (Satyanarayana et al., 2017). The silica bionanomaterials offer unique qualities that allow for the application in medication delivery includes small dimension, enormous surface area, ameliorated absorption proportion, adjustable pore size, and volume, and highly biocompatible (Zhou et al., 2017). Because of their tiny size, large surface area, improved absorption percentage, changeable pore size and volume, and excellent biocompatibility, silica bionanomaterials have special properties that make them suitable for use in drug administration (Zhou et al., 2017). The pharmaceutical industry has shown a great deal of interest in mesoporous silica nanoparticles compared to other types of silica nanoparticles because of their high structural stability, cellular uptake, and formulation for biosensing molecules. They also have a large loading capacity and a distinctive porous honeycomb-like structure (Shen et al., 2017; Yamamoto and Kuroda, 2016). The ceramic bionanomaterial is extremely resistant to environmental changes, ceramic has a distinct solid core that contains a mixture of metals and non-metals, and it is formed by the application of pressure and heat. Silica, alumina, zirconia, nitride, and various other metals, metal oxides, and metal sulphides comprise the bulk of ceramic bionanomaterials. Moreover, the solid core of ceramic is helpful to formulate bionanomaterials with a variety of sizes, shapes, and porosities, and these types of bionanomaterials are generally formulated to avoid the reticuloendothelial system by typically changing their size and surface makeup (Singh et al., 2016). They gave a variety of crucial traits, including higher resistance, excellent chemical stability, strong biocompatibility, and less density (Szczęsny et al., 2022). Nanocrystalline hydroxyapatite has demonstrated several uses in the engineering of tissues by improving titanium alloy biocompatibility and serving for an antibody delivering agent for various infections in bone which produces injectable pastes which are utilized as a replacement for creating the structure of bone since they have excellent osteoconductive capabilities (Sitharaman, 2016). Bio-nanocomposites contained biological material or biopolymers, such as polysaccharides, polypeptides, and aliphatic polyesters due to these materials bio-nanocomposites exhibited their utilization in the pharmaceutical field (Arora et al., 2018). Their size ranges between 1–100nm, and their preparation technique and peculiar features are generally determined by their structure; moreover, their different characteristics, including thermal stability, biocompatibility, biodegradability, and its solubility in water, are helpful in deciding on their preparation techniques and applications (Naik et al., 2020). Research started focusing on the adjoining other materials with bio-nanocomposites to be utilized in drug delivery, and bone regeneration such as starch-based bio-nanocomposites. Starch exhibits multiple properties: low toxicity, higher biocompatibility and biodegradability, and mechanical properties. Several other materials are also utilized by researchers, such as carbon nanotubes, and oleyl phosphate also utilized in the drug delivery (Karki et al., 2021).

Other than polymer and organic-based bionanomaterials, biologically based bionanomaterial has also gained a lot of interest from researchers. Two categories of biologically based bionanomaterials exist: biogenic-based and natural-based bionanomaterials. Biogenic-based bionanomaterials include DNA, RNA, Lipoproteins, and peptides-based bionanomaterials; on the other hand, natural-based bionanomaterials include plants, bacteria, fungi, and viral-based bionanomaterials. Lipoprotein contained a wide range of lipid ratios in each sub-class (HDL, LDL, VLDL, IDL). It comes in enormous categories of dimensions as well as densities. It undergoes creation through the liver and intestines in the human body, and comprises intrinsically stable and non-immunogenic materials. Its shapes are typically spherical but can change to a discoidal form during synthesis and hepatic catabolism. It also facilitates their transport in blood circulation: these are biocompatible and biodegradable. Thus, these traits of lipoprotein are utilized for delivery of drugs and may be potentially used for bionanomaterials (Busatto et al., 2020). Similarly, peptide-based bionanomaterials can be utilized in the administration of drugs because of their varied qualities such as a biocompatible and their easy way of targeting due to the availability of the specific functions in their sequence. Altering the structure of peptide-based bionanomaterials resulted in the modification in their structure such as icosahedron either by self-assembly or interacting with other nanomaterials (He et al., 2021; Yadav et al., 2020). Another biogenic-based bionanomaterial is DNA-based bionanomaterial. DNA, generally called deoxyribonucleic acid, consists of base pairs that carry genetic information, and is considered a versatile biomolecule (Beissenhirtz and Willner, 2006). According to Watson and Crick, base pairing allows the biopolymer DNA to assemble into a structure with two helices, and further is stacked together through p-p stacking, hydrogen bonds, and hydrophobic interactions. In drug delivery, DNA nanostructures exhibited various properties shapes, sizes, surface chemistry, and the capacity for controlled release (Angell et al., 2016). These traits can be useful for cancer treatment, which is one of the leading causes of death worldwide, and its treatment with radiation or chemotherapy, both have a variety of adverse health impacts. Thus, it has become crucial to develop some alternative cancer treatments (Akram et al., 2017). In this, DNA-based bionanomaterials are helpful in recognizing carcinoma cells and help for controlled passage of the anti-cancer drug at the site (Jiang et al., 2016).

Another biological-based bionanomaterial is the natural-based bionanomaterial, now receiving more attention in nanotechnology. Certain yeasts, fungi, plants, and viruses are utilized to create the bionanomaterials because they have excellent cytocompatibility, low toxicity, and are environmentally benign (Bhattacharya et al., 2021). The pharmaceutical field has discovered the possibilities for natural-based bionanomaterial. Viruses are now used in biotechnology since they have been shown to have potential applications in several scientific domains including bionanomaterial, bioimaging, and drug delivery (Zrazhevskiy et al., 2010). There are several viruses that can be employed for these kinds of applications, including Bromovirus, cowpea mosaic virus, hepatitis B virus, vault nanocapsules, M13, and MS2 bacteriophages. The relaxivity of the paramagnetic Gd3+ bound nanoparticles was found to be greater than that of bound protein GD 3+ Chelates (Manchester and Singh, 2006). And as a result, they are presently utilized for small animal in-vivo MRI. Canine parvovirus occurs in dogs that carry a gene delivery system that has been used to treat cancers (Sayed-Ahmed et al., 2018). Inorganic bionanomaterials refer to a variety of bionanomaterials that are natural material analogues including peptide nucleic acids, aptamers, and xeno nucleic acids (Musa et al., 2023). The DNA analogues in peptide nucleic acids are N-(2-aminoethyl)glycine moieties, which stand in for the units of sugar-phosphate to join the nucleobases. These molecules can use both standards, and because of the base pairing between Watson and Crick, which produces stable duplexes, and the hydrogen bonding between Watson and Crick and Hoogsteen, which is coupled with the formation of triplehelical adducts with their counterpart to ultimately cause dsDNA strand invasion, these molecules improve protein binding for pertinent DNA and RNA (Volpi et al., 2020). Peptide nucleic acid has demonstrated distinctive chemical and physical features, which leads to resistance for breakdown of enzymes in live cells, which aids in its acceptance for use in various fields (Swenson and Heemstra, 2020). If the sugar-phosphate backbone removes from the peptide nucleic acids it becomes neutral, which prevents electrostatic repulsion and aids in its ability to demonstrate a strong affinity for the target. On the basis of polyamide, peptide nucleic acid is a synthetic molecule that has greater chemical and pH stability than DNA and RNA (Gupta et al., 2017). This characteristic of Peptide nucleic acid enables it to attach to more particular sequences. Peptide nucleic acid is regarded as precursor of RNA since it typically contains adenine, cytosine guanine and uracil. Moreover, Peptide nucleic acid exhibits different ionic strength and thermal stability due to its capability to detect certain DNA or RNA sections that make use of the Watson-Crick hydrogen bonding configuration (Singh et al., 2020). Thus, it can be helpful in DNA genotyping, Nucleic acids regarded as evaluated utilizations for gene therapy as well as drug employing Peptide nucleic acids, hence it is also utilized in the biomedical area (Briones and Moreno, 2012). This family of nucleic acids with a modification with sugar moiety refers to xeno nucleic acids which considerably expanded the uses of nucleic acids in nanomedicine and biotechnology such as 2′-Omethyl(2′-OMe). RNA is a kind of RNA that modifies the native ribose's 2′-hydroxy by adding a methyl group in the presence of Ag2O, CH3I (Tu et al., 2022).

The most extensively researched elements in nanoparticles are numerous metal-based bionanomaterials and their oxides utilized in the pharmaceutical/biomedical field. Many nanomaterials have found major applications in numerous fields. The metals and their oxides nanoparticles are simple to assemble and define, and they exhibit exceptional qualities that set them apart from the bulk material. They are simple to make using physical, chemical, or biological methods, but in general, biologically generated bionanomaterials catch the researcher's interest since they are inexpensive, have no negative impact on the environment or living cells, and are devoid of toxic chemicals (Musa et al., 2023). Gold, zinc, silver, and iron are the most often utilized metallic bionanomaterials, but researchers are also beginning to concentrate on other metals like cerium, selenium, magnesium, and so forth. Metallic bionanomaterials have special qualities that make them the most often employed bionanomaterials in biomedical applications since they may be used for both therapeutic and diagnostic purposes (Ghosh et al., 2022a). These qualities include size, shape, conductivity, and others. These metal nanoparticle characteristics are improved when they are created with other chemicals, which in turn boosts their usefulness (Stankic et al., 2016). These bionanomaterial are therefore beneficial and have the potential to be applied in a variety of disciplines, such as drug delivery and biomedical applications.

1.5 CHARACTERIZATION OF BIONANOMATERIALS

Numerous characterization approaches are employed to comprehend the physicochemical characteristics of produced bionanomaterials. Identification of properties—such as size distribution, crystallinity, surface area, solubility, porosity, shape, aggregation state, charge on the surface, chemical composition of the surface, and so forth—are made possible by the characterization process. Proper characterization improves comprehension and increases the dependability of results (Jain et al., 2018). When synthesizing the bionanomaterials in some aspects, it is necessary to consider components such as solvent medium as well as the reducing agent and stabilising agent. The primary use of characterization techniques is to examine and enhance the chemical composition, distinct size, shape, solubility, molecular weight, and purity, as well as synthesized bionanomaterials stability. These characteristics play a crucial role while assessing the applications of these materials in a variety of fields, including biomedicine, sensing, agriculture and environment, and so forth (Biao et al., 2018). Characterization methods are typically not very useful for the characterization of bionanomaterials. This is due to the fact that the resolution that may be achieved is limited through diffracted light limit, which is significantly larger than vision fields of the majority of nano-biomaterials. Greater resolution ranges, around 50–100 nm, can be achieved using near-field scanning optical microscopy (NSOM) instruments; the method, however, still has drawbacks. The two categories of characterization approaches are chemical characterization and structural characterization (Figure 1.3).

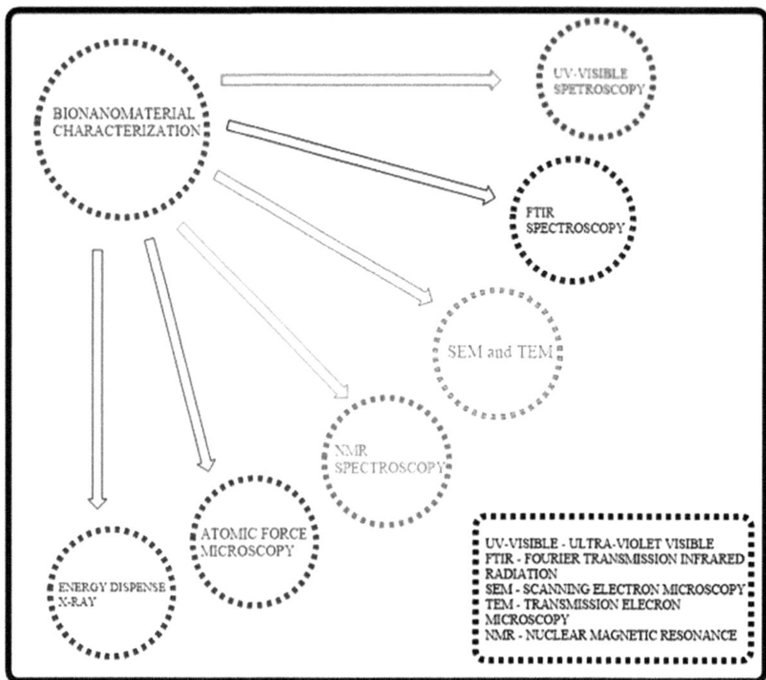

FIGURE 1.3 Characterization of Bionanomaterials. (Compiled by the Author.)

The chemical characterisation is used to identify numerous chemical atoms and surfaces as well as the spatial conjugation of the bionanomaterials, whereas the size, shape, and crystallinity of bionanomaterials are utilised to determine their structural morphology through structural characterisation. The characterisation methods make use of extremely advanced tools. The kind of bionanomaterials under investigation will determine which characterisation method is used.

1.5.1 Structural Characterization-

1.5.1.1 X-RAY Photoelectron Spectroscopy (XPS)

XPS is necessary for examining bionanomaterial physical and chemical properties and is also referred to as electron spectroscopy, utilized typically in chemical analysis (ESCA). XPS is frequently used for examining surface characteristics of bionanomaterials, which aids in understanding energy distribution and chemical state. It operates on the photoionization principle. According to research, XPS was used to examine how proteins from capsicum annuum interact with one another to produce silver nanoparticles (Li and Zhang, 2007).

1.5.1.2 Scanning Electron Microscopy (SEM) and Transmission Electron Microscopy (TEM)

These are the main prevalent instruments for assessing manufactured bionanomaterials due to their being simple and practical. They are employed to investigate the morphological characteristics of greenly generated bionanomaterials, such as size, shape, and size distribution. Compared to SEM, TEM has been viewed as a more advanced technology since it offers details on the material's chemical composition and produces an immediate picture with great resolution in space at the atomic scale (Jain et al., 2018). Apart from the specimen's thickness and the method for material collection of data, there are minor distinctions among SEM and TEM, which can be paired with X-ray spectroscopy to analyse bionanomaterials chemical composition in more detail. Other green bionanomaterials, such as those with ultra-thin electrically conductive coatings, can also be studied in this way (Sharma et al., 2018).

1.5.1.3 Atomic Force Microscopy (AFM)

Atomic force microscopy (AFM) is a less expensive tool that may be used to characterise bionanomaterials and has a number of advantages. A cantilever and a very thin probe are used to swivel across the surface of the sample. It gives 3D viewing together for X-Y resolutions in roughly 1 nm and geometric resolutions of less than 0.1 nm. It provides details on a number of physical parameters for each bionanomaterial. The ability to produce nano- and atomic-scaled surface images as well as perform the procedure in a wide range of environments, including air, vacuum, liquid, and so forth, making surface texture, size, roughness, and morphology one of the most cutting-edge, promising, and sophisticated technologies. This capability could be incredibly useful for biological studies (Jain et al., 2018). AFM is employed to identify and evaluate the distinctive qualities of numerous synthesized bionanomaterials (Bhattacherjee et al., 2018).

1.5.1.4 Energy Dispersive x-rAY (EDX)

Energy dispersive X-ray (EDX) analysis is one of the most cutting-edge techniques for determining the composition of bionanomaterials. The microstructure, elemental composition and spectra can be accurately detected when SEM and TEM are used in conjunction with EDX. By depicting the spatial distribution of components, EDX can provide qualitative, quantitative, or semi-quantitative analysis. Among the most effective methods for examining the many morphological, chemical, and compositional characteristics of produced bionanomaterials, such as gold and silver nanoparticles, is EDX (Gangadoo et al., 2015).

1.5.1.5 X-RAY Absorbance Spectroscopy

The elemental map and electrical structure of a particular substance can be accurately determined using X-ray spectroscopy. In general, it is possible to study any sort of substance, including crystal, gas or liquid. X-ray absorption gives a picture of the state corresponding to the unoccupied exterior electronic degrees of a specific atom. Details about the bond nature and bond's energy content can be obtained from both emission and absorption data (Liu et al., 2014).

1.5.1.6 X-ray Fluorescence (XRF)

The XRF technique follows the idea that atoms are regarded as substance which can be excited while being observed under a low-power X-ray tube. This energy from the created emission permits emission of secondary fluorescence that can be then discovered through a proportional detector after being reflected and measured by the crystal analyzer. Since practically all of the elements on the periodic table can be found in solid, liquid, powder, or film-based samples, XRF is incredibly versatile but also presents some difficulties (Bounakhla and Tahri, 2014).

1.5.1.7 X-ray Diffraction (XRD)

Because it determines the structure of crystal, crystallinity, defects of crystal, and crystalline material presentation in atomic spacing, the method is regarded as one of the key techniques applied in the nanotechnology field. Many environmentally friendly bionanoparticles, including silver NPs, are characterized using x-ray diffraction XRD (Titus et al., 2019).

1.5.2 Chemical Characterization

1.5.2.1 UV-Visible Spectroscopy

Spectroscopy is a fundamental method used to characterize all biological components on a quantitative and qualitative level. It is typically used for examining a variety of chemical as well as physical properties, including molecule concentration, size, and aggregation. It is used to research the structural conformation and optoelectronic characteristics of bionanomaterials. By using linear relationships between absorbance and concentrations, UV-visible spectroscopy depends on Beer-Lambert Law. There are numerous synthesised bionanomaterials, such as those made of silver, gold, copper, cerium, and selenium that have been studied for their structural characteristics (Biao et al., 2018).

1.5.2.2 Fourier Transform Infrared Spectroscopy (FTIR)

To characterize bionanomaterials, Fourier transform infrared spectroscopy (FT-IR) is thought to be the most accurate and speedy analysis technique. It operates by examining particular patterns of individual molecule vibration which assist in identifying functional group present in the bionanomaterials, such as aldehydes, ketones, alcohols, carboxylic acids, and terpenoids. As a result, it is used to analyze solid and liquid materials in both quantitative and qualitative ways. The structural composition, dynamics, stability, aggregation, and so forth, are also detected using this method. FT-IR has been used to investigate a number of produced bionanomaterials, including iron oxide NPs, silver NPs, gold NPs, and nickel NPs (Davar et al., 2009).

1.5.2.3 Nuclear Magnetic Resonance (NMR)–Spectroscopy

Nuclear magnetic resonance (NMR)(Sharma et al., 2018) is a commonly used spectroscopic method that uses electromagnetic radiation that magnetic field-affected nuclei absorb and then reemit to characterise the conformational dynamics of bionanomaterials at the atomic level. The different constituents of the sample are examined using a non-invasive technique. NMR is regarded as an effective analytical technique since it generates data with high output, excellent reproducibility, and ease of modification. The characteristics of diverse polymers, amorphous materials, biomolecules, and bionanomaterials are examined using NMR. Due to its non-invasiveness, enhanced flow, reduced preparation of sample, enhanced repeatability along with comfortably measurable data, the NMR-based nanoparticles evaluation has become more popular (Lupoi et al., 2014).

1.6 APPLICATION OF BIONANOMATERIALS

1.6.1 Drug Delivery

During their transition from their origin to their molecular location of action, all medicines must navigate a number of transport obstacles. Critical obstacles include transfer from the bloodstream to target cells inside tissues, fast kidney filtration, and clearance through reticulo-endothelial system (RES), especially for drugs that stay in the bloodstream for an extended period of time (Hubbell and Chilkoti, 2012). Numerous researchers have discussed that the use of nanomaterials as drug transporters is one strategy for resolving these drug delivery problems. Both the diagnosis and treatment of diseases as well as nano-biomedicine have greatly benefited from the use of nanotechnology (Li et al., 2018). The design, synthesis, and assembly of advanced nanostructures are the focus of the emerging topic of nanomedicine, which integrates elements of material science, chemistry, engineering, and medicine. These structures can have their chemical and physical characteristics altered through structural change, which enables the use of these materials in gene therapy delivery systems, medication delivery systems, and various theranostic techniques (Nasseri et al., 2018). A 1000 nm range for nanoscale materials and devices has been suggested for use in nanomedicine.

Nanomaterials can enter the body through a variety of channels, including the digestive system, respiratory system, and skin as well as medication injection therapy. Nanomaterials can then travel to various organs and have an impact on biological processes (such as inflammatory responses, oxidative damage, DNA damage and cell death) (Li et al., 2018). NPs have emerged as a crucial topic in drug delivery research due to their ability to carry and deliver an astounding array of medicines to organs or regions of the body. This allows them to provide targeted, regulated, and sustained therapeutic effects. NPs can include drugs or other bioactive compounds that are dissolved, trapped, encapsulated, adsorbed, or connected. Numerous studies have shown that employing NPs as carriers, biological macromolecules, proteins, hydrophobic and hydrophilic medicines, as well as vaccinations, can be successfully delivered (Teleanu et al., 2019).

1.6.2 Targeted Therapy

Targeted delivery occurs when the medicinal substance is effectively targeted and predominately accumulates in an intended location. Targeting a specific cell or tissue and releasing the encapsulated medicinal agents require the agent-laden system to be present inside the physiological system in the preferred amount for time period, to avoid the immune system, and achieve efficient targeted delivery (Davis et al., 2008). Targeted nanoparticle delivery in the treatment of cancer is currently being extensively researched. More than 20 per cent of the therapeutic nanoparticles that are currently being used in clinics or being tested in clinical trials were created for anti-cancer uses. Numerous anti-cancer substances can be included in nanoconjugates for the purpose of delivering drugs, which will release them and cause them to act therapeutically on cancer cells. Nevertheless, damage can also be done to healthy tissues, thus a targeted strategy is a crucial component of combined therapy to avoid this issue (Hu et al., 2010). By employing tumour biological markers as docking locations to concentrate the curative effect at the tumours, one can improve therapy effectiveness while reducing widespread exposure and inaccurate consequences. These tumours have numerous surface proteins or receptors as molecular markers that are expressed at substantially higher or lower levels in cancer cells and tumour vasculature, distinguishing tumour masses from the surrounding healthy tissues (Silva et al., 2014).

1.6.3 Gene Delivery

Gene therapy aims to change or modify missing and/or damaged gene sequences to treat inherited and acquired illnesses such cancer, cardiovascular disease, and acquired immunodeficiency syndrome (AIDS). Until now symptomatic treatment, radiation, and chemotherapy have been used to treat terminal diseases that are primarily caused by missing, defective, or altered genetic material. Gene therapy, in contrast, offers a novel therapeutic approach for changing the genetic material in cells (Keles et al., 2016). For gene delivery applications, a wide range of nanomaterials composed of fatty acids, polymer compounds, carbon nanotubes, graphene (CNTs), nanospheres, mesoporous nanoparticles (NPs), and other inorganic NPs are employed. Each has strengths and weaknesses as a platform for the transport of genes, and by functionalizing these substances contain both organic and inorganic components; one might increase the efficacy of gene delivery while reducing cytotoxicity (Ditto et al., 2009). The benefit of transmitting DNA using vectors that are not viral is that they may be able to escape the immune response and carry larger payload. The latest advances in materials science, polymer engineering, and nanomaterials have revolutionised our understanding of nanoscale delivery. Through improved transfection efficiency and the ability to get past various biological barriers, these developments are assisting in overcoming many of the drawbacks of non-viral vectors (Parhiz et al., 2013). Regardless of the vector, therapeutic nucleic acids like antisense oligonucleotides (AONS), small interfering RNA (siRNA), short hairpin RNA (shRNA), and microRNA (miRNA) can change the genetic makeup of the cells being targeted. Delivering DNA to replace a patient's dysfunctional gene with a functional copy involved through gene therapy (siRNA) (Tomari and Zamore, 2005).

1.6.4 Detection and Diagnosis

In addition to their small size and large surface area, which become essential features in the nanoscale system, new developments in the last decade have made it possible to produce them in a variety of unique forms and regulate the structure of nanoparticles (NPs). Finding better methods is everyone's aim, whether it be for drug delivery, advanced imaging, or evaluation in areas like tumours and immunotherapy (Chou et al., 2011; Pelaz et al., 2017; Chowdhuri et al., 2016). The detection of specific biological agents in bodily fluids for illness, early-stage screening like cancer, such as tumour-associated antigens and other bimolecular indicators, are important components of biomedical science since both the survival and prognosis rates are poor and these illnesses can be asymptomatic until advanced stages. The study on the usage of antibodies, which has also gained an abundance of interest, has enabled the synthesis and refinement of particular antibodies—such monoclonal antibodies (mAbs)—towards the target antigen. This has made possible a wide variety of prospective uses in numerous research and health science sectors, particularly the area of medical diagnostics (De Palma and Hanahan, 2012; Daniels et al., 2006). Due to the aforementioned characteristics of NPs, the application of NPs for elisa applications is currently among the most prospective and often utilised identification approaches for the highly sensitive and specific immunological identification of clinically important biomarkers. Generally immobilised on the NP surfaces, the bimolecular recognition components, such as antibodies, engage for analytes (i.e., antigens or molecular biomarkers) to alter the signal by causing a physico-chemical reaction. As a result, careful engineering of nanoparticles must also be taken into account when using them for immune sensing due to the fact that most exchanges happen at the NP's contact with the biological surroundings (Nel et al., 2009). Additionally, NPs for immunosensing must fulfil certain requirements, including outstanding signalling effectiveness (i.e., the effective signal absorption from the antigens), extremely little (or no) toxic effects, and chemical resistance in difficult biological settings. In addition to plasmonic gold nanoparticles and fluorescent quantum dots, research for immunosensing by nanomaterials application has also grown to encompass silicon NPs and carbon-based NPs amongst a repertoire that is always growing (Holzinger et al., 2017).

1.6.5 Molecular Imaging

Many academics and doctors in a variety of fields have recently become interested in molecular imaging, which is quantitative in nature, non-intrusive and reproducible scanning of certain biomolecules and the tracking of corresponding biological reactions in human beings. Two distinct groups were once utilised to describe it. The first group includes imaging techniques for medicine like magnetic resonance imaging (MRI), computed tomography (CT), and ultrasound imaging that can show anatomic features. The second category includes molecular or functional imaging methods as optical imaging, single photon emission computed tomography (SPECT), and positron emission tomography (PET) (Jokerst and Gambhir 2011). According to the current discoveries of specific contrast compounds, such as nanoparticles, iodinated molecules, Gd chelates, and fluorescent dyes, particular biomarkers can be detected utilising conventional imaging techniques. Owing to their innately unique magnetic or optical properties, which allow for the use of these materials to various methods of imaging, nanoparticles stood out among them as potent tools for molecular imaging (Zhu et al., 2017; Lee and Hyeon, 2012). To employ the nanoparticles for molecular imaging purposes, an appropriate targeting ligand must be coupled onto their outermost layer. Because the ability to alter the surface of these nanoparticles has been effectively established, many targeted molecules, such as antibodies, peptides, and aptamers, are able to be conveniently coupled, utilising conventional biological conjugation techniques. Due to the possibility of conjugating several targeting ligands, the targeted nanomaterials usually have a higher affinity for biomolecules or cells (Jiang et al., 2008).

1.6.6 Tissue Engineering

There is a recently developed field of bioengineering called *tissue engineering* in combination with methods along with ideas of cell biology, material science, and engineering to produce artificial tissues which can mimic real tissues both physically as well as psychologically. This official definition of tissue engineering given through National Science Foundation (NSF) is this:

> Tissue engineering combines engineering and life sciences to explore how normal mammalian tissues work and how they can be replaced or enhanced. It involves using a variety of materials, such as polymers and ceramics, along with bioactive substances and cells, to stimulate tissue regeneration where there is damage or injury. This process aims to understand the relationship between tissue structure and function and to develop biological substitutes that can restore, maintain, or even enhance tissue function.
>
> (Harding et al., 2007)

The bulk of tissue engineering methods, however, are predicated on the notion that, given the proper bioreactor conditions, cells that were either implanted or induced into three-dimensional, or 3D, biocompatible scaffolding might assemble onto functioning components that mimic real tissues. Early artificial scaffolds tried to provide cells structural integrity at the level of the microscopic estimation, but they only proved moderately effective. Nowadays, it is well accepted that scaffolds must create a tissue-specific environment in order to maintain and control cell activity and function in order to replicate correct tissue functionality (Goldberg et al., 2007). Extracellular matrix (ECM), which is characterised by a natural web of hierarchically structured nanofibres, surrounds cells in tissues. As a result of cell-extracellular matrix interactions, this essential nanoarchitecture plays crucial role for supporting cells and controlling how they behave. Additionally, the ECM is essential for the preservation, discharge, and stimulation of a numerous biological components as well as for facilitating interactions between cells and with soluble biological factors. Therefore, the capacity to create biomaterials that nearly approximate the level of complexity and function of ECM will impact whether regeneration of tissue is successful (Shi et al., 2010).

1.7 CONCLUSION

Bionanomaterials are the combination of nanomaterials and biomolecules. The biomolecules are encapsulated or loaded into the nanomaterials. Aquatic, microalgae, and plants are among the numerous sources for biomolecules, and the nanomaterials are divided into multiple categories, including polymer, organic, biological and metal nanomaterials. The bionanomaterial can be used to target the disease via the active and passive targeting. For an analysis or characterization of bionanomaterials there are different instruments, such as UV-visible, FTIR, NMR spectroscopy, SEM and TEM, X-ray diffraction. Bionanomaterials can be utilized in drug delivery, tissue engineering, targeted therapy, molecular imaging, detection and diagnosis, and gene delivery. Thus, bionanomaterials are the potential nanotechnology to utilize.

REFERENCES

ABATE, G., ZHANG, L., PUCCI, M., MORBINI, G., MAC SWEENEY, E., MACCARINELLI, G., RIBAUDO, G., GIANONCELLI, A., UBERTI, D. & MEMO, M. 2021. Phytochemical analysis and anti-inflammatory activity of different ethanolic phyto-extracts of Artemisia annua L. *Biomolecules,* 11, 975.

ABD ELLAH, N. H. & ABOUELMAGD, S. A. 2017. Surface functionalization of polymeric nanoparticles for tumor drug delivery: approaches and challenges. *Expert Opinion on Drug Delivery,* 14, 201–214.
ABU-SERIE, M. M., HABASHY, N. H. & MAHER, A. M. 2019. In vitro anti-nephrotoxic potential of Ammi visnaga, Petroselinum crispum, Hordeum vulgare, and Cymbopogon schoenanthus seed or leaf extracts by suppressing the necrotic mediators, oxidative stress and inflammation. *BMC Complementary and Alternative Medicine,* 19, 1–16.
AKRAM, M., IQBAL, M., DANIYAL, M. & KHAN, A. U. 2017. Awareness and current knowledge of breast cancer. *Biological Research,* 50, 1–23.
ALAVI, M. & HAMIDI, M. 2019. Passive and active targeting in cancer therapy by liposomes and lipid nanoparticles. *Drug Metabolism and Personalized Therapy,* 34(1):20180032
ANGELL, C., XIE, S., ZHANG, L. & CHEN, Y. 2016. DNA nanotechnology for precise control over drug delivery and gene therapy. *Small,* 12, 1117–1132.
ARIGA, K., LI, J., FEI, J., JI, Q. & HILL, J. P. 2016. Nanoarchitectonics for dynamic functional materials from atomic-/molecular-level manipulation to macroscopic action. *Advanced Materials,* 28, 1251–1286.
ARORA, B., BHATIA, R. & ATTRI, P. 2018. Bionanocomposites: green materials for a sustainable future. In: *New Polymer Nanocomposites for Environmental Remediation*, 699–712. Elsevier.
BARSANTI, L. & GUALTIERI, P. 2005. *Algae: Anatomy, Biochemistry, and Biotechnology*, CRC Press.
BEISSENHIRTZ, M. K. & WILLNER, I. 2006. DNA-based machines. *Organic & Biomolecular Chemistry,* 4, 3392–3401.
BHATTACHARYA, T., RATHER, G., AKTER, R., KABIR, M. T., RAUF, A. & RAHMAN, M. 2021. Nutraceuticals and bio-inspired materials from microalgae and their future perspectives. *Current Topics in Medicinal Chemistry,* 21, 1037–1051.
BHATTACHERJEE, A., GHOSH, T. & DATTA, A. 2018. Green synthesis and characterisation of antioxidant-tagged gold nanoparticle (X-GNP). and studies on its potent antimicrobial activity. *Journal of Experimental Nanoscience,* 13, 50–61.
BIAO, L., TAN, S., MENG, Q., GAO, J., ZHANG, X., LIU, Z. & FU, Y. 2018. Green synthesis, characterization and application of proanthocyanidins-functionalized gold nanoparticles. *Nanomaterials,* 8, 53.
BOUNAKHLA, M. & TAHRI, M. 2014. X-ray fluorescence analytical techniques. *National Center for Energy Sciences and Nuclear Techniques (CNESTEN), Rabat, Morocco,* 1, 1–73.
BRIONES, C. & MORENO, M. 2012. Applications of peptide nucleic acids (PNAs). and locked nucleic acids (LNAs). in biosensor development. *Analytical and Bioanalytical Chemistry,* 402, 3071–3089.
BUSATTO, S., WALKER, S. A., GRAYSON, W., PHAM, A., TIAN, M., NESTO, N., BARKLUND, J. & WOLFRAM, J. 2020. Lipoprotein-based drug delivery. *Advanced Drug Delivery Review,* 159, 377–390.
CARUSO, G., FLORIS, R., SERANGELI, C. & DI PAOLA, L. 2020. Fishery wastes as a yet undiscovered treasure from the sea: Biomolecules sources, extraction methods and valorization. *Marine Drugs,* 18, 622.
CHAKI, R., GHOSH, N. & MANDAL, S. C. 2022. Phytopharmacology of herbal biomolecules. In: *Herbal Biomolecules in Healthcare Applications*, 101–119. Academic Press.
CHANMANGKANG, S., WANGTUEAI, S., PANSAWAT, N., TEPWONG, P., PANYA, A. & MANEEROTE, J. 2022. Characteristics and properties of acid-and pepsin-solubilized collagens from the tail tendon of Skipjack Tuna (Katsuwonus pelamis). *Polymers,* 14, 5329.
CHATURVEDI, V. K., SINGH, A., SINGH, V. K. & SINGH, M. P. 2019. Cancer nanotechnology: a new revolution for cancer diagnosis and therapy. *Current Drug Metabolism,* 20, 416–429.
CHOI, C., NAM, J.-P. & NAH, J.-W. 2016. Application of chitosan and chitosan derivatives as biomaterials. *Journal of Industrial and Engineering Chemistry,* 33, 1–10.
CHOU, L. Y., MING, K. & CHAN, W. C. 2011. Strategies for the intracellular delivery of nanoparticles. *Chemical Society Review,* 40, 233–245.
CHOWDHURI, A. R., BHATTACHARYA, D. & SAHU, S. K. 2016. Magnetic nanoscale metal organic frameworks for potential targeted anticancer drug delivery, imaging and as an MRI contrast agent. *Dalton Transactions,* 45, 2963–2973.
CLEMONS, T. D., SINGH, R., SOROLLA, A., CHAUDHARI, N., HUBBARD, A. & IYER, K. S. 2018. Distinction between active and passive targeting of nanoparticles dictate their overall therapeutic efficacy. *Langmuir,* 34, 15343–15349.
D. FRIEDMAN, A., E. CLAYPOOL, S. & LIU, R. 2013. The smart targeting of nanoparticles. *Current Pharmaceutical Design,* 19, 6315–6329.

DANIELS, T. R., DELGADO, T., RODRIGUEZ, J. A., HELGUERA, G. & PENICHET, M. L. 2006. The transferrin receptor part I: biology and targeting with cytotoxic antibodies for the treatment of cancer. *Clinical Immunology,* 121, 144–158.

DATTA, L. P., MANCHINEELLA, S. & GOVINDARAJU, T. 2020. Biomolecules-derived biomaterials. *Biomaterials,* 230, 119633.

DAVAR, F., FERESHTEH, Z. & SALAVATI-NIASARI, M. 2009. Nanoparticles Ni and NiO: synthesis, characterization and magnetic properties. *Journal of Alloys and Compounds,* 476, 797–801.

DAVIS, M. E., CHEN, Z. G. & SHIN, D. M. 2008. Nanoparticle therapeutics: an emerging treatment modality for cancer. *Nature Review Drug Discovery,* 7, 771–782.

DE PALMA, M. & HANAHAN, D. 2012. The biology of personalized cancer medicine: facing individual complexities underlying hallmark capabilities. *Molecular Oncology,* 6, 111–127.

DÍEZ-PASCUAL, A. M. 2022. Surface engineering of nanomaterials with polymers, biomolecules, and small ligands for nanomedicine. *Materials (Basel),* 15(9):3251

DITTO, A. J., SHAH, P. N. & YUN, Y. H. 2009. Non-viral gene delivery using nanoparticles. *Expert Opinion Drug Delivery,* 6, 1149–1160.

DONNO, D., MELLANO, M. G., CERUTTI, A. K. & BECCARO, G. L. 2016. Biomolecules and natural medicine preparations: analysis of new sources of bioactive compounds from Ribes and Rubus spp. Buds. *Pharmaceuticals,* 9, 7.

FAHMY, T. M., FONG, P. M., GOYAL, A. & SALTZMAN, W. M. 2005. Targeted for drug delivery. *Materials Today,* 8, 18–26.

GANGADOO, S., TAYLOR-ROBINSON, A. & CHAPMAN, J. 2015. Nanoparticle and biomaterial characterisation techniques. *Materials Technology,* 30, B44–B56.

GENTILE, P., CHIONO, V., CARMAGNOLA, I. & HATTON, P. V. 2014. An overview of poly (lactic-co-glycolic). acid (PLGA)-based biomaterials for bone tissue engineering. *International Journal of Molecular Sciences,* 15, 3640–3659.

GHERABLI, A., GRIMI, N., LEMAIRE, J., VOROBIEV, E. & LEBOVKA, N. 2023. Extraction of valuable biomolecules from the microalga haematococcus pluvialis assisted by electrotechnologies. *Molecules,* 28, 2089.

GHOSH, R., DAS, S., MALLICK, S. P. & BEYENE, Z. 2022a. A review on the antimicrobial and antibiofilm activity of doped hydroxyapatite and its composites for biomedical applications. *Materials Today Communications,* 31:103311.

GHOSH, S., SARKAR, T., PATI, S., KARI, Z. A., EDINUR, H. A. & CHAKRABORTY, R. 2022b. Novel bioactive compounds from marine sources as a tool for functional food development. *Frontiers in Marine Science,* 9, 76.

GMEINER, W. H. & GHOSH, S. 2014. Nanotechnology for cancer treatment. *Nanotechnology Reviews,* 3, 111–122.

GOLDBERG, M., LANGER, R. & JIA, X. 2007. Nanostructured materials for applications in drug delivery and tissue engineering. *Journal of Biomaterials Science Polymer Edition,* 18, 241–268.

GOMES, T. A., ZANETTE, C. M. & SPIER, M. R. 2020. An overview of cell disruption methods for intracellular biomolecules recovery. *Preparative Biochemistry & Biotechnology,* 50, 635–654.

GOSWAMI, R. K., AGRAWAL, K. & VERMA, P. 2022. Microalgae Dunaliella as biofuel feedstock and β-carotene production: an influential step towards environmental sustainability. *Energy Conversion and Management: X,* 13, 100154.

GUADAGNINI, R., HALAMODA KENZAOUI, B., WALKER, L., POJANA, G., MAGDOLENOVA, Z., BILANICOVA, D., SAUNDERS, M., JUILLERAT-JEANNERET, L., MARCOMINI, A., HUK, A., DUSINSKA, M., FJELLSBØ, L. M., MARANO, F. & BOLAND, S. 2015. Toxicity screenings of nanomaterials: challenges due to interference with assay processes and components of classic in vitro tests. *Nanotoxicology,* 9, 13–24.

GUPTA, A., MISHRA, A. & PURI, N. 2017. Peptide nucleic acids: advanced tools for biomedical applications. *Journal of Biotechnology,* 259, 148–159.

HARDING, K. G., DENNIS, J. S., VON BLOTTNITZ, H. & HARRISON, S. T. 2007. Environmental analysis of plastic production processes: comparing petroleum-based polypropylene and polyethylene with biologically-based poly-beta-hydroxybutyric acid using life cycle analysis. *Journal of Biotechnology,* 130, 57–66.

HE, L., ZENG, Y., QIN, X., PAN, L., CHEN, T., WANG, Q., MA, Y. & FANG, J. 2021. Delivery of ferulic acid with PEGylated diphenylalanine nanoparticles for pre-arthritis therapy. *Journal of Biomedical Nanotechnology,* 17, 357–368. https://doi.org/10.21203/rs.3.rs-140007/v1

HIRAYAMA, H., AMOLEGBE, S. A., ISLAM, M. S., RAHMAN, M. A., GOTO, N., SEKINE, Y. & HAYAMI, S. 2021. Encapsulation and controlled release of an antimalarial drug using surface functionalized mesoporous silica nanocarriers. *Journal of Materials Chemistry B,* 9, 5043–5046.

HOLZINGER, M., LE GOFF, A. & COSNIER, S. 2017. Synergetic effects of combined nanomaterials for biosensing applications. *Sensors (Basel),* 17(5), 1010.

HU, C. M., ARYAL, S. & ZHANG, L. 2010. Nanoparticle-assisted combination therapies for effective cancer treatment. *Therapeutic Delivery,* 1, 323–334.

HU, W., GLADUE, R., HANSEN, J., WOJNAR, C. & CHALMERS, J. J. 2007. The sensitivity of the dinoflagellate Crypthecodinium cohnii to transient hydrodynamic forces and cell-bubble interactions. *Biotechnology Progress,* 23, 1355–1362.

HUBBELL, J. A. & CHILKOTI, A. 2012. Nanomaterials for drug delivery. *Science,* 337, 303–305.

IM, D.-S. 2020. Pro-resolving effect of ginsenosides as an anti-inflammatory mechanism of Panax ginseng. *Biomolecules,* 10, 444.

JAIN, A., RANJAN, S., DASGUPTA, N. & RAMALINGAM, C. 2018. Nanomaterials in food and agriculture: an overview on their safety concerns and regulatory issues. *Critical Reviews in Food Science and Nutrition,* 58, 297–317.

JEEVANANDAM, J., LING, J. K. U., BARHOUM, A., CHAN, Y. S. & DANQUAH, M. K. 2022. Chapter 1 – bionanomaterials: definitions, sources, types, properties, toxicity, and regulations. In: BARHOUM, A., JEEVANANDAM, J. & DANQUAH, M. K. (eds.). *Fundamentals of Bionanomaterials.* Elsevier.

JIANG, D., ENGLAND, C. G. & CAI, W. 2016. DNA nanomaterials for preclinical imaging and drug delivery. *Journal of Controlled Release,* 239, 27–38.

JIANG, W., KIM, B. Y., RUTKA, J. T. & CHAN, W. C. 2008. Nanoparticle-mediated cellular response is size-dependent. *Nature Nanotechnology,* 3, 145–150.

JOKERST, J. V. & GAMBHIR, S. S. 2011. Molecular imaging with theranostic nanoparticles. *Accounts of Chemical Research,* 44, 1050–1060.

KARKI, S., GOHAIN, M. B., YADAV, D. & INGOLE, P. G. 2021. Nanocomposite and bio-nanocomposite polymeric materials/membranes development in energy and medical sector: a review. *International Journal of Biological Macromolecules,* 193, 2121–2139.

KAUR, N., POPLI, P., TIWARY, N. & SWAMI, R. 2023. Small molecules as cancer targeting ligands: shifting the paradigm. *Journal of Controlled Release,* 355, 417–433.

KELES, E., SONG, Y., DU, D., DONG, W.-J. & LIN, Y. 2016. Recent progress in nanomaterials for gene delivery applications. *Biomaterials Science,* 4, 1291–1309.

KHAN, K., SHAIFUL ISLAM, M., AWAL, A., KHAN, M. I. & ULLAH, A. A. 2021. Studies on performances of copper oxide nanoparticles from catharanthus roseus leaf extract. *Microelectronics, Circuits and Systems: Select Proceedings of 7th International Conference on Micro2020.* Springer, 179–190.

KIM, G. J. & NIE, S. 2005. Targeted cancer nanotherapy. *Materials Today,* 8, 28–33.

KOBAYASHI, K., WEI, J., IIDA, R., IJIRO, K. & NIIKURA, K. 2014. Surface engineering of nanoparticles for therapeutic applications. *Polymer Journal,* 46, 460–468.

KOVAČ, D. J., SIMEUNOVIĆ, J. B., BABIĆ, O. B., MIŠAN, A. Č. & MILOVANOVIĆ, I. L. 2013. Algae in food and feed. *Food and Feed Research,* 40, 21–32-21-32.

KUMARI, P., GHOSH, B. & BISWAS, S. 2016. Nanocarriers for cancer-targeted drug delivery. *Journal of Drug Targeting,* 24, 179–191.

KUNDU, B., RAJKHOWA, R., KUNDU, S. C. & WANG, X. 2013. Silk fibroin biomaterials for tissue regenerations. *Advanced Drug Delivery Reviews,* 65, 457–470.

LEE, N. & HYEON, T. 2012. Designed synthesis of uniformly sized iron oxide nanoparticles for efficient magnetic resonance imaging contrast agents. *Chemical Society Reviews,* 41, 2575–2589.

LEONOR PINZON-DAZA, M., CAMPIA, I., KOPECKA, J., GARZÓN, R., GHIGO, D. & RIGANT, C. 2013. Nanoparticle-and liposome-carried drugs: new strategies for active targeting and drug delivery across blood-brain barrier. *Current Drug Metabolism,* 14, 625–640.

LI, J., CAI, C., LI, J., LI, J., LI, J., SUN, T., WANG, L., WU, H. & YU, G. 2018. Chitosan-based nanomaterials for drug delivery. *Molecules*, 23(10), 2661.

LI, X.-Q. & ZHANG, W.-X. 2007. Sequestration of metal cations with zerovalent iron nanoparticles a study with High Resolution X-ray Photoelectron Spectroscopy (HR-XPS). *The Journal of Physical Chemistry C,* 111**,** 6939–6946.

LIU, X., YANG, W. & LIU, Z. 2014. Recent progress on synchrotron-based in-situ soft X-ray spectroscopy for energy materials. *Advanced Materials,* 26**,** 7710–7729.

LUPOI, J. S., SINGH, S., SIMMONS, B. A. & HENRY, R. J. 2014. Assessment of lignocellulosic biomass using analytical spectroscopy: an evolution to high-throughput techniques. *BioEnergy Research,* 7**,** 1–23.

MAITI, S. & SEN, K. K. 2017. Introductory chapter: drug delivery concepts. In: *Advanced Technology Delivery Therapy*. Intech Publishers, 1–2.

MAKADIA, H. K. & SIEGEL, S. J. 2011. Poly lactic-co-glycolic acid (PLGA). as biodegradable controlled drug delivery carrier. *Polymers,* 3**,** 1377–1397.

MALAFAYA, P. B., SILVA, G. A. & REIS, R. L. 2007. Natural–origin polymers as carriers and scaffolds for biomolecules and cell delivery in tissue engineering applications. *Advanced Drug Delivery Reviews,* 59**,** 207–233.

MANCHESTER, M. & SINGH, P. 2006. Virus-based nanoparticles (VNPs): platform technologies for diagnostic imaging. *Advanced Drug Delivery Reviews,* 58**,** 1505–1522.

MCMAHON, R. E., WANG, L., SKORACKI, R. & MATHUR, A. B. 2013. Development of nanomaterials for bone repair and regeneration. *Journal of Biomedical Materials Research Part B: Applied Biomaterials,* 101**,** 387–397.

MONDAL, D. S. 2018. *UNIT–I Biomolecules*. 10.13140/RG.2.2.20736.74241

MÜLLER, B. & LAIBACH, N. 2022. Chapter 5. Extraction of valuable components from waste biomass. In: *Waste to Food: Returning Nutrients to the Food Chain*, 123003.. Wageningen Academic Publishers.

MUSA, S., OGALA, H., OKOJI, C., MOSES, N., IBOYI, N. & OJEI, S. 2023. Applications of bionanoenabled materials and coatings for affordable, reliable and sustainable green energy production. In: *Bionanotechnology Towards Green Energy*, 77–102. CRC Press.

NAIK, S., TIRKEY, A. & JUJJAVARAPU, S. E. 2020. Synthesis and characterization of nanocomposites of animal origin. In: *Green Polymeric Nanocomposites*, 47–80. CRC Press.

NASSERI, B., SOLEIMANI, N., RABIEE, N., KALBASI, A., KARIMI, M. & HAMBLIN, M. R. 2018. Point-of-care microfluidic devices for pathogen detection. *Biosensors Bioelectronics,* 117**,** 112–128.

NEL, A. E., MÄDLER, L., VELEGOL, D., XIA, T., HOEK, E. M., SOMASUNDARAN, P., KLAESSIG, F., CASTRANOVA, V. & THOMPSON, M. 2009. Understanding biophysicochemical interactions at the nano-bio interface. *Nature Materials,* 8**,** 543–557.

NURILMALA, M., HIZBULLAH, H. H., KARNIA, E., KUSUMANINGTYAS, E. & OCHIAI, Y. 2020. Characterization and antioxidant activity of collagen, gelatin, and the derived peptides from yellowfin tuna (Thunnus albacares). skin. *Marine Drugs,* 18**,** 98.

NUZZO, D., CONTARDI, M., KOSSYVAKI, D., PICONE, P., CRISTALDI, L., GALIZZI, G., BOSCO, G., SCOGLIO, S., ATHANASSIOU, A. & DI CARLO, M. 2019. Heat-resistant aphanizomenon flos-aquae (AFA). extract (Klamin®). as a functional ingredient in food strategy for prevention of oxidative stress. *Oxidative Medicine and Cellular Longevity,* 2019(1)**,** 9481390.

ODJADJARE, E. C., MUTANDA, T. & OLANIRAN, A. O. 2017. Potential biotechnological application of microalgae: a critical review. *Critical Reviews in Biotechnology,* 37**,** 37–52.

OLUWASANU, A. A., OLUWASEUN, F., TESLIM, J. A., ISAIAH, T. T., OLALEKAN, I. A. & CHRIS, O. A. 2019. Scientific applications and prospects of nanomaterials: a multidisciplinary review. *African Journal of Biotechnology,* 18**,** 946–961.

OTIMENYIN, S. 2022. Herbal biomolecules acting on central nervous system. In: *Herbal Biomolecules in Healthcare Applications*, 475–523. Academic Press

PARHIZ, H., SHIER, W. T. & RAMEZANI, M. 2013. From rationally designed polymeric and peptidic systems to sophisticated gene delivery nano-vectors. *International Journal of Pharmacy,* 457**,** 237–259.

PELAZ, B., ALEXIOU, C., ALVAREZ-PUEBLA, R. A., ALVES, F., ANDREWS, A. M., ASHRAF, S., BALOGH, L. P., BALLERINI, L., BESTETTI, A., BRENDEL, C., BOSI, S., CARRIL, M., CHAN, W. C., CHEN, C., CHEN, X., CHEN, X., CHENG, Z., CUI, D., DU, J., DULLIN, C., ESCUDERO, A., FELIU, N., GAO, M., GEORGE, M., GOGOTSI, Y., GRÜNWELLER, A., GU, Z., HALAS, N. J., HAMPP, N., HARTMANN, R. K., HERSAM, M. C., HUNZIKER, P., JIAN, J., JIANG, X., JUNGEBLUTH, P., KADHIRESAN, P., KATAOKA, K., KHADEMHOSSEINI, A., KOPEČEK, J.,

KOTOV, N. A., KRUG, H. F., LEE, D. S., LEHR, C. M., LEONG, K. W., LIANG, X. J., LING LIM, M., LIZ-MARZÁN, L. M., MA, X., MACCHIARINI, P., MENG, H., MÖHWALD, H., MULVANEY, P., NEL, A. E., NIE, S., NORDLANDER, P., OKANO, T., OLIVEIRA, J., PARK, T. H., PENNER, R. M., PRATO, M., PUNTES, V., ROTELLO, V. M., SAMARAKOON, A., SCHAAK, R. E., SHEN, Y., SJÖQVIST, S., SKIRTACH, A. G., SOLIMAN, M. G., STEVENS, M. M., SUNG, H. W., TANG, B. Z., TIETZE, R., UDUGAMA, B. N., VANEPPS, J. S., WEIL, T., WEISS, P. S., WILLNER, I., WU, Y., YANG, L., YUE, Z., ZHANG, Q., ZHANG, Q., ZHANG, X. E., ZHAO, Y., ZHOU, X. & PARAK, W. J. 2017. Diverse applications of nanomedicine. *ACS Nano,* 11**,** 2313–2381.

PRASANTHI, N. L., ROY, H., JYOTHI, N. & VAJRAPRIYA, V. S. 2016. A brief review on chitosan and application in biomedical field. *American Journal of PharmTech Research,* 6**,** 41–51.

RASHID, M. M. U. & AHMAD, Q. Z. 2019. Trends in nanotechnology for practical applications. In: *Applications of Targeted Nano Drugs and Delivery Systems*, 297–325. Elsevier

ROCO, M. C., MIRKIN, C. A. & HERSAM, M. C. 2011. Nanotechnology research directions for societal needs in 2020: summary of international study. *Journal of Nanoparticle Research,* 13**,** 897–919.

ROSADO-DE-CASTRO, P. H., MORALES, M. D. P., PIMENTEL-COELHO, P. M., MENDEZ-OTERO, R. & HERRANZ, F. 2018. Development and application of nanoparticles in biomedical imaging. *Contrast Media & Molecular Imaging,* 2018**,** 1403826.

SAHAYARAJ, K. 2012. Bionanomaterials: synthesis and applications. *Proceedings of the First National Seminar on New Materials Research and Nanotechnology (NSNMRN'2012)*, 12–14.

SALAHPOUR ANARJAN, F. 2019. Active targeting drug delivery nanocarriers: ligands. *Nano-Structures & Nano-Objects,* 19**,** 100370.

SALEHI, B., ATA, A., V. ANIL KUMAR, N., SHAROPOV, F., RAMIREZ-ALARCON, K., RUIZ-ORTEGA, A., ABDULMAJID AYATOLLAHI, S., VALERE TSOUH FOKOU, P., KOBARFARD, F. & AMIRUDDIN ZAKARIA, Z. 2019. Antidiabetic potential of medicinal plants and their active components. *Biomolecules,* 9**,** 551.

SATYANARAYANA, K., RANGAN, A., PRASAD, V. & MAGALHÃES, W. L. E. 2017. Preparation, characterization, and applications of nanomaterials (cellulose, lignin, and silica) from renewable (lignocellulosic) resources. In: *Handbook of Composites from Renewable Materials***,** 1–66. John Wiley & Sons.

SAYED-AHMED, M. Z., MAKEEN, H. A., ELSHERBINI, M. M., SYED, N. K. & SHOEIB, S. M. 2018. Oncolytic viruses: a gene therapy for treatment of cancer in companion animals. *Health Science Journal,* 12**,** 1–9.

SHARMA, G., PANDEY, S., GHATAK, S., WATAL, G. & RAI, P. K. 2018. Chapter 3 – potential of spectroscopic techniques in the characterization of "green nanomaterials". In: TRIPATHI, D. K., AHMAD, P., SHARMA, S., CHAUHAN, D. K. & DUBEY, N. K. (eds.). *Nanomaterials in Plants, Algae, and Microorganisms.* Academic Press.

SHEN, S., WU, Y., LIU, Y. & WU, D. 2017. High drug-loading nanomedicines: progress, current status, and prospects. *International Journal of Nanomedicine,* 12**,** 4085.

SHI, J., VOTRUBA, A. R., FAROKHZAD, O. C. & LANGER, R. 2010. Nanotechnology in drug delivery and tissue engineering: from discovery to applications. *Nano Letters,* 10**,** 3223–3230.

SILVA, J., FERNANDES, A. R. & BAPTISTA, P. V. 2014. Application of nanotechnology in drug delivery. InTech. http://dx.doi.org/10.5772/58424

SINGH, D., SINGH, S., SAHU, J., SRIVASTAVA, S. & SINGH, M. R. 2016. Ceramic nanoparticles: recompense, cellular uptake and toxicity concerns. *Artificial Cells, Nanomedicine, and Biotechnology,* 44**,** 401–409.

SINGH, K. R., SRIDEVI, P. & SINGH, R. P. 2020. Potential applications of peptide nucleic acid in biomedical domain. *Engineering Reports,* 2**,** e12238.

SINGH, K. R. B., NAYAK, V. & SINGH, R. P. 2021. Introduction to bionanomaterials: an overview. In: *Bionanomaterials.* IOP Publishing, 1–16.

SITHARAMAN, B. 2016. *Nanobiomaterials Handbook*, CRC Press.

SRIDHAR, R., LAKSHMINARAYANAN, R., MADHAIYAN, K., BARATHI, V. A., LIM, K. H. C. & RAMAKRISHNA, S. 2015. Electrosprayed nanoparticles and electrospun nanofibers based on natural materials: applications in tissue regeneration, drug delivery and pharmaceuticals. *Chemical Society Reviews,* 44**,** 790–814.

STANKIC, S., SUMAN, S., HAQUE, F. & VIDIC, J. 2016. Pure and multi metal oxide nanoparticles: synthesis, antibacterial and cytotoxic properties. *Journal of Nanobiotechnology,* 14**,** 1–20.

STEVENS, M. M. & GEORGE, J. H. 2005. Exploring and engineering the cell surface interface. *Science,* 310, 1135–8.

SURI, S. S., FENNIRI, H. & SINGH, B. 2007. Nanotechnology-based drug delivery systems. *Journal of Occupational Medicine and Toxicology,* 2, 1–6.

SWENSON, C. S. & HEEMSTRA, J. M. 2020. Peptide nucleic acids harness dual information codes in a single molecule. *Chemical Communications,* 56, 1926–1935.

SZCZĘSNY, G., KOPEC, M., POLITIS, D. J., KOWALEWSKI, Z. L., ŁAZARSKI, A. & SZOLC, T. 2022. A review on biomaterials for orthopaedic surgery and traumatology: from past to present. *Materials,* 15, 3622.

TELEANU, D. M., NEGUT, I., GRUMEZESCU, V., GRUMEZESCU, A. M. & TELEANU, R. I. 2019. Nanomaterials for drug delivery to the central nervous system. *Nanomaterials (Basel),* 9(3), 371

TITUS, D., SAMUEL, E. J. J. & ROOPAN, S. M. 2019. Nanoparticle characterization techniques. In: *Green Synthesis, Characterization and Applications of Nanoparticles,* 303–319. Elsevier.

TOMARI, Y. & ZAMORE, P. D. 2005. Perspective: machines for RNAi. *Genes and Development,* 19, 517–529.

TU, T., HUAN, S., KE, G. & ZHANG, X. 2022. Functional xeno nucleic acids for biomedical application. *Chemical Research in Chinese Universities,* 38, 912–918.

VOLPI, S., CANCELLI, U., NERI, M. & CORRADINI, R. 2020. Multifunctional delivery systems for peptide nucleic acids. *Pharmaceuticals,* 14, 14.

WANG, M. D., SHIN, D. M., SIMONS, J. W. & NIE, S. 2007. Nanotechnology for targeted cancer therapy. *Expert Review of Anticancer Therapy,* 7, 833–837.

WANI, S. U. D., GAUTAM, S. P., QADRIE, Z. L. & GANGADHARAPPA, H. V. 2020. Silk fibroin as a natural polymeric based bio-material for tissue engineering and drug delivery systems-a review. *International Journal of Biological Macromolecules,* 163, 2145–2161.

WEBSTER, D. M., SUNDARAM, P. & BYRNE, M. E. 2013. Injectable nanomaterials for drug delivery: Carriers, targeting moieties, and therapeutics. *European Journal of Pharmaceutics and Biopharmaceutics,* 84, 1–20.

WENK, E., MERKLE, H. P. & MEINEL, L. 2011. Silk fibroin as a vehicle for drug delivery applications. *Journal of Controlled Release,* 150, 128–141.

YAAKOB, Z., ALI, E., ZAINAL, A., MOHAMAD, M. & TAKRIFF, M. S. 2014. An overview: biomolecules from microalgae for animal feed and aquaculture. *Journal of Biological Research-Thessaloniki,* 21, 1–10.

YADAV, S., SHARMA, A. K. & KUMAR, P. 2020. Nanoscale self-assembly for therapeutic delivery. *Frontiers in Bioengineering and Biotechnology,* 8, 127.

YAMAMOTO, E. & KURODA, K. 2016. Colloidal mesoporous silica nanoparticles. *Bulletin of the Chemical Society of Japan,* 89, 501–539.

YUAN, X.-L., MAO, X.-X., DU, Y.-M., YAN, P.-Z., HOU, X.-D. & ZHANG, Z.-F. 2019. Anti-tumor activity of cembranoid-type diterpenes isolated from Nicotiana tabacum L. *Biomolecules,* 9, 45.

ZHAO, G.-X., YANG, X.-R., WANG, Y.-M., ZHAO, Y.-Q., CHI, C.-F. & WANG, B. 2019. Antioxidant peptides from the protein hydrolysate of spanish mackerel (Scomberomorous niphonius). muscle by in vitro gastrointestinal digestion and their in vitro activities. *Marine Drugs,* 17, 531.

ZHOU, J. & ROSSI, J. J. 2014. Cell-type-specific, aptamer-functionalized agents for targeted disease therapy. *Molecular Therapy-Nucleic Acids,* 3, e169.

ZHOU, M., SHEN, L., LIN, X., HONG, Y. & FENG, Y. 2017. Design and pharmaceutical applications of porous particles. *RSC Advances,* 7, 39490–39501.

ZHU, S., YANG, Q., ANTARIS, A. L., YUE, J., MA, Z., WANG, H., HUANG, W., WAN, H., WANG, J., DIAO, S., ZHANG, B., LI, X., ZHONG, Y., YU, K., HONG, G., LUO, J., LIANG, Y. & DAI, H. 2017. Molecular imaging of biological systems with a clickable dye in the broad 800- to 1,700-nm near-infrared window. *Proc Natl Acad Sci U S A,* 114, 962–967.

ZRAZHEVSKIY, P., SENA, M. & GAO, X. 2010. Designing multifunctional quantum dots for bioimaging, detection, and drug delivery. *Chemical Society Reviews,* 39, 4326–4354.

2 Synthesis of Bionanomaterials by Polymers

Tanima Bhattacharya, Pooja Mittal, Sanjukta Kar, Tanmoy Das, and Madhav Singla

2.1 INTRODUCTION

We are privileged to be alive in a century when nearly every medical specialty has made progress around the world. Due to the fact that individuals live longer nowadays, there is a need to widen the treatment spectrum and diagnostic capabilities in order to deal with whole new-complex disease traits. Diagnostic instruments, medication delivery nanocarriers, and imaging and sensor components have all entered an entirely novel phase of development because of nanotechnology. In 1974 Taniguchi coined the word "nanotechnology", which refers to the study and manipulation of materials on a molecular or atomic scale, typically on a scale of a few nanometres to one micron (Taniguchi, 1994). Exploring the applications of nanoscale materials in various areas such as healthcare and medicine can lead to the development of novel technologies with potential applications. The National Nanotechnology Initiative (NNI) gave the most credible definition of nanotechnology: the capacity to operate at the molecular level with the goal to design and fabricate nanomaterials with a functional structure, including essentially novel molecular characteristics (Sargent, 2013; Feynman, 1959). The emergence of examining the tunneling microscope and the development of cluster science gave a significant boost to the field of nanotechnology. Granqvist and Buhrman reported the first effective nanocrystal fabrication utilizing an inert-gas evaporation technique in 1976 (Drexler, 2004; Chopra et al., 2022).

A fascinating and significant advancement, resulting from the convergence of nanotechnology and biomedicine, involves the extensive utilization of engineered nanomaterials (NMs) in a wide range of technological, biotechnological, and biomedical applications. This cross-disciplinary breakthrough has paved the way for numerous innovations and advancements in various fields (Docter et al., 2015). They are man-made designed particles with very tiny dimensions to utilize their exclusive characteristics. The scientific community supports nanoparticles being classified as new chemicals and subject to the same laws and restrictions as these substances (Rakel et al., 2022). Growing interest in nanomaterials may be attributed to the growing need to modify specific molecules in the human body and the surrounding environment. The human hair is thousand of times larger than nanomaterials (Emam et al., 2014). Owing to their high conductivity and small size, they are perfect for enhancing the kinetics of analyte adsorption. They are already implemented into a vast array of commercially accessible products and services. Due to the novel design of NMs, physicochemical properties may be controlled in a stable environment, but not under physiological or natural situations (Rakel et al., 2022). Even though the toxicological information for a chemical in its bulk form is well researched, nanoparticles must always be treated as new compounds whose safety must be properly examined. This is especially true for drugs with the potential to cause cancer (Rakel et al., 2022; Emam et al., 2014). Polymeric NPs have garnered a lot of attention recently due to their minute size (Taniguchi, 1994; Sargent, 2013; Feynman, 1959). Using polymeric NPs as drug

 DOI: 10.1201/9781003432791-3

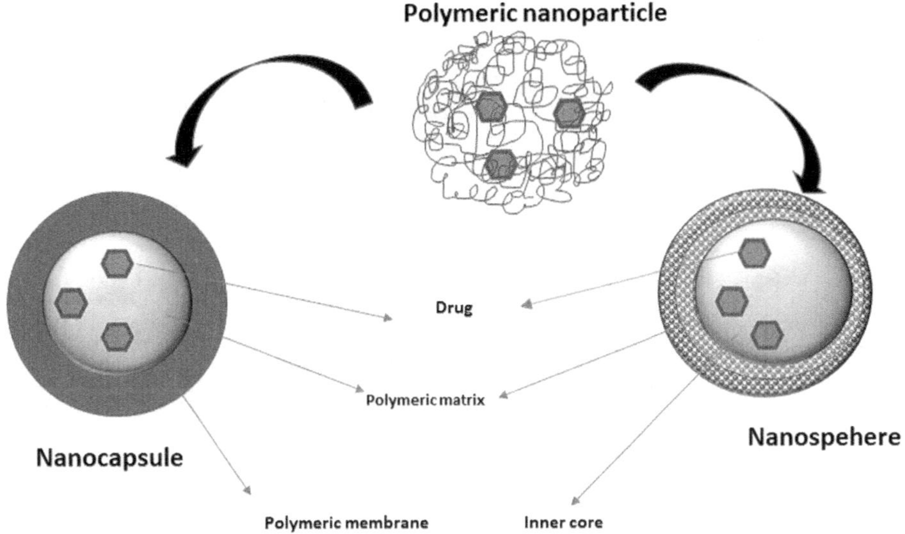

FIGURE 2.1 Structure of Polymeric Nanoparticles. (Figures Compiled by the Authors.)

carriers has a series of advantages, including the potential for sustained release, protecting the drug from degradation or instabilities, and increasing the bioavailability and therapeutic index of the drug (Taniguchi, 1994; Drexler, 2004). The two types of nanoparticles—nanospheres and nanocapsules—are included under the umbrella term "nanoparticle" (Docter et al, 2015). A nanocapsule's polymeric coating regulates the dissolution profile of the drug from the viscous core, where it is usually incorporated. Due to their uniform polymeric matrix, nanospheres can encapsulate the drug or have it deposited on their surface (Docter et al., 2015; Rakel et al., 2022; Emam et al., 2014). Figure 2.1 shows two kinds of polymer-based nanoparticles: A reservoir system and a matrix system (nanosphere) (Feynman, 1961).

2.2 POLYMERIC NANOPARTICLE PRODUCTION TECHNIQUES

Several techniques can be employed to create the particles depending on the kind of medicine that is going to encapsulated in polymeric NPs and how it needs to be administered (Jawahar et al., 2012). Generally, the polymerization of monomers or the premade dispersions of polymers are being used (Reis et al, 2006; Amgoth et al., 2019). The most popular methods are listed in a schematic way (Figure 2.2). Initially, organic solvents are employed to mix polymer in the majority of the approaches that call for the use of already made polymers (Amgoth et al., 2019). The solvents being used in the synthesis can cause the toxicity and environmental risk issues. The finished product also needs to be cleaned of any remaining solvents. Using methodologies that depend on the monomer polymerization process, chemicals can be incorporated into polymeric NP more efficiently and in a single reaction phase (Kamaly et al., 2016). The final product of the majority of synthesis techniques is an aqueous suspension of colloidal particles (Jawahar et al., 2012).

2.2.1 Solvent Evaporation:

A solvent evaporation method was used initially for the synthesis (Desgouilles et al., 2003). Figure 2.3 depicts the entire procedure. Initially, a polar organic solvent is utilized to create an organic phase, into which the polymer is immersed prior to the addition of the active component.

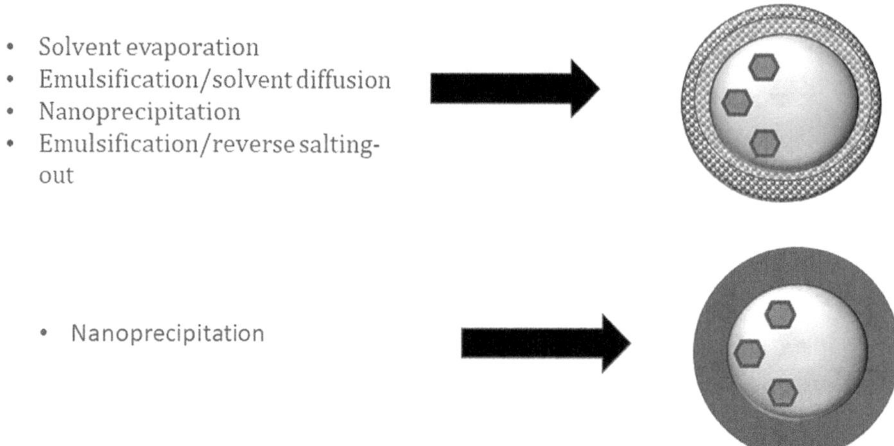

FIGURE 2.2 Preparation Techniques for Polymeric Panoparticles. (Figures Compiled by the Authors.)

Chloroform and dichloromethane are frequently utilized, albeit more so in the past (Grumezescu, 2017). Ethyl acetate now replaces them (Bohrey et al., 2016), which is safer and better suited for use in biomedical applications (Christine et al., 2017). A typical aqueous-phase formulation also contains a surfactant, such as polyvinyl acetate (Grumezescu, 2017). The surfactant will make an emulsion of the organic solution in the water phase and the mixture will get ultrasonification to be mixed instantly (Sharma et al., 2016). The resultant nanodroplets will be prepared and, after the evaporation of the organic solvent, the NPs become suspended in the aqueous solution. Purification, centrifugation, and freeze-drying of the NPs are all possible after solvent evaporation to increase shelf life. This method can be utilized to produce nanospheres (Szczęch et al., 2020).

2.2.2 Solvent Diffusion

An oil–water emulsion is produced by mixing a fluid that contains polymers and drugs and is only partially miscible with water with an aqueous solution that contains surfactants (Kumar et al., 2012; Souto et al., 2019). To ensure that both phases of the emulsion are initially in thermodynamic equilibrium at room temperature, the internal phase of the emulsion comprises a water-saturated organic solvent with modest hydro miscibility, such as benzyl alcohol or ethyl acetate (Souto et al., 2012). Colloidal particles are created when the solvent diffuses from the molecules that are dispersed to the outside and the mixture is then diluted with water. This process is frequently used to create nanospheres, but it can also create nanocapsules if a tiny amount of oil is added (Guterres et al., 2007). Figure 2.3 displays the schematic for this approach. Although there is a chance that the hydrophilic drug will penetrate the aqueous medium and the liquid must be stopped in large amounts, this method is widely used to make polymeric NPs (Quintanar-Guerrero et al., 1998; Vasile., 2018).

2.2.3 Reverse Salting Out

The aforementioned emulsification/solvent diffusion technique is regarded as a variation of the emulsification/reverse salting-out technique. The salting-out method uses salting-out action to separate a hydromiscible solvent from an aqueous solution, which could lead to the production of nanospheres (Wang et al., 2016). The main distinctions are the creation of the o/w emulsion, which uses a water-miscible polymer solvent such as acetone or ethanol, and the addition of a

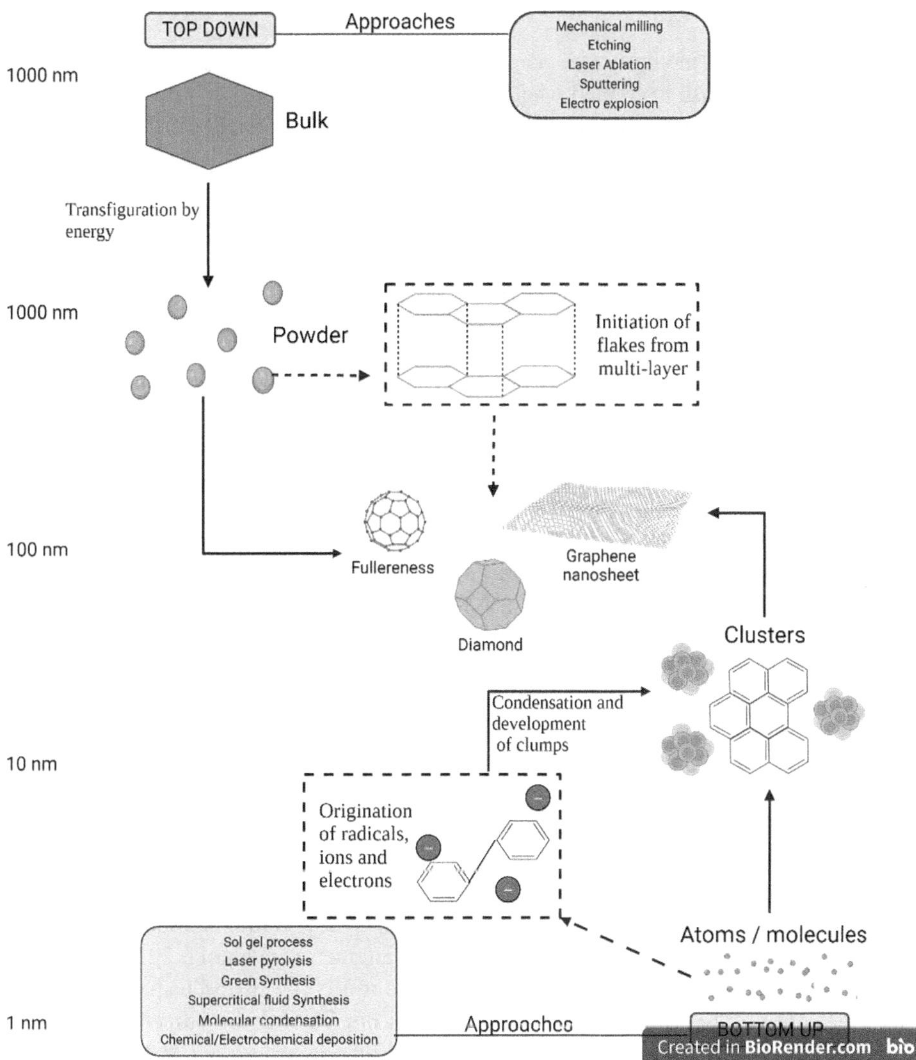

FIGURE 2.3 Techniques for the Synthesis of Nanomaterials. (Figures Compiled by the Authors.)

gel, the salting-out agent, and a colloidal stabilizer to the aqueous phase (Lim, 2018). Salting-out chemicals can include electrolytes like sucrose and non-electrolytes like magnesium chloride, calcium chloride, and magnesium acetate (Pal et al., 2011). It is feasible to diminish the miscibility of acetone and water by saturating the aqueous phase, allowing the other miscible phases to come together to form an o/w emulsion (Vauthier and Bouchemal, 2009; Zielińska et al., 2020). The o/w emulsion is created by forcefully combining the two components at room temperature. In order for the polymer to settle and form nanospheres, the organic solvent must first diffuse to the exterior phase of the emulsion. Thereafter, the emulsion is diluted with an aqueous solution. Cross-flow filtering is used to get rid of the remaining solvent and salting-out agent. The execution process is made simpler by perfect miscibility between the organic solvent and water, even if it is not required (Wang et al., 2016; Krishnamoorthy et al., 2015). With this method, nanospheres with a diameter of 170 to 900 nm were created; moreover the average particle size can be changed by varying the polymer concentration in the interior phase to the external phase (Crucho, 2017).

2.2.4 Nanoprecipitation

The solvent displacement method is the name given to this procedure, which calls for two miscible solvents. The internal phase is created when a polymer is submerged in a miscible organic solvent like acetone or acetonitrile (Canadas et al., 2016; Sánchez-López et al., 2018). They are immiscible in water, which makes them simple to evaporate away. This method emphasizes the stacking of a polymer on an interface and is based on the displacement of the organic solvent from the lipophilic solution to the aqueous phase. The polymer solution in a water-miscible solvent with a medium level of polarity is added to an aqueous solution while being agitated, either drop by drop or in a predetermined order. NPs immediately develop as a means of avoiding the water molecules when the polymer solution travels fast into the water phase on its own (Salatin et al., 2017, He et al., 2012). When the solvent is released from the nanodroplets, the polymer congeals into nanocapsules or nanospheres. Typically, the organic phase is introduced after the aqueous phase, although the procedure can alternatively be carried out backward without compromising the ability to create nanoparticles (Rivas et al., 2017). Surfactants can be used during the process to keep the colloidal suspension stable, but they are not required for the creation of nanoparticles. In comparison to emulsification solvent evaporation, this process typically produces nanoparticles that are more consistent in size and shape (Bilati et al., 2005). The process of nanoprecipitation allows for the generation of nanospheres or nanocapsules in addition to polymeric NPs with dimensions of around 170 nm (Chidambaram, 2014; Guterres et al., 2007). Nanospheres are created when the active component is mixed with the polymeric solution. By merging the internal and external components of an emulsion, nano capsules are created by dispersing the medication in an oil and then emulsifying it in an organic polymeric solution (Salatin et al., 2017; Rivas et al., 2017).

2.3 NANOMATERIALS

As soon as nanostructures were found in ancient meteorites, people became more interested in nanomaterials. Gold nanoparticles (NPs) were made for the first time in 1857. They were used in colloidal solutions to colour glass. Fume-silica NPs were made in the 1940s and subsequently commercialized in the United States (Feynman, 1961). Nanomaterials are materials with dimensions on the nanoscale, typically less than 100 nanometres (Ventola, 2012). They have also succeeded in getting enough funding, with over $7 billion being invested in this segment every year. They are currently employed in a broad spectrum of business goods and processes. Some nanomaterials come from nature, while others are made in a lab to serve a specific purpose. They can capture and move other molecules as they possess unique and useful attributes, such as being small, having strange physical properties and causing the process to speed up. They can also perk up drug delivery, create more efficient solar panels, and enhance the performance of electronic devices (Bhushan, 2017).

2.3.1 Synthesis of Nanomaterials

Consideration of the feasibility of synthesizing nanoparticles goes back quite some time. Until now, the majority of NMs have been developed as conceptual proof, and it has been challenging to bring them from the lab to clinical use (Mclaughlin et al., 2016). At the end of 2017, the FDA authorized 50 nanoparticle-containing drugs. Many nanomaterials may be manufactured using different techniques (Ventola, 2017). They consist of colloidal NPs, nanoclusters, nano powders, nanotubes, nanorods, nanowires, and so on. With a few modifications, nanomaterials can be manufactured using existing processes. Nanomaterials may currently be manufactured by a vast array of physical, chemical, biological, and hybrid techniques. The needed quantity, the targeted size of 0D, 1D, or 2D nanomaterials, and the synthesis technique all have a role (Khan et al., 2022). Typically, nanoengineered materials are synthesized using either a bottom-up or a top-down technique.

- **Bottom up**

  The production of nanostructures is the result of the bottom-up process, which starts with the atomic-level reduction of material and ends with self-assembly. This approach has benefited both quantum dots and NPs created during epitaxial growth and colloidal dispersion, respectively.

  (Khan et al., 2022)

(a) Chemical vapor deposition
This technique includes coating of a substrate with a thin layer of vapor-phase precursors. Consideration is given to the acceptability of precursors based on their volatility, chemical purity, evaporation stability, cost, and safety (Jones et al., 2009). Primarily, carbon-containing gas is supplied gradually. High temperature is implied to break down the gas to liberate carbon atoms that reassemble on the substrate to create carbon nanotubes (Shah and Tali, 2016). Because of its efficiency and dependability in the production of high-quality nanomaterials, this method is commonly employed to manufacture two-dimensional nanoparticles (Machac et al., 2020).

(b) Solvothermal and hydrothermal methods
The hydrothermal process is among the most often employed procedures for developing NM, together with the solvothermal method. In the hydrothermal process, a heterogeneous reaction is conducted under controlled circumstances in an aqueous medium near the critical point, resulting in the generation of nanostructured materials (Baig et al., 2021). Compared to the hydrothermal method, this process will be done in a non aqueous medium and all other aspects are identical (Chen et al., 2010). Both are often performed in enclosed spaces. These techniques for producing nanowires, nanorods, nanosheets, and nanospheres are both intriguing and useful (Dong et al., 2020; Chai et al., 2018).

(c) The sol–gel method
This is said to be a common wet chemical method for the production of nanomaterials. During the synthesis of metallic nanoparticles, the liquid material will be first converted into a sol which will be converted to a gel-like structure encoring a solid core inside. To produce sol, hydrogenation of the sol is produced by using alcohol or water (Danks et al., 2016; Tseng et al., 2010). Drying will be performed to remove the organic solvent from the gel which will reduce the porosity and increase the distance between the colloidal particles which is followed by the final stem, that is, calcinations (Parashar et al., 2020). Using this method, a wide range of excellent nanomaterials made from metal oxides are produced (Danks et al., 2016).

- Top down

  Top down permits external control of the macrostructure during the fabrication of the required nanostructures. Techniques such as mask-breaking etching, ball milling, and severe plastic deformation fall under this group. The imprecise surface structure is a significant drawback of this method. Because of the high aspect ratio of this approach, surface defects may impact the physical and surface properties of the NPs (Khan et al., 2022).

(d) Mechanical milling
The mechanical milling of mass-to-nanoscale materials is a cost-effective approach. It is an effective method for producing nanocomposites since it allows the mixing of various phases (Zhuang et al., 2016). It is used to produce a broad range of nano-materials, such as aluminium alloys reinforced with oxides and carbides. Carbon nanoparticles produced by ball milling represent a new class of nanomaterials with potential uses (Yadav et al., 2012; Wender et al., 2013).

(e) Electrospinning
Electrospinning is the most commonly used top-down method for creating nanostructured materials. Although a variety of materials can be used to create nanofibers, polymers are

the most common (Ostermann et al., 2011). One of the most important advancements in the history of electrospinning is coaxial electrospinning. It creates core-shell nanoarchitectures in an electric field using a spinneret made of two coaxial capillaries filled with two viscous liquids. It is a quick and efficient way to mass-produce extremely thin core-shell fibres. These incredibly thin nanomaterials can have diameters of a few centimetres (Kumar et al., 2014; Baig et al., 2021).

2.3.2 Classification of Nanomaterials

Two characteristics that may be used to classify nanomaterials are size and shape. There are four distinct dimensions—zero, one, two, and three—in which nanomaterials may be categorized (Singh et al., 2021).

1. Zero dimensional
   0d nanomaterials are NPs with a diameter of between 1 and 50 nm. In addition to cube and polygon shapes, 0D nanomaterials have also been found in cube and polygonal forms. Nanoparticles composed of precious metals (gold, palladium, platinum, and silver) or semiconductors (quantum dots) are examples of 0D nanomaterials.
2. One dimensional
   The one-dimensional (1D) nanomaterial has a size range of 1–100 nanometres. One-dimensional nanomaterials including nano- wires, fibres, rods, and tubes are common. Many materials, including metals (Au, Ag, Si, etc.), metal oxides (ZnO, TiO_2, CeO_2, etc.), and others, may be found in quantum dots.
3. Two dimensional
   This category has nanoscale dimension and 1D possesses macroscale dimensions. It may have a consistent nanometre-scale thickness and a surface area of few micrometres square.
4. Three dimensional
   There are no nanoscale dimensions in 3D nanomaterials; all dimensions are macroscale. Bulk materials are 3D nanomaterials made of distinct blocks with sizes ranging from the nanometre scale (1–100 nm) to much greater sizes.

2.3.3 Properties of Nanomaterials

Nanostructured materials are utilized in a wide range of industries, and their use is expanding quickly owing to their outstanding qualities and phenomenon structure-related properties (Edelstein, 1998; Bakirhan et al., 2017).

(a) Structural properties
- Clusters
  The crystal structure of clusters with a maximum of 20 atoms is unique from that of larger clusters. The lattice constant is reduced or a metastable structure is observed in bigger crystals. By balancing the surface energy and elastic energy, the free energy is minimized, and as a result, the lattice constant is decreased.
- Nanocrystalline solids
  The structure of the interfaces dictates the properties of nanocrystalline solids, which feature a high fraction of interface atoms. The structure of nanocrystalline interfaces has been outlined from two perspectives. A theory postulated that atoms at interfaces are "gas-like", indicating that they are more disorganized and possess lesser density than atoms in the same position in micrometre-sized polycrystalline materials. Another view is that the grain boundaries resemble those of a coarser-grained material. Recent studies imply that

the grain boundaries of nanomaterials are like those of coarse-grained materials, with key processing-specific details separating them.

(b) Thermal diffusion
They may have thermal diffusion rates several times quicker than bulk materials. Diffusion is improved by increasing voids and grain boundaries. The elimination of voids during sintering significantly lowers diffusion.

(c) Physical
The physical characteristics of nanomaterials are different from those of bulk materials, and these properties are both spectacular and unique.
- Colour
When materials are processed into different forms, they may acquire a new hue. For instance, when gold is reduced to nanoscale form, its colour changes to crimson. The relationship of AuNPs with light is significantly determined by particle diameter.
- Melting point
As particles are reduced to the nanoscale region, the melting point drops significantly. In addition, enhanced ratio of surface and volume of nanoscale materials reduces their melting point by hundreds of degrees in comparison with bulk materials.

(d) Chemical
The chemical composition aids in understanding how NPs interact with one another. The surface properties of the nanoparticles may vary due to the higher surface area, small size which leads to the production of highly charged species with imperfections. Chemical properties of nanomaterials are listed below (Edelstein, 1998; Bakirhan et al., 2017).
- As nanomaterials have a larger concentration of surface atoms than bulk materials, their atoms possess more energy.
- The interactions between nanoparticles (NPs) and their surfaces are adjustable.
- Due to their huge surface area, nanomaterials can be efficient catalysts.
- On non-metallic substrates, heterogeneous catalytic processes are often carried out by utilizing metal or metal oxide particles. If the reaction is excessively violent, there will be a propensity for particles to coalesce, since the mobility of the atoms result in the rise of particle size.

For instance, platinum nano crystallites are at their highest excited phase for oxygen reduction on carbon at around 3 nm in size. Catalytic particles are prevented from clumping by microporous scaffolds, for example, aluminosilicate zeolites.

(e) Optical properties
Nanomaterials have widely divergent optical characteristics from their composites. Particles' primary optical behaviours typically involve light scattering and its rate of absorption. The phenomenon of surface plasmon resonance is fully accountable for this, as unpaired electrons within the conduction band get excited coherently, resulting in in-phase oscillation. It depends on the size of the particle because it gets activated if the particle's size is smaller than the wavelength of radiation, making it responsible for the change in colour of metallic nanoparticles (Bakirhan et al., 2017). Due to its excellent light-scattering capabilities, TiO_2 has found widespread use as a nanomaterial in a wide variety of applications, including inks, films, fibres, and so forth.

2.4 TOXICITY OF NANOMATERIALS

Over time, nanotechnology has developed substantially, and, constantly, new uses for nanomaterials are discovered. The great bulk of research is devoted to the production and fabrication of novel nanomaterials, with less than 1% of studies concentrating on the destiny and biological consequences of nanoparticles (Mahmoudi et al., 2011). Despite all the hoopla around nanomaterials, it is still unclear how nanoparticles affect biological organisms physicochemically (Maynard et al., 2006; Khan et al., 2016). There are numerous ways for these nanostructured materials to enter and stay in the body. The size, shape, surface area and surface charge can be responsible for the toxicity of nanoparticles. They stay for a long time, enabling movement through epithelial and endothelial cells as well as other macromolecules (Aillon et al., 2009; Papp et al., 2008). The drugs contained in NPs form a clump-free, sustained, and hydrophilic material due to their extraordinarily high surface-to-volume ratio. To prevent the oxidative damage which is connected with the pathophysiology of neurological disorders, like Alzheimer's disease, Parkinson's disease, Huntington's disease, and Wilson's disease, nanocarrier-based medicine delivery methods are currently being studied (Re et al., 2012, Bansod et al., 2017). These cells live in the circulation because macrophages are unable to engulf them. The lack of toxicological data may impede the safe development of nanomaterials. However, the majority of intracellular and in vivo nanomaterial toxicity is caused by the production of too many reactive oxygen species (ROS), which can occur in the body through a number of different mechanisms (Nel et al., 2006; Moller et al., 2010; Unfried et al., 2007). Due to its high concentration of easily peroxidized unsaturated fatty acids, the central nervous system (CNS) is particularly susceptible to the toxicity of reactive oxygen species (Adibhatla et al., 2010). The apoptosis, autophagy, and necrosis pathways in the nanotoxicology profile are associated with mammalian cell death (Kroemer et al., 2009). Testing for toxicity can be done on living organisms like fish, mice, or rats as well as cell cultures (in vitro) (Kumar et al., 2017; Hadrup et al., 2019). Nanotoxicity has been evaluated in vitro, in vivo, and in silico; however, due to a lack of proper correlation, it is challenging to analyse the toxicity of nanomaterials. Employees must be sufficiently safeguarded from breathing nanoparticles because the long-term effects of exposure to nanomaterials during the creation of nano-based products are still unknown. Therefore, a range of physicochemical characteristics can be changed to significantly modify the toxicity of nanomaterials.

2.5 CHARACTERIZATION, PROPERTIES, AND MECHANISM OF NANOMATERIALS

Given the close relationship between morphological characterization and experimental execution evaluation, the former is clearly simpler than the latter. Wide-angle X-ray scattering (WAXS), which is straightforward to utilize, is commonly used in these kinds of investigations. However, these evaluations are not measurable and are prone to error (Davis et al., 2003; Morgan et al., 2003). In nanocomposites platelet separation, or d-spacing, is represented by a single peak; numerous reflections, as anticipated by Bragg's law, may result in the development of additional peaks. The drawbacks of this production process are illustrated by the incomplete exfoliation of injection-molded polyolefin nanocomposites. The finished product can be seen in X-ray scans as having tactoid peaks. It is possible that repeated readings of the milled surface towards the centre of the bar will not reveal a peak because the tactoids in these samples grow more irregular as they go closer to the as-moulded surface (H.S. Lee et al., 2005; Kim et al., 2007). Finding the peak, however, primarily requires a more accurate scan. According to the WAXS study, the peak position may change in relation to the organoclay in its pure form. Many people view the peak, which can change to smaller angles or wider d-spacing, as evidence that polymers have been "intercalated" into the galleries (LeBaron et al., 1999; Mai et al., 2006; Manias et al., 2001). It is also possible for the reverse to occur, which is often attributed to the removal of free surfactant from the gallery or to its degradation (Yoon, 2003; Paul, 2008). Perception is challenging because various phenomena could take place at the same time. The efficacy of these useful materials does not seem to be affected by intercalation. Small-angle X-ray scattering (SAXS), according to some scientists, may yield increasing amounts of quantitative data (Lincoln et al., 2001; Acosta-Torres et al., 2011; Schaefer, 2007). Only a small number of laboratories, most likely because they lacked the necessary SAXS equipment or skilled employees to analyse the results, have extensively used this approach. Several methods have been used to better understand clay dispersion, including neutron scattering and solid-state nuclear magnetic resonance (Bourbigot et al., 2004; Davis et al., 2003). Transmission electron microscopy (TEM) clearly reveals the nanocomposite material's electron structure. Although time-consuming and labor-intensive, this method may yield quantifiable data. A number of scientists are skeptical of TEM due to the limited observing area that displays the morphology. One way to do this is to take pictures from many different angles until they find a good one that shows the shape of the object correctly. The biggest problem with getting good TEM pictures is getting uniformly microtomized pieces that are big enough to show the shape. As the clay's elemental constitution is distinguishable from the polymer matrix's, no staining is required. In the case of nylon 6 and the w1 nm thick clay platelets, exfoliation is nearly complete at this point and may be seen as black lines when platelets are cut perpendicularly. Image analysis is employed to quantify the length distribution of platelets, but it takes analyzing several hundreds of particles to generate meaningful data (Fornes, 2003; Chavarria, 2004; Cui, 2007). The observed measurement indicate a disorganized slicing of an uneven platelet, therefore the maximum diameter is rarely observed (Corté and Leibler, 2005). Particles comprising two or more platelets may be seen even in the finest nylon-6 nanocomposites since exfoliation almost never proceeds in its entirety (Fornes, 2003). This platelet distribution could appear asymmetrical when compared to another (Chavarria, 2004). This may make certain platelets seem much longer than they are. These factors should be taken into account when assessing quantitative evaluations of a particle's aspect ratio and comparing actual performance to that anticipated by the composite concept (Fornes, 2003). Nanocomposites made of polyolefins and other polymers are observed to be thicker than single clay platelets, which results in a reduced rate of exfoliation (Stretz et al., 2005; Stretz et al., 2006; Kim et al., 2007; Cui et al., 2007). When the polymer-organoclay affinity is enhanced by providing a compatibilizer like PP-g-MA or by increasing the amount of a polar comonomer like vinyl acetate, the clay particles become narrower and shorter in length (D. Lee et al., 2005; Kim et al., 2007; Cui et al., 2007; Chavarria et al., 2007). A higher aspect ratio is typically associated with improved performance, if that particle's width decreases at a faster rate than its length. This may occur under extreme conditions, although falling particle sizes do not always reflect processing errors or the addition of clay platelets (Chavarria et al., 2007). Contrary to popular belief, tactoids do not have platelets of the same length that register together. There have been statements showing a more accurate picture of a tactoid, in which the particle length may be noticeably greater than that of the original one and there is development with growing dispersion (Kim et al., 2007; Cui et al., 2007). When particles are grouped by length and thickness, it is harder to figure out what their average aspect ratio is. To proceed, determine a weighted average of the distribution's numbers (Fornes., 2003; Cui et al., 2007). Furthermore, before computing the aspect ratio, one may compute the means of the lengths and the thicknesses separately (Cui et al., 2007). There isn't any conceptual guidance that it is either a better predictor of performance or a tool for composite modeling (Kim et al., 2007; Cui et al., 2007).

Clay platelets may help the mechanical, thermal, or barrier properties of nanocomposites by keeping them straight so that they can be as strong as possible. The way the test sample was made—by extrusion, injection molding, or some other method—has an effect on how the particles are arranged. Another problem that is often encountered during the mixing process is the level of exfoliation. Measurements taken from specimens with a high aspect ratio but not precisely oriented or flattened by processes like compression molding occasionally reduce the performance. TEM evaluates particle alignment and shape, which can be potentially integrated into the appropriate models to realize how these attributes impact performance (Yoon et al., 2003; Stretz., 2006; Weon et al., 2005; Masenelli-Varlot., 2007).

2.6 SURFACE CHEMISTRY OF NANOMATERIALS

The significance of the surface chemistry of nanomaterials has come to the forefront of nanoscience. Nanocarriers have been produced using a variety of materials. Certain sizes, shapes, compositions, surface charges, and biocompatibility are established for extended release (Schaefer, 2010). Different nano impacts on nanomaterials are provided by size and surface variables. The fundamentals of creating metal nanocrystals with regulated forms have been thoroughly addressed in a number of recent papers (Navya et al., 2019). Researchers have discovered that the stability ratio and the beginning of a doublet are significantly influenced by the size and volume fraction of nanoparticles. The size of the nanomaterials influences how certain tissues are treated as well as how cells take up medications. Their stability, rate of drug release, and dosage of loaded medication are also impacted (Karakoti et al., 2006). When it comes to the distribution of medications, shape is just as crucial as size. The synthesis of metal nanocrystals via shape control has attracted increasing scholarly interest because of their shape-dependent optical characteristics. The cubic crystal form with the face in the middle is typical of valuable metals. It is possible to modify the shape of metal nanocrystals through the surface coordination of tiny ligands, which aids in the creation of metal nanocomposites by lowering their surface energy (Liu et al., 2017). The surface charges of the nanoparticles have a significant role in how they interact with one another at the nano-bio interface. Nanomaterials' capacity to penetrate cells is controlled by their surface charge (Navya, 2016). Within the human body, phospholipid membranes are where particles are found most frequently. It has been found that NPs can traverse the blood-brain barrier, with the BBB's stability and ability to let substances into the brain being affected by the surface charge (Karakoti et al., 2006). Additionally, in vivo studies show that liver cells absorbed positively charged NP, whereas cancer sites collected negatively charged NP. Additionally, surface area significantly affects how materials interact with living cells (He et al., 2010). NPs are more reactive than their bigger counterparts due to their huge surface area to volume ratio (Karakoti et al., 2006). Its physicochemical properties have an impact on how well medications are distributed throughout the body and absorbed by cells. The absorption and dispersion of nanomaterials inside cells are also significantly influenced by a variety of additional characteristics. Surface-to-volume ratio, crystal planes, stability, nanoparticle-protein interactions, chemical composition, accumulation phases, cell types, and cell treatment may all influence the uptake and distribution of nanoparticles by cells.

2.7 APPLICATIONS OF NANOMATERIALS

2.7.1 Application of Nanotechnology in Food

Nanotechnology has become a captivating technology in many industrial sectors including food, health, textile, and so forth, in the last few decades. With a scale of roughly 1–100nm, this nanometer-scale technology deals with macromolecules, molecules, or atoms to produce food materials with unique characteristics (Singh et al., 2017). The increasing consciousness of consumers about the quality and health benefits of foods has created a need for foods with enhanced/escalated quality while disrupting the nutritional value as low as possible (Pathkoti et al., 2017). Nanotechnology has been seen offering various food solutions from manufacturing to processing to packaging, but its use in the food sector is mainly categorized into food nanosensing and nanostructured components (Ezhilarasi et al., 2013).

Food nanostructured components or ingredients in the food processing industry are usually utilized as food additives, nutrient carriers, antimicrobial agents, anti-caking agents, the increased mechanical strength of foods, and packaging material durability (Afzaal et al., 2022). Food nanosensing is utilized to attain better food safety and quality (Jagtiani, 2022). Its application helps enhance the shelf-life of different food materials addressing food wastage issues in India since India's food processing sector is underdeveloped and lacks the use of higher preservation and storage techniques (Pradhan et al., 2015).

2.7.2 Nanotechnology Application in Food Processing

Because it appears to stabilize food ingredients and reduce waste, the use of nanotechnology in the food processing industry is warranted (Pradhan et al., 2015). Nanocarriers are essential for transferring food additives in food products without altering the fundamental shape of the food, as noted by Ezhilarasi et al. (2013). Since particle size can affect a bioactive compound's transport to various locations within the body in some specific cell lines, submicron nanoparticles are an alternative (He et al., 2019). In comparison to the conventional encapsulation method, nanoparticles exhibit improved features for the efficiency of encapsulation and release (Rivas et al., 2017). Protecting food components from heat and moisture, controlling the release of active agents, regulating interactions between active components within the food matrix, and ensuring the availability of a certain rate and a target duration are all achieved through nanoencapsulation (Singh et al., 2017).

Currently, amorphous silica, silicon dioxide (SiO_2), titanium oxide (TiO_2), and other nanoparticles have been widely employed as food additives (Hayes and Sahu, 2017). Inorganic, water-insoluble titanium dioxide (TiO_2) has been manufactured as a white powder for more than a century and is used as a white pigment in paints, varnishes, printing inks, polymers, cosmetics, and other products (Gazsó and Pavlicek, 2020). Since 1966 and 1969, respectively, it has been used as colouring in cake frosting, chewing gum, cheese, candles, and powdered sugar on doughnuts (Westervelt, 2015; Blaznik et al, 2021).

Different delivery systems which are protein-based or lipid or carbohydrate-based have been used for improved nutrient bioavailability and enhanced stability of the nanoparticles (Pateiro et al., 2021). The addition of nanoparticle additives in food products considers different criteria like the synthesis pathway, structure of the nanoparticle, amounts, purity, and toxicity absence. Currently, the rutile crystalline phase of TiO_2 has been synthesized through chloride and sulfate processes, however, the former is more sustainable as it provides more narrow-sized particles (Ropers et al., 2017).

However, it is important to note that TiO_2 has no physiological or nutritional benefit and can only be taken with a maximum concentration of 1 per cent of the total food weight (Gazsó and Pavlicek, 2020). "Synthetic amorphous silica (SAS)" generally meets the specifications to be used as food additives but its constituent particles are larger than 100 nm (Fruijtier-Pölloth, 2016). SAS is produced in precipitated or pyrogenic forms of nanomaterials and used in food processing; however, its oral intake seems to pose a health risk (Aureli et al., 2020). Liu et al. (2020) and Yu et al. (2015) noted that the application of Nano-ZnO on Green soybean prevents the growth of microorganisms in coliforms, bacteria, molds, and yeast.

ZnO nanoparticles have been employed as a biocompatible material and can control food-borne pathogens (Sridhar et al., 2021). ZnO nanoparticles seem to increase cucumber quality by 36 per cent (Astafurova et al., 2017). The production of ZnO nanoparticles takes vapor phase and solution-based approaches, as the structure of nanoparticles can be manipulated by different materials, solvents, and reaction conditions. Microemulsion, hydrothermal, solvothermal, precipitation, and sol-gel processes have been used as solution-based approaches for the formation of NPs (Ong et al., 2018; Zorraquı´n-Pen˜a et al., 2020; Li, et al., 2020; He et al., 2019).

The combination of Cu+Se nanoparticles seems to enhance the chlorophyll content and overall yield of tomatoes by altering the enzymatic activity of the plant of tomatoes (Hernández-Hernández et al., 2019).

- **Encapsulation**

Bioactive compounds can be observed in vegetables, cereals, fruits and pulses, however, they suffer from the limitation of being chemically unstable, and vulnerable to oxidative degradation due to light exposure (Hu et al., 2020). Encapsulation technology has been highly utilized in the pharmaceutical and food-processing industry to entrap the micronutrients, polyphenols, enzymes, and antioxidants so as to protect them from harsh degradation conditions (Devi et al., 2017). They help in controlling the release, increasing bioavailability, and targeting precise bioactive compounds, and Shishir et al. (2018) and de Souza Simões et al. (2017) have noted that they further contribute to reducing undesirable flavors, volatility loss or evaporation, forming solid particles, and improving biological activity, physical stability, and bioactive compounds' shelf life.

Synthesis of nanoparticles encompasses two techniques, the "top-down" technique which utilizes energy for nanonization, and the "bottom-up" technique which works on physicochemical aggregation or association of monomers, molecules, ions, or atoms (Pateiro et al., 2021). "Emulsification", "High-Pressure Homogenization", and "Solvent Evaporation" processes come under the "top-down" techniques which focus on the reduction of participle size by encapsulation. In emulsification two immiscible liquids blend together with a surfactant with the formation of droplets (nano/ultrafine/mini/submicron emulsions) whose size (20–200 nm) can be manipulated by the synthesis technique, components, emulsions, and so forth (Paredes et al., 2016, Homar et al., 2007).

In contrast, the "bottom-up" technique works on particle size accretion by processes like coacervation, nanoprecipitation, supercritical fluids, and inclusion complexation. Phase separation of polyelectrolyte from a liquid-liquid solution in the coacervation process results in the deposition of a new coacervate phase (dense polymer-rich phase) whose strength is further increased by enzymatic or chemical cross-linking (Paredes et al., 2016). The process of nanoprecipitation involves solvent displacement by continuous emulsification of polymer and organic solvent containing internal phase into an external aqueous phase. The process is accompanied by polymer precipitation and organic solvent diffusion from organic solution and in an aqueous medium respectively (Pineda-Reyes et al., 2021). The use of solvent media for encapsulation can cause physicochemical fluctuation, toxicity, and microbial contagion, and are not economic, therefore the solvents are removed by lyophilization or spray, or freeze drying.

The application of nanotechnology can be observed in a sugar-based amphiphilic scorpion along with star-like nanomaterials that have been designed as a core-shell micelle appropriate for drug encapsulation (Bhardwaj and Kaushik, 2017). These bioactive shells seem to possess inherent targeting qualities which can be utilized to attain the targeted drug delivery to prevent scavenger receptors from limiting Parkinson's, atherosclerosis, and other diseases and treating cancer (Lewis et al., 2016).

2.7.3 Applications in Food Packaging

Nanomaterials in the food packaging sector provide several benefits including better mechanical barriers, improved nutrient bioavailability, and identification of microbial contamination (Hayes and Sahu, 2017). The application of ZnO and MgO NPs can be observed in food packaging along with Amorphous silica. ZnO, MgO, TiO_2, and SiO_2 are used in food packaging for their higher optical, antibacterial, and UV shielding abilities (Swaroop and Shukla, 2019; Garcia et al., 2018).

Nanomaterials provide a wide scope for food packaging applications, by active, smart, improved, and bio-based packaging. Improved packaging focuses on enhancing the physicomechanical properties of packaging like flexibility, mechanical strength, humidity, temperature resistance, and properties of gas barriers. Composites of clay nanoparticles reduce 80–90 per cent of CO_2 and O_2 permeability thereby ameliorating barrier properties. Nano laminates, nano-coating, and nano clay materials are widely used for improved packaging.

The amorphous areas of polymers entail free volume that enables the transmission of microscopic contents like microorganisms, atmospheric gases, and water vapor, which in turn promotes the degradation of organoleptic or nutritional characteristics of packed food because of oxidation (Swaroop and Shukla, 2019). A greater reinforcement of the MgO NPs results in the deterioration of qualities regarding water vapor barriers. It has been further noted that MgO NC films are capable of lowering the transmission of oxygen molecules to protect the quality of food products. Hoseinnejad et al. (2018) have referred to the application of TiO_2 in food packaging due to its durability, abundance, and cost-efficiency. However, TiO_2 is graded as GRAS material by FDA as it can cause cytotoxicity and genotoxicity in human cells as noted in toxicology studies (Hoseinnejad et al., 2018).

The use of silver nanoparticles cannot be ignored as it helps in reducing infection rates and is used in commercial products due to high thermal resistance, a variety of antimicrobial performances, and usage safety (Jaiswal et al., 2010). Some companies use silver nanoparticles in their packaging to improve shelf-life and prevent microbial growth in perishable food products (Wesley et al., 2014). Emamifar and Bavaisi (2020) have referred to the application of a bio-nanocomposite coating with nano-ZnO and sodium alginate on strawberries. Nano-ZnO increases moisture barriers in strawberries to reduce weight loss as the study outcome reveals that nanoparticles coated fruits have lesser weight loss than uncoated fruits (Jafarzadeh et al., 2021). Ti and Ag NPs further help in improving the water barrier characteristics in PLA films (Chi et al., 2019). Further, it has been noted that nano-TiO_2/nano-Ag and essential oils in mangoes help in reducing weight loss of mangoes managed with nano packaging film among all the other packaging (Jaiswal et al., 2019).

2.7.4 Biomedical Applications of Nanotechnology

The application of nanomaterials can be noted in drug delivery systems, diagnosis, implants, and prostheses (Ramos et al., 2017). Nanoscale materials are integrated into biomedical devices since the majority of biological systems are nanosized. These materials are generally used by the biomedical sector to generate nanotechnology products in the form of carbon nanotubes, liposomes, metallic surfaces, and inorganic, as well as, metal nanoparticles (Liu et al., 2016). Bhardwaj and Kaushik (2017) noted that transition metals like silver, platinum, gold, iron, cobalt, lanthanide, titanium, and technetium have special electrical, magnetic, and optimal properties among inorganic nanomaterials making them a better choice for multipurpose biomedical applications in electrical and optical sensing, diagnosis, optogenetic, and photo-thermal therapy. It has been further noted that nanotechnology and nanomaterials in association with stem cell biotechnology entail a better impact on regenerative medicine (Perán et al., 2012).

- **Applications in Diagnosis**

Effective disease prevention and treatment need up-to-date diagnostics along with the monitoring of the treatment efficacy with better detection of pathogens (Qasim et al., 2014). Now molecular diagnostics methods are being used widely to diagnose and control infections like HCV and HIV/AIDS since they have higher sensitivity and specificity than any other ELISA-based diagnostics (Naidoo et al., 2016). The term "nanodiagnostics" refers to the application of techniques and methods of nanotechnology and its key characteristics for diagnostic purposes (Jackson et al., 2017). Nanotechnology-based biochips have grown extremely popular for different types of nanodiagnostics that enable DNA analysis and are particularly designed to associate with the high-level specificity of cellular components (Alimardani et al., 2021). Point-of-care techniques and imaging are the most particular areas that could benefit from NPs application. A contrast agent could be used in terms of imaging to associate tumors with the bonding of NP surface ligands with the cancer biomarker which can be used to diagnose (Bayford et al., 2017). Zhang et al. (2019) have discussed the benefit and opportunities in utilizing nanotechnology to diagnose cancer since early diagnose is the only key factor in treating the disease. Nanotechnology helps in overcoming the limitation associated with conventional diagnostic procedures with better specificity, sensitivity, and multiplexed measurement capacity. However, nanotechnology has yet to be used clinically in cancer diagnosis despite having a variety of medical screening and tests taking place in the area like the use of gold NPs in pregnancy tests at home (Zhou et al., 2015).

The key technique in detecting cancer cells depends on the efficiency of the bindings of NP and the molecules or aptamers attached to the cancer cells. CTCs are further used later to determine the size, density, deformability, and physical properties of cancer cells (Zhou et al., 2015).

Nanotechnology in MRI has been noted where gold-based NPs have been used to diagnose diseases since they don't affect the blood or tissue contrast like iron. Pérez-Campaña et al. (2013) have provided evidence on how "radiolabeled positron-emitting isotopic" NPs can be utilized for detecting biological events in tumor enzyme activities and receptor levels (Bayford et al., 2017). Karimi et al. (2016) have suggested the implication of nanotechnology in the diagnosis and therapy of coronary artery disease (CAD). Atherosclerotic plaque amasses the coronary artery lining in CAD, generating narrowing of the artery lumen, lowering vessel wall compliance, and suddenly causing blood loss in the myocardium (Pennington and DeAngelis, 2016).

- **Applications in Drug Delivery**

The blood-brain barrier (BBB) is inside the brain and restricts the entry of the antigens or pollutants to the brain (Saeedi et al., 2019). However, nanoparticles can get past a number of obstacles and take several paths to brain neurons. Lipid NPs, by virtue of their lipophilic properties, let them permeate the BBB (Gao, 2016). By coating or conjugating ligands to transit through the BBB by receptor-mediated transcytosis, NPs can reach out to specific cells (Saeedi et al., 2019).

Specific ligands, including as lactoferrin, insulin, transferrin, and others, combine to the NP surface in the transcytosis receptor-mediated mechanism, which depends on endothelial receptors on the cell surface (Kim and Feldman, 2012). Lipoprotein is one more ligand molecule that can bind to receptors (De Jong and Borm, 2008). The brain's low-density lipoprotein receptors transport the key serum proteins apolipoprotein E, apolipoprotein A, and apolipoproteins (Saraiva et al., 2016).

The contribution of nanotechnology in terms of drug encapsulation cannot be ignored as it surpasses the limitation of traditional dosage characteristics and forms like stability, odor, taste, and bioavailability (Shishir et al., 2018; Iqbal et al., 2015). The "Biopharmaceutical classification system (BCS)", has noted that 40 per cent of total market drug molecules currently used show a poor soluble rate while 90 percent of drug molecules in researched drug development lists are also recognized as poorly soluble (Loftsson and Brewster, 2010; Rivas et al., 2017). The nanoprecipitation technique, in this context, has been utilized as an alternative for drug carrier formulations in agricultural research and pharmaceutical and is based on precipitation processing (Rivas et al., 2017). Polymer precipitation takes place after the inclusion of a non-solvent in a solution of polymer in four stages including supersaturation, nucleation, rise by condensation, and, last, development by coagulation to generate polymer nanoparticles . It is mainly applied for hydrophobic encapsulation or encapsulation of hydrophobic compounds (Smith, et al., 2017; Song et al., 2001).

Lammari et al. (2020) have provided an example of using a nanoprecipitation technique for the encapsulation of essential oils as they are crucial to agriculture, cosmic, food, and pharmaceutical areas. Carvacrol, a key component of various essential oil, is known for its antimicrobial activity, limiting the development of preformed biofilms and interfering with the formation of biofilms (Liolios et al., 2009). Encapsulation of Carvacrol is conducted using "poly (lactide-co-glycolide) nanocapsules" along with the solvent displacement mechanism (Iannitelli, et al., 2011). The developed NP has been observed to have less elasticity and mechanical stability and functionality of "*Staphylococcus epidermidis*" preformed biofilms (Lammari et al., 2020).

Correia et al. (2017), have examined the scope of the microfluidic method of liposome creation which is used for drug release, encapsulation efficacy, and stability. Since liposomes provide an amphipathic environment to enable hydrophilic and hydrophobic drug incorporation and provide an effective medium for drug delivery framework (Pattni et al., 2015). Microfluidic systems in liposome creation enhance the opportunities for drug delivery with unique features which enable controlling the particle stability and size for the final liposome products.

- **Prostheses**

The application of nanotechnology has been noted in traditional prosthetic acrylic resins to improve their performance efficiency. It has been noted that major dental restorations are generated from acrylic resins dependent on heat-cured "Poly(Methyl Methacrylate) (PMMA)". The application of additional metal NPs to organic materials seems to enhance surface hydrophobicity and lower biomolecules' adherence. TiO_2 and Fe2O3 NPs have been observed to simultaneously improve the coloring and the antimicrobial characteristics of MMA resins. Liu et al. (2022), have noted the scope of nanotechnology in developing a smaller and more efficient "cortical visual prosthesis" (CVP) that is more biocompatible than traditional CVP.

- **Implants**

Microbial infection entails a serious threat to biomedical implants in modern medicine. A biofilm infection has been recognized as the major reason for biomaterial implants since all medical devices are highly susceptible to microbial infection colonization (Velu et al., 2019): 60–70 per cent of hospital-acquired infections take place due to medical devices among critically ill patients (Bryers, 2008). Bacterial biofilms could colonize the surface of both the implanted device and tissues, resulting in tissue destruction, device dysfunction, and pathogen systemic dissemination to cause serious illness. Gram-positive (*"S. aureus", "Enterococcus faecalis" "S. epidermidis", "Streptococcus viridans"*) and Gram-negative (*"Klebsiella pneumoniae", "E. coli", "Proteus mirabilis", "P. aeruginosa"*) along with yeast are responsible for the formation of biofilm on devices (Chen et al., 2013; Veerachamy et al., 2014; He and Hwang, 2016).

The application of nanotechnology has the capability to prevent the formation of biofilm. Strategies like antimicrobial coatings, QS quenchers, nanostructured coatings, polymer coatings, biosurfactants, and surface modifications are the most popular methods recognized by scholars in preventing bacterial adherence (Veerachamy et al., 2014). Nano-silver coating can be considered one of the most effective strategies in preventing the formation of biofilm since silver seems to possess antimicrobial activity and silver NPs can be integrated into implant coatings either by encapsulation or directly with cement in joint replacement applications (D. Lee et al., 2005; Veerachamy et al., 2014). Velu et al. (2019) have suggested the application of metallic nanofillers including gold or silver NPs to improve antimicrobial properties and the electrical conductivity of tissues. The integration of gold NPs in "extracellular matrixes (ECM)" seems to improve cellularity while simultaneously reducing inflammation for engineering applications of musculoskeletal tissue.

2.8 FUTURE PERSPECTIVES

Nanoparticles amalgamate different characteristics of solids including optical, mechanical, magnetic, and radiation with molecule mobility. These characteristics open up a path for possibilities for consumer products, and industrial and medicinal advancements (Singh et al., 2017). The future scope of nanotechnology has been noticed in skeletal and dental surgery by improving the chemical reaction of "nanoparticulate calcium-based biomaterials" and the generation of bioconjugated nanocrystals to promote the development of medical imaging (Liu et al., 2022; Castaño, 2011). Liu et al. (2022) have noted the scope of nanotechnology growth on neurotechnology to gain a better understanding of brain functions like speech processing, motor control, and synthesis. It is important to integrate large brain coverage areas with higher sensor density in terms of medicinal and neuroscientific applications (Anumanchipalli et al., 2019). Nanomaterial technology might be capable of improving the sensor amount; as a result, advancement in CVP occurs (Stringer et al., 2019). "Carbon nanotubes (CNTs)", "silicon nanowires (Si NWs)", and "nanocrystalline diamonds (NCDs)" can be used for the development of CVP (David-Pur et al., 2014).

Pathkoti et al. (2017) have acknowledged the significance of assessing the potential risks nanotechnology poses to human health before making them commercially available. But there is a significant lack of universal guidelines for the safety assessment of food NPs. EU guidelines highlighted the fact that food ingredients generated from nanotechnology must conduct safety assessments prior to commercial use (Cubadda et al., 2013). "The Department of Science and Technology" is mostly responsible for the research and development of nanotechnology in India and "The Food Safety and Standard Authority of India (FSSAI)" has the authority to oversee the effects of any adverse effects of nanomaterials on human health, environment, and animals . Future studies must concern with food regulations that address the impact of NPs on environmental and human health (Weir et al., 2012; Lockman et al., 2004).

2.9 CONCLUSION

Many studies have been conducted to help comprehend the physicochemical nature of polymeric NPs; however, one of the key challenges faced in their assessment is their microsize. In this chapter, we took a comprehensive look at the many potential options for bionanomaterial development. Although there are two possible approaches to getting NM, we have focused on bottom-up synthesis throughout this chapter. These methods are more flexible, allowing for the manufacture of many nanomaterials. Thus, biocompatibility is primarily sought after for biomedical applications since it is associated with synthetic techniques. We also cover other relevant topics, such as the technique behind exfoliated clay-based nanocomposites, potential biological and electrical/electronic/optoelectronic applications, and more. The constant development of effective nanocarriers that pose no threat to environmental or human health necessitates the study of nanoecotoxicology. With chemical metrology in mind specifically, quantitative and qualitative measurements may be obtained to facilitate the establishment of concepts, practices, and guidelines that ensure analytically sound outcomes.

REFERENCES

Acosta-Torres, L. S., López-Marín, L. M., Nunez-Anita, R. E., Hernández-Padrón, G., & Castaño, V. M. (2011). Biocompatible metal-oxide nanoparticles: nanotechnology improvement of conventional prosthetic acrylic resins. *Journal of Nanomaterials,* 1–8. https://doi.org/10.1155/2011/941561

Adibhatla, R. M., & Hatcher, J. F. (2010). Lipid oxidation and peroxidation in CNS health and disease: from molecular mechanisms to therapeutic opportunities. *Antioxidants & Redox Signaling*, *12*(1), 125–169.

Afzaal, M., Saeed, F., Ahmed, A., Islam, F., Khalid, M. A., Hussain, M., ... & Afzal, A. (2022). Anticaking agents in food nanotechnology. In *Application of Nanotechnology in Food Science, Processing and Packaging*, 141–151. Springer International Publishing. https://doi.org/10.1007/978-3-030-98820-3_9

Aillon, K. L., Xie, Y., El-Gendy, N., Berkland, C. J., & Forrest, M. L. (2009). Effects of nanomaterial physicochemical properties on in vivo toxicity. *Advanced Drug Delivery Reviews*, *61*(6), 457–466.

Alimardani, V., Abolmaali, S. S., & Tamaddon, A. M. (2021). Recent advances on nanotechnology-based strategies for prevention, diagnosis, and treatment of coronavirus infections. *Journal of Nanomaterials*, *2021*(1), 1–20. https://doi.org/10.1155/2021/9495126

Amgoth, C., Phan, C., Banavoth, M., Rompivalasa, S., & Tang, G. (2019). Polymer properties: functionalization and surface modified nanoparticles. In *Role of Novel Drug Delivery Vehicles in Nanobiomedicine.* IntechOpen.

Anumanchipalli, G. K., Chartier, J., & Chang, E. F. (2019). Speech synthesis from neural decoding of spoken sentences. *Nature*, *568*(7753), 493–498. https://doi.org/10.1038/s41586-019-1119-1

Astafurova, T. P., Burenina, A. A., Suchkova, S. A., Zotikova, A. B. P., Kulizhskiy, S. P., & Morgalev, Y. (2017). Influence of ZnO and Pt nanoparticles on cucumber yielding capacity and fruit quality. In *Nano Hybrids and Composites* (Vol. 13, pp. 142–148). Trans Tech Publications Ltd. https://doi.org/10.4028/www.scientific.net/NHC.13.142

Aureli, F., Ciprotti, M., D'Amato, M., do Nascimento da Silva, E., Nisi, S., Passeri, D., ... & Cubadda, F. (2020). Determination of total silicon and SiO2 particles using an ICP-MS based analytical platform for toxicokinetic studies of synthetic amorphous silica. *Nanomaterials*, *10*(5), 888. https://doi.org/10.3390/nano10050888

Baig, N., Kammakakam, I., & Falath, W. (2021). Nanomaterials: a review of synthesis methods, properties, recent progress, and challenges. *Materials Advances*, *2*(6), 1821–1871.

Bakirhan, N. K., Uslu, B., & Ozkan, S. A. (2017). Sensitive and selective assay of antimicrobials on nanostructured materials by electrochemical techniques. In *Nanostructures for Antimicrobial Therapy*, 55–83. Elsevier.

Bansod, B., Kumar, T., Thakur, R., Rana, S., & Singh, I. (2017). A review on various electrochemical techniques for heavy metal ions detection with different sensing platforms. *Biosensors and Bioelectronics*, *94*, 443–455.

Bayford, R., Rademacher, T., Roitt, I., & Wang, S. X. (2017). Emerging applications of nanotechnology for diagnosis and therapy of disease: a review. *Physiological Measurement*, *38*(8), R183. https://eprints.mdx.ac.uk/21772/6/Revised_manuscript_clean_versionAAM.pdf

Bhardwaj, V., & Kaushik, A. (2017). Biomedical applications of nanotechnology and nanomaterials. *Micromachines*, *8*(10), 298. https://doi.org/10.3390/mi8100298

Bhushan, B. (2017). Introduction to nanotechnology. In *Springer Handbook of Nanotechnology*, 1–19. Springer.

Bilati, U., Allémann, E., & Doelker, E. (2005). Nanoprecipitation versus emulsion-based techniques for the encapsulation of proteins into biodegradable nanoparticles and process-related stability issues. *Aaps Pharmscitech*, *6*, E594–E604.

Blaznik, U., Krušič, S., Hribar, M., Kušar, A., Žmitek, K., & Pravst, I. (2021). Use of food additive titanium dioxide (E171) before the introduction of regulatory restrictions due to concern for genotoxicity. *Foods*, *10*(8), 1910. https://doi.org/10.3390/foods10081910

Bohrey, S., Chourasiya, V., & Pandey, A. (2016). Polymeric nanoparticles containing diazepam: preparation, optimization, characterization, in-vitro drug release and release kinetic study. *Nano Convergence*, *3*(1), 1–7.

Bourbigot, S., Vanderhart, D. L., Gilman, J. W., Bellayer, S., Stretz, H., & Paul, D. R. (2004). Solid state NMR characterization and flammability of styrene–acrylonitrile copolymer montmorillonite nanocomposite. *Polymer*, *45*(22), 7627–7638.

Bryers, J. D. (2008). Medical biofilms. *Biotechnology and Bioengineering*, *100*(1), 1–18. https://doi.org/10.1002/bit.21838

Cañadas, C., Alvarado, H., Calpena, A. C., Silva, A. M., Souto, E. B., García, M. L., & Abrego, G. (2016). In vitro, ex vivo and in vivo characterization of PLGA nanoparticles loading pranoprofen for ocular administration. *International Journal of Pharmaceutics*, *511*(2), 719–727.

Chai, B., Xu, M., Yan, J., & Ren, Z. (2018). Remarkably enhanced photocatalytic hydrogen evolution over MoS2 nanosheets loaded on uniform CdS nanospheres. *Applied Surface Science*, *430*, 523–530.

Chavarria, F.,& Paul, D. R. (2004). Comparison of nanocomposites based on nylon 6 and nylon 66. *Polymer*, *45*(25), 8501–8515.

Chen, A., & Holt-Hindle, P. (2010). Platinum-based nanostructured materials: synthesis, properties, and applications. *Chemical Reviews*, *110*(6), 3767–3804.

Chen, M., Yu, Q., & Sun, H. (2013). Novel strategies for the prevention and treatment of biofilm related infections. *International Journal of Molecular Sciences*, *14*(9), 18488–18501. https://doi.org/10.3390/ijms140918488.

Chi, H., Song, S., Luo, M., Zhang, C., Li, W., Li, L., & Qin, Y. (2019). Effect of PLA nanocomposite films containing bergamot essential oil, TiO2 nanoparticles, and Ag nanoparticles on shelf life of mangoes. *Scientia Horticulturae*, *249*, 192–198. https://doi.org/10.1016/j.scienta.2019.01.059

Chidambaram, M., & Krishnasamy, K. (2014). Modifications to the conventional nanoprecipitation technique: an approach to fabricate narrow sized polymeric nanoparticles. *Advanced Pharmaceutical Bulletin*, *4*(2), 205.

Christine, V., & Ponchel, G. (2017). Polymer nanoparticles for nanomedicines. A guide for their design. *Anticancer Research*, *37*, 1544.

Chopra, H., Bibi, S., Kumar, S., Khan, M. S., Kumar, P., & Singh, I. (2022). Preparation and evaluation of chitosan/PVA based hydrogel films loaded with honey for wound healing application. *Gels*, *8*(2), 111.

Correia, M. G. S., Briuglia, M. L., Niosi, F., & Lamprou, D. A. (2017). Microfluidic manufacturing of phospholipid nanoparticles: stability, encapsulation efficacy, and drug release. *International Journal of Pharmaceutics*, *516*(1–2), 91–99. https://doi.org/10.1016/j.ijpharm.2016.11.025

Corté, L., & Leibler, L. (2005). Analysis of polymer blend morphologies from transmission electron micrographs. *Polymer*, 46(17), 6360–6368.

Crucho, C. I., & Barros, M. T. (2017). Polymeric nanoparticles: a study on the preparation variables and characterization methods. *Materials Science and Engineering: C, 80*, 771–784.

Cubadda, F., Aureli, F., D'Amato, M., Raggi, A., & Mantovani, A. (2013). Conference: Nanomaterials in the food sector: new approaches for safety assessment. *Rome, Instituto Superiore Di Sanità, September 27, 2013: Proceedings.*

Cui, L., Ma, X., & Paul, D. R. (2007). Morphology and properties of nanocomposites formed from ethylene-vinyl acetate copolymers and organoclays. *Polymer, 48*(21), 6325–6339.

Danks, A. E., Hall, S. R., & Schnepp, Z. J. M. H. (2016). The evolution of 'sol–gel'chemistry as a technique for materials synthesis. *Materials Horizons, 3*(2), 91–112.

David-Pur, M., Bareket-Keren, L., Beit-Yaakov, G., Raz-Prag, D., & Hanein, Y. (2014). All-carbon-nanotube flexible multi-electrode array for neuronal recording and stimulation. *Biomedical Microdevices, 16*, 43–53. https://doi.org/10.1007/s10544-013-9804-6

Davis, R. D., Gilman, J. W., & VanderHart, D. L. (2003). Processing degradation of polyamide 6/montmorillonite clay nanocomposites and clay organic modifier. *Polymer Degradation and Stability, 79*(1), 111–121.

De Jong, W. H., & Borm, P. J. (2008). Drug delivery and nanoparticles: applications and hazards. *International Journal of Nanomedicine, 3*(2), 133–149. https://www.tandfonline.com/doi/full/10.2147/ijn.s596

de Souza Simões, L., Madalena, D. A., Pinheiro, A. C., Teixeira, J. A., Vicente, A. A., & Ramos, Ó. L. (2017). Micro-and nano bio-based delivery systems for food applications: in vitro behavior. *Advances in Colloid and Interface Science, 243*, 23–45. https://doi.org/10.1016/j.cis.2017.02.010

Desgouilles, S., Vauthier, C., Bazile, D., Vacus, J., Grossiord, J. L., Veillard, M., & Couvreur, P. (2003). The design of nanoparticles obtained by solvent evaporation: a comprehensive study. *Langmuir, 19*(22), 9504–9510.

Devi, N., Sarmah, M., Khatun, B., & Maji, T. K. (2017). Encapsulation of active ingredients in polysaccharide–protein complex coacervates. *Advances in Colloid and Interface Science, 239*, 136–145. https://doi.org/10.1016/j.cis.2016.05.009

Docter, D., Strieth, S., Westmeier, D., Hayden, O., Gao, M., Knauer, S. K., & Stauber, R. H. (2015). No king without a crown–impact of the nanomaterial-protein corona on nanobiomedicine. *Nanomedicine, 10*(3), 503–519.

Dong, Y., Du, X. Q., Liang, P., & Man, X. L. (2020). One-pot solvothermal method to fabricate 1D-VS4 nanowires as anode materials for lithium ion batteries. *Inorganic Chemistry Communications, 115*, 107883.

Drexler, K. E. (2004). Nanotechnology: from Feynman to funding. *Bulletin of Science, Technology & Society, 24*(1), 21–27.

Edelstein, A. S., & Cammaratra, R. C. (Eds.) (1998). *Nanomaterials: Synthesis, Properties and Applications.* CRC press.

Emam, A. N., Girgis, E., Khalil, W. K., & Mohamed, M. B. (2014). Toxicity of plasmonic nanomaterials and their hybrid nanocomposites. In *Advances in Molecular Toxicology* (Vol. 8, pp. 173–202). Elsevier.

Emamifar, A., & Bavaisi, S. (2020). Nanocomposite coating based on sodium alginate and nano-ZnO for extending the storage life of fresh strawberries (Fragaria× ananassa Duch.). *Journal of Food Measurement and Characterization, 14*(2), 1012–1024. https://doi.org/10.1007/s11694-019- 00350-x.

Ezhilarasi, P. N., Karthik, P., Chhanwal, N., & Anandharamakrishnan, C. (2013). Nanoencapsulation techniques for food bioactive components: a review. *Food Bioprocessing Technology,* 6, 628–647. https://doi.org/10.1007/s11947-012-0944-0

Feynman, R. (1959, December). There is plenty of room at the bottom: An invitation to enter a new field of physics. In *Lecture at American Physical Society Meeting* (Vol. 29). American Physical Society.

Feynman, R. P. (1961). There's plenty of room at the bottom. In *Miniaturization*, 282–296 (HD Gilbert ed). Reinhold.

Fruijtier-Pölloth, C. (2016). The safety of nanostructured synthetic amorphous silica (SAS) as a food additive (E 551). *Archives of Toxicology, 90*, 2885–2916. https://doi.org/10.1007/s00204-016-1850-4

Gao, H. (2016). Progress and perspectives on targeting nanoparticles for brain drug delivery. *Acta Pharmaceutica Sinica B, 6*(4), 268–286. https://doi.org/10.1016/j.apsb.2016.05.013

Garcia, C. V., Shin G. H., & Kim, J. T. (2018). Metal oxide-based nanocomposites in food packaging: applications, migration, and regulations. *Trends Food Science Technology. 82*, 21–31. https://doi.org/10.1016/j.tifs.2018.09.021.

Gazsó, A., & Pavlicek, A. (2020). Titanium dioxide as a food additive. https://www.researchgate.net/profile/Anna-Pavlicek/publication/346801661_Titanium_Dioxide_as_a_Food_Additive/links/5fd0fe1045851568d1507209/Titanium-Dioxide-as-a-Food-Additive.pdf?_sg%5B0%5D=started_experiment_milestone&origin=journalDetail

Grumezescu, A. M. (Ed.) (2017). *Design and Development of New Nanocarriers*. William Andrew.

Guterres, S. S., Alves, M. P., & Pohlmann, A. R. (2007). Polymeric nanoparticles, nanospheres and nanocapsules, for cutaneous applications. *Drug Target Insights*, *2*, 117739280700200002.

Hadrup, N., Loeschner, K., Mandrup, K., Ravn-Haren, G., Frandsen, H. L., Larsen, E. H., ... & Mortensen, A. (2019). Subacute oral toxicity investigation of selenium nanoparticles and selenite in rats. *Drug and Chemical Toxicology*, *42*(1), 76–83.

Hayes, A. W. & Sahu, S. C. (2017). Nanotechnology in the food industry: a short review. *Food Safety Magazine*, *23*(1), 18–19. https://www.food-safety.com/articles/5193-nanotechnology-in-the-food-industry-a-short-review#References

He, C., Hu, Y., Yin, L., Tang, C., & Yin, C. (2010). Effects of particle size and surface charge on cellular uptake and biodistribution of polymeric nanoparticles. *Biomaterials*, *31*(13), 3657–3666.

He, X., & Hwang, H. M. (2016). Nanotechnology in food science: functionality, applicability, and safety assessment. *Journal of Food and Drug Analysis*, *24*(4), 671–681. https://doi.org/10.1016/j.jfda.2016.06.001

He, X., Deng, H., & Hwang, H. M. (2019). The current application of nanotechnology in food and agriculture. *Journal of Food and Drug Analysis*, *27*(1), 1–21. https://doi.org/10.1016/j.jfda.2018.12.002

Hernández-Hernández, H., Quiterio-Gutiérrez, T., Cadenas-Pliego, G., Ortega-Ortiz, H., Hernández-Fuentes, A. D., Cabrera de la Fuente, M., ... & Juárez-Maldonado, A. (2019). Impact of selenium and copper nanoparticles on yield, antioxidant system, and fruit quality of tomato plants. *Plants*, *8*(10), 355. https://doi.org/10.3390/plants8100355

Homar, M., Ubrich, N., Ghazouani, F. E., Kristl, J., Kerč, J., & Maincent, P. (2007). Influence of polymers on the bioavailability of microencapsulated celecoxib. *Journal of Microencapsulation*, 24(7), 621–633. https://doi.org/10.1080/09637480701497360

Hoseinnejad, M., Jafari, S. M., & Katouzian, I. (2018). Inorganic and metal nanoparticles and their antimicrobial activity in food packaging applications. *Critical Reviews in Microbiology*, *44*(2), 161–181.

Hu, W., Ye, X., Chantapakul, T., Chen, S., & Zheng, J. (2020). Manosonication extraction of RG-I pectic polysaccharides from citrus waste: optimization and kinetics analysis. *Carbohydrate Polymers*, *235*, 115982. https://doi. org/10.1016/j.carbpol.2020.115982

Iannitelli, A., Grande, R., Di Stefano, A., Di Giulio, M., Sozio, P., Bessa, L. J., ... & Cellini, L. (2011). Potential antibacterial activity of carvacrol-loaded poly (DL-lactide-co-glycolide)(PLGA) nanoparticles against microbial biofilm. *International Journal of Molecular Sciences*, *12*(8), 5039–5051. https://doi.org/10.3390/ijms12085039

Iqbal, M., Zafar, N., Fessi, H., & Elaissari, A. (2015). Double emulsion solvent evaporation techniques used for drug encapsulation. *Int ernationl Journal Pharmacy*, *496*, 173–190. https://doi.org/10.1016/j.ijpharm.2015.10.057

Jackson, T. C., Patani, B. O., & Ekpa, D. E. (2017). Nanotechnology in diagnosis: a review. *Advances in Nanoparticles*, *6*(3), 93–102. https://doi.org/10.4236/anp.2017.63008

Jafarzadeh, S., Nafchi, A. M., Salehabadi, A., Oladzad-Abbasabadi, N., & Jafari, S. M. (2021). Application of bio-nanocomposite films and edible coatings for extending the shelf life of fresh fruits and vegetables. *Advances in Colloid and Interface Science*, *291*, 102405. https://doi.org/10.1016/j.cis.2021.102405

Jagtiani, E. (2022). Advancements in nanotechnology for food science and industry. *Food Frontiers*, *3*(1), 56–82. https://doi.org/10.1002/fft2.104

Jaiswal, L., Shankar, S., & Rhim, J. W. (2019). Applications of nanotechnology in food microbiology. In *Methods in Microbiology* (Vol. 46, pp. 43–60). Academic Press. https://doi.org/10.1016/bs.mim.2019.03.002

Jaiswal, S., Duffy, B., Jaiswal, A. K., Stobie, N., & McHale, P. (2010). Enhancement of the antibacterial properties of silver nanoparticles using β-cyclodextrin as a capping agent. *International Journal of Antimicrobial Agents*, *36*(3), 280–283. https://doi.org/10.1016/j.ijantimicag.2010.05.006

Jawahar, N., & Meyyanathan, S. N. (2012). Polymeric nanoparticles for drug delivery and targeting: a comprehensive review. *International Journal of Health & Allied Sciences*, *1*(4), 217.

Jones, A. C., & Hitchman, M. L. (2009). Overview of chemical vapour deposition. *Chemical Vapour Deposition: Precursors, Processes and Applications*, *1*, 1–36.

Kamaly, N., Yameen, B., Wu, J., & Farokhzad, O. C. (2016). Degradable controlled-release polymers and polymeric nanoparticles: mechanisms of controlling drug release. *Chemical Reviews*, *116*(4), 2602–2663.

Karakoti, A. S., Hench, L. L., & Seal, S. (2006). The potential toxicity of nanomaterials—the role of surfaces. *Jom*, *58*, 77–82.

Karimi, M., Zare, H., Bakhshian Nik, A., Yazdani, N., Hamrang, M., Mohamed, E., … & Hamblin, M. R. (2016). Nanotechnology in diagnosis and treatment of coronary artery disease. *Nanomedicine*, *11*(5), 513–530. https://www.ncbi.nlm.nih.gov/pmc/articles/PMC4794112/pdf/nnm-11-513.pdf

Khan, H. A., Ibrahim, K. E., Khan, A., Alrokayan, S. H., Alhomida, A. S., & Lee, Y. K. (2016). Comparative evaluation of immunohistochemistry and real-time PCR for measuring proinflammatory cytokines gene expression in livers of rats treated with gold nanoparticles. *Experimental and Toxicologic Pathology*, *68*(7), 381–390.

Khan, Y., Sadia, H., Ali Shah, S. Z., Khan, M. N., Shah, A. A., Ullah, N., … & Khan, M. I. (2022). Classification, synthetic, and characterization approaches to nanoparticles, and their applications in various fields of nanotechnology: a review. *Catalysts*, *12*(11), 1386.

Kim, B., & Feldman, E. L. (2012). Insulin resistance in the nervous system. *Trends in Endocrinology & Metabolism*, *23*(3), 133–141. https://doi.org/10.1016/j.tem.2011.12.004

Kim, D. H., Fasulo, P. D., Rodgers, W. R., & Paul, D. R. (2007). Effect of the ratio of maleated polypropylene to organoclay on the structure and properties of TPO-based nanocomposites. Part I: Morphology and mechanical properties. *Polymer*, *48*(20), 5960–5978.

Kim, D. H., Fasulo, P. D., Rodgers, W. R., & Paul, D. R. (2007). Structure and properties of polypropylene-based nanocomposites: effect of PP-g-MA to organoclay ratio. *Polymer*, *48*(18), 5308–5323.

Krishnamoorthy, K., & Mahalingam, M. (2015). Selection of a suitable method for the preparation of polymeric nanoparticles: multi-criteria decision making approach. *Advanced Pharmaceutical Bulletin*, *5*(1), 57–67.

Kroemer, G., Galluzzi, L., Vandenabeele, P., Abrams, J., Alnemri, E. S., Baehrecke, E. H., … & Melino, G. (2009). Classification of cell death: recommendations of the Nomenclature Committee on Cell Death 2009. *Cell Death & Differentiation*, *16*(1), 3–11.

Kumar, P. S., Sundaramurthy, J., Sundarrajan, S., Babu, V. J., Singh, G., Allakhverdiev, S. I., & Ramakrishna, S. (2014). Hierarchical electrospun nanofibers for energy harvesting, production and environmental remediation. *Energy & Environmental Science*, *7*(10), 3192–3222.

Kumar, S., Dilbaghi, N., Saharan, R., & Bhanjana, G. (2012). Nanotechnology as emerging tool for enhancing solubility of poorly water-soluble drugs. *Bionanoscience*, *2*, 227–250.

Kumar, V., Sharma, N., & Maitra, S. S. (2017). In vitro and in vivo toxicity assessment of nanoparticles. *International Nano Letters*, *7*(4), 243–256.

Lammari, N., Louaer, O., Meniai, A. H., & Elaissari, A. (2020). Encapsulation of essential oils via nanoprecipitation process: overview, progress, challenges and prospects. *Pharmaceutics*, *12*(5), 431. https://doi.org/10.3390/pharmaceutics12050431

LeBaron, P. C., Wang, Z., & Pinnavaia, T. J. (1999). Polymer-layered silicate nanocomposites: an overview. *Applied Clay Science*, *15*(1–2), 11–29.

Lee, D., Cohen, R. E., & Rubner, M. F. (2005). Antibacterial properties of Ag nanoparticle loaded multilayers and formation of magnetically directed antibacterial microparticles. *Langmuir*, *21*(21), 9651–9659. https://doi.org/10.1021/la0513306

Lee, H. S., Fasulo, P. D., Rodgers, W. R., & Paul, D. R. (2005). TPO based nanocomposites. Part 1. Morphology and mechanical properties. *Polymer*, *46*(25), 11673–11689.

Lewis, D. R., Petersen, L. K., York, A. W., Ahuja, S., Chae, H., Joseph, L. B., … & Moghe, P. V. (2016). Nanotherapeutics for inhibition of atherogenesis and modulation of inflammation in atherosclerotic plaques. *Cardiovascular Research*, *109*(2), 283–293. https://doi.org/10.1093/cvr/cvv237

Li, M., Liu, H. L., Dang, F., Hintelmann, H., Yin, B., & Zhou, D. (2020). Alteration of crop yield and quality of three vegetables upon exposure to silver nanoparticles in sludge-amended soil. *ACS Sustainable Chemistry & Engineering*, *8*(6), 2472–2480. https://doi.org/10.1021/acssuschemeng.9b06721

Lim, K., & Hamid, Z. A. (2018). Polymer nanoparticle carriers in drug delivery systems: Research trend. In *Applications of Nanocomposite Materials in Drug Delivery* (pp. 217–237). Woodhead Publishing.

Lincoln, D. M., Vaia, R. A., Wang, Z. G., & Hsiao, B. S. (2001). Secondary structure and elevated temperature crystallite morphology of nylon-6/layered silicate nanocomposites. *Polymer*, *42*(4), 1621–1631.

Liolios, C. C., Gortzi, O., Lalas, S., Tsaknis, J., & Chinou, I. (2009). Liposomal incorporation of carvacrol and thymol isolated from the essential oil of Origanum dictamnus L. and in vitro antimicrobial activity. *Food Chemistry*, *112*(1), 77–83. https://doi.org/10.1016/j.foodchem.2008.05.060

Liu D, Yang F, Xiong F, & Gu N. (2016). The smart drug delivery system and its clinical potential. *Theranostics*, *6*, 1306–1323. https://doi.org/10.7150/thno.14858

Liu, P., Qin, R., Fu, G., & Zheng, N (2017). Surface coordination chemistry of metal nanomaterials. *Journal of the American Chemical Society*, *139*(6), 2122–2131.

Liu, W., Zhang, M., & Bhandari, B (2020). Nanotechnology–A shelf life extension strategy for fruits and vegetables. *Critical Reviews in Food Science and Nutrition*, *60*(10), 1706–1721. https://doi.org/10.1080/10408398.2019.1589415

Liu, X., Chen, P., Ding, X., Liu, A., Li, P., Sun, C., & Guan, H. (2022). A narrative review of cortical visual prosthesis systems: the latest progress and significance of nanotechnology for the future. *Annals of Translational Medicine*, *10*(12). https://dx.doi.org/10.21037/atm-22-2858

Lockman, P. R., Koziara, J. M., Mumper, R. J., & Allen, D. D. (2004). Nanoparticle surface charges alter blood–brain barrier integrity and permeability. *Journal of Drug Targeting*, *12*(9–10), 635–641.

Loftsson, T., & Brewster, M.E. (2010). Pharmaceutical applications of cyclodextrins: basic science and product development. *Journal of Pharmacy and Pharmacology*, *62*, 1607–1621. https://doi.org/10.1111/j.2042- 7158.2010.01030.x

Machac, P., Cichon, S., Lapcak, L., & Fekete, L. (2020). Graphene prepared by chemical vapour deposition process. *Graphene Technology*, *5*, 9–17.

Mahmoudi, M., Azadmanesh, K., Shokrgozar, M. A., Journeay, W. S., & Laurent, S. (2011). Effect of nanoparticles on the cell life cycle. *Chemical Reviews*, 111(5), 3407–3432.

Mai, Y. W., & Yu, Z. Z (2006). *Polymer nanocomposites*. CRC Press.

Manias, E., Touny, A., Wu, L., Strawhecker, K., Lu, B., & Chung, T. C. (2001). Polypropylene/montmorillonite nanocomposites. Review of the synthetic routes and materials properties. *Chemistry of Materials*, *13*(10), 3516–3523.

Masenelli-Varlot, K., Vigier, G., Vermogen, A., Gauthier, C., & Cavaillé, J. Y. (2007). Quantitative structural characterization of polymer–clay nanocomposites and discussion of an "ideal" microstructure, leading to the highest mechanical reinforcement. *Journal of Polymer Science Part B: Polymer Physics*, *45*(11), 1243–1251.

Maynard, A. D., Aitken, R. J., Butz, T., Colvin, V., Donaldson, K., Oberdörster, G., … & Warheit, D. B. (2006). Safe handling of nanotechnology. *Nature*, *444*(7117), 267–269.

Mclaughlin, S., Podrebarac, J., Ruel, M., Suuronen, E. J., McNeill, B., & Alarcon, E. I. (2016). Nano-engineered biomaterials for tissue regeneration: what has been achieved so far?. *Frontiers in Materials*, *3*, 27.

Møller, P., Jacobsen, N. R., Folkmann, J. K., Danielsen, P. H., Mikkelsen, L., Hemmingsen, J. G., … & Loft, S. (2010). Role of oxidative damage in toxicity of particulates. *Free Radical Research*, *44*(1), 1–46.

Morgan, A. B., & Gilman, J. W. (2003). Characterization of polymer-layered silicate (clay) nanocomposites by transmission electron microscopy and X-ray diffraction: a comparative study. *Journal of Applied Polymer Science*, *87*(8), 1329–1338.

Naidoo, A., Parboosing, R., & Moodley, P. (2016). Real-time polymerase chain reaction optimised for hepatitis C virus detection in dried blood spots from HIV-exposed infants, KwaZulu-Natal, South Africa. *African Journal of Laboratory Medicine*, *5*(1), 1–6. https://journals.co.za/doi/abs/10.4102/ajlm.v5i1.269

Navya, P. N., & Daima, H. K. (2016). Rational engineering of physicochemical properties of nanomaterials for biomedical applications with nanotoxicological perspectives. *Nano Convergence*, *3*, 1–14.

Navya, P. N., Kaphle, A., Srinivas, S. P., Bhargava, S. K., Rotello, V. M., & Daima, H. K. (2019). Current trends and challenges in cancer management and therapy using designer nanomaterials. *Nano Convergence*, *6*, 1–30.

Nel, T., Xia, L. Madler and Li, N. (2006). Toxic potential of materials at the nanolevel. *Science*, *311*, 622–627.

Ong, C. B., Ng, L. Y., & Mohammad, A. W. (2018). A review of ZnO nanoparticles as solar photocatalysts: synthesis, mechanisms and applications. *Renewable and Sustainable Energy Reviews*, *81*, 536–551. https://doi.org/10.1016/j.rser.2017.08.020

Ostermann, R., Cravillon, J., Weidmann, C., Wiebcke, M., & Smarsly, B. M. (2011). Metal–organic framework nanofibers via electrospinning. *Chemical Communications*, *47*(1), 442–444.

Pal, S. L., Jana, U., Manna, P. K., Mohanta, G. P., & Manavalan, R. (2011). Nanoparticle: an overview of preparation and characterization. *Journal of Applied Pharmaceutical Science*, *30*, 228–234.

Parashar, M., Shukla, V. K., & Singh, R. (2020). Metal oxides nanoparticles via sol–gel method: a review on synthesis, characterization and applications. *Journal of Materials Science: Materials in Electronics*, *31*, 3729–3749.

Paredes, A. J., Asensio, C. M., Llabot, J. M., Allemandi, D. A., & Palma, S. D. (2016). Nanoencapsulation in the food industry: manufacture, applications and characterization. https://ri.conicet.gov.ar/handle/11336/64263

Pateiro, M., Gómez, B., Munekata, P. E., Barba, F. J., Putnik, P., Kovačević, D. B., & Lorenzo, J. M. (2021). Nanoencapsulation of promising bioactive compounds to improve their absorption, stability, functionality and the appearance of the final food products. *Molecules, 26*(6), 1547. https://doi.org/10.3390/molecules26061547

Pathkoti, K., Manubolu, M., & Hwang, H. M. (2017). Nanostructures: current uses and future applications in food science. *Journal of Food and Drug Analysis, 25*(2), 245–253. https://doi.org/10.1016/j.jfda.2017.02.004

Pattni, B. S., Chupin, V. V., & Torchilin, V. P. (2015). New developments in liposomal drug delivery. *Chemical Reviews*, *115*(19), 10938–10966. https://doi.org/10.1021/acs.chemrev.5b00046

Paul, D. R., & Robeson, L. M. (2008). Polymer nanotechnology: nanocomposites. *Polymer*, *49*(15), 3187–3204.

Pennington, K. L., & DeAngelis, M. M. (2016). Epidemiology of age-related macular degeneration (AMD): associations with cardiovascular disease phenotypes and lipid factors. *Eye and Vision*, *3*(1), 1–20. https://doi.org/10.1186/s40662-016-0063-5

Perán, M., García, M. A., López-Ruiz, E., Bustamante, M., Jiménez, G., Madeddu, R., & Marchal, J. A. (2012). Functionalized nanostructures with application in regenerative medicine. *International Journal of Molecular Sciences*, *13*(3), 3847–3886. https://doi.org/10.3390/ijms13033847

Pérez-Campaña, C., Gómez-Vallejo, V., Puigivila, M., Martín, A., Calvo-Fernández, T., Moya, S. E., … & Llop, J. (2013). Biodistribution of different sized nanoparticles assessed by positron emission tomography: a general strategy for direct activation of metal oxide particles. *ACS Nano*, *7*(4), 3498–3505. https://doi.org/10.1021/nn400450p

Pineda-Reyes, A. M., Delgado, M. H., de la Luz Zambrano-Zaragoza, M., Leyva-Gómez, G., Mendoza-Munoz, N., & Quintanar-Guerrero, D. (2021). Implementation of the emulsification-diffusion method by solvent displacement for polystyrene nanoparticles prepared from recycled material. *RSC Advances*, *11*(4), 2226–2234. https://doi.org/10.1039/D0RA07749F

Pradhan, N., Singh, S., Ojha, N., Srivastava, A., Barla, A., Rai, V., et al. (2015). Facets of nanotechnology as seen in food processing, packaging, and preservation industry. *BioMed Research International, 2015*, 365672. https://doi.org/10.1155/2015/365672

Qasim, M., Lim, D. J., Park, H., & Na, D. (2014). Nanotechnology for diagnosis and treatment of infectious diseases. *Journal of Nanoscience and Nanotechnology*, *14*(10), 7374–7387. https://doi.org/10.1166/jnn.2014.9578

Quintanar-Guerrero, D., Allémann, E., Doelker, E., & Fessi, H. (1998). Preparation and characterization of nanocapsules from preformed polymers by a new process based on emulsification-diffusion technique. *Pharmaceutical Research*, 15, 1056–1062.

Rakel, D., & Minichiello, V. (Eds.) (2022). *Integrative Medicine, E-Book*. Elsevier Health Sciences.

Ramos, A. P., Cruz, M. A., Tovani, C. B., & Ciancaglini, P. (2017). Biomedical applications of nanotechnology. *Biophysical Reviews*, *9*(2), 79–89. DOI 10.1007/s12551-016-0246-2

Re, F., Gregori, M., & Masserini, M. (2012). Nanotechnology for neurodegenerative disorders. *Maturitas*, 73(1), 45–51.

Reis, C. P., Neufeld, R. J., Ribeiro, A. J., & Veiga, F (2006). Nanoencapsulation I. Methods for preparation of drug-loaded polymeric nanoparticles. *Nanomedicine: Nanotechnology, Biology and Medicine*, *2*(1), 8–21.

Rivas, C. J. M., Tarhini, M., Badri, W., Miladi, K., Greige-Gerges, H., Nazari, Q. A., … & Elaissari, A. (2017). Nanoprecipitation process: from encapsulation to drug delivery. *International Journal of Pharmaceutics*, *532*(1), 66–81.

Rivas, C. J. M., Tarhini, M., Badri, W., Miladi, K., Greige-Gerges, H., Nazari, Q. A., … & Elaissari, A. (2017). Nanoprecipitation process: from encapsulation to drug delivery. *International Journal of Pharmaceutics*, *532*(1), 66–81. https://doi.org/10.1016/j.ijpharm.2017.08.064

Ropers, M. H., Terrisse, H., Mercier-Bonin, M., & Humbert, B. (2017). Titanium dioxide as food additive. *Application of Titanium Dioxide*, *10*. https://doi.org/10.5772/intechopen.68883

Saeedi, M., Eslamifar, M., Khezri, K., & Dizaj, S. M. (2019). Applications of nanotechnology in drug delivery to the central nervous system. *Biomedicine & Pharmacotherapy*, *111*, 666–675. https://doi.org/10.1016/j.biopha.2018.12.133

Salatin, S., Barar, J., Barzegar-Jalali, M., Adibkia, K., Kiafar, F., & Jelvehgari, M. (2017). Development of a nanoprecipitation method for the entrapment of a very water soluble drug into Eudragit RL nanoparticles. *Research in Pharmaceutical Sciences*, *12*(1), 1.

Sánchez-López, E., Egea, M. A., Davis, B. M., Guo, L., Espina, M., Silva, A. M., … & Cordeiro, M. F. (2018). Memantine-loaded PEGylated biodegradable nanoparticles for the treatment of glaucoma. *Small*, *14*(2), 1701808.

Saraiva, C., Praça, C., Ferreira, R., Santos, T., Ferreira, L., & Bernardino, L. (2016). Nanoparticle-mediated brain drug delivery: overcoming blood–brain barrier to treat neurodegenerative diseases. *Journal of Controlled Release*, *235*, 34–47. https://doi.org/10.1016/j.jconrel.2016.05.044

Sargent Jr, J. F. (2013, August). *The National Nanotechnology Initiative: Overview, Reauthorization, and Appropriations Issues*. Library of Congress Washington DC Congressional Research Service.

Schaefer, D. W., & Justice, R. S. (2007). How nano are nanocomposites? *Macromolecules*, *40*(24), 8501–8517.

Schaefer, H. E (2010). *Nanoscience: the Science of the Small in Physics, Engineering, Chemistry, Biology and Medicine*. Springer Science & Business Media.

Shah, K. A., & Tali, B. A. (2016). Materials science in semiconductor processing synthesis of carbon nanotubes by catalytic chemical vapour deposition: a review on carbon sources, catalysts and substrates. *Materials Science in Semiconductor Processing*, *41*, 67–82.

Sharma, N., Madan, P., & Lin, S. (2016). Effect of process and formulation variables on the preparation of parenteral paclitaxel-loaded biodegradable polymeric nanoparticles: a co-surfactant study. *Asian Journal of Pharmaceutical Sciences*, *11*(3), 404–416.

Shishir, M. R. I., Xie, L., Sun, C., Zheng, X., & Chen, W. (2018). Advances in micro and nano-encapsulation of bioactive compounds using biopolymer and lipid-based transporters. *Trends in Food Science & Technology*, *78*, 34–60. https://doi.org/10.1016/j.tifs.2018.05.018

Singh, R. P., & Singh, K. R. (Eds.) (2021). *Nanomaterials in Bionanotechnology: Fundamentals and Applications*. CRC Press.

Singh, T., Shukla, S., Kumar, P., Wahla, V., Bajpai, V. K., & Rather, I. A. (2017). Application of nanotechnology in food science: perception and overview. *Frontiers in Microbiology*, *8*, 1501. https://doi.org/10.3389/fmicb.2017.01501

Smith, S. E., Snider, C. L., Gilley, D. R., Grant, D. N., Sherman, S. L., Ulery, B. D., … & Grant, S. A. (2017). Homogenized porcine extracellular matrix derived injectable tissue construct with gold nanoparticles for musculoskeletal tissue engineering applications. *Journal of Biomaterials and Nanobiotechnology*, *8*(2), 125. https://doi.org/10.4236/jbnb.2017.82009

Song, W., So, S. K., & Cao, L. (2001). Angular-dependent photoemission studies of indium tin oxide surfaces. *Applied Physics A: Materials Science and Processing*, *72*(3), 361–365.

Souto, E. B., Severino, P., & Santana, M. H. A. (2012). Preparação de nanopartículas poliméricas a partir da polimerização de monômeros: parte I. *Polímeros*, *22*, 96–100.

Souto, E. B., Souto, S. B., Campos, J. R., Severino, P., Pashirova, T. N., Zakharova, L. Y., … & Santini, A. (2019). Nanoparticle delivery systems in the treatment of diabetes complications. *Molecules*, *24*(23), 4209.

Sridhar, A., Ponnuchamy, M., Kumar, P. S., & Kapoor, A (2021). Food preservation techniques and nanotechnology for increased shelf life of fruits, vegetables, beverages and spices: a review. *Environmental Chemistry Letters*, *19*, 1715–1735. https://doi.org/10.1007/s10311-020-01126-2

Stretz, H. A., & Paul, D. R (2006). Properties and morphology of nanocomposites based on styrenic polymers. Part I: styrene-acrylonitrile copolymers. *Polymer*, *47*(24), 8123–8136.

Stretz, H. A., Paul, D. R., & Cassidy, P. E. (2005). Poly (styrene-co-acrylonitrile)/montmorillonite organoclay mixtures: a model system for ABS nanocomposites. *Polymer*, *46*(11), 3818–3830.

Stringer, C., Pachitariu, M., Steinmetz, N., Reddy, C. B., Carandini, M., & Harris, K. D. (2019). Spontaneous behaviors drive multidimensional, brainwide activity. *Science, 364*(6437), eaav7893. https://doi.org/10.1126/science.aav7893

Swaroop, C., & Shukla, M. (2019). Development of blown polylactic acid-MgO nanocomposite films for food packaging. *Composites Part A: Applied Science and Manufacturing, 124*, 105482.

Szczęch, M., & Szczepanowicz, K. (2020). Polymeric core-shell nanoparticles prepared by spontaneous emulsification solvent evaporation and functionalized by the layer-by-layer method. *Nanomaterials, 10*(3), 496.

Papp, T., Schiffmann, D., Weiss, D., Castranova, V., Vallyathan V., and Rahman, Q. (2008). Human health implications of nanomaterial exposure. *Nanotoxicology, 2*, 9–27.

Taniguchi, N. (1994). 1993 ASPE distinguished lecturer. *Precision Engineering, 16*(1), 5–24.

Toumey, C. (2008). Reading Feynman Into Nanotechnology: A Text for a New Science. *Techne: Research in Philosophy & Technology, 12*(3).

Tseng, T. K., Lin, Y. S., Chen, Y. J., & Chu, H. (2010). A review of photocatalysts prepared by sol-gel method for VOCs removal. *International Journal of Molecular Sciences, 11*(6), 2336–2361.

Unfried, K., Albrecht, C., Klotz, L. O., Von Mikecz, A., Grether-Beck, S., & Schins, R. P. (2007). Cellular responses to nanoparticles: target structures and mechanisms. *Nanotoxicology, 1*(1), 52–71.

Vasile, C (2018). *Polymeric Nanomaterials in Nanotherapeutics*. Elsevier.

Vauthier, C., & Bouchemal, K. (2009). Methods for the preparation and manufacture of polymeric nanoparticles. *Pharmaceutical Research, 26*, 1025–1058.

Veerachamy, S., Yarlagadda, T., Manivasagam, G., & Yarlagadda, P. K. (2014). Bacterial adherence and biofilm formation on medical implants: a review. *Proceedings of the Institution of Mechanical Engineers, Part H: Journal of Engineering in Medicine, 228*(10), 1083–1099. https://doi.org/10.1177/0954411914556137

Velu, R., Calais, T., Jayakumar, A., & Raspall, F. (2019). A comprehensive review on bio-nanomaterials for medical implants and feasibility studies on fabrication of such implants by additive manufacturing technique. *Materials, 13*(1), 92. https://doi.org/10.3390/ma13010092

Ventola, C. L (2012). The nanomedicine revolution. *Pharmacology &. Therapeutics, 40*, 525.

Ventola, C. L (2017). Progress in nanomedicine: approved and investigational nanodrugs. *Pharmacy and Therapeutics, 42*(12), 742.

Wang, Y., Li, P., Truong-Dinh Tran, T., Zhang, J., & Kong, L. (2016). Manufacturing techniques and surface engineering of polymer based nanoparticles for targeted drug delivery to cancer. *Nanomaterials, 6*(2), 26.

Weir, A., Westerhoff, P., Fabricius, L., Hristovski, K., & Von Goetz, N. (2012). Titanium dioxide nanoparticles in food and personal care products. *Environmental Science & Technology, 46*(4), 2242–2250. https://doi.org/10.1021/es204168d

Wender, H., Migowski, P., Feil, A. F., Teixeira, S. R., & Dupont, J. (2013). Sputtering deposition of nanoparticles onto liquid substrates: recent advances and future trends. *Coordination Chemistry Reviews, 257*(17–18), 2468–2483.

Weon, J. I., & Sue, H. J. (2005). Effects of clay orientation and aspect ratio on mechanical behavior of nylon-6 nanocomposite. *Polymer, 46*(17), 6325–6334.

Wesley, S. J., Raja, P., Raj, A. A., & Tiroutchelvamae, D (2014). Review on-nanotechnology applications in food packaging and safety. *International Journal of Engineering Research, 3*(11), 645–651. https://maken.wikiwijs.nl/userfiles/b/bf23d059784ef52881d2f6cb5e0cda1d5871b245.pdf

Westervelt, A (2015). Dunkin'Donuts to remove titanium dioxide from donuts. *The Guardian, 11*. https://chemistry.beloit.edu/classes/nanotech/food/Dunkin%20Donuts%20to%20remove%20titanium%20dioxide%20from%20donuts%20%7C%20Guardian%20Sustainable%20Business%20%7C%20The%20Guardian.pdf

Yadav, T. P., Yadav, R. M., & Singh, D. P. (2012). Mechanical milling: a top down approach for the synthesis of nanomaterials and nanocomposites. *Nanoscience and Nanotechnology*, 2(3), 22–48.

Yoon, P. J., Hunter, D. L., & Paul, D. R (2003). Polycarbonate nanocomposites: Part 2. Degradation and color formation. *Polymer, 44*(18), 5341–5354.

Yu, N., Zhang, M., Islam, N., Lu, L., Liu, Q.,& Cheng, X (2015). Combined sterilizing effects of nano-ZnO and ultraviolet on convenient vegetable dishes. *LWT – Food Science and Technology, 61*(2), 638–643. https://doi.org/10.1016/j.lwt.2014.12.042.

Zhang, Y., Li, M., Gao, X., Chen, Y., & Liu, T. (2019). Nanotechnology in cancer diagnosis: progress, challenges and opportunities. *Journal of Hematology & Oncology*, *12*(1), 1–13. https://doi.org/10.1186/s13045-019-0833-3

Zhou, W., Gao, X., Liu, D., & Chen, X. (2015). Gold nanoparticles for in vitro diagnostics. *Chemical Reviews*, *115*(19), 10575–10636. https://doi.org/10.1021/acs.chemrev.5b00100

Zhuang, S., Lee, E. S., Lei, L., Nunna, B. B., Kuang, L., & Zhang, W. (2016). Synthesis of nitrogen-doped graphene catalyst by high-energy wet ball milling for electrochemical systems. *International Journal of Energy Research*, *40*(15), 2136–2149.

Zielińska, A., Carreiró, F., Oliveira, A. M., Neves, A., Pires, B., Venkatesh, D. N., … & Souto, E. B (2020). Polymeric nanoparticles: production, characterization, toxicology and ecotoxicology. *Molecules*, *25*(16), 3731.

Zorraquı´n-Pen˜a I., Cueva C., Bartolome´ B., & Moreno-Arribas M. V (2020). Silver nanoparticles against foodborne bacteria. Effects at intestinal level and health limitations. *Microorganisms*, *8*, 132. https://doi.org/10.3390/microorganisms8010132.

3 Characterization, Properties, and Application of Bionanomaterials Extracted from Fish Scales

S.C. Onwubu, S.C. Mkhize, Z. Obiechefu, T.H. Mokhothu, P.S. Mdluli, and A.K. Misra

3.1 INTRODUCTION

Bionanomaterials are a new class of materials derived from natural sources that exhibit unique properties and functionalities at the nanoscale. The development of these materials has been driven by the growing demand for materials with specific properties and functionalities that are not available from synthetic materials. In the past few years, there has been an increasing fascination with the extraction of bionanomaterials from bio-waste, such as the waste produced by the fishing industry. One particularly prevalent type of waste generated by the fishing industry is bio-waste from fish scales (Lionetto and Esposito Corcione, 2021). It is a rich source of biologically active compounds such as collagen, chitin, and calcium carbonate, which have the potential to be transformed into bionanomaterials with unique properties and functionalities (Debeaufort, 2021). The extraction of bionanomaterials from fish scales bio-waste not only provides a solution for the management of this waste but also helps to reduce the environmental impact of the fishing industry.

Furthermore, the study of bionanomaterials extracted from fish scales bio-waste has been a rapidly growing field of research. These materials have been found to have excellent mechanical strength, biocompatibility, and biodegradability, making them suitable for use in tissue engineering and biomedical applications (Halim et al., 2021). Additionally, they have excellent antioxidant activity and good adsorption properties, making them useful for environmental applications such as water purification and the removal of heavy metals from contaminated water (Saleem and Zaidi, 2020). The extraction of bionanomaterials from fish scales bio-waste is a promising solution for the management of this waste and has the potential to contribute to the development of new and improved materials with unique properties and functionalities. The purpose of this chapter is to provide a comprehensive overview of bionanomaterials extracted from fish scales bio-waste. It aims to highlight their unique properties and functionalities and to discuss the potential applications in different fields. The chapter is intended for researchers, scientists, and students who are interested in learning about this new class of materials and their potential for contributing to the development of new and improved materials with unique properties and functionalities.

3.2 CLASSIFICATION OF BIONANOMATERIALS FROM BIO-WASTE

Bionanomaterials from bio-waste can be broadly categorized into two main types: polysaccharides and proteins. Polysaccharides, also known as carbohydrates, are a class of biopolymers that are widely found in bio-waste, such as food and agricultural waste. Examples of polysaccharides

DOI: 10.1201/9781003432791-4

commonly extracted from bio-waste include cellulose, chitin, and starch (Guo et al., 2019). These materials have a high molecular weight and a highly branched structure, which makes them attractive for use in various applications due to their mechanical strength, biocompatibility, and biodegradability (Bilal et al., 2021). Proteins are another class of biopolymers commonly found in bio-waste, such as waste from the food and pharmaceutical industries. These materials are highly versatile and can be used for a range of applications due to their ability to self-assemble into various structures and their biocompatibility. Examples of proteins commonly extracted from bio-waste include gelatin and silk. In addition to polysaccharides and proteins, other biopolymers such as lipids, nucleic acids, and polyphenols may also be present in bio-waste and have potential as bionanomaterials.

The specific type of bionanomaterial extracted from bio-waste will depend on the source of the waste and the desired properties of the material. For instance, when it comes to applications demanding superior mechanical strength, polysaccharides might be appropriate, whereas proteins could be more suitable for applications requiring biocompatibility and biodegradability. The choice of bionanomaterial will also depend on the available extraction methods and the desired production scale. Figure 3.1 illustrates various examples of bionanomaterials that can be derived from fish scales. These are further elaborated below:

- Chitosan
  Chitosan is a biopolymer that is extracted from fish scales using a process of deacetylation (Takarina et al., 2017). It has a range of properties that make it useful for various applications, including its biocompatibility, biodegradability, and antimicrobial properties (Martău et al., 2019).

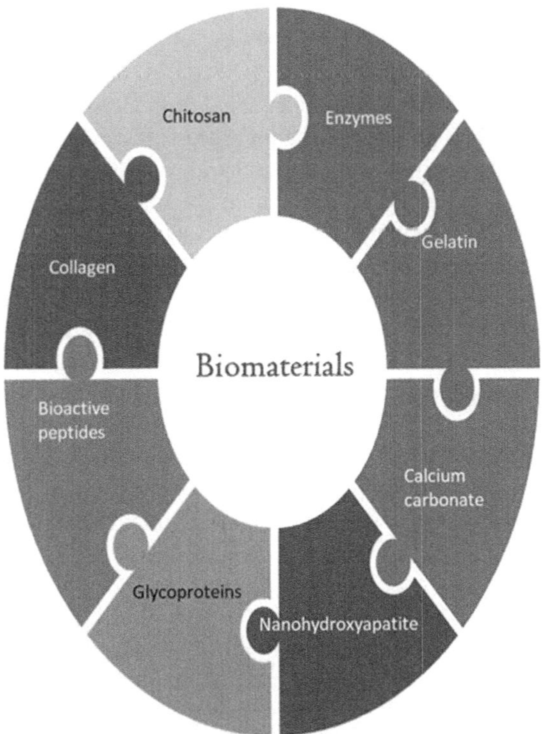

FIGURE 3.1 Types of biomaterials that can be extracted from fish scales. (Compiled by the authors.)

- Collagen
  Collagen, a fibrous protein abundant in fish scales, plays a vital role in providing strength and elasticity to the scales. Extracting collagen from fish scales offers potential applications in the fields of tissue engineering and wound healing (Jafari et al., 2020).
- Calcium Carbonate
  Fish scales are composed mainly of calcium carbonate, which can be extracted and used as a bio-based material. Calcium carbonate has potential applications as a filler in various products, including paper, plastics, and rubber (Saulat et al., 2020).
- Enzymes
  Fish scales possess a variety of enzymes, including transglutaminase, which can be extracted and employed in diverse applications such as food processing and industrial biotechnology (Zheng et al., 2022).
- Peptides
  Fish scales contain various bioactive peptides that have potential applications in biomedical and cosmetic fields. These peptides have anti-inflammatory, antioxidant, and antimicrobial properties (Ahmed et al., 2020).
- Glycoproteins
  Fish scales contain various glycoproteins that have potential applications in the biomedical field. These glycoproteins have properties such as anticoagulant and anti-inflammatory activities (Tabasum et al., 2017).

It should be emphasized that the mentioned information does not encompass all possible bionanomaterials, and there is a possibility of discovering new ones in the future. Extensive research is necessary to comprehensively explore the potential of these materials, as well as to develop novel extraction techniques and applications.

3.2.1 Chitosan

Chitosan, a biopolymer derived from the exoskeletons of crustaceans like shrimp and crab, can also be extracted from fish scales through a deacetylation process. It consists of β-D-glucosamine and N-acetyl-D-glucosamine units, forming a linear polysaccharide. One of the unique properties of chitosan is its biocompatibility, which makes it useful in various biomedical applications (Iber et al., 2022). Chitosan is a biodegradable and non-toxic material that is biocompatible with human tissues and does not elicit an adverse immune response (Islam et al., 2017). This makes it an ideal material for use in medical devices, such as wound dressings, drug delivery systems, and tissue engineering scaffolds.

Another important property of chitosan is its antimicrobial activity (Yan et al., 2021). Chitosan has been shown to inhibit the growth of various microorganisms, including bacteria, fungi, and viruses. This makes it a useful material for use in food packaging, wound dressings, and other applications where antimicrobial properties are required (Meng et al., 2021). Chitosan also has excellent water solubility, making it useful for various applications, including drug delivery systems and wound healing. In addition, chitosan has good film-forming properties, making it useful for use in food packaging, cosmetics, and other applications where film formation is required (Priyadarshi and Rhim, 2020).

3.2.2 Collagen

Collagen, a fibrous protein found abundantly in various tissues, including fish scales, plays a crucial role in the strength and elasticity of tissues as a key component of the extracellular matrix. Extracted from fish scales, collagen holds great potential in diverse fields, such as tissue engineering and wound healing (Rezvani Ghomi et al., 2021). In tissue engineering, collagen can serve as a scaffold to facilitate cell growth and tissue regeneration. By mimicking the natural extracellular matrix, collagen scaffolds create a supportive environment for cell growth and tissue regeneration. Furthermore, collagen exhibits favourable biocompatibility and biodegradability, making it an appropriate material for tissue engineering applications (Rezvani Ghomi et al., 2021).

In the context of wound healing, collagen has been employed as a wound dressing to enhance healing and prevent infection. Collagen dressings have demonstrated the ability to expedite wound healing by providing a moist environment conducive to wound healing, stimulating cell growth and tissue regeneration, and reducing the risk of infection. Collagen dressings can also help to reduce scarring and improve the appearance of the wound after healing (Lin et al., 2019). In addition to tissue engineering and wound healing, collagen has potential applications in other fields, such as drug delivery and cosmetics. Collagen can be used as a drug delivery vehicle, as it can complex with various drugs and promote their controlled release (Furtado et al., 2022). In cosmetics, collagen has been used as an ingredient in skincare products to improve skin elasticity and reduce the appearance of wrinkles (Lupu et al., 2020).

3.2.3 Calcium Carbonate

Calcium carbonate (CaCO3) is a common bio-based material that can be extracted from fish scales (Harikrishna et al., 2017). Fish scales are composed mainly of calcium carbonate, which makes them an attractive source of this material. Calcium carbonate has a range of properties, including its hardness, chemical stability, and low cost, which make it useful for various applications (Saulat et al., 2020). One of the most important applications of calcium carbonate is as a filler in various products, including paper, plastics, and rubber. In paper production, calcium carbonate is used as a filler to improve the brightness and opacity of the paper. It also increases the mechanical strength of the paper, making it more durable. In the plastics and rubber industries, calcium carbonate is used as a filler to reduce the cost of the final product and to improve its mechanical properties (Saulat et al., 2020).

In addition to its use as a filler, calcium carbonate has potential applications in various fields, such as construction, agriculture, and environmental protection. In construction, calcium carbonate can be used as a building material, because it is strong, durable, and abundant. In agriculture, cal cium carbonate is used as a soil amendment to neutralize soil acidity and to improve soil fertility. In environmental protection, calcium carbonate can be used as a neutralizing agent to reduce the acidity of wastewater and to prevent soil and water pollution (Yadav et al., 2021).

3.2.4 Enzymes

Enzymes are biological molecules that play a crucial role in many biological processes. Fish scales are known to contain a range of enzymes, including transglutaminase, which can be extracted and used for various applications. Transglutaminase is an enzyme that cross-links proteins by forming covalent bonds between the amino acids of two proteins. This property makes transglutaminase useful for food processing, as it can be used to improve the texture and flavour of various food products (Zheng et al., 2022). For example, transglutaminase can be used to bind proteins in meat products, such as sausages and ham, which results in a more cohesive and flavourful product (Alves et al., 2021). In industrial biotechnology, transglutaminase can be used in various applications, including the production of bio-based materials. For example, transglutaminase can be used to cross-link proteins from renewable sources, such as plants and microorganisms, to produce bio-based materials with improved mechanical and thermal stability (Lamp et al., 2022).

3.2.5 Bioactive Peptides

Fish scales are a source of various bioactive peptides, which are small chains of amino acids (Ishak and Sarbon, 2018). These peptides have been found to have a range of beneficial properties, making them useful for various applications in the biomedical and cosmetic fields (Felician et al., 2018). One of the main properties of these peptides is their anti-inflammatory activity. Inflammation is a key factor in many diseases, including cardiovascular disease, arthritis, and cancer. Bioactive peptides extracted from fish scales have been found to have anti-inflammatory properties, making them potential candidates for the development of new anti-inflammatory drugs (Abril et al., 2022).

Furthermore, it has been discovered that peptides derived from fish scales possess antioxidant properties that safeguard cells against harm induced by reactive oxygen species. These properties make these peptides useful for the development of new cosmetics and skincare products, as they may help to reduce the signs of aging and protect the skin from damage. Finally, some fish scale peptides have been found to have antimicrobial properties, making them useful for the development of new antibiotics and other antimicrobial agents (Abril et al., 2022). These peptides have the potential to be used as alternative or complementary treatments for various infections caused by bacteria, viruses, and other pathogens (Ghosh et al., 2019).

3.2.6 Glycoproteins

Glycoproteins are complex biomolecules that consist of a protein backbone and carbohydrates attached to it (Shivatare et al., 2022). Fish scales are rich in glycoproteins, and they can be extracted and used for various applications in the biomedical field (Qin et al., 2022). Glycoproteins in fish scales have shown anticoagulant activities, which means they can help prevent blood clots (Nandish et al., 2020). This makes them useful in the development of drugs for treating cardiovascular diseases and stroke. Additionally, some glycoproteins from fish scales have anti-inflammatory properties, which make them useful for the treatment of various inflammatory diseases such as arthritis and inflammatory bowel disease (Tabasum et al., 2017).

Another potential application of glycoproteins from fish scales is in wound healing. These glycoproteins have been shown to promote cell proliferation and wound closure, making them useful for the development of wound healing products (Serra et al., 2017). Overall, the extraction and characterization of glycoproteins from fish scales bio-waste hold great potential for various applications in the biomedical field. Further research is needed to fully understand their properties and functionalities, as well as to develop new and improved products based on these materials.

3.2.7 Nanohydroxyapatite

Nanohydroxyapatite is a nanoscale form of the mineral hydroxyapatite, which is the main mineral component of fish scales and bones (see Figure 3.2). Nanohydroxyapatite can be extracted from fish scales using various methods, including acid extraction and precipitation (Sricharoen et al., 2020). One of the unique properties of nanohydroxyapatite is its high biocompatibility, making it useful for various biomedical applications (Zainol, 2021). For example, nanohydroxyapatite has been used as a scaffold material for tissue engineering, due to its ability to support cell growth and promote tissue regeneration (Ma et al., 2021). It has also been used as a filler material for dental and bone repair, due to its ability to integrate with living tissues (Du et al., 2021). In addition, nanohydroxyapatite has been shown to have antimicrobial properties, making it useful for the development of antimicrobial materials. It has also been used as a drug delivery system, due to its ability to slowly release drugs over time (Lowe et al., 2019).

3.3 CHARACTERIZATION OF BIONANOMATERIALS FROM FISH SCALES BIO-WASTE

The characterization of bionanomaterials extracted from fish scales bio-waste is a crucial step in understanding their properties and functionalities. This information is essential for the development

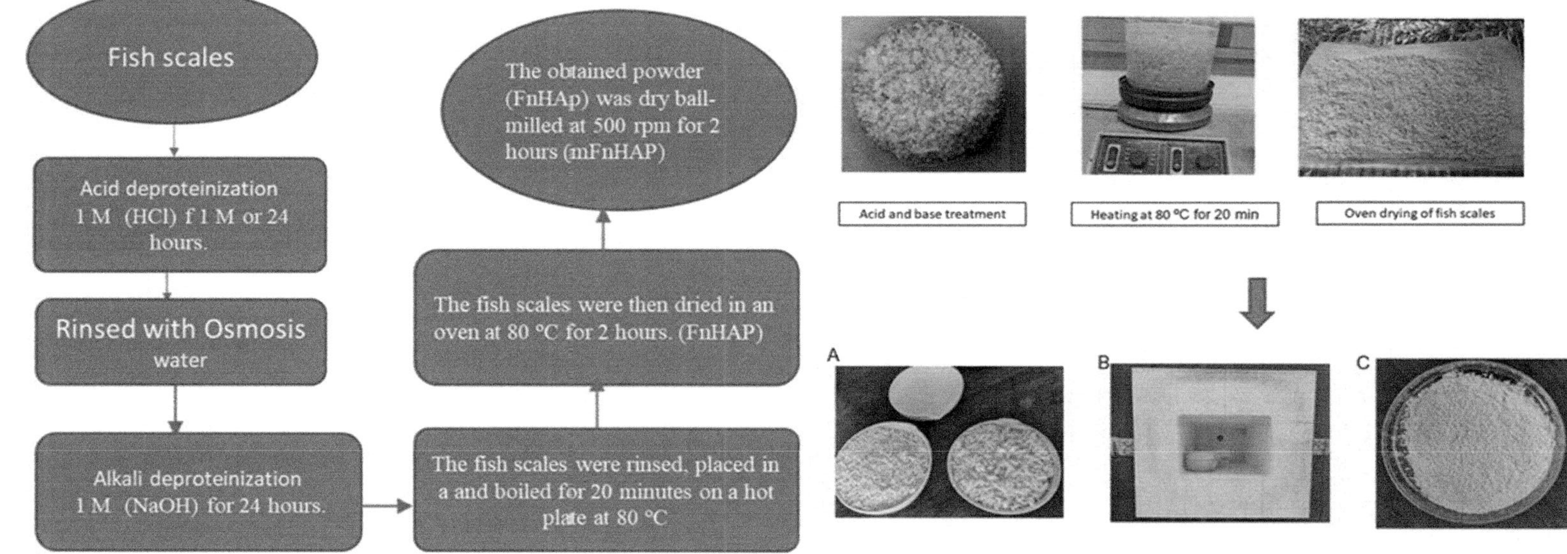

FIGURE 3.2 Preparation and extraction of nHAp from fish scales (A=oven dried fish scales, B=calcination of the fish scales, C=extracted nHAp). (Compiled by the authors.)

of new applications for these materials and for optimizing their use in existing applications. The characterization of bionanomaterials extracted from fish scales bio-waste can be carried out using a variety of techniques, including spectroscopy, microscopy, and X-ray diffraction (Surya et al., 2021).

- Spectroscopy is a powerful tool for characterizing bionanomaterials. It provides information about the chemical composition and structure of these materials, including their functional groups, chemical bonds, and molecular weight. Fourier-transform infrared spectroscopy (FTIR), ultraviolet-visible spectroscopy (UV-vis), and Raman spectroscopy are commonly employed spectroscopy techniques for the characterization of bionanomaterials obtained from fish scales bio-waste (Kontomaris and Stylianou, 2017). For instance, the FTIR spectrum in Figure 3.3 provides the functional group composition of nanohydroxyapatite from three samples labelled A-C. The absorption bands typical of hydroxyapatite characteristics were observed around ~1000 cm^{-1} and ~560 cm^{-1} for all the samples (A-C), respectively.
- Microscopy is another important tool for the characterization of bionanomaterials. It provides information about the shape, size, and distribution of these materials at the nanoscale. Transmission electron microscopy (TEM), scanning electron microscopy (SEM), and atomic force microscopy (AFM) are widely utilized microscopy techniques for the characterization of bionanomaterials derived from fish scales bio-waste (Wonorahardjo, 2019). Figure 3.4 illustrates the particle size and elemental analysis of nanohydroxyapatite biomaterial extracted from fish scales. The image in Figure 3.4A is obtained using TEM while the image in Figure 3.4B is obtained using SEM.
- X-ray diffraction is a powerful technique for the characterization of the crystalline structure of bionanomaterials. It provides information about the arrangement of the atoms in the material, including the size and shape of the crystalline domains, the crystalline phase, and the degree of crystallinity (Figure 3.5). This information is essential for understanding the mechanical properties and other functionalities of the material (Rabiei et al., 2020).

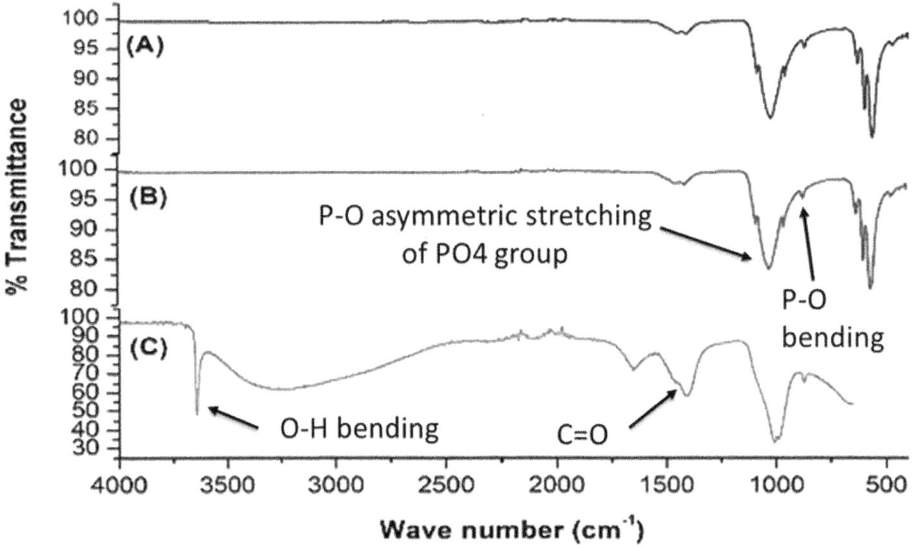

FIGURE 3.3 FTIR spectra of bionanomaterials extracted from biowaste (A and B are spectrum of nanohydroxyapatite extracted from fish scales while C is the spectrum of nanohydroxyapatite from eggshells). (Compiled by the authors.)

FIGURE 3.4 Microscopy and elemental characterization of bionanomaterials (Note that image A is obtained using TEM, B is obtained using SEM while C is the energy dispersive spectroscopy of the nHAp). (Compiled by the authors.)

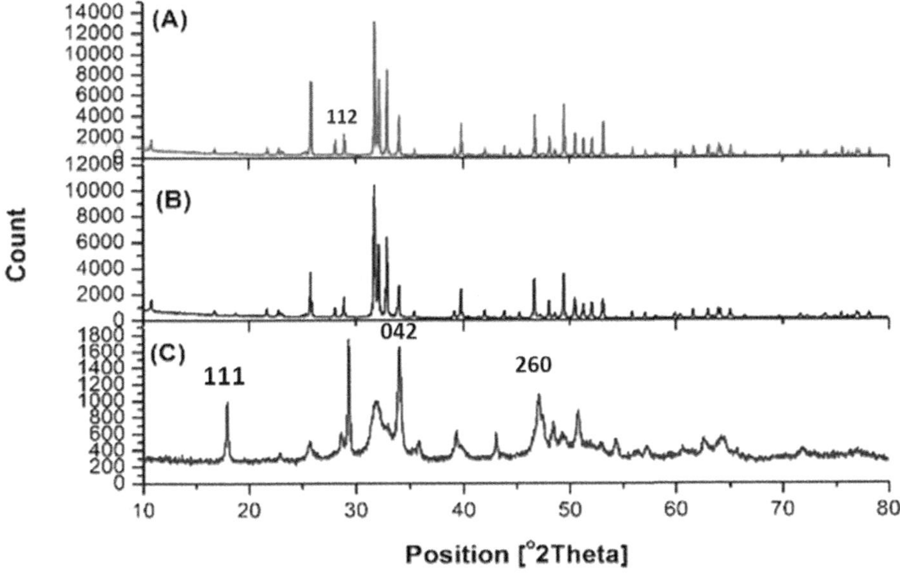

FIGURE 3.5 XRD pattern of bionanomaterials extracted from biowaste (A and B are pattern of nanohydroxyapatite extracted from fish scales while C is the spectrum of nanohydroxyapatite from eggshells). (Compiled by the authors.)

In summary, the characterization of bionanomaterials extracted from fish scales bio-waste is a crucial step in understanding their properties and functionalities. A combination of spectroscopy, microscopy, and X-ray diffraction can provide a comprehensive understanding of these materials, which is essential for their development and optimization for use in various applications.

3.3.1 Properties of Bionanomaterials from Fish Scales Bio-Waste

Bionanomaterials extracted from fish scales bio-waste have unique properties and functionalities that make them useful for various applications in different fields. These properties are due to the presence of biologically active compounds such as collagen, chitin, and calcium carbonate in the fish scales (Sathiyavimal et al., 2020). The unique properties of biomaterials include:

- **Mechanical properties**

One of the most notable properties of bionanomaterials from fish scales bio-waste is their mechanical strength. These materials have been found to have excellent tensile strength and toughness, making them suitable for use in tissue engineering and biomedical applications (Sathiyavimal et al., 2020). The mechanical properties can be characterized using techniques such as tensile testing and compression testing (Kim et al., 2019).

- **Biocompatibility and biodegradability**

Additionally, bionanomaterials have good biocompatibility, making them suitable for use in direct contact with biological tissues. Another important property of bionanomaterials from fish scales bio-waste is their biodegradability (Aswathi et al., 2022). These materials are biodegradable and therefore do not pose a long-term environmental hazard. This property makes them suitable for use in environmental applications such as water purification and the removal of heavy metals from contaminated water. Bionanomaterials from fish scales bio-waste have excellent antioxidant activity, making them useful for a variety of biomedical applications. Antioxidants can help to protect cells from oxidative damage, which is a major factor in the development of many diseases (Kim et al., 2019). Additionally, bionanomaterials from fish scales bio-waste have good adsorption properties, making them useful for the removal of heavy metals and other pollutants from contaminated water (Pai et al., 2020).

- **Size and shape**

Bionanomaterials can vary in size and shape, depending on the type of bio-waste and the extraction method used. The size and shape of bionanomaterials can be characterized using techniques such as transmission electron microscopy (TEM), scanning electron microscopy (SEM), and dynamic light scattering (DLS). The size and shape of bionanomaterials can affect their properties and potential applications (Aswathi et al., 2022). For example, smaller and more uniform bionanomaterials may be more suitable for biomedical applications, while larger and more irregular bionanomaterials may be more suitable for use in composite materials.

- **Surface area and porosity**

Bionanomaterials can have a high surface area and porosity, which can affect their properties and potential applications (Saleh, 2020). The surface area and porosity can be characterized using techniques such as Brunauer-Emmett-Teller (BET) analysis and mercury intrusion porosimetry (MIP). High surface area and porosity can make bionanomaterials suitable for use in adsorbents, catalysts, and drug delivery systems. Bionanomaterials can have a high crystallinity and a specific crystal structure, which can affect their properties and potential applications (Eichhorn et al., 2022). The crystallinity and crystal structure can be characterized using techniques such as X-ray diffraction (XRD) and Fourier transform infrared spectroscopy (FTIR). High crystallinity and a specific crystal structure can make bionanomaterials suitable for use in composite materials and biomedical applications.

- **Chemical composition**

The chemical composition of bionanomaterials can have a significant impact on their properties and potential applications. Bionanomaterials are typically made up of biological molecules, such as proteins, nucleic acids, lipids, and carbohydrates, which can be characterized using a variety of techniques (Nagamune, 2017). Some common techniques for characterizing the chemical composition of bionanomaterials include:

1. Spectroscopy: Spectroscopic techniques such as infrared (IR), Raman, and nuclear magnetic resonance (NMR) spectroscopy can be used to identify and quantify the chemical groups present in bionanomaterials.
2. Chromatography: Chromatographic techniques such as high-performance liquid chromatography (HPLC) and gas chromatography (GC) can be used to separate and analyze the components of complex bionanomaterials.
3. Mass spectrometry: Mass spectrometry (MS) can be used to determine the molecular weight and composition of bionanomaterials by ionizing and analyzing their constituent molecules.
4. X-ray crystallography: X-ray crystallography can be used to determine the three-dimensional structure of bionanomaterials and their constituent molecules, which can provide insight into their chemical composition and properties.

By using these techniques, researchers can gain a better understanding of the chemical composition of bionanomaterials and how it affects their properties and potential applications.

In summary, the unique properties and functionalities of bionanomaterials from fish scales biowaste make them useful for a variety of applications in different fields. Their mechanical strength, biocompatibility, biodegradability, antioxidant activity, and adsorption properties make them suitable for use in tissue engineering, biomedical applications, and environmental applications.

3.3.2 Applications of Bionanomaterials from Fish Scales Bio-Waste

Bionanomaterials extracted from fish scales bio-waste have a wide range of potential applications in different fields due to their unique properties and functionalities. In this section, we will discuss the most promising applications of these materials in three main areas: tissue engineering, biomedical, and environmental.

- **Tissue engineering**

Bionanomaterials from fish scales bio-waste have excellent mechanical strength and biocompatibility, making them ideal for use in tissue engineering applications. They can be used as scaffolds for the growth of new tissue and can also be used to reinforce tissue, improving its strength and stability. Additionally, the biodegradable nature of these materials makes them a promising alternative to traditional synthetic materials in tissue engineering applications (Sharifianjazi et al., 2021).

- **Biomedical**

Bionanomaterials from fish scales bio-waste have good antioxidant activity and biocompatibility, making them ideal for use in biomedical applications (Sathiyavimal et al., 2020). They can be used as a drug delivery system, as they can effectively encapsulate and release drugs in a controlled manner. Additionally, they can be used to remove heavy metals and other pollutants from the body, making them useful for the treatment of heavy metal poisoning and other environmental

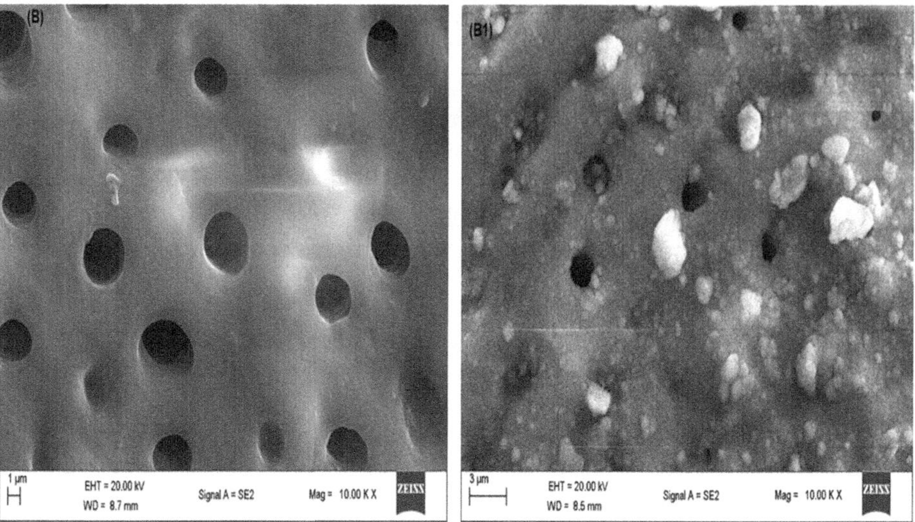

FIGURE 3.6 The occluding characteristics of nHAp extracted from fish scales (B) before treatment; (B1) after agitation treatment. (Compiled by the authors.)

exposure-related diseases (Pai et al., 2020). In dentistry, nHAp extracted from fish scales can be use in toothpaste formulation in the management of dentine hypersensitivity. Preliminary in vitro investigation carried provide evidence of the remineralization capabilities of fish scale nHAp (Figure 3.6).

- **Environmental**

Bionanomaterials from fish scales bio-waste have excellent adsorption properties and are biodegradable, making them ideal for use in environmental applications. They can be used to remove heavy metals and other pollutants from contaminated water, making them useful for water purification (Pai et al., 2020). Additionally, their biodegradable nature makes them a promising alternative to traditional synthetic materials in environmental applications, as they do not pose a long-term environmental hazard.

In conclusion, bionanomaterials obtained from fish scales bio-waste exhibit diverse properties and functionalities, offering a broad spectrum of potential applications in various fields. Their utilization in tissue engineering, biomedical, and environmental contexts showcase their versatility and potential as an alternative to conventional materials.

3.4 LIMITATIONS AND ADVANTAGES TO THE EXTRACTION OF BIONANOMATERIALS FROM FISH SCALES BIO-WASTE

3.4.1 Advantages

- **Sustainability**

The use of bio-waste from fish scales as a source of bionanomaterials provides a sustainable solution for managing waste and reducing environmental impact.

- Cost-effectiveness
  The extraction of bionanomaterials from fish scales bio-waste is relatively low-cost compared to the synthesis of materials from synthetic sources.

- Unique Properties
  Fish scales bio-waste is a rich source of materials with unique properties and functionalities, such as chitosan, collagen, calcium carbonate, enzymes, peptides, and glycoproteins, which have potential applications in various fields (Coppola et al., 2021).

3.4.2 Disadvantages

- **Lack of standardization**

The extraction process of bionanomaterials from fish scales bio-waste is still in its early stages, and there is a lack of standardization in the process, which can result in variations in the quality and properties of the extracted materials.

- **Limited production**

The production of bionanomaterials from fish scales bio-waste is currently limited and may not be able to meet the demand for these materials on a large scale.

3.4.3 Limitations

- **Limited availability**

Fish scales bio-waste is a limited resource, and the availability of this waste material may become a constraint for the production of bionanomaterials in the future.

- **Technical challenges**

The extraction of bionanomaterials from fish scales bio-waste can be technically challenging and requires further research to optimize the process and improve the yield and quality of the extracted materials. In summary, the extraction of bionanomaterials from fish scales bio-waste has great potential, but there are also several limitations and challenges that need to be addressed through further research and development.

3.5 DISCUSSION AND IMPLICATIONS

The study of bionanomaterials extracted from fish scales bio-waste has significant implications for various fields, including materials science, biomedical, and environmental (Aswathi et al., 2022). These materials have the potential to serve as a promising alternative to traditional synthetic materials, providing unique properties and functionalities that can be utilized in a variety of applications. In the field of materials science, the exploration of bionanomaterials derived from fish scales bio-waste holds promise for advancing the development of novel materials boasting distinct properties and functionalities. This is important as the demand for materials with specific properties and functionalities continues to grow, and new sources of these materials are needed. By studying bionanomaterials from fish scales bio-waste, valuable insights can be gained to guide the creation of innovative materials and potentially uncover new materials with exceptional characteristics.

Within the biomedical domain, the utilization of bionanomaterials sourced from fish scales bio-waste has the potential to revolutionize disease treatment approaches. Their remarkable antioxidant activity and biocompatibility render them highly suitable for biomedical applications like drug delivery and heavy metal removal, with the potential to significantly enhance patient outcomes and improve overall quality of life (Serati-Nouri et al., 2020). The use of these materials in biomedical applications has the potential to significantly improve patient outcomes and quality of life. In the environmental field, the use of bionanomaterials from fish scales bio-waste has the potential to significantly reduce the negative impact of traditional synthetic materials on the environment. Their biodegradable nature and excellent adsorption properties make them ideal for use in environmental applications, such as water purification and heavy metal removal. Utilizing these materials in environmental applications has the potential to considerably enhance environmental quality and safeguard public health.

3.6 CONCLUSION AND SUMMARY

In conclusion, the study of bionanomaterials extracted from fish scales bio-waste has significant implications for various fields, including materials science, biomedical, and environmental. Bionanomaterials have the potential to serve as a promising alternative to traditional synthetic materials, providing unique properties and functionalities that can be utilized in a variety of applications. The importance of this field cannot be overstated, and continued research is necessary to fully explore its potential and bring these materials to their full potential. One advantage of using bionanomaterials extracted from fish scales is that they are derived from a renewable and abundant source. Fish scales are readily available and can be obtained from a waste stream, reducing the need for virgin materials and reducing waste. Additionally, the extraction process can be sustainable and environmentally friendly, as it does not rely on harsh chemicals or pollutants. However, there are also certain disadvantages to consider. The extraction process for bionanomaterials from fish scales can be complex and may require specialized equipment and expertise. Additionally, the purity of the extracted bionanomaterials may vary depending on the source and processing methods used, which can affect their properties and functionalities.

Considering the limitations, there is a need for further research and development to fully understand the properties and functionalities of bionanomaterials extracted from fish scales. This includes a deeper understanding of the structure–property relationships of these materials and the optimization of their extraction and purification methods. Overall, the use of bionanomaterials extracted from fish scales has great potential, but it is important to carefully consider the advantages, disadvantages, and limitations of these materials. Further research and development is needed to fully unlock their potential for various applications.

3.7 RECOMMENDATIONS AND FUTURE DIRECTIONS

The study of bionanomaterials extracted from fish scales bio-waste is a promising field with significant potential to contribute to the development of new materials with unique properties and functionalities. However, despite the promising results, further research is necessary to fully explore and realize the potential of these materials. One of the main areas for future research is the development of new extraction methods for bionanomaterials from fish scales bio-waste. The current extraction methods have limitations, and the development of new methods that are more efficient, cost-effective, and environmentally friendly is necessary.

Another area for future research is the improvement of the properties of bionanomaterials from fish scales bio-waste. The current materials have excellent properties, but further research is necessary to improve their properties and functionalities, making them more useful for a wider range of applications. Finally, the development of new applications for bionanomaterials from fish scales bio-waste is a crucial area for future research. The current applications are promising, but further research is necessary to fully explore the potential of these materials and develop new applications in fields such as tissue engineering, biomedical, and environmental.

In conclusion, bionanomaterials extracted from fish scales bio-waste are a promising source of materials for various applications due to their unique properties and functionalities. Further research is needed to fully explore their potential and to develop new and improved applications for these materials. The sustainable and environmentally friendly extraction of bionanomaterials from fish scales bio-waste will provide a solution for the management of this waste and help to reduce the environmental impact of the fishing industry.

REFERENCES

ABRIL, A. G., PAZOS, M., VILLA, T. G., CALO-MATA, P., BARROS-VELÁZQUEZ, J. & CARRERA, M. 2022. Proteomics characterization of food-derived bioactive peptides with anti-allergic and anti-inflammatory properties. *Nutrients,* 14, 4400.

AHMED, M., VERMA, A. K. & PATEL, R. 2020. Collagen extraction and recent biological activities of collagen peptides derived from sea-food waste: a review. *Sustainable Chemistry and Pharmacy,* 18, 100315.

ALVES, M. C., PAULA, M. M. D. O., COSTA, C. G. C. D., SALES, L. A., LAGO, A. M. T., PIMENTA, C. J. & GOMES, M. E. D. S. 2021. Restructured fish cooked ham: effects of the use of carrageenan and transglutaminase on textural properties. *Journal of Aquatic Food Product Technology,* 30, 451–461.

ASWATHI, V., MEERA, S., MARIA, C. A. & NIDHIN, M. 2022. Green synthesis of nanoparticles from biodegradable waste extracts and their applications: a critical review. *Nanotechnology for Environmental Engineering,* 7, 1–21.

BILAL, M., GUL, I., BASHARAT, A. & QAMAR, S. A. 2021. Polysaccharides-based bio-nanostructures and their potential food applications. *International Journal of Biological Macromolecules,* 176, 540–557.

COPPOLA, D., LAURITANO, C., PALMA ESPOSITO, F., RICCIO, G., RIZZO, C. & DE PASCALE, D. 2021. Fish waste: from problem to valuable resource. *Marine Drugs,* 19, 116.

DEBEAUFORT, F. 2021. Active biopackaging produced from by-products and waste from food and marine industries. *FEBS Open Bio,* 11, 984–998.

DU, M., CHEN, J., LIU, K., XING, H. & SONG, C. 2021. Recent advances in biomedical engineering of nano-hydroxyapatite including dentistry, cancer treatment and bone repair. *Composites Part B: Engineering,* 215, 108790.

EICHHORN, S. J., ETALE, A., WANG, J., BERGLUND, L. A., LI, Y., CAI, Y., CHEN, C., CRANSTON, E. D., JOHNS, M. A. & FANG, Z. 2022. Current international research into cellulose as a functional nanomaterial for advanced applications. *Journal of Materials Science,* 57, 5697–5767.

FELICIAN, F. F., XIA, C., QI, W. & XU, H. 2018. Collagen from marine biological sources and medical applications. *Chemistry & Biodiversity,* 15, e1700557.

FURTADO, M., CHEN, L., CHEN, Z., CHEN, A. & CUI, W. 2022. Development of fish collagen in tissue regeneration and drug delivery. *Engineered Regeneration,* 3, 154–168.

GHOSH, C., SARKAR, P., ISSA, R. & HALDAR, J. 2019. Alternatives to conventional antibiotics in the era of antimicrobial resistance. *Trends in Microbiology,* 27, 323–338.

GUO, Z., YAN, N. & LAPKIN, A. A. 2019. Towards circular economy: integration of bio-waste into chemical supply chain. *Current Opinion in Chemical Engineering,* 26, 148–156.

HALIM, N. A. A., HUSSEIN, M. Z. & KANDAR, M. K. 2021. Nanomaterials-upconverted hydroxyapatite for bone tissue engineering and a platform for drug delivery. *International Journal of Nanomedicine,* 16, 6477.

HARIKRISHNA, N., MAHALAKSHMI, S., KIRAN KUMAR, K. & REDDY, G. 2017. Fish scales as potential substrate for production of alkaline protease and amino acid rich aqua hydrolyzate by Bacillus altitudinis GVC11. *Indian Journal of Microbiology,* 57, 339–343.

IBER, B. T., KASAN, N. A., TORSABO, D. & OMUWA, J. W. 2022. A review of various sources of chitin and chitosan in nature. *Journal of Renewable Materials,* 10, 1097.

ISHAK, N. & SARBON, N. 2018. A review of protein hydrolysates and bioactive peptides deriving from wastes generated by fish processing. *Food and Bioprocess Technology,* 11, 2–16.

ISLAM, S., BHUIYAN, M. R. & ISLAM, M. 2017. Chitin and chitosan: structure, properties and applications in biomedical engineering. *Journal of Polymers and the Environment,* 25, 854–866.

JAFARI, H., LISTA, A., SIEKAPEN, M. M., GHAFFARI-BOHLOULI, P., NIE, L., ALIMORADI, H. & SHAVANDI, A. 2020. Fish collagen: extraction, characterization, and applications for biomaterials engineering. *Polymers,* 12, 2230.

KIM, D.-K., WOO, W., KIM, E.-Y. & CHOI, S.-H. 2019. Microstructure and mechanical characteristics of multi-layered materials composed of 316L stainless steel and ferritic steel produced by direct energy deposition. *Journal of Alloys and Compounds,* 774, 896–907.

KONTOMARIS, S. & STYLIANOU, A. 2017. Atomic force microscopy for university students: applications in biomaterials. *European Journal of Physics,* 38, 033003.

LAMP, A., KALTSCHMITT, M. & DETHLOFF, J. 2022. Options to improve the mechanical properties of protein-based materials. *Molecules,* 27, 446.

LIN, Z., WU, T., WANG, W., LI, B., WANG, M., CHEN, L., XIA, H. & ZHANG, T. 2019. Biofunctions of antimicrobial peptide-conjugated alginate/hyaluronic acid/collagen wound dressings promote wound healing of a mixed-bacteria-infected wound. *International Journal of Biological Macromolecules,* 140, 330–342.

LIONETTO, F. & ESPOSITO CORCIONE, C. 2021. Recent applications of biopolymers derived from fish industry waste in food packaging. *Polymers,* 13, 2337.

LOWE, B., HARDY, J. G. & WALSH, L. J. 2019. Optimizing nanohydroxyapatite nanocomposites for bone tissue engineering. *ACS Omega,* 5, 1–9.

LUPU, M. A., GRADISTEANU PIRCALABIORU, G., CHIFIRIUC, M. C., ALBULESCU, R. & TANASE, C. 2020. Beneficial effects of food supplements based on hydrolyzed collagen for skin care. *Experimental and Therapeutic Medicine,* 20, 12–17.

MA, P., WU, W., WEI, Y., REN, L., LIN, S. & WU, J. 2021. Biomimetic gelatin/chitosan/polyvinyl alcohol/nano-hydroxyapatite scaffolds for bone tissue engineering. *Materials & Design,* 207, 109865.

MARTĂU, G. A., MIHAI, M. & VODNAR, D. C. 2019. The use of chitosan, alginate, and pectin in the biomedical and food sector—biocompatibility, bioadhesiveness, and biodegradability. *Polymers,* 11, 1837.

MENG, Q., SUN, Y., CONG, H., HU, H. & XU, F.-J. 2021. An overview of chitosan and its application in infectious diseases. *Drug Delivery and Translational Research,* 11, 1340–1351.

NAGAMUNE, T. 2017. Biomolecular engineering for nanobio/bionanotechnology. *Nano Convergence,* 4, 9.

NANDISH, S. K. M., KENGAIAH, J., RAMACHANDRAIAH, C., SHIVAIAH, A., SHANKAR, R. L. & SANNANINGAIAH, D. 2020. Purification and characterization of non-enzymatic glycoprotein (NEGp) from flax seed buffer extract that exhibits anticoagulant and antiplatelet activity. *International Journal of Biological Macromolecules,* 163, 317–326.

PAI, S., KINI, S. M., SELVARAJ, R. & PUGAZHENDHI, A. 2020. A review on the synthesis of hydroxyapatite, its composites and adsorptive removal of pollutants from wastewater. *Journal of Water Process Engineering,* 38, 101574.

PRIYADARSHI, R. & RHIM, J.-W. 2020. Chitosan-based biodegradable functional films for food packaging applications. *Innovative Food Science & Emerging Technologies,* 62, 102346.

QIN, D., BI, S., YOU, X., WANG, M., CONG, X., YUAN, C., YU, M., CHENG, X. & CHEN, X.-G. 2022. Development and application of fish scale wastes as versatile natural biomaterials. *Chemical Engineering Journal,* 428, 131102.

RABIEI, M., PALEVICIUS, A., MONSHI, A., NASIRI, S., VILKAUSKAS, A. & JANUSAS, G. 2020. Comparing methods for calculating nano crystal size of natural hydroxyapatite using X-ray diffraction. *Nanomaterials,* 10, 1627.

REZVANI GHOMI, E., NOURBAKHSH, N., AKBARI KENARI, M., ZARE, M. & RAMAKRISHNA, S. 2021. Collagen-based biomaterials for biomedical applications. *Journal of Biomedical Materials Research Part B: Applied Biomaterials,* 109, 1986–1999.

SALEEM, H. & ZAIDI, S. J. 2020. Developments in the application of nanomaterials for water treatment and their impact on the environment. *Nanomaterials,* 10, 1764.

SALEH, T. A. 2020. Nanomaterials: classification, properties, and environmental toxicities. *Environmental Technology & Innovation,* 20, 101067.

SATHIYAVIMAL, S., VASANTHARAJ, S., LEWISOSCAR, F., SELVARAJ, R., BRINDHADEVI, K. & PUGAZHENDHI, A. 2020. Natural organic and inorganic–hydroxyapatite biopolymer composite for biomedical applications. *Progress in Organic Coatings,* 147, 105858.

SAULAT, H., CAO, M., KHAN, M. M., KHAN, M., KHAN, M. M. & REHMAN, A. 2020. Preparation and applications of calcium carbonate whisker with a special focus on construction materials. *Construction and Building Materials,* 236, 117613.

SERATI-NOURI, H., JAFARI, A., ROSHANGAR, L., DADASHPOUR, M., PILEHVAR-SOLTANAHMADI, Y. & ZARGHAMI, N. 2020. Biomedical applications of zeolite-based materials: a review. *Materials Science and Engineering: C,* 116, 111225.

SERRA, M. B., BARROSO, W. A., SILVA, N. N. D., SILVA, S. D. N., BORGES, A. C. R., ABREU, I. C. & BORGES, M. O. D. R. 2017. From inflammation to current and alternative therapies involved in wound healing. *International Journal of Inflammation,* 2017, 1–17

SHARIFIANJAZI, F., ESMAEILKHANIAN, A., MORADI, M., PAKSERESHT, A., ASL, M. S., KARIMI-MALEH, H., JANG, H. W., SHOKOUHIMEHR, M. & VARMA, R. S. 2021. Biocompatibility and mechanical properties of pigeon bone waste extracted natural nano-hydroxyapatite for bone tissue engineering. *Materials Science and Engineering: B,* 264, 114950.

SHIVATARE, S. S., SHIVATARE, V. S. & WONG, C.-H. 2022. Glycoconjugates: synthesis, functional studies, and therapeutic developments. *Chemical Reviews,* 122, 15603–15671.

SRICHAROEN, P., LIMCHOOWONG, N., NUENGMATCHA, P. & CHANTHAI, S. 2020. Ultrasonic-assisted recycling of Nile tilapia fish scale biowaste into low-cost nano-hydroxyapatite: ultrasonic-assisted adsorption for Hg2+ removal from aqueous solution followed by "turn-off" fluorescent sensor based on Hg2+-graphene quantum dots. *Ultrasonics Sonochemistry,* 63, 104966.

SURYA, P., NITHIN, A., SUNDARAMANICKAM, A. & SATHISH, M. 2021. Synthesis and characterization of nano-hydroxyapatite from Sardinella longiceps fish bone and its effects on human osteoblast bone cells. *Journal of the Mechanical Behavior of Biomedical Materials,* 119, 104501.

TABASUM, S., NOREEN, A., KANWAL, A., ZUBER, M., ANJUM, M. N. & ZIA, K. M. 2017. Glycoproteins functionalized natural and synthetic polymers for prospective biomedical applications: a review. *International Journal of Biological Macromolecules,* 98, 748–776.

TAKARINA, N. D., NASRUL, A. A. & NURMARINA, A. 2017. Degree of deacetylation of chitosan extracted from white snapper (Lates sp.) scales waste. *International Journal of Pharma Medicine and Biological Sciences,* 6, 16–19.

WONORAHARDJO, S. 2019. Spectroscopy for characterization and application of silica-cellulose biomaterials, a brief review article about analytical methods. *IOP Conference Series: Materials Science and Engineering*, IOP Publishing, 012102.

YADAV, V. K., YADAV, K. K., CABRAL-PINTO, M. M., CHOUDHARY, N., GNANAMOORTHY, G., TIRTH, V., PRASAD, S., KHAN, A. H., ISLAM, S. & KHAN, N. A. 2021. The processing of calcium rich agricultural and industrial waste for recovery of calcium carbonate and calcium oxide and their application for environmental cleanup: a review. *Applied Sciences,* 11, 4212.

YAN, D., LI, Y., LIU, Y., LI, N., ZHANG, X. & YAN, C. 2021. Antimicrobial properties of chitosan and chitosan derivatives in the treatment of enteric infections. *Molecules,* 26, 7136.

ZAINOL, I. 2021. Mechanical properties improvement of epoxy composites by natural hydroxyapatite from fish scales as fillers. *Journal of Polymer Research, 28, 1–12.*

ZHENG, H., ZHAO, M., DONG, Q., FAN, M., WANG, L. & LI, L. 2022. Extruded transglutaminase-modified gelatin–beeswax composite packaging film. *Food Hydrocolloids,* 132, 107849.

Part 2

Bionanomaterials in Food Applications

4 Bionanomaterials in Food Packaging

Duygu Aslan Türker

4.1 INTRODUCTION

Packaging system protects the product inside from external factors. The primary role of food packaging is to provide food safety and quality throughout transportation and storage as well as to prevent factors and conditions such as pathogenic microorganisms, air, light, oxygen, and humidity from spoiling the food (Siegrist et al., 2008). Food packages should not absorb moisture into the product and should prevent moisture loss from the product. Packaging materials should preserve the food from microbial contamination and serve as a barrier to the entry of other volatile components such as carbon dioxide, water vapor, flavor, and oxygen cinto the food (Tornuk et al., 2015). As customers have grown more conscientious, there has been a surge in interest in solutions of packaging for food where all of these negatives are avoided, and greater quality and safe food preservation are achievable. It is possible to preserve the physical and chemical structures of foods with applications such as non-thermal, PEF (Pulsed Electric Field) and UV. However, these methods have some disadvantages such as not being able to inactivate spores of spore-producing bacteria. All these problems have made it inevitable to increase the tendency towards new generation technologies and food packaging materials such as modified atmosphere packaging, edible film, bioplastic material technologies, and coating materials, active and smart packaging systems. Nanotechnology is used to make the attributes of the polymer used in food packaging much stronger. Among these applications, the usage of packaging materials made from biodegradable bionanomaterials, which are not harmful to health, should become widespread. Therefore, this chapter compiles the rapidly novel applications of bionanomaterials in food packaging. In this chapter, the usage of bionanomaterials in food packaging, their manufacturing and characterization procedures as well as the food safety issues are investigated.

4.2 NANOTECHNOLOGY APPLICATIONS IN FOOD PACKAGING

Freshness is the most fundamental quality indicator of the food, and this characteristic is an essential parameter used by customers to assess whether the food will be purchased. Fish, for example, is one of the most difficult items to keep fresh. When purchasing fish, you may be able to tell its freshness based on its eyes, aliveness, and color; however, determining the freshness of the product in cut and packed fish is quite difficult. While packaging producers attempt to keep food fresher for longer periods of time, customers request to view the freshness of the product without having to open the package. Nanotechnology and bionanomaterials offers options that provide these advantages in foods.

DOI: 10.1201/9781003432791-6

Nanotechnology is a discipline of technology and science that attempts to develop products and systems with novel physical, chemical, and biological features, such as the manufacture, processing, characterization, design, and modeling of substances spanning in size from 1 to 100 nm (Silvestre et al., 2011). When the particle size is decreased to these dimensions, the resulting material has drastically varying chemical and physical characteristics than do macro-materials consisting of the similar constituent (Duncan, 2011). In nano dimensions, improving the "surface area/volume" ratio makes the material considerably more effective, causing it to interact differently with the surrounding molecules. Nanotechnological applications have increased rapidly in recent years, and today there are many specialized companies producing nano-sized materials (Roco et al., 2011).

Nanotechnological packaging provides many advantages in the food industry. Therefore, nanotechnology advancements are becoming an increasingly crucial topic for the food sector in both emerging and industrialized countries. Although local priorities are important, the benefits that it delivers in general are as follows:

- Improved techniques for food production/less pesticide use
- Food and feed processing that is more hygienic
- Foods with a longer shelf life
- Innovative, lightweight, strong and functional packaging (Chaudhry et al., 2008).

Besides, nanotechnology improves product durability while decreasing manufacturing costs. While it keeps food fresh for a long period, its environmental impact is less than that of petroleum-based polymers. Therefore, research on these newly designed packages has intensified. Nevertheless, their use is not yet prevalent. However, nanotechnology tends to produce innovative materials in the food packaging industry. The mechanical, barrier, and thermal qualities will be strengthened with the inclusion of nanoparticles suited for acceptable foods. The consequences of nanotechnology-based foods and packaging materials remain mostly unknown. Future research will eliminate legal and scientific deficiencies.

Many sectors, including materials, electronics, computers, medicine, pharmaceuticals, textiles, the environment, energy, biotechnology, agriculture, and food, can benefit from nanotechnology. As be seen in Figure 4.1, nanotechnology is used in the food products such as nano-emulsions, nanoparticles, in the field of food security and safety, food materials, and food processing.

Food packaging science and technology have the broadest application area in nanotechnology. Fillers are used to offer the necessary qualities of food packaging when plastic or biodegradable materials are manufactured. The extra infill material is classified into three types based on its form.

1. Equidimensional nanoparticles: They are nanometer-scale fillers with three dimensions.
2. Nanotubes: These are elongated structures having two nanoscale dimensions and a third larger dimension.
3. Polymer-layered crystal structures: Nano-scale composites (Alexandre et al., 2000; De Azeredo, 2009).

Applications for nanotechnology in the food sector include developing food packaging systems for quality and reliable food production, ensuring the traceability of foods using biosensors, and identifying bacteria by developing active and smart packaging systems (Vermeiren et al., 1999). Food nanotechnology is primarily concerned with food additives, nutritional supplements and antimicrobial food and beverage containers. Nanoparticles used in food packaging systems ensure that the packaging becomes more durable, flexible, and resistant to light and gases in adverse environmental conditions (Gupta, 2021). Figure 4.2 summarizes possible uses of nanotechnology in the food and food packaging industry. Food nanotechnology applications include: (a) processing of

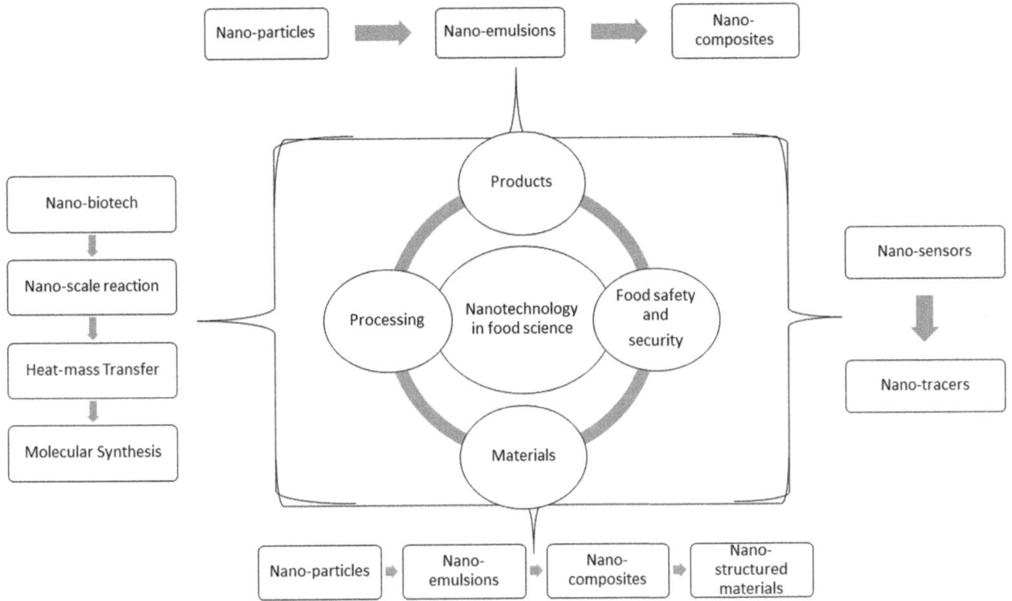

FIGURE 4.1 The usage of nanotechnology in foods. (Nile et al., 2020.)

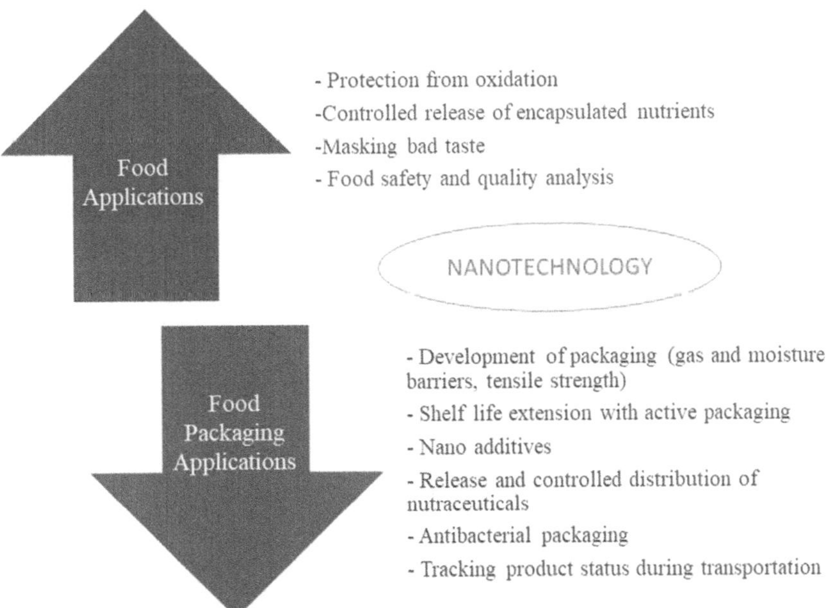

FIGURE 4.2 Nanotechnology's potential applications in the food and food packaging industries. (Compiled by the author.)

food and functional product development, (b) transportation and controlled release of biologically active compounds, (c) pathogen detection and increasing food safety, (d) packaging system development (Erdem et al., 2015).

Packaging of food is another branch where nanotechnology is used in the field of food. The primary purpose of nano packaging has been to increase the product shelf-life by improving the barriers of the package, reducing gas and moisture exchange, and reducing food products' contact with UV rays. In this regard, studies on food packaging produced using nanotechnology and bionanomaterials have increased considerably in recent years. In the previous study, various nanoparticles such as titanium and silver oxide were added to the packaging materials, modifying the permeability of the product, creating an anaerobic environment by adding oxygen to the packaging surface that is in contact with the food, thereby creating antimicrobial and antifungal surfaces (ElAmin, 2005; Roach, 2006). Another study found that employing different nanocomposites reduced the oxygen and carbon dioxide permeability of packing materials, prevented undesirable smells, conserved product freshness, and enhanced shelf life (Sherman, 2004).

The usage of nanoparticle materials provides flexibility, durability, heat/moisture stability, excellent barrier qualities against gases, light resistance, and strong mechanical and thermal performance to food packaging. It is possible to guarantee that the packing material is light, tear-proof, and resistant to high temperatures, while also enhancing its mechanical qualities. Other applications include increasing the permeability of the material by adding different nanoparticles to food packaging, creating an anaerobic environment by providing oxygen absorption on the packaging surface that comes into contact with food, and providing an antimicrobial and antifungal surface (Mahalik et al., 2010). In the literature, nanomaterials with properties such as antibiotic carriers (Gu et al., 2003) and killing agents (Huang et al., 2005) was studied for antimicrobial activity for packaging of food. The packaging film produced by coating with nano titanium dioxide (TiO_2) is expected to be capable of reducing *E.coli* contamination on the fresh vegetables and fruits surface (Theinsathid et al., 2011). Antimicrobial peptides such as nisin as nano-sized coating material were used to create antimicrobial films (De Azeredo, 2009). Nanocomposite films (nanocomposites) made with nanoparticles are widely used in the fruit and vegetables, meat and meat products, seafood, and confectionary industries. Nanocomposites have a wide range of features, such as reinforcement of composite, barrier qualities, resistance to flame, improved electro-optical features, antibacterial capabilities, and cosmetic applications (D.R. Paul et al., 2008). As a result, current research indicates that both are conceivable with nanotechnology.

Bionanomaterials can also be found in food packaging as nanosensors or nanobiosensors. Some nanotechnology applications in food packaging have concentrated on films with nanoparticles or nanosensors incorporated in packaging material that may detect the presence of harmful bacteria. These sensors can detect signals electrically, optically, thermally, or chemically. Nanosensors are rapidly detecting microorganisms that cause food poisoning or create toxins based on the temperature, mass, and color changes. There has been an upsurge in the studies on this topic recently. As an example of these investigations, nano TiO_2 and nano SnO_2 were used as oxygen sensors in MAP-applied food packaging. As the oxygen concentration of the package increases owing to microbial activity, these nanoparticles will sensitize the redox dyes in the polymer surroundings to light, producing bleaching in the sensor color of the package (De Azeredo, 2009; Mills et al., 2009). Biosensors are described by the International Union of Pure and Applied Chemistry (IUPAC) as "devices that convert the biological response to a chemical compound into optical, thermal or electrical signals". Biosensors consist of two principal parts as transducer and bioreceptor (ligand). The bioreceptor is the biological binding site that captures the target chemical. The converter, on the other hand, is the part that is formed as a result of the binding and converts biochemical/physicochemical interactions into electronic signals (Nagel et al., 1992). In recent years, nanobiosensors have been improved by integrating nanotechnology with biosensing methods. These sensors' great sensitivity and rapid reaction times are owing to the use of small-sized nanostructures with exceptionally large surface areas, such as nanotubes, nanofibers, and nanoparticles. Microorganisms in meat and dairy products, toxic compounds in meat and fruit juices, and food components such as glucose and carbohydrates, are easily and quickly detected by means of nanosensors and nanobiosensors (Palaniappan et al., 2013; M. Yang et al., 2008; Zhang et al., 2014). Systems that detect and respond automatically when there is any external stimulus effect or change on the packaging are called bioswitch systems. In these systems, which are triggered by environmental variations including pH, light, and temperature, if there are microorganisms in the environment the antimicrobial substance is released in the package. Starch-based packaging film containing antimicrobial agent nanocapsules can be given as an example (Huff, 2008; Muhammad et al., 2005). The conclusion to be derived from this and the study conducted, show that the desired properties can be controlled to a greater extent with nanoparticles added to packaging materials compared to other methods.

4.3 CLASSIFICATION OF BIONANOMATERIALS

Bionanomaterials are categorized as carbon-based organic and inorganic nanoparticles due to their chemical structures.

4.3.1 Organic Nanoparticles

Organic polymer-based nanoparticles are of natural origin (gelatin, chitosan, alginate, nanocrystalline cellulose) and synthetic-based polycaprolactone (PCL), polylactic glycolic acid (PLGA), and cyanoacrylate. Non-polymer-based nanoparticles are carbon-based carriers liposomes and solid-lipid nanoparticles (Lai et al., 2014). The most used in the food industry are gelatin, chitosan, and liposome. Gelatin nanoparticles are produced by desolvation and water-in-oil emulsification. Chitosan is produced by ionic gelation, polyelectrolyte complex, drying processes, and emulsion system (Shoueir et al., 2021). Liposomes are surface-active (usually phospholipids) spherical structures capable of spontaneous bilayer formation in aqueous solutions that have two layers of nonpolar molecules facing another (McClements, 2014). Thin film hydration, solvent injection, detergent removal, heating, homogenization, supercritical liquids, and high pressure homogenization methods are used in its production (Subramani et al., 2020).

The use of weak (non-covalent) interactions in molecular design aid in the conversion of organic nanoparticles into desirable forms such as liposomes, dendrimers, polymer nanoparticles, and micelles. These nanoparticles degrade naturally and are safe. The hollow core of certain particles, such as liposomes and micelles, which are often referred to as nanocapsules, makes them susceptible to electromagnetic and thermal radiation such as light and heat (Tiwari et al., 2008).

Moreover, it is possible to synthesize herbal nanoparticles from plants. Nanoparticles synthesized from plants are noted for being rapid, stable, and affordable due to their widely accessible and broad properties. From the beginning of the twentieth century, it has been recognized that plant extracts may decrease metal ions, although it is still unclear how these reducing agents work naturally. Medicinal plants such as *Camelia sinensis Allium sativum*, *Acalypha indica*, *Calotropis procera*, and *Boswellia ovalifoliolata* are some of the plants used in the production of silver nanoparticles (Yavuz et al., 2021). It has been reported that silver nanoparticles obtained from plants with strong content in terms of phytochemicals such as quinones and protein remain more stable (Beykaya et al., 2016). It has been stated that the nanoparticles obtained by synthesizing silver nanoparticles (AgNP) from *Salvia limbata* extract, which is among the endemic plants in Iran, eliminate the toxic effect and protect the environment (Sugandha et al., 2014). The rapid use of powdered silver nanoparticle extracts in biological syntheses makes it environmentally friendly as well as being practical and efficient. However, apart from being used in different ways in biomedical studies, being economical, and predisposed to medical and medical studies, it also has the potential to be a commercial product. Similar research found that nanoparticles made from the seed extracts of *Argyeria nervosa* have a strong antagonistic impact on bacteria and fungus (Thombre et al., 2014). Another study found that silver nanoparticles produced with *Ficus benghalensis* leaf extract have environmental friendly properties, and that the amino groups of proteins has a key role in the produced particles stability (Saware et al., 2014). In recent years, the capacities of plants to decrease metal salts have been used. The conformity of the structural properties of silver nanoparticle and its antimicrobial functionality, as well as their capacity to decrease toxicity in the cell, are unique to these particles. The antibacterial activity of silver nanoparticles developed with environmentally friendly green nanotechnology against *Pseudomonas aeruginosa, Escherichia coli, Proteus* species, and *Staphylococcus epidermidis* was proven in a study by Oyagi et al. (2014) using *Argyreia cymosa* rather than Rooibos tea.

4.3.2 Inorganic Nanoparticles

Non-carbon-based nanoparticles are referred to as inorganic nanoparticles. Inorganic nanoparticles are typically categorized as those made of metal and metal oxide. These nanoparticles are created synthetically as metal nanoparticles, such as Ag or Au, and MeO nanoparticles, such as ZnO and Ti_2O (Jeevanandam et al., 2018). Inorganic nanoparticles are obtained from metals and trace elements with magnetic properties, including silver, copper, titanium, iron, and zinc. Metallic nanoparticles bind seamlessly to DNA, so they are utilized in medical applications (Mohammed, 2016). Moreover, due to the superior attributes of metal nanoparticles, they are utilized in the development of solar cells, sensors and trace matter detection, photoimaging, quantum lasers and computers, dye, optical limiting devices, textile industry, energy storage, energy, environment, food packaging, drug delivery, and as wound healing agents (Sana et al., 2021). Small size and large surface areas are characteristics of metal-based nanoparticles. Metal-based nanoparticles are those created by either constructive or destructive techniques from metals down to nanometric sizes. Nanoparticles may be created from almost any metal (Ealia et al., 2017). The toxicity mechanism of metal nanoparticles can be variable. Most of the metals are toxic for all cell types, nevertheless, this activity might vary among bacteria. Metal-based nanoparticles studied for antibacterial activity and reported activities contain aluminum oxide, iron, silver, CuO, FeO, Ti_2O, ZnO, gallium, and gold nanoparticles (Aderibigbe, 2017). The metal oxides, which contain both low- and high-melting point metals, are produced when oxygen and metal are combined. Depending on their various electronic structures, metal oxides demonstrate semiconducting, metallic, and insulating properties. Metal oxide nanoparticles, which range in size between 1 and 100 nm and come in a variety of forms and sizes, are one type of these nanomaterials. The improved reactivity and yields of metal oxide nanoparticles are the key reasons they are manufactured. As compared to their metal counterparts, these nanoparticles exhibit an exceptional characteristic (Gangwar et al., 2016).

4.3.3 Carbon-Based Nanoparticles

Carbon-based nanoparticles, include carbon and in general can take the form of hollow tubes, ellipsoids, or spheres. Carbon-based nanomaterials combine the particular properties of sp^2 hybridized carbon bonds with the uncommon properties of nanoscale physics and chemistry (Mauter et al., 2008). Carbon nanotubes are tubular forms of carbon that has a diameter as tiny as 1 nm and a length ranging from few nm to several microns. The atomic arrangement of the carbon atoms that make up the nanotube determines whether it is metallic or semiconductor in nature. As a result, the electrical, mechanical, optical, and thermal characteristics are all unique. Therefore, carbon nanotubes are mostly benefitted in different applications (Purohit et al., 2014). The structure of fullerene shows unique chemical, optical, and structural properties (Mauter and Elimelech, 2008). After graphite and diamond, fullerene is regarded as the third allotropic form of carbon (Siqueira et al., 2017). The monoatomic, two-dimensional carbon sheet known as graphene is distinguished by its exceptional electrical, chemical, physical, and optical characteristics. Recently, according to certain research, nanoparticle-graphene hybrid structures might work in concert to provide a variety of special physicochemical features. These graphene-nanoparticle complexes are of special interest because they can display additional qualities that improve the selectivity and sensitivity possible utilizing various detection modalities in addition to exhibiting the specific capabilities of graphene and nanoparticles (Yin et al., 2013).

4.4 FOOD PACKAGING WITH BIONANOMATERIALS

In food packaging, nanotechnology is investigated in three groups: (a) nanocomposite packaging materials, (b) biodegradable nanocomposite packaging materials, (c) antimicrobial active and smart nano packaging.

4.4.1 Bionanocomposite Packaging Materials

Many material applications already make use of a significant quantity of naturally occurring polymers made from renewable resources, including starch, cellulose, and rubber. Natural polymers could be classified in different ways. For instance: while starch and cellulose are divided into different groups according to their physical characteristics, both are polysaccharides according to chemical classification. Figure 4.3 provides a list of some natural polymers. These natural polymers fulfill their unique functions in different ways. For example, the function of polysaccharides is in intra-cellular junctions and membranes; lipids serve as energy reservoirs and proteins serve as catalysts and building blocks. Nature provides an effective arrangement of polymers that potentially find applications in films, thermosetting, and thermoplastics resins, gels, foams, fibers, coatings, and adhesives (Yu et al., 2006).

Water is employed as a plasticizer, for dispersion and as a solvent in the creation of many natural polymer blends since the most natural polymers are soluble-in-water. Polysaccharides and proteins are the primary components of natural polymers, their interactions with each other and water give the relationships of structural properties in these materials. In natural polymers, the examination of thermal profile and glass transition temperature best illustrates the role of water (Yuet al.,

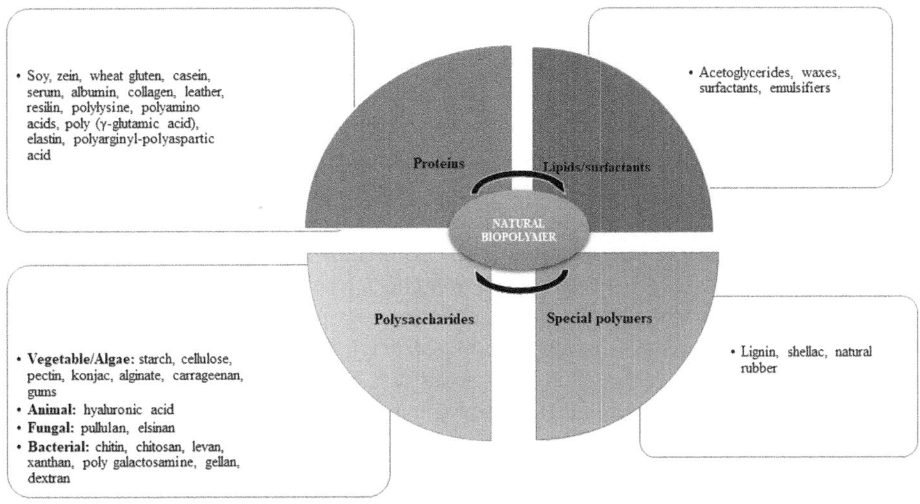

FIGURE 4.3 Natural biopolymers. (Compiled by the author.)

2006). It has been shown that bio-based nanocomposite films are stronger than artificial polymeric films and edible films in terms of barrier qualities, particularly mechanical attributes. Bio-based nanocomposites could be utilized to prolong the shelf-life of fresh food such as vegetables and fruits by regulating the respiration cycle. By delaying moisture loss, improving product appearance, minimizing discoloration and lipid oxidation, and lowering fat uptake from coating and paneling during frying, it can also enhance the quality of fresh, processed, frozen poultry, meat, and seafood (Akbari et al., 2007). These bionanocomposites exhibit many advantages as nanometer-sized particles are obtained by dispersion. Some of these can be listed as follows:

- Reduces packing weight, waste and volume.
- Extends shelf-life and overall quality of unpackaged components.
- Controls internal components.
- Provides tiny food particles in separate packing, such as raisins and nuts.
- Antimicrobial and antioxidant functions as carriers for agents.
- Provides regulated release of the active substances.
- Resources that can be renewed annually. Thus, as a cutting-edge packaging and processing technology; bionanocomposite packaging materials offer enormous promise for enhancing food safety, stability and quality.
- Packaging made of natural biopolymers has a special benefit, allowing the development of novel products, including individual packaging for particulate foods, transport of nutritional additives and active compounds.
- It is biodegradable.
- Improves organoleptic properties of products, such as smell, appearance, and flavor (Zhao et al., 2008).

4.4.1.1 Polysaccharide-based Bionanocomposites

4.4.1.1.1 Cellulose-based Bionanocomposites

Cellulose is the most prevalent and naturally occurring biopolymer in the world. It is made up of 1,4-D glucosidic linkages that connect linear, unbranched chains of D-glucose molecules. It has a high potential to be used in the production of biopackaging materials as it is biocompatible, renewable and biodegradable. Naturally, it is a high molecular weight polymer because it is very highly crystalline. It is transformed into derivatives to make it more practical because it is not soluble. Being the genuine thermoplastic resin among the cellulose ethers, hydroxypropyl cellulose (HPC) may be extruded into films that are formed in the molten state. Nowadays, cellulose acetate is employed in high-scale applications like thermoplastic injection molding, films, and fibers. In recent years, the improvement of the permeability and thermal/mechanical attributes of cellulose acetate films has raised interest in the creation of nanocomposites using cellulosic materials to produce useful materials (Rhim et al., 2007).

The use of cellulose nanoparticle reinforcement improves the moisture barrier capabilities of polymer films. Crystalline fibers are supposed to make materials more curved, which results in a slower diffusion process and decreased permeability. If the infill is less permeable, evenly distributed across the matrix, and has a high aspect ratio, barrier characteristics are improved. It has been stated that nanosized cellulose fibrils also improve the thermal properties of polymers. As compared to comparable polymer clusters, the thermal stability of polymers in nanocomposites containing cellulose crystals improved. Wu et al. (2009) observed that when they developed polyurethane with cellulose nanofibrils, its elongation was reduced by conventional micro-sized cellulose filler. These variations might be explained by various levels of matrix–cellulose interactions. In accordance with the Jordan et al. (2005) study, the addition of weakly interacting nanoreinforcements with matrix causes elongation and decrease in material strength, while on the other hand, it is determined that the modulus is not dependent on these interactions.

4.4.1.1.2 Starch-based Bionanocomposites

Starch is a common carbohydrate found in nature. It is the carbon source of many plants. Starch consists of glucose units linked together to form two structurally different fractions, amylose (linear), and amylopectin (branched). Besides being an important food ingredient, starch is a biopolymer frequently used by the paper industry to improve the strength of paper. Depending on the derivatization method, it can be dissolved in hot water up to a concentration of 20–30 per cent by weight and can be applied on fiber-based packaging materials by the coating method. As a result of the structure of starch and its interaction with nanoparticles, complex nanocomposite systems are formed. Starch exhibits multilayer structures from macroscale to molecular scale: (a) starch granules (mm), (b) semicrystalline and amorphous multimolecular structures (100 μm), (c) crystalline regions (10 nm), and (d) branched amylopectin (nm) chains with linear amylose. In this way, starch-based nanocomposites with many different application areas are obtained by adding nanoparticles into starch, which is a complex multi-level structure (Söğüt et al., 2017). However, many applications are limited by poor mechanical properties such as lack of water barrier properties and film shine caused by high intermolecular forces (Rouf et al., 2018). In the literature, there are many starch-based nanocomposite studies produced using different nano-fillers and different preparation techniques (Gao et al., 2012; He et al., 2012; Raquez et al., 2011).

4.4.1.1.3 Chitin/Chitosan-based Bionanocomposites

One of the most prevalent biopolymers in nature, along with cellulose, is chitosan. Chitosan is generated through partial deacetylation of the natural polysaccharide chitin (Casettari et al., 2012). It is biodegradable, biocompatible, and non-toxic (Çabuk et al., 2011). Chitosan has many functions such as moisture adsorption, precipitation, film formation, antimicrobial effect, and enzyme immobilization. Chitosan also has properties such as high viscosity, gel formation, and water binding capacity. Similar to dietary fiber, it cannot be hydrolyzed by digestive enzymes. It has antibacterial properties, and this effect varies according to the bacteria type and the chitosan molecular weight (No et al., 2002). There are many studies on chitosan-based nanocomposites in the literature (Chang et al., 2010; Ramirez et al., 2017; Zamudio-Flores et al., 2010). Clay nanoparticles (Grigoriadi et al., 2015), zinc oxide nanoparticles (Youssef et al., 2015), silver and gold nanoparticles (Youssef et al., 2014), multilayer carbon nanotubes (Hernández-Vargas et al., 2014), and graphene oxide nanosheets (Ahmed, Mulla, et al., 2017) were added to chitosan films to improve thermal and mechanical attributes. Furthermore, according to L. Sun et al. (2017), bioactive packaging materials including chitosan films and apple polyphenols can extend the shelf life of foods.

4.4.1.1.4 Bionanocomposites from Other PoSysaccharide sources

Pectins, which are among the natural polymers, are amorphous, white, and complex carbohydrates found in some vegetables and ripe fruits. Pectin is a byproduct of the manufacture of sugar, sunflower oil, and juices. Given that it is a byproduct of the food-processing sector, it is a prime choice for ecologically friendly biodegradable materials. However, the use of pectins in materials produced from natural sources is limited due to their weak water resistance and low strength. A different approach to solving this issue is to change natural polymers by adding inorganic fillers, which broadens the scope of their use in more intricate or critical circumstances. High amylase starch or poly(vinyl alcohol) and pectin blends are plasticized to produce water-soluble bags for detergents and sturdy, flexible films that can be employed as medicinal delivery systems or pesticides (Mangiacapra et al., 2006).

The yeast *Aureobasidium pullulans*, which resembles a fungus, produces the extracellular microbial polysaccharide known as pullulan. It dissolves in water and produces colorless, flexible, and transparent films. Pullulan plasticized with sorbitol has been used to create bionanocomposite materials, and as the reinforcing phase, pure granules of waxy corn starch were subjected to acid hydrolysis at 35°C to create starch nanocrystals. Increasing starch nanocrystalline content in turn increased the crystallinity of composite biopolymer films. The water absorption of starch -pullulan nanocomposites reduced when the filler content was increased; nonetheless, water vapor permeability consistently remained over 20 per cent (w/w); however, it drastically dropped with the addition of nanocrystals (Kristo et al., 2007).

4.4.1.2 Protein-based Bionanomaterials

Proteins are heterogeneous polymers whose functions mostly depend on their amino acid content. Usually, protein film creation requires the physical and/or chemical partial denaturation of the polypeptide chain, followed by the organization and orientation of the peptide chains and the formation of new bonds between the partly denatured peptide chains. As a result, the protein matrix is formed (Luecha et al., 2010). Milk proteins produce transparent, tasteless, and flexible films. These films can be a vehicle for food additives including antimicrobial agents, colorants, and antioxidants. These edible films have a higher oxygen barrier when relative humidity is low or medium, but they still have poor water vapor permeability (Mohamed et al., 2020). Caseins and whey proteins are two kinds of milk proteins. Whey protein is produced by precipitating casein protein (Mohamedet al., 2020). Whey protein isolate (WPI), the purest form of whey protein (90% protein), was used for the development of edible films for food packaging (Mohamedet al., 2020; Ramos et al., 2012). WPI, low density polyethylene (LDPE), vinyl alcohol, high density polyethylene, polyvinylidene chloride (PVDC), and ethylene vinyl alcohol have promising mechanical properties as well as good oxygen barriers and moderate moisture permeability as compared to cellophane and polyester (Ramoset al., 2012). Wakai et al. (2015) found that the incorporation of MMT to WPI films reduced the solids migration and modified the film characteristics.

In contrast to other protein films, zein, the prolamin fraction of maize protein that is produced from maize gluten meal during starch production, is moderately hydrophobic, has poor mechanical properties because of its relatively high content of non-polar amino acids, and forms an excellent barrier to oxygen. Many hybrid approaches, such as blending with biopolymers or synthetic polymers, the lamination of zein with other lipids and proteins, incorporation of various nanofillers and crosslinking approach have been developed in addition to the usage of different plasticizers to enhance the film flexibility (Vahedikia et al., 2019).

In a previous work, corn zeine nanocomposite coatings on PP films were studied as an alternative to synthetic polymer barrier layers. The insertion of organomodified montmorillonite into the zein matrix by solution intercalation greatly enhanced the oxygen and water vapor barriers of the coated polypropylene films (Ozcalik et al., 2013).

Gelatin, which is generated from collagen hydrolysis, is one of the animal proteins utilized in the food sector. Gelatin is a mixture of polypeptides and proteins inexpensively obtained from various animals, with no distinctive color or taste. It is a product of mild thermal denaturation of collagen in an alkaline or acidic environment (Z. Yang et al., 2021). According to Farahnaky et al. (2014), while the tensile strength of the films is related to the clay content, the addition of clay decreased the elongation and permeability of water vapor of the gelatin–nanoclay films. Several mechanical and physical characteristics of gelatin biofilms were enhanced by nanoclay.

Three different forms of soy protein products are made from soybean processing; soy protein concentrate (65–72% protein), soy protein isolate (≥90% protein), and soy flour (54% protein). Soy protein isolate-based films are more flexible, smoother, and clearer than other vegetable protein-based films. It has stronger gas barrier characteristics than films made of polysaccharides and lipids (Song et al., 2011). However, the developed soy protein isolate-based films' inferior mechanical qualities and significant moisture sensitivity have restricted their practical uses (Liu et al., 2017). Echeverría et al. (2014) studied the influence of the montmorillonite (MMT) addition at various concentrations up to 10 g/100 g soy protein on the physicochemical attributes of the obtained nanocomposites. The MMT-soy film enhanced mechanical deformation resistance, protein solubility in water, and water vapour permeability without compromising appearance. The higher the clay content, the better results were obtained. Lee et al. (2010) created nanocomposite films using ontmorillonite MMT as the nanofilling material at weight ratios of 0.3,6,9,12, and 15 per cent based on the dry weight of soy protein isolate. MMT concentrations were determined to be between 3 and 12 for improving the functional qualities of composite films.

4.4.1.3 Poly(lactic acid)-based Bionanocomposites

Poly(lactic acid) (PLA) is a biodegradable, thermoplastic polymer synthesized from cyclic lactic dimers or lactic acid monomers. During the fermentation process, natural carbohydrates like corn and wheat or food/agricultural industry waste may be produced. Today, PLA used in commercial packaging applications has a very high yield and is manufactured by the polymerization of lactic acids through lactide production (Auras et al., 2005). Since PLA is not water soluble, the use of this polymer as a coating material is limited. However, the coating obtained by dissolving PLA in chloroform greatly reduces the water absorption values and increases the surface hydrophobicity of the materials utilized in food packaging with high moisture content or frozen foods. In a study using coating solutions, the upper limit for the solubility of PLA in chloroform was 5 per cent by weight (Rhim and Ng, 2007). However, it is not appropriate to use a toxic solvent in packaging materials that come into contact with food. Therefore, extrusion coating is a more suitable method in PLA-containing studies. According to the studies, the brittleness of PLA polymer is high (Paul et al., 2003), and this situation causes bigger problems when nanoparticles are added (Plackett et al., 2006). The microfibrillation procedure was used by Kakroodi et al. (2017) to develop various features of PLA films, such as gas permeability.

4.4.1.4 Polyhydroxy Alkanate-based Bionanocomposites

Biopolymers known as polyhydroxy alkonates (PHA) are created by microorganisms as a source of carbon and energy. Copolymers of poly(hydroxybutyrate-covalerate) (PHB/V) hydroxybutyrate (HB), polyhydroxybutyrate (PHB), and hydroxyvalerate (HV) can be given as examples of this group. PHB generates very crystalline films and therefore films produced from copolymers including longer alkyl chains of hydroxyvalerate are highly brittle despite excellent hardness and strength. Since they are highly hydrophobic, they show water-resistant coating and film-forming properties. PHA films are very suitable as packaging materials for food since they have values close to the water vapor permeability values of low density polyethylene (LDPE) (Söğüt and Seydim, 2017). The disadvantage of PHAs is their high production costs and poor gas barrier. In addition, PHB polymer exhibits thermal disorder and rapid degradation at high temperatures limits the commercial use of PHB (Ray et al., 2005). Dasan et al. (2017) showed that the oxygen barrier properties were improved in PLA/PHBV polymer films prepared using nanocrystalline cellulose.

4.4.1.5 Polycaprolactone-based Bionanocomposites

Polycaprolactone (PCL),- a linear aliphatic polyester with high crystallization and hydrophobic character, produced ε-caprolactone polymerization via ring-opening process. It is produced synthetically. The starting material for the synthesis of ε-caprolactone originates from petroleum. Despite its synthetic origin, PCL is partially biodegradable and has nominal tensile strength and very elevated elongation at break values. PCL has high compatibility with other polymers and is utilized as a coating material directly as well as in composite barrier films. Elen et al. (2012) reported that the gas permeability and mechanical attributes of PCL films containing ZnO nanoparticles were improved.

4.4.2 Biodegradable Bionanocomposite Packaging Materials

Another packaging area that benefits from nanotechnology is biobased and biodegradable nanocomposites. Biodegradable bionanocomposite packaging is defined as the integration of food-grade protein, oil- and lipid-based nanocomposite biopolymers obtained as a result of natural, synthetic, or microbial fermentation into the packaging film (Lamabam et al., 2018). Biopolymers could be classified as natural biopolymers (vegetable carbohydrates like agar, alginate, starch, carrageenan and cellulose as well as animal or vegetable proteins like gluten, gelatin, soy protein, casein collagen, zein and whey protein); synthetic biodegradable polymers (polylactide, polyglycolic acid, polycaprolactone, polybuty polyesters); polysaccharides (pullulan, chitosan) based on the origin and manufacturing procedures of raw materials. These biopolymers have poor barrier and mechanical characteristics such as brittle structure, elevated gas permeability, low thermal bending temperature, and nominal melt viscosity. These properties need to be developed so that they can be used instead of traditional non-degradable plastics. Biodegradable nano food packaging not only causes the least harm to the environment, but also has a positive impact on consumer health by providing safer and healthier packaged food. The nanoparticles dispersed on the plastic prevent the passage of oxygen, carbon dioxide and moisture. Nanoplates also make plastics lighter in color, stronger and more heat resistant (Cerqueira et al., 2018; Lamabam and Thangjam, 2018; Salgado et al., 2019).

In response to customer demand for high-quality food products, natural biopolymers like proteins and polysaccharides that are yearly renewable have been employed to make biodegradable packaging materials. These materials use biopolymers rather than petrochemicals, which have non-biodegradable plastic packaging materials that contribute to environmental waste problems (Rhim and Ng, 2007). Biodegradable, that is, plastics that degrade in nature; synthesized from natural polymers such as protein, cellulose, and starch. They pollute the environment less because they can degrade in nature, and they are preferred because they reduce the waste problem (Dursun et al., 2010). Biopolymer-based packaging materials offer certain advantages such as increasing shelf-life by reducing microbial growth in the product and improving food quality. These materials function not only as barriers to gases, moisture, solutes and water vapor, but also as transporters of some active compounds. Also, they work well for adding a variety of additives, including as antifungal agents, antimicrobials, food colorants, antioxidants, and other food ingredients. Natural biopolymers are superior than synthetic polymers because they are edible, biodegradable, and renewable (Rhim and Ng, 2007). The potential use of biopolymer-based antimicrobial films for a variety of food groups, such as fish, meat, and poultry products, cereal, cheese, fruit, and vegetable products, has increased attention in the food sector. However, the usage of combined biopolymer films is limited, especially in humid environments, due to its sensitivity to water and its relatively low stiffness and strength. Reducing hydrophilicity and enhancing mechanical qualities have been the main focus of several studies aimed at enhancing the physical characteristics of biopolymer-based films. The moisture barrier capabilities of biopolymer films have been improved by the incorporation of hydrophobic substances such waxes, fatty acids, or neutral lipids (Rhim et al., 2006). Biopolymer films slow down the loss of volatile components and moisture migration, lower the respiration rate and postpone modifications to textural attributes. These films are utilized for coating a diverse variety of products like cherry, tangerine, and strawberry. In comparison to typical synthetic films, they have a high sensitivity gas permeability ratio (CO_2/O_2) and are effective oil barriers (Casariego et al., 2009). The incorporation of nanotechnology into these polymers creates new opportunities for property enhancement as well as low-cost effectiveness (Sorrentino et al., 2007). By integrating nanoparticles into biodegradable plastics and strengthening, the negative properties of materials are improved and new materials with radically different characteristics are developed (Dursunet al., 2010). Recently, aliphatic polyesters, including polylactic acid, starch, and its derivatives, polyhydroxybutyrate, polycaprolactone, and poly(butylene succinate), have recently been researched as the most effective biodegradable nanocomposites for packaging applications (Sorrentinoet al., 2007). The formation ways of bionanocomposites was given in Figure 4.4. Bionanocomposites made from starch, soybean oil, and polylactic acid (PLA) have improved in barrier and mechanical characteristics.

Several methods have been developed to improve the functional qualities of biopolymers so that they can compete with tougher and more formable commercial polymers like polypropylene or

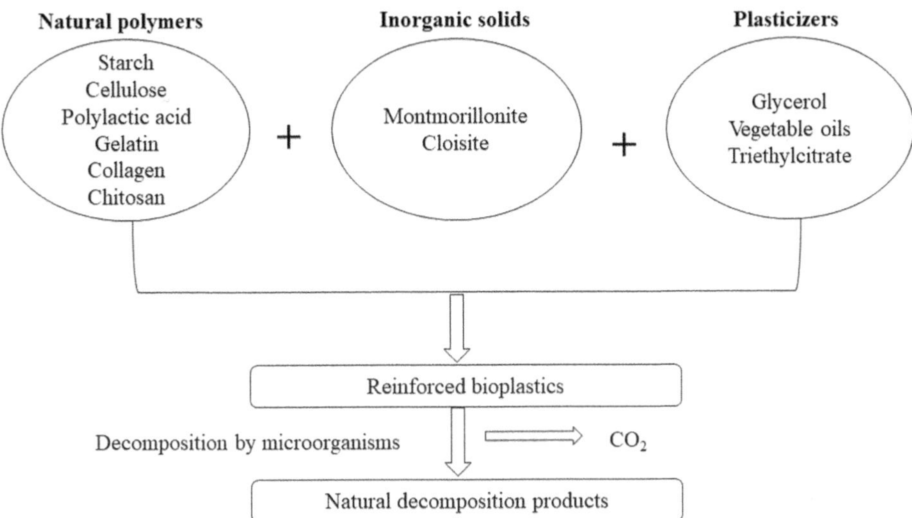

FIGURE 4.4 Formation of bionanocomposites. (Dursunet al., 2010.)

polyethylene. The use of polymer-layered silicate nanocomposite technology has been successful in enhancing qualities, including barrier performance, mechanical strength, and thermal stability. While there is more interest in polymer/clay nanocomposites, there is relatively little interest in biopolymer/clay nanocomposites such as clay/polylactide, clay/cotton, clay/poly(butylene succinate), and clay/vegetable oil nanocomposites (Wang et al., 2007). Nanostructured substances with enhanced gas barrier, thermal, and mechanical qualities are called bionanocomposites. Bionanocomposites are used in food packaging to preserve food and extend shelf-life while lowering the quantity of plastic needed in packaging, resulting in a more environmentally friendly option. However, since biodegradable films, today's alternative packaging material, exhibit poor barrier and mechanical attributes, it is necessary to greatly improve their properties before they may be substituted by conventional polymers, and accordingly aid the waste problem of the world (Sozer et al., 2009). By incorporating inorganic particles like clay into the biopolymer matrix, as well as by modifying layered silicates with surfactants, it is possible to increase a packaging material's biodegradability. The usage of inorganic particles also reveals the multi-functionality that aids in the dispersion of micronutrients in edible capsules (Sozer and Kokini, 2009).

4.4.3 Antimicrobial Smart and Active Bionano Packaging Materials

Studies for enhancing food shelf-life can only achieve concrete results with nanotechnology. Smart packaging includes those that use nano-sized materials. This new technology has led to the development of odor-, air-tight, and light food packages that preserve freshness for a long time (Sürengil et al., 2011). Table 4.1 lists various active packaging systems and application areas in foods. Smart packaging provides great prospects for increasing food quality, convenience, and safety in the food field. Smart packaging takes advantage of the packaging's communication capacity to aid decision making (Dursun et al., 2009).

Antimicrobial characteristics are demonstrated in nanocomposite materials created by incorporating certain nano metals or metal oxides into polymers. Nanoparticles' antibacterial capabilities are used. These compounds inhibit the growth of microorganisms, resulting in a longer shelf-life. Furthermore, the condition of food may be checked with nanosensors placed in plastic (Chaudhryet al., 2008). Nanosensor research is also ongoing, with the goal of detecting infections

TABLE 4.1
Active/intelligent packaging systems and their use in foods

Active packaging system	Structure	Food applications
CO_2 capture systems	Iron, metal/acid, enzyme	Bread, cake, pizza, cheese, cured meat and fish, dry foods, coffee, etc.
O_2	$FeO/Ca(OH)_2$, ascorbate/ $NaHCO_3$, CaO/activated carbon	Coffee, fresh fish and meat, oil seeds (peanuts, etc.)
Ethylene capture systems	$KMnO_4$, activated carbon	Fruit/vegetables and cereal products
Ethanol capture systems	Capsulated ethanol	Pizza, cake, bread, biscuits, fish
Dehumidifiers	Silica gel, minerals	Fish, meat and meat products, cereals, dried foods, vegetables and fruits
Taste/odor traps	Cellulose acetate, citric acid, activated carbon/zeolite	Juices, fish, farm products, dairy products and fruit
Preservative compound spreaders	Organic acids, plant extracts, BHA/BHT, vitamin E	Cereals, meat and fish, bread, cheese, fruit and vegetables

Source: Table compiled by the author.

in food by color change. Sensors placed in food packages show the product's freshness and whether there is microbial deterioration in the products, oxidative bitterness, and temperature-related changes (Sürengil and Kılınç, 2011). Recently, the use of chemical sensor and biosensor technologies in packaging applications of food has been increasing rapidly.

Antimicrobial packages produced by nanotechnology are produced by adding synthetic and natural antimicrobial agents into a polymer or by coating the surface of the package or leaving a small bag inside the package. Antimicrobial packaging materials prolong the logarithmic growth period of microorganisms and suppress microorganism growth to enhance shelf-life by ensuring food safety. It is emphasized that the materials obtained by forming nano-sized composites of antimicrobial films, coatings, and edible films with nanoparticles or nanofillers will provide more effective protection in fishery products (Dursunet al., 2010). Moreover, the principles of nanoscience provide not only a packaging system that improves food stability, but also an encapsulation system that protects sensitive bioactives in foods from the environment, and from unwanted interactions during storage and processing, without reducing food quality and at the same time allowing the food to be released to the target region in the body after digestion (İlyasoğlu et al., 2010).

4.5 BIONANOMATERIALS WITH ADDITIVE PROPERTIES

4.5.1 Antioxidant Properties

Reactive nanoparticles can be included into the packaging material. Including oxygen scavengers in food packaging is a particularly helpful application for maintaining a low oxygen level in the packaging environment. Nano-sized titanium oxide, for example, has been found to be an effective oxygen scavenger and is appropriate for packaging oxygen-sensitive foods (Tulsyan et al., 2017). The most significant drawback of titanium oxide is that it requires UVA light. Nanosensors are capable of responding to environmental changes (temperature and humidity change in the storage room, increase in oxygen level), degradation products or microbial contamination (Nathalie et al., 2017). Chemical components, infections, and poisons found in food packaging may be identified using nanosensors. The development of non-reversible and non-toxic oxygen sensors to assess the presence of oxygen in oxygen-free packaging systems, such as nitrogen or vacuum packing, is gaining momentum. Mihindukulasuriya et al. (2013) created an oxygen sensor employing titanium oxide nanoparticles that are activated by UV and can detect the presence of oxygen in their investigation. The active ingredients, titanium oxide, glycerol, and methylene blue, were electro-spun into poly(ethylene oxide) fibers and employed as a carrier. López-Córdoba et al. (2017) showed that adding rosemary nanoparticles to starch films increased their mechanical and antioxidative capabilities. By mixing nano clay and blueberry extract with antioxidant activity, Gutiérrez et al. (2017) created active packaging while also improving the thermoplastic qualities of the packaging material.

4.5.2 Antimicrobial Activity

The properties of some nanoparticles or nanocomposites, such as their ability to inhibit growth (Ahmed, Hiremath, et al., 2017), can be used as an antimicrobial agent carrier (Bi et al., 2011) or form films with direct antimicrobial action (Reesha et al., 2015). In nanocomposite antimicrobial systems, the surface reactivity of the nano-sized antimicrobial agent provides a more effective inhibition of microorganisms (Ben-Sasson et al., 2014). Natural biopolymers (such as chitosan), metal ions (such as Ag, Cu, Au, Pd), MeOs (such as Ti_2O, ZnO and MgO), enzymes (such as lysozyme and peroxidase), and synthetic antimicrobials (propionic and benzoic acid EDTA) are frequently utilized in nanocomposites. Silver nanoparticles are the most often used for adding functionality to polymers. This is because it has benefits such as optical, electrical, catalytic, thermal stability, and antibacterial characteristics (Dallas et al., 2011). Copper ions are antimicrobial and antiviral, and they are required for life as an enzyme component. Ben-Sassonet et al. (2014) developed a nanocomposite membrane with antimicrobial surface properties containing copper nanoparticles, but it is not used in the food packaging industry because copper is toxic when in contact with food and accelerates biochemical deterioration in foods due to its catalytic effect on oxidation. MeOs such as Ti_2O, ZnO, and MgO are used in the preparation of antimicrobial films because of their high stability and activity. By collecting energy from a light source and producing catalytic activity, these metal oxides are utilized as photocatalysts. When a photocatalyst is exposed to UV light, it forms highly reactive oxygen species, which is one of the antimicrobial action mechanisms (Dong et al., 2011). There are research on the use of mineral clays as biocide carriers that predominantly contained inorganic biocides such as Au, Cu, Zn, and Mg. Biocidal metals are added to the clay structure to gain ion charge through ion exchange. Magnesium phyllosilicate has been demonstrated to efficiently inhibit *Candida albicans*, *Staphylococcus aureus*, and *Escherichia coli* species (Chandrasekaran et al., 2011). The mechanism of inhibiting microbial growth of the nanoclay used is the disruption of the cell's membrane integrity due to the charge interaction of these groups with the amino groups and the inability to use the essential components by the cell.

4.5.3 Nano-Sized Enzyme Immobilization Systems

In the food industry, enzymes are applied in a wide range of processes. Since enzymes are sensitive to processing conditions or components that may induce inhibition, their direct usage is restricted in several circumstances (in which case the enzyme is inactivated or the enzyme is shorter-lived). The resistance of enzymes to pH and temperature fluctuations, other components that induce denaturation, and adaptability to ambient conditions, reusability, or controlled release can be enhanced when immobilized using appropriate carriers (Mihindukulasuriya and Lim, 2013). The addition of enzymes such as cholesterol reductase or lactase to packaging materials responds to the problems of consumers with enzyme-related health problems (Fernández et al., 2008). Clays, for example, have a high affinity for adsorption of protein and were proven in experiments to be a good enzyme carrier (Benucci et al., 2018; J. Sun et al., 2017). Silica nanoparticles were modified to lactate dehydrogenase and immobilize glutamat, and the immobilized enzymes showed unique activity, allowing the use of modified silica nanoparticles in biosensor applications (Qhobosheane et al., 2001). Caseli et al. (2007) used the mutual adsorption approach using counter-ion-loaded materials to immobilize glucose oxidase onto chitosan films and demonstrated that the activities of enzyme was similar to the solution. Electrospinning is a quick and simple technology for creating nanofibers from a variety of substances. The high specific surface areas and porous structure of the produced fibers make these nanofibers an unique enzyme support and increase the catalysis ability of immobilized enzymes (Ren et al., 2006).

4.6 PRODUCTION AND CHARACTERIZATION OF BIONANOMATERIALS

4.6.1 Production of Nanomaterials

Nanomaterials are basically obtained by two different production techniques called "top-down" or "bottom-up". In the "top-down" production technique, which is mainly used in commercial scale production, nanoparticles are developed by reducing the macromolecule to nano size by physical fragmentation methods such as crushing, grinding, etching, or lithography. In the "bottom-up" production technique, which is a newer technique, nanoparticles are obtained as nano-sized, multi-molecular structures with the self-arranging of each atom or molecule (Moraru et al., 2003; Sanguansri et al., 2006). In the first approach, which is called top-down, the process starts with the whole material and the material is divided into small pieces. In this main approach, the structural dimensions of microscopic elements are reduced to the nanometer scale, with special processing and chemical etching techniques, lithography, and extremely flawless surface shaping. In the bottom-up production approach, the material is synthesized as a result of the growth of atoms and molecules in size through chemical reactions. Atomic and molecular elements are brought together in a controlled manner to form larger systems, clusters, organic lattices, multimolecular structures, and synthesized macromolecules (Kellar, 2006). The surface areas of the nano-sized materials developed by these methods increase, as does their usefulness. Increased surface areas in nanotechnology products have the ability to give several functional qualities such as greater water absorption, higher adhesion and release of taste and color compounds, and increased reaction catalysis rate. Furthermore, the standard size distributions that can be obtained are of great importance in terms of quality control as well as the functionality of the product (Sanguansri and Augustin, 2006). Several spectrophotometric, microscopic, and spectroscopic approaches may be used to characterize nanomaterials, including parameters like as mass, structure, size, composition, shape, and charge (Luykx et al., 2008). Atomic force microscopy (AFM) and scanning electron microscopy (SEM), particularly transmission electron microscopy (TEM), are often employed for imaging nanostructures and getting qualitative information about their dimensions and morphological features (Figure 4.5). With mass spectrometry (MS), nanostructures can be defined in terms of mass and composition.

In the mass characterization of carbohydrate, protein-peptide and lipid-based nanostructures, mass spectrometry with various analysis characteristics, such as Ion Mobility-Mass Spectroscopy (IM-MS), Electrospray Ionization (ESI-MS), Matrix-Assisted Laser Desorption Ionization (MALDIMS), and Desorption Electrospray Ionization (DESI-S) also plays a significant role. Another standardized technique for figuring out nanoparticle diameters and size distributions is the dynamic light scattering (DLS) approach. On the other hand, Small-Angle X-Ray Scattering (SAXS) analysis may be used to identify the structural, shape and size, characteristics of particles smaller than 100 nm. Another approach for analyzing the atomic structures of nanoparticles is X-ray Diffraction (XRD) (Tarhan et al., 2010).

4.6.2 Production of Nanofibers

Throughout history, fibers have existed in nature as continuous filaments or elongated objects. Spiders, for example, utilize their fiber webs to catch their prey. Meanwhile, silkworms are very skilled at producing silk cocoons, which are wonders of fiber. Numerous natural systems like as

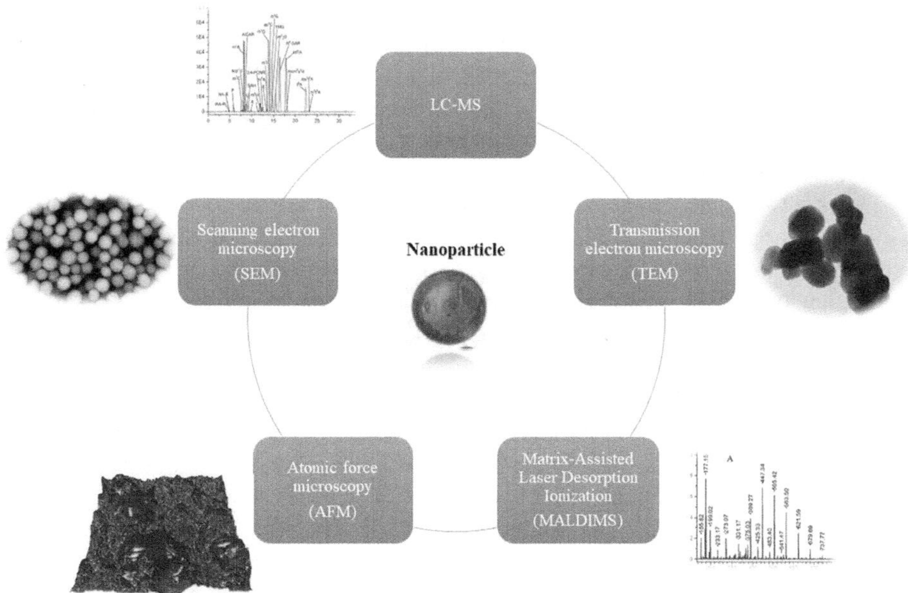

FIGURE 4.5 The methods for the characterization of nanomaterials. (Compiled by the author.)

these have inspired people to think about the historical and current role of man-made fibers (Xue et al., 2019). The fibers shrink to micro size, allowing for a massive increase in specific surface area of up to 1000 m^2/g. The increase in surface area by reducing the diameter of the fibers from 10 μm to 10 nm greatly affects the chemical, biological reactivity and electroactivity of the fibers. With the recognition of the potential impact of this shift in fibers, there has been a worldwide surge in nanofiber research (Barhoum et al., 2018). Nanofibers are described as nanomaterials with a diameter of up to 100 nm. Nanofibers, which are considered as one of the most interesting nanomaterials to study in the world of science and industry, are subjected to different chemical and physical reactions during and after manufacturing and, thanks to these modifications, they offer wide possibilities to create many new products with new properties. In this way, it brings solutions to our basic problems in order to make our lives easier in areas such as energy, environment, medicine, and food (Ko and Wan, 2014). Nanofibril and nanofibers can be categorized based on their size (pore size length and diameter) and composition (metals and their oxides, polymers, ceramics, hybrid and carbon, hollow, non-porous, core-shell, mesoporous, multicomponent and biological component) (Barhoumet al., 2018). Nanofibers can be produced from almost any polymer. However, the application areas and properties vary according to the polymer type. Their diameters also vary according to the polymer type and the production techniques. Researchers have developed various methods to produce nanofibers. These methods include drawing, electrospinning, self-assembly, template, synthesis and phase separation (Asmatulu et al., 2018). The three most significant techniques for creating nanofibers are self-assembly, electrospinning, and phase separation since template synthesis cannot create continuous fibers and can only be utilized with viscoelastic materials that can withstand the forces exerted during the drawing process (Eatemadi et al., 2016).

4.6.2.1 Electrospinning Method

Electrospinning technique is the method of collecting polymer fibers from polymer melts or solutions on films with the help of a certain electric force in order to increase the performance of these fibers with various additives—for example, as reinforcement material or to increase their barrier properties (Topuz et al., 2020). Its active use has been accepted as the simplest, most versatile, and most effective technique to fabricate both electrospraying and conventionally fibrous materials from tens of nanometers (Kahraman et al., 2018; Wu et al., 2012). One of the primary benefits of the electrospinning method is its versatility. Several morphologies might be obtained with different parameters and configurations of the electrospinning method (Alberti et al., 2017). Taken together, the same characteristics may be utilized in a diverse variety of fields, including the creation of diverse smart packaging materials for the food sector, high-performance smart textiles, biosensors, and tissue engineering (Coelho et al., 2018). As the use of packaged foods increases, packaging technology plays a universal role, especially in the stages of storage and delivery to the consumer. As the tendency to use biopolymers as packaging material increases, interest in electrospinning has also increased by researchers and the food industry (Hottle et al., 2013).

4.6.2.2 Drawing Method

This method is also known as the molecular application of the dry drawing method. In the manufacturing process, long, continuous, and single nanofibers are employed. The most significant benefit of this method is that by producing a single nanofiber, its characteristics and applications may be studied (Asmatulu and Khan, 2018; Ko and Wan, 2014). The most important material required for the application stages of this method is a pointed tip or micropipette. The resulting liquid is deposited as a drop from a polymer solution onto a surface. With the help of a micromanipulator, the micropipette is immersed in this droplet. The immersed micropipette is gently pulled through the droplet at a certain speed. The diameters and dimensions of the fibers obtained in this method depend on the drawing speed, which is the basic parameter, and the solution viscosity. With the drawing of the micropipette, the solvent in the polymer solution, which has a high surface area, evaporates, and fiber is formed as the viscosity increases (Alghoraibi et al., 2018; Asmatulu and Khan, 2018).

4.6.2.3 Template Synthesis Method

Template synthesis technique is used to synthesize nanofibers, nanotubes, and nanowires of controllable length and diameter. In this method, a template or mold containing cylindrical nanopores is used to produce nanofibers of uniform diameter. Thanks to the uniform pores, the diameters of the nanofibers to be formed are controlled (Asmatulu and Khan, 2018). Extrusion is provided by passing the polymer solution through nano-sized cylindrical pores with water under high pressure being applied to one surface of the polymer solution. When the solution passing through the pores comes together with the solidifying solution, solidification occurs in the polymer and fibers with nano pore diameters being formed. Different diameters of fibers can be obtained by using different templates. This approach, however, cannot produce continuous nanofibers (Alghoraibi and Alomari, 2018; Ko and Wan, 2014; Ramakrishna, 2005).

4.6.2.4 Phase Separation (Sol-Gel) Method

The phase separation technique is also defined as the sol-gel method. The sol solution progressively evolves into a gel-like structure comprising both liquid and solid phases in this process (Ko and Wan, 2014). It is fundamentally based on the separation of phases caused by temperature differences and is based on thermodynamics. There are basically four phases in this process:

- Solvent dissolving of polymer at high or room temperature.
- Gelation, which is the most challenging phase for controlling the morphological pore size of the formed nanofiber. Gelation time varied according to gelation temperature and polymer concentration.
- Water extraction of the solvent from the generated gel.
- Freeze-drying or freeze-drying under vacuum (Alghoraibi and Alomari, 2018).

Phase separation method is a cost-effective and low-temperature method. In this respect, it is suitable to be preferred, but it is not a frequently used method in industrial terms due to the long duration of the processes (Eatemadiet al., 2016).

4.7 FOOD SAFETY ASPECTS OF BIONANOMATERIALS

The term 'nanofood', which began to be pronounced with the advancement of nanotechnology in the field of food, is defined as food products that are harvested, processed, produced, and packaged using nanotechnology techniques and tools, or obtained by adding nanomaterials such as nanoparticles like zinc, iron, and nanocapsules containing active ingredients (Mousavi et al., 2011). Nanotechnology applications in food systems are developing rapidly. Although the potential benefits of new food products produced with this technology are intensely emphasized, there are concerns about the influence on public health, as there is no knowledge about the safety of these products yet. However, several scientists have lately begun to investigate the harmful impacts of nanotechnology on biological systems and have drawn attention to some of its possible consequences (Nel et al., 2006; Oberdörster et al., 2005). Nanomaterials such as Au, Mg and ZnO incorporated into packaging materials may contaminate food products in direct contact and cause serious health risks when these products are consumed. Protein, carbohydrate, and fat-based systems, or nutritional support products that can be used in the transport of food additives may cause negative effects by exposing the consumer to the transported substance at high doses (Bouwmeester et al., 2009).

Deeper study should be conducted before using nano-sized materials in food packaging, and the associated health and ethical responsibilities should be assumed. It is important to remember that all dietary components, from proteins and carbohydrates, which are assumed to be innocuous, to nanometer-sized particles, can have harmful consequences at sufficient proportions. The packaging material should be examined for potential toxicities. Appropriate regulations should be checked before using any nanomaterial to be applied as food packaging material. All materials in contact with food are subject to a general regulation (Regulation (EC) No. 1935/2004). The European Commission's relevant regulations for active nanomaterials (Regulation (EC) 450/2009 European Commission, 2009) and smart packaging and plastics in contact with food in its revised form (Regulation (EU) 2016/1416 European Parliament and Council, 2011) provide a more specific explanation (Söğüt and Seydim, 2017). If the nanoparticle is explicitly included in the regulations, this means that the nanomaterial can be used. The only material specified as a nanoparticle in these regulations is titanium nitride, which is used as a polymer additive. Furthermore, although not specified as nanoparticles, silicon dioxide and zinc oxide, whose dimensions are given as 100 nm and below, are also included in the regulation. It is required to review the food additive regulations for components released from active packaging to food (e.g. the use of nano silver as an antibacterial agent; Commission, 2009).

4.7.1 Effects of Bionanomaterials on Human Health

Migration into food may occur from packaging materials prepared with bionanomaterials. It is of great importance to know the potential health effects of bionanomaterial packaging materials, since foods are the direct consumption material of humans. According to several researchers, nanoparticles can have toxicological impacts on biological systems (Mao et al., 2016). The high volume-to-area ratio of nanomaterials makes them more reactive and hazardous. They will rapidly react with other components during disposal and recovery since they are more reactive. According to certain settings, nanoparticles may be new allergens, create new hazardous effects, and be absorbed faster by the environment. It has been demonstrated that titanium oxide nanoparticles contained in meals may interact with living cells (Weir et al., 2012). Exposure to nanomaterials in food packaging material can occur in three different ways: skin contact, respiratory tract, or ingestion (Alger et al., 2014). It is stated that the ability of nanoparticles to enter the body and cells can cause toxic substances to spread inside the body, resulting in cell and tissue damage and malfunctions in the defense mechanism. Furthermore, it is reported that inhalation of nanosized materials might cause lung diseases in humans and mammals (Moore, 2006). Furthermore, nanoparticles may have been dispersed into the environment and subsequently contaminated food. According to Silvestre et al. (2011), certain free nanoparticles pass through the cell wall and cause oxidative and inflammatory responses. On the other hand, what happens to these particles after they are ingested is not well understood. The two main organs that ensure the nanoparticle distribution in the body—after the nanoparticle reaches the blood circulation system—are the spleen and the liver. The hydrophilic nature of nanoparticles increases their circulation rate drastically, and their surfaces are positively charged (Silvestreet al., 2011). Insoluble nanoparticles are also predicted to collect in secondary target organs.

4.7.2 Risk of Contamination and Migration from Bionanomaterial Packaging

Direct contact between packaging material and food may cause ions and molecular components from the packaging to migrate unintentionally to the food, resulting in both loss of quality in the food and health problems for the consumer (Brown et al., 2003). Small residues of monomers or oligomers (such as lactic acid) that react at the end of the polymerization process, such as plasticizers or other formulation additives, which are usually used to increase flexibility, are compounds that have the potential to migrate through the packaging material (Plackettet al., 2006). Adhesives in closing materials that are not in direct touch with food yet, are in close proximity to the packed food, which is another cause of compound migration (Brown and Williams, 2003). Components in the environment can be absorbed by the packaging material and eventually will migrate to the food. The presence of migration can be determined either visually or by chemical investigation. According to Echegoyen et al. (2013), the migration values of silver nanoparticles in three different food containers were between 1.66 and 31.46 ng/cm^2.

4.8 CONCLUSIONS

Nanotechnological packaging provides many advantages in the food industry. Nanotechnology improves product durability while decreasing manufacturing costs. While it keeps food fresh for a long period, its environmental impact is less than that of petroleum-based polymers. Therefore, research on these newly designed packages has intensified. However, their use is not yet prevalent. Yet, nanotechnology tends to produce innovative materials in the field of packaging. The thermal barrier and mechanical attributes will be stronger with the addition of nanoparticles suitable for appropriate foods. Bionanomaterials are potential substances made from a variety of biological components, including nucleic acids, peptides, bacteria, fungus, and plants. The use of bionanomaterials in biomedicine has attracted considerable attention because these materials are biosynthesized and biocompatible. Bionanomaterials exhibit extraordinary properties due to their miniature size, making them widely used in various field. However, studies and projects on foods and packaging of processed products show that the use of bionanomaterials in this sector will develop in the future. Furthermore, there is growing interest in foods enriched with newly developed bionanoparticles, and the research being conducted is progressing in this direction. Food packaging has been the most commonly used application of nanotechnology in the food industry. Nanotechnology or bionanoparticles can be employed to improve innovative food packaging. Food packaging materials with thermal, mechanical, antibacterial, and antioxidant capabilities can be created by adding functional qualities to polymers via bionanomaterials. Thus, this chapter summarized the fundamentals of bionanomaterials and their potential applications in food packaging.

REFERENCES

Aderibigbe, B. A. (2017). Metal-based nanoparticles for the treatment of infectious diseases. *Molecules, 22*(8), 1370.

Ahmed, J., Hiremath, N., & Jacob, H. (2017). Antimicrobial efficacies of essential oils/nanoparticles incorporated polylactide films against L. monocytogenes and S. typhimurium on contaminated cheese. *International Journal of Food Properties, 20*(1), 53–67.

Ahmed, J., Mulla, M., & Arfat, Y. A. (2017). Mechanical, thermal, structural and barrier properties of crab shell chitosan/graphene oxide composite films. *Food Hydrocolloids, 71*, 141–148.

Akbari, Z., Ghomashchi, T., & Moghadam, S. (2007). Improvement in food packaging industry with biobased nanocomposites. *International Journal of Food Engineering, 3*(4), 1–24.

Alberti, T., S Coelho, D., Voytena, A., Pitz, H., de Pra, M., Mazzarino, L., ... Veleirinho, B. (2017). Nanotechnology: A promising tool towards wound healing. *Current Pharmaceutical Design, 23*(24), 3515–3528.

Alexandre, M., & Dubois, P. (2000). Polymer-layered silicate nanocomposites: Preparation, properties and uses of a new class of materials. *Materials Science and Engineering: R: Reports, 28*(1–2), 1–63.

Alger, H., Momcilovic, D., Carlander, D., & Duncan, T. V. (2014). Methods to evaluate uptake of engineered nanomaterials by the alimentary tract. *Comprehensive Reviews in Food Science and Food Safety, 13*(4), 705–729.

Alghoraibi, I., & Alomari, S. (2018). Different methods for nanofiber design and fabrication. In *Handbook of nanofibers*, 1–46. Springer.

Asmatulu, R., & Khan, W. S. (2018). *Synthesis and applications of electrospun nanofibers*: Elsevier.

Auras, R. A., Singh, S. P., & Singh, J. J. (2005). Evaluation of oriented poly (lactide) polymers vs. existing PET and oriented PS for fresh food service containers. *Packaging Technology and Science: An International Journal, 18*(4), 207–216.

Barhoum, A., Rasouli, R., Yousefzadeh, M., Rahier, H., & Bechelany, M. (2018). Nanofiber technology: History and developments. In *Handbook of nanofibers*, 1–42. Springer International Publishing AG.

Ben-Sasson, M., Zodrow, K. R., Genggeng, Q., Kang, Y., Giannelis, E. P., & Elimelech, M. (2014). Surface functionalization of thin-film composite membranes with copper nanoparticles for antimicrobial surface properties. *Environmental Science & Technology, 48*(1), 384–393.

Benucci, I., Liburdi, K., Cacciotti, I., Lombardelli, C., Zappino, M., Nanni, F., & Esti, M. (2018). Chitosan/clay nanocomposite films as supports for enzyme immobilization: An innovative green approach for winemaking applications. *Food Hydrocolloids, 74*, 124–131.

Beykaya, M., & Caglar, A. (2016). An investigation on synthesis of silver-nanoparticles (AgNP) and their antimicrobial effectiveness by using herbal extracts. *Afyon Kocatepe University Journal of Sciences and Engineering, 16*(3), 631–641.

Bi, L., Yang, L., Narsimhan, G., Bhunia, A. K., & Yao, Y. (2011). Designing carbohydrate nanoparticles for prolonged efficacy of antimicrobial peptide. *Journal of Controlled Release, 150*(2), 150–156.

Bouwmeester, H., Dekkers, S., Noordam, M. Y., Hagens, W. I., Bulder, A. S., De Heer, C., ... Sips, A. J. (2009). Review of health safety aspects of nanotechnologies in food production. *Regulatory Toxicology and Pharmacology, 53*(1), 52–62.

Brown, H., & Williams, J. (2003). Packaged product quality and shelf life. In *Food Packaging Technology*, 65–94. Blackwell Publishing.

Casariego, A., Souza, B., Cerqueira, M., Teixeira, J., Cruz, L., Díaz, R., & Vicente, A. (2009). Chitosan/clay films' properties as affected by biopolymer and clay micro/nanoparticles' concentrations. *Food Hydrocolloids, 23*(7), 1895–1902.

Caseli, L., dos Santos Jr, D. S., Foschini, M., Gonçalves, D., & Oliveira Jr, O. N. (2007). Control of catalytic activity of glucose oxidase in layer-by-layer films of chitosan and glucose oxidase. *Materials Science and Engineering: C, 27*(5–8), 1108–1110.

Casettari, L., Vllasaliu, D., Castagnino, E., Stolnik, S., Howdle, S., & Illum, L. (2012). PEGylated chitosan derivatives: Synthesis, characterizations and pharmaceutical applications. *Progress in Polymer Science, 37*(5), 659–685.

Cerqueira, M. A., Vicente, A. A., & Pastrana, L. M. (2018). Nanotechnology in food packaging: Opportunities and challenges. *Nanomaterials for Food Packaging Materials, Processing Technologies, and Safety Issues*, 1–11. Elsevier.

Chandrasekaran, G., Han, H.-K., Kim, G.-J., & Shin, H.-J. (2011). Antimicrobial activity of delaminated aminopropyl functionalized magnesium phyllosilicates. *Applied Clay Science, 53*(4), 729–736.

Chang, P. R., Jian, R., Yu, J., & Ma, X. (2010). Fabrication and characterisation of chitosan nanoparticles/plasticised-starch composites. *Food Chemistry, 120*(3), 736–740.

Chaudhry, Q., Scotter, M., Blackburn, J., Ross, B., Boxall, A., Castle, L., … Watkins, R. (2008). Applications and implications of nanotechnologies for the food sector. *Food Additives and Contaminants, 25*(3), 241–258.

Coelho, D. S., Veleirinho, B., Alberti, T., Maestri, A., Yunes, R., Dias, P. F., & Maraschin, M. (2018). Electrospinning technology: Designing nanofibers toward wound healing application. In *Nanomaterials-Toxicity, Human Health and Environment*, 1–19. IntechOpen.

Çabuk, M., Yavuz, M., & Hlavac, J. (2011). Biodegradable and anti-carcinogenic chitosan/benzaldehyde modification and preparation of its nanocomposite. *Erciyes University Journal of the Institute of Science and Technology, 27*(3), 247–251.

Dallas, P., Sharma, V. K., & Zboril, R. (2011). Silver polymeric nanocomposites as advanced antimicrobial agents: classification, synthetic paths, applications, and perspectives. *Advances in Colloid and Interface Science, 166*(1–2), 119–135.

Dasan, Y., Bhat, A., & Ahmad, F. (2017). Polymer blend of PLA/PHBV based bionanocomposites reinforced with nanocrystalline cellulose for potential application as packaging material. *Carbohydrate Polymers, 157*, 1323–1332.

De Azeredo, H. M. (2009). Nanocomposites for food packaging applications. *Food Research International, 42*(9), 1240–1253.

Dong, C., Song, D., Cairney, J., Maddan, O. L., He, G., & Deng, Y. (2011). Antibacterial study of Mg (OH) 2 nanoplatelets. *Materials Research Bulletin, 46*(4), 576–582.

Duncan, T. V. (2011). Applications of nanotechnology in food packaging and food safety: barrier materials, antimicrobials and sensors. *Journal of Colloid and Interface Science, 363*(1), 1–24.

Dursun, S., & Erkan, N. (2009). The use of edible protein films in seafood. *Journal of Fisheries Sciences, 3*(4), 352.

Dursun, S., Erkan, N., & Yesiltas, M. (2010). Application of natural biopolymer based nanocomposite films in seafood. *Journal of Fisheries Sciences, 4*(1), 50.

Ealia, S. A. M., & Saravanakumar, M. (2017). *A review on the classification, characterisation, synthesis of nanoparticles and their application.* Paper presented at the IOP conference series: materials science and engineering. Vol. 263. No. 3. IOP Publishing.

Eatemadi, A., Daraee, H., Zarghami, N., Melat Yar, H., & Akbarzadeh, A. (2016). Nanofiber: Synthesis and biomedical applications. *Artificial Cells, Nanomedicine, and Biotechnology, 44*(1), 111–121.

Echegoyen, Y., & Nerín, C. (2013). Nanoparticle release from nano-silver antimicrobial food containers. *Food and Chemical Toxicology, 62*, 16–22.

Echeverría, I., Eisenberg, P., & Mauri, A. N. (2014). Nanocomposites films based on soy proteins and montmorillonite processed by casting. *Journal of Membrane Science, 449*, 15–26.

ElAmin, A. (2005). Nanotechnology targets new food packaging products. May, 18, 2011. Available from www.foodproductiondailyusa.com/news/ng.asp?id¼63147

Elen, K., Murariu, M., Peeters, R., Dubois, P., Mullens, J., Hardy, A., & Van Bael, M. (2012). Towards high-performance biopackaging: Barrier and mechanical properties of dual-action polycaprolactone/zinc oxide nanocomposites. *Polymers for Advanced Technologies, 23*(10), 1422–1428.

Erdem, S., & Gülel, G. T. (2015). Nanotechnology applications in the food industry. *Etlik Journal of Veterinary Microbiology, 26*(2), 52–57.

European Commission. (2009). Commission Regulation (EC) No 450/2009 of 29 May 2009 on active and intelligent materials and articles intended to come into contact with food., L135 C.F.R. *Official Journal of the European Union*, 135, 3–11.

Farahnaky, A., Dadfar, S. M. M., & Shahbazi, M. (2014). Physical and mechanical properties of gelatin–clay nanocomposite. *Journal of Food Engineering, 122*, 78–83.

Fernández, A., Cava, D., Ocio, M. J., & Lagarón, J. M. (2008). Perspectives for biocatalysts in food packaging. *Trends in Food Science & Technology, 19*(4), 198–206.

Gangwar, J., Gupta, B. K., & Srivastava, A. K. (2016). Prospects of emerging engineered oxide nanomaterials and their applications. *Defence Science Journal, 66*(4), 323–340.

Gao, W., Dong, H., Hou, H., & Zhang, H. (2012). Effects of clays with various hydrophilicities on properties of starch–clay nanocomposites by film blowing. *Carbohydrate Polymers, 88*(1), 321–328.

Grigoriadi, K., Giannakas, A., Ladavos, A. K., & Barkoula, N.-M. (2015). Interplay between processing and performance in Chitosan-based clay nanocomposite films. *Polymer Bulletin, 72*, 1145–1161.

Gu, H., Ho, P., Tong, E., Wang, L., & Xu, B. (2003). Presenting vancomycin on nanoparticles to enhance antimicrobial activities. *Nano Letters, 3*(9), 1261–1263.

Gupta, N. (2021). 2 Nanotechnology in agrifood sector: Ethical, regulatory, and governance landscape in EU. *Ethics in Nanotechnology: Social Sciences and Philosophical Aspects*, 25, 25–58.

Gutiérrez, T. J., Ponce, A. G., & Alvarez, V. A. (2017). Nano-clays from natural and modified montmorillonite with and without added blueberry extract for active and intelligent food nanopackaging materials. *Materials Chemistry and Physics, 194*, 283–292.

He, Y., Kong, W., Wang, W., Liu, T., Liu, Y., Gong, Q., & Gao, J. (2012). Modified natural halloysite/potato starch composite films. *Carbohydrate Polymers, 87*(4), 2706–2711.

Hernández-Vargas, J., González-Campos, J. B., Lara-Romero, J., Prokhorov, E., Luna-Bárcenas, G., Aviña-Verduzco, J. A., & González-Hernández, J. C. (2014). Chitosan/MWCNTs-decorated with silver nanoparticle composites: Dielectric and antibacterial characterization. *Journal of Applied Polymer Science, 131*(9), 40214.

Hottle, T. A., Bilec, M. M., & Landis, A. E. (2013). Sustainability assessments of bio-based polymers. *Polymer Degradation and Stability, 98*(9), 1898–1907.

Huang, L., Li, D.-Q., Lin, Y.-J., Wei, M., Evans, D. G., & Duan, X. (2005). Controllable preparation of Nano-MgO and investigation of its bactericidal properties. *Journal of Inorganic Biochemistry, 99*(5), 986–993.

Huff, K. (2008). *Active and intelligent packaging: innovations for the future*: *Department of Food Science & Technology. Virginia Polytechnic Institute and State University*, 1–13.

İlyasoğlu, H., & El, S. N. (2010). Nanoemulsions: Formations, structures and application areas as collodial delivery system in food sector. *Journal of Food, 35*(2), 143–150.

Jeevanandam, J., Barhoum, A., Chan, Y. S., Dufresne, A., & Danquah, M. K. (2018). Review on nanoparticles and nanostructured materials: History, sources, toxicity and regulations. *Beilstein Journal of Nanotechnology, 9*(1), 1050–1074.

Jordan, J., Jacob, K. I., Tannenbaum, R., Sharaf, M. A., & Jasiuk, I. (2005). Experimental trends in polymer nanocomposites—A review. *Materials Science and Engineering: A, 393*(1–2), 1–11.

Kahraman, H. T., Yar, A., Avcı, A., & Pehlivan, E. (2018). Preparation of nanoclay incorporated PAN fibers by electrospinning technique and its application for oil and organic solvent absorption. *Separation Science and Technology, 53*(2), 303–311.

Kakroodi, A. R., Kazemi, Y., Nofar, M., & Park, C. B. (2017). Tailoring poly (lactic acid) for packaging applications via the production of fully bio-based in situ microfibrillar composite films. *Chemical Engineering Journal, 308*, 772–782.

Kellar, J. J. (2006). *Functional fillers and nanoscale minerals: New markets/new horizons*: SME.

Ko, F., & Wan, Y. (2014). *Introduction to nanofibre materials*: Cambridge University Press. https://doi.org/10.1107/S2052520615000165

Kristo, E., & Biliaderis, C. G. (2007). Physical properties of starch nanocrystal-reinforced pullulan films. *Carbohydrate Polymers, 68*(1), 146–158.

Lai, P., Daear, W., Löbenberg, R., & Prenner, E. J. (2014). Overview of the preparation of organic polymeric nanoparticles for drug delivery based on gelatine, chitosan, poly (d, l-lactide-co-glycolic acid) and polyalkylcyanoacrylate. *Colloids and Surfaces B: Biointerfaces, 118*, 154–163.

Lamabam, S. D., & Thangjam, R. (2018). Progress and challenges of nanotechnology in food engineering. In *Impact of Nanoscience in the Food Industry*, 87–112. Elsevier.

Lee, J. E., & Kim, K. M. (2010). Characteristics of soy protein isolate-montmorillonite composite films. *Journal of Applied Polymer Science, 118*(4), 2257–2263.

Liu, X., Song, R., Zhang, W., Qi, C., Zhang, S., & Li, J. (2017). Development of eco-friendly soy protein isolate films with high mechanical properties through HNTs, PVA, and PTGE synergism effect. *Scientific Reports, 7*(1), 44289.

López-Córdoba, A., Medina-Jaramillo, C., Piñeros-Hernandez, D., & Goyanes, S. (2017). Cassava starch films containing rosemary nanoparticles produced by solvent displacement method. *Food Hydrocolloids, 71*, 26–34.

Luecha, J., Sozer, N., & Kokini, J. L. (2010). Synthesis and properties of corn zein/montmorillonite nanocomposite films. *Journal of Materials Science, 45*, 3529–3537.

Luykx, D. M., Peters, R. J., van Ruth, S. M., & Bouwmeester, H. (2008). A review of analytical methods for the identification and characterization of nano delivery systems in food. *Journal of Agricultural and Food Chemistry, 56*(18), 8231–8247.

Mahalik, N. P., & Nambiar, A. N. (2010). Trends in food packaging and manufacturing systems and technology. *Trends in Food Science & Technology, 21*(3), 117–128.

Mangiacapra, P., Gorrasi, G., Sorrentino, A., & Vittoria, V. (2006). Biodegradable nanocomposites obtained by ball milling of pectin and montmorillonites. *Carbohydrate Polymers, 64*(4), 516–523.

Mao, X., Nguyen, T. H., Lin, M., & Mustapha, A. (2016). Engineered nanoparticles as potential food contaminants and their toxicity to Caco-2 cells. *Journal of Food Science, 81*(8), T2107–T2113.

Mauter, M. S., & Elimelech, M. (2008). Environmental applications of carbon-based nanomaterials. *Environmental Science & Technology, 42*(16), 5843–5859.

McClements, D. J. (2014). *Nanoparticle-and microparticle-based delivery systems: Encapsulation, protection and release of active compounds*: CRC press.

Mihindukulasuriya, S. D., & Lim, L.-T. (2013). Oxygen detection using UV-activated electrospun poly (ethylene oxide) fibers encapsulated with TiO 2 nanoparticles. *Journal of Materials Science, 48*, 5489–5498.

Mills, A., & Hazafy, D. (2009). Nanocrystalline SnO2-based, UVB-activated, colourimetric oxygen indicator. *Sensors and Actuators B: Chemical, 136*(2), 344–349.

Mohamed, S. A., El-Sakhawy, M., & El-Sakhawy, M. A.-M. (2020). Polysaccharides, protein and lipid-based natural edible films in food packaging: A review. *Carbohydrate Polymers, 238*, 116178.

Mohammed, A. M. (2016). Fabrication and characterization of gold nano particles for DNA biosensor applications. *Chinese Chemical Letters, 27*(5), 801–806.

Moore, M. (2006). Do nanoparticles present ecotoxicological risks for the health of the aquatic environment? *Environment International, 32*(8), 967–976.

Moraru, C. I., Panchapakesan, C. P., Huang, Q., Takhistov, P., Liu, S., & Kokini, J. L. (2003). Nanotechnology: A new frontier in food science understanding the special properties of materials of nanometer size will allow food scientists to design new, healthier, tastier, and safer foods. *Nanotechnology, 57*(12), 24–29.

Mousavi, S. R., & Rezaei, M. (2011). Nanotechnology in agriculture and food production. *Journal of Applied Environmental and Biological Sciences, 1*(10), 414–419.

Muhammad, I., Razali, F., & Liza, M. (2005). *Study of an active antimicrobial system using a bio-switch concept*. Project Report. Universiti Teknologi Malaysia (Unpublished).

Nagel, B., Dellweg, H., & Gierasch, L. (1992). Glossary for chemists of terms used in biotechnology (IUPAC Recommendations 1992). *Pure and Applied Chemistry, 64*(1), 143–168.

Nathalie, G., Stéphane, P., Jose, M. L., Yolanda, E., & Carole, G. (2017). Nanotechnologies for active and intelligent food packaging: Opportunities and risks. In *Nanotechnology in Agriculture and Food Science*, 177–196. Wiley.

Nel, A., Xia, T., Madler, L., & Li, N. (2006). Toxic potential of materials at the nanolevel. *Science, 311*(5761), 622–627.

Nile, S. H., Baskar, V., Selvaraj, D., Nile, A., Xiao, J., & Kai, G. (2020). Nanotechnologies in food science: Applications, recent trends, and future perspectives. *Nano-Micro Letters, 12*(1), 1–34.

No, H. K., Park, N. Y., Lee, S. H., & Meyers, S. P. (2002). Antibacterial activity of chitosans and chitosan oligomers with different molecular weights. *International Journal of Food Microbiology, 74*(1–2), 65–72.

Oberdörster, G., Maynard, A., Donaldson, K., Castranova, V., Fitzpatrick, J., Ausman, K., … Lai, D. (2005). Principles for characterizing the potential human health effects from exposure to nanomaterials: Elements of a screening strategy. *Particle and Fibre Toxicology, 2*(1), 1–35.

Oyagi, M., Michira, I. N., Guto, P., Baker, P., Kamau, G., & Iwuoha, I. (2014). *Polydisperse low diameter 'non-toxic' silver nanoparticles encapsulated by rooibos tea templates.* Paper presented at the Nano Hybrids.

Ozcalik, O., & Tihminlioglu, F. (2013). Barrier properties of corn zein nanocomposite coated polypropylene films for food packaging applications. *Journal of Food Engineering, 114*(4), 505–513.

Palaniappan, A., Goh, W., Fam, D., Rajaseger, G., Chan, C., Hanson, B., … Liedberg, B. (2013). Label-free electronic detection of bio-toxins using aligned carbon nanotubes. *Biosensors and Bioelectronics, 43*, 143–147.

Paul, D. R., & Robeson, L. M. (2008). Polymer nanotechnology: Nanocomposites. *Polymer, 49*(15), 3187–3204.

Paul, M.-A., Alexandre, M., Degée, P., Henrist, C., Rulmont, A., & Dubois, P. (2003). New nanocomposite materials based on plasticized poly (L-lactide) and organo-modified montmorillonites: Thermal and morphological study. *Polymer, 44*(2), 443–450.

Plackett, D. V., Holm, V. K., Johansen, P., Ndoni, S., Nielsen, P. V., Sipilainen-Malm, T., … Verstichel, S. (2006). Characterization of l-polylactide and l-polylactide–polycaprolactone co-polymer films for use in cheese-packaging applications. *Packaging Technology and Science: An International Journal, 19*(1), 1–24.

Purohit, R., Purohit, K., Rana, S., Rana, R., & Patel, V. (2014). Carbon nanotubes and their growth methods. *Procedia Materials Science, 6*, 716–728.

Qhobosheane, M., Santra, S., Zhang, P., & Tan, W. (2001). Biochemically functionalized silica nanoparticles. *Analyst, 126*(8), 1274–1278.

Ramakrishna, S. (2005). *An introduction to electrospinning and nanofibers*: World scientific.

Ramirez, O., Bonardd, S., Saldías, C., Radic, D., & Leiva, Á. (2017). Biobased chitosan nanocomposite films containing gold nanoparticles: Obtainment, characterization, and catalytic activity assessment. *ACS Applied Materials & Interfaces, 9*(19), 16561–16570.

Ramos, Ó. L., Silva, S. I., Soares, J. C., Fernandes, J. C., Poças, M. F., Pintado, M. E., & Malcata, F. X. (2012). Features and performance of edible films, obtained from whey protein isolate formulated with antimicrobial compounds. *Food Research International, 45*(1), 351–361.

Raquez, J. M., Nabar, Y., Narayan, R., & Dubois, P. (2011). Preparation and characterization of maleated thermoplastic starch-based nanocomposites. *Journal of Applied Polymer Science, 122*(1), 639–647.

Ray, S. S., & Bousmina, M. (2005). Poly (butylene sucinate-co-adipate)/montmorillonite nanocomposites: Effect of organic modifier miscibility on structure, properties, and viscoelasticity. *Polymer, 46*(26), 12430–12439.

Reesha, K., Panda, S. K., Bindu, J., & Varghese, T. (2015). Development and characterization of an LDPE/chitosan composite antimicrobial film for chilled fish storage. *International Journal of Biological Macromolecules, 79*, 934–942.

Ren, G., Xu, X., Liu, Q., Cheng, J., Yuan, X., Wu, L., & Wan, Y. (2006). Electrospun poly (vinyl alcohol)/glucose oxidase biocomposite membranes for biosensor applications. *Reactive and Functional Polymers, 66*(12), 1559–1564.

Rhim, J.-W., Hong, S.-I., Park, H.-M., & Ng, P. K. (2006). Preparation and characterization of chitosan-based nanocomposite films with antimicrobial activity. *Journal of Agricultural and Food Chemistry, 54*(16), 5814–5822.

Rhim, J.-W., & Ng, P. K. (2007). Natural biopolymer-based nanocomposite films for packaging applications. *Critical Reviews in Food Science and Nutrition, 47*(4), 411–433.

Roach, S. (2006). Nanotechnology passes first toxicity hurdle. *Available from Error.*

Roco, M. C., Mirkin, C. A., & Hersam, M. C. (2011). Nanotechnology research directions for societal needs in 2020: Summary of international study. *Journal of Nanoparticle Research, 13*, 897–919. Available from www.foodproductiondaily-usa.com/news/ ng.asp?id¼69557

Rouf, T. B., & Kokini, J. L. (2018). Natural biopolymer-based nanocomposite films for packaging applications. In *Bionanocomposites for Packaging Applications*, 149–177. Springer.

Salgado, P. R., Di Giorgio, L., Musso, Y. S., & Mauri, A. N. (2019). Bioactive packaging: Combining nanotechnologies with packaging for improved food functionality *Nanomaterials for Food Applications* (pp. 233–270): Elsevier.

Sana, S. S., Li, H., Zhang, Z., Sharma, M., Usmani, Z., Hou, T., … Gupta, V. K. (2021). Recent advances in essential oils-based metal nanoparticles: A review on recent developments and biopharmaceutical applications. *Journal of Molecular Liquids, 333*, 115951.

Sanguansri, P., & Augustin, M. A. (2006). Nanoscale materials development–A food industry perspective. *Trends in Food Science & Technology, 17*(10), 547–556.

Saware, K., Sawle, B., Salimath, B., Jayanthi, K., & Abbaraju, V. (2014). Biosynthesis and characterization of silver nanoparticles using Ficus benghalensis leaf extract. *International Journal of Research Engineering and Technology, 3*, 867–874.

Sherman, L. M. (2004). Chasing nanocomposites. *Plastics Technology, 50*(11), 56–61.

Shoueir, K. R., El-Desouky, N., Rashad, M. M., Ahmed, M., Janowska, I., & El-Kemary, M. (2021). Chitosan based-nanoparticles and nanocapsules: Overview, physicochemical features, applications of a nanofibrous scaffold, and bioprinting. *International Journal of Biological Macromolecules, 167*, 1176–1197.

Siegrist, M., Stampfli, N., Kastenholz, H., & Keller, C. (2008). Perceived risks and perceived benefits of different nanotechnology foods and nanotechnology food packaging. *Appetite, 51*(2), 283–290.
Silvestre, C., Duraccio, D., & Cimmino, S. (2011). Food packaging based on polymer nanomaterials. *Progress in Polymer Science, 36*(12), 1766–1782.
Siqueira, J., & Oliveira, O. (2017). *Carbon-based nanomaterials. Nanostructures*: Elsevier.
Song, F., Tang, D.-L., Wang, X.-L., & Wang, Y.-Z. (2011). Biodegradable soy protein isolate-based materials: a review. *Biomacromolecules, 12*(10), 3369–3380.
Sorrentino, A., Gorrasi, G., & Vittoria, V. (2007). Potential perspectives of bio-nanocomposites for food packaging applications. *Trends in Food Science & Technology, 18*(2), 84–95.
Sozer, N., & Kokini, J. L. (2009). Nanotechnology and its applications in the food sector. *Trends in Biotechnology, 27*(2), 82–89.
Söğüt, E., & Seydim, A. C. (2017). Bio-based nanocomposites and food packaging applications. *Journal of Food, 42*(6), 821–833.
Subramani, T., & Ganapathyswamy, H. (2020). An overview of liposomal nano-encapsulation techniques and its applications in food and nutraceutical. *Journal of Food Science and Technology, 57*(10), 3545–3555.
Sugandha, V., Babita, K., & Shrivastava, J. (2014). Green synthesis of silver nanoparticles using single cell protein of Spirulina platensis. *International Journal of Pharma and Bio Sciences, 5*(2), 458–464.
Sun, J., Yendluri, R., Liu, K., Guo, Y., Lvov, Y., & Yan, X. (2017). Enzyme-immobilized clay nanotube–Chitosan membranes with sustainable biocatalytic activities. *Physical Chemistry Chemical Physics, 19*(1), 562–567.
Sun, L., Sun, J., Chen, L., Niu, P., Yang, X., & Guo, Y. (2017). Preparation and characterization of chitosan film incorporated with thinned young apple polyphenols as an active packaging material. *Carbohydrate Polymers, 163*, 81–91.
Sürengil, G., & Kılınç, B. (2011). Nanotechnology applications in foods-packagings and importance for seafoods. *Journal of Fisheries Sciences, 5*(4), 317–325.
Tarhan, Ö., Gökmen, V., & Harsa, Ş. (2010). Applications of nanotechnology in food science and technology. *The Journal of Food, 35*(3), 219–225.
Theinsathid, P., Visessanguan, W., Kingcha, Y., & Keeratipibul, S. (2011). Antimicrobial effectiveness of biobased film against Escherichia coli 0157: h7, Listeria monocytogenes and Salmonella typhimurium. *Advance Journal of Food Science and Technology, 3*(4), 294–302.
Thombre, R., Parekh, F., & Patil, N. (2014). Green synthesis of silver nanoparticles using seed extract of Argyreia nervosa. *International Journal of Pharmacy and Biological Sciences, 5*(1), 114–119.
Tiwari, D. K., Behari, J., & Sen, P. (2008). Application of nanoparticles in waste water treatment. *World Applied Sciences Journal, 3*(3), 417–433.
Topuz, F., & Uyar, T. (2020). Antioxidant, antibacterial and antifungal electrospun nanofibers for food packaging applications. *Food Research International, 130*, 108927.
Tornuk, F., Hancer, M., Sagdic, O., & Yetim, H. (2015). LLDPE based food packaging incorporated with nanoclays grafted with bioactive compounds to extend shelf life of some meat products. *LWT-Food Science and Technology, 64*(2), 540–546.
Tulsyan, G., Richter, C., & Diaz, C. A. (2017). Oxygen scavengers based on titanium oxide nanotubes for packaging applications. *Packaging Technology and Science, 30*(6), 251–256.
Vahedikia, N., Garavand, F., Tajeddin, B., Cacciotti, I., Jafari, S. M., Omidi, T., & Zahedi, Z. (2019). Biodegradable zein film composites reinforced with chitosan nanoparticles and cinnamon essential oil: Physical, mechanical, structural and antimicrobial attributes. *Colloids and Surfaces B: Biointerfaces, 177*, 25–32.
Vermeiren, L., Devlieghere, F., van Beest, M., de Kruijf, N., & Debevere, J. (1999). Developments in the active packaging of foods. *Trends in Food Science & Technology, 10*(3), 77–86.
Wakai, M., & Almenar, E. (2015). Effect of the presence of montmorillonite on the solubility of whey protein isolate films in food model systems with different compositions and pH. *Food Hydrocolloids, 43*, 612–621.
Wang, B., & Sain, M. (2007). Isolation of nanofibers from soybean source and their reinforcing capability on synthetic polymers. *Composites Science and Technology, 67*(11–12), 2521–2527.
Weir, A., Westerhoff, P., Fabricius, L., Hristovski, K., & Von Goetz, N. (2012). Titanium dioxide nanoparticles in food and personal care products. *Environmental Science & Technology, 46*(4), 2242–2250.

Wu, H., Liu, C., Chen, J., Chang, P. R., Chen, Y., & Anderson, D. P. (2009). Structure and properties of starch/α-zirconium phosphate nanocomposite films. *Carbohydrate Polymers, 77*(2), 358–364.

Wu, J., Wang, N., Wang, L., Dong, H., Zhao, Y., & Jiang, L. (2012). Electrospun porous structure fibrous film with high oil adsorption capacity. *ACS Applied Materials & Interfaces, 4*(6), 3207–3212.

Xue, J., Wu, T., Dai, Y., & Xia, Y. (2019). Electrospinning and electrospun nanofibers: Methods, materials, and applications. *Chemical Reviews, 119*(8), 5298–5415.

Yang, M., Kostov, Y., & Rasooly, A. (2008). Carbon nanotubes based optical immunodetection of Staphylococcal Enterotoxin B (SEB) in food. *International Journal of Food Microbiology, 127*(1–2), 78–83.

Yang, Z., Chaieb, S., & Hemar, Y. (2021). Gelatin-based nanocomposites: A review. *Polymer Reviews, 61*(4), 765–813.

Yavuz, İ., & Yılmaz, E. Ş. (2021). Nanoparticules with biological systems. *Gazi University Journal of the Faculty of Science, 2*(1), 93–108.

Yin, P. T., Kim, T.-H., Choi, J.-W., & Lee, K.-B. (2013). Prospects for graphene–nanoparticle-based hybrid sensors. *Physical Chemistry Chemical Physics, 15*(31), 12785–12799.

Youssef, A. M., Abdel-Aziz, M. S., & El-Sayed, S. M. (2014). Chitosan nanocomposite films based on Ag-NP and Au-NP biosynthesis by Bacillus subtilis as packaging materials. *International Journal of Biological Macromolecules, 69*, 185–191.

Youssef, A. M., Abou-Yousef, H., El-Sayed, S. M., & Kamel, S. (2015). Mechanical and antibacterial properties of novel high performance chitosan/nanocomposite films. *International Journal of Biological Macromolecules, 76*, 25–32.

Yu, L., Dean, K., & Li, L. (2006). Polymer blends and composites from renewable resources. *Progress in Polymer Science, 31*(6), 576–602.

Zamudio-Flores, P. B., Torres, A. V., Salgado-Delgado, R., & Bello-Pérez, L. A. (2010). Influence of the oxidation and acetylation of banana starch on the mechanical and water barrier properties of modified starch and modified starch/chitosan blend films. *Journal of Applied Polymer Science, 115*(2), 991–998.

Zhang, W., Zhang, X., Zhang, L., & Chen, G. (2014). Fabrication of carbon nanotube-nickel nanoparticle hybrid paste electrodes for electrochemical sensing of carbohydrates. *Sensors and Actuators B: Chemical, 192*, 459–466.

Zhao, R., Torley, P., & Halley, P. J. (2008). Emerging biodegradable materials: starch-and protein-based bionanocomposites. *Journal of Materials Science, 43*, 3058–3071.

5 Bionanomaterials in Increasing Shelf Life of Food

Jeyakumar Saranya Packialakshmi, Duraisamy Elango, Arumugam Priyadharsan, and Jun Tae Kim

5.1 INTRODUCTION

The term shelf life refers to the commercial status of the product and can be defined as a period of time when the food remains safe to eat under defined storage conditions. During the shelf-life period, the food retains its desired chemical, physical, sensory, functional, and microbiological characters. Shelf life is an important requirement that ensures food safety. In addition to ensuring food safety, shelf life also conceptualizes the reduction in wastage of food. The US Food and Drug Administration (FDA) states that about 1.3 billion tons—that accounts for one-third of the total food—is disposed of or otherwise wasted. Food may be spoiled by several physical, chemical, enzymatic, and microbial factors. Foods past their shelf life lose tone, appearance, consistency, nutrition, palatability, and delectability. Consuming foods beyond their shelf life leads to several food borne-illnesses, even to death in severe cases (Nyachuba, 2010). Extension of shelf life of food products importantly focuses on managing the enzymatic changes and rendering protection from pathogenic microbial infections. Over the decades, several techniques have been adopted for altering the storage life of food. From simple approaches of olden times to modern molecular methods, the list is boundless. The nature of the food determines the shelf-life extension technique adopted. Heating/drying, chilling, freezing, chemical treatment, modified atmospheric packaging (MAP) and controlled atmospheric packaging (CAP), vacuum sealing, shrink wrapping—these are the techniques used for food preservation (Amit et al., 2017): scientific and technological advancements change the course of shelf-life extension techniques. The modernization of the farming sector shares a similarity with the modernization of medicine distribution and pharmaceuticals (Arshad et al., 2021). Nanomaterials are one addition to the long, evolving list of techniques for the extension of shelf life. The nanomaterials synthesized from various chemical sources possessed a higher surface-to-volume ratio, exclusive morphology, and related functional properties. Yet their utility was limited by the cell cytotoxicity of foods they worked with. To overcome these issues, bionanomaterials were created.

Bionanomaterials are molecular materials composed of partial/non-partial biological molecules like proteins, antibiotics, enzymes, nucleic acids, lipids, oligo/polysaccharides, viruses, secondary metabolites, and so forth. Simply put, bionanomaterials are biologically derived nanomaterials (Singh et al., 2021b). The bionanomaterials display exceptional properties of biodegradability, biocompatibility, and non-toxicity compared to conventional nanomaterials. These bionanomaterials are employed in several fields, such as healthcare, electronics, textiles, aerospace engineering, environmental protection, food packaging, and so forth (Wang et al., 2017a). In the food industry,

 DOI: 10.1201/9781003432791-7

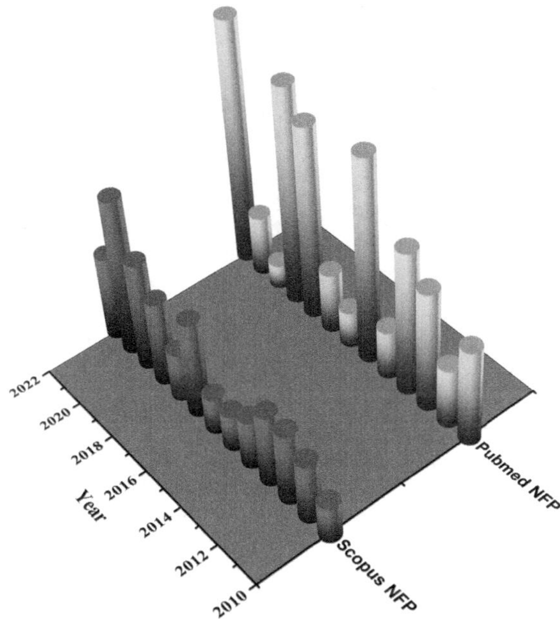

FIGURE 5.1 Research Trends in Nano Food Packaging (NFP), as Estimated by the Number of Pubmed and Scopus Publications Hits from 2010 to 2022. (www.ncbi.nlm.nih.gov/pubmed; www.scopus.com)

bionanoparticles are utilized for the shelf-life extension of food. Food's shelf-life extension is mostly accomplished by packing the food with bionanomaterials based packaging films. The majority of research (apart from that of the sample population) shows that nanoparticles are well accepted by consumers as packaging materials rather than direct application on the food itself (Bhardwaj et al., 2019).

Application of nanomaterials/bionanomaterials in food systems has resulted in a wide range of unique products with improved food qualities such as texture, taste, sensory aspects, stability, and so on (Pateiro et al., 2021). Figure 5.1 depicts the increased interest in this topic, as evidenced by the growing number of publications since 2010.

5.2 TRADITIONAL SHELF-LIFE EXTENSION TECHNIQUES AND THEIR PITFALLS

Extension of shelf life of food basically depends inhibiting/delaying microbial growth. This may involve change in physiological and storage conditions (Gould 1996). Heating or drying involves removal of moisture from the food. Removal of water reduces the water activity, the moisture available for microbial growth. During drying/heating, the moisture of the food becomes lower than that required for microbial growth. But in the process color, flavor, and texture of the food is subjected to alterations; further there occurs modification/losses of nutrients, which reduces the bioavailability of the phytochemicals (Rahman and Perera, 2007). Chilling/freezing foods to extend their shelf life was a practice adopted by common people by means of refrigeration. Refrigeration over years has become a mainstay for food preservation. Refrigerated food retains high quality, good safety, and market value. Freezing is the advanced form of chilling where smaller-size ice crystals form, reduces the water activity, and the product quality is retained (James and James, 2023). But the cold preservation processes often reduce the dietary benefits and brings about textural changes. Development of rancidity and color changes also occur. Frozen fruits are said to have fewer antioxidants, vitamins

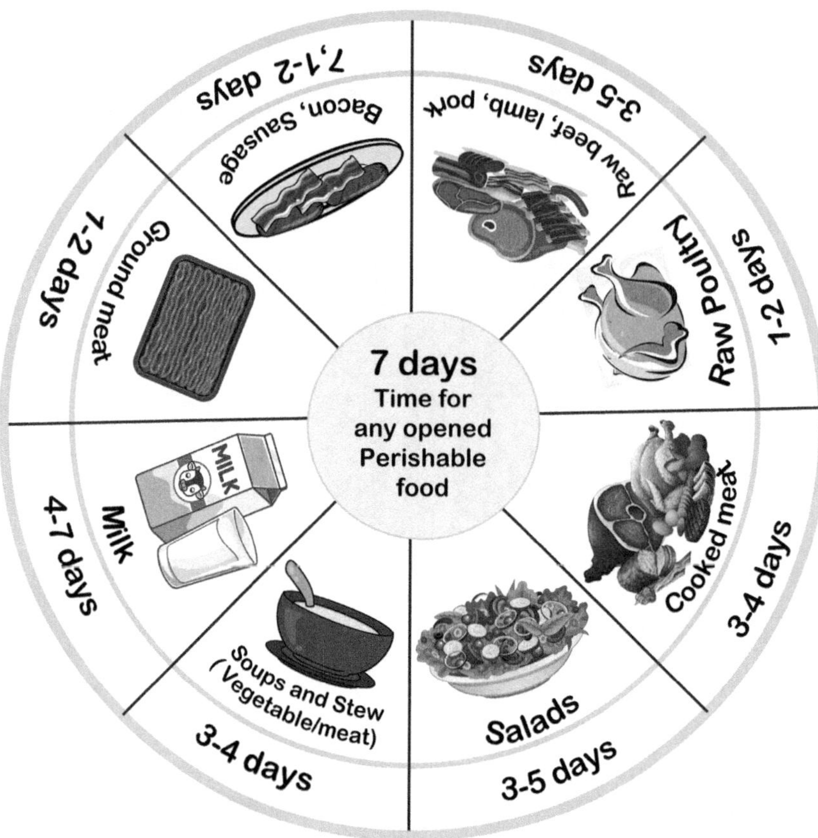

FIGURE 5.2 Shelf Life of Foods Under Refrigerated Conditions. (Compiled by the Authors.)

B and C compared to fresh fruits. Freezer burn, formation of damaged ice crystals, catching up of odor from other foods, emission of CFC—these are the major drawbacks of the cold preservation techniques. The chemical techniques hamper their own disadvantages. There is a chance of formation of high volume of hazardous materials like O_3, H_2O_2, Cl_2, and so forth. Certain chemical preservatives when taken for extended periods of time, causes severe illness and chronic life conditions (Abdulmumeen et al., 2012). The advanced packaging technologies such as modified atmosphere, controlled atmosphere packaging technologies, incurs extra costs in addition to the alteration of the physiology of the food due to the gases in contact with the food materials. These above mentioned are some of the pitfalls existing in the shelf-life extension techniques. The shelf life of different food under refrigerated conditions is shown in Figure 5.2.

5.3 POSSIBILITY OF NOVEL TECHNOLOGIES IN EXTENDING THE STORAGE LIFE OF FOODS

Over the past era, quick advancement of nanotechnology has altered numerous areas in food science, particularly those related to food packaging and many other aspects in processing of fresh produce (Kumar et al., 2022a; Jayakumar et al., 2022; Sagar et al., 2022). Nanotechnology is regarded as an ecofriendly, and nascent, technique in the recent decade. Nanomaterials are produced either through bottom up or top-down approaches (Hamad et al., 2018; Hermes et al., 2020; Shweta et al., 2021). The agriculture and food processing sectors have recently made extensive utilization of nanomaterials for increasing food shelf life. Nanopackaging enhance the shelf life of food which in turn reduces wastage of food. (Nurfatihah and Siddiquee, 2019; Neme et al., 2021; Sagar et al., 2022; Haris et al., 2023). Improvements in mechanical, thermic, moisture and gas barriers, and chemical resistance properties in food packaging films could be made with the help of nanotechnology (Mustafa and Andreescu, 2020). The main characteristics displayed during the transformation of materials from macro to nano scale enhances the excellent optical properties, hardness, and inimitable barrier (Pasricha and Sachdev 2017; Tiwari et al., 2022). With nanomaterials in hand, a suitable environment for extension of shelf life can be provided by food packaging. This mitigates the risk of water loss, gas distribution, microbial degradation, and destruction of organoleptic properties of food products by retardation of the metabolic and enzymatic activities (Ma et al., 2017; Jafarzadeh et al., 2021; Yadav et al., 2022). Nano packaging protects foods from both internal and external factors by acting as a thermal or mechanical barrier along with their antimicrobial properties (Singh et al., 2017; Kuswandi and Moradi 2019; Rai et al., 2019; Jalgaonkar. et al., 2022). The following sections discuss the key characteristics of nano-packaging in terms for the extension of shelf-life.

5.3.1 Thermal Properties

Thermal properties are vital for processing fresh produce and vegetables (Jalgaonkar et al., 2022). Thermal properties were improved by the rearrangements of nanoparticles and matrix molecules into a higher energy state (Smith et al., 2019). Thermal conductivity, thermal diffusivity, specific heat and density, and packaging materials are the most important properties to consider when analyzing thermal processes (Basumatary et al., 2022). In food packaging, thermal property degradation is proportional to polymer shear force, molecular weight, structure, time, and temperature (Shojaeiarani et al., 2019; Jafarzadeh et al., 2021; Xu et al., 2021). Nanomaterials are the prime options for enhancing the thermal stability of food packaging systems to overcome thermal degradation (Fahmy et al., 2020). The most commonly used nanofillers to improve the thermal stability are nanoparticles, nanotubes, nanofibrils, and nanorods (Yu et al., 2017; Sagar et al., 2022). The temperature profile of the nano-packaging can be investigated through differential scanning calorimetry (DSC) and thermal gravimetric analysis (TGA). To determine a sample transition temperature, DSC investigates the glass transition temperature (Tg), crystallization temperature (Tc), and melting temperature (Tm), whereas TGA illustrates the loss of mass (Kumari et al., 2021; Shen et al., 2022). Recent researches indicates that the nanoparticles increased the thermal stability of the nano-packaging materials.

5.3.2 Mechanical Properties

In general, favorable interactions between nanoparticles and matrix, as well as nanoparticles with nanoparticles, improve the mechanical characteristics of nanocomposite materials in food packaging application. The effect of addition of cellulose nanoparticles (CN) to alginate biopolymer films was characterized in terms of the mechanical, physical and optical properties. The outcomes displayed that upon increment from 0 to 5 per cent of CN content, the tensile strength of the films increased evincing the CN ability to affect the mechanical properties (Abdollahi et al., 2013). However, the possible potential of aggregation of the nanoparticles may happen at higher nanoscale filler material concentrations, which would reduce the reinforcing effect. Additionally, Cheng et al. (2016) found that adding nanocrystalline cellulose to guar gum, nanocrystalline cellulose films can significantly enhance their mechanical characteristics, such as hardness and elastic modulus. According to a different study, combining cellulose nanoparticles with alginate film enhanced the mechanical properties of biopolymer matrix.

5.3.3 Barrier Characteristics

Barrier properties were enhanced by using nanomaterials in food packaging due to their finite "quantum" properties. When compared to conventional preservation materials, nanomaterials have more advantages (Sagar et al., 2022). Although nanoparticles possess elevated surface area per unit volume than the pure polymer alone, only a minimum of addition (approximately 5 per cent, w/w) is necessary for the polymer to establish effective surface interactions, giving it longer shelf life (Wang et al., 2017b; Muller et al., 2017; Liu et al., 2022a). Nanoscale fillers being impermeable, the gases pocketed inside move out of the package by strenuous diffusion path amidst the nanoscale fillers rather than the pure polymers in case of gas permeation (Rhim et al., 2013; Cui et al., 2015). Researchers developed a nano-crystalline cellulose (NCC) film amended with guar gum (GG), and determined that the addition of nano-crystalline cellulose altered the film's pore morphology and increased the capacity of the film to obstruct oxygen (Gupta et al., 2018; Dubey et al., 2023). Additionally, researchers created the concise torsional model shown below to address the lengthening effect of nanoscale fillers on diffusion path.

$$\frac{P_s}{P_p} = \frac{1-\phi}{1+\frac{L}{2W\phi}} \tag{1}$$

Where, Ps is the permeability of nanoscale filler and Pp is the permeability of original polymer, the length and width of the nanocomposites are designated by L and W. But scientists have already developed an alternative model, to predict the dispersal effect of clay in semi-dilute and dilute composites.

$$\frac{P_s}{P_p} = \frac{1}{1+\mu\left(\frac{L}{2W}\right)^2 \varphi^2} \tag{2}$$

Where, μ = л2/16 ln (L/2W)2, and Ps, Pp, L and W in equation (2) are same as in equation (1).

5.3.4 Antimicrobial Properties

Antimicrobial properties of nanoparticles and nanocomposites can sustain the freshness of the produce by retarding the micro biogenesis, inhibiting the growth of the individual microorganism (Liu et al., 2020a; Kumar et al., 2022). In general, two different types of antimicrobial nanoparticles were currently emerged, such as, metal ion type and oxide photocatalytic type. Metal ionic antimicrobial nanoparticles including silver, copper, cobalt, nickel, iron, zinc, and aluminium with natural or synthetic nanocarriers were utilized in food packaging application (Abd Elkodous et al., 2019; Kumar et al., 2022b; Chouke et al., 2022; Sagar et al., 2022). Porous materials with large surface volume ratio like layered clay, silicate, phosphate have extraordinary adsorption properties, less cytotoxicity and have the additional benefit as a carrier for antimicrobial nanomaterials (Ali et al., 2021; Virmani and Pathak, 2022). On the other hand, antibacterial abilities of photocatalytic materials primarily depend on the ROS generation ability (Wang et al., 2017b; Ahmad et al., 2020; Shi et al., 2021). Additionally, sulfhydryl (-SH) and/or amine (-NH) functional groups from proteins and genetic materials can react with antimicrobial metal ions, which results in the denaturation of functional proteins and cleavage of nucleic acids (Nain et al., 2020; Guo et al., 2021). Antimicrobial metal ions can damage the membrane, get into the bacterial cells, and interact with DNA and/or enzymes to retard the normal functions of the pathogens (Koechler et al., 2015; Salem and Fouda, 2021). The contact reaction theory postulates that when cationic antibacterial nanoparticles come in contact with anionic bacterial cells and there is increased interaction between one another the result is increasing effective penetration into the bacterial cell membrane and denaturation of the protein (Kaur and Liu, 2016; Imani et al., 2020; Tu et al., 2021). Ceased respiration and metabolism due to the nanoparticles results in the death of the bacterial cells.

5.3.5 Photocatalytic Properties

Photocatalytic nanomaterials have a broad application range due to their unique properties, which include high surface area, reactivity, pore volume, electrostatic interactions, and hydrophilic and hydrophobic interactions (Yamashita et al., 2018; Wen et al., 2019). The band gaps existing between the valence band and conduction band having contrasting energy levels a low energy valence band and a high energy conduction band with band gaps between is responsible for the photocatalytic activity of the nanoparticles (Lin et al., 2019; Nasir et al., 2021). On exposure to light, valence-bond electrons are excited into the conduction band with a higher energy than the bandgap; this causes them to simultaneously produce a series of holes (h^+) that are positively charged and an array of exceedingly active electrons (e^-) on the conduction band (Liu et al., 2022b; Jilani et al., 2022; Zilberg et al., 2022; Das et al., 2022). Ultimately high-activity electron-hole pairs generated on the semiconductor surface (Valenzuela et al., 2020). In order to transfer energy, these charged holes and active electrons interact with the water molecules and dissolved oxygen that adsorb on the surface of the particles. As a result, highly reactive hydroxyl radicals (OH) and powerful oxidizing superoxide radicals were formed ($O_2^{\cdot}$) (Liu et al., 2020b; Padmanabhan et al., 2021; Feng et al., 2022). The primary reactions are as follows:

$$OH^- + h^+ \longrightarrow .OH \quad (3)$$

$$H_2O + h^+ \longrightarrow .OH + H^+ \quad (4)$$

$$e^- + O_2 \longrightarrow O_{2-} \quad (5)$$

These active groups can interact with organic compounds (Eq. 6&7).

$$OH + \text{organics} \longrightarrow CO_2 + H_2O \quad (6)$$

$$O_2 + \text{organics} \longrightarrow CO_2 + H_2O \quad (7)$$

The entire reaction can be pronounced responsible for ethylene degradation and microbe inhibition. As a result, photocatalytic nanoparticles impact the food preservation for a longer time.

5.4 NANOPARTICLE FORMS UTILIZED IN FOOD PACKAGING

Commercially, various kinds of nanomaterials are being utilized for fabricating innovative structural and functional packaging (Table 5.1). Nanotechnology aids in the preservation of food product quality and enhances the following qualities: (1) product features; (2) purpose; and (3) nutritional and sensory qualities (Idumah et al., 2020; McClements, 2020; Liu et al., 2022c; Cheng et al., 2022). Numerous nanostructured food ingredients have unique properties, such as improved texture, consistency, and taste. Excellent colloidal properties are present in the nanomaterials, and certain types of organic materials and natural (colloids) particles have a significant impact on these properties (Shafiq et al., 2020; Sahoo et al., 2021; Aguilar-Perez et al., 2021). Functional nanomaterials are derived from various organic and inorganic materials. Several types of organic and inorganic nanomaterials have been utilized for shelf-life extension (Lombardo et al., 2019; Basumatary et al., 2022). Ability to encapsulate active substances, improved functionality, stability, bioavailability,

TABLE 5.1
Summary of Recent Research Works on Positive Impact of Nano-Edible Coatings on Food Product Quality

Form of nanomaterial	Constituents	Test Sample	Beneficial impact	Reference
Nanoparticles	Ag	Papaya	Delayed ripening, Negligible loss in soluble solids and mass.	(Wu et al., 2019)
		Pumpkin and Tomato	Delayed microbial growth with more shelf life.	(Mohsen et al., 2020)
	Cu	Pear and Apple	Superior textural properties.	(Singh and Sahareen 2017)
	ZnO	Carrot	Less surface microbial population.	(Sarojini et al., 2019)
		Mango	Lesser deterioration after a storage period of 33 days.	(Sapelli et al., 2021)
		Black Grapes	Extended storage time up to 9 days with fresh appearance	(Indumathi et al., 2019)
	TiO_2	Banana	Sufficient O_2 and CO_2 exchange, preserving the fruits up to 12 days in storage.	(Kaur et al., 2017)
		Cherry Tomatoes	Storage time increased to 20 days owing to UV induced ethylene degradation.	(Meindrawan et al., 2018)
		Grapes and Apples	Decreased oxidation effect, moisture loss, and microbial contamination.	(Mohsen et al., 2020)
Nanocomposites	Chitosan, alginate and ZnO nanoparticles	Guavas	Storage time increased to 20 days as opposed to the uncoated ones which could be stored only for 7 days.	(Arroyo et al., 2020)
		Fresh-cut papaya	Reduced microbial infection on the surface.	(Lavinia et al., 2019)
	Cassava-starch and starch nanocrystals	Pears	Inhibition of browning enzyme activities. Color, texture and permeability remains unchanged.	(Dai et al., 2020)
Nano emulsions	Alginate and basil oil	Okra	Lesser fungal infection on storage. Increased freshness and overall acceptance.	(Gundewadi et al., 2018)
	Chitosan, carboxymethyl cellulose and citral	Fresh-cut melons	Prolongation of storage life to 10 days due to reduced microbial contamination.	(Arnon-Rips et al., 2019)
	Sodium alginate and sweet orange essential oil	Tomatoes	Decreased total mesophilic bacterial count. Lesser weight loss due to retained firmness.	(Das et al., 2020)
	Carnauba wax	mandarins and tangors	Gloss retained throughout the storage, less ethylene production and dehydration.	(Miranda et al., 2021)
		Papaya	Delayed ripening.	(Zucchini et al., 2021)
Polymeric nanoparticles	Chitosan	Bananas	Megre expression of genes, *MaACS1* and *MaACO* for the coated bananas delaying the ripening.	(Lustriane et al., 2018)
		Table grapes	Retention of firmness, total soluble solids and total sugar even after longer storage. Increased table time.	(Castelo Branco Melo et al., 2018)
		Bell pepper (Fresh cut)	Stabilization of mass reduction and sensory properties upto12 days at 5°C	(Hu et al., 2020)
Nanofibers	Cellulose	Strawberry	Delayed senescence due to the restriction of the cell respiration.	(Kwak et al., 2021)
		Grapes	Less reduction in weight retaining the firmness and juiciness.	(Silva-Vera et al., 2018)
	Whey protein isolate nanofibers, glycerol and carvacrol	Salted duck egg yolk	Maintenance of weight and lustrous shell throughout the storage.	(Wang et al., 2020)

lesser cytotoxicity, easy handling, and so forth, makes nanomaterials prominent candidates for food packaging application.

This use of nanotechnology offers the packaging materials for food by altering the way foils penetrate, enhancing the effects of mechanical, chemical, and microbial obstacles, and improving heat resistance (Sahoo et al., 2022). Bio-based nano packaging has a number of benefits over traditional materials. They are safe, environmentally friendly, and biodegradable (Primozic et al., 2021; Singha et al., 2022). The biodegradable nano-packaging is composed of three types of biopolymers: naturally occurring (polysaccharides and proteins), microbially derived (Xanthan and so forth), and manufactured biopolymers (polylactic acid, polypropylene, polystyrene and polyglycolic acid). In addition, polymers like gelatin, carrageenan, cellulose, chitosan, pectin, pullulan, and zein are also used in the production of edible food packaging films (Maciel et al., 2020; Shahidi and Hossain, 2022; Atta et al., 2022; Kumar et al., 2022; Gupta et al., 2022; Aayush et al., 2022). Due to their biodegradable nature, these nano-films are less/non-toxic and environment friendly as well as fresh produce (Sagar et al., 2022). In order to ensure consumer safety, the current trend in packaging includes nano-reinforcement, nanocomposite, nanosensing, and biodegradability (Chausali et al., 2022). Nanotechnology produces biodegradable packaging, which reduces environmental pollution.

5.4.1 Nanoparticles

Nanoparticles possess special optical, catalytic, magnetic and biological characteristics, such as antimicrobial and antioxidant functions, and are widely utilized in the food science sector lately (Nile et al., 2020). Nanoparticles added to edible films enhance their physicochemical characteristics, such as strength, mechanical characteristics, and flexibility (Bahrami et al., 2019). It enhances the food's flavor, color, and texture. In order to improve the characteristics of packaging and provide beneficial properties like antimicrobial activity, these nanoparticles are added to polymer matrices (Paidari et al., 2023). To enhance the heat tolerance, mechanical, and antimicrobial functions of the food packaging materials, metallic nanoparticles (copper, gold, zinc oxide, silver, and titanium oxide) are typically preferred (Ballesteros et al., 2022; Jadhav et al., 2023). Compared to organic nanomaterials, the addition of inorganic nanoparticles to food packaging systems exhibits excellent antimicrobial properties (Suvarna et al., 2022). They can interact with organic material in food and cause cell damage as a result of oxidized cell parts, secondary metabolite production by the interactions with microbial cells (Stambulska and Bayliak, 2020; Aguirre-Becerra et al., 2021). For extension of storage time of food and feed products, magnesium oxide (MgO), silicon dioxide (SiO_2), ferric oxide (Fe_3O_4), and other biocompatible nanoparticles are used (Barciela et al., 2023).

5.4.2 Nano-Sized Emulsions

Nano-emulsions are gaining popularity in a variety of industries, particularly the food industry. Nano emulsions (NEs) are a type of emulsion with extremely small dispersed phase droplet sizes ranging from 10 to 100 nm (Singh and Pulikkal, 2022). Owing to their large surface area, translucent appearance, long-lasting stability, and tunable rheology, nano emulsions have attracted a lot of attention in a variety of industrially significant fields (Sharma et al., 2020; Kumar et al., 2021; Gauthier and Capron, 2021). The most widely used techniques for creating nano emulsions include low-energy techniques like spontaneous emulsification, phase inversion composition, phase inversion temperature, and emulsion inversion point, as well as high-energy techniques like HPVH, micro fluidizers, and ultrasonic homogenization (Safaya and Rotliwala, 2020; K. and Kumar, 2022). Nano-emulsion are used to encapsulate, preserve, improve bioavailability, and target release sensitive functional compounds (Islam et al., 2023). Naturally occurring antioxidant-rich and antimicrobial-rich oils—such as lemon grass oil, thyme oil, mandarin oil, and cinnamon oil—are being incorporated into chitosan polymer for the development of nano-emulsion based edible films (Xing et al., 2016; Tian and Liu, 2020; Ghoshal, 2021; Bhandari et al., 2022).

5.4.3 Polymers

The term "polymer" refers to a substance with a large molecular structure made entirely of rings or chains of numerous unique repeating units (Peterson and Choi 2020). Numerous natural and synthetic polymers were widely utilized for food packaging applications. Over the past decade, there seems to be an increasing demand for biopolymers for functional food packaging (Panda et al., 2022). Owing to their desirable qualities, such as strength, pliability, durability and producibility, polymers are now a necessary component of modern life (Mangaraj et al., 2019; Priyadarshi et al., 2021). Several proteins, including gelatin, keratin, and casein, contain fascinating polymer properties like flexural, shear strength, and tensile modulus in addition to exceptional material qualities like toughness, strength, and elasticity (Zhang et al., 2023). Protein polymer films are used in the food packaging industry as edible films that can be consumed with the food, for example, gelatin, wheat gluten, corn, zein, soy proteins, milk proteins, and egg white (Mihalca et al., 2021; Gupta et al., 2022). Therefore, new biodegradable polymers can be created using these proteins for a variety of industrial uses.

5.5 BIONANOMATERIALS AND SHELF-LIFE EXTENSION BY FOOD PACKAGING

The expanding food sector is under increasing demand to fulfil shoppers' desire for hygienic, nutritious, and fresh food, as well as to meet current severe food safety requirements. Foods are very reactive to spoiling, rendering them unsuitable for human consumption. Consumers, desire for uncured, less processed, nutritional, affordable, and ready-to-eat food products with explicit labelling has increased as a result of quick access to technology. Food packaging is a vital factor in appropriate food processing and quality preservation. The four essential roles of traditional food packaging are protection or preservation, confinement, convenience, and communication. To ensure food and nutrition security and honesty all over the food production chain, food suppliers, distributors, consumers, and food regulatory bodies pursue a unique, profitable, quick, and appropriate tool for evaluating the quality of packaged foods; doing so more efficaciously than the conventional passive barrier approach, as packing is a critical activity of almost all section of the food industry (Siracusa, 2012; Joardder and Masud, 2019). Enhancements have been made to these fundamental functions in order to provide better, more active, and smart packaging (Majid et al., 2018). Conventional or standard food packaging is being replaced with active or creative food packaging that incorporates nanotechnology to create intelligent, interactive, and responsive food packaging with enhanced functions (Ashfaq et al., 2022).

Bionanomaterials are the class of nanomaterials that are formulated from biological materials or nanomaterials encapsulated in a biomaterial. The bionanomaterials are isolated from any biological sources like plants, animals, and insects, and also products, agriculture and marine wastes, microorganisms and their derivative materials, and so forth (Jeevanandam et al., 2022). Fabrication of bionanomaterial technique includes the fabrication of structures, devices, and components with the least dimension of 1–100 nm size through development, characterization, and modification of materials. The resulting material has physicochemical characteristics that differ significantly from the capabilities of macroscale material consisting of the same ingredient when the particle size is reduced below this limit (Fanfair et al., 2007).

5.5.1 Antibacterial Food Packaging

Food packaging is an important activity performed in the food industry, as it permits the storage of food materials for a considerable time, without loss in nutrition and palatability. Packing not only keeps the food fresh but also protects delicate bioactive chemicals from severe circumstances, both physical and environmental. Nanotechnology contributes to the development of enhanced packaging like active and smart packaging (Ashfaq et al., 2022). Bionanomaterials-based active

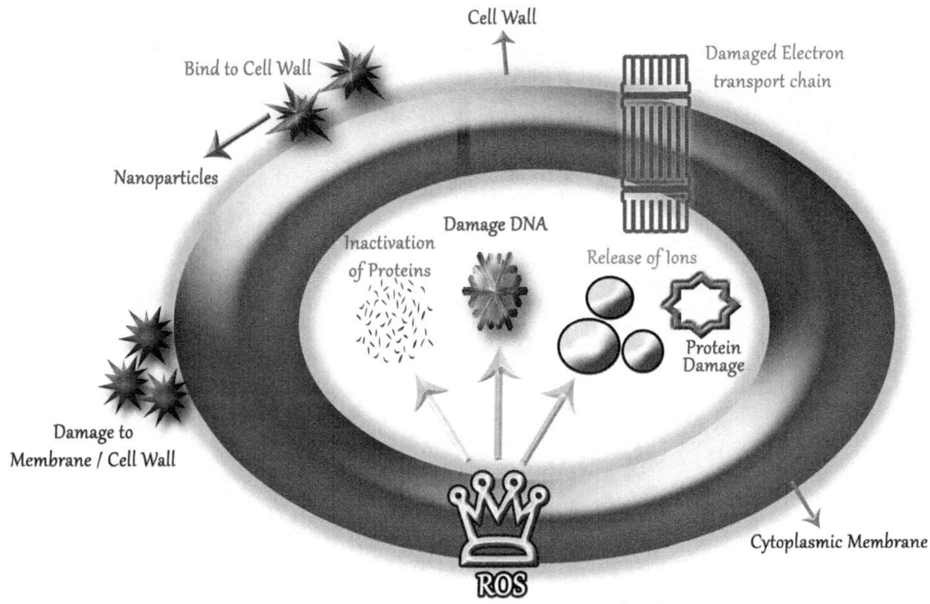

FIGURE 5.3 Schematic Diagram of Nanoparticle Toxicity and Antibacterial Mechanisms. (Compiled by the Authors.)

packaging involves addition of active compounds into food, outstretching the storage life of food on reaction with internal and external variables. Active packaging is a revolutionary way of preserving foods by reducing bacterial infection, moisture gain, oxidation, over-ripening, and other undesirable effects (Figure 5.3). To improve the characteristics of the packaging material, functional materials like oxygen scavengers, ethylene absorbers, antioxidants, carbon dioxide emitters, and antibacterial agents are purposely incorporated or inserted in the container headspace in active packaging (Day, 2008).

Antibacterial agents inhibit the cell wall synthesis or damage the cell wall, and interfere with the bacterial metabolism (Romling and Balsalobre, 2012). Antimicrobial packaging materials reduce the infection of microorganisms in food, hence halting food from spoiling and preventing foodborne disease transmission. It can be believed that, in the long term, the active packaging will become extra-active packaging, having broad spectrum antimicrobial activity (inhibiting bacterial infection and bacterial multiplication), atmospheric control, gas, light, UV protection, tampering, spoilage indication. The facet of maintaining the safety standards of the food and food packaging material improves over time, and some active packages maintain the food shelf life by releasing distinct chemicals, and there is also smart packaging, which warns customers if the food is contaminated. There are also antibacterial packets containing probiotics, a novel form of bio preservation (Rahman et al., 2022). Bionanoparticles induces the formation of free radicals, which interferes with replication of nucleic acid and the development of antibacterial packing, resulting in cell deterioration or death (Hamida et al., 2020). Many antimicrobial nanostructures are utilized in food packaging because they work in different ways and with varied implementations.

5.5.2 Antioxidant Food Packaging

Nanomaterials with antioxidant properties are a fascinating choice of additives to the food packaging materials owing to the dismissive impact of the free radicals on the biological systems (Losada-Barreiro and Bravo-Diaz, 2017). One recognized method by which antioxidants reduce

FIGURE 5.4 Process and Socio-Economic Benefits of Developing Antioxidant Food Packaging. (Compiled by the Authors.)

oxidation cell membrane and other biological entities is by free radical scavenging (Banerjee et al., 1999; Sanchez-Moreno et al., 1999). But, several common food products, such as meat, poultry, fish, and baked items, are wrapped in plastic trays without any head-on contact with the top layer of the packing material (sealed closure film). Thus, removing the free radicles as soon as they are generated considerably delays future oxidation by limiting the lipid peroxidation. Predominantly, employment of antioxidant incorporated packaging eliminates the requirement for both high-barrier and vacuum packing. Active packing with synthetic antioxidants (such as butylated hydroxy anisole) can preserve food against oxidative rancidity. Furthermore, customers, desire for less processed goods is driving an increase in the usage of natural antioxidants as artificial antioxidant alternatives (Singh et al., 2021a). Figure 5.4 depicts a schematic representation of the notion of combining active natural components and biopolymers for food preservation.

To address this issue, researchers began with a simple but effective technique, including natural antioxidants, and as a result, plant extracts and essential oils have largely replaced synthetic additives (Basavegowda and Baek, 2021; Christaki et al., 2021; Al-Maqtari et al., 2022). This has an additional benefit, which is the economic development of agro-industries as several byproducts become a cost-effective and practical source of powerful antioxidants (Reguengo et al., 2022). Since many essential oils have a strong odor, it is critical to identify the quantity of essential oil that is used, in case it may significantly affect the organoleptic properties of the food (Sanchez-Gonzalez et al., 2011). This unique notion has piqued the attention of both the research scientists and the commercial sectors in the last two decades, and so the investigation and identification of novel antioxidant sources, characterization, standardization and employment in food packaging offers tremendous potential to replace traditional packaging.

5.5.3 Bio Nano-Sensor Packaging for Food Protection

As some individuals are concerned about the hazards of nanotechnology, nanoparticles have the possibility to be employed as a nano sensor to alert the user of the food quality. For example, barcodes are used on milk bottles for newborns; they are safe options as they do not interfere with the food but informs the user of all the required safety data. Rather, it informs the customer on the state of the product. Sensors can possibly be developed employing bionanomaterials. The bionanomaterials can be used for the fabrication of calorimetric, fluorescence, electrochemical sensors, which can be utilized for sensing of the pathogens, toxins, drug residues, heavy metals, pesticides, and so forth, in food and food products (Qiao et al., 2022). The usage of bionanoparticles offers both advantages and disadvantages. People who are afflicted with sickness or face health concerns should avoid eating foods containing nanoparticles. As a result, bionanoparticles are increasingly being employed in packaging to reduce health risks. Bionanomaterials are both cost-effective and environmental friendly for use in food packaging. But cost remains the primary determining factor. The normal costs and techniques of manufacturing, as well as better application, determines the use of bionanomaterials in shelf-life extension of food (Sekhon, 2014).

Monitoring food safety is critical everywhere in the food manufacturing process as well as the supply chain. Assessment of food quality is now done in centralized laboratories using standard sophisticated instruments, with restrictions in sample size. Development of assays that are quick, cheap, and portable allows food producers and customers to check quality of the food product during production, shipping, and usage. The bionanomaterials based sensors can enable monitoring the food from the source. The sensors rely on the chemical transduction processes, which are typically used to measure volatile substances, bimolecular components, biological receptors for better sensitivity. The sensors comprise a biological receptor (enzymes, aptamers, cells, antibodies, and so on) which detects a specific target, while the process of binding is transduced by electrochemical, optical, mass, or thermal processes bringing changes in color and voltage. Pathogenic bacteria, contaminants, for example, heavy metals, residual veterinary antibiotics, residual pesticides, and illegal additives (Velasco-Garcia and Mottram, 2003), mycotoxins, bisphenol A, and so forth, allergens, and nutrients (e.g., antioxidants) have been detected using biosensors, overcoming the restrictions of costly and time-consuming research methods. Nanomaterials frequently used in these technologies includes Au and Ag bionanoparticles, which have surface platform resonance (SPR) properties and high conductivity, enhancing their usage in sensors; Magnetic NPs such as Fe_2O_3 nanoparticles obtained from biological sources, which brings about paramagnetic alteration functions for the efficient separation and enrichment of the targeted analytes.

Intelligent packaging materials could be identified by the European Commission in 2004 as materials that monitor the state of packed food or the environment surrounding the food (Pocas et al., 2008). Smart packaging is also referred to as intelligent packaging because it includes smart devices such as barcodes, Radio Frequency Identification Tags (RFID), sensors, and indicators that communicate, monitor, sense, record, track, and indicate information about food safety, quality, and history, throughout the supply chain.

Nano sensors, nanotechnology's eminent applications, are utilized for observing the conditions of fresh and processed food items. Most foods are perishable and have the ability to degrade with storage; several indicators such as time-temperature indicators, pH indicators, and leak indicators are designed for monitoring the food standards. Time and temperature sensors show time past opening by a perceptible color difference and display the time is remaining for safe consumption the meal.

5.6 ENVIRONMENTAL EFFECTS, SAFETY, AND TOXICITY OF BIONANOMATERIALS IN THE FOOD SECTOR

Degree of environmental impact by the nanoparticles depends on the specific application and concentration used. In the food industry both factors coincide, conducing higher exposure risk when the nanoparticles are used as food additives. Quality food production and supply are frequently linked to human progress and welfare, along with ecological preservation. Constantly increasing global population, existing environmental risks, climate change, a scarcity of resources, and cultivable land shrinking, contribute to the use of contemporary technologies for enhancing the supply of quality food products to consumers. Regardless its numerous advantages, the implementation of nanoscience in food poses potential environmental, social hazards because the nanoparticles enter the ecosphere via pesticide application for food production or application in processed food for packaging, raising toxicity concerns (Bumbudsanpharoke and Ko, 2015). The greater danger of nanoengineered particles is, the higher reactivity and increased bioavailability to human bodies, resulting in long-term degenerative repercussions. Nanomaterials enter the food chain directly by nanoparticle integration in new foods such as nano emulsions, nano capsules, and nano antimicrobial coatings. In another case, nanomaterials such as nanolaminates, nano sensors, and carbon nanotubes are being used in food manufacture, handling, preserving, and monitoring.

Nanotechnology offers several favorable implications into the food industry. The huge potential and aptitude of this strong technology to tackle numerous challenges has already demonstrated triumphs in the agro-industrial sector. Several other potential applications are now being researched but face less commercialization. Utilization of bionanoparticles for the identification of crop diseases, smart delivery of bioactive materials, sensing food quality during packaging and transport, detection of food contamination, and adulteration have revolutionized the food industry globally. Numerous research studies have found that the public reaction to nanotechnology knowledge is neutral. Thompson et al. (2011) investigated the general publics perception of the acceptance of application of nanotechnology in food, finding that 30 per cent of the answerers preferred utilization of nanomaterials over the risks, 44 per cent thought the risks are coincidental with benefits, and 26 per cent thought the risks outweighed the overall benefits. Existing option or choice, as well as understanding of the science underlying its manufacture, dictate consumer adoption of each unique food item. Employment of nanotechnology for sustainable environment and society is commendable, but is not complimentary to the health hazards associated with its usage. People view nanotechnology dangers as they do genetically engineered foods, lowering their use of such foods. Individual responses are also influenced by government laws and customer preferences.

Essentially, contamination of food items is caused by environmental exposure, improper food processing, and low-quality packaging. Most toxicity investigations have been carried out at considerably greater doses than the practical dose (Satin, 2002). As Paracelsus correctly stated, "the dose creates the poison", and these tests demonstrate this adage. Most hazardous compounds are innocuous in tiny amounts and dangerous only when eaten in large quantities. Exact measurement of bionanoparticle release into the environment and occupational exposure is difficult. Based on modelling studies, the half-life and life cycle of nanomaterials have been documented. This research requires refinement since information pertaining to the industrialized production of nanoparticles have not yet been widely included in research.

When bionanomaterials are discharged into the environment, they can take on distinct forms owing to their strikingly different physical and chemical features, contradicting the bulk parent molecules. These unique adoptive characteristics of the nanomaterials influence the way they interact with the biological components (Mishra et al., 2018). Figure 5.5 shows that the toxicity of bionanoparticles is governed by several factors. Bionanoparticles in the environment rarely persist in its native state due to the extremely reactive surfaces. The reactive chemical groups present on the nanoparticle surface determine the method by which cellular uptake, accumulation, and clearance will occur. Furthermore, the nanoparticles, interaction with the biological membrane might be physical or chemical. Physical interactions primarily cause damage to membranes and their functionality, protein folding, and different transport channels and their activities. Chemical interactions, on the other hand, primarily induce the formation of reactive oxygen species (ROS) and oxidative damage (Zhang et al., 2021). Environmental interactions complicate the assessment of bionanoparticle toxicity.

5.7 CONCLUSION

Early humans got their food by hunting and gathering. Food became the major reason for societal development. With time, human civilization acquired the knowledge of preserving the food by

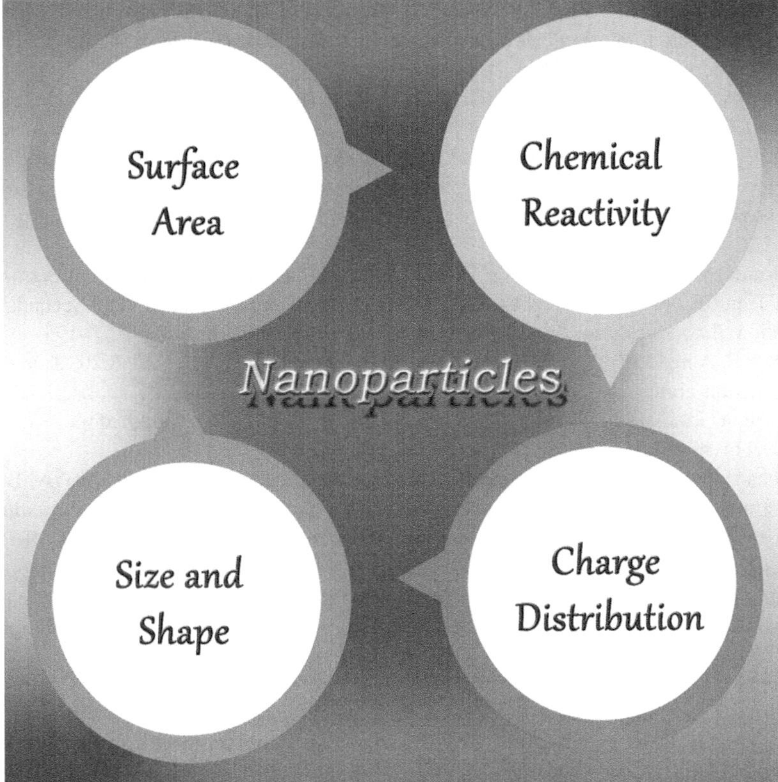

FIGURE 5.5 Factors Affecting the Overall Toxicity of a Bionanoparticle. (Compiled by the Authors.)

extending their shelf life. But to date, extension of storage life without compromising the native food quality poses a great challenge. As a recent innovation, bionanomaterials are proposed to expand the storage life of food products. Though there is a recent surge in research employing bionanomaterials, the studies are still in the exploratory lab stages. Regardless of the fact that addition of bionanomaterials is so encouraging in food packaging, numerous imperfections such as toxicity, migration, and biomagnification in the environmental forefront must be addressed. Large-scale industrial production technologies and facilities are still an unachieved need. Considering the scientific data available, the above sought data should be revealed prior to the utilization ensuring the quality assurance in the food industry. When it comes to food ingested by people and inadequate knowledge on the safety of various forms of nanoparticles, many organizations choose to reinforce nanostructures into the food lower than the detection limits. Addressing these issues of utility are important. Extension of shelf life with bionanomaterials will increase the marginable profit reducing the wastage of food.

REFERENCES

Aayush K, McClements DJ, Sharma S, Sharma R, Singh GP, Sharma K, Oberoi K (2022) Innovations in the development and application of edible coatings for fresh and minimally processed Apple. *Food Control* 141:109188. https://doi.org/10.1016/j.foodcont.2022.109188

Abd Elkodous M, El-Sayyad GS, Abdelrahman IY, El-Bastawisy HS, Mosallam FM, Nasser HA, Gobara M, Baraka A, Elsayed MA, El-Batal AI (2019) Therapeutic and diagnostic potential of nanomaterials for enhanced biomedical applications. *Colloids Surf B Biointerfaces* 180:411–428. https://doi.org/10.1016/j.colsurfb.2019.05.008

Abdollahi M, Alboofetileh M, Behrooz R, Rezaei M, Miraki R (2013) Reducing water sensitivity of alginate bio-nanocomposite film using cellulose nanoparticles. *Int J Biol Macromol* 54:166–173. https://doi.org/10.1016/j.ijbiomac.2012.12.016

Abdulmumeen HA, Risikat AN, Sururah AR (2012) Food: Its preservatives, additives and applications. *Int J Chem Biochem Sci* 1:36–47

Aguilar-Perez KM, Ruiz-Pulido G, Medina DI, Parra-Saldivar R, Iqbal HM (2023) Insight of nanotechnological processing for nano-fortified functional foods and nutraceutical-opportunities, challenges, and future scope in food for better health. *Crit Rev Food Sci Nutr* 63:1–18. https://doi.org/10.1080/10408398.2021.2004994

Aguirre-Becerra H, Vazquez-Hernandez MC, Saenz de la O D, Alvarado-Mariana A, Guevara-Gonzalez RG, Garcia-Trejo JF, Feregrino-Perez AA (2021) Role of Stress and Defense in Plant Secondary Metabolites Production. *Bioactive natural products for pharmaceutical applications* pp. 151–195

Ahmad A, Ullah S, Ahmad W, Yuan Q, Taj R, Khan AU, Rahman AU, Khan UA (2020) Zinc oxide-selenium heterojunction composite: Synthesis, characterization and photo-induced antibacterial activity under visible light irradiation. *J Photochem Photobiol B* 203:111743. https://doi.org/10.1016/j.jphotobiol.2019.111743

Ali SS, Al-Tohamy R, Koutra E, Moawad MS, Kornaros M, Mustafa AM, Mahmoud YA, Badr A, Osman ME, Elsamahy T, Jiao H (2021) Nanobiotechnological advancements in agriculture and food industry: Applications, nanotoxicity, and future perspectives. *Sci Total Environ* 792:148359. https://doi.org/10.1016/j.scitotenv.2021.148359

Al-Maqtari QA, Rehman A, Mahdi AA, et al (2022) Application of essential oils as preservatives in food systems: Challenges and future prospectives - A review. *Phytochem Rev* 21:1209–1246. https://doi.org/10.1007/s11101-021-09776-y

Amit SK, Uddin MdM, Rahman R, et al (2017) A review on mechanisms and commercial aspects of food preservation and processing. *Agric Food Secur* 6:51. https://doi.org/10.1186/s40066-017-0130-8

Arnon-Rips H, Porat R, Poverenov E (2019) Enhancement of agricultural produce quality and storability using citral-based edible coatings; the valuable effect of nano-emulsification in a solid-state delivery on fresh-cut melons model. *Food Chem* 277:205–212. https://doi.org/10.1016/j.foodchem.2018.10.117

Arroyo BJ, Bezerra AC, Oliveira LL, et al (2020) Antimicrobial active edible coating of alginate and chitosan add ZnO nanoparticles applied in guavas (Psidium guajava L.). *Food Chem* 309:125566. https://doi.org/10.1016/j.foodchem.2019.125566

Arshad R, Gulshad L, Haq I, et al (2021) Nanotechnology: A novel tool to enhance the bioavailability of micronutrients. *Food Sci Nutr* 9:3354–3361. https://doi.org/10.1002/fsn3.2311

Ashfaq A, Khursheed N, Fatima S, et al (2022) Application of nanotechnology in food packaging: Pros and cons. *J Agric Food Res* 7:100270. https://doi.org/10.1016/j.jafr.2022.100270

Atta OM, Manan S, Shahzad A, Ul-Islam M, Ullah MW, Yang G (2022) Biobased materials for active food packaging: A review. *Food Hydrocoll* 125:107419. https://doi.org/10.1016/j.foodhyd.2021.107419

Bahrami A, Mokarram RR, Khiabani MS, Ghanbarzadeh B, Salehi R (2019) Physico-mechanical and antimicrobial properties of tragacanth/hydroxypropyl methylcellulose/beeswax edible films reinforced with silver nanoparticles. *Int J Biol Macromol* 129:1103–1112. https://doi.org/10.1016/j.ijbiomac.2018.09.045

Ballesteros LF, Lamsaf H, Sebastian CV, Cerqueira MA, Pastrana L, Teixeira JA (2022) Active packaging systems based on metal and metal oxide nanoparticles. In: J Parameswaranpillai, RE Krishnankutty, A Jayakumar, SM Rangappa, S Siengchin, (Eds.), *Nanotechnology – Enhanced Food Packaging*. Wiley, pp. 143–181. https://doi.org/10.1002/9783527827718.ch7

Banerjee BD, Seth V, Bhattacharya A, Pasha ST, Chakraborty AK (1999) Biochemical effects of some pesticides on lipid peroxidation and free-radical scavengers. *Toxicol Lett* 107:33–47. https://doi.org/10.1016/S0378-4274(99)00029-6

Barciela P, Carpena M, Li NY, Liu C, Jafari SM, Simal-Gandara J, Prieto MA (2023) Macroalgae as biofactories of metal nanoparticles; biosynthesis and food applications. *Adv Colloid Interface Sci* 311:102829. https://doi.org/10.1016/j.cis.2022.102829

Basavegowda N, Baek KH (2021) Synergistic antioxidant and antibacterial advantages of essential oils for food packaging applications. *Biomolecules* 11:1267. https://doi.org/10.3390/biom11091267

Basumatary IB, Mukherjee A, Katiyar V, Kumar S (2022) Biopolymer-based nanocomposite films and coatings: recent advances in shelf-life improvement of fruits and vegetables. *Crit Rev Food Sci Nutr* 62:1912–1935. https://doi.org/10.1080/10408398.2020.1848789

Bhandari N, Bika R, Subedi S, Pandey S (2022) Essential oils amended coatings in citrus postharvest management. *J Agric Food Res* 10:100375. https://doi.org/10.1016/j.jafr.2022.100375

Bhardwaj A, Alam T, Talwar N (2019) Recent advances in active packaging of agri-food products: a review. *J Postharvest Technol* 07:33–62

Bumbudsanpharoke N, Ko S (2015) Nano-food packaging: An overview of market, migration research, and safety regulations. *J Food Sci* 80:R910-R923. https://doi.org/10.1111/1750-3841.12861

Melo NF, de MendoncaSoares BL, Diniz KM, Leal CF, Canto D, Flores MA, da Costa Tavares-Filho JH, Galembeck A, Stamford TL, Stamford-Arnaud TM, Stamford TC (2018) Effects of fungal chitosan nanoparticles as eco-friendly edible coatings on the quality of postharvest table grapes. *Postharvest Biol Technol* 139:56–66. https://doi.org/10.1016/j.postharvbio.2018.01.014

Chausali N, Saxena J, Prasad R (2022) Recent trends in nanotechnology applications of bio-based packaging. *J Agric Food Res* 7:100257. https://doi.org/10.1016/j.jafr.2021.100257

Cheng H, Chen L, McClements DJ, Xu H, Long J, Zhao J, Xu Z, Meng M, Jin Z (2022) Recent advances in the application of nanotechnology to create antioxidant active food packaging materials. *Crit Rev Food Sci Nutr* 1–16. https://doi.org/10.1080/10408398.2022.2128035

Cheng S, Zhang Y, Cha R, Yang J, Jiang X (2016) Water-soluble nanocrystalline cellulose films with highly transparent and oxygen barrier properties. *Nanoscale* 8:973–978. https://doi.org/10.1039/C5NR07647A

Chouke PB, Shrirame T, Potbhare AK, Mondal A, Chaudhary AR, Mondal S, Thakare SR, Nepovimova E, Valis M, Kuca K, Sharma R (2022) Bioinspired metal/metal oxide nanoparticles: A road map to potential applications. *Mater Today Adv* 16:100314. https://doi.org/10.1016/j.mtadv.2022.100314

Christaki S, Moschakis T, Kyriakoudi A, Biliaderis CG, Mourtzinos I (2021) Recent advances in plant essential oils and extracts: Delivery systems and potential uses as preservatives and antioxidants in cheese. *Trends Food Sci Technol* 116:264–278. https://doi.org/10.1016/j.tifs.2021.07.029

Cui Y, Kumar S, Rao Kona B, van Houcke D (2015) Gas barrier properties of polymer/clay nanocomposites. *RSC Adv* 5:63669–63690. https://doi.org/10.1039/C5RA10333A

Dai L, Zhang J, Cheng F (2020) Cross-linked starch-based edible coating reinforced by starch nanocrystals and its preservation effect on graded Huangguan pears. *Food Chem* 311:125891. https://doi.org/10.1016/j.foodchem.2019.125891

Das HT, Dutta S, Beura R, Das N (2022) Role of polyaniline in accomplishing a sustainable environment: recent trends in polyaniline for eradicating hazardous pollutants. *Environ Sci Poll Res* 29:49598–49631. https://doi.org/10.1007/s11356-022-20916-5

Das S, Vishakha K, Banerjee S, Mondal S, Ganguli A (2020) Sodium alginate-based edible coating containing nanoemulsion of Citrus sinensis essential oil eradicates planktonic and sessile cells of food-borne pathogens and increased quality attributes of tomatoes. *Int J Biol Macromol* 162:1770–1779. https://doi.org/10.1016/j.ijbiomac.2020.08.086

Day, Brian PF (2008) Active packaging of food. *Smart packaging technologies for fast moving consumer goods*, 1–17

Dubey N, Chitranshi S, Dwivedi SK, Sharma A (2023) Postharvest physiology, value chain advancement, and nanotechnology in fresh-cut fruits and vegetables. In: MR Goyal, SK Mishra, S Kumar (Eds.), *Nanotechnology Horizons in Food Process Engineering: Volume 3: Trends, Nanomaterials, and Food Delivery*. CRC Press, pp. 76–99. https://doi.org/10.1201/9781003305408-6

Fahmy HM, Eldin RE, Serea ES, Gomaa NM, AboElmagd GM, Salem SA, Elsayed ZA, Edrees A, Shams-Eldin E, Shalan AE (2020) Advances in nanotechnology and antibacterial properties of biodegradable food packaging materials. *RSC Adv* 10:20467–20484. https://doi.org/10.1039/D0RA02922J

Fanfair D, Salil D, Christopher K (2007) The early history of nanotechnology. *Connexions* 6:1–15

Feng C, Wu ZP, Huang KW, Ye J, Zhang H (2022) Surface modification of 2D photocatalysts for solar energy conversion. *Adv Mater* 34:2200180. https://doi.org/10.1002/adma.202200180

Gauthier G, Capron I (2021) Pickering nanoemulsions: An overview of manufacturing processes, formulations, and applications. *JCIS Open* 4:100036. https://doi.org/10.1016/j.jciso.2021.100036

Ghoshal G (2021) Recent advancements in edible and biodegradable food packaging. In: MR Goyal, P Birwal, M Sharma (Eds.), *Handbook of Research on Food Processing and Preservation Technologies*. Apple Academic Press, pp. 165–199. https://doi.org/10.1201/9781003153221

Gould GW (1996) Methods for preservation and extension of shelf life. *Int J Food Microbiol* 33:51–64. https://doi.org/10.1016/0168-1605(96)01133-6

Gundewadi G, Rudra SG, Sarkar DJ, Singh D (2018) Nanoemulsion based alginate organic coating for shelf life extension of okra. *Food Packag Shelf Life* 18:1–12. https://doi.org/10.1016/j.fpsl.2018.08.002

Guo Y, Sun Q, Wu FG, Dai Y, Chen X (2021) Polyphenol containing nanoparticles: Synthesis, properties, and therapeutic delivery. *Adv Mater* 33:2007356. https://doi.org/10.1002/adma.202007356

Gupta A, Pal AK, Patwa R, Dhar P, Katiyar V (2018) Green composites with excellent barrier properties. In: AN Netravali (Ed.), *Advanced Green Composites*. John Wiley & Sons, pp. 321–356. https://doi.org/10.1163/156855107782325230

Gupta V, Biswas D, Roy S (2022) A comprehensive review of biodegradable polymer-based films and coatings and their food packaging applications. *Materials* 15:5899. https://doi.org/10.3390/ma15175899

Hamad AF, Han JH, Kim BC, Rather IA (2018) The intertwine of nanotechnology with the food industry. *Saudi J Biol Sci* 25:27–30. https://doi.org/10.1016/j.sjbs.2017.09.004

Hamida RS, Ali MA, Goda DA, Khalil MI, Al-Zaban MI (2020) Novel biogenic silver nanoparticle-induced reactive oxygen species inhibit the biofilm formation and virulence activities of Methicillin-Resistant *Staphylococcus aureus* (MRSA) strain. *Front Bioeng Biotechnol* 8. https://doi.org/10.3389/fbioe.2020.00433

Haris M, Hussain T, Mohamed HI, Khan A, Ansari MS, Tauseef A, Khan AA, Akhtar N (2023) Nanotechnology - A new frontier of nano-farming in agricultural and food production and its development. *Sci Total Environ* 857:159639. https://doi.org/10.1016/j.scitotenv.2022.159639

Hermes PH, Gabriela MP, Denisse VG, Ileana VR, Fabian FL (2020) Edible crop production by nanotechnology. In: C Keswani (Ed.), *Intellectual Property Issues in Nanotechnology*. 1st ed. CRC Press, pp. 11–40. https://doi.org/10.1201/9781003052104-4

Hu X, Saravanakumar K, Sathiyaseelan A, Wang MH (2020) Chitosan nanoparticles as edible surface coating agent to preserve the fresh-cut bell pepper (*Capsicum annuum* L. var. grossum (L.) Sendt). *Int J Biol Macromol* 165:948–957. https://doi.org/10.1016/j.ijbiomac.2020.09.176

Idumah CI, Zurina M, Ogbu J, Ndem JU, Igba EC (2020) A review on innovations in polymeric nanocomposite packaging materials and electrical sensors for food and agriculture. *Compos Interfaces* 27:1–72. https://doi.org/10.1080/09276440.2019.1600972

Imani SM, Ladouceur L, Marshall T, Maclachlan R, Soleymani L, Didar TF (2020) Antimicrobial nanomaterials and coatings: Current mechanisms and future perspectives to control the spread of viruses including SARS-CoV-2. *ACS Nano* 14:12341–12369. https://doi.org/10.1021/acsnano.0c05937

Indumathi MP, Saral Sarojini K, Rajarajeswari GR (2019) Antimicrobial and biodegradable chitosan/cellulose acetate phthalate/ZnO nano composite films with optimal oxygen permeability and hydrophobicity for extending the shelf life of black grape fruits. *Int J Biol Macromol* 132:1112–1120. https://doi.org/10.1016/j.ijbiomac.2019.03.171

Islam F, Saeed F, Afzaal M, Hussain M, Ikram A, Khalid MA (2023) Food grade nanoemulsions: promising delivery systems for functional ingredients. *J Food Sci Technol* 60:1461–1471. https://doi.org/10.1007/s13197-022-05387-3

Jadhav R, Pawar P, Choudhari V, Topare N, Raut-Jadhav S, Bokil S, Khan A (2023) An overview of antimicrobial nanoparticles for food preservation. *Mater Today Proc* 72:204–216. https://doi.org/10.1016/j.matpr.2022.07.045

Jafarzadeh S, Nafchi AM, Salehabadi A, Oladzad-Abbasabadi N, Jafari SM (2021) Application of bionanocomposite films and edible coatings for extending the shelf life of fresh fruits and vegetables. *Adv Colloid Interface Sci* 291:102405. https://doi.org/10.1016/j.cis.2021.102405

Jalgaonkar K, Mahawar MK, Bibwe B, Kannaujia P (2022) Postharvest profile, processing and waste utilization of dragon fruit (*Hylocereus Spp* .): A review. *Food Rev Int* 38:733–759. https://doi.org/10.1080/87559129.2020.1742152

James SJ, James C (2023) Chilling and freezing. In: S Clark, S Jung, B Lamsal (Eds.), *Food Safety Management*. Elsevier, pp. 453–474. https://doi.org/10.1002/9781118846315.ch5

Jayakumar A, Radoor S, Kim JT, et al (2022) Recent innovations in bionanocomposites-based food packaging films - A comprehensive review. *Food Packag Shelf Life* 33:100877. https://doi.org/10.1016/j.fpsl.2022.100877

Jeevanandam J, Ling JK, Barhoum A, San Chan Y, Danquah MK (2022) Bionanomaterials: Definitions, sources, types, properties, toxicity, and regulations. In: A Barhoum, J Jeevanandam, MK Danquah (Eds.), *Fundamentals of Bionanomaterials*. Elsevier, pp. 1–29. https://doi.org/10.1016/B978-0-12-824147-9.00001-7

Jilani A, Ansari MO, ur Rehman G, Shakoor MB, Hussain SZ, Othman MH, Ahmad SR, Dustgeer MR, Alshahrie A (2022) Phenol removal and hydrogen production from water: Silver nanoparticles decorated on polyaniline wrapped zinc oxide nanorods. *J Ind Eng Chem* 109:347-358. https://doi.org/10.1016/j.jiec.2022.02.021

Joardder MUH, Masud MH (2019) Challenges and mistakes in food preservation. In. MUH Joardder, MH Masud (Eds.), *Food Preservation in Developing Countries: Challenges and Solutions.* Springer, pp. 175–198. https://doi.org/10.1007/978-3-030-11530-2_7

K. S, Kumar A (2022) Nanoemulsions: Techniques for the preparation and the recent advances in their food applications. *Innovat Food Sci Emerg Technol* 76:102914. https://doi.org/10.1016/j.ifset.2021.102914

Kaur M, Kalia A, Thakur A (2017) Effect of biodegradable chitosan-rice-starch nanocomposite films on post-harvest quality of stored peach fruit. *Starch -Starke* 69:1600208. https://doi.org/10.1002/star.201600208

Kaur R, Liu S (2016) Antibacterial surface design - Contact kill. *Prog Surf Sci* 91:136–153. https://doi.org/10.1016/j.progsurf.2016.09.001

Koechler S, Farasin J, Cleiss-Arnold J, Arsene-Ploetze F (2015) Toxic metal resistance in biofilms: Diversity of microbial responses and their evolution. *Res Microbiol* 166:764–773. https://doi.org/10.1016/j.resmic.2015.03.008

Kumar L, Ramakanth D, Akhila K, Gaikwad KK (2022a) Edible films and coatings for food packaging applications: A review. *Environ Chem Lett* 20:875–900. https://doi.org/10.1007/s10311-021-01339-z

Kumar L, Ramakanth D, Akhila K, Gaikwad KK (2022b) Edible films and coatings for food packaging applications: A review. *Environ Chem Lett* 20:875–900. https://doi.org/10.1007/s10311-021-01339-z

Kumar N, Verma A, Mandal A (2021) Formation, characteristics and oil industry applications of nanoemulsions: A review. *J Pet Sci Eng* 206:109042. https://doi.org/10.1016/j.petrol.2021.109042

Kumar V A, Hasan M, Mangaraj S, et al (2022) Trends in edible packaging films and its prospective future in food: a review. *Appl Food Res* 2:100118. https://doi.org/10.1016/j.afres.2022.100118

Kumari A, Wasnik P, Patel SS, Kumar S (2021) Characterization, applications, and safety aspects of nanomaterials in the food and dairy industries. In: MR Goyal, P Birwal, M Sharma (Eds.), *Handbook of Research on Food Processing and Preservation Technologies.* Academic Press, pp. 233–256. https://doi.org/10.1201/9781003305378_4

Kuswandi B, Moradi M (2019) Improvement of food packaging based on functional nanomaterial., In: S Siddiquee, GJH Melvin, MdM Rahman (Eds.), *Nanotechnology: Applications in Energy, Drug and Food.* Springer, pp. 309–344. https://doi.org/ 10.1007/978-3-319-99602-8_16

Kwak H, Shin S, Kim J, Kim J, Lee D, Lee H, Lee EJ, Hyun J (2021) Protective coating of strawberries with cellulose nanofibers. *Carbohydr Polym* 258:117688. https://doi.org/10.1016/j.carbpol.2021.117688

Lavinia M, Hibarturrahman SN, Harinata H, Wardana AA (2019) Antimicrobial activity and application of nanocomposite coating from chitosan and ZnO nanoparticle to inhibit microbial growth on fresh-cut papaya. *Food Res* 4:307–311. https://doi.org/10.26656/fr.2017.4(2).255

Lin J, Hu Z, Li H, Qu J, Zhang M, Liang W, Hu S (2019) Ultrathin nanotubes of Bi_5O_7I with a reduced band gap as a high-performance photocatalyst. *Inorg Chem* 58:9833–9843. https://doi.org/10.1021/acs.inorgchem.9b00858

Liu F, Li M, Wang Q, Yan J, Han S, Ma C, Ma P, Liu X, McClements DJ (2022) Future foods: Alternative proteins, food architecture, sustainable packaging, and precision nutrition. *Crit Rev Food Sci Nutr* 1–22. https://doi.org/10.1080/10408398.2022.2033683

Liu W, Zhang M, Bhandari B (2020) Nanotechnology - A shelf life extension strategy for fruits and vegetables. *Crit Rev Food Sci Nutr* 60:1706–1721. https://doi.org/10.1080/10408398.2019.1589415

Lombardo D, Kiselev MA, Caccamo MT (2019) Smart nanoparticles for drug delivery application: Development of versatile nanocarrier platforms in biotechnology and nanomedicine. *J Nanomater* 2019:1–26. https://doi.org/10.1155/2019/3702518

Losada-Barreiro S, Bravo-Diaz C (2017) Free radicals and polyphenols: The redox chemistry of neurodegenerative diseases. *Eur J Med Chem* 133:379–402. https://doi.org/10.1016/j.ejmech.2017.03.061

Lustriane C, Dwivany FM, Suendo V, Reza M (2018) Effect of chitosan and chitosan-nanoparticles on post harvest quality of banana fruits. *J Plant Biotechnol* 45:36–44. https://doi.org/10.5010/JPB.2018.45.1.036

Ma L, Zhang M, Bhandari B, Gao Z (2017) Recent developments in novel shelf life extension technologies of fresh-cut fruits and vegetables. *Trends Food Sci Technol* 64:23–38. https://doi.org/10.1016/j.tifs.2017.03.005

Maciel VBV, Contini LRF, Yoshida CMP, Venturini AC (2020) Application of edible biopolymer coatings on meats, poultry, and seafood. In: *Biopolymer Membranes and Films*. Elsevier, 515–533

Majid I, Ahmad Nayik G, Mohammad Dar S, Nanda V (2018) Novel food packaging technologies: Innovations and future prospective. *J Saudi Soc Agri Sci* 17:454–462. https://doi.org/10.1016/j.jssas.2016.11.003

Mangaraj S, Yadav A, Bal LM, et al (2019) Application of biodegradable polymers in food packaging industry: A comprehensive review. *J Packag Technol Res* 3:77–96. https://doi.org/10.1007/s41783-018-0049-y

Mary Isabella Sonali J, Subhashree S, Senthil Kumar P, Veena Gayathri K (2022) New analytical strategies amplified with carbon-based nanomaterial for sensing food pollutants. *Chemosphere* 295:133847. https://doi.org/10.1016/j.chemosphere.2022.133847

McClements DJ (2020) Nanotechnology approaches for improving the healthiness and sustainability of the modern food supply. *ACS Omega* 5:29623–29630. https://doi.org/10.1021/acsomega.0c04050

Meindrawan B, Suyatma NE, Wardana AA, Pamela VY (2018) Nanocomposite coating based on carrageenan and ZnO nanoparticles to maintain the storage quality of mango. *Food Packag Shelf Life* 18:140–146. https://doi.org/10.1016/j.fpsl.2018.10.006

Mihalca V, Kerezsi AD, Weber A, et al (2021) Protein-based films and coatings for food industry applications. *Polymers (Basel)* 13:769. https://doi.org/10.3390/polym13050769

Miranda M, Sun X, Ference C, et al (2021) Nano- and micro- carnauba wax emulsions versus shellac protective coatings on postharvest citrus quality. *J Am Soc Horticult Sci* 146:40–49. https://doi.org/10.21273/JASHS04972-20

Mishra RK, Ha SK, Verma K, Tiwari SK (2018) Recent progress in selected bio-nanomaterials and their engineering applications: An overview. *J Sci: Adv Mater Devices* 3:263–288. https://doi.org/10.1016/j.jsamd.2018.05.003

Mohsen Z, Almasi H, Dardmeh N (2020) Investigating the effect of the package containing titanium dioxide and zinc oxide nanoparticles on shelf life and quality changes of apples and grapes. *Innov Food Technol* 8:63–82. https://doi.org/ 10.22104/JIFT.2020.4206.1974

Muller K, Bugnicourt E, Latorre M, et al (2017) Review on the processing and properties of polymer nanocomposites and nanocoatings and their applications in the packaging, automotive and solar energy fields. *Nanomaterials* 7:74. https://doi.org/10.3390/nano7040074

Mustafa F, Andreescu S (2020) Nanotechnology-based approaches for food sensing and packaging applications. *RSC Adv* 10:19309–19336. https://doi.org/10.1039/D0RA01084G

Nain A, Tseng YT, Lin YS, et al (2020) Tuning the photoluminescence of metal nanoclusters for selective detection of multiple heavy metal ions. *Sens Actuators B Chem* 321:128539. https://doi.org/10.1016/j.snb.2020.128539

Nasir SNS, Mohamed NA, Tukimon MA, et al (2021) Direct extrapolation techniques on the energy band diagram of BiVO4 thin films. *Physica B Condens Matter* 604:412719. https://doi.org/10.1016/j.physb.2020.412719

Neme K, Nafady A, Uddin S, Tola YB (2021) Application of nanotechnology in agriculture, postharvest loss reduction and food processing: Food security implication and challenges. *Heliyon* 7:e08539. https://doi.org/10.1016/j.heliyon.2021.e08539

Nile SH, Baskar V, Selvaraj D, Nile A, Xiao J, Kai G (2020) Nanotechnologies in food science: Applications, recent trends, and future perspectives. *Nanomicro Lett* 12:45. https://doi.org/10.1007/s40820-020-0383-9

Nurfatihah Z, Siddiquee S (2019) Nanotechnology: Recent trends in food safety, quality and market analysis. In: S Siddiquee, GJH Melvin, MdM Rahman (Eds.), *Nanotechnology: Applications in Energy, Drug and Food*. Springer International Publishing, pp. 283–293

Nyachuba DG (2010) Foodborne illness: is it on the rise? *Nutr Rev* 68:257–269. https://doi.org/10.1111/j.1753-4887.2010.00286.x

Padmanabhan NT, Thomas N, Louis J, Mathew DT, Ganguly P, John H, Pillai SC (2021) Graphene coupled TiO2 photocatalysts for environmental applications: A review. *Chemosphere* 271:129506. https://doi.org/10.1016/j.chemosphere.2020.129506

Paidari S, Ahari H, Pasqualone A, Anvar A, Allah Yari Beyk S, Moradi S (2023) Bio-nanocomposites and their potential applications in physiochemical properties of cheese: An updated review. *J Food Measure Charact* 17:2595–2606. https://doi.org/10.1007/s11694-022-01800-9

Panda PK, Sadeghi K, Seo J (2022) Recent advances in poly (vinyl alcohol)/natural polymer based films for food packaging applications: A review. *Food Packag Shelf Life* 33:100904. https://doi.org/10.1016/j.fpsl.2022.100904

Pasricha R, Sachdev D (2017) Biological characterization of nanofiber composites. In: M Ramalingam, S Ramakrishna (Eds.), *Nanofiber Composites for Biomedical Applications*. Elsevier, pp. 157–196. https://doi.org/ 10.1016/B978-0-08-100173-8.00005-3

Pateiro M, Gomez B, Munekata PE, Barba FJ, Putnik P, Kovacevic DB, Lorenzo JM (2021) Nanoencapsulation of promising bioactive compounds to improve their absorption, stability, functionality and the appearance of the final food products. *Molecules* 26:1547. https://doi.org/10.3390/molecules26061547

Peterson GI, Choi TL (2020) Cascade polymerizations: Recent developments in the formation of polymer repeat units by cascade reactions. *Chem Sci* 11:4843–4854. https://doi.org/10.1039/D0SC01475C

Pocas MFF, Delgado TF, Oliveria FAR (2008) Smart packaging technologies for fruits and vegetables. In: J Kerry, P Butler (Eds.), *Smart Packaging Technologies for Fast Moving consumer goods*. John Wiley and Sons, pp. 151–165. https://doi.org/ 10.1002/9780470753699

Primozic M, Knez Z, Leitgeb M (2021) (Bio) nanotechnology in food science-Food packaging. *Nanomaterials* 11:292. https://doi.org/10.3390/nano11020292

Priyadarshi R, Ezati P, Rhim JW (2021) Recent advances in intelligent food packaging applications using natural food colorants. *ACS Food Sci Technol* 1:124–138. https://doi.org/10.1021/acsfoodscitech.0c00039

Qiao X, He J, Yang R, Li Y, Chen G, Xiao S, Huang B, Yuan Y, Sheng Q, Yue T (2022) Recent advances in nanomaterial-based sensing for food safety analysis. *Processes* 10:2576. https://doi.org/10.3390/pr10122576

Rahman M, Islam R, Hasan S, Zzaman W, Rana MR, Ahmed S, Roy M, Sayem A, Matin A, Raposo A, Zandonadi RP (2022) A comprehensive review on bio-preservation of bread: An approach to adopt wholesome strategies. *Foods* 11:319. https://doi.org/10.3390/foods11030319

Rahman MS, Perera CO (2007) Drying and food preservation. In: M Shafiur Rahman (Ed.), *Handbook of Food Preservation*. CRC Press, pp. 421–450. https://doi.org/ 10.1201/9781420017373

Rai M, Ingle AP, Gupta I, Pandit R, Paralikar P, Gade A, Chaud MV, dos Santos CA (2019) Smart nanopackaging for the enhancement of food shelf life. *Environ Chem Lett* 17:277–290. https://doi.org/10.1007/s10311-018-0794-8

Reguengo LM, Salgaco MK, Sivieri K, Marostica Junior MR (2022) Agro-industrial by-products: Valuable sources of bioactive compounds. *Food Res Int* 152:110871. https://doi.org/10.1016/j.foodres.2021.110871

Rhim JW, Park HM, Ha CS (2013) Bio-nanocomposites for food packaging applications. *Prog Polym Sci* 38:1629–1652. https://doi.org/10.1016/j.progpolymsci.2013.05.008

Romling U, Balsalobre C (2012) Biofilm infections, their resilience to therapy and innovative treatment strategies. *J Intern Med* 272:541–561. https://doi.org/10.1111/joim.12004

Safaya M, Rotliwala YC (2020) Nanoemulsions: A review on low energy formulation methods, characterization, applications and optimization technique. *Mater Today Proc* 27:454–459. https://doi.org/10.1016/j.matpr.2019.11.267

Sagar NA, Kumar N, Choudhary R, Bajpai VK, Cao H, Shukla S, Pareek S (2022) Prospecting the role of nanotechnology in extending the shelf-life of fresh produce and in developing advanced packaging. *Food Packag Shelf Life* 34:100955. https://doi.org/10.1016/j.fpsl.2022.100955

Sahoo M, Panigrahi C, Vishwakarma S, Kumar J (2022) A review on nanotechnology: Applications in food industry, future opportunities, challenges and potential risks. *J Nanotechnol Nanomater* 3. https://doi.org/10.33696/Nanotechnol.3.029

Sahoo M, Vishwakarma S, Panigrahi C, Kumar J (2021) Nanotechnology: Current applications and future scope in food. *Food Front* 2:3–22. https://doi.org/10.1002/fft2.58

Salem SS, Fouda A (2021) Green synthesis of metallic nanoparticles and their prospective biotechnological applications: An overview. *Biol Trace Elem Res* 199:344–370. https://doi.org/10.1007/s12011-020-02138-3

Sanchez-Gonzalez L, Vargas M, Gonzalez-Martínez C, Chiralt A, Chafer M (2011) Use of essential oils in bioactive edible coatings: A review. *Food Eng Rev* 3:1–16. https://doi.org/10.1007/s12393-010-9031-3

Sanchez-Moreno C, A. Larrauri J, Saura-Calixto F (1999) Free radical scavenging capacity and inhibition of lipid oxidation of wines, grape juices and related polyphenolic constituents. *Food Res Int* 32:407–412. https://doi.org/10.1016/S0963-9969(99)00097-6

Sapelli KS, Borba KR, Miranda M, Spricigo PC, Bresolin JD, Foschini MM, Correa DS, Ferreira MD (2021) Postharvest quality of papaya fruit wrapped with polyvinyl chloride film added with silver. *Acta Hortic* 265–272. https://doi.org/10.17660/ActaHortic.2021.1325.38

Sarojini SK, Indumathi MP, Rajarajeswari GR (2019) Mahua oil-based polyurethane/chitosan/nano ZnO composite films for biodegradable food packaging applications. *Int J Biol Macromol* 124:163–174. https://doi.org/10.1016/j.ijbiomac.2018.11.195

Satin M (2002) Use of irradiation for microbial decontamination of meat: Situation and perspectives. *Meat Sci* 62:277–283. https://doi.org/10.1016/S0309-1740(02)00129-8

Sekhon B (2014) Nanotechnology in agri-food production: An overview. *Nanotechnol Sci Appl* 31. https://doi.org/10.2147/NSA.S39406

Shafiq M, Anjum S, Hano C, Anjum I, Abbasi BH (2020) An overview of the applications of nanomaterials and nanodevices in the food industry. *Foods* 9:148. https://doi.org/10.3390/foods9020148

Shahidi F, Hossain A (2022) Preservation of aquatic food using edible films and coatings containing essential oils: A review. *Crit Rev Food Sci Nutr* 62:66–105. https://doi.org/10.1080/10408398.2020.1812048

Sharma S, Loach N, Gupta S, Mohan L (2020) Phyto-nanoemulsion: An emerging nano-insecticidal formulation. *Environ Nanotechnol Monit Manage* 14:100331. https://doi.org/10.1016/j.enmm.2020.100331

Shen Y, Gong W, Li Y, Deng J, Shu X, Wu D, Pellegrini N, Zhang N (2022) The physiochemical and nutritional properties of high endosperm lipids rice mutants under artificially accelerated ageing. *LWT* 154:112730. https://doi.org/10.1016/j.lwt.2021.112730

Shi J, Wang J, Liang L, Xu Z, Chen Y, Chen S, Xu M, Wang X, Wang S (2021) Carbothermal synthesis of biochar-supported metallic silver for enhanced photocatalytic removal of methylene blue and antimicrobial efficacy. *J Hazard Mater* 401:123382. https://doi.org/10.1016/j.jhazmat.2020.123382

Shojaeiarani J, Bajwa DS, Rehovsky C, Bajwa SG, Vahidi G (2019) Deterioration in the physico-mechanical and thermal properties of biopolymers due to reprocessing. *Polymers (Basel)* 11:58. https://doi.org/10.3390/polym11010058

Shweta, Sood S, Sharma A, Chadha S, Guleria V (2021) Nanotechnology: A cutting-edge technology in vegetable production. *J Hortic Sci Biotechnol* 96:682–695. https://doi.org/10.1080/14620316.2021.1902864

Silva-Vera W, Zamorano-Riquelme M, Rocco-Orellana C, Vega-Viveros R, Gimenez-Castillo B, Silva-Weiss A, Osorio-Lira F (2018) Study of spray system applications of edible coating suspensions based on hydrocolloids containing cellulose nanofibers on grape surface (Vitis vinifera L.). *Food Bioproc Tech* 11:1575–1585. https://doi.org/10.1007/s11947-018-2126-1

Singh AK, Ramakanth D, Kumar A, Lee YS, Gaikwad KK (2021) Active packaging technologies for clean label food products: A review. *J Food Meas Charact* 15:4314–4324. https://doi.org/10.1007/s11694-021-01024-3

Singh IR, Pulikkal AK (2022) Preparation, stability and biological activity of essential oil-based nano emulsions: A comprehensive review. *Open Nano* 8:100066. https://doi.org/10.1016/j.onano.2022.100066

Singh KR, Nayak V, Singh RP (2021b) Introduction to bionanomaterials: An overview. In: RP Singh, KRB Singh (Eds.), *Bionanomaterials*. IOP Publishing, pp. 1–16.

Singh M, Sahareen T (2017) Investigation of cellulosic packets impregnated with silver nanoparticles for enhancing shelf-life of vegetables. *LWT* 86:116–122. https://doi.org/10.1016/j.lwt.2017.07.056

Singh T, Shukla S, Kumar P, Wahla V, Bajpai VK, Rather IA (2017) Application of nanotechnology in food science: perception and overview. *Front Microbiol* 8. https://doi.org/10.3389/fmicb.2017.01501

Singha K, Regubalan B, Pandit P, Maity S, Ahmed S (2022) Introduction to nanotechnology - Enhanced food packaging industry. In: J. Parameswaranpillai, RE Krishnankutty, A Jayakumar, SM Rangappa, S Siengchin (Eds.), *Nanotechnology-Enhanced Food Packaging*. Wiley, pp. 1–17. https://doi.org/10.1002/9783527827718.ch1

Siracusa V (2012) Food packaging permeability behaviour: A report. Int *J Polym Sci* 2012:1–11. https://doi.org/10.1155/2012/302029

Smith SJ, Hou R, Lau CH, Konstas K, Kitchin M, Dong G, Lee J, Lee WH, Seong JG, Lee YM, Hill MR (2019) Highly permeable Thermally Rearranged Mixed Matrix Membranes (TR-MMM). *J Memb Sci* 585:260–270. https://doi.org/10.1016/j.memsci.2019.05.046

Stambulska UYa, Bayliak MM (2020) Legume-rhizobium symbiosis: Secondary metabolites, free radical processes, and effects of heavy metals. In: J-M Merillon, KG Ramawat (Eds.), *Co-evolution of Secondary Metabolites*. Springer, pp. 291–322. https://doi.org/ 10.1007/978-3-319-96397-6_43

Suvarna V, Nair A, Mallya R, Khan T, Omri A (2022) Antimicrobial nanomaterials for food packaging. *Antibiotics* 11:729. https://doi.org/10.3390/antibiotics11060729

Thompson S (2011) News about NANOTECHNOLOGY: A longitudinal framing analysis of newspaper reporting on nanotechnology. *Thesis (Doctoral)*, Bournemouth University.

Tian B, Liu Y (2020) Chitosan-based biomaterials: From discovery to food application. *Polym Adv Technol* 31:2408–2421. https://doi.org/10.1002/pat.5010

Tiwari SK, Pandey R, Wang N, Kumar V, Sunday OJ, Bystrzejewski M, Zhu Y, Mishra YK (2022) Progress in diamanes and diamanoids nanosystems for emerging technologies. *Adv Sci* 9:2105770. https://doi.org/ 10.1002/advs.202105770

Tu Y, Li P, Sun J, Jiang J, Dai F, Li C, Wu Y, Chen L, Shi G, Tan Y, Fang H (2021) Remarkable antibacterial activity of reduced graphene oxide functionalized by copper ions. *Adv Funct Mater* 31:2008018. https:// doi.org/10.1002/adfm.202008018

Valenzuela L, Iglesias-Juez A, Bachiller-Baeza B, Faraldos M, Bahamonde A, Rosal R (2020) Biocide mechanism of highly efficient and stable antimicrobial surfaces based on zinc oxide-reduced graphene oxide photocatalytic coatings. *J Mater Chem B* 8:8294–8304. https://doi.org/10.1039/D0TB01428A

Velasco-Garcia MN, Mottram T (2003) Biosensor technology addressing agricultural problems. *Biosyst Eng* 84:1–12. https://doi.org/10.1016/S1537-5110(02)00236-2

Virmani R, Pathak K (2022) Consumer nanoproducts for cosmetics. In: S Mallakpour, CM Hussain (Eds.), *Handbook of Consumer Nanoproducts*. Springer Nature, pp. 931–961.

Wang J, Li H, Tian L, Ramakrishna S (2017a) Nanobiomaterials: State of the art. In: XM Wang, M Ramalingam, X Kong, L Zhao (Eds.), *Nanobiomaterials*. Wiley, pp. 3–35.

Wang L, Hu C, Shao L (2017b) The antimicrobial activity of nanoparticles: present situation and prospects for the future. *Int J Nanomedicine* 12:1227–1249. https://doi.org/10.2147/IJN.S121956

Wang Q, Liu W, Tian B, Li D, Liu C, Jiang B, Feng Z (2020) Preparation and characterization of coating based on protein nanofibers and polyphenol and application for salted duck egg yolks. *Foods* 9:449. https://doi. org/10.3390/foods9040449

Wen M, Li G, Liu H, Chen J, An T, Yamashita H (2019) Metal-organic framework-based nanomaterials for adsorption and photocatalytic degradation of gaseous pollutants: Recent progress and challenges. *Environ Sci Nano* 6:1006–1025. https://doi.org/10.1039/C8EN01167B

Wu Z, Deng W, Luo J, Deng D (2019) Multifunctional nano-cellulose composite films with grape seed extracts and immobilized silver nanoparticles. *Carbohydr Polym* 205:447–455. https://doi.org/10.1016/j.carb pol.2018.10.060

Xing Y, Xu Q, Li X, Chen C, Ma L, Li S, Che Z, Lin H (2016) Chitosan-based coating with antimicrobial agents: Preparation, property, mechanism, and application effectiveness on fruits and vegetables. *Int J Polym Sci* 2016:1–24. https://doi.org/10.1155/2016/4851730

Xu Y, Liu X, Jiang Q, Yu D, Xu Y, Wang B, Xia W (2021) Development and properties of bacterial cellulose, curcumin, and chitosan composite biodegradable films for active packaging materials. Carbohydr Polym 260:117778. https://doi.org/10.1016/j.carbpol.2021.117778

Yadav A, Kumar N, Upadhyay A, Sethi S, Singh A (2022) Edible coating as postharvest management strategy for shelf-life extension of fresh tomato (*Solanum lycopersicum* L.): An overview. J Food Sci 87:2256–2290. https://doi.org/10.1111/1750-3841.16145

Yamashita H, Mori K, Kuwahara Y, Kamegawa T, Wen M, Verma P, Che M (2018) Single-site and nano-confined photocatalysts designed in porous materials for environmental uses and solar fuels. Chem Soc Rev 47:8072–8096. https://doi.org/10.1039/C8CS00341F

Yu HY, Zhang H, Song ML, Zhou Y, Yao J, Ni QQ (2017) From Cellulose Nanospheres, Nanorods to Nanofibers: Various Aspect Ratio Induced Nucleation/Reinforcing Effects on Polylactic Acid for Robust-Barrier Food Packaging. ACS Appl Mater Interfaces 9:43920–43938. https://doi.org/10.1021/ acsami.7b09102

Zhang Q, Liu Y, Yang G, Kong H, Guo L, Wei G (2023) Recent advances in protein hydrogels: From design, structural and functional regulations to healthcare applications. Chemical Engineering Journal 451:138494. https://doi.org/10.1016/j.cej.2022.138494

Zhang W, Chen L, Xiong Y, Panayi AC, Abududilibaier A, Hu Y, Yu C, Zhou W, Sun Y, Liu M, Xue H (2021) Antioxidant Therapy and Antioxidant-Related Bionanomaterials in Diabetic Wound Healing. Front Bioeng Biotechnol 9:. https://doi.org/10.3389/fbioe.2021.707479

Zilberg S, Stekolshik Y, Palii A, Tsukerblat B (2022) Controllable Electron Transfer in Mixed-Valence Bridged Norbornylogous Compounds: *Ab Initio* Calculation Combined with a Parametric Model and Through-Bond and Through-Space Interpretation. J Phys Chem A 126:2855–2878. https://doi.org/10.1021/acs.jpca.1c09637

Zucchini NM, Florencio C, Miranda M, Borba KR, Oldoni FC, Oliveira Filho JG, Bonfim NS, Rodrigues KA, de Oliveira RM, Mitsuyuki MC, Hubinger SZ (2021) Effect of carnauba wax nanoemulsion coating on postharvest papaya quality. Acta Hortic 199–206. https://doi.org/10.17660/ActaHortic.2021.1325.29

6 Bionanomaterials in Nutraceuticals

Neeraj Narwat, Nitu Rani, Babita Thakur, and Sukhminderjit Kaur

6.1 INTRODUCTION

Nanobiotechnology is the rising field that works on the principles of two vast fields named nanotechnology and biotechnology. This field mainly involves the tools and application of microtechnology and biotechnology, which help in the development and manufacturing of certain devices and substances that have a nanoscale dimension of at least 1–100nm in length, known as nanomaterials (Durejaet al., 2003). These materials have some unique properties, such as large surfacearea-to-volume, and have high mechanical strength, which ensures their application in various target specifics and biological approaches. By employing nanoscale materials, our biological systems can be modulated through the investigation of intrinsic biological phenomena, facilitating various applications including cellular drug delivery, disease diagnosis, molecular imaging, and the development of diverse beneficial products. These products span agriculture, medicine, and food supplements such as nutraceuticals, as well as environmental applications for energy production, among others. According to Richard W. Siegel, nano biomaterials are classified into four-dimensional nanostructures based on various modulatory dimensionalities as Zero dimension (contains 0D atomics spheres and clusters), one dimension (1D multilayer nanowires, fibers, and rods), two dimensions (2D crystal clear overlayers of nanoplates and films), three dimension (3D nanophase materials). In recent years bionanomaterials fabricate intense interest in themselves because of their unusual physiochemical properties. Some examples of nanoparticles are inorganic-based nanomaterials that contain metals and their oxides (like silver, gold, copper, aluminum, zinc oxides, copper oxides, silica oxides, and so forth), carbon-based nanomaterials like graphene, carbon nanotubes, carbon fibers, and so forth. Organic-based nanomaterials that contain organic/carbon materials like dendrimers, liposomes, micelles, and so forth. Composite-based nanomaterials (these materials are very complex substances mainly composed of metals, metal oxides, and organic carbon materials.

With changing times our lifestyle is also changing due to which our way of consuming food and food nutrients are also changing; that's why these days people are more focused on those ingredients which are more target specific and have various health benefits. Again, nanobiotechnology comes into play by applying various new technologies to provide a wide range of improved and healthy beneficial food products and supplements like nutraceuticals. It also helps in the processing, packaging, and enhancing of food flavor, color, texture, and so forth. The techniques that are majorly involved in the production of target-specific ingredients and food supplements use different types of nano biomaterials such as the incorporation of nutraceuticals, nanoencapsulation, nano-emulsions, nutrient delivery, minerals, and vitamins fortification, and so forth.

Nutraceuticals are derived from the combination of two terms, "nutrition and pharmaceuticals". These are the bioactive proteins product that gives essential nutrients and nourishment to the body and imparts many health benefits, like providing good immunity and strength against various diseases, a

DOI: 10.1201/9781003432791-8

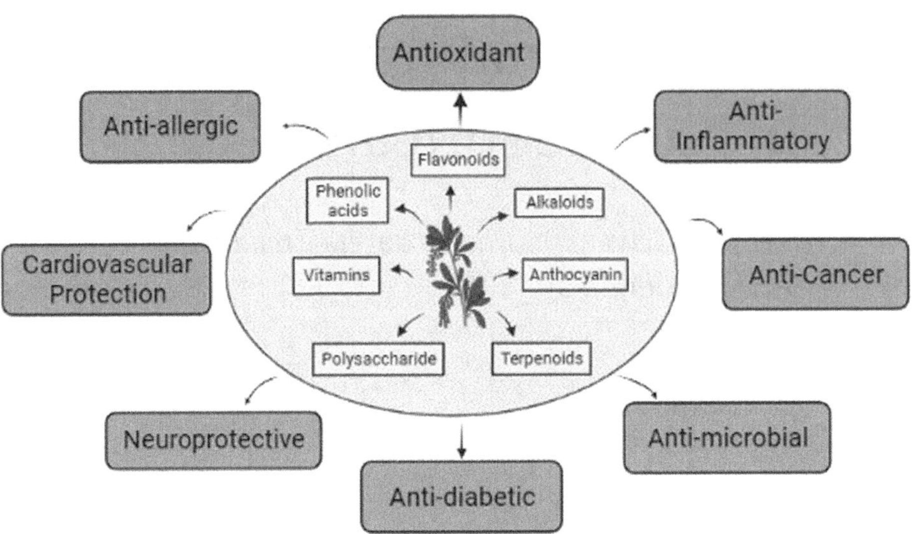

FIGURE 6.1 Nutraceuticals and their Applications. (Figures Created by the Authors.)

natural antioxidant, better bioavailability and long half-life, and fewer side effects than drugs (Trottier et al., 2010) as shown in Figure 6.1. Due to many advantages and fewer side effects the demand for the nutraceutical is increasing each day. These are nutritious food supplements that are routinely consumed in the daily diet, but some nutraceuticals have low solubility, as they are not easily absorbed in the human gut. Encapsulation of these bioactive compounds can be done to increase the bioavailability of the nutraceuticals by changing their pharmacokinetics and biodistribution (Huang et al., 2009). Nutraceuticals are of two main types: (i) Dietary supplements such as protein supplements derived from food products like probiotics and prebiotics, (ii) Functional food (basic food nutrient with various specific ingredients that provide health benefits to the body such as: beta carotene from carrot, lycopene from tomato, omega 3 acids from salmon oil. These nutrients help in decreasing the risk of getting chronic diseases and provide endless benefits to the body), and phytochemicals (Kalra, 2003; Neethirajan and Jayas, 2011). The biological activity of these supplements is increased by decreasing their particle size, which makes them available to cross the intestine to enhance their solubility and make them available for delivery properties (Chen et al., 2006, Shegokar and Muller, 2010).

6.2 NUTRACEUTICALS FROM PLANTS

According to the global medical organization, over 80 per cent of the world's population relies on traditional plant-based medical systems such as phytochemicals, herbal supplements, or functional food (Kasbia, 2005). The productive output and accessibility of nutraceuticals is a highly desirable goal to enhance the well-being of a nation's population, particularly those who are poor. Currently, the development of nutraceuticals quality and quantity is a major focus of ongoing biotechnological research.

6.2.1 Phytochemicals

These are biological chemicals that are naturally found in plants and help them to provide colour, flavour, taste, and texture (Brouns, 2002). Glucosinolates in green vegetables, limonoids in citrus fruits, flavonols in flaxseed, lycopene in tomatoes, and catechins in tea are some examples of naturally occurring phytochemicals. Various phytochemicals have been found in different plants, and their therapeutic properties have been elaborated in Table 6.1. Each has a different bioactive compound possessing therapeutic applications and can be used to prevent various diseases (Bagchi, 2006).

TABLE 6.1
List of Phytochemicals in Plants with Medicinal Properties

Nutraceuticals	Plant Sources	Properties	References
Methoxyflavones	*Kaempferiaparviflora*	Anti-melanogenic activity	(Chen et al., 2023)
Isoquercitin	leafy vegetables, fruits, grapes, tea	Helps against oxidative stress, inflammation	(Shabir et al., 2022)
Zeaxanthin	Paprika, corn, saffron, wolfberries	Reduces the effects of Age-related macular degeneration disease and is useful to maintain eye health	(Khoo et al., 2019)
Saponins	*Aralia elata, Agavaceae, Dioscoreaceae and Liliaceae*	Plays a vital role in diabetes, hepatitis, stomach spasms, neurasthenia and also act as an anti-tumour compound	(Cheng et al., 2021, Li F. et al., 2019)
Lignan	Carrots, soybean, broccoli, kale spinach	Used as an antioxidant, anti-arthritis, anti-inflammatory, hepatoprotective, cardioprotective	(Fuad et al., 2020)
Deoxyelephantopin	*Elephantopuscarolinianus, Elephantopusscaber*	Has antitumor and apoptosis-inducing properties, help in preventing and treatment of uterine leiomyoma	(Pandey et al., 2020, Kabeer et al., 2019)
Xanthone	*Garcinia onlongifolia,C. caledonicum,Gentiana*	Used as anticarcinogenic, antimicrobial, neuroprotective, antidiabetic, and cardioprotective	(Kovacevic et al., 2019, Li P. et al., 2016)
Angelicin	*Angelica archangelica, PimpinellinsaxifragaL.,* Heracleum*spondylium*L.	Used in phototherapy, show anticancer activities, and shows multiple applications in therapies such as osteoporosis Sickle cell anemia	(Oliveira et al., 2019)
Polyphenolic Compounds	*Menyanthestrifoliata, Salvia chinensis*	They have antioxidant, antimicrobial, and reduce oxidative damage	(Kowalczyk et al., 2019)
Ascorbic acid and chlorogenic acid	*Morusniger, Tomato, organge*	Used in food and beverages, benefits for skin and personal care	(Turan et al., 2017)
Lutein	*Lycopersicanesculentum, Tropaeolummajus, T. minus*	Help in cardiovascular and eye health	(Lockwood, 2016)
Caffeic acid	*Fruits, vegetables, citrus*	Antioxidant-like activities may reduce the risk of degenerative diseases, heart diseases & eye diseases	(Pellegrino et al., 2016)
Proanthocyanidins	Cranberries, cranberry products, cocoa, chocolate	May improve urinary tract health; may reduce the risk of CVD	(Pellegrino et al., 2016)
Lycopene	Tomato (*Lycopersiconesculentum*), spinach (*Spinaceaoleracea*)	Reduces risk of prostate cancer	(Pellegrino et al., 2016)
B-sitosterol	*Nitrariaretusa, Bacillariophyceae, Asterionellaglacialis*	Used for heart disease, hypercholesterolemia, prevent cancer, hair loss, and rheumatoid arthritis.	(Saeidnia et al., 2014)
Betaine	(Trimethyl Glycine) from green vegetables and germinated grains	Decrease harmful homocysteine accumulation	(Pulliainen et al., 2010)
Ellagic Acid	Strawberries and Raspberries	Used to fight against cancer in humans	(Vattem and shetty 2005)
Carnitine or L-carnitine	*Asparagus,* wheat seed, cabbage head leafs	Used to carry a long-chain fatty acid group into the mitochondria.	(Steiber et al., 2004)

6.2.2 Dietary Supplements

These supplements are abundant sources of necessary nutrients that individually or in combination have a nutritious or therapeutic effect (Bickford et al., 2006, Devi and Rehman, 2002). They include those products that a customer can buy without any recommendation or prescription. Antioxidant use in the form of dietary intake or supplementation has been linked to a wide range of possible advantages as this may hopefully minimize tumor and cerebrovascular illness (Brouns, 2002). Additionally, we can enhance or supplement a person's diet by adding food additives like vitamins, minerals, phytochemicals, and proteins, although these supplements are not advised for consumption alone or in place of any food or medication.

6.2.3 Functional Food

These are consumed as a part of the regular diet and serve to enhance the body's requirements for vital nutrients. They might also support activities that promote expansion and improvement, and they can be taken as regular foods or as foods enriched with bioactive substances to reduce the risk of disease.

Nutraceuticals come from a variety of sources, such as fundamental human and mammalian metabolites, dietary elements derived from plants and animals, synthetic ingredients, and plant secondary metabolites. They are also increasingly made by microbial fermentation. The majority of nutraceuticals come from plants and are consumed as whole plant foods like flaxseed or as single, purified components like resveratrol, multi-component products like pycnogonid, or single, pure components like resveratrol. The most studied plant-based nutraceuticals are those made from soy and tea, but many scientific and clinical articles discuss the polyphenolic compounds found in many plants, including grapes, wine, and many more. A list of commercially accessible single-component nutraceuticals include lycopene, lutein, zeaxanthin, and Y-linolenic acid, and so forth, which frequently occur in a variety of plants, along with their plant sources and medicinal properties. From particular plants, a variety of pure multi-component nutraceuticals is also obtained, for example GSPE (Grape Seed Procyanidin Extract), which constitutes either catechin or epicatechin monomer (Figure 6.2).

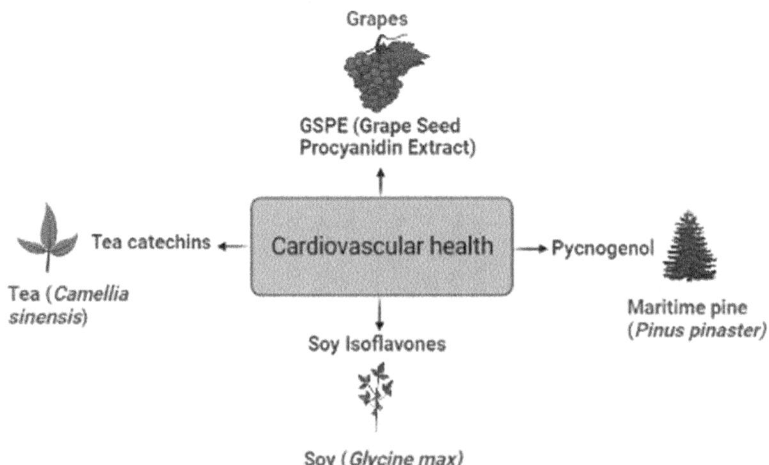

FIGURE 6.2 Multicomponent Nutraceuticals from Plants. (Figures Created by the Authors.)

6.3 NUTRACEUTICALS FROM ANIMALS

Animal products are utilized in medications as both active and inactive components. Inactive components include binders, carriers, stabilizers, fillers, and colorants. Additionally, animals and their by-products play crucial roles in pharmaceutical production even without being included in the final product. The most vital animal components that are used as nutraceuticals or supplements to provide various health benefits are as follows:

6.3.1 Lanolin

This is the most prominent source of vitamin D, obtained from the boiled wool of sheep. It is collected by centrifugal separation or solvent extraction after rinsing the wool in hot water and detergent. It is frequently utilized in supplements as it is cheaper than Vegan vitamin D3 which is extracted from algae.

6.3.2 Collagen

This is a quite well-liked supplement that is present in many skins, and hair-and-nail care products, boiling cow bone and tissue yields collagen which is then dried and made into a powdered supplement.

6.3.3 Glucosamine and Chondroitin

These are human fibers tissue, derived from shellfish and shark cartilage, and mainly used to manage arthralgia, protecting cartilage, and are also applied in cosmetics.

6.3.4 Carotene

Beta carotene, a reddish-orange pigment found particularly in colorful vegetables, as well as in other plants and fruits, is converted by our body into vitamin A. Another source for extracting carotene is animal tissue.

6.3.5 Choline

It is a complex nutritious molecule mainly derived from animals including fish, chicken, duck, and so forth. Choline is used for healthy fat digestion to decrease the risk of hepatic steatosis (Sherriff et al., 2016).

6.4 NUTRACEUTICALS FROM BACTERIA

Many microorganisms produce some compounds that show or offer various therapeutic benefits for the gut and immunological health: these microorganisms are known as probiotics. Probiotic bacteria are widely used in the food industry, including in fruits and vegetables, dairy, and animal products. Nevertheless, they must be consumed in adequate amounts to benefit the host (Kehinde et al., 2020). These compounds' beneficial effects are related to a particular microbial strain, not just a particular species (Zielińska et al., 2018). When taken in sufficient amounts, these probiotic microbial strains are said to provide various positive effects on human health (Sanders et al., 2018). Functional foods that contain beneficial microorganisms display assessments of the vitality of the microbial cells about factors like the circumstances of food preparation and preservation and the ph of the polymeric matrix (de britoalves et al., 2020, Frakolaki et al., 2021).

6.5 NANOBIOTECHNOLOGY IN NUTRACEUTICALS

Nanobiotechnology is a study that can transform, quantify, analyze, regulate, and modify certain devices or biological substances at the scale of around 1–100nm, where innovative interfacial

TABLE 6.2
Most Common Bioactive Compounds Obtained from Bacteria

Nutraceuticals	Microorganism	Function	References
Beta-Carotene	*Sphingomonas sp.,*Blakesleat *rispora,Erwiniauredovora*	Prevention of breast cancer and macular degeneration	(Ram et al., 2020)
Astaxanthin	*Dunaliellasalina and Haematococcuspluvialis*	Act as an antioxidant helps in protecting the damaged cells, improves the immune system response, and involve in Alzheimer's disease, athletic performance	(Park et al., 2010)
Omega-6 fatty acids	*Mortierellaalpina, Mortierellaalliacea, and Rhodotorulamucilaginosa*	Antiviral, antioxidant, antitumor, anti-inflammatory	(Patel et al., 2019)
EPA (Eicosapentaenoic acid)	*Rhodopseudomonasfaecalis*	Neural development, prevention of cardiovascular disease, anti-inflammatory	(Brinton et al., 2017)
Lycopene	*Rhodospirillum rubrum, Escherichia coli*	Nutrient supplements used as antioxidants, anti-cancer, and anti-inflammatory	(Wang et al., 2016)
Lycogen	Rhodobactersphaeroides	Anti-inflammatory, anti-oxidative, and glucose homeostasis effects	(Wang et al., 2016)
Probiotic	*Bifidobacteria, Lactobacillus*	Anticancer, anti-inflammatory immunomodulating	(Lin et al., 2014)
Alkaloids (Benzylisoquinoline alkaloids (BIAs), tyrosine, monoterpene-indole alkaloids (MIAs)	*E.coil and S. cerevisiae*	Act as anti-cancer, and antimalarial, have therapeutic values	(Hawkins et al., 2008)
Poly-€-L-Lysine	*Streptomyces sp., M-Z18*	Protection of cancer, defence against retinopathy, and defense against Alzheimer's disease	(Chen et al., 2011)
DHA (Docosahexaenoic acid)	*Crypthecodiniumcohnii, Pavlova salina, Isochrysisgalbana*	Crohn's disease and cystic fibrosis	(Guedes et al., 2011, Borowitzka, 2013)

properties bring new functionalities. This unique potential of nanotechnology gives rise to a wide range of new advancements that influence almost every area of research science, business, market, environment, and even the daily life of living beings as shown in Table 6.2. (Baeumner, 2004). The nanoparticles that are commonly used for nutraceutical delivery are of two types, organic and inorganic.

Organic nanoparticles are small, solid particles that are made up of lipids and polymers of phospholipids, for example, liposomes, polymersomes, polymer construct, dendrimeric, micelles, and so forth, for sensing and therapeutic gene and drug delivery (Qiu et al., 2006). Inorganic nanoparticles are of a size between 1–100 nanometres, for example, gold, silver, magnesium, zinc, manganese dioxide (MnO_2), iron-platinum nanoparticles (FePt), titanium, cassiterite, and so forth (Figure 6.3).

Several nanoscale phenomena have been used in the formulation and production of nutraceuticals and functional foods such as carotenoids, CoQ10, lycopene, and so forth. New ideas of nanotechnology have been developed to increase the bioavailability and functionality of

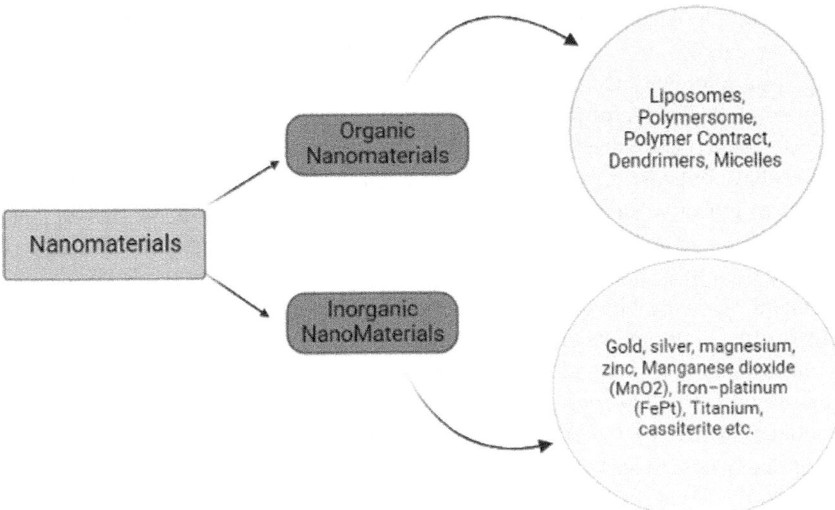

FIGURE 6.3 Types of Nanomaterials. (Figures Created by the Authors.)

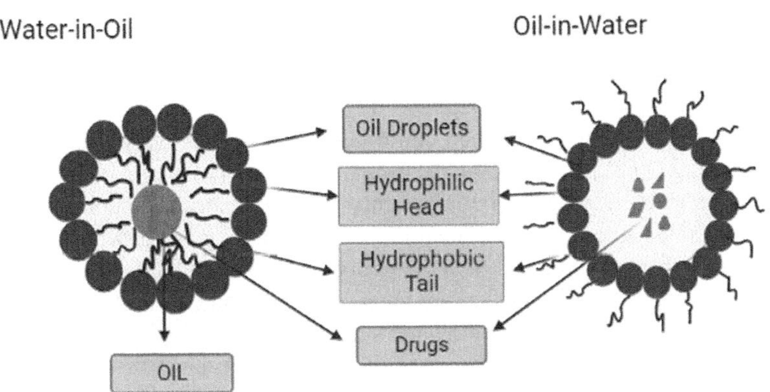

FIGURE 6.4 Nanoemulsion(O/W) and (W/O). (Figures Created by the Authors.)

nutraceutical products. A few of these nanotechnology-based innovations with wide interfacial areas like nanoemulsion, dispersion, bilayer structured fluid, and so forth, are used. Greater knowledge of these structures' functionality and a better ability to visualize them in nanometer precision are both made possible by recently gained capabilities in nanoscale characterization (Tolles and Rath, 2003).

6.6 NANOPARTICULATE DELIVERY SYSTEM

6.6.1 Nanoemulsion

Nanoemulsions are very small oil-in-water (O/W) or water-in-oil (W/O) emulsions having a droplet diameter of about 50–200nm(Figure 6.4.). Emulsions are mixtures of two fully or partially non-miscible fluids such as oil and water in which one solvent is scattered throughout the other and referred to as droplets. Examples like milk, mayonnaise, sauces, and so forth. Nanoemulsions are quite sufficient to not disperse light in the visible range of the spectrum, which gives them an optically transparent look rather than optically opaque look; also, due to their diminutive size, they do not show a creaming process. Techniques like high-pressure homogenizers and microfluidic channels are used to create macro-emulsion. It is stated that nanoemulsions are thermodynamically stable over a broad range of pH values to preserve flavor and insulate it from heat, oxidation, enzymes process, and breakdowns (Samal, 2017). Because of excellent transparency and small size, nanoemulsions have outstanding penetrating characteristics that assure the quick transport of high concentrations of bioactive substances to cell membranes. For example, the delivery of lipophilic active compounds in nanoemulsions can significantly increase their bioavailability (Nakajima, 2005). They are used to confine active ingredients and functional compounds such as antioxidants and nutraceuticals, also used by entrapping bioactive components in the disperse phase, which are self-stabilizers due to surfactants and biopolymers in the continuous phase (Krishna et al., 2022), such as lutein, which is trapped under poly (lactic-co-glycolic acid, PLGA) nanoparticles resulting in high bioavailability and aqueous solubility of lutein to enter the circulatory system and targeted organs (Bodoki et al., 2019). Curcuminorganogel has been recently developed for the emulsification of curcumin. Tween 20 was also chosen as the emulsifier to increase the bioaccessibility of the curcumin (Yu et al., 2012).

6.6.2 Nanoencapsulation

This is the process of encapsulating core or active solid, liquified, and gaseous nanoparticles inside the matrix or shell of the secondary substance to create nanocapsules. The active compounds inside the core of the nanocapsules (drugs, fragrances, biocontrols, vitamins, and so forth) are covered and sealed by the shell to protect it from the surrounding environment. This shielding can be temporary or permanent depending upon the core released by diffusion or by some trigger factors like stress, pH, or enzyme action and, thus, allow their controlled and timely delivery to a specified site (Desai and Park. 2005, Jyothi et al., 2010). Nanoencapsulation covers flavor and odor, helps in regulating the interactions of biologically active components with the matrix component, and controls the delivery of active substances, showing flexibility with various molecules in the system (Ubbink and Kruger, 2006; Weiss et al., 2006). With the help of nanoencapsulation, it is possible to preserve bioactive substances such as vitamins, antioxidants, proteins, and so forth, for the manufacturing of food products with improved functioning and stability and also used in the drug delivery system based on the chemical and physical characteristics of the substance that is used to encapsulate (Paredes et al., 2016). For example, Saponin is efficiently encapsulated by polyelectrolyte complex nanoparticles (PEC NPs) using negatively charged polysaccharide, TLH-3, and positively charged sodium caseinate (Zhang et al., 2022); Beta-carotene is cored around ethylcellulose and zein NPs to increase the bioavailability (Afonso et al., 2020). 2,2-diphenyl-1-picrylhydrazyl (DPPH) method increases the antioxidant activity of lutein microspheres of lutein-alginate produced by the calcium chloride gelation process (Huang et al., 2019).

6.6.2.1 Micelles

Micelles are spherical nanosized particles with an average diameter of about 5–100nm. They are readily formed by dissolving the chemical compound into the water up to the concentration above a then critical level known as "Critical micelle concentration" (CMC). This thermodynamically controlled self-build mechanism minimizes contacts of the non-polar tail (hydrophobic group) of surfactants with water, whereas maximizing contacts of the polar head (hydrophilic group) with water (Kronberg et al., 1995). As a result, micelle viability is frequently retained for a long time under a particular combination of environmental factors (pH, temperature, osmotic pressure). Micelles have some unique characteristics: they can encapsulate non-water-soluble compounds and nonpolar compounds like triglycerides, antibacterial/antiviral drugs, flavonoids, and vitamins (Chen et al., 2006). Although micelles are widely used in pharmaceutical chemicals nowadays, they are also used in food industries. For example, certain bioactive compounds like loteolin, lutein, omega-3 fatty acids, and so forth are emulsified inside the micelles for effective nutrition delivery inside the body system. Xanthones, bioaccessibility is increased by the amalgamation of bile salt mixed with micelles (Bumrungpert et al., 2009). Casein micelles exhibit extraordinary surface activity to stabilize nanoemulsion and have a self-assembling tendency to integrate bioactive and the capacity to bind hydrophobic compounds when heated (Sadiq et al., 2021), and polymeric micelles have properties including reduced particle size and greater drug loading capacity used in novel drug delivery systems (Gong et al., 2012).

6.6.2.2 Liposomes

Liposomes are small lipid bilayer vesicles that are made up of polar lipids readily available in the environment, like phospholipids from plant-based food (such as soy) and egg. These are aggregates of many molecules with bi-layered shell structured and hydrophilic and hydrophobic cores as shown in Figure 6.5. So, like micelles, these lipid vesicles also include a wide range of different functional elements and help in enclosing both lipids as well as water-soluble components. Based on the technique of formation liposomes can be single or multi-lamellar, with one or several bilayer shells. Liposomes are variable in size from 20nm to a few hundred micrometers (see Figure 6.6). These are efficiently used in the encapsulation of proteins and give an interior microenvironment that allows the protein to continue functioning irrespective of the outer environment (Taylor et al., 2005). There are various food applications of liposomes, as they help in extending the shelf life of food products by carrying components such as antibacterial polypeptide nisin Z and bactericidal glycoprotein lactoferrin, encapsulate vitamin C and maintained the 50 per cent of its activity compared to ascorbic

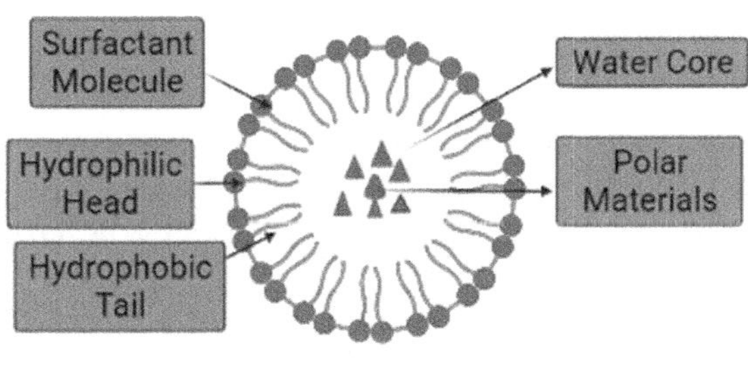

FIGURE 6.5 Structure of Micelles. (Figures Created by the Authors.)

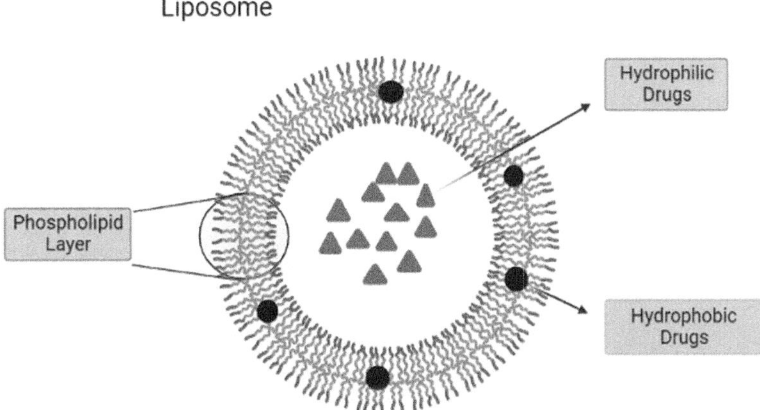

FIGURE 6.6 Structure of Liposomes. (Figures Created by the Authors.)

acid and also inhibit lipid oxidation inside the dairy products by encapsulating phosvitin (Gaysinksy et al., 2005, Were et al., 2003). PEG derivative vitamin E and carotenoid are encapsulated inside the lipid bilayer of liposome (Sercombe et al., 2015) and drugs like doxorubicin, nystatin, amphotericin are carried out by liposomal carriers (Xia et al., 2015). Chemicals like phytocannabinoids obtained from cannabis plants recently used in therapeutics are encapsulated inside liposomes to increase the bioavailability of cannabinoids (Assadpour et al., 2023).

6.6.2.3 Biopolymericnanoparticles

These particles are the biopolymers of protein and polysaccharides that are joined together chemically or intermolecularly to produce solid particles. A single biopolymer can make up a nanoparticle, or it could have a core-shell structure. These particles have quickly emerged as the most promising nanoscale delivery systems in the biomedical and cosmetics industry sector due to their flexibility in terms of the compounds that can be encapsulated, the degree to which these particles can be engineered, and the degree to which surface properties can be tailored. To encapsulate and distribute substances a broad range of polymers, both natural and artificial, has been utilized: among these are the artificial polymers l- d- and dl-polylactic acid (PLA) polyglycolic acid (PGA) and polycaprolactone acid, and the naturally occurring antibacterial and antioxidant polymer like chitosan, which is extracted from the crustaceans' shells. Also becoming more popular are copolymers produced by combining the monomers lactidegalactide and caprolactone (Shahidi and Abuzaytoun, 2005). Biopolymeric nanoparticles like chitosan, starch alginate, cellulose, and so forth, obtained from the ionic gelation method have established effective encapsulation by maintaining the bioactivity of the implanted molecules (Liu et al., 2020).

6.7 CONCLUSION

In the last ten years, considerable advancements have been made in the formulation and manufacturing of better food, nutritional supplement, and new drugs. Nanobiotechnology can enhance food flavor, color, texture, and so forth by changing the taste, making it healthier and more nourishable as well as providing new food supplies, packaging, and preservations. Moreover, the majority of implementations are still at an early stage and, at least initially, are more focused on slightly upgraded, elevated products. The application of microtechnology in the food sector is either majorly successful or it fails, based on customer belief, faith, and appreciation, just like any other modern technology.

In the present chapter, we discussed various nutraceuticals of plant, animal, and microbe origin. The various types of nanomaterials used in nutraceuticals are organic nanoparticles, for example, liposomes, polymersomes, micelles, and so forth, and inorganic nanoparticles, for example, Mg, gold, silver. The application of nanoscience in maintaining and controlling the quality of nutraceuticals is widely used and consumed by customers, as some low-soluble nutraceuticals are not easily absorbed by the body, and hence encapsulation of these bioactive compounds is done to increase the bioavailability of the nutraceuticals by changing the pharmacokinetics (PK) and biodistribution (BD). With this changing physiology, nutraceuticals are easily absorbed and solubilized inside the gastrointestinal tract to give essential nutrients and nourishment to the body and impart many health benefits like providing good immunity and strength against various diseases, a natural antioxidant, better bioavailability, and long half-life as well as fewer side effects than drugs.

REFERENCES

Afonso, B. S., Azevedo, A. G., Gonçalves, C., Amado, I. R., Ferreira, E. C., Pastrana, L. M., & Cerqueira, M. A. (2020). Bio-based nanoparticles as a carrier of β-Carotene: Production, characterisation and in vitro gastrointestinal digestion. *Molecules (Basel, Switzerland)*, *25*(19), 4497. https://doi.org/10.3390/molecules25194497

Assadpour, E., Rezaei, A., Das, S. S., Krishna Rao, B. V., Singh, S. K., Kharazmi, M. S., Jha, N. K., Jha, S. K., Prieto, M. A., & Jafari, S. M. (2023). Cannabidiol-loaded nanocarriers and their therapeutic applications. *Pharmaceuticals (Basel, Switzerland)*, *16*(4), 487. https://doi.org/10.3390/ph16040487

Baeumner, A. (2004). Nanosensors identify pathogens in food. *Food Technology,58*(8), 51–55.

BagchiD. (2006). Nutraceuticals and functional foods regulations in the United States and around the world. *Toxicology*, *221*(1), 1–3. https://doi.org/10.1016/j.tox.2006.01.001

Bickford, P. C., Tan, J., Shytle, R. D., Sanberg, C. D., El-Badri, N., & Sanberg, P. R. (2006). Nutraceuticals synergistically promote proliferation of human stem cells. *Stem Cells and Development*, *15*(1), 118–123. https://doi.org/10.1089/scd.2006.15.118

Bodoki, E., Vostinaru, O., Samoila, O., Dinte, E., Bodoki, A. E.,Swetledge, S., ...& Sabliov, C. M. (2019). Topical nanodelivery system of lutein for the prevention of selenite-induced cataract. *Nanomedicine: Nanotechnology, Biology and Medicine*, *15*(1), 188–197.

Borowitzka, M. (2013).High-value products from microalgae – their development and commercialisation. *Journal of Applied Phycology*, 25(3): 743–756.

Brinton, E. A., & Mason, R. P. (2017). Prescription omega-3 fatty acid products containing highly purified eicosapentaenoic acid (EPA). *Lipids in Health and Disease*, *16*(1), 23.https://doi.org/10.1186/s12944-017-0415-8

Brouns, F.(2002). Soya isoflavones: a new and promising ingredient for the health foods sector. *Food Research International, 35*, 187–193.

Bumrungpert, A., Kalpravidh, R. W., Suksamrarn, S., Chaivisuthangkura, A., Chitchumroonchokchai, C., & Failla, M. L. (2009). Bioaccessibility, biotransformation, and transport of alpha-mangostin from *Garcinia mangostana* (Mangosteen) using simulated digestion and Caco-2 human intestinal cells. *Molecular Nutrition & Food Research*, *53 Suppl 1*, S54–S61. https://doi.org/10.1002/mnfr.200800260

Chen, H., Weiss, J., & Shahidi, F. Nanotechnology in nutraceuticals and functional foods. *Food Technology*, *60*, 30–36.

Chen,H., Lee, S., Yoo, M. J., Lee, B. S., Jang, Y. S., Kim, H. K.,Lee, S., Han, Y. B., & Ki, H. K. (2023). Methoxyflavones from black Ginger (*Kaempferiaparviflora* Wall. ex-Baker) and their inhibitory effect on melanogenesisin B16F10mouse melanoma cells. https://doi.org/10.3390/plants12051183

Cheng, Y., Liu, H., Tong, X., Liu, Z., Zhang, X., Chen, Y., Wu, F., Jiang, X., & Yu, X. (2021). Effects of shading on triterpene saponin accumulation and related gene expression of *Aralia elata* (Miq.)Seem. *Plant Physiology and Biochemistry: PPB*, *160*, 166–174. https://doi.org/10.1016/j.plaphy.2021.01.009

de Brito Alves, J. L., de Oliveira, Y., de Sousa, V. P., & de Souza, E. L. (2020). Probiotics for humans: current status and future prospects. *New and Future Developments in Microbial Biotechnology and Bioengineering*, *14*, 243–254.

Desai,K. G. H., & Park, H. J. (2005). Recent developments in microencapsulation of food ingredients. *Drying Technology*, 23(7), 1361–1394.

Devi,V.K., & Rehman, F. (2002). Nutraceutical antioxidants-an overview. *Indian Journal of Pharmaceutical Education*, *36*(1), 3–8.

Frakolaki, G., Giannou, V., Kekos, D., & Tzia, C. (2021). A review of the microencapsulation techniques for the incorporation of probiotic bacteria in functional foods. *Critical Reviews in Food Science and Nutrition*, *61*(9), 1515–1536. https://doi.org/10.1080/10408398.2020.1761773

Fuad,N.,Sekar,M., Gan, S.H., & Lum, P.T. (2020). Lutein: a comprehensive review on its chemical, biological activities and therapeutic potentials. https://doi.org/10.5530/pj.2020.12.239.

Gaysinsky, S., Davidson, P. M., Bruce, B. D., & Weiss, J. (2005). Stability and antimicrobial efficiency of eugenol encapsulated in surfactant micelles as affected by temperature and pH. *Journal of Food Protection*, *68*(7), 1359–1366. https://doi.org/10.4315/0362-028x-68.7.1359

Gong, J., Chen, M., Zheng, Y., Wang, S., & Wang, Y. (2012). Polymeric micelles drug delivery system in oncology. *Journal of Controlled Release: Official Journal of the Controlled Release Society*, *159*(3), 312–323. https://doi.org/10.1016/j.jconrel.2011.12.012

Guedes, A. C., Amaro, H. M., & Malcata, F. X. (2011). Microalgae as sources of high added-value compounds a brief review of recent work. *Biotechnology Progress*, *27*(3), 597–613.

Hawkins, K. M., & Smolke, C. D. (2008). Production of benzylisoquinoline alkaloids in Saccharomyces cerevisiae. *Nature Chemical Biology*, *4*(9), 564–573. https://doi.org/10.1038/nchembio.105

Huang, J., Bai, F., Wu, Y., Ye, Q., Liang, D., Shi, C., & Zhang, X. (2019). Development and evaluation of lutein-loaded alginate microspheres with improved stability and antioxidant. *Journal of the Science of Food and Agriculture*, *99*(11), 5195–5201. https://doi.org/10.1002/jsfa.9766

Huang, Q., & Given, P. (2009). *Micro/Nano Encapsulation of Active Food Ingredients*. American Chemical Society.

Jyothi, N. V. N., Prasanna, P. M., Sakarkar, S. N., Prabha, K.S., Ramaiah, P.S., Srawan, G. Y., et al. (2010). Microencapsulation techniques, factors influencing encapsulation efficiency. *Journal of Microencapsulation,27*, 187–197.

Kabeer, F. A., Rajalekshmi, D. S., Nair, M. S., & Prathapan, R. (2019). In vitro and in vivo antitumor activity of deoxyelephantopin from a potential medicinal plant Elephantopusscaber against Ehrlich ascites carcinoma. *Biocatalysis and Agricultural Biotechnology*, *19*, 101106.https://doi.org/10.1016/j.bcab.2019.101106

Kalra E. K. (2003). Nutraceutical--definition and introduction. *AAPS Pharmsci*, *5*(3), E25. https://doi.org/10.1208/ps050325

KasbiaG. S.(2005). Functional foods and nutraceuticals in the management of obesity. *Nutrition and Food Science*, 35,344–351.

Kehinde, B. A., Panghal, A., Garg, M. K., Sharma, P., & Chhikara, N. (2020). Vegetable milk as probiotic and prebiotic foods. *Advances in Food and Nutrition Research*, *94*, 115–160. https://doi.org/10.1016/bs.afnr.2020.06.00

Kovacevic G. T., NevenaZogovic, D. K.-M. (2020). Secondary metabolites from endangered *Gentiana*, *Gentianella*, *Centaurium*, and *Swertia* species (Gentianaceae): promising natural biotherapeutics. https://doi.org/10.1016/B978-0-12-819541-3.00019-0

Kowalczyk, T., Sitarek, P., Skała, E., Toma, M., Wielanek, M., Pytel, D., Wieczfińska, J., Szemraj, J., & Śliwiński, T. (2019). Induction of apoptosis by in vitro and in vivo plant extracts derived from *Menyanthestrifoliata L.* in human cancer cells. *Cytotechnology*, *71*(1), 165–180. https://doi.org/10.1007/s10616-018-0274-9

Krishna, A. R., Gurumoorthy, S., Elayappan, P., Sakthivadivel, P., Kumaran, S., & Pushparaj, P. (2022). A review on the application of nanotechnology in food industries. *Current Research in Nutrition & Food Science*, *10*(3).

Khoo, H. E., Ng, H. S., Yap, W. S., Goh, H. J. H., & Yim, H. S. (2019). Nutrients for prevention of macular degeneration and eye-related diseases. *Antioxidants (Basel, Switzerland)*, *8*(4), 85. https://doi.org/10.3390/antiox8040085

Kronberg, B., Costas, M., & Silveston, R. (1995). Thermodynamics of the hydrophobic effect in surfactant solutions: micellization and adsorption. *Pure and Applied Chemistry*, *67*(6), 897–902.

Li, F., Wang, W., & Xiao, H. (2021). The evaluation of anti-breast cancer activity and safety pharmacology of the ethanol extract of *Aralia elata* Seem leaves. *Drug and Chemical Toxicology*, *44*(4), 427–436. https://doi.org/10.1080/01480545.2019.1601211

Li, P., A. Senthilkumar, H., Wu, S. B., Liu, B., Guo, Z. Y., Fata, J. E., Kennelly, E. J., & Long, C. L. (2016). Comparative UPLC-QTOF-MS-based metabolomics and bioactivities analyses of Garcinia oblongifolia. *Journal of Chromatography. B, Analytical Technologies in the Biomedical and Life Sciences*, *1011*, 179–195. https://doi.org/10.1016/j.jchromb.2015.12.061

Lin, C. S., Chang, C. J., Lu, C. C., Martel, J., Ojcius, D. M., Ko, Y. F., Young, J. D., & Lai, H. C. (2014). Impact of the gut microbiota, prebiotics, and probiotics on human health and disease. *Biomedical Journal*, *37*(5), 259–268. https://doi.org/10.4103/2319-4170.138314

Liu, Q., Cai, W., Zhen, T., Ji, N., Dai, L., Xiong, L., & Sun, Q. (2020). Preparation of debranched starch nanoparticles by ionic gelation for encapsulation of epigallocatechin gallate. *International Journal of Biological Macromolecules*, *161*, 481–491. https://doi.org/10.1016/j.ijbiomac.2020.06.070

Lockwood, G.B. (2016). https://basicmedicalkey.com/plant-nutraceuticals/

Nakajima, M. (2005). Development of nanotechnology and materials for innovative utilization of biological functions. *Proceedings of the 34th United States and Japan Natural Resources (UJNR) Food and Agriculture Panel, Susono, Japan*.

Neethirajan, S., & Jayas, D. S. (2011). Nanotechnology for the food and bioprocessing industries. *Food and Bioprocess Technology*, *4*, 39–47.

Oliveira, C. R., Spindola, D. G., Garcia, D. M., Erustes, A., Bechara, A., Palmeira-Dos-Santos, C., Smaili, S. S., Pereira, G. J. S., Hinsberger, A., Viriato, E. P., Cristina Marcucci, M., Sawaya, A. C. H. F., Tomaz, S. L., Rodrigues, E. G., & Bincoletto, C. (2019). Medicinal properties of Angelica archangelica root extract: cytotoxicity in breast cancer cells and its protective effects against in vivo tumor development. *Journal of Integrative Medicine*, *17*(2), 132–140. https://doi.org/10.1016/j.joim.2019.02.001

Pandey, V., Tripathi, A., Rani, A., & Dubey, P. K. (2020). Deoxyelephantopin, a novel naturally occurring phytochemical impairs growth, induces G2/M arrest, ROS-mediated apoptosis and modulates lncRNA expression against uterine leiomyoma. *Biomedecine & Pharmacotherapie*, *131*, 110751.https://doi.org/10.1016/j.biopha.2020.110751

Park, J. S., Chyun, J. H., Kim, Y. K., Line, L. L., & Chew, B. P. (2010). Astaxanthin decreased oxidative stress and inflammation and enhanced immune response in humans. *Nutrition & Metabolism*, *7*, 18. https://doi.org/10.1186/1743-7075-7-18

Paredes, A. J., Asensio, C. M., Llabot, J. M., Allemandi, D. A., & Palma, S. D. (2016). Nanoencapsulation in the food industry: manufacture, applications and characterization. *Journal of Food Bioengineering and Nanoprocessing, 1*(1), 56–79.

Patel, A., Rova, U., Christakopoulos, P. (2019). Simultaneous production of DHA and squalene from *Aurantiochytrium* sp. grown on forest biomass hydrolysates. *Biotechnol Biofuels, 12*, 255.https://doi.org/10.1186/s13068-019-1593-6

Pellegrino, D. (2016). Antioxidants and cardiovascular risk factors. *Diseases* (Basel, Switzerland), *4*(1), 11. https://doi.org/10.3390/diseases4010011

Pulliainen, K., Nevalainen, H., Väkeväinen, H., Jutila, K., & Gummer, C. L. (2010). An analytical method for the determination of betaine (trimethylglycine) from hair. *International Journal of Cosmetic Science*, *32*(2), 135–138. https://doi.org/10.1111/j.1468-2494.2009.00554.x

Qiu, L. Y., & Bae, Y. H. (2006). Polymer architecture and drug delivery. *Pharmaceutical Research*, *23*(1), 1–30. https://doi.org/10.1007/s11095-005-9046-2

Ram, S., Mitra, M., Shah, F., Tirkey, S. R., & Mishra, S. (2020). Bacteria as an alternate biofactory for carotenoid production: a review of its applications, opportunities and challenges. *Journal of Functional Foods, 67, 103867*.https://doi.org/10.1016/j.jff.2020.103867

Sadiq, U., Gill, H., & Chandrapala, J. (2021). Casein micelles as an emerging delivery system for bioactive food components. *Foods (Basel, Switzerland)*, *10*(8), 1965.https://doi.org/10.3390/foods10081965

Saeidnia, S., Manayi, A., Gohari, A.R., Abdollahi, M. (2014). The story of beta-sitosterol- a review. https://doi.org/10.9734/EJMP/2014/7764

Samal,D. (2017). Use of nanotechnology in food industry: a review. *International Journal of Environment, Agriculture and Biotechnology*,*2*(4), 2270–2278.

Sanders, M. E., Merenstein, D., Merrifield, C. A., & Hutkins, R. (2018). Probiotics for human use. *Nutrition Bulletin, 43*(3), 212–225.doi:10.1111/nbu.12334

Sercombe, L., Veerati, T., Moheimani, F., Wu, S. Y., Sood, A. K., & Hua, S. (2015). Advances and challenges of liposome assisted drug delivery. *Frontiers in Pharmacology*, *6*, 286. https://doi.org/10.3389/fphar.2015.00286

Shahidi, F., & Abuzaytoun, R. (2005). Chitin, chitosan, and co-products: chemistry, production, applications, and health effects. *Advances in Food and Nutrition Research*, *49*, 93–135. https://doi.org/10.1016/S1043-4526(05)49003-8

Shabir, I., Kumar Pandey, V., Shams, R., Dar, A. H., Dash, K. K., Khan, S. A., Bashir, I., Jeevarathinam, G., Rusu, A. V., Esatbeyoglu, T., & Pandiselvam, R. (2022). Promising bioactive properties of quercetin for potential food applications and health benefits: a review. *Frontiers in Nutrition*, *9*, 999752. https://doi.org/10.3389/fnut.2022.999752

Shegokar, R., & Müller, R. H. (2010). Nanocrystals: industrially feasible multifunctional formulation technology for poorly soluble actives. *International Journal of Pharmaceutics*, *399*(1-2), 129–139. https://doi.org/10.1016/j.ijpharm.2010.07.044

Sherriff, J. L., O'Sullivan, T. A., Properzi, C., Oddo, J. L., & Adams, L. A. (2016). Choline, its potential role in nonalcoholic fatty liver disease, and the case for human and bacterial genes. *Advances in nutrition*, *7*(1), 5–13.

Steiber, A., Kerner, J., & Hoppel, C. L. (2004). Carnitine: a nutritional, biosynthetic, and functional perspective. *Molecular Aspects of Medicine*, *25*(5-6), 455–473. https://doi.org/10.1016/j.mam.2004.06.006

Taylor, T. M., Davidson, P. M., Bruce, B. D., & Weiss, J. (2005). Liposomal nanocapsules in food science and agriculture. *Critical Reviews in Food Science and Nutrition*, *45*(7–8), 587–605. https://doi.org/10.1080/10408390591001135

Tolles, W. M., & Rath, B. B. (2003). Nanotechnology, a stimulus for innovation. *Current Science, 85*, 1746–1759.

Trottier, G., Boström, P.J., Lawrentschuk, N., & Fleshner, N. E. (2010). Nutraceuticals and prostate cancer prevention: a current review. *Nature Reviews: Urology*, *7*(1), 21–30. https://doi.org/10.1038/nrurol.2009.234

Turan, I., Demir,S., Kilinc, K., Burnaz, N. A., Yaman, S. O., Akbulut, K., Mentese, A., Aliyazicioglu, Y., & Deger, O. (2017). Antiproliferative and apoptotic effect of *Morusnigra* extract on human prostate cancer cells. *Saudi Pharmaceutical Journal, 25*(2), 241–248. https://doi.org/10.1016/j.jsps.2016.06.002

Ubbink,–J., & Kruger,J. (2006). Physical approaches for the delivery of active ingredients in foods. *Trends Food ScienceTechnology*, 17, 244–254. https://doi.org/10.1016/j.tifs. 01.007.

Vattem, D. A., & Shetty, K. (2005). Biological functionality of ellagic acid: a review. *Journal of Food Biochemistry,* 29(3), –234–266. https://doi.org/10.1111/j.1745-4514.2005.00031.x

Wang, C. C., Ding, S., Chiu, K. H., Liu, W. S., Lin, T. J., & Wen, Z. H. (2016). Extract from a mutant *Rhodobactersphaeroides* as an enriched carotenoid source. *Food & Nutrition Research*, *60*, 29580. https://doi.org/10.3402/fnr.v60.29580

Weiss, J., Takhistov, P., & McClements, D. J. (2006). Functional materials in food nanotechnology. *Journal of Food Science*, *71*(9), 107–116.

Were, L. M., Bruce, B. D., Davidson, P. M., & Weiss, J. (2003). Size, stability, and entrapment efficiency of phospholipid nanocapsules containing polypeptide antimicrobials. *Journal of Agricultural and Food Chemistry*, *51*(27), 8073–8079. https://doi.org/10.1021/jf0348368

Xia, S., Tan, C., Zhang, Y., Abbas, S., Feng, B., Zhang, X., & Qin, F. (2015). Modulating effect of lipid bilayer-carotenoid interactions on the property of liposome encapsulation. *Colloids and Surfaces. B, Biointerfaces*, *128*, 172–180. https://doi.org/10.1016/j.colsurfb.2015.02.004

Yu, H., & Huang, Q. (2012). Improving the oral bioavailability of curcumin using novel organogel-based nanoemulsions. *Journal of Agricultural and Food Chemistry*, *60*(21), 5373–5379. https://doi.org/10.1021/jf300609p

Zhang, D., Jiang, Y., Xiang, M., Wu, F., Sun, M., Du, X., & Chen, L. (2022). Biocompatible polyelectrolyte complex nanoparticles for lycopene encapsulation attenuate oxidative stress-induced cell damage. *Frontiers in Nutrition,* 9, 902208.

Zielińska, D., Ołdak, A., Rzepkowska, A., & Zieliński, K. (2018). Enumeration and identification of probiotic bacteria in food matrices. *Advances in Biotechnology for Food Industry*, 167–196. https://doi.org/10.1016/b978-0-12-811443-8.00006-2

7 Bionanomaterials in Food Applications and their Risk Assessment

Shilpa Borehalli Mayegowda, Hemavathi Brijesh, H. Dheeraj, Y. Harinath Reddy, and N.G. Manjula

7.1 INTRODUCTION

The advent of the modern era in food industrialization has helped in the emergence of food nanotechnology to improvise the food properties like solubility, diffusivity, mechanical resistance being formulated for helping to processes, store, package, and transport food products. A variety of techniques has efficiently helped to increase the shelf life of varied food products like the use of mild heat treatments, modified atmospheric gas composition, and low temperatures for storage. However, some of the following treatments are found to be less effective and also to lower efficiency rates (Saravanakumar et al., 2020). The resulting outcome led to the use of nanoparticles (NPs), that have a key role to extend food product shelf life, along with preservation and maintenance of the quality. The most important aspect is to assist in reducing the post-harvest loss by using active elements in packaging to help enhance the gas permeation and mechanical rigidity leading to maintaining the greater quality of agricultural foods. The basic strategy is to maintain the permeable factor with respect to the atmospheric gases, mechanical rigidity and antimicrobial properties (Borehalli et al., 2022; Kumar et al., 2022; Yadav et al., 2021). These green synthesized NPs have been used to assist the developing microfluid devices, or nanosensors, for varied food analysis in industries (Couto and Almeida., 2022).

The manufacture of high-quality food requires sophisticated processing, packaging, and long-term storage techniques. Nanotechnology has a significant impact on the food business by enhancing food quality by enriching the flavour and texture (Patra et al., 2018). Consumers may get information on the condition of the food in the package and also on the nutritional factors through nanomaterials and nanosensors, which are improvised for security by detecting food-borne pathogens. Since most food-derived bioactive compounds are hydrophobic in nature and have low bioavailability and stability, their mode of action may be to fight against various disease-causing agents. However, nanotechnology-based delivery methods have been able to increase the bioavailability of food bioactive compounds and transport the active portion of them to specific locations (Wang et al., 2021). Government and industry face enormous hurdles as a result of the meals based on nanotechnology, but this also ensures that consumers will embrace and believe in the nanofoods that are now available in the market. It has lately been extensively reported that nano colloidal particles are actively being used in several food areas, commercial and marketable, including food quality, safety, nutrition, processing, and packaging. While creating meals that are safer, healthier, and have higher quality and sustainability, it is crucial to consider the characteristics and behaviour of colloidal particles. They have gained noteworthy consideration in recent decades with their probable potentiality with applications in various fields like in electronics, energy, and biotechnology. Synthetic NPs are divided based on the varying composition and synthesis method, such as synthetic polymers—man-made and mimicking properties of natural materials of proteins and

DOI: 10.1201/9781003432791-9

polysaccharides (polyethyleneimine (PEI) and polyethylene glycol (PEG) (Ossipov et al., 2007). The inorganic NPs are basically the metal and metal oxides with an extensive variety of potential applications, including imaging, sensing, and drug delivery. While lipid-based NPs are used in gene therapy, drug delivery, imaging and dendrimers that are highly branched, with defined, uniform structures used in varied applications.

Many areas of food industries and production can benefit from nanotechnology. The foremost applications remain in the food processing units that form the interactive foods and food preservation. To provide nutrients, boost nutrient absorption by the body, and lengthen product shelf life, NPs can be added to already prepared foods. The assistances of using these NPs in food processing include the capacity to regulate flavour release, create novel tastes and sensations, encapsulate food ingredients or additives, and improve the bioavailability of nutritional components. On the other hand, consumer acceptability and the examination of regulatory concerns will be key factors in determining if these improvements are successful. With the use of nanotechnology, food producers and manufacturers may significantly improve the safety of their products, and customers would similarly also benefit (Durán et al., 2013).

However, nanotechnology has its pros and cons that need to be verified for their enhanced utilization of various chemical synthesized NPs that have major environmental and ecological issues. In order to help overcome the aforementioned issues with physical and chemical synthesis, green synthesized NPs using plants and microbes have gained immense popularity worldwide, eliminating toxic chemicals and organic solvents used as reducing agents (Zhao et al., 2021). These biological systems produce a wide range of alkaloids, terpenoids, polyphenols and flavonoids that are not only excellent stabilizers but also reducing agents (Rotti et al., 2023; Ananda et al., 2022; Lavanya et al., 2024). In this emerging area, both pros and cons have impacts on the applications processing and packing food products. There are numerous safety concerns with respect to human health, and therefore complete data on the safety and risks associated with use of NPs in food is of high concern. With respect to these remarks and the recent development in the production and utilization of green synthesized NPs for their major role in food industry have been taken into account.

7.2 CLASSIFICATION OF NANOMATERIALS AND THEIR FOOD APPLICATIONS

The emerging potential to help activate the new features to be included in food and packaging items, nanomaterials are being thoroughly investigated. The major objectives for food packaging are to ensure food quality and its preservation, protection from contaminants, and keeping food fresh, with minimal food waste. The four main NPs subcategories used include inorganic-based NPs, carbon-based, organic-based NPs, and composite-based nanomaterials (see Figure 7.1).

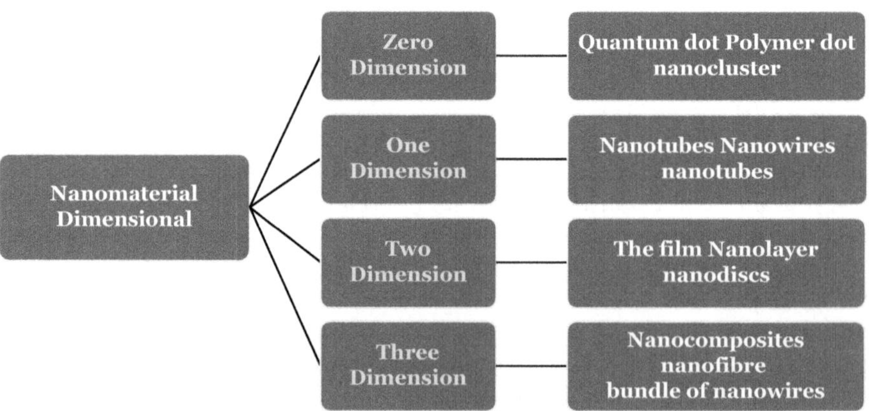

FIGURE 7.1 Dimensional classification of nanomaterials. (Compiled by the authors.)

7.2.1 Inorganic Nanoparticles

The NPs categorization of inorganic includes gold nanoparticles (AuNPs) and silver nanoparticles (AgNPs), and metal oxide used as different packaging materials (Hoseinnejad et al., 2018).

7.2.1.1 Metal-based NPs

AgNPs are well denoted for their antimicrobial properties and with hydroxypropyl methylcellulose polymer incorporated with Ag NPs demonstrated effective inhibition on different type of bacteria. While AuNPs may be effective against food-derived enzymatic reactions related to respiratory pathogenic bacteria, these enzymes act by changing their conformational leading to the damage of cells by disrupting DNA transcription, delaying the unwinding of DNA, and ultimately leading to cell death (Hoseinnejad et al., 2018).

7.2.1.2 Metal Oxide-based NPs

Due to their smaller size and large surface area, these are used in a range of applications, as biosensors, bio-nanotechnology, and nanomedicine. Copper oxide (CuO), titanium oxide (TiO_2), and silicon dioxide (SiO_2) are examples of metal oxide-based inorganic NPs. It has been noted that the lack of immunizations, pathogen mutation with rise in antibiotic resistance has posed significant health risks to people (Jadimurthy et al., 2022; Manjula et al., 2022; Adarsha et al., 2022). Antibacterial chemical development to battle against pathogens present in food—*Salmonella typhi, Clostridium perfringens,* and *Pseudomonas aeruginosa*—has significant goals in present research. Zinc oxide is sized down to the micron and nanometre range that acts as an antibacterial agent. Bactericidal effects are produced by ZnO-NPs that interact with the surface and nucleus of bacterial cells. These NPs have strong photochemical and catalytic activity, as well as antibacterial and antifungal characteristics (Webster and Seil, 2012).

With exposure to UV light, titanium dioxide (TiO_2) exhibits antimicrobial activity against a diversity of foodborne pathogens such as *Listeria monocytogenes*, *Salmonella sps*, and *Vibrio parahaemolyticus*. It has vast possible efficiency for reducing food safety risks in the food processing sector by forming reactive oxygen species (ROS) produced after absorbing light energy, aiding in microbial eradication. Specialists have proven the potential of TiO_2 NPs for application in food packaging and composite materials by changing their physical, chemical, and biological activities by inserting them into films (Zhu et al., 2018). Copper oxide (CuONPs) with their extensive and powerful antibacterial capabilities, inhibit bacterial, viral, and fungal growth. Due to these properties, it is the utmost capable metal oxide for use in packaged food items in industrial sectors. Silicon dioxide (SiO_2-NPs) is used for its antibacterial activity in packaging material when compared to other stored samples, as in shrimp packaged with LDPE-SiO_2 (Luo et al., 2015). Others include cans, bags, and bottles coated with silicon to make them non-stick (SBM Mayegowda et al., 2022).

7.2.2 Carbon Nanomaterials

There are several NPs made up of carbon including Graphene (GR), Carbon Dots (CDs), single- and multi-walled carbon nanotubes, Ordered Mesoporous Carbon (OMC).

7.2.2.1 Graphene

GR is used in development of novel food safety sensors applicable with enhanced polymers. GR composites have various features and functions combined with polymeric materials, with excellent performance for a variety of food safety applications. GR may produce more stable composites with PDA (polydopamine) and other biopolymers such as polyallylamine, polychitosan, and others (Ramakrishnappa et al., 2022; SB Mayegowda et al., 2022; Manjula et al., 2022a). Molecularly imprinted biomimetic sensors' performance is further improvised, and a wider range of applications for food safety are opened up by combining MIPs and ionic liquids in graphene. By coating AuNPs with reduced GR oxides, a novel electrochemical sensor with molecular impingement polymers (MIPs) for the detection of carbofuran with high efficiency and adsorption capabilities is being generated. Consequently, the AuNP were added to boost the precise surface area to improve the probe signal when using with immobilized molecularly weighted biosensor for the tebuconazole detection in food samples of fruits and vegetables (PB-AuNP) (Bahadır et al., 2016).

7.2.2.2 Carbon Nanotubes

Single-walled CNTs (SWCNTs) and multi-walled CNTs are two different forms of CNTs that can be distinguished based on how the graphene cylinders are arranged (MWCNTs). Catalytic activity, stability, electrical conductivity, and biocompatibility are just a few of the many electrochemical characteristics of SWCNTs and MWCNTs that have crucial applications for creating biological and chemical sensors for food safety concerns. These types of electrochemical sensors based on acetylcholinesterase (AChE) that are sensitive and reasonably priced in order to measure of pesticides in food and environmental specimen (Finegan, 1989). However, SWCNT is used for efficient detection of multiplex foodborne pathogen. MIPs have lately drawn a great deal of interest for their prospective role in the food matrix filtration, isolation, and analysis, medicinal and environment sampling (Barsan et al., 2016. In order to create composite NPs that enhance other nanomaterials, NPs of metals and their transition complexes can be effectively changed on the CNT surfaces to detect the biomimetic sensors in samples of food processing products (Chen et al., 2008).

7.2.2.3 Carbon Dots

Low toxicity, environmental friendliness, affordability, and an easy synthetic pathway are just a few of the advantageous features of CD. Fluorescence quenching of CDs is used to identify impurities. One crucial aspect of food safety is the identification and measurement of microorganisms. Regarding the creation of luminescence-based MIPs, food safety tests using certain CDs and MIPs have two major findings (Zhang et al., 2018).

7.2.2.4 Ordered Mesoporous Carbon

The large surface area with specificity, high absorbency, and variable pore size, tuneable substance and structure of the pore, aggregate simplicity and lack of physiological toxicity are extraordinary characteristics of carbon neutral materials (Ryoo et al., 1999).

Bionanomaterials being a diverse group can be categorized into natural and synthetic substances, based on their origins. Natural bionanomaterials are derived from living organisms, while synthetic bionanomaterials are produced through chemical or biological synthesis. Within these broad categories, they are further divided according to composition, structure, and possessions, as well as their potential applications. The classification of natural bionanomaterials is based on their composition, such as proteins and natural polymers composed of amino acids, and they have a broad spectrum of biological functions like collagen, silk, along with elastin, while polysaccharides include cellulose, chitin, and alginate and lipids are various naturally occurring groups of molecules contains hydrocarbons, oils, liposomes and phospholipids (Bao et al., 2013).

7.3 CLASSIFIED BASED ON THEIR PROPERTIES AND POTENTIAL APPLICATIONS

7.3.1 Biocompatible Bionanomaterials

These are designed to interact with living systems in a safe and non-toxic way and have attracted substantial attention in the medical arena considering their possibilities in many different applications, like drug delivery systems, biomedical implants and diagnostic tools. Polymers that are biocompatible, such as polyethylene glycol (PEG) and polyvinylpyrrolidone (PVP), are used to create a variety of materials like films and fibres, as well as scaffolds. These materials are non-toxic and non-immunogenic and are applied to create hydrogels that imitate the natural extracellular pattern and provide a supportive environment for cells (Reddy et al., 2021). Biocompatible ceramics, such as hydroxyapatite and bioactive glasses used to create scaffolds for tissue engineering, can be designed to mimic the natural mineral composition of bone. These materials can also be used to create coatings for biomedical implants to improve integration with the surrounding tissue (Fiume et al., 2021). Liposomes are spherical structures made of lipids that can encapsulate drugs or other materials and have been demonstrated to be biocompatible and non-toxic when employed as medication delivery devices, in biomedical imaging and as biosensors. Protein-based materials made from proteins like collagen and gelatin are used to create scaffolds in tissue engineering and wound healing (Chen et al., 2015). These materials can be broken down by enzymes, which are not harmful and are often produced from natural sources, for example cow hide, pig skin, or fish skin (Bozzuto et al., 2015)

AuNPs are biocompatible and, along with biomolecules such as antibodies or peptides, can be used to target specific cells or tissues. They can be used for imaging as contrast agents and as drugs carriers, and can be easily visualized by various imaging techniques (Nune et al., 2009). Biocompatible NPs are crucial in various applications as they minimize the risk of adverse reactions and support the integration of materials with host tissue. Moreover, they can aid in tissue repair and regeneration by creating scaffolds that replicate the natural extracellular matrix and promote the creation of novel tissue (Mostafavi et al., 2020). It is vital to remember that biocompatibility is a complex and multi-factorial property, and not all of them are inherently biocompatible, as it depends on the materials properties and the system in which it is used. Therefore, extensive testing and evaluation are needed to assure that they are suitable for use in medical applications and do not produce dangerous and harmful responses (Witika et al., 2020).

7.3.2 Biodegradable Bionanomaterials

These include bionanomaterials that can be broken down by natural processes such as enzymes, microbes, or by physical means such as UV-light or heat. These materials are designed to be more environmentally friendly than traditional materials as they are eliminated after use, reducing the potential for pollution and harming the environment (Ahmann and Dorgan, 2007). Biodegradable polymers such as polylactic acid (PLA) and polyhydroxyalkanoates (PHAs) are examples of these and are used to create a variability of constituents as films, fibres, and scaffolds. These materials are made from renewable resources, like maize starch, and are degraded by microbes in the presence of oxygen (Nduko et al., 2021). Liposomes used as drug delivery systems, biomedical imaging, biosensors while, protein-based materials are used for wound healing and tissue engineering (Bozzuto et al., 2015; Chattopadhyay et al., 2014).

Chitosan, a biodegradable polymer produced from chitin, a crustacean shell component, has been utilized to make scaffolds for tissue engineering, wound healing, and drug delivery (Thandapani et al., 2017). Being the primary component of plant cell walls, cellulose has been modified to create biodegradable nanocellulose, a 3D structured material with high mechanical strength with a wide range of applications. These have gained attention as potential benefits over traditional materials with varied applications such as biomedical implants and packaging, and broken down in the environment without creating pollution or harmful effects (Norizan et al., 2022).

7.3.3 Functional Bionanomaterials

These constitute a subset with specific properties and functions, such as mechanical, electrical, optical, or magnetic properties. For example, quantum dots offer distinctive optical features, like highly structured fluorescence and strong-colour tunability, for use in biomedical imaging, sensing and diagnostics. Quantum dots have been employed in-vivo to track cell mobility and identify the presence of certain biomolecules. Small hollow tubes made of carbon atoms are carbon nanotubes (CNTs) with excellent electrical, mechanical, and thermal characteristics and a variety of applications, including computerized electronics, energy storage as well as production, and biotechnology. CNTs have been used to create lightweight, strong, and flexible/elastic materials, to enhance the performance of batteries, solar cells, and fuel cells, and to create new biosensors and imaging agents. Graphene, a 2-D substance consisting of only one layer of carbon atoms with exceptional properties, is widely studied for potential applications in electronics, and energy. Graphene has been used to create new types of transistors, solar cells, and batteries, and has potential for biosensing and drug delivery applications (Butler et al., 2013).

Magnetic NPs allow them to respond to magnetic fields and are used in medicine, and environmental science to create new contrast agents for imaging, to deliver drugs specifically to cancer cells, and to remove pollutants from water and soil (Flores-Rojas et al., 2022). AuNPs are tiny particles made of gold with distinctive optical and electronic characteristics that are employed in biological applications such as diagnostics and drug delivery. Functionalized biomolecules can bind to specific targets, such as cancer cells, and used as contrast mediators in imaging process, and as carriers for drugs (Wang et al., 2016).

a. **Functional bionanomaterials:** These could have a potential impact on atmosphere as well as on mankind, thus it is significant to conduct extensive safety tests before use in applications. Additionally, these materials are often synthesized using complex chemical methods, which could pose environmental risks due to the production and disposal of the chemicals used (Gupta et al., 2018).
b. **Medical bionanomaterials:** Designed to mimic or interact with biological systems at the nanoscale. A lot of interest has been observed in the medical field because of their potential towards revolutionizing the way we diagnose and treat disease. It is used to create extremely sensitive and specific symptomatic tests, considering a wide range of diseases. AuNPs may be implemented through specific biomolecules, like antibodies that bind to disease markers and generate a signal that can be easily detected. This approach can be used to create rapid and low-cost diagnostic tests for conditions like infectious diseases and cancer. NPs are used to deliver medications directly to the location of an illness. They are made of lipids or polymers loaded with drugs to specific target cells or tissues using biomolecules such as antibodies or peptides with increased drugs efficacy and reduce their side effects by ensuring that they are delivered directly to the cells that need them (Nune et al., 2016). Also used to create scaffolds that support growth of new tissue by developing nanofiber scaffolds resembling the native extracellular matrix as well as provide supportive conditions for cells to grow and differentiate. These scaffolds can be used to repair or regenerate damaged tissue, such as bone, cartilage, or skin (Borehalli et al., 2022, a).

   NPs created visualizing agents reveal molecular and cellular mechanisms of disease at nanometer scale. AuNPs and quantum dots can be functionalized with biomolecules that bind to specific targets, such as proteins or cells, used to track these targets in vivo using techniques such as fluorescence imaging or X-ray computed tomography (Wang et al., 2016). Despite its potential, there are significant safety problems with its usage as well. Since these are so tiny, they may readily pass through biological membranes and penetrate cells, potentially causing toxicity. Therefore, it is important to conduct extensive safety testing before use in medical applications. Overall, an encouraging field of study with the potential to transform the way we diagnose and treat diseases with the development of new NPs and their application is an active and exciting field of research expected to produce significant medical advancements in the near future (Sabourian et al., 2020).

c. **Environmental bionanomaterials**: The field of environmental science has gained immense popularity by the use of NPs because of their ability to handle a variety of environmental challenges. Heavy metals, microbes, and pesticides are among the contaminants removed from water treatment by them. Researchers have developed NPs such as Fe2O3, TiO_2, or ZnO, that adsorb pollutants from water and that can be effortlessly recovered and reprocessed (Archana et al., 2021; Shilpa et al., 2022). They are used to remove pollutants from air, such as volatile organic compounds and particulate matter. Developed catalytic NPs, such as titanium dioxide, degrade pollutants in the air when exposed to sunlight. NPs help to remove pollutants from soil, for example, biological compounds and heavy metals, cleaning of contaminated sites, promotes growth of microbes that can break down pollutants such as oils and chlorinated compounds (Mayegowda et al., 2023). The sensors and monitoring environmental pollutants are engineered to bind to specific pollutants generating signals to detect heavy metals or organic compounds (Willner and Vikesland, 2018). NPs use raises certain safety issues as there is little understanding on the long-term fate and effects of their persistence, accumulation and toxicity. To ensure safety in the environment, understand their potential impacts and to develop regulations and guidelines is necessary (Ray et al., 2009).

7.4 BIONANOMATERIALS IN FOOD APPLICATIONS

Nanotechnology in food-related applications has provided several advantages to consumers, including potential depletion in the usage of salt, preservatives, surfactants, and fat in food items, as well generate novel or superior flavors, textures, and oral experiences through nanometre scale processing of food. As compared to their bulk equivalents, nano-formulations have enhanced mineral and supplement uptake, absorption, and bioavailability in the cells (Jafari and McClements, 2017). Polymer composites produced from NPs provide new, lightweight, yet robust food packaging solutions that can keep food goods secure throughout transit, fresher for extended periods of storage, and free of microbiological infections.

Antimicrobial nano-coatings over food preparation items may help keep food clean throughout processing, while "smart" labeling can help safeguard the food item's safety and validity throughout the supply network/chain. The anticipated advantages and presence of nanotechnology in food and associated areas are still an innovative, developing application in most countries, and despite a constant increase in the number of accessible items, the vast majority of new breakthroughs are currently in research and development or just near-market phases.

7.4.1 Food Processing

The present and near-term expected nanotechnology uses in the food industry have been recognized in recent years with assessments. The primary areas include the food packaging of the food items that have ingredients and additives which are nanosized or nanoencapsulated (Figure 7.2). The food industries are concerned with the ingredients, packaging, and food analysis techniques (Balassa et al., 1971). Few of the aspects have already been working with the NPs applications of greater potential that will help for better processing, packaging, and storage of food (Sing et al., 2017). The fact that at the nanoscale, a variety of food structures are typically present, nanofood has indeed been a component of food preparation for ages. The goals with nanofood are to increase food safety, enhance nutrients and flavour, and lower costs. The present nanotechnology applications in arena of food science allow for the speedy, sensitive, and labor-efficient detection of food pathogens using nanosensors.

In light of the aforementioned considerations, it might be obvious that nanotechnology can help ensure a rather more reliable supply of premium food for the entire world's inhabitants.

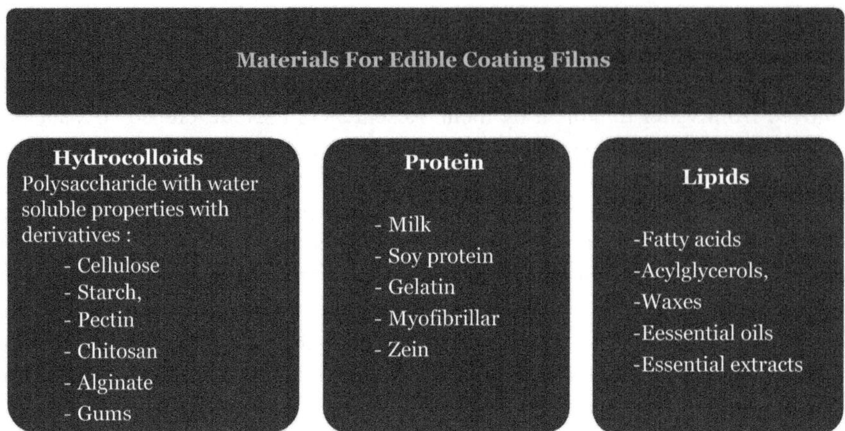

FIGURE 7.2 Different materials used for coating food as edible films. (Compiled by the authors.)

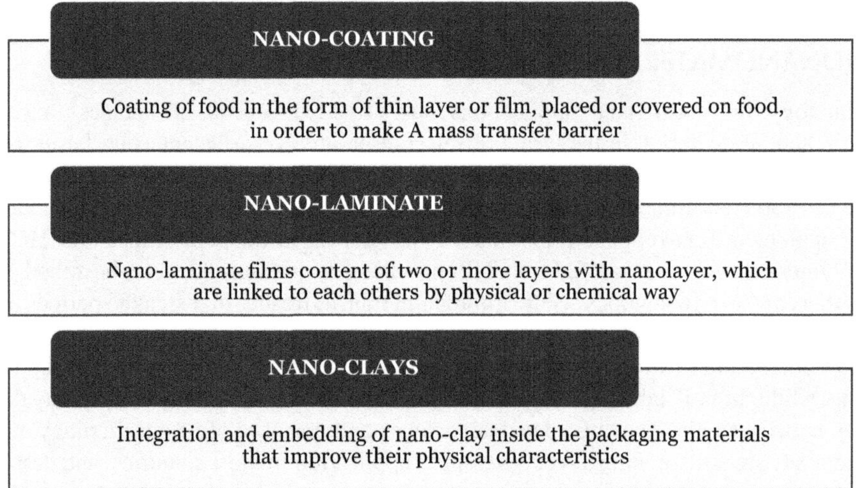

FIGURE 7.3 Nanoparticles in different forms used for preservation and packaging of food. (Compiled by the authors.)

Food preparation and consumption, however, encompass much than simply the logical provision of sufficient nutrition (Maleta, 2006). The propensity of molecular building blocks are specially formed accumulates is frequently used in nanotechnology that includes the amphiphilic molecules' ability to self-assemble into micelles at low quantities and lamellar mesophases at high levels of certain proteins molecules. The encapsulation process has helped to transport nutrients to specific areas of the gastrointestinal walls in a great number of other instances as NPs technique allows for the accurate assembly of a structure from individual components, molecules, and it is obvious that they can be used in the preparation of these capsules. These encapsulated NPs are typically 1–10 m in size, with a large interfacial area, making them difficult to effectively enclose. On the other hand, probiotics are made up of bacterial cells. Vesicles are not a food matrix in and of themselves, but they can be used as capsules for certain substances to be mixed into a cuisine. Figure 7.3 shows nanoparticles in different forms used for preservation and packaging of food.

7.4.2 Edible Nanocoating

These serve as a protective container whether or not they are a component of the food products. Edible food coatings can be applied directly to the product to enhance food with good taste, colour, enzymes, antioxidants, and anti-browning substances (Nile et al., 2020). Edible nano-coatings of 5 nm thin with the addition of chemicals like antioxidants and anti-browning are generally used in the meat processing, cheese, baking and agricultural sectors to protect fruit and vegetable products and so forth. Additionally, nanoclays for their mechanical, thermal, barrier abilities as well as inexpensiveness, are widely accepted and investigated for casting of food. The components for biodegradable coatings films fabrication include polysaccharides like chitosan, pullulan, cellulose, alginate, starch, carrageenan, pectin, and kefiran that are primarily investigated to create coating films that are effective barriers against the passage of O_2 and CO_2 (Cazón et al., 2017). Spraying, dipping, or rubbing edible nanocoating on food are simple ways to apply them, and these are often eco-friendly and do not need to be removed from food before consumption (Zambrano-Zaragoza et al., 2018).

7.4.3 Biosensors

Biosensors created from the fusion have the ability to process food safely reducing the health risks and financial burden of foodborne illness. They can quickly identify microorganisms, potential contaminants, and chemical/biological substances in case of a possible spread. (Pérez-López and Merkoçi, 2011). In smart packaging, food pathogens as well as chemicals are screened for using polymers and nanosensors screening, food storage and transportation operations. They are capable of detecting environmental deviations, like temperature, humidity and so forth, produced by the development of microorganisms and waste products from food deterioration. *Bacillus cereus*, *Vibrio parahaemolyticus*, and *Salmonella spp.* are among the foodborne organisms that can be found and identified by polymer nanocomposites made of carbon black and polyaniline (Arshak et al., 2007).

Nanosensors can also be attached to packages to identify microbes that contaminate food. Instead of sending packaged food to a lab for sampling, the sensor shows the nutritional content, where the consumers directly interpret built on colour differences. Nanosensors detect food contaminants and pathogens in food are most potential application of nanotechnology (Thiruvengadam et al.,2018). Food packaging considered to be the main application in the food industry focuses more on using advanced techniques like nano casting which can be engineered for enzyme release, flavours, antimicrobials, antioxidants, and nutritional supplements to extend the food shelf life. Bioactive packaging materials support food oxidation and, hence, halt the emergence of off-flavours also unfavourable textures by NP-based utilising biosensor for food monitoring. For quicker electron transfer and greater specificity from biomolecules, electrode surfaces have been modified with novel NPs in recent years (Pérez et al., 2009). This benefit has inspired research to combine NPs-based biosensors with biomolecules with advantages among various fields like food quality, clinical investigation, and environmental control.

Biosensors certainly offer the following advantages with reduced overpotentials for guaranteeing reversibility of certain redox reactions that are irreparable with unaltered electrodes, many analytically significant electrochemical reactions, and opens up new labelling possibilities, including multi-detection capabilities. Food analysis of pathogens like *Salmonella* and *E. coli*, pesticides, sugar, and so forth can be detected. Optical biosensors are important sensing instruments with a range of uses in agricultural industry, biomedical, healthcare, pharmaceutical, and outdoor surveillance. NPs such as Au NPs, Ag NPs, Fe3O4 MNPs, and QDs possess exceptional optical properties, making them excellent optical labels to enhance the optical detector surfaces' sensitivity in biosensors. Optical converters are particularly attractive for the development of robust gadgets that are simple to use, portable, and possibly inexpensive analytical systems (Narayanaswamy and Wolfbeis, 2004).

7.5 REGULATION AND SAFETY MANAGEMENT: RISK ASSESSMENT

Bionanomaterials are made from biological components, such as proteins or DNA, and have nano-scale dimensions with unique properties and possible uses in the areas of medicine, energy as well as other disciplines. The advancements in the domains of enhancing food quality, extending food shelf life, alter the structure, composition of the organic compounds, and producing edible films in fruits and veggies are all examples of food preservation techniques. Nanotechnology is extensively employed in the food industry (Barhoum et al., 2022). Since organic chemicals have no negative effects on people or the environment, encapsulating bioactive substances will boost the product's stability to control the pesticides. The best outcomes for extending food's shelf life came from the liposomes applied and used in the food industry. Additionally, it gives the food good flavour and nutrients (Zabot et al., 2022). However, their small size and unique properties also raise concerns about their safety and potential risks. Nanotechnology fosters innovation for better products, processes, preservation, and food analysis. However, given the presence of toxicity, concerns about health aspects and their impact on the environment may arise. Future efforts should be made to ensure that nanotechnology is used safely and effectively (Sing et al., 2017).

A crucial component of ensuring the security and responsible growth of biological NPs is their regulation. In the United States, bionanomaterials are currently governed by numeral governmental bodies, like the Occupational Safety and Health Administration (OSHA), the Environmental Protection Agency (EPA), and the Food and Drug Administration (FDA). These agencies have their own specifications and regulative guidelines that apply on biological-derived NPs (Poirot-Mazères, 2011). The FDA is responsible for regulations to be used in medical applications, such as diagnostic tests and therapeutic products with several regulations in place for medical products that contain or are made from NPs, in addition to the Public Health Service Act (PHSA), the Federal Food, Drug, and Cosmetic Act (FD&C Act) as well as the Medical Device Amendments of 1976 (*Biotechnology Law Report*, FDA, 2011). These regulations require that products containing or made from bionanomaterials are safe and effective, and can be manufactured and distributed under good manufacturing practices (GMPs).

The Environmental Protection Agency (EPA) is accountable for modifiable products and regulates the NPs used in pesticides and other environmental petitions under the Federal Insecticide, Fungicide, and Rodenticide Act (FIFRA) and the Pesticide Registration Improvement Act (PRIA). These regulations require the ecosystem and human health be unaffected by bionanomaterials that might be present or constitutes the pesticides, and that before they can be disseminated or sold, they must be registered with the EPA (EPA, 1989).

Bionanomaterials in the workplace are governed by the Occupational Safety and Health Act (OSHA), including ensuring the safety of workers who may be exposed to these materials. OSHA has several regulations for use and handling in the workplace that are secure, including the OSHA and the Hazard Communication Standard (HCS). These regulations require employers to offer their employees a secure and healthy work atmosphere, including protecting them from hazards associated with these bionanomaterials. Safety management is crucial for assuring the responsible development and usage of these biomaterials that involves a combination of laboratory practices, risk assessments, and regulations (Gergely et al., 2010). The proper laboratory practices are essential to the safe handling, storage, and disposal of bionanomaterials that include using appropriate personal protective equipment, handling materials safely, and disposing of materials properly. This also includes ensuring the laboratory is equipped with the necessary safety equipment, such as fume hoods, eye wash stations, and emergency showers in the US National Research Council Committee on Prudent Practices in the Laboratory (American Chemical Society, 2011).

Risk assessment is a major element of the safety administration to evaluate the potential hazards related with these ingredients and identify measures to control or mitigate those hazards. The risk assessment process for biological NPs typically involves several key steps, including hazard identification, exposure assessment, and risk characterization (Warheit, 2018). Identifying potential risks associated with the bionanomaterial is the first stage in the risk-assessment process, such as its toxicity, reactivity, and flammability. This step also includes an evaluation of the material's qualities that can affect its potential for toxicity and other risks include its size, form, and surface chemistry. The next step is exposure assessment, identifying the routes of exposure that could lead to harm, such as inhalation, ingestion, and dermal contact also includes an evaluation of the potential for exposure, taking into account factors such as the intended use of the bionanomaterial, the population that may be exposed, and the environment in which the exposure may occur. The final step is the risk-characterization process, which involves the evaluation of possible dangers connected to the bionanomaterial, taking into account the information gathered during the steps for identifying hazards and evaluating exposure (Warheit, 2018). It also includes an evaluation of the uncertainties associated with the risk assessment, such as lacking information on the characteristics of the substance or the potential for exposure. It is important to note that bionanomaterials are a relatively new field, and the risk assessment of these materials needs to be addressed (Laux et al., 2018). The method of assessing risk is uncertain and lacking in data. In addition, the risk assessment process may also involve other activities, such as the development of risk management strategies, implementation of risk communication plans, monitoring, and evaluation of the material's safety over time (Isigonis et al., 2019).

Nanomaterials are mostly used in environmental remediation to remove contaminants from water and make it pure and drinkable, which is a significant problem in society. Investigators have begun to inspect saltwater as a significant supply of drinking water, however desalinating seawater is a costly procedure. In this sense, CNT-based membranes may lower desalination costs, and CNT nanofilters may be the most effective at removing impurities from surface or ground water. Nanosensors are also capable of precisely detecting pollution. Air pollution is the next significant environmental issue for which nanomaterials are utilized as a remedy. Here, different nanomaterials were employed to filter the air using a variety of approaches. Similar to this, nanofibers are employed in industry smokestacks or automotive tailpipes to isolate pollutants and stop them from entering the open air. The extremely slight leaking of dangerous gases can also be found using nanosensors (Rahman et al., 2022). Nanomaterials have several exciting applications in air filtration to remove the contaminant: related techniques including degradation and sequestration are utilized. Last, but not least, due to their superior quality and enhanced analytical capabilities, the nanoproducts need to be classified and named in order to answer the needs of consumers and logistical issues, which will eventually result in a speedy financial clearance. Topics must often schedule special meetings to examine the use of nanotechnology that is conceivable; this effort will intensify in the future.

7.6 FUTURE PERSPECTIVE

Although NPs are produced all over the world, very few nations have the requisite regulatory framework in place to allow for the use in food items. The applications lie on the efficacy of biological nanosystems that is challenging and needs further scientific investigations. As food additives, these NPs can have serious health issues when used in food packaging. Further, with a constant risk of entering through food chain, that can result in DNA damage, disruption of cell membranes, and cell death. A vivo research needs to be carried to determine on nanofoods affecting both human and animal health. In order to promote customer acceptance, proper labelling and restrictions should be recommended for the marketing and needs to be handled, regulated properly for significant contribution in enhancing food processing and product quality. While toxicity research is still in its early stages, research on green synthesised NPs are to be carried for their potentiality, methodologies and biotransformation. Future efforts should focus on technical verification as well as ensuring the sensor's constancy and permanent receptacle under environmental restrictions. Another problem to be better understood is the movement of NPs from food containers through the environment. In addition, consumers are reluctant to employ nanofoods due to the lack of regulations governing their use in sustenance. NPs used as a nanosensor warn a customer if a product is no longer suitable for consumption. For instance, most people find barcodes for guaranteed food security to be beneficial, since they have no negative effects on a person's health when used for antibacterial milk bottles for infants (Singh et al., 2023). Several nations have a regulatory structure for managing dangers related as a result of the rising regulatory issues. Legal nanotechnological uses require comprehensive government regulations and laws as well as strict toxicological screening procedures that are critically necessary to establish an international regulatory framework for the control of nanoparticle use in the food business.

7.7 CONCLUSION

Green synthesized NPs, a rapidly growing area of research, has the potential to revolutionize many different areas, including biomedicine with drug transport, tissue engineering, and biomedical devices, food industries, and environmental bioremediation. These materials are designed to interact with living systems in a safe and non-toxic way, minimizing the risk of adverse reactions, and supporting the integration of the materials with host tissue. However, as with any material intended for use in the human body, biocompatibility testing is necessary to guarantee these products' effectiveness and safety. These nanomaterials that are non-toxic and do not elicit an adverse immune response when in contact with living tissues. Polyethylene glycol (PEG) and polyvinyl alcohol (PVA) are a few examples. In conclusion, green synthesized NPs are a rapidly growing area of research that has the ability to transform a variety of fields, from electronics, renewable energy, and biomedicine. As this field of study advances, scientists are working to create new functional bionanomaterials and to find new ways to use existing materials to deal with some of the most urgent issues that the community is currently facing, including genetically engineered bionanomaterials to perform specific functions, such as delivering drugs or sensing specific molecules need to be addressed.

REFERENCES

Adarsha JR, Ravishankar TN, Ananda A, Manjunatha CR, Shilpa BM, Ramakrishnappa T (2022) Hydrothermal synthesis of novel heterostructured Ag/TiO_2/$CuFe_2O_4$ nanocomposite: Characterization, enhanced photocatalytic degradation of methylene blue dye, and efficient antibacterial studies. *Water Environment Research* 94(6). https://doi.org/10.1002/wer.10744

Ahmann D, Dorgan JR (2007) Bioengineering for pollution prevention through development of biobased energy and materials state of the science report. *Industrial Biotechnology* 3(3):218–259. https://doi.org/10.1089/ind.2007.3.218

American Chemical Society (2011) Book listing of prudent practices in the laboratory: Handling and management of chemical hazards. *Journal of the American Chemical Society*, 133(39):15795–15795. https://doi.org/10.1021/ja208310g

Ananda A, Ramakrishnappa T, Archana S, Reddy Yadav LS, Shilpa BM, Nagaraju G, Jayanna BK (2022) Green synthesis of MgO nanoparticles using Phyllanthus emblica for Evans blue degradation and antibacterial activity. *Materials Today: Proceedings* 49:801–810. https://doi.org/10.1016/j.matpr.2021.05.340

Archana S, Jayanna BK, Ananda A, B.M S, Pandiarajan D, Muralidhara HB, Kumar KY (2021) Synthesis of nickel oxide grafted graphene oxide nanocomposites – A systematic research on chemisorption of heavy metal ions and its antibacterial activity. *Environmental Nanotechnology, Monitoring & Management* 16:100486. https://doi.org/10.1016/j.enmm.2021.100486

Arshak K, Adley C, Moore E, Cunniffe C, Campion M, Harris J (2007) Characterisation of polymer nanocomposite sensors for quantification of bacterial cultures. *Sensors and Actuators B: Chemical* 126(1):226–231. https://doi.org/10.1016/j.snb.2006.12.006

Bahadır EB, Sezgintürk MK (2016) Applications of graphene in electrochemical sensing and biosensing. *TrAC Trends in Analytical Chemistry* 76:1–14. https://doi.org/10.1016/j.trac.2015.07.008

Balassa LL, Fanger GO, Wurzburg OB (1971) Microencapsulation in the food industry. *C R C Critical Reviews in Food Technology* 2(2):245–265. https://doi.org/10.1080/10408397109527123

Bao Ha TL, Minh T, Nguyen D, Minh D (2013) Naturally Derived Biomaterials: Preparation and Application. In: Andrades JA (ed) *Regenerative Medicine and Tissue Engineering*. InTech. https://doi.org/10.5772/55668

Barhoum A, García-Betancourt ML, Jeevanandam J, Hussien EA, Mekkawy SA, Mostafa M, Omran MM, S. Abdalla M, Bechelany M (2022) Review on natural, incidental, bioinspired, and engineered nanomaterials: History, definitions, classifications, synthesis, properties, market, toxicities, risks, and regulations. *Nanomaterials* 12(2):177. https://doi.org/10.3390/nano12020177

Bozzuto G, Molinari A (2015) Liposomes as nanomedical devices. *International Journal of Nanomedicine* 975. https://doi.org/10.2147/IJN.S68861

Butler SZ, Hollen SM, Cao L, Cui Y, Gupta JA, Gutiérrez HR, Heinz TF, Hong SS, Huang J, Ismach AF, Johnston-Halperin E, Kuno M, Plashnitsa VV, Robinson RD, Ruoff RS, Salahuddin S, Shan J, Shi L, Spencer MG, Terrones M, Windl W, Goldberger JE (2013) Progress, challenges, and opportunities in two-dimensional materials beyond graphene. *ACS Nano* 7(4):2898–2926. https://doi.org/10.1021/nn400280c

Cazón P, Velazquez G, Ramírez JA, Vázquez M (2017) Polysaccharide-based films and coatings for food packaging: A review. *Food Hydrocolloids* 68:136–148. https://doi.org/10.1016/j.foodhyd.2016.09.009

Chattopadhyay S, Raines RT (2014) Collagen-based biomaterials for wound healing. *Biopolymers* 101(8):821–833. https://doi.org/10.1002/bip.22486

Chen H, Zuo X, Su S, Tang Z, Wu A, Song S, Zhang D, Fan C (2008) An electrochemical sensor for pesticide assays based on carbon nanotube-enhanced acetycholinesterase activity. *Analyst* 133(9):1182. https://doi.org/10.1039/b805334k

Chen P-H, Liao H-C, Hsu S-H, Chen R-S, Wu M-C, Yang Y-F, Wu C-C, Chen M-H, Su W-F (2015) A novel polyurethane/cellulose fibrous scaffold for cardiac tissue engineering. *RSC Adv* 5(9):6932–6939. https://doi.org/10.1039/C4RA12486C

Couto C, Almeida A (2022) Metallic nanoparticles in the food sector: A mini-review. *Foods* 11(3):402. https://doi.org/10.3390/foods11030402

Durán N, Marcato PD (2013) Nanobiotechnology perspectives. Role of nanotechnology in the food industry: A review. *Int J Food Sci Technol* 48(6):1127–1134. https://doi.org/10.1111/ijfs.12027

EPA (1989) *Risk assessment guidance for superfund: Pt. A. Human health evaluation manual*. Office of Emergency and Remedial Response, U.S. Environmental Protection Agency.

Finegan PA (1989) FIFRA Lite: A regulatory solution or part of the pesticide problem? *Pace Environmental Law Review* 6(2):615. https://doi.org/10.58948/0738-6206.1505

Fiume E, Magnaterra G, Rahdar A, Verné E, Baino F (2021) Hydroxyapatite for biomedical applications: A short overview. *Ceramics* 4(4):542–563. https://doi.org/10.3390/ceramics4040039

Flores-Rojas GG, López-Saucedo F, Vera-Graziano R, Mendizabal E, Bucio E (2022) Magnetic nanoparticles for medical applications: Updated review. *Macromology* 2(3):374–390. https://doi.org/10.3390/macromol2030024

Gergely A, Chaudhry Q, Bowman DM (2010) Regulatory Perspectives on Nanotechnologies in Foods and Food Contact Materials. In: Hodge GA. Bowman DMJ.S, Maynard A.D. (eds) *International Handbook on Regulating Nanotechnologies*. Edward Elgar.

Gupta R, Xie H (2018) Nanoparticles in daily life: Applications, toxicity and regulations. *Journal of Environmental Pathology Toxicology and Oncology* 37(3):209–230. https://doi.org/10.1615/JEnvironPatholToxicolOncol.2018026009

Hoseinnejad M, Jafari SM, Katouzian I (2018) Inorganic and metal nanoparticles and their antimicrobial activity in food packaging applications. *Critical Reviews in Microbiology* 44(2):161–181. https://doi.org/10.1080/1040841X.2017.1332001

Isigonis P, Hristozov D, Benighaus C, Giubilato E, Grieger K, Pizzol L, Semenzin E, Linkov I, Zabeo A, Marcomini A (2019) Risk governance of nanomaterials: Review of criteria and tools for risk communication, evaluation, and mitigation. *Nanomaterials* 9(5):696. https://doi.org/10.3390/nano9050696

Jadimurthy R, Mayegowda SB, Nayak SC, Mohan CD, Rangappa KS (2022) Escaping mechanisms of ESKAPE pathogens from antibiotics and their targeting by natural compounds. *Biotechnology Reports* 34:e00728. https://doi.org/10.1016/j.btre.2022.e00728

Jafari SM, McClements DJ (2017) Nanotechnology Approaches for Increasing Nutrient Bioavailability. In: Toldrá F (ed) *Advances in Food and Nutrition Research,* Volume 81. Elsevier, pp 1–30.

Laux P, Tentschert J, Riebeling C, Braeuning A, Creutzenberg O, Epp A, Fessard V, Haas K-H, Haase A, Hund-Rinke K, Jakubowski N, Kearns P, Lampen A, Rauscher H, Schoonjans R, Störmer A, Thielmann A, Mühle U, Luch A (2018) Nanomaterials: Certain aspects of application, risk assessment and risk communication. *Archives of Toxicology* 92(1):121–141. https://doi.org/10.1007/s00204-017-2144-1

Luo Z, Xu Y, Ye Q (2015) Effect of nano-SiO2-LDPE packaging on biochemical, sensory, and microbiological quality of Pacific white shrimp Penaeus vannamei during chilled storage. *Fish Science* 81(5):983–993. https://doi.org/10.1007/s12562-015-0914-3

Maleta K (2006) Undernutrition. *Malawi Medical Journal* 18(4):189–205

Manjula NG, Sarma G, Shilpa BM, Suresh Kumar K (2022) Environmental Applications of Green Engineered Copper Nanoparticles. In: Shah MP, Roy A (eds) *Phytonanotechnology*. Springer Nature Singapore, Singapore, pp 255–276.

Mayegowda SBM, Bhoomika S, Nagshetty K, Manjula NG (2022) Environmental Adequacy of Green Polymers and Biomaterials. In: Agarwal P., Tripathy DB, Gupta A, Kuanr BK (eds) *Polymeric Biomaterials*, 193–214. CRC Press.

Mayegowda SB, Ng M, Alghamdi S , Atwah B, Alhindi Z, Islam F (2022) Role of antimicrobial drug in the development of potential therapeutics. *Evidence-Based Complementary and Alternative Medicine* 2022:1–17. https://doi.org/10.1155/2022/2500613

Mayegowda SB, Sarma G, Gadilingappa MN, Alghamdi S, Aslam A, Refaat B, Almehmadi M, Allahyani M, Alsaiari AA, Aljuaid A, Al-Moraya IS (2023) Green-synthesized nanoparticles and their therapeutic applications: A review. *Green Processing and Synthesis* 12(1):20230001. https://doi.org/10.1515/gps-2023-0001

Mayegowda SB, Sureshkumar K, Yashaswini R, Ramakrishnappa T (2022b) Phytonanotechnology for the removal of pollutants from the contaminated soil environment. In: Shah MP, Roy A (eds) *Phytonanotechnology*. Springer Nature Singapore, Singapore, pp 319–336.

Mostafavi E, Medina-Cruz D, Kalantari K, Taymoori A, Soltantabar P, Webster TJ (2020) Electroconductive nanobiomaterials for tissue engineering and regenerative medicine. *Bioelectricity* 2(2):120–149. https://doi.org/10.1089/bioe.2020.0021

Narayanaswamy R, Wolfbeis OS (2004) *Optical sensors: Industrial, environmental, and diagnostic applications*. Springer, Berlin; New York.

Nduko JM, Taguchi S (2021) Microbial production of biodegradable lactate-based polymers and oligomeric building blocks from renewable and waste resources. *Frontiers in Bioengineering and Biotechnology* 8:618077. https://doi.org/10.3389/fbioe.2020.618077

Nile SH, Baskar V, Selvaraj D, Nile A, Xiao J, Kai G (2020) Nanotechnologies in food science: Applications, recent trends, and future perspectives. *Nano-Micro Letters* 12(1):45. https://doi.org/10.1007/s40820-020-0383-9

Norizan MN, Shazleen SS, Alias AH, Sabaruddin FA, Asyraf MRM, Zainudin ES, Abdullah N, Samsudin MS, Kamarudin SH, Norrrahim MNF (2022) Nanocellulose-based nanocomposites for sustainable applications: A review. *Nanomaterials* 12(19):3483. https://doi.org/10.3390/nano12193483

Nune KC, Misra RDK (2016) Biological activity of nanostructured metallic materials for biomedical applications. *Materials Technology* 31(13):772–781. https://doi.org/10.1080/10667857.2016.1225148

Nune SK, Gunda P, Thallapally PK, Lin Y-Y, Laird Forrest M, Berkland CJ (2009) Nanoparticles for biomedical imaging. *Expert Opinion on Drug Delivery* 6(11):1175–1194. https://doi.org/10.1517/17425240903229031

Ossipov DA, Brännvall K, Forsberg-Nilsson K, Hilborn J (2007) Formation of the first injectable poly(vinyl alcohol) hydrogel by mixing of functional PVA precursors. *Journal of Applied Polymer Science* 106(1):60–70. https://doi.org/10.1002/app.26455

Patra JK, Shin H-S, Paramithiotis S (2018) Editorial: Application of nanotechnology in food science and food microbiology. *Frontiers in Microbiology* 9: 714. https://doi.org/10.3389/fmicb.2018.00714

Pérez S, Farré Marinel la, Barceló D (2009) Analysis, behavior and ecotoxicity of carbon-based nanomaterials in the aquatic environment. *TrAC Trends in Analytical Chemistry* 28(6):820–832. https://doi.org/10.1016/j.trac.2009.04.001

Pérez-López B, Merkoçi A (2011) Nanomaterials based biosensors for food analysis applications. *Trends in Food Science & Technology* 22(11):625–639. https://doi.org/10.1016/j.tifs.2011.04.001

Poirot-Mazères I (2011) Chapitre 6. Legal aspects of the risks raised by nanotechnologies in the field of medicine. *Journal International de Bioéthique* 22(1):99. https://doi.org/10.3917/jib.221.0099

Rahman S, Bozal-Palabiyik B, Unal DN, Erkmen C, Siddiq M, Shah A, Uslu B (2022) Molecularly imprinted polymers (MIPs) combined with nanomaterials as electrochemical sensing applications for environmental pollutants. *Trends in Environmental Analytical Chemistry* 36 :e00176. https://doi.org/10.1016/j.teac.2022.e00176

Ramakrishnappa SBM, Sureshkumar K, Sahana M, Ramakrishnapp T (2022) pH and Thermo-Responsive Systems. In: Agarwal P., Tripathy DB, Gupta A, Kuanr BK (eds) *Polymeric Biomaterials*, 63–83. CRC Press. https://doi.org/10.1201/9781003240884-4.

Ray PC, Yu H, Fu PP (2009) Toxicity and environmental risks of nanomaterials: challenges and future needs. *Journal of Environmental Science and Health, Part C* 27(1):1–35. https://doi.org/10.1080/10590500802708267

Reddy MSB, Ponnamma D, Choudhary R, Sadasivuni KK (2021) A comparative review of natural and synthetic biopolymer composite scaffolds. *Polymers* 13(7):1105. https://doi.org/10.3390/polym13071105

Rotti RB, Sunitha DV, Manjunath R, Roy A, Mayegowda SB, Gnanaprakash AP, Alghamdi S, Almehmadi M, Abdulaziz O, Allahyani M, Aljuaid A, Alsaiari AA, Ashgar SS, Babalghith AO, Abd El-Lateef AE, Khidir EB (2023) Green synthesis of MgO nanoparticles and its antibacterial properties. *Frontiers of Chemistry* 11:1143614. https://doi.org/10.3389/fchem.2023.1143614

Ryoo, R., Joo, S. H., & Kim, J. M. (1999). Energetically favored formation of mcm-48 from cationic–neutral surfactant mixtures. *The Journal of Physical Chemistry B*, 103(35): 7435–7440. https://doi.org/10.1021/jp9911649

Sabourian P, Yazdani G, Ashraf SS, Frounchi M, Mashayekhan S, Kiani S, Kakkar A (2020) Effect of physico-chemical properties of nanoparticles on their intracellular uptake. *IJMS* 21(21):8019. https://doi.org/10.3390/ijms21218019

Saravanakumar K, Hu X, Chelliah R, Oh D-H, Kathiresan K, Wang M-H (2020) Biogenic silver nanoparticles-polyvinylpyrrolidone based glycerosomes coating to expand the shelf life of fresh-cut bell pepper (Capsicum annuum L. var. grossum (L.) Sendt). *Postharvest Biology and Technology* 160:111039. https://doi.org/10.1016/j.postharvbio.2019.111039

Shilpa BM, Rashmi R, Manjula NG, Sreekantha A (2022) Bioremediation of Heavy Metal Contaminated Sites Using Phytogenic Nanoparticles. In: Shah MP, Roy A (eds) Phytonanotechnology. Springer Nature Singapore, Singapore, pp 227–253.

Singh T, Shukla S, Kumar P, Wahla V, Bajpai VK, Rather IA (2017) Application of nanotechnology in food science: Perception and overview. *Frontiers in Microbiology* 8:1501. https://doi.org/10.3389/fmicb.2017.01501

Thandapani G, Prasad S, Sudha PN, Sukumaran A (2017) Size optimization and in vitro biocompatibility studies of chitosan nanoparticles. *International Journal of Biological Macromolecules* 104:1794–1806. https://doi.org/10.1016/j.ijbiomac.2017.08.057

Thiruvengadam M, Rajakumar G, Chung I-M (2018) Nanotechnology: current uses and future applications in the food industry. *3 Biotech* 8(1):74. https://doi.org/10.1007/s13205-018-1104-7

Wang C, Zhang H, Zeng D, San L, Mi X (2016) DNA nanotechnology mediated gold nanoparticle conjugates and their applications in biomedicine. *Chinese Journal of Chemistry* 34(3):299–307. https://doi.org/10.1002/cjoc.201500839

Wang M, Li S, Chen Z, Zhu J, Hao W, Jia G, Chen W, Zheng Y, Qu W, Liu Y (2021) Safety assessment of nanoparticles in food: Current status and prospective. *Nano Today* 39:101169. https://doi.org/10.1016/j.nantod.2021.101169

Warheit DB (2018) Hazard and risk assessment strategies for nanoparticle exposures: How far have we come in the past 10 years? *F1000Res* 7:376. https://doi.org/10.12688/f1000research.12691.1

Webster TJ, Seil I (2012) Antimicrobial applications of nanotechnology: Methods and literature. *IJN* :2767. https://doi.org/10.2147/IJN.S24805

Willner MR, Vikesland PJ (2018) Nanomaterial enabled sensors for environmental contaminants. *Journal of Nanobiotechnology* 16(1):95. https://doi.org/10.1186/s12951-018-0419-1

Witika BA, Makoni PA, Matafwali SK, Chabalenge B, Mwila C, Kalungia AC, Nkanga CI, Bapolisi AM, Walker RB (2020) Biocompatibility of biomaterials for nanoencapsulation: Current approaches. *Nanomaterials* 10(9):1649. https://doi.org/10.3390/nano10091649

Yadav LSR, Shilpa BM, Suma BP, Venkatesh R, Nagaraju G (2021) Synergistic effect of photocatalytic, antibacterial and electrochemical activities on biosynthesized zirconium oxide nanoparticles. *European Physical Journal Plus* 136(7):764. https://doi.org/10.1140/epjp/s13360-021-01606-6

Zabot GL, Schaefer Rodrigues F, Polano Ody L, Vinícius Tres M, Herrera E, Palacin H, Córdova-Ramos JS, Best I, Olivera-Montenegro L (2022) Encapsulation of bioactive compounds for food and agricultural applications. *Polymers* 14(19):4194. https://doi.org/10.3390/polym14194194

Zambrano-Zaragoza M, González-Reza R, Mendoza-Muñoz N, Miranda-Linares V, Bernal-Couoh T, Mendoza-Elvira S, Quintanar-Guerrero D (2018) Nanosystems in edible coatings: A novel strategy for food preservation. *IJMS* 19(3):705. https://doi.org/10.3390/ijms19030705

Zhang Y, Gao Z, Zhang W, Wang W, Chang J, Kai J (2018) Fluorescent carbon dots as nanoprobe for determination of lidocaine hydrochloride. *Sensors and Actuators B: Chemical* 262:928–937. https://doi.org/10.1016/j.snb.2018.02.079

Zhao X, Wang K, Ai C, Yan L, Jiang C, Shi J (2021) Improvement of antifungal and antibacterial activities of food packages using silver nanoparticles synthesized by iturin A. *Food Packaging and Shelf Life* 28:100669. https://doi.org/10.1016/j.fpsl.2021.100669

Zhu Z, Cai H, Sun D-W (2018) Titanium dioxide (TiO 2) photocatalysis technology for nonthermal inactivation of microorganisms in foods. *Trends in Food Science & Technology* 75:23–35. https://doi.org/10.1016/j.tifs.2018.02.018

FDA (2011) Guidance for industry considering whether an FDA-Regulated product involves the application of nanotechnology. *Biotechnology Law Report* 30(5):613–616. https://doi.org/10.1089/blr.2011.9814

8 Bionanomaterials in Improving Food Quality and Safety

B.K. Nithin Gowda, Bhagya Venkanna Rao, and Shilpa Borehalli Mayegowda

8.1 INTRODUCTION

Bionanomaterials are defined as the nanomaterial synthesised from biomolecules such as proteins, lipids, nucleic acid, polysaccharides and oligosaccharides, secondary metabolites, and so forth (Singh, 2011). Bionanomaterials are fabricated by using biomolecules or the traditional nanomaterial is encapsulated or immobilized in a biomolecule. The sources of biomolecules utilized for the synthesis of bionanomaterials are obtained from live sources, for example, plants, microorganisms, and some animal sources (Jeevanandam et al., 2022). These are the nanosized entities produced by virtue of biomolecules, which include amino acids, proteins, and enzymes (Mishra et al., 2018). The size of the bionanomaterials becomes a major factor as nanosized materials can be easily and efficiently made to interact with host tissues and this helps in determining nanomaterials', final destiny. The bionanomaterials exhibit exemplary mechanical properties due to its dimensions and dramatic effects of bionanomaterials (Khalil et al., 2017. Various types of bionanomaterials include organic, biological and synthetic bionanomaterials. Organic bionanomaterials consists of nanoparticles composed of metals, for example, zinc, copper, titanium oxide nanoparticles (NPs), and so forth (Manjula et al., 2022). Silica bionanomaterials, polymeric, such as chitosan, silk fibronin, and decomposable biopolymers like (poly lactic acid-co-glycolic acid) PLGA (Manjula et al., 2022; Mayegowda et al., 2024a, 2024b; Borehalli Mayegowda et al., 2022, 2023) and carbon-based bionanomaterials. The inorganic NPs of are basically the metal and metal oxides with an extensive variety of potential implications, including sensing, tomographical understanding like imaging and drug delivery. While lipid-based NPs are used in gene therapy, drug delivery, imaging, and dendrimers that are highly branched, nanoscale polymers with defined, uniform structures have been used in a varied range of applied fields. Biological bionanomaterials are of green synthesis from plant, bacteria, and algae sources. Green synthesis of nanocompounds mainly involves the cell components like its biomolecules DNA, lipids, proteins, RNA and in viruses, its entire structure (Manjula et al., 2023; Lavanya et al., 2024). Among which bionanomaterials used the components of peptide, polypetides, nucleic acid, xeno nucleic acid, and so forth.

Various techniques can be selected for the fabrication of bionanomaterials. This includes physical, chemical, and biological methods. A biological approach, or green synthesis, is preferred in recent days since it is easy, eco-friendly, involves less usage of toxic chemicals, and is cost effective. The biological approach for synthesis uses make use of bacteria, fungi, plants, and some animals (Ananda et al., 2022; Archana et al., 2021; Adarsha et al., 2022). Many bio-synthesis process also utilizes natural substances such as starch, glucose, honey, and so forth, which are derived from

DOI: 10.1201/9781003432791-10

biological sources. The key advantage of green synthesis is that it allows for the monitoring of bionanomaterials' adaptability and preservation of their natural origin. (Yadav et al., 2021; Mittal et al., 2013). These bionanomaterials include nanoparticles, nanotubes, nanowires, nanofibers, and so forth. Even though bionanomaterials are used in several areas of the biomedical field, the exact activity of bionanomaterial with cells, tissues, and organs of host is still unclear. Bionanomaterials should meet some specifications like biocompatibility, biodegradability, and non-toxicity to be of use in the nanomedical field (Ananada et al., 2022; Archana et al., 2021; Adarsha et al., 2022).

8.2 SYNTHESIS OF BIONANOMATERIALS

Commonly two types of process are employed in the nanoparticle's synthesis:

8.2.1 Top-Down Process

This process involves usage of bulk materials that are broken down into nano-scale objects. This technique makes use of a microfabrication technique wherein materials are broken down, milled, and shaped into the anticipated or planned structure with the help of machines. In this method metallic NPs are synthesized by milling the contents mechanically, etching, laser cutting out, splattering, and electro explosions.

8.2.2 Bottom-Up Process

This method involves fabricating each atom, molecule, or self-organization to assemble a defined structure. In this method, the self-assembling capability of the single molecule is used to build complex nanoscale conformation. This method involves synthesis of NPs by laser pyrolysis, chemical reduction, molecular condensation and most importantly by green synthesis (Khanna et al., 2019).

Bionanomaterials are synthesised via bio-reduction methods whereby reducing sugar, proteins, enzymes, and phenolic compounds cause the reduction.

Biogenesis is environmentally benign procedure for the generation of nanoparticles. The steps involved in biogenesis are:

1. The choice of solvent medium
2. Choosing the apt reducing agent, which is benign
3. Choosing nanoparticle stabilizer, which is non-toxic.

The green NPs, generation from a plant and its parts or microbial whole cell or free extract is taken and the metal salts are added to the extract. Nanoparticles are synthesized due to the reaction between metal salts and enzymes and other components in the extract. The successful creation of natural-based NPs depends more on the amide of peptide chains, plasma membrane enzymes, microbial cell walls, phytochelatins, oxidoreductases, quinines, chitosan, terpenoids, polyphenols, and other components found in plants (Mayegowda et al., 2022a, 2022b).

8.3 BIOLOGICAL BASES FOR THE GENERATION OF BIONANOMATERIALS

The common biological sources for the biosynthesis of bionanomaterials are plant derivatives, microbes, viruses, polysaccharides, and proteins. These sources are discussed in brief below.

8.3.1 Plant Derivatives

The bio-active compound obtained after removal of plant tissues using the solvent system is plant extract. The extract contains phytochemicals, active components, and macro-molecules, which act as a necessary source of bioactive compound various application in different fields of nanotechnology. Extracts from the plants have been considered as the most dependable for bionanomaterial synthesis as they are environment-friendly, plentiful, provide better control over nanoparticle's morphology, and reduce the usage and production of toxic substances. Since plant extracts have so many advantages, they are employed to modify many nanomaterials, especially metal nanoparticles. The plant extract contains ketones, alkaloids, phenolics, carbonyl, tannins, and other functional groups that encourage nanoparticle synthesis by acting as a reductant, which reduces metallic ions into atoms (Rajeshkumar et al., 2018; Venkat Kumar and Rajeshkumar, 2018, Vijayaraghavan and Ashokkumar, 2017). Various plant excerpts are employed in the formulation of NPs: *Euphorbia prostrata* (Zahir et al., 2015), *Sargassum algae* (Momeni et al., 2015), *Ginkgo biloba* (Nasrollahzadeh et al., 2015), *Cymbopogon citratus* (Murugan et al., 2015), *Azadirachta indica* (Rotti et al., 2023), *Nigella sativa* (Amooaghaie et al., 2015), *Cocus roseus* (Kalaiselvi et al., 2015), *Artocarpus gomezianus*, commonly known as breadfruit (Suresh et al., 2015), and so forth, are used in the production of nanoparticles.

8.3.2 Microorganisms

Microbes are ubiquitous in nature. Several studies have shown that microbes can be utilized for synthesizing bionanomaterials, which have applications in biosensor, antimicrobial, anticancer, medical diagnosis, and antibiofouling (Salunke et al., 2016). Microbes including bacteria, actinomycetes, yeast, lichens, algae and fungi are utilized for the synthesis of bionanomaterials. Microbial mediated synthesis is advantageous and possesses several advantages, such as synthesized nanomaterials will be of definite morphology, dimensions, and chemical content, simple to maintain and harvest the microbial cells, can be scaled up, and can readily adapt to a wide range of environmental conditions (Jacob et al., 2021). Microbial synthesis can be differentiated into two methods, that is, intracellular and extracellular. The intracellular approach is a process in which the metal ions are internalized inside the cell. In extracellular approach, the metal ions settle, or adsorb, on the cell surface. In both the approaches oxidoreductase enzymes are significantly high contributing in reducing metallic ions either outside or inside the cell is microbial enzyme catalyzed, which is reasonable for the synthesis of metallic NPs (Salunke et al., 2016).

8.3.2.1 Bacteria

Microorganisms, including *Pseudomonas deceptionensis*, *Weissella oryzae*, *Bacillus methylotrophicus*, and so forth have shown the capability of generating excellent NPs when combined with silver and gold metallic oxides. Several diverse genera of bacteria exhibited the production of potential metal nanoparticle from *Bacillus sps*, *Pseudomonas sps*, *Klebsiella sps*, *Escherichia coli*, *Enterobacter sps*, *Aeromonas sps*, *Corynebacterium sps*, *Lactobacillus sps*, *Rhodobacter*, *Streptomyces*, *Trichoderma*, *Sargassum*, *Pyrobalcum* and others (Jo et al., 2015; Wang et al., 2016; Li et al., 2011).

8.3.2.2 Fungi

Mycosynthesis is the process of using fungi instead of bacteria for the synthesis of NP (Alghutaymi et al, 2015). Fungal colonies such as the species of *Penicillium* and *Fusarium oxysporum*, contains enzymes nitrate reductase and α-NADPH-dependent reductase is known for its pivotal role NPs synthesis in combination with fungal extracts (Golinska et al, 2014).

8.3.2.3 Viruses

The common size of most viral capsids is between 20–500 nm, hence, they are known to be a natural nanoparticle because of their nanoscale dimension. Viruses are considered as prefabricated nanoparticles that are used in the field of nanoscience (Jeevanandham et al., 2019). The plant viruses, or phytophages, have been identified as befitting the production of bionanomaterials, that relate to various environmental conditions, such as nanoscale dimensions, simple functionalisation, lack of covering, and spectacular solidity. Most of the phytophages facilitates 3D building blocks for nanoscale structures assembly multidimensionally (Culver et al., 2015). Most commonly used plant viruses are helical or icosahedral shaped. Viruses like *Cowpea Chlorotic Mottle*, *Cowpea Mosaic Virus*, and *Brome Mosaic Virus* are the most frequently common icosahedral virus, whereas *Tobacco Mosaic Virus* is the most common rod-shaped virus (Zhang et al., 2018).

8.3.3 Polysaccharides

Polysaccharides are a type of carbohydrate composed of monomer units of sugar. They are polymers of naturally occurring sugar, hence they are called "Biopolymer". These polysaccharides are castoff to synthesize natural-based nanomaterials with particular functionalities that could be utilized for various applications because of the material's biodegradable nature and the biocompatibility of materials (Mayegowda et al., 2022; Mayegowda et al., 2023). Starch and cellulose, polysaccharides derived from plants are having more attention because of their abundance and profitability as well as simplicity of process (Zhou et al., 2020). Recently, starch-based material which had 3-D nanoporous network was prepared for the encapsulation of clove-based oil (CEO) which has a property of food preservative. This study showed that nanomaterial encapsulated CEO has the enhanced susceptibility to bacteria like *Bacillus subtilis*, *Staphylococcus aureus*, and *Escherichia coli* with comparison to pure CEO, exhibiting excellent antibacterial activity (Fang et al., 2020).

8.3.4 Proteins

Proteins showcase amino acids that are chemically diverse, hence they are excellent blocks that have well-defined binding characteristics. The binding sites aid in the synthesis of novel bionanomaterials by enabling self-organisation so that it can direct into one, two or three arrays that could be used as bio-templates. Nanomaterials to a protein base are generally classified as nanoparticles and nanofibers that have a widespread variety of biomedical application in several fields, including drug delivery, diagnostics, and biosensing. For example, a protein isolated from maize and known as zein, is widely used in nanomaterial synthesis as a bio template (Kasaai, 2018).

8.4 IMPORTANCE AND APPLICATIONS OF BIONANOMATERIALS

Bionanomaterials are used in various biomedical application which include tissue engineering, drug delivery, nanomedicines, nanopharmaceuticals, nanobiosensors and nanoimaging, and so forth (Singh et al., 2021).

8.4.1 Tissue Engineering

Tissue engineering is an emerging area of science in which a wide variety of biobased active molecules, scaffolds, and generation of cells preserve or progress the damaged tissue and/or organs are obtained. While constructing an ideal scaffold, it should possess some characteristics, such as, it should be capable of carrying active biomolecules, have the capacity to initiate physiological signals, and have mechanical properties consisting of a striking similarity to original tissue (Kumar et al., 20220). Bionanomaterials have a refined superfine structure and allow them to have interaction with cell or surface receptors of tissue. Thus, creating a microenvironment for tissue regeneration (An et al., 2013). The defects in the traditional vascular grafts are overcome by the vascular tissue engineering approach. The actual nanostructure in the vascular tissue is mimicked by the nanomaterials. Several bionanomaterials have been designed in such a way that it controls the vascular endothelial cells and SMCs function. It also helps to overcome connected difficulties like inflammation and thrombosis (Mayegowda et al., 2022a).

For nerve repair application various biomaterials have been used to prepare nerve grafts but all these have limitations (Chiono et al., 2009). Bionanomaterials' extraordinary mechanical and electrical possessions and cytocompatibility provide the best chance to heal damaged nerves. Chitosan-heparin nanoparticles were conjugated with nerve-growth factors; this molecule was able to improve neuron outgrowth and could be utilized as potential drug delivery system for nerve repair (Gonçalves et al., 2012).

8.4.2 Nanomedicines

The discovery of nanomedicine made it possible to achieve a few hitherto impossible medical needs. It made possible integration of highly toxic but effective molecules into the system. It increased efficacy, bioavailability, dose toxicity decreases, increasing drug target, controlled and suit-specific drug release, and increased efficiency of transport through biological barriers (Figure 8.1). All the above-mentioned properties are due to the unique properties of bionanomaterials. Bionanomaterials aid in adsorption of biomolecules owing to their nanoscale size and high surface area-to-volume ratio, when they contact biological fluids. Synthetic bionanomaterials include polylactic acid, PLGA, polyethyleneimine, and so forth. Natural bionanomaterials such as PLGA conjugated with several anticancer drugs like cisplatin, doxorubicin, 5-fluorouracil, and so forth, which showed promising effect in delivering drug to cancer cells. A nanomaterial with substantial possibilities to use in several medicinal approaches is carbon nanotubes (CNTs). Nevertheless, due to its form, which mimics asbestos fibres, worries have been raised about its toxicity. Long CNTs have demonstrated to behave like fibres that cannot be digested and cause phagocytosis to fail and the creation of granulomas (Ali et al., 2020).

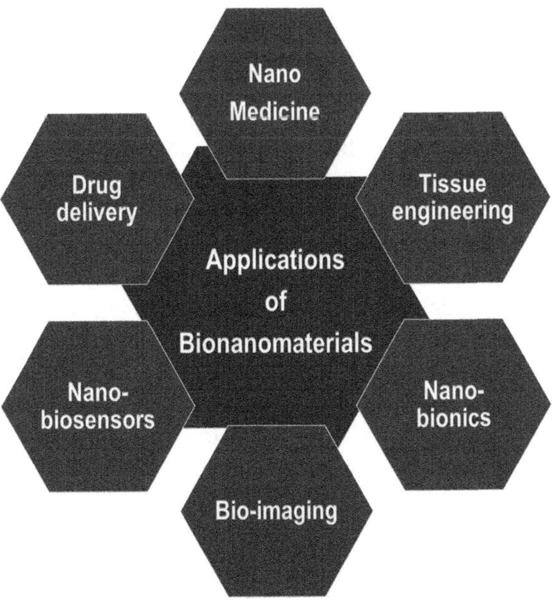

FIGURE 8.1 Bionanomaterials Are Being Used by Biological Sciences (Compiled by the Authors.)

8.4.3 Drug Delivery

Bionanomaterials used for drug delivery should be prepared by following certain criteria. It must have the capability to amplify the hydrophilic nature and solubility of hydrophobic drugs in water and controlled drug release and enhancement of degree of drug delivery. A lot of exploration is done on Carbon Nanotubes (CNTs) since they showcase the distinctively enclosed nanochannels, which makes them apt candidates for drug delivery application (Chen et al., 2014). In recent advancements three-dimensional (3D) technology is gaining importance as it effectively controls the release of drugs for up to four months and sometimes longer (Figure 8.1).

8.4.4 Nanobionics

The technology involving nanomaterials measuring 3-4 nm in size is appropriate for the creation of bionic nanoparticles and nanostructures. Bionic particles exhibit similar properties of inorganic materials and are capable of upregulating the chemical performance of supremely developed biomolecules mimicking the living system (Figure 8.1). One such well known combination is conjugation of cadmium telluride and cytochrome C to form super particles in a highly efficient process. These bionic nanoparticles make use of sunlight, and a series of chemical reactions are initiated; the nanomaterials also can be integrated with enzyme to convert pollutant nitrate into nitrite. The nanoparticles can also modify the plants in such a way that it converts sunlight into fuel. The bionic nanoparticles go around with natural processes, where working parts are constantly renewed through self-assembly.

8.5 FOOD QUALITY AND SAFETY

Nanotechnology is mainly employed in food industries for the following:

1. Using of bio-nanosensors to ensure food quality and safety;
2. Design devices to target the nutrient delivery nutrition therapy;
3. Developing systems to control nutrient release through nanoencapsulation, and
4. Development of new product and food fortification via an enzymatic reactor (Dwivedi et al., 2018).

Nanotechnology has been extensively applied in the food industry to extend the life of storage, enhancing food security, refining flavour, nutrient transfer, pathogen recognition, and serving functional food (He and Hwang, 2016). Food ingredients with nanostructures have been developed claiming to increase taste, improve texture and consistency. Nanotechnology is applied in several forms. These forms attribute to the effective delivery system. Various forms include encapsulation, biopolymer, solutions, emulsions, and colloids. The conventional materials in packaging of food are now being substituted by nano-polymers. The existence of contaminants, mycotoxins, and microorganisms can be detected by the use of nano-biosensors (Mayegowda et al., 2023). Nanoparticles offer improved properties and efficiency for encapsulation compared to a traditional encapsulation system. This delivery system makes delivery of active compounds to target sites very efficient, since they have the capacity to infiltrate deeply into the tissues with regard to their smaller size (Lamprecht et al., 2004) (Figure 8.2).

8.5.1 Food Processing

Food processing refers to the process and procedures in the preservation of food that converts raw food material into a usable, consumable, and palatable state (Pradhan et al., 2015). It is concerned with the transformation of resources into complete and semi-finished foodstuffs with a particular

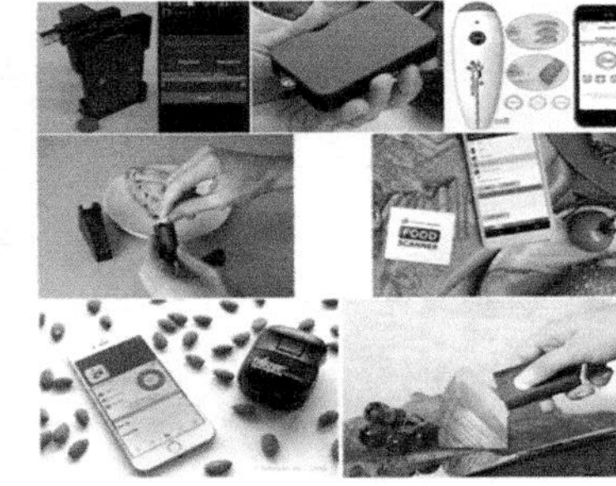

FIGURE 8.2 Bio-nanotechnology in Food Industries Includes Storage, Food Security, and Nutritive Value (Compiled by the Authors.)

set of techniques and methods (Monteiro et al., 2010). These procedures include techniques such as washing, slicing, cooking, pasteurization, and so forth. Food processing also comprises adding elements to lengthen the shelf life of food (Dwyer et al., 2012; Weaver et al., 2014).

Additional processing consists of toxin removal, pathogen prevention, food preservation, and distribution of nutrients will increase with respect to uniformity in the food. This process improves the marketing and distribution of food (Chellaram et al., 2014). Processed food has an advantage over unprocessed food since it keeps its freshness longer and is more convenient to transfer from the producer to the consumer over great distances. Bionanomaterials' advantage and role in bettering food processing are assessed in terms of the food's flavour, texture, appearance, nutritional value, and shelf life.

The best gastronomic balance is delivered by using the nanoencapsulation method, which also improves taste retention (Nakagawa et al., 2014). Anthocyanins, which are highly sensitive and unsteady to pigment from plants having a heightened spectrum of biological activity, have been enclosed in nanoparticles. The thermal stability and photostability of both recombinant soybean seed H-2 subunit ferritin rH-2 was enhanced by enclosing cyanidin-3-0 glucoside(C3G) molecules into the inner cavity. This type of multifunctional nanocarrier design can be utilized to conduct and safeguard bioactive molecules (Zhang et al., 2014). Due to their ease of synthesis using natural food ingredients and the capacity to improve water dispersion and bioavailability, nanoemulsions are frequently used for the administration of lipid-soluble bioactive chemicals (Ozturk et al., 2015). Due to their sub-cellular size, nanoparticles provide encouraging ways to increase the bioavailability of nutraceuticals, which results in more medication availability. Metallic oxides like Tio_2 and Sio_2 have been employed as coloring agents in the past. One of the most popular food nanomaterials for carrying flavours or scents in food products is SiO_2 (Dekkers et al., 2011).

Most of the bioactive compounds are vulnerable to chemical parameters like acidic due to pH, along with change on enzyme activity in the stomach and duodenum of the digestive tract. Such adverse conditions can be overcome by encapsulation of bioactive compounds. Encapsulation also aids in in assimilation of food products. Nanoparticles are formulated as capsules aiming to progress in medicine delivery, delivering vitamins and sensitive micronutrients in routine food to impart remarkable health benefits (Koo et al., 2005).

The bioactive compounds frequently become degraded and lead to loss of function in functional food as a consequence of incompatible conditions. Nanoencapsulation hinders the breaking down process of biologically active compounds, thereby accounting for the enhancement of serviceable life of the food by delaying the product degradation, until the product reaches the site of target. Nanocoatings which are edible in nature on various food product act as a hurdle the interchange of gases and moisture and also assist in delivering flavours, antioxidants, colours, enzymes, and anti-browning agents and it also could have a role in increasing the post opening shelf life of the food (Weiss et al., 2006).

8.5.2 Food Packaging

An immense surge has been seen in the use of bionanomaterials in various aspects of the food industry. They stirred the interest of lots of researchers owing to their superior mechanical and physiochemical properties over the traditionally used alloys. Moreover, green bionanomaterials are preferred because in the food industry as they are capable of blocking or resisting water and oxygen in packaging materials. Bionanomaterials are used as an efficient delivery system of highly unstable nutrients. The encapsulation of these nutrients with the nanomaterials makes them more stable and able to be efficiently delivered to the target site. Biopolymers such as starch, cellulose, chitosan, maltodextrin, sodium alginate, albumin, zein protein, and so forth are extensively studied for encapsulation. Nanomaterials packaging of food have an advantage, like microbial contamination detection, enhanced mechanical barriers, and possible enhancement of bioavailability of nutrients. This is the popular implementation of nanomaterials in industries associated with food (Bradley et al., 2011). Numerous nanocomposites, polymer consisting of nanoparticles are applied the for food packing and contact materials used in food processing (Llorens et al., 2012). ZnO and MgO nanoparticles have been found to have a potential application in food packaging (Gerloff et al., 2009). Amorphous silica has a potential application in packaging of food and food containers (Uboldi et al., 2012). Engineered water nanostructures generated in the form of aerosols are effective in killing food-borne pathogens such as *Escherichia coli*, *Listeria* and *Salmonella* over food utilisation surface (Pyrgiotakis et al., 2015) (Figure 8.2). So, it can be concluded that food contact substances containing nanomaterials have the ability to migrate into food substance from packaging material.

8.5.2.1 Food Packaging Is Categorized as Several Types:

1. **Improved packaging**—Packaging of food bar can be raised by using functional nanomaterials which have improved physico-chemical properties like temperature stability, durability, flexibility, mechanical strength, and so forth.
2. **Active packaging**—Nanomaterials with an active function also helps to upgrade the packaging standard of food. For example, addition of nanoparticles with antimicrobial, antioxidative and UV protective properties.
3. **Smart packaging**—This packaging involves the addition of nano-sensors having smart and brilliant application in the recognition of gases, traces of organic molecules, active stage and identification of primary and secondary metabolites.
4. **Bio-based packaging**—Bio-based packaging, also known as biodegradable packaging, biocompatibility, low-waste packaging and eco-friendly packaging are all improved by incorporating bionanomaterials into the packaging materials. Polymer nanocomposites, also known as bio-nanocomposites, are a fusion of nano-structured ingredients with improved gas barrier, thermal, and mechanical properties. Bio-nanocomposites are said to be a more environmentally friendly choice because they lessen the demand for plastic packaging materials. The packaging materials can be made biodegradable by adding inorganic material such as clay to the matrix of biopolymers. Surfactant utilised in the alteration of layered silicate can be used to manage them as well (Bratovčić et al., 2022).

When it comes to packaging food, nanomaterials have a number of advantages over traditional materials. The most popular nanotechnology implication is in enhancing the qualities of food packaging by nano-coating. Food can be covered with variety of food coatings, such as thin layer coatings or films to help provide a barrier against mass transfer. The coatings can also be prepared using edible nanoparticles. In order to give the products flavour, colour, enzymes, antioxidants, and anti-browning compounds, edible nano-coatings that are less than 5 nm thin coatings are applied in the meat-processing industry and for protecting fruit, vegetable products, cheese, and bakery industry, and so forth. Hydrocolloids such as extracts of polymers like cellulose, pectin, alginate and starch are used as materials for edible coating (Nile et al., 2020). The usage of lipids and their derivatives, specifically fatty acids, waxes, essential oils, and so forth. Edible coatings use proteins including gelatin, myofibrillar proteins, milk and soy proteins, zein, and others.

8.5.3 Food Nano-sensor

Nanomaterials are employed in nano-sensors to control the food environment and detect contamination. They are effective at detecting the presence of food and microbiological pollutants. Because of this, nano-sensors have a potential application in food and packaging manufacturing. We can keep an eye on food quality as it is being transported and stored (Bouwmeester et al., 2009; Buzby et al., 2010). These can be used to detect deficiency of nutrients in edible plants and signals dispensers, which contain nutrients, to provide nutrition to the plants (Figure 8.2). Henceforth, nanomaterials can be trusted in nano-sensors and nano-tracers, providing unlimited potential to the food industries.

Nanomaterial is utilized in building the biosensors as it provides high sensitivity and other new applications. Nano-sensors and nano-biosensors are found to be beneficial in food industries for the identification of disease-causing organisms in food constituents, quantifying the available constituents in food, which is important to maintain food's the safety standards (Thakur et al., 2002). In food-storage applications, nano-sensors detect the environment conditions like temperature, humidity, contamination by microbes and degradation of products (Bouwmeester et al., 2009).

Enormous nanostructures have been tested to determine their potential for use in biosensors, including thin films, nanorods, quantum dots, and nanofibers (Jianrong et al., 2004). Highly sensitive detection systems are the result of thin-film built optical immunosensor for the detection of microorganisms or cells. The aforementioned immunosensors immobilise specific antibodies, antigens, or protein molecules on thin nanofilms or sensor chips so that they can generate signals when they detect target ligands (Viswanathan and Radecki, 2008).

A dimethylsiloxane microfluidic immunosensor is joined with a specific antibody that is immobilised on an alumina nanoporous membrane with an electrochemical impedance spectrum for the quick recognition of food-borne pathogenic microbes like *Escherichia coli* O157:H7 and *Staphylococcus aureus* (Tan et al., 2011). Bio-nanotechnology has tremendous insight in terms of a tracking, tracing, and monitoring chain for food quality, along with the identification traces of pesticides in the food products (Mayegowda et al., 2024a, 2024b), infections (Stephen Inbaraj and Chen, 2015), and toxins (Palchetti and Mascini, 2008).

8.5.4 Food Safety

Even though there has been tremendous advancement in the usage of nanoparticles in food industry, still much less investigation is done about nanoparticle toxicity. The major concern while using nanomaterials is the release of allergens and heavy metal (Shilpa et al., 2022). Without the requisite knowledge and regulations, nanomaterials are already being employed in foodstuffs at a quicker rate, harming the surroundings and mankind (Ranjan et al., 2014). We can produce nutritious food stuff if we possess complete insight into the properties of nanomaterials such as solubility, size, composition, and surface chemistry. The unique characteristic of nanomaterials makes them appealing for use in many fields, but in the case of food application, this could be debatable and pose possible risk to human health (Ameta et al., 2020).

Even though bio-nanotechnology has extensive applications and advantages in the food industry, we cannot neglect the safety issues related with nanomaterial. The likelihood of nanoparticle migration into food released from packaging material and its influence on human health have been highlighted in several platform discussions of safety concerns associated with nanomaterials (Bradley et al., 2011). Even when a substance is GRAS—normally observed as safe—more thorough research must be done to assess the risk of the nano-dimensional structures since they possess distinct physiochemical properties from those in microstate. The risk of bioaccumulation in the tissues and organs of the host increases with decreasing nanomaterial size (Savolainen et al., 2010). To assure product quality and health, regulatory agencies must step up and create rules for the application of nanomaterials in commercial products (Figure 8.2).

Even though several studies suggest that migration of nanoparticles from packaging to food is negligible, it cannot be neglected and a doubt prevails in the concerned consumer's mind. Government have asked the public for regulation of nanomaterial migration limit since no specific regulation exists. Various media outlets and non-governmental organizations (NGOs) raised this issue through their communication channels. These organisations focus on two major things:

1. To make people aware about the usage of nanomaterials in daily life with uncertain safety regulation;
2. To stimulate government to regulate the use of nanomaterials in food packaging.

In current EU regulation, only three nanomaterials have been approved for use instead of plastic food packaging (Reig et al., 2014). On the other hand, the USFDA has given GRAS status to aluminium, nano clay, and ZnO, but it does not cover all the materials that have already been applied for packaging of food.

In 2009, the United Nations (UN) Food and Agriculture Organization (FAO) and the World Health Organization (WHO) held a conference titled "Nanotechnologies in the Food and Agriculture Sectors: Potential Food Safety Implications". Experts from 13 countries attended this meeting. Experts at the meeting discussed existing and emerging applications of nanotechnologies, data regarding migration of nanoparticles into food packaging, any possible threat of nanotechnology, and the existing capability to evaluate such risks.

In February 2014, a draft regulation mandating the labelling nanomaterials synthesized to be engineering used in food was proposed to the European Union (EU) Parliament by the European Commission and the European Consumer Organization (BEUC). This proposal was created based on the concept that the safety of nanoparticles is still ambiguous; consumers have the right to know when their food contains nanoparticles and to make their own decisions regarding use of the products (European Union, 2023). However, this bill was rejected by the lawmakers in the EU Parliament's Committee for Environment, Public Health and Food Safety. The reason for this denial was that this new proposal would contradict the existing nanofood additives that already have been approved.

8.6 CONCLUSION

Nanomaterials have been found to have a potential application in sensing of contaminants and pathogens via nano-biosensors. The high sensitivity of nanomaterials aids in the detection of even a tiny number of contaminants and small deviation in the storage conditions. Bionanomaterials also have antimicrobial activity, so we can capitalize on bionanomaterials in food preservation. Furthermore study needs to be conducted on the impact of nanomaterial on the consumers and its toxicity. When used as antimicrobial agents, the bacterial resistance towards the nanoparticles has to be taken into account. When it comes to safety, people are more concerned about biodegradable products, particularly those that are edible.

The safety of consumers and environmental impact of bionanomaterials have to be prioritized while developing or studying the bionanomaterials. There is no official regulation for nanomaterial in any countries. A few nanoparticles have been approved as a food contact material in some countries, whereas some countries assume nanomaterials are toxic.

REFERENCES

Adarsha JR, Ravishankar TN, Ananda A, Manjunatha CR, Shilpa BM, Ramakrishnappa T (2022) Hydrothermal synthesis of novel heterostructured Ag/TiO_2/$CuFe_2O_4$ nanocomposite: Characterization, enhanced photocatalytic degradation of methylene blue dye, and efficient antibacterial studies. *Water Environment Research* 94(6). https://doi.org/10.1002/wer.10744

Alghuthaymi MA, Almoammar H, Rai M, Said-Galiev E, Abd-Elsalam KA (2015) Myconanoparticles: Synthesis and their role in phytopathogens management. *Biotechnology & Biotechnological Equipment* 29(2):221–236. https://doi.org/10.1080/13102818.2015.1008194

Ali I, Alsehli M, Scotti L, Tullius Scotti M, Tsai S-T, Yu R-S, Hsieh MF, Chen J-C (2020) Progress in polymeric nano-medicines for theranostic cancer treatment. *Polymers* 12(3):598. https://doi.org/10.3390/polym12030598

Ameta SK, Rai AK, Hiran D, Ameta R, Ameta SC (2020) Use of Nanomaterials in Food Science. In: Ghorbanpour M, Bhargava P, Varma A, Choudhary DK (eds) *Biogenic Nano-Particles and their Use in Agro-Ecosystems*. Springer Singapore, Singapore, pp 457–488.

Amooaghaie R, Saeri MR, Azizi M (2015) Synthesis, characterization and biocompatibility of silver nanoparticles synthesized from Nigella sativa leaf extract in comparison with chemical silver nanoparticles. *Ecotoxicology and Environmental Safety* 120:400–408. https://doi.org/10.1016/j.ecoenv.2015.06.025

An J, Chua CK, Yu T, Li H, Tan LP (2013) Advanced nanobiomaterial strategies for the development of organized tissue engineering constructs. *Nanomedicine* 8(4):591–602. https://doi.org/10.2217/nnm.13.46

Ananda A, Ramakrishnappa T, Archana S, Reddy Yadav LS, Shilpa BM, Nagaraju G, Jayanna BK (2022) Green synthesis of MgO nanoparticles using Phyllanthus emblica for Evans blue degradation and antibacterial activity. *Materials Today: Proceedings* 49:801–810. https://doi.org/10.1016/j.matpr.2021.05.340

Archana S, Jayanna BK, Ananda A, B.M S, Pandiarajan D, Muralidhara HB, Kumar KY (2021) Synthesis of nickel oxide grafted graphene oxide nanocomposites – A systematic research on chemisorption of heavy metal ions and its antibacterial activity. *Environmental Nanotechnology, Monitoring & Management* 16:100486. https://doi.org/10.1016/j.enmm.2021.100486

Borehalli Mayegowda, S., Bhoomika, S., Nagshetty, K., & Manjula, N. G. (2022). Environmental adequacy of green polymers and biomaterials. In P. Agarwal, D. B. Tripathy, A. Gupta, & B. K. Kuanr, *Polymeric Biomaterials* (1st ed., pp. 193–214). CRC Press. https://doi.org/10.1201/9781003240884-10

Borehalli Mayegowda, S., Roy, A., N. G., M., Pandit, S., Alghamdi, S., Almehmadi, M., Allahyani, M., Awwad, N. S., & Sharma, R. (2023). Eco-friendly synthesized nanoparticles as antimicrobial agents: An updated review. Frontiers in Cellular and Infection Microbiology, 13, 1224778. https://doi.org/10.3389/fcimb.2023.1224778

Bouwmeester H, Dekkers S, Noordam MY, Hagens WI, Bulder AS, De Heer C, Ten Voorde SECG, Wijnhoven SWP, Marvin HJP, Sips AJAM (2009) Review of health safety aspects of nanotechnologies in food production. *Regulatory Toxicology and Pharmacology* 53(1):52–62. https://doi.org/10.1016/j.yrtph.2008.10.008

Bradley, E. L., Castle, L., & Chaudhry, Q. (2011). Applications of nanomaterials in food packaging with a consideration of opportunities for developing countries. *Trends in Food Science & Technology*, 22(11): 604–610. https://doi.org/10.1016/j.tifs.2011.01.002

Bratovcic, A. (2022). Bio- and synthetic nanocomposites for food packaging. In: Ameta SC, Ameta R, (eds) *The Science of Nanomaterials* (1st ed.). Apple Academic Press, pp. 303–334. https://doi.org/10.1201/9781003283126-11

Buzby JC (2010) Nanotechnology for food applications: More questions than answers. *Journal of Consumer Affairs* 44(3):528–545. https://doi.org/10.1111/j.1745-6606.2010.01182.x

Chellaram C, Murugaboopathi G, John A A, Sivakumar R, Ganesan S, Krithika S, Priya G (2014). Significance of nanotechnology in food industry. *APCBEE Procedia*, 8: 109–113. https://doi.org/10.1016/j.apcbee.2014.03.010

Chen J, Shi M, Liu P, Ko A, Zhong W, Liao W, Xing MMQ (2014) Reducible polyamidoamine-magnetic iron oxide self-assembled nanoparticles for doxorubicin delivery. *Biomaterials* 35(4):1240–1248. https://doi.org/10.1016/j.biomaterials.2013.10.057

Chiono V, Vozzi G, Vozzi F, Salvadori C, Dini F, Carlucci F, Arispici M, Burchielli S, Scipio FD, Geuna S, Fornaro M, Tos P, Nicolino S, Audisio C, Perroteau I, Chiaravalloti A, Domenici C, Giusti P, Ciardelli G (2009) Melt-extruded guides for peripheral nerve regeneration. Part I: Poly(ε-caprolactone). *Biomedical Microdevices* 11(5):1037

Culver JN, Brown AD, Zang F, Gnerlich M, Gerasopoulos K, Ghodssi R (2015) Plant virus directed fabrication of nanoscale materials and devices. *Virology* 479–480:200–212. https://doi.org/10.1016/j.virol.2015.03.008

Dekkers S, Krystek P, Peters RJB, Lankveld DPK, Bokkers BGH, Van Hoeven-Arentzen PH, Bouwmeester H, Oomen AG (2011) Presence and risks of nanosilica in food products. *Nanotoxicology* 5(3):393–405. https://doi.org/10.3109/17435390.2010.519836

Dwivedi C, Pandey I, Misra V, Giulbudagian M, Jungnickel H, Laux P, Luch A, W. Ramteke P, Vikram Singh A (2018) The prospective role of nanobiotechnology in food and food packaging products. *Integrative Food and Nutrition Metabolism* 5(6). https://doi.org/10.15761/IFNM.1000237

Dwyer JT, Fulgoni VL, Clemens RA, Schmidt DB, Freedman MR (2012). Is "processed" a four-letter word? The role of processed foods in achieving dietary guidelines and nutrient recommendations. *Advances in Nutrition,* 3(4): 536–548. https://doi.org/10.3945/an.111.000901

The European Union, 2023, European Parliament opposes European Commission proposal on nanomaterials in food. https://www.beuc.eu/press-releases/european-parliament-opposes-european-commission-proposal-nanomaterials-food. Accessed 21 May 2023

Application of polymer nanocomposite materials in food packaging. *Croatian Journal of Food Science and Technology* 7(2):86–94. https://doi.org/10.17508/CJFST.2015.7.2.06

Fang Y, Fu J, Liu P, Cu B (2020) Morphology and characteristics of 3D nanonetwork porous starch-based nanomaterial via a simple sacrifice template approach for clove essential oil encapsulation. *Industrial Crops and Products* 143:111939. https://doi.org/10.1016/j.indcrop.2019.111939

Gerloff K, Albrecht C, Boots AW, Förster I, Schins RPF (2009) Cytotoxicity and oxidative DNA damage by nanoparticles in human intestinal Caco-2 cells. *Nanotoxicology* 3(4):355–364. https://doi.org/10.3109/17435390903276933

Golinska P, Wypij M, Ingle AP, Gupta I, Dahm H, Rai M (2014) Biogenic synthesis of metal nanoparticles from actinomycetes: Biomedical applications and cytotoxicity. Applied Microbiology and Biotechnology 98(19):8083–8097. https://doi.org/10.1007/s00253-014-5953-7

Gonçalves NP, Oliveira H, Pêgo AP, Saraiva MJ (2012) A novel nanoparticle delivery system for *in vivo* targeting of the sciatic nerve: Impact on regeneration. *Nanomedicine* 7(8):1167–1180. https://doi.org/10.2217/nnm.11.188

He X, Hwang H-M (2016) Nanotechnology in food science: Functionality, applicability, and safety assessment. *Journal of Food and Drug Analysis* 24(4):671–681. https://doi.org/10.1016/j.jfda.2016.06.001

Jacob JM, Ravindran R, Narayanan M, Samuel SM, Pugazhendhi A, Kumar G (2021) Microalgae: A prospective low cost green alternative for nanoparticle synthesis. *Current Opinion in Environmental Science & Health* 20:100163. https://doi.org/10.1016/j.coesh.2019.12.005

Jeevanandam, Jaison, Siaw Fui Kiew, Stephen Boakye-Ansah, Sie Yon Lau, Ahmed Barhoum, Michael K. Danquah, and João Rodrigues. "Green Approaches for the Synthesis of Metal and Metal Oxide Nanoparticles Using Microbial and Plant Extracts." *Nanoscale* 14, no. 7 (February 17, 2022): 2534–71. https://doi.org/10.1039/D1NR08144F

Jianrong C, Yuqing M, Nongyue H, Xiaohua W, Sijiao L. (2004). Nanotechnology and biosensors. *Biotechnology Advances*, 22(7): 505–518. https://doi.org/10.1016/j.biotechadv.2004.03.004

Jo Y-K, Cromwell W, Jeong H-K, Thorkelson J, Roh J-H, Shin D-B (2015). Use of silver nanoparticles for managing *Gibberella fujikuroi* on rice seedlings. *Crop Protection*, 74: 65–69. https://doi.org/10.1016/j.cropro.2015.04.003

Kalaiselvi A, Roopan SM, Madhumitha G, Ramalingam C, Elango G (2015) Synthesis and characterization of palladium nanoparticles using Catharanthus roseus leaf extract and its application in the photo-catalytic degradation. *Spectrochimica Acta Part A: Molecular and Biomolecular Spectroscopy* 135:116–119. https://doi.org/10.1016/j.saa.2014.07.010

Kasaai MR (2018) Zein and zein -based nano-materials for food and nutrition applications: A review. *Trends in Food Science & Technology* 79:184–197. https://doi.org/10.1016/j.tifs.2018.07.015

Khanna P, Kaur A, Goyal D (2019) Algae-based metallic nanoparticles: Synthesis, characterization and applications. *Journal of Microbiological Methods* 163:105656. https://doi.org/10.1016/j.mimet.2019.105656

Koo OM, Rubinstein I, Onyuksel H (2005) Role of nanotechnology in targeted drug delivery and imaging: A concise review. *Nanomedicine: Nanotechnology, Biology and Medicine* 1(3):193–212. https://doi.org/10.1016/j.nano.2005.06.004

Kumar KM, Ram S, Manjula NG, Mayegowda SB (2022). Biopolymers and their applications in biomedicine. In Agarwal P, Tripathy DB, Gupta A., Kuanr BK (eds) *Polymeric Biomaterials* (1st ed., pp. 35–62). CRC Press. https://doi.org/10.1201/9781003240884-3

Lamprecht A, Saumet J-L, Roux J, Benoit J-P (2004) Lipid nanocarriers as drug delivery system for ibuprofen in pain treatment. *International Journal of Pharmaceutics* 278(2):407–414. https://doi.org/10.1016/j.ijpharm.2004.03.018

Li X, Xu H, Chen Z-S, Chen G (2011) Biosynthesis of nanoparticles by microorganisms and their applications. *Journal of Nanomaterials* 2011:1–16. https://doi.org/10.1155/2011/270974

Llorens A, Lloret E, Picouet PA, Trbojevich R, Fernandez A (2012) Metallic-based micro and nanocomposites in food contact materials and active food packaging. *Trends in Food Science & Technology* 24(1):19–29. https://doi.org/10.1016/j.tifs.2011.10.001

Manjula NG, Sarma G, Shilpa BM, Suresh Kumar K (2022) Environmental Applications of Green Engineered Copper Nanoparticles. In: Shah MP, Roy A (eds) *Phytonanotechnology*. Springer Nature, Singapore, pp 255–276

Manjula NG, Tajunnisa, Mamani V, Meghana CA, Mayegowda SB (2023). Fungal-based synthesis to generate nanoparticles for nanobioremediation. In Policarpo Tonelli FM, Roy A, Ananda Murthy HC (Eds.), *Green Nanoremediation: Sustainable Management of Environmental Pollution* (. Springer, 83–108. https://doi.org/10.1007/978-3-031-30558-0_4

Mayegowda SB, Bhoomika S, Nagshetty K, Manjula NG (2022a). Environmental adequacy of green polymers and biomaterials. In Agarwal P, Tripathy DB, Gupta A., Kuanr BK (eds) *Polymeric Biomaterials* (1st ed.). CRC Press, 193–214. https://doi.org/10.1201/9781003240884-10

Mayegowda SB, Sureshkumar K, Yashaswini R, Ramakrishnappa T (2022b) Phytonanotechnology for the Removal of Pollutants from the Contaminated Soil Environment. In: Shah MP, Roy A (eds) *Phytonanotechnology*. Springer Nature, Singapore, pp 319–336

Mayegowda SB, Sarma G, Gadilingappa MN, Alghamdi S, Aslam A, Refaat B, Almehmadi M, Allahyani M, Alsaiari AA, Aljuaid A, Al-Moraya IS (2023) Green-synthesized nanoparticles and their therapeutic applications: A review. *Green Processing and Synthesis* 12(1):20230001. https://doi.org/10.1515/gps-2023-0001

Mayegowda, S. B., Chikkud, V., Barua, S., & Manjula, N. G. (2024a). 13—Heavy metal detection by nanotechnology-based sensors. In F. M. Policarpo Tonelli, A. Roy, M. Ozturk, & H. C. A. Murthy (Eds.), Nanotechnology-based Sensors for Detection of Environmental Pollution (pp. 237–263). Elsevier. https://doi.org/10.1016/B978-0-443-14118-8.00013-9

Mayegowda, S. B., Nithin Gowda, B. K., Chandan Gowda, U., Joshi, V., & Manjula, N. G. (2024b). 27—Sustainability and green nanomaterials on nanotechnology-based sensors. In F. M. Policarpo Tonelli, A. Roy, M. Ozturk, & H. C. A. Murthy (Eds.), Nanotechnology-based Sensors for Detection of Environmental Pollution (pp. 553–572). Elsevier. https://doi.org/10.1016/B978-0-443-14118-8.00027-9

Mishra RK, Ha SK, Verma K, Tiwari SK (2018) Recent progress in selected bio-nanomaterials and their engineering applications: An overview. *Journal of Science: Advanced Materials and Devices* 3(3):263–288. https://doi.org/10.1016/j.jsamd.2018.05.003

Mittal AK, Chisti Y, Banerjee UC (2013) Synthesis of metallic nanoparticles using plant extracts. *Biotechnology Advances* 31(2):346–356. https://doi.org/10.1016/j.biotechadv.2013.01.003

Momeni S, Nabipour I (2015) A simple green synthesis of palladium nanoparticles with sargassum alga and their electrocatalytic activities towards hydrogen peroxide. *Applied Biochemistry and Biotechnology* 176(7):1937–1949. https://doi.org/10.1007/s12010-015-1690-3

Monteiro CA, Levy RB, Claro RM, Castro IRRD, Cannon G (2010) A new classification of foods based on the extent and purpose of their processing. *Cad Saúde Pública* 26(11):2039–2049. https://doi.org/10.1590/S0102-311X2010001100005

Murugan K, Benelli G, Panneerselvam C, Subramaniam J, Jeyalalitha T, Dinesh D, Nicoletti M, Hwang J-S, Suresh U, Madhiyazhagan P (2015) Cymbopogon citratus-synthesized gold nanoparticles boost the predation efficiency of copepod Mesocyclops aspericornis against malaria and dengue mosquitoes. *Experimental Parasitology* 153:129–138. https://doi.org/10.1016/j.exppara.2015.03.017

Nakagawa K (2014) Nano- and Microencapsulation of Flavor in Food Systems. In: Kwak H-S (ed) *Nano- and Microencapsulation for Foods*. John Wiley & Sons, Ltd, Chichester, UK, pp 249–271

Nasrollahzadeh M, Mohammad Sajadi S (2015) Green synthesis of copper nanoparticles using Ginkgo biloba L. leaf extract and their catalytic activity for the Huisgen [3 + 2] cycloaddition of azides and alkynes at room temperature. *Journal of Colloid and Interface Science* 457:141–147. https://doi.org/10.1016/j.jcis.2015.07.004

Nile SH, Baskar V, Selvaraj D, Nile A, Xiao J, Kai G (2020) Nanotechnologies in food science: Applications, recent trends, and future perspectives. *Nano-Micro Letters* 12(1):45. https://doi.org/10.1007/s40820-020-0383-9

Ozturk B, Argin S, Ozilgen M, McClements DJ (2015) Formation and stabilization of nanoemulsion-based vitamin E delivery systems using natural biopolymers: Whey protein isolate and gum Arabic. *Food Chemistry* 188:256–263. https://doi.org/10.1016/j.foodchem.2015.05.005

Palchetti I, Mascini M (2008). Electroanalytical biosensors and their potential for food pathogen and toxin detection. *Analytical and Bioanalytical Chemistry*, 391(2): 455–471. https://doi.org/10.1007/s00216-008-1876-4

Pradhan N, Singh S, Ojha N, Shrivastava A, Barla A, Rai V, Bose S (2015) Facets of nanotechnology as seen in food processing, packaging, and preservation industry. *BioMed Research International* 2015:1–17. https://doi.org/10.1155/2015/365672

Pyrgiotakis G, McDevitt J, Gao Y, Branco A, Eleftheriadou M, Lemos B, Nardell E, Demokritou P (2014). Mycobacteria inactivation using engineered water nanostructures(Ewns). *Nanomedicine: Nanotechnology, Biology and Medicine*, 10(6): 1175–1183. https://doi.org/10.1016/j.nano.2014.02.016

Rajeshkumar S, Kumar SV, Ramaiah A, Agarwal H, Lakshmi T, Roopan SM (2018). Biosynthesis of zinc oxide nanoparticles usingMangifera indica leaves and evaluation of their antioxidant and cytotoxic properties in lung cancer (A549) cells. *Enzyme and Microbial Technology*, 117: 91–95. https://doi.org/10.1016/j.enzmictec.2018.06.009

Ranjan S, Dasgupta N, Chakraborty AR, Melvin Samuel S, Ramalingam C, Shanker R, Kumar A (2014). Nanoscience and nanotechnologies in food industries: Opportunities and research trends. *Journal of Nanoparticle Research*, 16(6): 2464. https://doi.org/10.1007/s11051-014-2464-5

Reig CS, Lopez AD, Ramos MH, Ballester VAC (2014). Nanomaterials: A map for their selection in food packaging applications. *Packaging Technology and Science*, 27(11), 839–866. https://doi.org/10.1002/pts.2076

Rotti RB, Sunitha DV, Manjunath R, Roy A, Mayegowda SB, Gnanaprakash AP, Alghamdi S, Almehmadi M, Abdulaziz O, Allahyani M, Aljuaid A, Alsaiari AA, Ashgar SS, Babalghith AO, Abd El-Lateef AE, Khidir EB (2023) Green synthesis of MgO nanoparticles and its antibacterial properties. *Frontiers in Chemistry* 11:1143614. https://doi.org/10.3389/fchem.2023.1143614

Salunke BK, Sawant SS, Lee S-I, Kim BS (2016) Microorganisms as efficient biosystem for the synthesis of metal nanoparticles: Current scenario and future possibilities. *World Journal of Microbiology and Biotechnology* 32(5):88. https://doi.org/10.1007/s11274-016-2044-1

Savolainen K, Pylkkänen L, Norppa H, Falck G, Lindberg H, Tuomi T, Vippola M Alenius H, Hämeri K, Koivisto J, Brouwer D, Mark D, Bard D, Berges M, Jankowska E, Posniak M, Farmer P, Singh R, Krombach F, Bihari P, Kasper G, Seipenbusch M (2010) Nanotechnologies, engineered nanomaterials and occupational health and safety – A review. *Safety Science* 48(8):957–963. https://doi.org/10.1016/j.ssci.2010.03.006

Shilpa BM, Rashmi R, Manjula NG, Sreekantha A (2022) Bioremediation of Heavy Metal Contaminated Sites Using Phytogenic Nanoparticles. In: Shah MP, Roy A (eds) Phytonanotechnology. Springer Nature, Singapore, pp 227–253.

Singh KR, Nayak V, Singh J, Singh AK, Singh RP (2021) Potentialities of bioinspired metal and metal oxide nanoparticles in biomedical sciences. *RSC Adv* 11(40):24722–24746. https://doi.org/10.1039/D1RA04273D

Singh RP (2011) Prospects of nanobiomaterials for biosensing. *International Journal of Electrochemistry* 2011:1–30. https://doi.org/10.4061/2011/125487

Stephen Inbaraj B, Chen BH (2015) Nanomaterial-based sensors for detection of foodborne bacterial pathogens and toxins as well as pork adulteration in meat products. *Journal of Food and Drug Analysis* 24(1):15–28. https://doi.org/10.1016/j.jfda.2015.05.001

Suresh D, Shobharani RM, Nethravathi PC, Pavan Kumar MA, Nagabhushana H, Sharma SC (2015) Artocarpus gomezianus aided green synthesis of ZnO nanoparticles: Luminescence, photocatalytic and antioxidant properties. *Spectrochimica Acta Part A: Molecular and Biomolecular Spectroscopy* 141:128–134. https://doi.org/10.1016/j.saa.2015.01.048

Tan F, Leung PHM, Liu Z, Zhang Y, Xiao L, Ye W, Zhang X, Yi L, Yang M. (2011). A PDMS microfluidic impedance immunosensor for E. coli O157:H7 and Staphylococcus aureus detection via antibody-immobilized nanoporous membrane. *Sensors and Actuators B: Chemical*, 159(1): 328–335. https://doi.org/10.1016/j.snb.2011.06.074

Uboldi C, Giudetti G, Broggi F, Gilliland D, Ponti J, Rossi F (2012) Amorphous silica nanoparticles do not induce cytotoxicity, cell transformation or genotoxicity in Balb/3T3 mouse fibroblasts. *Mutation Research/Genetic Toxicology and Environmental Mutagenesis* 745(1–2):11–20. https://doi.org/10.1016/j.mrgentox.2011.10.010

Venkat Kumar S, Rajeshkumar S (2018) Plant-Based Synthesis of Nanoparticles and Their Impact. In: Tripathi DK, Ahmad P, Sharma S, Chauhan DK, Dubey NK (eds) *Nanomaterials in Plants, Algae, and Microorganisms*. Academic Press, pp 33–57. https://doi.org/10.1016/B978-0-12-811487-2.00002-5

Vijayaraghavan K, Ashokkumar T (2017) Plant-mediated biosynthesis of metallic nanoparticles: A review of literature, factors affecting synthesis, characterization techniques and applications. *Journal of Environmental Chemical Engineering* 5(5):4866–4883. https://doi.org/10.1016/j.jece.2017.09.026

Wang, C., Kim, Y. J., Singh, P., Mathiyalagan, R., Jin, Y., & Yang, D. C. (2016). Green synthesis of silver nanoparticles by Bacillus methylotrophicus , and their antimicrobial activity. Artificial Cells, Nanomedicine, and Biotechnology, 1–6. https://doi.org/10.3109/21691401.2015.1011805

Weaver CM, Dwyer J, Fulgoni VL, King JC, Leveille GA, MacDonald RS, Ordovas J, Schnakenberg D (2014) Processed foods: Contributions to nutrition. *The American Journal of Clinical Nutrition* 99(6):1525–1542. https://doi.org/10.3945/ajcn.114.089284

Weiss J, Takhistov P, McClements DJ (2006) Functional materials in food nanotechnology. *Journal of Food Science* 71(9):R107–R116. https://doi.org/10.1111/j.1750-3841.2006.00195.x

Yadav LSR, Shilpa BM, Suma BP, Venkatesh R, Nagaraju G (2021) Synergistic effect of photocatalytic, antibacterial and electrochemical activities on biosynthesized zirconium oxide nanoparticles. *European Physical Journal Plus* 136(7):764. https://doi.org/10.1140/epjp/s13360-021-01606-6

Zahir AA, Chauhan IS, Bagavan A, Kamaraj C, Elango G, Shankar J, Arjaria N, Roopan SM, Rahuman AA, Singh N (2015) Green synthesis of silver and titanium dioxide nanoparticles using euphorbia prostrata extract shows shift from apoptosis to G_0/G_1 arrest followed by necrotic cell death in Leishmania donovani. *Antimicrobial Agents and Chemotherapy* 59(8):4782–4799. https://doi.org/10.1128/AAC.00098-15

Zhang T, Lv C, Chen L, Bai G, Zhao G, Xu C (2014) Encapsulation of anthocyanin molecules within a ferritin nanocage increases their stability and cell uptake efficiency. *Food Research International* 62:183–192. https://doi.org/10.1016/j.foodres.2014.02.041

Zhang Y, Dong Y, Zhou J, Li X, Wang F (2018) Application of plant viruses as a biotemplate for nanomaterial fabrication. *Molecules* 23(9):2311. https://doi.org/10.3390/molecules23092311

Zhou R, Zhao L, Wang Y, Hameed S, Ping J, Xie L, Ying Y (2020) Recent advances in food-derived nanomaterials applied to biosensing. *TrAC Trends in Analytical Chemistry* 127:115884. https://doi.org/10.1016/j.trac.2020.115884

9 Bionanomaterials as Edible Food Packaging Materials

Anindita Ray, Prishila Dutta, Saptak Bhattacharjee, Tulika Mukhopadhyay, and Shakeel Ahmed

9.1 INTRODUCTION

In the present era, people are largely dependent on packaged food. Food is packaged to protect it from chemical, biological, and physical changes. Food packaging has constantly been a censorious factor wherein proper coping with and maintenance of high quality of foods are required because foods are noticeably at risk of spoilage, which can make it unacceptable to its consumers. Active and innovative food packaging using nanotechnology by providing mechanical and barrier properties, detecting pathogens, has taken the place of traditional packaging methods to process smart, interactive, and reactive food packaging with progressed functionalities (Ashfaq, et al., 2022).

In recent decades, the food sector has fundamentally changed due to the advancing technique of nanotechnology. To revamp the functional behaviour of food and conceive nanoparticles from food ingredients, food nanotechnology applies various nanomaterials with sizes ranging from 1 to 100 nanometers. As consumers are increasingly concerned about food exorcism and various health benefits, researchers are tracking down the way to improve the quality of foods with minimal disruption to the products' nutritive value. There has been an increased demand for nanoparticle-based materials in the food industry, as many of them contain essential elements and have also been found to be innocuous. They were also found to be stable at a high temperature and pressure (Singh, et al., 2017). This development in nanotechnology, supported by increased global investment, has fueled nano-packaging markets worldwide in recent years. Edible coatings loaded with nanomaterials and nanoparticles are more beneficial than traditional packaging materials as they can better conserve and maintain the quality of food standard. The functional properties of packaging materials are enhanced by organic natural substances such as proteins, carbohydrates, and fats. Inorganic products such as metals and metal oxides, and combinations of these two (nanoclay) nanoparticles, are commonly made by integrating them into polymer matrices. Improving vapor and gas barrier properties through the manifestation of nanofillers, antimicrobial properties of nanocomposite films, and smart packaging based on nanosensors are some of the procedures that nanotechnology applies to maintain the food quality and the food standard through bundling (Ashfaq, et al., 2022).

Previously, the most commonly used food packaging materials were plastic, metals, glass, paper, and so forth, but researchers have found that these materials have many hazards and limitations. Plastics like polyethylene terephthalate (PET), polypropylene, polyvinyl chloride (PVC), polyvinylidene chloride (PVDC), polyamide, and so forth have been commonly used in food packaging. But these polythenes are not environmentally safe and can manifest many environmental hazards. Nowadays biopolymer materials for food packaging have become more popular, as they are produced from edible biomolecules such as cellulose, starches, proteins, and lipids; and are less toxic, are environment-safe, biodegradable, and recyclable; so, the manufacturers use these

DOI: 10.1201/9781003432791-11

bio-based, environment-friendly food packaging materials, keeping in mind the consumer's consciousness about health and environment. Operationalized nano particles have, moreover, been utilized to provide defense against overwhelming metals, the potential sources of harmful explosions, and have also been utilized to avoid bio-film development arrangement on food items (Dey, et al., 2022). Nanosensors, which are nanoscale devices, utilise nanoparticles to inform consumers about food items that are unsafe for consumption, as there are concerns over the risks associated with nanotechnology and its possible effects on human health. For example, barcodes for antibacterial baby milk bottles to ensure food safety are acceptable for most people because they do not affect a person's well-being (Chadha, et al., 2022). Nanoparticles are framed as an outgrowth of processing techniques identical to homogenization and milling, and there is also the possibility that the combination of ingredients instinctively self-assembles into micelles, nanofibers, and so on. Because it can lead to renovation in the texture, taste, processability, and stability of foods throughout their shelf life, nanotechnology has the capacity to mutate the food system and have a valuable influence on food science (Krishna, et al., 2022). As these materials possess cohesive structures, so are they able to form a protective coating to the outer covering of the food, which prolongs the storage life of the food without affecting the food standard through several mechanisms (Otoni, et al., 2017; Mihindukulasuriya et al., 2014).

1. Providing an obstruction between food and environment to anticipate the exchanges of condensation, gases and lipid from the environment to the food materials and vice versa.
2. Providing the antimicrobial effects, thus preventing contamination and maintaining food safety.
3. Preventing the evaporation of different volatile compounds, flavoured compounds and loses of nutrients from the food.

These edible materials can also be accustomed in the form of parcels, containers, and so forth.

9.2 NANOSYSTEMS AS COMPONENTS OF EDIBLE COATINGS

With the advancement in technologies, it has become more feasible to find out more about the functional modifications of edible coatings and acceptability of the ingredients that will be infused in the food products. The compounds that are generally used in this process of producing edible packaging are nanofibers, nanoemulsions, polymeric nanoparticles, solid lipid nanoparticles, nanostructured lipid carriers, nanotubes, nanocrystals, or the blend of inorganic and organic nano-sized components. These nanosystems are incorporated into protein or polysaccharide matrices which are commonly known as 'nanocomposites' (Zambrano et al., 2018).

The chief function of edible packaging and coating is to regulate a huge transfer between the food and the ambient environment. They act as the barrier to an external environment and a suitable biopolymer with proper permeability is used to do the job. They act as barrier against the following:

1. **Moisture:** The final finished food products, if exposed to humidity can increase their water activity by absorbing water from its surroundings, thereby changing in its texture, increasing microbial growth and aggravating undesirable chemical and enzymatic reactions. Water vapour permeability values of the films are an important parameter for determining the efficiency of the films. It has been found that WVP values of hydrocolloidal based films are quite high compared to edible wax and plastic films. This is due to the low water affinity, low polarity and dense molecular matrix of the lipid or hydrophobic compound-based films. If edible coatings can be applied to the foods before osmotic dehydration, then it can prevent the loss of valuable water-soluble compounds. They also help in selective dehydration.

2. **Oxygen:** Presence of oxygen and its contact with the food can lead to oxidation of lipids, enzymatic browning of the fresh cut fruit and vegetable products, microbial growth and discolouration of the cut raw meat. These problems can be solved by using edible packaging with low oxygen permeability. This helps extend the storage life and condition of oxygen-sensitive foods.
3. **Aroma:** Volatile organic compounds are the chemistry behind these mouth-watering aromas and flavors of the foods, which are susceptible to losses. Along with this, off-flavors may also migrate to the foods during storage and distribution, which also needs to be prevented. Here, barrier-efficient edible packaging plays a role. Edible films have to select in such a way that the migrating compounds have low affinity and diffusibility through the film material and matrices. Protein and polysaccharide based edible films are hydrophilic in nature, and this property makes them an appropriate barrier against non-polar aromatic compounds. So, based on this property, flavors and aroma are encapsulated in a protein and polysaccharide based edible films for their better delivery.
4. **Oil:** Hydrocolloid based films help to retain moisture and reduce fat intake in deep-fried foods. Moreover, hydrophilic property of proteins and polysaccharide-based films are believed to give grease resistance to the lipid containing foods (Janjarasskul et al., 2010).

9.3 EDIBLE FILM MAKING PROCEDURE

Preparation of edible packaging is one of the most important things to be done properly, as it decides the possibility of the packaging material being successful (see Figure 9.1). First of all, the film-forming solution is to be produced by mixing all the components in a proper way and in a proper amount. It is suggested to stir the mixture slowly for better results. It is then homogenized in a proper way. Next, the film-forming formulation is degassed, that is, the air micro bubbles are removed. Vacuum degassing and ultrasonic degassing are mainly used to remove the excess air. The next step is the formation of the edible film and drying it so that it can be applied to the food item. Finally at room temperature, it is cooled and then placed on the roller. Generally edible films

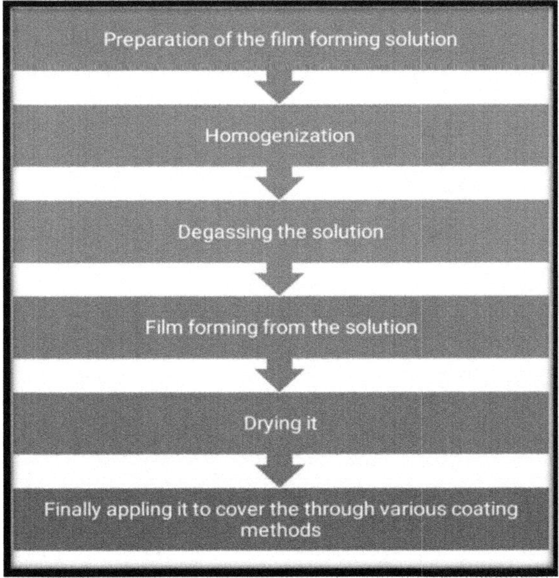

FIGURE 9.1 Different steps in making edible packaging. (Figure created by authors.)

are applied to food by spraying the solution on food or by immersing the food into the solution, by multilayer coating of the solution on the food, spreading the solution, and so forth (Otono, et al., 2017; Hernalsteens, et al., 2020).

9.4 BIOFILMS IN EDIBLE PACKAGING

Edible packaging can be applied to food items through so many ways and, among those, are two common examples: wrapping and coating. Coating is directly added to food products, whereas wrapping is a procedure by which the packaging is done through wrapping the food product. Edible films are generally produced by two methods, one is wet and the other is dry. Apart from that, dipping, spraying, panning, and so forth are performed so that the food item can hold the coating in a proper way. The wet method is known as casting and the dry one is known as extraction (Mahela, et al., 2021).

There are varieties of organic nanomaterials which are used as edible biofilms (see Figure 9.2). The notable ones are as follows:

9.4.1 Polysaccharides

Monosaccharides and disaccharides are condensed by glycosidic linkage to form polysaccharides, present in nature abundantly, which is edible and safe. As polysaccharides are tightly packed so they act as a barrier of oxygen, moistures, volatile compounds, by creating H-bonds with added agents. Polysaccharides are hydrophilic or water loving in nature; they can absorb water and swell up. The hydrophilic properties of polysaccharides are obstructed with the inclusion of hydrophobic substances like wax, lipids and so forth. The advantages of using polysaccharides and their derivatives are as follows:

- They are present in ample amount, cheap, and handy.
- They have a wide extent of plating solution viscosities.
- They provide an excellent mechanical and gas, oils, and lipids barrier properties.

But they barely provide any resistance against water migration.

There are different types of polysaccharides (animal origin, plant origin, marine origin) used in food packaging like cellulose, starch, pectin, agar, chitosan, chitin and so forth.

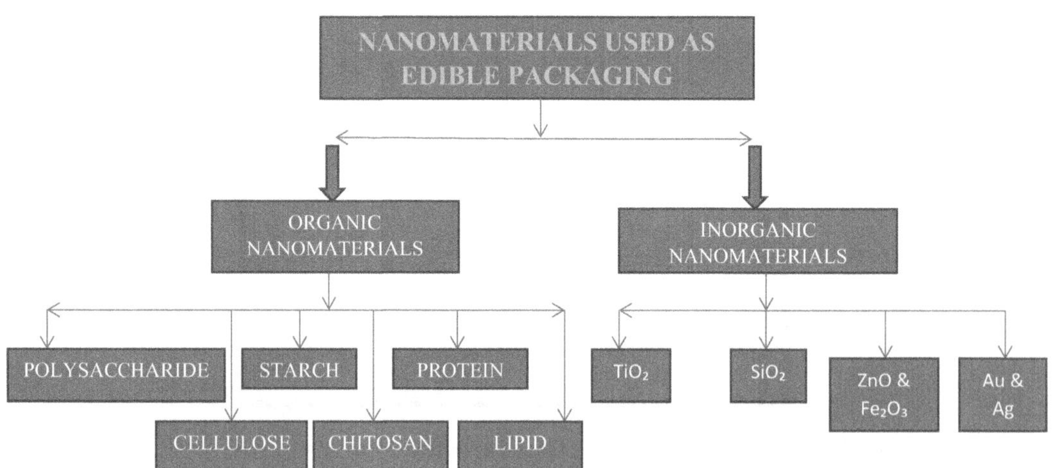

FIGURE 9.2 Various nano materials used in making edible packaging. (Figure created by authors.)

9.4.2 Cellulose

Cellulose is plant origin polysaccharides which are formed by ß-1, 4 linkages, condensing to glucose monomer. It absorbs water and then becomes a gel-like substance. Cellulose is often used as coating materials in various fruits. The polymer chains in the cellulose are compacted together, and they have a high crystalline form due to its regular structure and an ordered arrangement of hydroxyl groups, so they are water insoluble but their solubility can be enhanced by esterification. The common commercially used water soluble cellulose is methyl cellulose, hydroxypropyl cellulose, hydroxypropylmethyl cellulose and carboxymethyl cellulose. Edible coatings made from these cellulose ethers are used to provide barriers from oil, moisture, and oxygen (Janjarasskul, et al., 2010).

9.4.3 Nano Cellulose

Cellulose is a most abundantly found polysaccharide in cell walls of plants. Nano cellulose is a nano-structured cellulose consisting of alternating crystalline and amorphous strings <100 nm in diameter (Sharma, et al., 2019). The crystalline structure is made of a hydrogen bond making the polymer relatively stable. The hydrogen bond system makes the cellulose axial chain highly stiff, providing excellent barrier properties (Eichhorn, et al., 2010). Even if the nano-cellulose is present in a smaller percentage then it can also elevate the effectiveness and rigidity of the polymer because of its high aspect ratio. Cellulose nano-whiskers are prepared by delignification of coconut fibers by hydrolysis (Rosa, et al., 2010). Nano-cellulose polymers are used in paper and composite industries to improve strength, mechanical properties, uniformity, and biodegradability (Naseer, et al., 2018). Similar traits have been shown by the cellulose microfibers even when they were isolated from the different functional cells of the same plant (Oun, et al., 2016). Cellulose nano-particles isolated from grain straws have a huge potential in the food packaging industry. It is reported that the cellulose nano-particles were separated from wheat straw by the acid hydrolysis method, and the carboxymethyl cellulose composite membrane was prepared by different modification treatment. Thus, modified composite tensile strength increased by 45.7 per cent, and the water vapor permeability decreased by 26.3 per cent. Because of its degradability, superior mechanics, and water vapor permeability, this type of modified cellulose nano-particles composite material is now potentially being used in the food packaging industry (Wu, et al., 2020). After removing the hemicellulose and lignin by pre-treatment from the plant cell to retain the cellulose as a skeleton material, nano-cellulose is further obtained through oxidation treatment, acid treatment, and alkali treatment combined with mechanical treatment. Table 9.1 shows various methods of extraction of cellulose from plants.

The first step of the methodology for the isolation of nano-cellulose crystals from ethanol residues is that the bio-residues should be purified by the removal of extractives using a Soxhlet apparatus (Oksman, et al., 2011).

TABLE 9.1
Different Methods of Cellulose Extraction from Plant Sources

Category	Pretreatment	Treatment	Experiment	Reference
Sweet potato vine	3 mm cut in length	Acid-alkali treatment	$NaClO_2$ (10g/L), acetic acid (2ml/L), bleaching (70°C), 2 ml of acetic acid should be added every three hours for 6 times; 6 wt% of KOH solution to be added at 80 °C for 3 hours. Then they should be washed.	[Nakagaito et al., 2018)
Onion and garlic	The stalks of the onion, garlic should be washed, dried and pulverized.	Acid-alkali treatment	The mixed solution of toluene and ethanol (2:1) was reflux for 6 hrs; 0.7% (w/v) $NaClO_2$, 100°C, 2 hrs, fiber to solution mass ratio 1:50.	[Reddy et al., 2018)
Buckwheat and rice	They should be immersed in acetone solution at room temperature for 24 hrs.	Acid-alkali treatment	They should be bleached for four times at 75°C with $NaClO_2$ under pH 4–5; using 4% (w/w) NaOH solution they should be treated for 1 hr.	[Nakamura et al., 2019)

Further processes are as follows:

Extraction of the samples at 150°C for 6 hours using the mixture of toluene and ethanol in the ratio of 2:1

↓

Then, they should be allowed to cool

↓

The sample has to be placed in a Buckner funnel and for the removal of excess solvent they should be vacuum dried at room temperature

↓

They should be dried using ethanol for the removal of the toluene traces

↓

This same process is repeated for the re-extraction after drying

↓

For the removal of any traces of remaining solvent residues, the sample has to be vacuum-dried for 24 hours at room temperature

↓

After reweighing the sample, it is now ready for the bleaching step

↓

700 ml deionized water should be preheated to 70°C for 12 hours in a conical flask, then to it 5 ml of acetic acid and 6.7 g of NaCl to be added and then the samples have to kept for 12 hours at 70°C

↓

For any further removal of residual chemicals by centrifugation, deionized water to be added to the mixture after 24 hours.

FIGURE 9.3 Purification and extraction of nano cellulose. (Figure created by authors.)

The procedure of separating nano-cellulose from reject cellulose (Tashiro, et al., 1991) is as follows:

At first the rejected cellulose should be placed in a 65% sulfuric acid solution at 40°C with a constant mechanical stirring for 30 mins.

↓

Deionized water should be added to dilute the suspension and centrifuged at 6,000 rpm for several times in cycles of 5 mins each.

↓

Supernatant should be removed which should be replaced by deionized water and mixed.

↓

After at least five times of washing or after the formation of turbidity in the supernatant, the centrifugation process should stop. Now this turbid supernatant has to be collected to go through the process of dialysis against deionized water till it obtains a constant ph.

↓

To avoid overheating the samples have to be sonicated in an ice bath for 2 mins.

FIGURE 9.4 Bondeson's method of purification and extraction of nano cellulose. (Figure created by authors.)

Both crystals have similar lengths, approximately 300 nm for nano-cellulose crystals from ethanol residues and 377 nm for nano-cellulose crystals from reject cellulose; though, the length distribution of nano-cellulose crystals from ethanol residues ranges of 375–449 nm and nano-cellulose crystals from reject cellulose between 300 and 374 nm (Herrera, et al., 2012).

9.4.4 Starch: Nano Starch

Starch is the foremost potential raw material because it could be a natural, ecological innocuous polysaccharide broadly available from plants; currently, starch nanoparticles are gaining more interest for improved quality and wide applications involved in food, paper making, pharmaceuticals, packaging, and rubber due to its low cost (Gonera, et al., 2002). Starches are abundantly found in roots of plants, staple crops, and cereals such as rice, maize, wheat, barley, corn, tapioca, potato, and others. Starch happens in the configuration of discrete and mostly crystalline granules of amylose and amylopectin. Starches that are used as nano-particles have at least one dimension smaller than 1000 nm but are bigger than a single molecule, and are utilized as nano-fillers in composites to progress strength, adaptability, biodegradability, water impermeability, temperature, and barrier properties (Dularia, et al., 2019; Campelo et al., 2020). Starch-based carriers are widely used in the encapsulation and control release of bioactive compounds for applications in food and pharmaceutical formulations due to their high biocompatibility and their small size, which offers a high surface:volume ratio and high surface:weight ratio when hollow nano-particles are synthesized (Caldonazo, et al., 2021). Amylose and hydroxypropyl amylose films are used as coatings and encapsulating agents for providing protection against oxygen, lipids, and to enhance texture. They are generally used in confectionaries, bakery, butters, and meat products.

There are various techniques for starch nanoparticle preparations, including acid hydrolysis, enzymes, or combination of two, regeneration and mechanical treatments using extrusion, irradiation, ultrasound, or precipitation by co-solvent (Ahmad, et al., 2019). The easy and cost-effective methods for starch and starch derivative nanoparticles are always of paramount importance. Among various such methods, nano-precipitation and ultra-sonication are very simple and reliable methods for nanoparticle production with desired size.

Starch nano-particles are now obtained from the breakdown of the starch granules using different methods, classified into two processes, top-down and bottom-up depending on the material used for the synthesis (Wang, 2004; Sant, 2012). The top-down processes include methods of synthesis where the reduced size SNPs are obtained and the bottom-up processes cover all the synthesis obtained from the assembly of molecules or macro-molecules in the form of small primary nuclei that are growing. The following are the methods:

A. Nano-precipitation
B. Micro emulsion
C. Emulsion cross-linking
D. Dialysis
E. Sacrificial templates
F. Ball Milling
G. Acid Hydrolysis (with or without)
H. Ultrasonic Atomization

9.4.5 Chitosan

Chitosan is produced by fusion of chitin with alkali chitosan coatings with semi permeability and is best for fresh fruits to enhance post-harvest life. The antimicrobial and antioxidant properties provided by chitosan are due to its cationic properties. They are also able to slow release functional compounds. It has been found they slow down enzymatic browning in fresh cut fruits.

TABLE 9.2
Proteins Films and their Sources

Types of protein	Examples of food
1. Whey protein isolated film / coating	sliced Kasar cheese, soft rennet curd cheese
2. Collagen edible film	pork
3. Gelatin edible film	fruits and vegetables
4. Soy protein isolate film	shrimp

Source: Table created by Authors.

Other polysaccharides used are pectin, alginate, carrageenan (complex mixture of several polysaccharides), exudate gums, seed gum, and gellan gum (see Figure 9.4).

9.4.6 Protein

Protein is the polymeric form of amino acids and is formed by peptide bonds. Structurally, protein is mainly of two types—globular and fibrous. Researchers have shown that fibrous protein such as soy protein, whey protein, and globular protein such as collagen have good potential to give film and coating on food and increase shelf life of the food by creating a barrier to lipophilic compounds like fats and oils. (Kumar, et al., 2022). Other research has shown that protein extracted films and coating has good efficacy to provide antioxidant as well as antimicrobial effects (Saklani, et al., 2019). The various sources of protein has been elaborated in Table 9.2 and their properties in Table 9.3.

9.4.7 Lipid

Lipids are not a polymeric form of any molecule. Lipids are hydrophobic in nature, so a good barrier for water and moisture content, and they are more efficient packaging material than carbohydrate and protein. A study has shown that wax film is more efficient than other lipid-based films. Waxes can be obtained from plants as well as animals. Edible waxes are esters of long chain fatty acids and alcohol. Wax also manifests reduction of the vaporization rate of water from food. Studies have shown that aloe-vera gel coating on fruits has a good efficacy to create a barrier for moisture content (Kumar, et al., 2022).

In the presence of water there are mainly four different structures formed: micelle, inverted micelles, bilayer, and bilayer vesicles (Mouradian, 2021). Generally, wax, paraffin and glycerides are used, as they help to avoid drying or dehydration of edible films. Waxes are totally insoluble in water and thus most effective. Whereas in the case of triglycerides, long chain triglycerides are partially soluble in water, although short chain triglycerides are not, but rather are hydrophobic. Some monoglycerides are solvent in hexane, helping to increase adhesion between two different parts with different hydrophobicity. For example, it enhances adhesion between the edible film and food items and even sometimes between lipidic layer and the hydrocolloid layer of the edible film. In case of fruits and vegetables, paraffin and natural wax are used to store these for long time. According to some studies, respiration of strawberries was reduced successfully by using beeswax into chitosan-based coating. Lipid-based edible films are also proved to be helpful to preserve meat in some studies, though it has some disadvantages too in case of meat, reported by some studies (Díaz-Montes, et al., 2021). Along with the advantages of lipid based edible packaging materials, some disadvantages are also seen such as in case of wax the changes in taste, texture, even the changes in surface and potential rancidity are observed when implemented (Saklani, et al., 2019).

TABLE 9.3
Different Proteins Used as Edible Films and their Properties

NAME OF PROTEINS	CONSTITUENT COMPOUNDS	PROPERTIES AND USAGE
Wheat gluten	Gliadins, Glutenins	Acts as moisture, gas and solute barrier; can be used for active packaging, drug delivery and modified environment packaging.
Corn zein	Group of Prolamins	Used as finishing agents as it provides surface gloss; acts as moisture, oxygen, lipid barrier to the nuts, candies, confectionary products.
Soy protein isolate		They can be applied to precooked meat product to control lipid oxidation and to restrict surface moisture loss.
Collagen and Gelatin	Glycine, Hydroproline, Proline Gelatin is hydrolyzed product of Collagen	Collagen based films are used on processed meats to reduce shrink loss, increase juiciness, absorbing fluid exudates for a variety of cooked meats. They can reduce oxygen, moisture and oil migration and can even carry bioactive compounds. So, they are used as an encapsulating agent in hard and soft gel capsules for low moisture and oil-based food ingredients, dietary supplements and pharmaceuticals.
	MILK PROTEINS	
Caseins	Phosphoproteins	They act as moisture barrier of water-soluble pouches, fresh products, dried fruits and frozen foods; they are efficient oxygen barrier of casein-based films at low water activity and thus reduce lipid oxidation of nuts.
Whey proteins	Predominant protein is beta-lactoglobulin	Reduce enzymatic browning in fresh cut fruits especially apples.

Source: Table created by Authors.

9.4.8 Additives

Apart from polysaccharides, proteins, and lipids some additives are also used to bring positive changes in edible packaging. For example, plasticizers (glycerol, aloe, raisins), chaotropic agents (urea), and polyphenols are used to improve the packaging in different aspects. These additives show different characteristics which supports the whole process of packaging. For example, plasticizers help in reduction of intermolecular force along with the melting temperature of the mixture. It also modifies viscosity and theological rheological properties. On the other hand, chaotropic agent increases the solubility of polymers in water. Polyphenols are helpful to protect the products and also works as a stabilizer (Díaz-Montes, et al., 2021). Along with that, some emulsifiers, antimicrobials like nisin, chitosan, plant extracts like essential oil extracts from different plants like cinnamon, clove, thyme, onion, and so forth, organic acids and their salts, bacteriocins, antioxidants like butylatedhydroxyanisole, butylatedhydroxytoluene, tertiary butylhydroquinone, citric acid, ascorbic acid, tartaric acid, and so forth, and even some enzymes like lysozyme and lactoperoxidase are also used for better results (Saklani, et al., 2019).

The examples of the additives used in edible films for improvisation are as follows:

1. **Plasticizers:** Generally, plasticizers are the agents used in edible films to improve the mechanical properties of the edible films by increasing durability along with flexibility. These properties are among the important things to be maintained while preparing an edible film

for better results. Plasticizers are generally hydrophilic and non-volatile in nature, and their molecular weight is small. Some mono saccharomyces, di-saccharides, and even oligosaccharides are used as plasticizer. Sometimes, polyols, lipids, and surfactants are also used as plasticizer in edible films. For example, glucose, sucrose, fructose-glucose syrups, glycerol, some glycerol derivatives, polyethylene glycols, sorbitol, derivatives of lipids along with lipids, and so forth, are generally used (Otoni, et al., 2017).

2. **Emulsifiers:** Like plasticizers, emulsifiers are also used as additives to the edible films. Emulsifiers are needed to modify the film by the modification of interfacial energy of interface, where two different systems meet and interact, for example, the interface between water and lipid, the interface between water and air, and so on. Emulsifiers ensure the proper coverage of the surface along with so many additional advantages; these are employed to the edible films for improvement (Saklani et al., 2019).
3. **Antimicrobials:** To prolong the shelf life of food goods, antifungal and antibacterial properties are essential traits anticipated in edible films, as these films or coatings serve as protective coverings for the food. Along with the natural antimicrobials, some synthetic antimicrobial agents are used in edible packaging to safeguard the food item from any microbiological contamination and control its growth. Some examples of antimicrobials used to in edible packaging are nisin, some organic acids, chitosan, plant extracts, and so forth are generally used (Saklani, et al., 2019).
4. **Chitosan:** The most important characteristic of chitosan is its antimicrobial property. Chitosan is a polysaccharide used in edible films as an additive. It also helps in chelating trace metals, and this is how it plays a vital important role in controlling a microorganism's growth along with preventing toxin production (Otono, et. al., 2017).
5. **Organic acids and their salts:** Some organic acids and their salts are important additives that are employed with edible films for so many reasons. For example, sorbic acid is effective against lactic acid bacteria along with yeasts and molds, benzoic acid and its salts are also known for its efficacy as a preservative effective against molds and bacteria. Apart from these two acids, acetic acid, lactic acid, propionic acid and fumaric acid are also used as additives (Saklani, et al., 2019).
6. **Plant extracts:** Some plant extracts are also used here as an additive like chitosan, organic and so on. Essentials oils collected from several plants that are rich sources of phenolic compounds are used as additives in edible films. These have so many biological effects like antioxidant properties, antimicrobial properties, and so on. Some basic examples of the plants from which extracts are collected are onion, grapefruit seed, garlic, radish, clove, oregano, mustard, and so forth (Saklani, et al., 2019).
7. **Bacteriocins:** Like other additives, bacteriocins are also used for betterment of the edible films. Bacteriocins are generally produced by various bacteria like colicins, nisin, lacticins, and pediocins. These are also effective against pathogenic bacteria and thus contribute to enhance the shelf life and control the spoilage of the food item (Saklani, et al., 2019).
8. **Antioxidants:** Antioxidants are the substances generally applied to edible food packaging or edible films to modify the quality of the film. Antioxidants do help by delaying the oxidation rate and this it is added to the films. There are so many good sources of antioxidants which can be used here. Among those the antioxidants used are butylatedhydroxytoluene, tertiary butylhydroquinone, butylhydroquinone, citric acid, ascorbic acid, tartaric acid, and so forth (Saklani, et al., 2019).
9. **Enzymes:** Some enzymes are also used as additives in edible films. These are also effective against bacteria and thus contribute to enhance the shelf life of the food product. These are collected from so many natural sources like milk. For example, lysozyme, lactoperoxidase, and so forth are used as additives (Saklani, et al., 2019).

9.5 CHARACTERISTICS OF FOOD PACKAGING

Food packaging is a very important method for a food item, as it decides the success rate of a food item along with its acceptance. Hence, so many researchers are going on to explore more in this field. There are many different types of food packaging which are active and smart (Survana, et al., 2022).

9.5.1 Active Packaging

Active packaging is a type in which the packaging is done with better functionalities using innovative materials. Generally, active packaging plays a vital role in controlling spoilage of foods by fighting against the foodborne pathogens and thus ensuring food safety and increasing shelf life. Petroleum-based polymers are used as they have so many superior functionalities that are actually so beneficial, but this leads to so many environmental problems. Nowadays, biopolymer-based materials are employed to produce a sustainable packaging having superior functionalities. These are generally eco-friendly and sometimes these are even edible too.

Active agents for food packaging include O_2 scavengers (iron, palladium, ascorbic acid, pyrogallol, gallic acid, glucose oxidase, laccase), ethylene scavengers ($KMnO_4$, metal oxides, activated carbon, MOFs, titanium dioxide), antimicrobials (metals, nisin, essential oils, lactoferrin, lysozyme, chitosan), CO_2 emitters (sodium bicarbonate, citric acid, ferrous carbonate), and antioxidants (phenolic compounds, lignin, essential oils, plant extracts, α-Tocopherol), which help in active packaging of food items by influencing the physiological processes such as respiration, ripening, and transpiration in fresh food items, the physical processes such as powders caking and bread staling, the chemical processes such as oxidation of oils and fats, and so forth. They also protect the food items from microbiological spoilage caused by bacteria, fungi and yeast (Survana, et al., 2022).

9.5.2 Smart Packaging

Smart packaging ensures and gives consumers information regarding the safety of the food items and its sustainability. For example, it informs the customer about the condition of the food and its environment like temperature, pH, and so forth. Thus, it detects unsafe food easily with so many benefits. Research is going on to explore more in this field. It is actually designated to track the internal and external quality of a food product so that it can assure a consumer about the quality of food and its safety. This technology is mainly used in dairy products, the meat and seafood industry, and bakery, and confectionery products (Survana, et al., 2022).

9.6 PACKAGING MATERIALS

The packaging materials are of following types:

1. Petroleum-based plastic polymer

Petroleum-based plastic polymer is one of the satisfactory food packaging materials. When crude oil and natural gas is used to prepare the plastic then that plastic is known as petroleum-based plastic polymer. The good qualities that made it acceptable as packaging material are as follows:

- They are economical
- Their processing method is trouble-free
- Their weight is low
- They are resistant from oil and chemical substances
- They provide an effective resistivity from gas and water vapour
- They are easily reusable and recyclable

Some of the most commonly used examples of plant-based plastics and petroleum-based plastics that are used in food packaging industries are polyethylene terephthalate, polyvinyl chloride, polystyrene, polypropylene, low-density polyethylene, and high-density polyethylene. The negative traits of these plastics are non-biodegradability, non-renewability and non-compostability (Survana, et al., 2022).

2. Biodegradable Polymers

Most petroleum-based plastic polymer is non-biodegradable, which is an environmental hazard. In recent years, to reduce the harmful effects on the environment, interests developed to increase the formation and production of biodegradable polymers from renewable sources, which would be the least threat to the environment. Biodegradable polymers do not get accumulated in the environment because they can be decomposed by the action of enzymatic catalysis by the microorganisms like bacteria and fungi if they are kept in bioactive habitats like landfills. Other than enzymatic catalysis, there are other mechanisms like chemical hydrolysis, which is a non-enzymatic breakdown of polymer chains. Another benefit of using biodegradable polymer is that they generally produce methane, water, carbon dioxide biomass and other natural substances as the end products, which help in balancing greenhouse emissions. Thus, biodegradable or edible packaging, plant extracts and nanocomposite materials are preferred over synthetic or non-biodegradable packaging as we can see that they confer less hazardous impact on the environment.

Commonly used biodegradable polymers for edible packaging are proteins and polysaccharides with their derivatives because they are easily available and present in ample amount and moreover, they have the capacity to form brittle and hard materials when they are polymerized. Some of the approaches that lead to the formation of polysaccharides and protein films that include the casting process, extrusion and molding. There are certain physical, chemical, and mechanical characteristics of synthetic biodegradable polymers that make them perfect as a packaging material. There is a way of biodegrading the synthetic materials, which includes slow chemical hydrolysis in an aqueous environment, and this process can be supported by enzymatic catalysis. The varieties of compounds that are conventionally used as antimicrobial packaging materials are polyhydroxyesters which includes polyglycolic acid; polylactic acid (PLA) and its co-polymers polyactideglycolide. Plasticizers including glycerol, polyethylene glycol, sorbitol, propylene glycol, ethylene glycol, and others are blended with biodegradable polymers because they have been found to improve the flexibility, extensibility, and moisture sensitivity of packaging films. Biodegradable films incorporated with nanoparticles shown to have greater potential for active packaging with an extended shelf life and effective storage of packaged food (Survana, et al., 2022).

3. Paper

The most prominent use of paper in packaging is in food labelling. But what are the properties that make paper suitable as a packaging material? The answer follows:

- Paper is cheap, which makes it affordable and economical.
- It is very light in weight. So, if it is used in packaging, it would not result in a heavy weight to the package and ultimately the whole product.
- It can be printed on easily, and attractive printing will enhance the appeal of the product.
- Paper has satisfactory mechanical qualities.

In spite of having all these good qualities, there are two cons to using paper as packaging material: their sensitivity to humidity: that is paper can absorb moisture easily and promptly. This can interfere with the integrity and barrier quality of the packaging and can degrade the food quality. In the process of manufacturing various additives may get added to paper, paperboard, and recycled paper, which can lead to hazardous health effects. Coating the surface with biopolymer increases the hydrophobicity, which can keep the barrier qualities of the paper intact and cuts the toxic additives from being added (Survana et al., 2022).

4. Examples of Some Edible Packaging

1. Milk packaging: Casein based films can be used replacing plastic wraps which will reduce non-recyclable/non-biodegradable waste and thus will effectively prevent oxygen from migrating inside the packaging and spoiling food.
2. Package-less froyo: Its objective is to reduce the amount of packaging in their products, so, in this technology organic fruit skins are used to keep moisture in and contaminants and oxygen out.
3. KFC's edible coffee cup: these cups are made from water coated with sugar paper lined with heat-resistant white chocolate.
4. Monosol's edible pouches: They dissolve away in hot water without making any difference in the taste of the food.
5. Bob's burger edible wrapper.
6. Tomatoes transformed into edible food containers.
7. KFC's edible bowl: This tortilla bowl was introduced in Bangalore, India, to replace plastic bowls, bags, plates, napkins and so forth (Mouradian 2017).

9.7 CONCLUSION

With the advancement of technologies, the ways of improving and maintaining human health is also improving. Moreover, we are also concerned with our environment and people are trying to diminish the utilization of plastics and non-biodegradable substances. So, a new technology of developing edible food packaging developed that is now a very important field of interest.

The penetrable nature of the food bundling material(s) is the major disadvantage of the packaging technique. Currently, no packaging material offers absolute protection to atmospheric gases, water vapor, and food and packaging materials. Available options include organic polymeric materials: polypropylene, polyethylene, polyethylene terephthalate, polystyrene, and polyvinyl chloride continue to be the materials of choice in the food packaging industry, favored for their low cost, ease of processing, and light weight. However, the main complication lies in their inherent permeability to gases and other small molecules.

REFERENCES

Ahmad, M. et al. Nano encapsulation of catechin in starch nanoparticles: Characterization; release behavior and bioactivity retention during in-vitro digestion. *Food Chem.* 2019a, 270, 95–104.

Ashfaq A., Khursheed N., Fatima S., Anjum Z., Younis K. Application of nanotechnology in food packaging: pros and Cons. *J Agri Food Res.* 2022 March, 7, 100270.

Caldonazo A., Almeida S.L., Bonetti A.F., Lazo R.E.L., Mengarda M., Murakami F.S. Pharmaceutical applications of starch nanoparticles: a scoping review. *Int J Biol Macromol.* 2021, 181, 697–704.

Campelo P.H., Sant'Ana A.S., Pedrosa Silva Clerici M.T. Starch nanoparticles: production methods, structure, and properties for food applications. *Curr Opin Food Sci.* 2020, 33, 136–140.

Chadha U., Bhardwaj P., Selvaraj S.K., et al. Current trends and future perspectives of nano materials in food packaging application. [JOURNAL OF NANOMATERIALS2022, 2022, Article ID 2745416.

Dey A., Pandey G., Rawtani D.. Functionalized nano materials driven antimicrobial food packaging: a technological advancement in food science. *Food Control.* 2022 January, 131, 108469.

Díaz-Montes E, Castro-Muñoz R. Edible films and coatings as food-quality preservers: an overview. *Foods.* 2021, 10(2), 249.

Dularia C., Sinhmar A., Thory R., Pathera A.K., Nain V. Development of starch nanoparticles based composite films from non-conventional source – water chestnut (Trapa bispinosa). *Int J Biol Macromol.* 2019, 136, 1161–1168.

Eichhorn S.J., Dufresne A, Aranguren M., et al. Review: current international research into cellulose nanofibers and nanocomposites. *J Mater Sci.* 2010, 45, 1–33.

Janjarasskul T, Krochta J.M. Edible packaging materials. *Annu Rev Food Sci Technol.* 1, 415–448.[2010]

Gonera A., Cornillon P. Gelatinization of starch/gum/sugar systems studied by using DSC, NMR, and CSLM. *Starch-Starke.* 2002, 54(11), 508–516.

Herrera, M.A., Mathew, A.P., Oksman, K. Characterization of cellulose nanowhiskers: a comparison of two industrial bio-residues. *IOP Conf Ser Mater Sci Eng.* 2012, 31, 012006.

Hernalsteens S., de Moraes M.A., da Silva C.F., Vieira R.S. Chapter 24 – Edible films and coatings made up of fruits and vegetables. *Biopolymer Membranes and Films.* Elsevier. 2020, pp. 575–588.

Krishna A.R., Gurumoorthy S, Elayappan P, Sakthivadivel P, Kumaran S, Pushparaj P. A review on the application of nanotechnology in food industries. *Cur Res Nut Food Sci.* 2022. https://dx.doi.org/10.12944/CRNFSJ.10.3.5;

Kumar L., Ramakanth D., Akhila K., Gaikwad K.K. Edible films and coatings for food packaging applications: a review. *Environ Chem Lett.* 2022, 20, 875–900.

Mahela U., Rana D.K., Joshi U., Tariyal Y.S. Nanoedible coatings and their applications in food preservation. *J Postharvest Technol.* 2021, 08(4), 52–63.

Mihindukulasuriya S.D.F., Lim L.-T. Nanotechnology development in food packaging: a review. *Trends Food Sci Technol.* 2014 December, 40(2), 149–167.

Mouradian N. 2017. 11 Examples of packaging you can actually eat. *Sustainable Packaging*, Oct. 19, 2017. https://thedieline.com/11-examples-of-packaging-you-can-actually-eat/

Nakagaito A.N., Takagi H. Easy cellulose nanofiber extraction from residue of agricultural crops. *Int J Mod Phys B.* 2018, 32(19), 1840080.

Nakamura Y., Ono Y., Saito T., Isogai, A. Characterization of cellulose microfibrils, cellulose molecules, and hemicelluloses in buckwheat and rice husks. *Cellulose.* 2019, 26, 6529–6541.

Naseer B., Srivastava G., Qadri O.S., et al. Importance and health hazards of nanoparticles used in the food industry. *Nanotechnol Rev.* 2018, 7 (6), 623–641.

Oksman K., Etang J., Mathew A., Jonoobi M. Cellulose nanowhiskers separated from a bio-residue from wood bioethanol production. *Biomass Bioenergy.* 2011, 35, 146–52.

Otoni C.G., Avena-Bustillos R.J., Azeredo H.M.C., Lorevice M.V., Moura M.R., Mattoso L.H.C., McHugh T.H. Recent advances on edible films based on fruits and vegetables—a review. *Compr Rev Food Sci Food Saf.* 2017, 16, 1151–1169.

Oun A.A., Rhim J.W. Isolation of cellulose nanocrystals from grain straws and their use for the preparation of carboxymethyl cellulose-based nanocomposite films. *Carbohydr Polym.* 2016, 150, 187–200.

Reddy J.P., Rhim J.-W. Extraction and characterization of cellulose microfibers from agricultural wastes of onion and garlic. *J Nat Fibers.* 2018, 15, 465–473.

Rosa M.F., Medeiros E.S., Malmonge J.A., et al. Cellulose nanowhiskers from coconut husk fibers: effect of preparation conditions on their thermal and morphological behaviour. *Carbohyr Polym.* 2010, 81, 83–92.

Saklani P.S., Nath S., Kishor Das S., Singh S.M. A review of edible packaging for foods. *Int J Curr Microbiol Appl Sci.* 2019, 8(7), 2885–2895.

Sant, S.B. Nanoparticles: from theory to applications. *Mater Manuf Process.* 2012, 27, 1462–1463.

Sharma A., Thakur M., Bhattacharya M., Mandal T., Goswami S. Commercial Application of cellulose nanocomposites – a review. *Biotechnology Reports*, vol. 21. Elsevier B.V. 2019.

Singh T., Shukla S., Kumar P., Wahla V., Bajpai V.K., Rather I.A. Application of Nanotechnology in food science: perception and overview. *Front Microbiol.* 2017 August, 8

Suvarna V., Nair A., Mallya R., Khan T., Omri A. Antimicrobial nanomaterials for food packaging. *Antibiotics* 2022, 11, 729.

Wang, R. The chemistry of nanomaterials—Rao C. N. R., Müller A., Cheetham A. K.(eds), WILEY-VCH Verlag GmbH & Co. KGaA, Weinheim 2004. ISBN 3-527-30686-2, 741 pages. *Colloid Polym Sci.* 2004, 283, 234.

Wu G., Sun J., Huang C., Ren H., Zhao R. Research progress on mechanical properties of tenon-mortise joints in traditional Chinese wood structures. *J For Eng.* 2020, 5, 29–37.

Zambrano-Zaragoza M.L., González-Reza R., Mendoza-Muñoz N., et al. Nanosystems in edible coatings: a novel strategy for food preservation. *Int J Mol Sci.* 2018, 19(3), 705.

Part 3

Bionanomaterials in Biomedical Applications

10 Applications of Biomimetics and Bionanomaterials in Dentistry

Zahra Vahedi Azad, Morteza Banakar, Seyyed Mojtaba Mousavi, Ashkan Badkoobeh, and Chin Wei Lai

10.1 INTRODUCTION

Bionanomaterials and biomimetics have a strong relationship in the field of dentistry. Bionanomaterials are materials that are composed of both biological and nanoscale components, and they can mimic the structures and functions of natural materials. Biomimetics, on the other hand, is the field of design and technology that seeks to imitate the principles and mechanisms of nature. In dentistry, bionanomaterials and biomimetics have the potential to revolutionize the way that dental treatments are approached. The study of materials on a nanoscale is known as *nanotechnology*. This use of nanotechnology in healthcare is known as *nanomedicine*. Although some relevant techniques have previously been put into practice, researchers are always looking into new diagnostic and therapeutic approaches leveraging nanoscale particles and technologies. This has led to the introduction of nanotechnology and the development of nanobiomaterials in many dental specialties (Table 10.1). Researchers face a significant clinical challenge in finding innovative therapy modalities to help repair damaged tissue. Developing natural biomaterials that may be used to repair and regenerate bone and other hard tissues is critical. Significant prior work has been done in the area of structural optimization. Current synthetic biomaterials are rudimentary in comparison to nanobiomaterials. Despite their importance in restoring hard tissues, typical biomaterials like ceramics, bio-glass, and bone cement cannot induce self-tissue regeneration, which is needed for healthy bones to function properly. Therefore, creating novel biomaterials that may be used as templates for hard tissue formation is a huge challenge (Jafari et al. 2020; Komasa and Okazaki 2022). Over time, it has become easier to design synthetic materials using natural components.

Material that mimics the structure of real tissue has been demonstrated to have a more profound effect on tissue targeted in recent studies (Qasim et al. 2020). Because of desired features, including biocompatibility, reversible disintegration, mechanical strength, and configurable functional behavior, self-assembled hydrogels such as peptides, nucleic acids, and carbohydrates have recently garnered increased attention (Abolmaali et al. 2014). Otto Schmitt, a biophysicist, and biomedical engineer, developed the term "biomimetic" in the 1950s. Restorative dentistry employing biomimetic techniques includes using bioinspired peptides to induce remineralization to restore defects in teeth, as well as using bioactive and biomimetic materials to promote regeneration. These approaches mimic natural processes to repair and restore damaged dental tissue (Figure 10.1). Recent advances have led to improvements in the characteristics of restoration materials, including adhesive restorative materials, have increased our understanding of the interaction between biomaterials and tissue at the nanoscale and microscale. These developments can potentially enhance the effectiveness and performance of restorative materials in dental applications (Zafar et al. 2020).

DOI: 10.1201/9781003432791-13

TABLE 10.1
Nanomaterials in Dentistry

Nanomaterial	Characteristics	Usage in dentistry	References
Chitosan	Nontoxic, biocompatible, biodegradable, hydrating and antimicrobial properties	Gels, microparticles, microspheres, nanoparticles, scaffolds, beads, nanofibers, and sponges, drug carrier, a bioactive and biocompatible coating for orthopedic and craniofacial implant devices, mouthwash, chewing gums	(Sanap, Hegde et al. 2020)
Gelatin	Safe nature, ingestible, nontoxic, biodegradable, low cost, availability	Drug delivery, gelatin-based nanoformulation as a new tool in common dental issues	(Foox and Zilberman 2015, Shahi, Sharifi et al. 2022)
Gelatin hydrogels as a nanocarrier of fibroblast growth factor-2 (FGF-2)	Short half-life due to rapid degradation via oral enzymes, uncontrolled release mechanism	Induces dentinogenesis	(Kikuchi, Kitamura et al. 2007)
Chitosan-gelatin bioglass scaffold as ananocarrier for Hydroxyapatite	Water retention ability	Remineralization of tooth enamel	(Chen, Lee et al. 2021)
Lyposomes	Hydrophobic and hydrophilic	Drug delivery Preservations Dental restorations	(Nguyen, Hiorth et al. 2011, Melling, Colombo et al. 2018)
Silver nanoparticles	Well-tolerated tissue response, low toxicity profile in human cells, and effective antimicrobial and inhibitory properties without inducing microbial resistance	Prostheses, surgical devices, wound dressing applications, oral hygiene products in the dental field, such as toothpaste, antigingivitis agents, and restorative materials, antimicrobial agents	(Noronha, Paula et al. 2017)
Poly Lactide-co-Glycolic Acid (PLGA)	One of the most biodegradable and biocompatible synthetic polymer, local delivery of medicaments	Periodontal infections, Osteomyelitis treatment, Endodontic treatment, Alveolar bone defects treatment	(Chachlioutaki, Karavasili et al. 2022)
Zinc oxide (ZnO) nanoparticles	Versatile inorganic material, antibacterial and antifungal,	Dental infection prophylaxis (prevent antimicrobial resistance by inhibiting rapid microbial adaptation and transformation), anti-biofilm in infected root canals, a stable interface between bone and implant, In Orthodontics: (1) rolling effects, which enable two surfaces to slide on each other, and (2) functioning as spacers to prevent contact of two opposing surfaces	(Moradpoor, Safaei et al. 2021)

TABLE 10.1 (Continued)
Nanomaterials in Dentistry

Nanomaterial	Characteristics	Usage in dentistry	References
Titanium Dioxide Nanoparticles	Antibacterial effects, have been observed to improve the performance of dental resins, increase the tensile strength, shear bond strength, containing 0.1%- 0.25% of TiO_2NP can simulate the opalescence of human enamel	Used in dental material (acrylic resins based on heat-cured poly(methylmethacrylate) and GIC) TiO_2NP incorporated with the dental material can be used as a drug delivery system for long-term antibacterial activity, Glass Ionomers, Orthodontic adhesive	(Liu, Chen et al. 2022, Mansoor, Khan et al. 2022)
Nanoemulsion-Based Approach	Thermodynamically stable, optically isotropic, transparent, and homogeneous/heterogeneous mixtures of oil and water, where the nanodroplets of the dispersing medium are stabilized by the thin layer of surfactant, stable, biocompatible, safe, inhibit biofilm formation	Periodontic diseases, improving oral hygiene	(Monika, Sharma et al. 2021)

Reconstructive biomaterials and tissue engineering techniques can be utilized to repair missing or damaged dental tissues. Implant therapy has been a common feature of prosthetic care in recent years, and it may be administered in several different oral environments. The nanostructure of the material's surface has the potential to alter how bone marrow cells initially respond, which may have implications for various applications, including early osseointegration. There are situations in which implant treatment is not possible without the use of bone grafting. However, bone-filling materials that encourage bone development may allow for minimally invasive and risk-free implant therapy. Further study may lead to the creation of dental materials with improved qualities, including the use of nanoparticles, which can help in early bone building. Nanoscale devices have also been used in dental diagnosis and therapy with dental implants. It is expected that the use of biodegradable and bio-friendly materials will be of great importance in the future of green dentistry.

Chitosan is used in various dentistry fields, including periodontology, oral surgery, restorative dentistry, conservative dentistry, oral pathology, and more. Its versatility makes chitosan a useful material in various dental applications (Banakar and Sijanivandi 2022). Green nanomaterials, which are special compounds that are clean and safe, can be used with green chemistry or advanced materials in nano dimensions with high-performance applications, resistant and effective structures, and are sustainable in terms of application and production in the environment. These materials have various uses, including in drug delivery and the production of antimicrobial and antiviral materials (Banakar et al. 2022a). This chapter discusses how bionanomaterials might affect contemporary dental care (Figure 10.2).

10.2 BIOMIMETIC MATERIALS BASED ON THE TEETH

Additive manufacturing is paving the way for creating biomimetic materials thanks to its rapid growth. To increase ceramic performance, graded structures have emerged, such as those seen in enamel, DEJ, and dentin. According to an assessment of this technology, it is possible to make

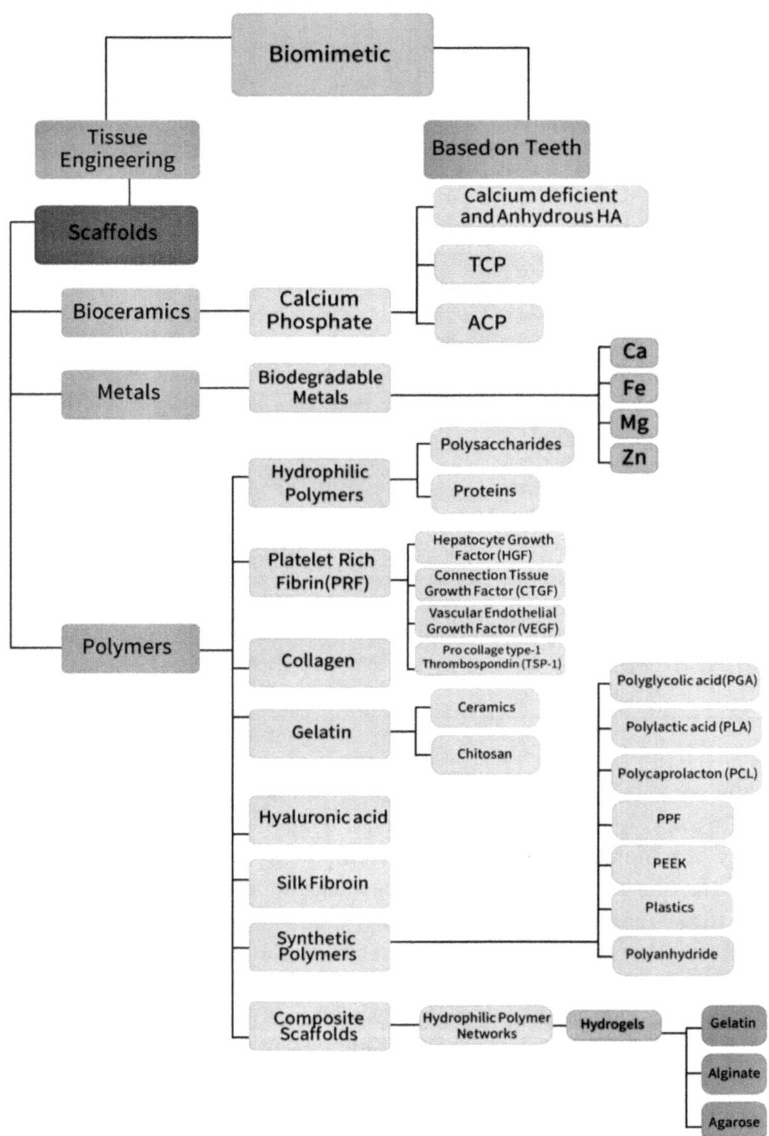

FIGURE 10.1 Biomaterial Used in Dentistry. (Figure Created by Authors.)

graded materials that utilize powder bed 3D printing processes (Chen et al. 2019; Galante et al. 2019). Hexagonally tightly packed discontinuous rods mimicking the arrangement of prisms in teeth were harder and more rigid than random or off-axis-oriented configurations of the same rods (de Obaldia et al. 2015). Changing the rods' aspect ratio, volume percentage, and orientation anticipated the observed indentation deformations and cracking from their FEA models. Using 3D-printed biomimetic Bouligand structures, this team has also examined the structure–property correlations (Huang et al. 2019). Rheology is an important part of 3D extrusion printing ceramics and cement.

Open architecture specimens including intricate forms, such as Bouligand structures with enhanced specific energy absorption, may be 3D printed utilising a 400m nozzle with a nano clay additive (Sajadi et al. 2019). A 1.36 mm nozzle was once the sole option for printing Portland cement Bouligand structures (Moini et al. 2018). New analog composite architectures reveal the decussation

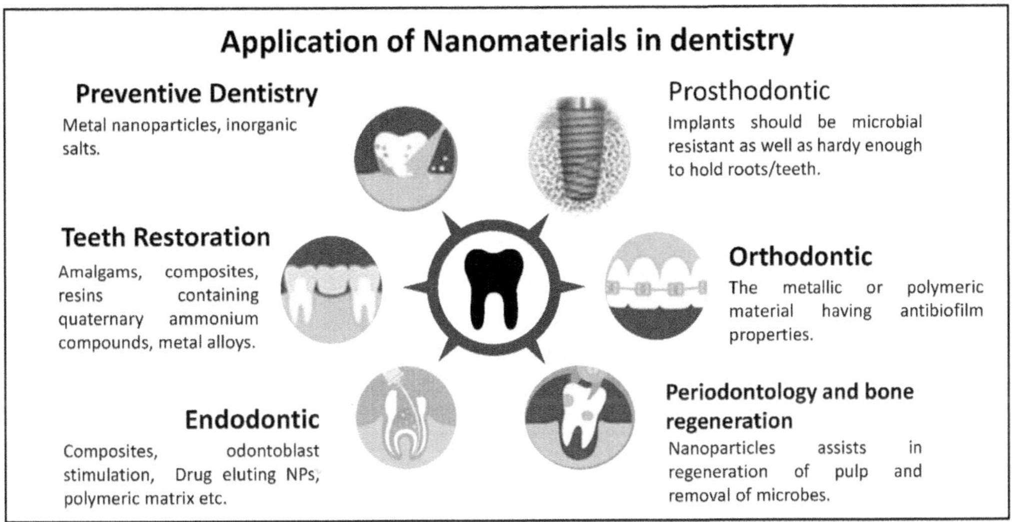

FIGURE 10.2 Pictorial Representation of Applications of Nanomaterial in Dentistry. (Sreenivasalu et al. 2022.)

of enamel prisms and the hierarchical organization of HA nano-crystallites. Hydroxyapatite (HA) has a chemical composition and structure like natural bone, making it suitable for various biomedical applications. Nanomaterials made of HA (HA-NMs) have been the subject of much research, leading to the development of new techniques of manufacturing as well as functionalization and characterization. Due to their high biocompatibility, these nanomaterials might find usage in a variety of medical settings.

One of the main advantages of HA-NMs is their ability to form strong chemical bonds with tissues and cells, which makes them suitable for use as drug carriers. These materials can be functionalized with a wide range of biomolecules and drugs, which can be released in a controlled manner over a specific time. In addition, the surface-to-volume ratio of HA-NMs is rather high (Figure 10.3), which allows them to interact with many different kinds of tissues and organs, leading to improved drug-delivery efficiency. Due to their ability to absorb and scatter light, HA-NMs can also be used in imaging applications. These materials can be functionalized with contrast agents and used for imaging techniques such as magnetic resonance imaging (MRI), computed tomography (CT), and ultrasound. Overall, HA-NMs have significant potential for use in various biomedical applications due to their biocompatibility, ability to form strong chemical bonds with tissues and cells, and high ratio of surface area to total volume (Figure 10.4) (Munir et al. 2022).

When extruded via a printer nozzle, a mixture of alumina nanoplatelets and epoxy may be organized in a circular pattern. This allows for the precise positioning of the nanoplatelets within the matrix, which can influence the material's properties and performance. When cured and tested, a Bouligand structure made from these prisms will exhibit R-curve behavior (Feilden et al. 2017). Aligned fiber-reinforced for enamel prism hierarchical structures may now be advantageous (Lewicki et al. 2017; Blok et al. 2018). Using multi-nozzle print heads, it is possible to print complex material structures with graded characteristics. There are still uncertainties about how various materials interact with one another when printing with numerous nozzles at once. A coaxial nozzle, or many coaxial nozzles, may be an answer. Printing multi-metal structures using coaxial nozzles on a microscale has been accomplished (Reiser et al. 2019). To do this, a single multichannel nozzle is used to swap and combine two metals "on-the-fly". They claim to be able to print chemical features of a certain size using electrohydrodynamic redox printing.

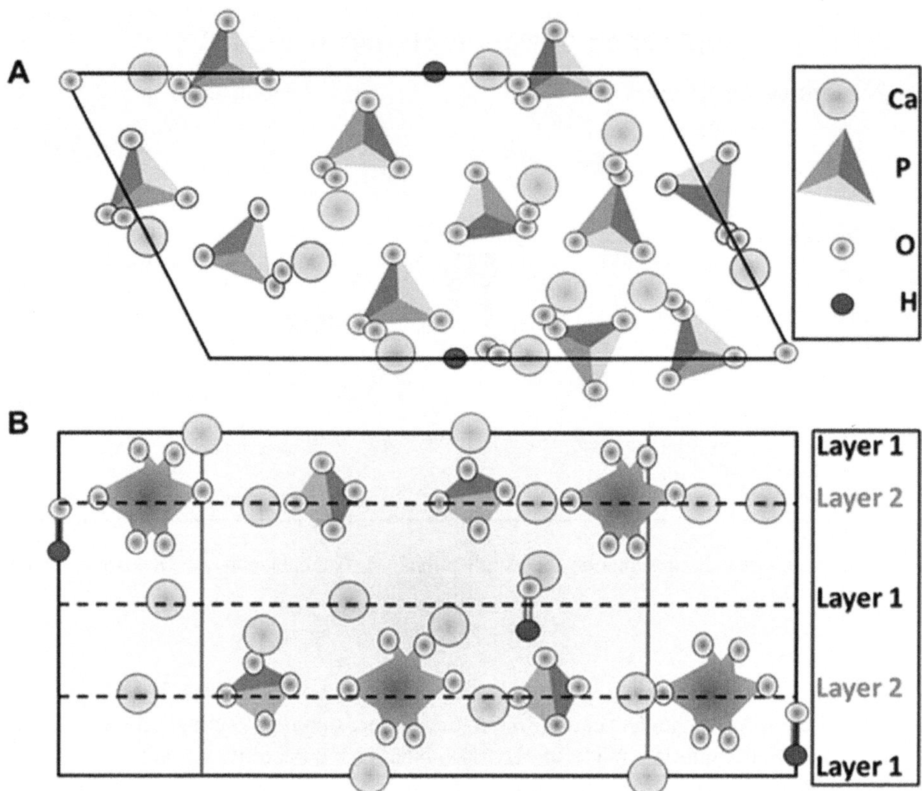

FIGURE 10.3 Hydroxyapatite Crystal Structure, (A) Top View and (B) Side View. Dashed Lines Represent One Line. (Munir et al. 2022.)

These days, demineralization and loss of enamel can be treated. Chen et al. attempted to develop a technique using pulsed laser deposition (PLD) and an erbium-doped yttrium aluminum garnet (Er:YAG) laser to directly produce a hydroxyapatite (HAp) layer on enamel surfaces. The coatings produced using this method were almost entirely hydrolyzed into HAp within two days, as demonstrated by X-ray diffraction peaks and scanning electron microscopy images showing the transformation of alpha-tricalcium phosphate (APC) powder into microparticles upon irradiation. The micro-Vickers hardness test also showed that the coatings restored nearly all the hardness lost during decalcification. These findings indicate that the Er:YAG-PLD technique has promise for usage in clinical settings to restore enamel abnormalities (Chen et al. 2021).

10.2.1 Biomimetic Tissue-Engineering with Nanomaterials Aspects

Since then, the discipline of tissue engineering has grown tremendously to demonstrate its ability to regenerate practically every organ and tissue (Langer 1993). Biomimetic tissue engineering aims to build or generate a biological system from scratch to treat or improve diseased or damaged tissues. Tissue engineering based on biomimicry may achieve its goals by integrating knowledge from various domains, including physics and biology (Nör 2006). Three key concepts are utilized while working with tissues in tissue engineering (Kottoor 2013; Zafar et al. 2015b). Cellular attachment to the surrounding tissues produces a new matrix while delivering growth factors to support and endorse cell functions. Biomimetic scaffolds are implanted to help cells differentiate, proliferate, and synthesize (Ikada 2006).

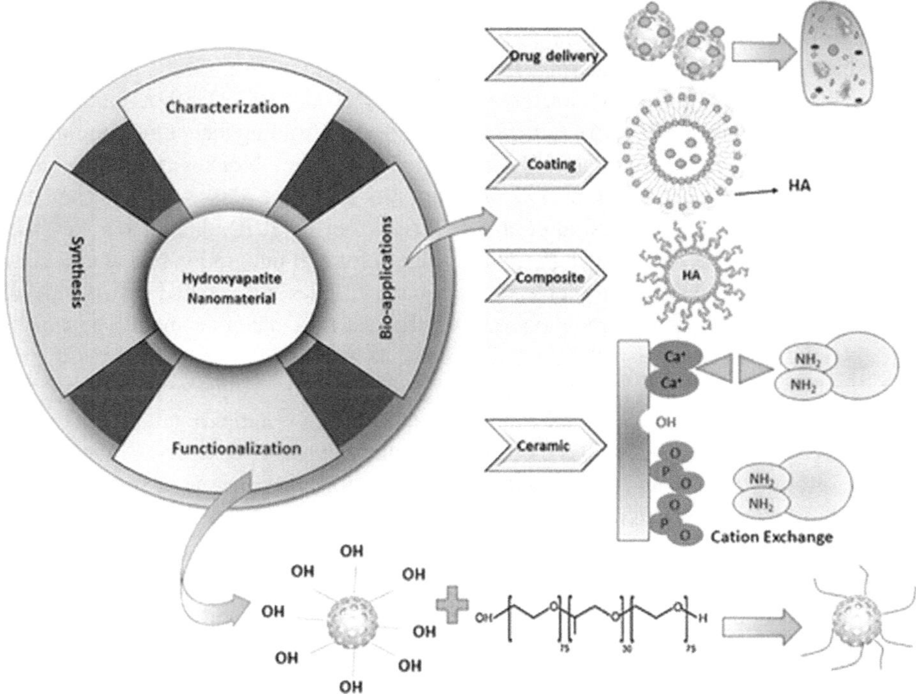

FIGURE 10.4 A Review of Hydroxyapatite Nanomaterials and Future Prospects. (Munir et al. 2022.)

Multiple obstacles exist for bone repair, utilizing mesenchymal stem cells. Adipose-derived dedifferentiated fat (DFAT) cells were investigated by Nakano et al. They tested their cell proliferation rate, osteoblast differentiation, and bone repair capabilities when combined with activated platelet-rich plasma (aPRP). After isolating aPRP and rat DFATs by centrifugation and ceiling culture, they found that aPRP substantially boosted cell proliferation. On day 21, after induction was initiated, a 9 mm critical hole had nearly completely healed, as evidenced by positive alizarin red staining and significantly higher Runt-related transcription factor 2 (Runx2) and osteocalcin expression than in the controls (60%). Because of this, bone regeneration using DFAT cells and aPRP may be an effective option (Tanaka et al. 2017; Nakano et al. 2020).

10.2.2 Scaffolds for Biomimetic Tissue Engineering

Biomimetic scaffolds may be used to generate a three-dimensional microenvironment. There are both good and damaged tissues as cells multiply, differentiate, and heal themselves (Kemppainen and Hollister 2010). Through diverse mechanisms, such as scaffold design, biomimetic scaffolds imitate the extracellular matrix's (ECM's) architectural and compositional constituent. Bioresorbable and biomimetic materials have been thoroughly investigated worldwide to build scaffolds with certain desirable features (Bowlin 2011). Making a biomimetic scaffold with the right mix of physical, mechanical, and biological qualities to aid in tissue regeneration is a difficult undertaking. When using biomimetic scaffolds, the materials and their breakdown products must be biocompatible and toxin-free. The original tissue ECM should be used as a standard for mechanical and physical qualities. For bone regeneration, scaffolds should have similar physical and mechanical qualities as bone tissues. Collagen and other components of the ECM organic matrix are crucial in controlling cell behaviour and organization (Kadler 2004; Roseti et al. 2017). Typically, nanofibers have diameters between 50 to 500 nanometers (Elsdale and Bard 1972; Kadler et al. 1996). The extreme porosity and interconnection of the material's holes make tissue regeneration, cell growth, differentiation, and proliferation simpler (Yu et al. 2018). The biomimetic scaffold's mechanical qualities are also critical for biomimetic tissue engineering. If regenerative mechanics are to be used, they must be designed or modified to fit those characteristics (Henkel et al. 2013; Jaschouz and Mehl 2014; Kumbar et al. 2014; Bouët et al. 2015). Functionally graded scaffolding techniques imitate the shape of real tissues, which can be beneficial (Qasim et al. 2020). The "bioactivity" of the biomimetic scaffold is an additional critical feature. The fabrication of bioactive glass scaffolds can be done in various ways (Fu et al. 2010; Khurshid et al. 2019; Sajadi et al. 2019) and encourage host tissue neoformation via cell differentiation and migration suited for cell integration (Ge et al. 2008).

Additionally, biodegradable, and bioresorbable characteristics are deemed vital for developing biomimetic scaffolds. The biodegradation rate of the scaffold material should be tunable and regulated, simulating tissue regrowth (Ge et al. 2008). Transplanting synthetic, three-dimensional (3D) bone graft replacements is a promising strategy for repairing extensive bone loss. Even with abundant cells and cell cytokines, significant bone defects may require a suitable 3D scaffold to sustain and accelerate bone healing. Nanogel cross-linked porous-freeze-dry (NanoCliP-FD) gel was created for bone tissue engineering, and its in-vivo performance was assessed by Adachi et al. Spectroscopic imaging of bone tissue produced in vivo after applying NanoCliP-FD gel demonstrated this gel scaffold's potential as a therapeutic method for bone diseases characterized by major bone defects (Adachi et al. 2020).

10.2.3 Materials for Biomimetic Scaffold Fabrication

People can profit from replacing or repairing damaged and diseased human tissues using biomimetic materials during tissue regeneration. There are a variety of tissue-engineering applications in restorative dentistry that make use of biomimetic biomaterials (Bouët et al. 2015). Many materials have been used to mimic biomimicry, including polymers, ceramics, and metal.

10.2.3.1 Polymers

All natural biomaterials have the potential to be both viable and biocompatible. These include polysaccharides and proteins (Schmidt and Baier 2000; Ige et al. 2012). Their self-assembly or cross-linking characteristics make them ideal for dissolving cell membranes, which are called hydrophilic polymers (Rios et al. 2011). Natural polymers are not commonly used in these applications because of their inability to withstand heavy loads and their lack of mechanical qualities (Nigam and Mahanta 2014). The following is a list of natural polymers dentists use to make biomimetic scaffolds. Platelet-rich fibrin (PRF) is a biodegradable scaffold that is formed by a combination of hepatocyte growth factor (HGF), connective tissue growth factor (CTGF), vascular endothelial growth factor (VEGF), and procollagen type I thrombospondin 1 (TSP-1). The activity of certain markers, such as alkaline phosphatase (ALP), osteocalcin, and bone sialoprotein, decreases in the presence of PRF, while the expression of collagen-I and CRP23 increases. This suggests that PRF may be able to promote the formation of new tissue and support the healing process (Zhao et al. 2013). Wound healing and other tissue regeneration applications have both been linked to platelet-rich fibrin (Femminella et al. 2016; Miron et al. 2017), periodontal defects (Ajwani et al. 2015; Agarwal et al. 2016; Najeeb et al. 2017), and bone regeneration (Junxian et al. 2023; Bansal and Bharti 2013; Mathur et al. 2015; Shah et al. 2015; Cortese et al. 2016).

Bio-C sealer is a newly developed material with high flexibility based on calcium silicate. Calcium hydroxide and polyethylene glycol, a dispersion agent, were detected by Fourier-transform infrared (FTIR) spectroscopy. In an in vitro assay employing the V-79 cell line, the cytotoxicity of Bio-C Sealer at 1:4 dilutions were lower than that of Calcipex II. The histological response of the peri-radicular tissue around the beagle dog roots was assessed after 90 days. Seventeen out of 18 roots evaluated for peri-radicular inflammation were filled with Bio-C Sealer. These findings corroborate the potential usefulness of the Bio-C sealer as a root-end filling material, supporting the conclusions of previous research (Okamura et al. 2020).

There are several uses for collagen in biomimetic tissue engineering scaffolds (Yeo et al. 2008). Enzymatic biodegradation and mechanical qualities are the primary advantages of this material (Asti and Gioglio 2014). Inorganic chemicals, such as HA, may be used to cross-link collagen, changing its characteristics (Turnbull et al. 2018). It was discovered that mesenchymal stem cells extracted from the bone marrow of mice might be used to repair a calvarial lesion. Collagen–HA scaffolds were degraded after three weeks, resulting in wound healing. Human-derived bone MSCs and collagen–HA scaffolds can cure osteochondral lesions satisfactorily (Turnbull et al. 2018). The electrospun collagen ECM mimics had greater osteoblastic differentiation on titanium than the real thing (Iafisco et al. 2013).

It is possible to make gelatin out of a natural substance that is both biocompatible and biodegradable. It is mainly employed for wound and medication delivery because of its low mechanical characteristics (Wu et al. 2010). A combination of this scaffold material and others may be used to modify the mechanical properties for bone and cartilage healing applications (Asti and Gioglio 2014). In addition to gelatin, ceramics and chitosan are often used as gelatin-based scaffolds (Kim et al. 2009; Jiang et al. 2010; Peter et al. 2010; Hunter and Ma 2013; Jing et al. 2013). The antibacterial and biocompatibility properties of chitosan make it an ideal biomaterial for various biomedical applications, including medication administration and dental tissue regeneration (Hamilton et al. 2015; Luo et al. 2015; Xu et al. 2015; Lotfi et al. 2016; Tang et al. 2016; Ali et al. 2017, 2020; Konovalova et al. 2017; Sarwar et al. 2020). Using chitosan scaffolds for dental pulp stem cells increased cell differentiation, adhesion, and proliferation because of the 3D matrix. Researchers created composites using biomimetic materials like chitosan and other biomaterials and increased the scaffold qualities (Chen and Fan 2007; Kim et al. 2009). Blood vessels formed more quickly in the body when chitosan and tricalcium phosphate were used together. When chitosan and collagen are stacked together, they may generate a biomimetic material that can change its mechanical properties by layering HAT-7 dental epithelial cells and DSCs on top of it (Liao et al. 2010; Ravindran et al. 2010). Chitosan and hydroxyapatite composite scaffolds have been described for use in bone regeneration (Chavanne et al. 2013; Liu et al. 2013). Other biomaterials containing chitosan were also used.

The substance's strong biocompatibility, biodegradability, and viscoelasticity of hyaluronic acid are well-known properties (Asti and Gioglio 2014). Tissue engineering cannot use this material due to its low mechanical strength and quick rate of disintegration. Biomaterials like chitosan and alginate are widely used to enhance hyaluronic acid's physical properties and cell and mechanical capabilities (Ganesh et al. 2013; Miranda et al. 2016). Cell adhesion, proliferation, contact, and cell growth were enhanced when hydrogel and the RGD peptide were mixed with hyaluronic acid. Endodontics and pulp regeneration have been studied together in the lab (Collins and Birkinshaw 2013). Hydrogels containing heparin-modified hyaluronic acid increased MSC adhesion and proliferation in biomimetic biomedical settings (Gwon et al. 2017). However, due to its low mechanical strength, it has a restricted range of applications. Calcium chloride may be used as a cross-linking agent to boost mechanical strength (Ehrbar et al. 2004).

In Matsumoto's study (Matsumoto et al. 2020), the adherence of protein and bone marrow cells to titanium (Ti) surfaces was measured in real-time using a quartz–crystal microbalance (QCM). The findings of the quantitative cell microscopy (QCM) experiments demonstrated that surface treatment enhanced the adhesion of sticky proteins and bone marrow cells to the implant surfaces. QCM proved to be a useful tool for studying the earliest stages of implant attachment, and both UV and atmospheric-pressure plasma treatments led to improved osseointegration behavior and reduced levels of reactive oxidant species in the implants. The overall performance was highest for samples treated with atmospheric-pressure plasma, making them the best candidates for clinical usage (Metavarayuth et al. 2021).

Silkworms are the primary source of natural silk, a polymeric biomaterial mostly made from silk (Zarkoob et al. 2004). Fibrin, the primary structural component of silkworm silk, is hydrophobic and resistant to most solvents, including water (Hardy et al. 2008). The protein material known as silk fibroin has a heavier hydrophobic heavy chain and a lighter hydrophobic light chain than most other proteins (Shimura et al. 1976; Zafar et al. 2015a). This silk fibroin's primary and secondary confirmation is mostly of -sheet repeats, with glycine and serine rounding out the amino acid makeup (Zhang et al. 2007). The crystallinity, hydrophobicity, and strength qualities of silk fibroin are due to the primary and secondary conformation of silk fibroin. Silk fibroin has exhibited advantageous qualities for various biomimetic biomedical applications, such as high biocompatibility, biodegradability, mechanical capabilities, and the ability to work under changing humidity and temperature conditions (Sheu et al. 2004; Kim et al. 2005; Hakimi et al. 2007). Because of this, silk has been studied for various biological uses, such as drug delivery scaffolds for tissue engineering and dental crowns (Soffer et al. 2008; Wenk et al. 2008; Mandal et al. 2009; Lammel et al. 2010; Jindal et al. 2014). Depending on the application, silk biomaterials may be treated into a wide variety of micron or even nanoscale morphologies, such as film and foamed hydrogels, non-woven silk fibers, and hydrogel coatings (Hardy et al. 2008). Another way to create composites with custom-tailored qualities is by mixing natural silk along with natural, manmade, and biological components (Chen et al. 2012). On the other hand, natural silk-based biomaterials offer a promising future in tissue regeneration. Synthetic polymers, in addition to natural ones, have shown improvements in cell adhesion, the capacity to transfer soluble compounds, a controlled breakdown rate, and the ability to construct diverse forms in addition to natural polymers (Asti and Gioglio 2014). These include PPF, PEEK, polyanhydride, PEEK, polycaprolactone (PCL), polylactic acid (PLA), and polyglycolic acid (PGA), among the synthetic polymers (PGA).

More durable synthetic plastics can be produced in large quantities at a reduced cost, making them ideal for large-scale production. The mechanical and physical properties of synthetic polymers allow them to be used to replace both soft and hard tissues. Lab-made polymers have biocompatibility issues since they are less capable of interacting with cells (Dhandayuthapani et al. 2011). Composite scaffolds are the greatest approach to overcoming these drawbacks (Nigam and Mahanta 2014). Hydrogels, or hydrophilic polymer networks, are a crucial class of polymers in bone-tissue engineering due to their capacity to control how new tissues develop in the body. Hydrogels, such as gelatin, alginate, and agarose, may be both natural and artificial. Cell differentiation, adhesion, and proliferation are made possible by the water-absorbent properties of these materials. In several tissue engineering applications, hydrogels are employed to transport bioactive compounds and replicate the topography of the extracellular matrix (ECM) (Matassi et al. 2011).

10.2.3.2 Bioceramics

Ceramics made from calcium phosphate (CaP) are well-known for their bioactivity and are often utilized to make synthetic bone grafts (Erol-Taygun et al. 2013; Khurshid et al. 2019; Skallevold et al. 2019). There are many types of calcium phosphates, such as amorphous calcium phosphate (ACP), tricalcium phosphate (TCP), and calcium-deficient and anhydrous hydroxyapatite (Wang et al. 2014). Many hard tissues have been regenerated using these bioceramics owing to the biocompatibility, high osteo-inductivity, and osteo-conductivity of these materials. Using a magnesium-enhanced nanocrystalline ceramic HA to repair an ameloblastoma-related bone defect was recently effective (Grigolato et al. 2015). Bioglasses (SiO_2Na_2O-CaO-P_2O_5) are commonly utilized in endodontics because of their strong bioactivity. These bioceramics are limited in their use for tissue engineering because of their poor mechanical properties, high stability, and its hard-to-shape nature.

In an effort to promote bone development, Yamaguchi created and tested a biphasic bone replacement coated with hydroxyapatite (HA) crystals. Bone replacement granules with a biphasic structure were made by dissolving HA granules in a supersaturated calcium phosphate solution and combining them with five different medical injectable solutions. Analysis of the resultant silt and testing of the biphasic HA granules' biological activity were performed in vitro and in vivo. Precipitated calcium phosphate crystals appeared visibly as needle-shaped crystals on the surface of the HA granules, and the results demonstrated that low-crystalline HA was isolated from these crystals. Bone formation and cell division were both greatly aided by this. Bone mineral density, volume, and area were all considerably boosted in animal trials when treated with biphasic hydroxyapatite granules. These results indicate that this substance has the potential as a bone replacement for use in dental implant therapy (Yamaguchi et al. 2021).

10.2.3.3 Metals

Scaffolds made of biomaterials that mimic natural bone and tissue may be 3D printed out of metals, which are seen as promising materials for this purpose. Materials that might be employed in the 3D printing of scaffolds include chromium and cobalt alloys and nitinol and stainless-steel-based materials (Chou et al. 2013). The term "biodegradable metals" (BMs) refers to metallic biomaterials that degrade over time. These materials have recently been introduced for use in biomimetic tissues. According to Zheng and co-workers, metal corrosion products delivered in vivo elicited an appropriate host response. These items facilitate tissue regeneration since they leave no implant remains behind (Zheng et al. 2014). Iron, calcium, magnesium, and zinc are among the elements that make up this group. Mg- and Fe-based BMs have lately been used in scaffold fabrication. However, their potential is still unknown because of a lack of published data on their biocompatibility and cell survival statistics (Zhuang et al. 2008; Vorndran et al. 2011; Chou et al. 2013). To increase the variety of materials available for biomimetic 3D printing scaffolds for bone regeneration, it is suggested that more studies be done employing BMs. Titanium (Ti) is a preferred metal for bone replacement in dental implants due to its high biocompatibility, corrosion resistance, mechanical strength, and other physical properties. Increased osseointegration and bone ingrowth at the implant interface are two other benefits of this chemical (Gong et al. 2015).

Graphene's strong electrical and thermal conductivity, as well as its mechanical resilience, make it a promising material for use in electronics. This technique is used in medication administration devices made of biomaterials, and another is in structural materials used to assist tissue regeneration (Liu et al. 2008). Graphene has good antibacterial properties because of its well-ordered structure, increasing its oxidized state (GO). Oxygen and carbon atoms are organized in a honeycomb arrangement in GO, giving it a two-dimensional appearance. While GO's edges are made up of carbonyl and carboxyl groups, its base comprises epoxide and hydroxyl groups (Raslan et al. 2020). Because of its hydrophilicity and negative charge, GO can bind to osteoblasts effectively (Foroutan et al. 2018). Because of the GO coating, bacteria experience oxidative and membrane stress due to the implant's antibacterial action and chemical functionalization (Pinto et al. 2013). Oxidation and membrane stress contribute to antimicrobial capabilities. After cells have deposited on graphene-based materials, they will come into touch with the material's sharp nanosheets, which will cause membrane stress and ultimately lead to the antimicrobial process of anion-independent superoxide oxidation (Stankovich et al. 2007). Compared to Nb2O5 coating, graphene-coated titanium alloy significantly increased corrosion resistance (Kalisz et al. 2015). It has been shown that GO may be used as a coating material for implants and bone scaffolds. Collagen membranes are used in oral surgery to correct bone abnormalities. These collagen membranes prevent soft tissue from entering the bone as it grows. Graphene oxide, a graphene derivative, is applied to collagen membranes to enhance bone and soft tissue biocompatibility. Radunovic et al. have shown promising results in collagen membrane biocompatibility after coating with GO utilizing DPSCs. The DPSCs' development into bone cells was hastened by the GO-coated collagen membrane's high biocompatibility. Bone growth and clinical outcomes will improve if GO cannot take the role of conservative membranes (Radunovic et al. 2017).

Key reasons for oral dysfunction include peri-implantitis and the resulting shortening of implant lifespans. Removing oral biofilms and avoiding future infections require both frequent self-care and expert mechanical teeth cleaning (PMTC). So, using agar particles, Sato devised a novel blasting cleansing method that is gentle on the implant surface. When white alumina (WA) abrasive grains and CaCO3 particles were used for blasting, the replicated spots were almost entirely eliminated, and the roughness, which was at first believed to be produced by tiny scratches and thus increasing the roughness, transformed into a satin layer. Most synthetic marks can be removed by surface blasting the sample surface with glycine and agar granules. The sample's shine was preserved despite a slight rise in surface roughness following cleaning (Sato et al. 2021).

10.3 METHODS FOR BIOMIMETIC SCAFFOLD MANUFACTURING

Copying the extracellular matrix's activities and architecture has been the traditional method of replicating these processes. There have been several developments in creating porous 3D biomimetic scaffolds in recent years. Using these tiny scaffolds, tissue regeneration may be regulated precisely. 3D scaffold manufacturing processes have been studied extensively to find new and better ways to make structures. One kind of biomaterial, porogen, is the particle leached from the biomaterial during freeze-drying (Nigam and Mahanta 2014).

10.4 THERAPEUTIC USE OF DENTAL STEM CELLS FOR BIOMIMICRY

Dental stem cells (DSCs) are a vital component of tissue regeneration and engineering techniques (Yen and Sharpe 2008; Rodríguez-Lozano et al. 2011; Rodríguez-Lozano et al. 2012). The term "stem cells" refers to young, undifferentiated cells that can specialize in a wide range of cell types and replace themselves (Conrad and Huss 2005). These cells, which form a "stem cell niche", may be found in various human tissues. Mesenchymal stem cells and multipotent mesenchymal stem cells may be found in many cell types in the mesenchyme (MSCs). As a result of its potential for regenerative applications, doctors and researchers are very interested in multipotent stem cells (MSCs), which can differentiate into a wide range of cells (Chagastelles and Nardi 2011). The utilization of cells with the right regeneration capacity may be administered through stem cell therapy, a sophisticated approach for treating deteriorated tissues. Post-natal dental stem cells come in various forms and origins (Garcia-Godoy and Murray 2006). Differentiated cells can be transformed into embryonic channels by reprogramming or somatic cell nuclear transfer, two novel stem cell techniques that sidestep the immunological rejection that is typically an issue with embryonic cells (Horch 2012). Stem cell-based tissue-engineering techniques have the potential to restore oral structures that have been destroyed (Zafar et al. 2015a). Understanding the fundamental molecular principles of stem cell destiny is essential to guarantee the safe and successful use of stem cell therapies (Witkowska-Zimny 2011).

10.5 NANOTECHNOLOGY IN ENDODONTICS

The success of endodontic treatment has been consistently high, with rates often surpassing 96 per cent, particularly in cases where apical periodontitis is absent. This trend indicates that the procedure has become increasingly reliable over time. The essential components of a successful endodontic treatment include biomechanical manipulation, disinfection, and the creation of a three-dimensional seal in the root canal system through obturation techniques. Root canal therapy has a high success rate, but it can still fail if the intricate architecture of the canal system is not properly manipulated or if there is microleakage in the sealing material. Modern endodontic materials have been known to have various issues, such as shrinkage, solubility, and sensitivity to moisture, which can reduce the longevity of treatment success. As such, researchers are developing materials with improved biomechanical and sealing properties to ensure the long-term effectiveness of endodontic therapy (Bucchi et al. 2022).

Nanotechnology research has advanced to the point where materials that enhance therapeutic outcomes are being developed. Many studies in the endodontology field aim to improve various aspects of clinical treatment, including files and filling materials. The antibacterial capabilities of certain nanoparticles can improve the efficacy of endodontic materials, irrigation solutions, and intracanal medicaments by penetrating deep into the intricate anatomical structures of root canal systems. A lot of effort has been put into developing "nanomodified" materials. The sealability of obturation and sealer materials used in root restoration and root-end filling processes may be enhanced by incorporating these particles into cutting-edge materials (Alenazy et al. 2018).

10.5.1 Endodontic Treatment Devices Utilizing Nanotechnology

Nanotechnology-based devices for endodontic treatment have been developed as a means of improving the effectiveness of the procedure. These devices utilize nanoscale materials and techniques to enhance various aspects of treatment, such as the efficiency of materials, irrigation solutions, and intracanal medicaments. Endodontic devices that use nanotechnology may lead to better patient outcomes due to their increased precision and powerful antibacterial effects during root canal therapy. Research on these devices is ongoing and holds promise for the future of endodontic treatment. Nickel-titanium (NiTi) rotary files are widely used among the various tools employed in endodontics due to their desirable properties, including excellent corrosion resistance and super elasticity, which enables them to retain their shape memory. With these data, an endodontist can assess the intricate root canal architecture and determine whether the planned treatment would be effective. It has been discovered that applying a cobalt coating on Ni-Ti files that have been coated with fullerene-like WS2 nanoparticles greatly increases the files' fatigue resistance and breaking time (Alenazy et al. 2018).

10.5.2 Application of Nanotechnology to a Canal Disinfectant

Root canal systems can be disinfected using a range of irrigants and medicaments. In a clinical setting, medications are generally used in smaller quantities compared to irrigants, as they require more time to come into contact with the canal walls and have an effect on cleansing the canals. To better seal and sanitize the entire root canal system, it has been suggested that nanoparticles be incorporated into irrigants and medications (Sen et al. 2022).

10.5.3 Irrigants for Endodontic Canals

The biomechanical procedure aims to eliminate and destroy microbial biofilms within the root canal system. While mechanical instrumentation has received much attention, root canal irrigants are essential to the disinfection process within the canal system. This disinfecting property is especially helpful in canal anatomical systems like fins and isthmuses that might be hard to clean or neglected altogether when relying on equipment alone (Alenazy et al. 2018; Sen et al. 2022).

Choosing an irrigating fluid and procedure that can remove materials like the smear layer and biofilms from canal gaps without damaging normal tissues is a major difficulty in endodontics. Root canals can be manually or mechanically disinfected using various irrigants and procedures. The use of nanoparticle-based irrigants to enhance root canal cleaning has been the subject of recent investigations, with a focus on tissue reactions. Dental care has long made use of silver nanoparticles for their antimicrobial and antibacterial properties, and they have also found use in biotechnology and bioengineering. An animal model's tissue fared better after three months when treated with a sponge containing 47 or 23 ppm of silver nanoparticles as a dispersion material enclosed in a polyethylene tube as opposed to a control sponge containing 2.5 per cent sodium hypochlorite and fibrin. The authors suggest employing a silver nanoparticle dispersion material with a concentration of 23 ppm when dealing with tissue, as suggested by the research. However, some have raised concerns about the safety of silver nanoparticles due to reports of "inflammatory, oxidative, genotoxic, and cytotoxic outcomes" resulting from exposure to these particles. More research is needed to confirm the safety of using these nanoparticles in clinical medicine and dentistry settings. Antimicrobial photodynamic treatment using nanoparticles has been identified as a potential new approach for cleansing root canal areas and has shown promising early results. Using transmission electron microscopy and the photosensitizer methylene blue, Pagonis et al. evaluated the efficacy of polyacticcoglycolic acid (PLGA) nanoparticles synergized with light against Enterococcus faecalis (MB) in vitro. The researchers found that the nanoparticle component had a broad-spectrum effect on the cell walls of the microorganisms, resulting in a significant decrease in the CFU count. They concluded that antibacterial root canal therapy using PLGA nanoparticles encapsulated with protective medications (Pagonis et al. 2010). A study comparing the efficacy of various disinfectants found that a cationic photosensitizer effectively eliminated the bacterial content of biofilms formed by E. faecalis and disrupted the biofilm's structure. As an alternative to the standard irrigants used in endodontic therapy, these nanoparticle-based disinfectants show promise (Kishen et al. 2010).

10.5.4 Root Canal Therapy Sealers

Root canal sealers play a crucial role in endodontic treatment by filling in any small gaps or imperfections between the root canal space and obturating material, such as gutta-percha. This helps to create an impermeable seal, which is necessary to prevent the ingress of microbes. In addition, sealers can act as a microbial control agent in cases where some microbes remain in the dentinal tubules after root canal therapy (Komabayashi et al. 2020). Sealers also function as lubricants, enabling the tight packing of obturating material, which is essential for an effective seal. Some studies have shown that sealers based on nanotechnology, specifically calcium phosphate cement with nanohydroxyapatite crystals as the primary component, demonstrate enhanced adhesive properties to dentin tubules and improved effectiveness against various bacteria compared to traditional sealers (Chen et al. 2007). A previous investigation found that nanocrystalline tetracalcium phosphate demonstrated superior antibacterial activity in an agar diffusion test. One hypothesis proposed that settling amorphous calcium hydroxide in the agar gel created a zone of inhibition around the samples. Several commercially available root restoration products, such as Mineral Trioxide Aggregate (MTA-Fillapex), iRoot SP, and EndoSequence BC, contain nanosize particles which possess biomineralizing, antibacterial, and safe for human use properties. These materials also demonstrated a low level of toxicity that decreased after they were fully set (Salles et al. 2012; Alenazy et al. 2018).

10.5.5 Retrofilling and Root-Repair Materials

Recently, there has been a focus on using materials such as MTA and EndoSequence BC root healing material putty for retrograde root filling during periapical surgery. Wu et al. emphasized the importance of achieving a long-lasting seal for such fillings. Several studies have indicated that the success of surgical therapy may be compromised if an inadequate root canal filling is used, and clinical investigations examining the extent of healing after periradicular surgery have demonstrated the value of completing an adequate root canal filling prior to the procedure. Despite its lengthy handling and setting times, MTA has become the preferred material for retrograde fillings. However, researchers recently evaluated a nanomodified version of MTA with improved physiochemical characteristics in an effort to address these limitations. They found that incorporating nanodispersion into the MTA powder shortened the setting time and increased microhardness. This modification led to the faster setting of the MTA, but at the cost of reduced hardness after it had been set (Mutluay and Mutluay 2021). In addition to improving existing materials, some researchers are exploring developing novel materials. Polymer nanocomposites (PNCs) are polymeric materials that contain nanoparticles such as clays and carbon nanotubes at a concentration lower than traditional composites. Due to their high surface-to-volume ratio, the dispersed phase in PNCs is highly manipulable, and even at low filler content (between 0% and 5%), these materials have demonstrated significantly enhanced mechanical and thermal characteristics (Idumah et al. 2021).

Previous research has demonstrated that using nanocomposite polymers can significantly enhance the properties of heat resistance, dimensional stability, stiffness, decreased electrical conductivity, and drug elution. In a recent in vitro study, two of these nanocomposite polymers (NERP1 and NERP2) were compared to a conventional polymer-based compomer in terms of their ability to create an initial apical seal. The incorporation of NERP1 was found to reduce apical microleakage significantly (Chogle et al. 2011).

10.5.6 Development of Nano Applications for Regeneration of Dental Pulp

The field of endodontology, which focuses on the health and treatment of dental pulp, aims to ensure the complete healing of this tissue. To achieve this goal, researchers and clinicians in this field seek to influence the development of embryonic stem cells in order to restore or retain the original structures and functions of the dental pulp. The hard encasement walls of the tooth pulp, a low-compliance system in the human body, can make it difficult to access this tissue without inciting an inflammatory response (Sen et al. 2022). Therefore, it is crucial to understand the fundamental processes involved in the repair and regeneration of dental pulp before attempting to address damage, inflammatory, or necrotic tissue. The unique features of dental pulp, including its vast volume, lack of anastomoses, and microvascular terminal supplies, make it a rare and complex part of the human body. The arteries and veins that supply and drain dental pulp are the only vessels that enter and exit this tissue. One of the major challenges in repairing and regenerating dental pulp is that it is encased in hard tissues such as dentin, enamel, and cementum. Despite this, dental pulp has been shown to have a significant ability to self-repair (Gomez-Sosa et al. 2022).

Tooth caries is a major cause of stress to the dental pulp, leading to inflammation and microorganism invasion of the enamel and dentin, eventually reaching the pulp tissue. In many cases, pulpal damage and inflammation can be repaired if the underlying cause, such as caries, is treated early on. Recent studies have demonstrated that even in a low-compliance system like tooth pulp, repair and regeneration are possible (Lin et al. 2020).

However, the complexity of factors involved in tissue repair and regeneration, including stem and progenitor cell proliferation, differentiation, scaffold types, brain and vascular tissue regeneration, signaling pathways, and signaling proteins, makes it difficult to translate findings from animal studies to humans (Ganey and Temple 2023). Dental pulp stem cells (DPSCs) have been studied for their behavior on various scaffolds, including those made of poly/gelatin and nano-hydroxyapatite (NHA). These scaffolds have been found to promote DPSC adhesion, odontoblast proliferation, and odontoblast differentiation in vitro and to support DPSC differentiation into odontoblast-like cells in vivo. These findings significantly advance our understanding of how better to control cell activity in tissue repair and regeneration. DPSCs were implanted onto electrospun poly/gelatin scaffolds with or without NHA, and then subcutaneously into immunocompromised nude mice. Scaffolds with NHA but no DPSCs were used as the standard of comparison. Differentiation of DPSCs into odontoblast-like cells was enhanced in both in vitro and in vivo environments, as evidenced by DNA content analysis, alkaline phosphatase movement analysis, and osteocalcin measurement (Alenazy et al. 2018; Dogra et al. 2020).

One study evaluated the differentiation of human odontogenic dental pulp stem cells (DPSCs) on nanofiber poly L-lactic acid (PLLA) scaffolds. The investigation utilized bone morphogenic protein-7 (BMP-7) and dexamethasone (DXM) media in combination with highly permeable nanofiber PLLA scaffolds that resembled collagen type-I fibers. The combination of DXM and the growth factor BMP-7 was found to promote odontogenic DPSC differentiation than DXM alone more effectively. These results suggest that DPSCs are well-supported by the nano scaffold environment during the regeneration of tooth pulp, dentin, and enamel (Alenazy et al. 2018; Dogra et al. 2020; Sen et al. 2022).

10.6 NANOMATERIALS USAGE IN DENTAL IMPLANTS

One potential strategy to prevent bacteria from attaching to implants is to alter the nano topography of the surface. Silver nanoparticles (AgNPs) are promising for dental implant doping and restoration due to their strong antibacterial properties. AgNPs can bond electrostatically to bacteria and damage their cytoplasmic membrane and cell wall, leading to their destruction. In addition to their antibacterial effects, AgNPs have been shown to enhance titanium implants' osteogenic and soft-tissue integration properties. Using anodic spark deposition, AgNPs have been deposited on titanium implants to provide antibacterial activity against S. epidermidis, S. mutans, and E. coli, as well as osteogenic activity in human osteoblast-like cells and SAOS-2 cells. Cytotoxicity tests using AgNPs on sand-blasted, large grit, and acid-etched titanium implants showed that the AgNPs were not harmful to cells, although they did inhibit the growth of S. aureus and F. nucleatum. Overall, these results indicate that titanium implants coated with AgNPs have the potential to be both antibacterial and osteogenic, making them a promising option for therapeutic applications. AgNPs isolated from plants has also been found to have bactericidal activity against S. aureus and P. aeruginosa. Due to these unique features, there is significant interest in using AgNPs in dental implants (Nayar et al. 2011; Besinis et al. 2015; Zhang et al. 2021).

10.6.1 Synthesis Processes

In general, bottom-up and top-down strategies are used to synthesize silver nanoparticles (AgNPs). Figure 10.5 depicts a top-down method of producing AgNPs via a series of physical procedures that scale down the initial material from bulk to nanoscale (Foong et al. 2020). The surface arrangement of metal nanoparticles is a major factor in determining their physicochemical characteristics. However, this method requires high temperatures and pressures, which can be energy-intensive. Among the techniques that fall within this category are thermolysis, pyrolysis, radiation-induced synthesis, and lithography. Through a variety of biological and chemical processes, the final nanomaterial of the required size is assembled from synthesized nanoparticles using the bottom-up method, commonly known as the self-assembly technique. The chemical compositional uniformity of the AgNPs produced, and the number of surface flaws are likely to improve using this method, and the production cost will go down. However, the biological uses of these chemical processes are constrained by the necessity of nonpolar organic solvents, poisonous compounds, synthetic capping agents, and other stabilizing agents (Besinis et al. 2015; Zhang et al. 2021; Banakar et al. 2022a; Banakar and Sijanivandi 2022).

Alternatives to the usual chemical synthesis methods for generating silver nanoparticles are currently being developed by researchers (AgNPs). Due to their biocompatibility and little environmental effect, plant extracts have been used in the production of AgNPs. By reducing Ag+ to Ag0, the organism serves as a capping agent, reducing agent, or stabilizing agent in the biological synthesis of silver nanoparticles. Since they are inexpensive, abundant, and safe for both humans and the environment, plant-derived chemicals are increasingly being used in biological procedures (Besinis et al. 2015; Banakar et al. 2022d, b).

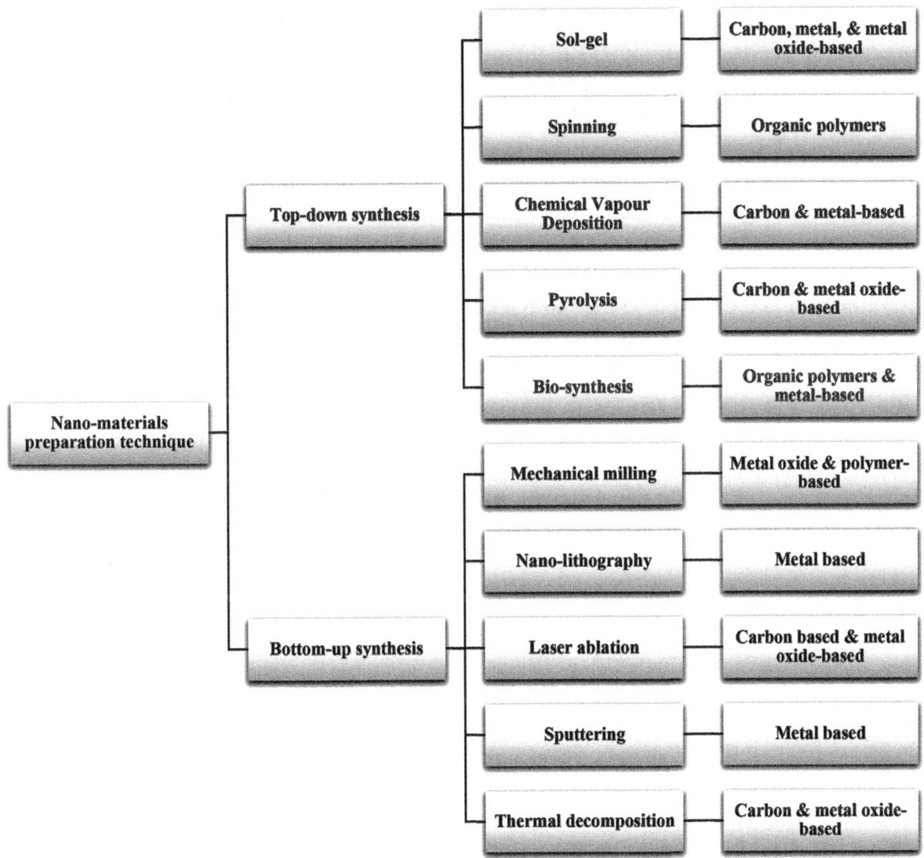

FIGURE 10.5 Conventional Synthetic Techniques for Nano Materials. (Foong et al. 2020.)

10.6.2 Chemical Method

There are several advantages to using chemical methods to produce silver nanoparticles. These methods often require relatively simple equipment, which can be obtained easily. It is well known that a reducing agent is necessary in order to convert silver ions into metallic silver nanoparticles. In this process, polyvinylpyrrolidone (PVP) is often used to regulate the size and act as a capping agent, while sodium borohydride and trisodium citrate are used to stabilize the nanoparticles. However, producing silver nanoparticles can be expensive and risky due to the need for various chemicals and compounds, including borohydride, 2-mercaptoethanol, citrate, and thioglycerol. In addition, producing uniformly sized silver nanoparticles and preventing them from clumping together can be challenging. The synthesis process also produces several potentially dangerous and toxic by-products. While the ability to synthesize nanoparticles with precise specifications is a major benefit of this technology, usage of harmful chemicals and severe reaction conditions, such as high temperatures, high pressures, and toxic by-products, have also contributed to environmental problems (Wang et al. 2002; Calderón-Jiménez et al. 2017; Hossain et al. 2022).

10.6.3 Physical Method

Nanoparticles can be synthesized using "top-down" physical processes, including ultrasonication, microwave irradiation, and electrochemical methods. Evaporation, condensation, and laser ablation are common physical methods for synthesizing nanoparticles. In evaporation, a material is evaporated using a pontoon, and the resulting gas is directed towards a burner. Condensation involves the formation of nanoparticles through the cooling and condensation of vapour. Laser ablation involves the use of a laser to vaporize a material, which is then condensed to form nanoparticles. These physical methods can be effective for synthesizing nanoparticles. However, they may also have limitations, such as the need for specialized equipment and the potential for producing particles with non-uniform sizes (Jung et al. 2006). It was found that when the surface temperature of the heater was kept constant, polydispersed nanoparticles could be generated. These silver nanoparticles were smooth and did not stick together. Physical methods for synthesizing nanoparticles, such as plasma catalysis and laser ablation, are examples of techniques that can be used. However, traditional methods for synthesizing nanoparticles, such as those based on physical processes, can be labour-intensive and costly due to the need for specialized equipment and trained individuals. Nanoparticles produced using these methods may be less stable and can be damaged by the high temperatures and bright lights used in the process. Physical methods have the advantage of being quick, chemical-free, and allowing the use of radiation as a reducing agent. However, these methods can also have limitations, such as inefficient energy usage, inefficient product distribution, and the potential for introducing unwanted solvents (Elsupikhe et al. 2015; Mikhailova 2020; Hossain et al. 2022).

10.6.4 Biosynthesis Method

Creating silver nanoparticles through chemical and physical processes can be wasteful and harmful to ecosystems. Therefore, it is important to develop a cost-effective and environmentally friendly system. Biological methods, which involve the use of organisms such as fungus, bacteria, and yeast, along with organic materials such as plant matter, can be used for various purposes in healthcare and offer a more sustainable alternative to traditional chemical and physical production techniques. However, biological methods also have their challenges, including the need for cell lysis for Ag-NP purification, the risk of infection due to microbial adhesion to the surface of the nanoparticles, and the need to identify and cultivate potent strains of microorganisms capable of producing Ag-NPs. There may also be a lack of control over pH and temperature during nanoparticle production when using these microbial strains for growth and maintenance, making the process more costly. There are a number of benefits to using plant material as a reducing agent in silver nanoparticle production, including convenience, affordability, low upkeep, and less impact on the environment (Fahmy and Mobarak 2011; Banakar et al. 2022c; Hossain et al. 2022).

10.6.5 Mechanism of Silver Nanoparticles Biosynthesis

Enzymes play a significant role in the synthesis of Ag-NPs, whether the process occurs within or outside of cells. Nitrate reductase, which requires the cofactor NADH, has been identified as a key enzyme in the synthesis of AgNPs. This enzyme transfers electrons from nitrate to the metal ion, facilitating the formation of nanoparticles by serving as an electron shuttle. A wide range of plants, including medicinal herbs, can be used to generate the "factories" that produce silver nanoparticles through mechanisms similar to those found in bacteria and enzymatic synthesis of AgNPs. Plant cells contain a variety of antioxidant metabolites that protect against deterioration due to oxidation, and the chemicals used to protect and encase the nanoparticles are also unique. The future practical uses of AgNPs will rely heavily on their anti-inflammatory, antioxidant, anticancer, and other benefits. Selective attachment of capping agents to certain nanoparticle crystal faces is possible. Changing the surface free energy and surface area ratio, which can avoid agglomeration, reduce toxicity, enhance antibacterial properties, and increase bacterial cell adhesion and activity of AgNPs. Silver nanoparticles (AgNPs) have potential applications in dental implants due to their antibacterial activity, biocompatibility, enhanced strength, and surface modification. The shape, size, and toxicity of nanoparticles are all important characteristics for AgNPs to function effectively. Researchers have been working to obtain AgNPs from a wide range of sources to achieve the desired shape and size to improve the material's activity in dental implants and obtain them less toxic manner. It has been shown that plant extracts can be a practical source for obtaining highly potent AgNPs. Despite the extensive literature available on this topic, further research is needed to improve plant-extracted silver nanoparticles' shape, size, and purity for use in dental implant applications (Besinis et al. 2015; Banakar and Sijanivandi 2022; Banakar et al. 2022a; Hossain et al. 2022).

10.7 TOXICITY OF NANOMATERIALS

The integration of nanomaterials into various industries and sectors has the potential to bring about significant technological advancements. However, the sustainability of green nanotechnology is uncertain and requires careful consideration. Although certain uses of nanomaterials may be benign to the environment and human health, it is nevertheless vital to be aware of the hazards and unforeseen effects that may arise from their widespread use. The need for caution in the use of nanomaterials is underscored by their pervasive presence in the environment as a result of their clandestine introduction into soil, water, and air through diverse human activities, including purposeful deposition during environmental treatments (Libralato et al. 2020). Toxic and damaging cellular consequences can be induced by magnetic nanoparticles, which are not normally seen in bigger micron-sized materials. It has been proven that these substances, when ingested or inhaled, can translocate to various organs and tissues within an organism, where they may cause damage. While there have been some studies on the toxic effects of nanomaterials on plant and animal cells, there has been limited research on the toxic effects of magnetic nanomaterials on plants. Aquatic creatures, including algae, bacteria, daphnia, and fish, are vulnerable to the toxic effects of dissolved silver due to the extensive usage of silver nanoparticles in consumer items that have been released into aquatic ecosystems (Ogunsona et al. 2020). The respiratory system is particularly vulnerable to the potential toxicity of nanomaterials due to the inhalation of these particles. Nanomaterials have a variety of applications in biosystems, but the long-term effects of human exposure to different concentrations of these materials on health are not yet fully understood. However, the study of the environmental impacts of nanomaterials is expected to receive increased attention in the future. The toxicity of nanomaterials can depend on various factors, including particle size, shape, surface characteristics, functionalized groups, charge, and free energy, which can influence the ability of these materials to interact with protein molecules. This interaction can have detrimental biological effects, such as protein unfolding, thiol cross-linking, fibrillation, and loss of enzymatic activity. In addition, some nanomaterials may release toxic ions when they dissolve in a biological environment or suspension medium under certain thermodynamic conditions (Abbasalipourkabir et al. 2015). Seawater and hard water are ideal environments for nanomaterial accumulation, while the presence of organic molecules or other natural particles (colloids) in fresh water might alter the behaviour of nanomaterials. How nanomaterials are dispersed can also affect their eco-toxicity, and several abiotic factors can influence this, including salinity, pH, and the presence of organic matter. It is important to carefully analyze these factors using eco-toxicological studies to understand their potential impacts (Fang et al. 2015).

10.8 CONCLUSION

Dentistry has seen a dramatic surge in technical and scientific advancements over the past several years. Nanotechnology, manufacturing techniques, and the functionalization of nanobiomaterials have all been studied in depth. Over the past decade, considerable advancements have been made in the capacity of biomimetic restorative materials to imitate the properties of natural tissues. However, biomimetic restorative material development is still in its early phases because of the complexity of dental tissues' structure and functional nature. Although several in vitro and in vivo studies have shown promising outcomes, few dental nanoformulations have been licensed for clinical use because of safety concerns. Another issue with mass-producing nanoformulations is that the manufacturing process is time-consuming. Toxicological profiles and other elements of nanoformulation have received little attention in the literature so far. Most research has been conducted on the release pattern and degradation profile. For the reasons stated above, it is challenging to go from animal studies to human studies, and it is hard to move from research to commercial manufacturing. Drug delivery via nanoparticles is unquestionably entering a new age, and the scientific community is also seeing encouraging results in employing nanoparticles to address dental problems. Further detailed studies should confirm the favourable outcomes of these investigations, and studies on nanoformulations should concentrate on the various characteristics of nanoformulations. There has been a tremendous expansion in biomimetic tissue engineering, which has gone from a theoretical phase to a rapidly evolving area in recent decades. However, converting these breakthroughs into practical and therapeutic applications still needs additional study. Restorative and periodontal soft tissue management techniques and dentin, enamel, and pulp regeneration are all possibilities. Biological regeneration will be demonstrated in the near future using these modalities to strengthen and complete the tooth structure. New insights into the industrial applications of nanoparticles for regenerating the periodontal apparatus may be revealed by the rapid advances in nanomaterials and nanotechnology. Dentine, cementum, periodontal ligaments, and bone all fall under this category. For example, the extracellular matrix can be mimicked by nanoparticles and tissue engineering triad-impregnated scaffolds to promote host tissue development. Due to their reduced toxicity, antibacterial properties, and enhanced protein-surface interactions, nanoparticles may be used in various dental applications. Combining traditional methods with clinical applications of nanotechnology has the potential to improve dental treatment.

REFERENCES

Abbasalipourkabir R, Moradi H, Zarei S, et al (2015) Toxicity of zinc oxide nanoparticles on adult male Wistar rats. *Food Chem Toxicol* 84:154–160.

Abolmaali SS, Tamaddon A, Najafi H, Dinarvand R (2014) Effect of l-Histidine substitution on Sol–Gel of transition metal coordinated poly ethyleneimine: Synthesis and biochemical characterization. *J Inorg Organomet Polym Mater* 24:977–987.

Adachi T, Boschetto F, Miyamoto N, et al (2020) In vivo regeneration of large bone defects by cross-linked porous hydrogel: A Pilot study in mice combining micro tomography, histological analyses, Raman spectroscopy and synchrotron infrared imaging. *Materials* 13:4275.

Agarwal A, Gupta ND, Jain A (2016) Platelet rich fibrin combined with decalcified freeze-dried bone allograft for the treatment of human intrabony periodontal defects: A randomized split mouth clinical trail. *Acta Odontol Scand* 74:36–43.

Ajwani H, Shetty S, Gopalakrishnan D, et al (2015) Comparative evaluation of platelet-rich fibrin biomaterial and open flap debridement in the treatment of two and three wall intrabony defects. *J Int Oral Health* 7:32.

Alenazy MS, Mosadomi HA, Al-Nazhan S, Rayyan MR (2018) Clinical considerations of nanobiomaterials in endodontics: A systematic review. *Saudi Endod J* 8:163.

Ali S, Sangi L, Kumar N, et al (2020) Evaluating antibacterial and surface mechanical properties of Chitosan modified dental resin composites. *Technol Health Care* 28:165–173.

Ali S, Sangi L, Kumar N (2017) Exploring antibacterial activity and hydrolytic stability of resin dental composite restorative materials containing chitosan. *Technol Health Care* 25:11–18.

Asti A, Gioglio L (2014) Natural and synthetic biodegradable polymers: Different scaffolds for cell expansion and tissue formation. *Int J Artif Organs* 37:187–205.

Banakar M, Shahbazi Z, Mousavi SM, et al (2022a) Antimicrobial and Antiviral Properties of Herbal Green Materials. In: Baskar C, Ramakrishna S, Daniela La Rosa A (eds) *Encyclopedia of Green Materials*. Springer Nature Singapore, Singapore, pp 1–10.

Banakar M, Shahbazi Z, Mousavi SM, et al (2022b) Herbal Green Nanomaterials and Their Applications. In: Baskar C, Ramakrishna S, Daniela La Rosa A (eds) *Encyclopedia of Green Materials*. Springer Nature Singapore, Singapore, pp 1–8.

Banakar M, Shahbazi Z, Mousavi SM, et al (2022c) Green Materials Sterilization Solutions. In: Baskar C, Ramakrishna S, Daniela La Rosa A (eds) *Encyclopedia of Green Materials*. Springer Nature Singapore, Singapore, pp 1–10.

Banakar M, Sijanivandi S, Mousavi SM, et al (2022d) Green Materials for 3D Printing in Dentistry. In: Baskar C, Ramakrishna S, Daniela La Rosa A (eds) *Encyclopedia of Green Materials*. Springer Nature Singapore, Singapore, pp 1–6.

Banakar M, Sijanivandi SS (2022) Green Dentistry. In: Baskar C, Ramakrishna S, Daniela La Rosa A (eds) *Encyclopedia of Green Materials*. Springer Nature Singapore, Singapore, pp 1–6.

Bansal C, Bharti V (2013) Evaluation of efficacy of autologous platelet-rich fibrin with demineralized-freeze dried bone allograft in the treatment of periodontal intrabony defects. *J Indian Soc Periodontol* 17:361

Besinis A, De Peralta T, Tredwin CJ, Handy RD (2015) Review of nanomaterials in dentistry: Interactions with the oral microenvironment, clinical applications, hazards, and benefits. *ACS Nano* 9:2255–2289.

Blok LG, Longana ML, Yu H, Woods BKS (2018) An investigation into 3D printing of fibre reinforced thermoplastic composites. *Addit Manuf* 22:176–186.

Bouët G, Marchat D, Cruel M, et al (2015) In vitro three-dimensional bone tissue models: from cells to controlled and dynamic environment. *Tissue Eng Part B Rev* 21:133–156.

Bowlin GL (2011) Enhanced porosity without compromising structural integrity: The nemesis of electrospun scaffolding. *J Tissue Sci Eng* 2:103e.

Bucchi C, Rosen E, Taschieri S (2022) Non-surgical root canal treatment and retreatment versus apical surgery in treating apical periodontitis: A systematic review. *Int Endod J 56*: 475-486.

Calderón-Jiménez B, Johnson ME, Montoro Bustos AR, et al (2017) Silver nanoparticles: Technological advances, societal impacts, and metrological challenges. *Front Chem* 5:6.

Chagastelles PC, Nardi NB (2011) Biology of stem cells: an overview. *Kidney Int Suppl* 1:63–67.

Chavanne P, Stevanovic S, Wüthrich A, et al (2013) 3D printed chitosan/hydroxyapatite scaffolds for potential use in regenerative medicine. *Biomed Eng /Biomed Tech* 58:000010151520134069.

Chen H, Fan M (2007) Chitosan/carboxymethyl cellulose polyelectrolyte complex scaffolds for pulp cells regeneration. *J Bioact Compat Polym* 22:475–491.

Chen J-P, Chen S-H, Lai G-J (2012) Preparation and characterization of biomimetic silk fibroin/chitosan composite nanofibers by electrospinning for osteoblasts culture. *Nanoscale Res Lett* 7:1–11.

Chen L, Hontsu S, Komasa S, et al (2021) Hydroxyapatite film coating by Er: YAG pulsed laser deposition method for the repair of enamel defects. *Materials* 14:7475.

Chen Z, Li Z, Li J, et al (2019) 3D printing of ceramics: A review. *J Eur Ceram Soc* 39:661–687.

Chen ZL, Wei W, Feng ZD, et al (2007) The development and in vitro experiment study of a bio-type root canal filling sealer using calcium phosphate cement. *Shanghai Kou Qiang Yi Xue* 16:530–533.

Chogle SMA, Duhaime CF, Mickel AK, et al (2011) Preliminary evaluation of a novel polymer nanocomposite as a root-end filling material. *Int Endod J* 44:1055–1060.

Chou D-T, Wells D, Hong D, et al (2013) Novel processing of iron–manganese alloy-based biomaterials by inkjet 3-D printing. *Acta Biomater* 9:8593–8603.

Collins MN, Birkinshaw C (2013) Hyaluronic acid based scaffolds for tissue engineering—A review. *Carbohydr Polym* 92:1262–1279.

Conrad C, Huss R (2005) Adult stem cell lines in regenerative medicine and reconstructive surgery. *J Surg Res* 124:201–208.

Cortese A, Pantaleo G, Borri A, et al (2016) Platelet-rich fibrin (PRF) in implant dentistry in combination with new bone regenerative technique in elderly patients. *Int J Surg Case Rep* 28:52–56.

de Obaldia EE, Jeong C, Grunenfelder LK, et al (2015) Analysis of the mechanical response of biomimetic materials with highly oriented microstructures through 3D printing, mechanical testing and modeling. *J Mech Behav Biomed Mater* 48:70–85.

Dhandayuthapani B, Yoshida Y, Maekawa T, Kumar DS (2011) Polymeric scaffolds in tissue engineering application: A review. *Int J Polym Sci* 2011: 1–19.

Dogra S, Gupta A, Goyal V, et al (2020) Recent Trends, Therapeutic Applications, and Future Trends of Nanomaterials in Dentistry. In: Kanchi S (ed) *Nanomaterials in Diagnostic Tools and Devices*. Elsevier, Amsterdam, pp 257–292.

Ehrbar M, Djonov VG, Schnell C, et al (2004) Cell-demanded liberation of VEGF121 from fibrin implants induces local and controlled blood vessel growth. *Circ Res* 94:1124–1132

Elsdale T, Bard J (1972) Collagen substrata for studies on cell behavior. *J Cell Biol* 54:626–637.

Elsupikhe RF, Shameli K, Ahmad MB, et al (2015) Green sonochemical synthesis of silver nanoparticles at varying concentrations of κ-carrageenan. *Nanoscale Res Lett* 10:1–8.

Erol-Taygun M, Zheng K, Boccaccini AR (2013) Nanoscale bioactive glasses in medical applications. *Int J Appl Glass Sci* 4:136–148.

Fahmy TYA, Mobarak F (2011) Green nanotechnology: A short cut to beneficiation of natural fibers. *Int J Biol Macromol* 48:134–136.

Fang Q, Shi X, Zhang L, et al (2015) Effect of titanium dioxide nanoparticles on the bioavailability, metabolism, and toxicity of pentachlorophenol in zebrafish larvae. *J Hazard Mater* 283:897–904.

Feilden E, Ferraro C, Zhang Q, et al (2017) 3D printing bioinspired ceramic composites. *Sci Rep* 7:1–9.

Femminella B, Iaconi MC, Di Tullio M, et al (2016) Clinical comparison of platelet-rich fibrin and a gelatin sponge in the management of palatal wounds after epithelialized free gingival graft harvest: A randomized clinical trial. *J Periodontol* 87:103–113.

Foong LK, Foroughi MM, Mirhosseini AF, et al (2020) Applications of nano-materials in diverse dentistry regimes. *RSC Adv* 10:15430–15460.

Foroutan T, Nazemi N, Tavana M, et al (2018) Suspended graphene oxide nanoparticle for accelerated multilayer osteoblast attachment. *J Biomed Mater Res A* 106:293–303.

Fu Q, Rahaman MN, Fu H, Liu X (2010) Silicate, borosilicate, and borate bioactive glass scaffolds with controllable degradation rate for bone tissue engineering applications. I. Preparation and in vitro degradation. *J Biomed Mater Res A* 95:164–171.

Galante R, Figueiredo-Pina CG, Serro AP (2019) Additive manufacturing of ceramics for dental applications: A review. *Dental Mater* 35:825–846.

Ganesh N, Hanna C, Nair S V, Nair LS (2013) Enzymatically cross-linked alginic–hyaluronic acid composite hydrogels as cell delivery vehicles. *Int J Biol Macromol* 55:289–294.

Ganey T, Temple HT (2023) Introduction to Regenerative Medicine. In: Hunter C W, Davis T T, DePalma M J (eds) *Regenerative Medicine*. Springer, New York pp 3–14.

Garcia-Godoy F, Murray PE (2006) Status and potential commercial impact of stem cell-based treatments on dental and craniofacial regeneration. *Stem Cells Dev* 15:881–887.

Ge Z, Jin Z, Cao T (2008) Manufacture of degradable polymeric scaffolds for bone regeneration. *Biomed Mater* 3:22001.

Gomez-Sosa JF, Cardier JE, Caviedes-Bucheli J (2022) The hypoxia-dependent angiogenic process in dental pulp. *J Oral Biosci* 64(4):381–391. https://doi.org/10.1016/j.job.2022.08.004

Gong T, Xie J, Liao J, et al (2015) Nanomaterials and bone regeneration. *Bone Res* 3:1–7.

Grigolato R, Pizzi N, Brotto MC, et al (2015) Magnesium-enriched hydroxyapatite as bone filler in an ameloblastoma mandibular defect. *Int J Clin Exp Med* 8:281.

Gwon K, Kim E, Tae G (2017) Heparin-hyaluronic acid hydrogel in support of cellular activities of 3D encapsulated adipose derived stem cells. *Acta Biomater* 49:284–295.

Hakimi O, Knight DP, Vollrath F, Vadgama P (2007) Spider and mulberry silkworm silks as compatible biomaterials. *Compos B Eng* 38:324–337.

Hamilton MF, Otte AD, Gregory RL, et al (2015) Physicomechanical and antibacterial properties of experimental resin-based dental sealants modified with nylon-6 and chitosan nanofibers. *J Biomed Mater Res B Appl Biomater* 103:1560–1568.

Hardy JG, Römer LM, Scheibel TR (2008) Polymeric materials based on silk proteins. *Polymer (Guildf)* 49:4309–4327.

Henkel J, Woodruff MA, Epari DR, et al (2013) Bone regeneration based on tissue engineering conceptions—A 21st century perspective. *Bone Res* 1:216–248.

Horch RE (2012) New developments and trends in tissue engineering: An update. *J Tissue Sci Eng* 3:110–114.

Hossain N, Islam MA, Chowdhury MA (2022) Synthesis and characterization of plant extracted silver nanoparticles and advances in dental implant applications. Heliyon 8:e12313.

Huang W, Restrepo D, Jung J, et al (2019) Multiscale toughening mechanisms in biological materials and bioinspired designs. *Adv Mater* 31:1901561.

Hunter KT, Ma T (2013) In vitro evaluation of hydroxyapatite–chitosan–gelatin composite membrane in guided tissue regeneration. *J Biomed Mater Res A* 101:1016–1025.

Iafisco M, Quirici N, Foltran I, Rimondini L (2013) Electrospun collagen mimicking the reconstituted extracellular matrix improves osteoblastic differentiation onto titanium surfaces. *J Nanosci Nanotechnol* 13:4720–4726.

Idumah CI, Obele CM, Ezeani EO (2021) Understanding interfacial dispersions in ecobenign polymer nano-biocomposites. *Polym Plast Technol Mater* 60:233–252.

Ige OO, Umoru LE, Aribo S (2012) Natural products: A minefield of biomaterials. *Int Sch Res Notices* 2012(2): 983062–983062. https://doi.org/10.5402/2012/983062

Ikada Y (2006) Challenges in tissue engineering. *J R Soc Interface* 3:589–601.

Jafari M, Abolmaali SS, Najafi H, Tamaddon AM (2020) Hyperbranched polyglycerol nanostructures for anti-biofouling, multifunctional drug delivery, bioimaging and theranostic applications. *Int J Pharm* 576:118959.

Jaschouz S, Mehl A (2014) Reproducibility of habitual intercuspation in vivo. *J Dent* 42:210–218.

Jiang T, Zhang Z, Zhou Y, et al (2010) Surface functionalization of titanium with chitosan/gelatin via electrophoretic deposition: Characterization and cell behavior. *Biomacromolecules* 11:1254–1260.

Jindal SK, Kiamehr M, Sun W, Yang XB (2014) Silk Scaffolds for Dental Tissue Engineering. In: Kundu C S (ed) *Silk Biomaterials for Tissue Engineering and Regenerative Medicine*. Elsevier, Amsterdam, pp 403–428.

Jing W, Chunxi Y, Yizao W, et al (2013) Laser patterning of bacterial cellulose hydrogel and its modification with gelatin and hydroxyapatite for bone tissue engineering. *Soft Mater* 11:173–180.

Jung JH, Oh HC, Noh HS, et al (2006) Metal nanoparticle generation using a small ceramic heater with a local heating area. *J Aerosol Sci* 37:1662–1670.

Junxian L, Mehrabanian M, Mivehchi H, et al (2023) The homeostasis and therapeutic applications of innate and adaptive immune cells in periodontitis. *Oral Dis* 29:2552–2564: https://doi.org/10.1111/odi.14360

Kadler K (2004) Matrix loading: assembly of extracellular matrix collagen fibrils during embryogenesis. *Birth Defects Res C Embryo Today* 72:1–11.

Kadler KE, Holmes DF, Trotter JA, Chapman JA (1996) Collagen fibril formation. *Biochem J* 316:1–11.

Kalisz M, Grobelny M, Mazur M, et al (2015) Comparison of mechanical and corrosion properties of graphene monolayer on Ti–Al–V and nanometric Nb2O5 layer on Ti–Al–V alloy for dental implants applications. *Thin Solid Films* 589:356–363.

Kemppainen JM, Hollister SJ (2010) Tailoring the mechanical properties of 3D-designed poly (glycerol sebacate) scaffolds for cartilage applications. *J Biomed Mater Res B Appl Biomater* 94:9–18.

Khurshid Z, Husain S, Alotaibi H, et al (2019) Novel Techniques of Scaffold Fabrication for Bioactive Glasses. In: Kaur G (ed) *Biomedical, Therapeutic and Clinical Applications of Bioactive Glasses*. Elsevier, Amsterdam, pp 497–519.

Kim K-H, Jeong L, Park H-N, et al (2005) Biological efficacy of silk fibroin nanofiber membranes for guided bone regeneration. *J Biotechnol* 120:327–339.

Kim NR, Lee DH, Chung P-H, Yang H-C (2009) Distinct differentiation properties of human dental pulp cells on collagen, gelatin, and chitosan scaffolds. *Oral Surg Oral Med Oral Pathol Oral Radiol Endodontol* 108:e94–e100.

Kishen A, Upadya M, Tegos GP, Hamblin MR (2010) Efflux pump inhibitor potentiates antimicrobial photodynamic inactivation of Enterococcus faecalis biofilm. *Photochem Photobiol* 86:1343–1349.

Komabayashi T, Colmenar D, Cvach N, et al (2020) Comprehensive review of current endodontic sealers. *Dent Mater J* 39:703–720.

Komasa S, Okazaki J (2022) Advances in dental bio-nanomaterials. *Materials* 15:2098. https://doi.org/10.3390/ma15062098

Konovalova M V, Markov PA, Durnev EA, et al (2017) Preparation and biocompatibility evaluation of pectin and chitosan cryogels for biomedical application. *J Biomed Mater Res A* 105:547–556.

Kottoor J (2013) Biomimetic endodontics: Barriers and strategies. *Health Sci* 2:7–12.

Kumbar S, Laurencin C, Deng M (2014) *Natural and synthetic biomedical polymers*. Elsevier, Amsterdam.

Lammel AS, Hu X, Park S-H, et al (2010) Controlling silk fibroin particle features for drug delivery. *Biomaterials* 31:4583–4591.

Langer R (1993) Vacanti JP: Tissue engineering. *Science (1979)* 260:920–926.

Lewicki JP, Rodriguez JN, Zhu C, et al (2017) 3D-printing of meso-structurally ordered carbon fiber/polymer composites with unprecedented orthotropic physical properties. *Sci Rep* 7:1–14.

Liao F, Chen Y, Li Z, et al (2010) A novel bioactive three-dimensional β-tricalcium phosphate/chitosan scaffold for periodontal tissue engineering. *J Mater Sci Mater Med* 21:489–496.

Libralato G, Lofrano G, Siciliano A, et al (2020) Toxicity Assessment of Wastewater After Advanced Oxidation Processes for Emerging Contaminants' Degradation. In: Sacco O, Vaiano V (eds) *Visible Light Active Structured Photocatalysts for the Removal of Emerging Contaminants*. Elsevier, Amsterdam, pp 195–211.

Lin LM, Ricucci D, Saoud TM, et al (2020) Vital pulp therapy of mature permanent teeth with irreversible pulpitis from the perspective of pulp biology. *Austr Endod J* 46:154–166.

Liu H, Peng H, Wu Y, et al (2013) The promotion of bone regeneration by nanofibrous hydroxyapatite/chitosan scaffolds by effects on integrin-BMP/Smad signaling pathway in BMSCs. *Biomaterials* 34:4404–4417.

Liu Z, Robinson JT, Sun X, Dai H (2008) PEGylated nanographene oxide for delivery of water-insoluble cancer drugs. *J Am Chem Soc* 130:10876–10877.

Lotfi G, Shokrgozar MA, Mofid R, et al (2016) Biological evaluation (in vitro and in vivo) of bilayered collagenous coated (nano electrospun and solid wall) chitosan membrane for periodontal guided bone regeneration. *Ann Biomed Eng* 44:2132–2144.

Luo Y, Teng Z, Li Y, Wang Q (2015) Solid lipid nanoparticles for oral drug delivery: Chitosan coating improves stability, controlled delivery, mucoadhesion and cellular uptake. *Carbohydr Polym* 122:221–229.

Mandal BB, Kapoor S, Kundu SC (2009) Silk fibroin/polyacrylamide semi-interpenetrating network hydrogels for controlled drug release. *Biomaterials* 30:2826–2836.

Matassi F, Nistri L, Paez DC, Innocenti M (2011) New biomaterials for bone regeneration. *Clin Cases Mineral Bone Metabol* 8:21.

Mathur A, Bains VK, Gupta V, et al (2015) Evaluation of intrabony defects treated with platelet-rich fibrin or autogenous bone graft: A comparative analysis. *Eur J Dent* 9:100–108.

Matsumoto T, Tashiro Y, Komasa S, et al (2020) Effects of surface modification on adsorption behavior of cell and protein on titanium surface by using quartz crystal microbalance system. *Materials* 14:97.

Metavarayuth K, Villarreal E, Wang H, Wang Q (2021) Surface topography and free energy regulate osteogenesis of stem cells: Effects of shape-controlled gold nanoparticles. *Biomaterials Translational* 2:165.

Mikhailova EO (2020) Silver nanoparticles: Mechanism of action and probable bio-application. *J Funct Biomater* 11:84.

Miranda DG, Malmonge SM, Campos DM, et al (2016) A chitosan-hyaluronic acid hydrogel scaffold for periodontal tissue engineering. *J Biomed Mater Res B Appl Biomater* 104:1691–1702.

Miron RJ, Fujioka-Kobayashi M, Bishara M, et al (2017) Platelet-rich fibrin and soft tissue wound healing: a systematic review. *Tissue Eng Part B Rev* 23:83–99.

Moini M, Olek J, Youngblood JP, et al (2018) Additive manufacturing and performance of architectured cement-based materials. *Adv Mater* 30:1802123.

Munir MU, Salman S, Ihsan A, Elsaman T (2022) Synthesis, characterization, functionalization and bio-applications of hydroxyapatite nanomaterials: An overview. *Int J Nanomedicine* 17:1903.

Mutluay M, Mutluay AT (2021) Sealing efficiency of MTA, accelerated MTA, Biodentine and RMGIC as retrograde filling materials. *Balkan J Dental Med* 25:159–165.

Najeeb S, Khurshid Z, Agwan MAS, et al (2017) Regenerative potential of platelet rich fibrin (PRF) for curing intrabony periodontal defects: A systematic review of clinical studies. *Tissue Eng Regen Med* 14:735–742.

Nakano K, Kubo H, Nakajima M, et al (2020) Bone regeneration using rat-derived dedifferentiated fat cells combined with activated platelet-rich plasma. *Materials* 13:5097.

Nayar S, Bhuminathan S, Muthuvignesh J (2011) Upsurge of nanotechnology in dentistry and dental implants. *Ind J Multidiscipl Dentistry* 1(5): 254–268.

Nigam R, Mahanta B (2014) An overview of various biomimetic scaffolds: Challenges and applications in tissue engineering. *J Tissue Sci Eng* 5:1.

Nör JE (2006) Buonocore memorial lecture: Tooth regeneration in operative dentistry. *Oper Dent* 31:633–642.

Ogunsona EO, Muthuraj R, Ojogbo E, et al (2020) Engineered nanomaterials for antimicrobial applications: A review. *Appl Mater Today* 18:100473.

Okamura T, Chen L, Tsumano N, et al (2020) Biocompatibility of a high-plasticity, calcium silicate-based, ready-to-use material. *Materials* 13:4770.

Pagonis TC, Chen J, Fontana CR, et al (2010) Nanoparticle-based endodontic antimicrobial photodynamic therapy. *J Endod* 36:322–328.

Peter M, Binulal NS, Nair S V, et al (2010) Novel biodegradable chitosan–gelatin/nano-bioactive glass ceramic composite scaffolds for alveolar bone tissue engineering. *Chem Eng J* 158:353–361.

Pinto AM, Goncalves IC, Magalhaes FD (2013) Graphene-based materials biocompatibility: A review. *Colloids Surf B Biointerfaces* 111:188–202.

Qasim SS Bin, Zafar MS, Niazi FH, et al (2020) Functionally graded biomimetic biomaterials in dentistry: An evidence-based update. *J Biomater Sci Polym Ed* 31:1144–1162.

Radunovic M, De Colli M, De Marco P, et al (2017) Graphene oxide enrichment of collagen membranes improves DPSCs differentiation and controls inflammation occurrence. *J Biomed Mater Res A* 105:2312–2320.

Raslan A, Del Burgo LS, Ciriza J, Pedraz JL (2020) Graphene oxide and reduced graphene oxide-based scaffolds in regenerative medicine. *Int J Pharm* 580:119226.

Ravindran S, Song Y, George A (2010) Development of three-dimensional biomimetic scaffold to study epithelial–mesenchymal interactions. *Tissue Eng Part A* 16:327–342.

Reiser A, Lindén M, Rohner P, et al (2019) Multi-metal electrohydrodynamic redox 3D printing at the submicron scale. *Nat Commun* 10:1–8.

Rios HF, Lin Z, Oh B, et al (2011) Cell-and gene-based therapeutic strategies for periodontal regenerative medicine. *J Periodontol* 82:1223–1237.

Rodríguez-Lozano FJ, Bueno C, Insausti CL, et al (2011) Mesenchymal stem cells derived from dental tissues. *Int Endod J* 44:800–806.

Rodríguez-Lozano F-J, Insausti C-L, Iniesta F, et al (2012) Mesenchymal dental stem cells in regenerative dentistry. *Med Oral Patol Oral Cir Bucal* 17:e1062

Roseti L, Parisi V, Petretta M, et al (2017) Scaffolds for bone tissue engineering: State of the art and new perspectives. *Mater Sci Eng C* 78:1246–1262.

Sajadi SM, Boul PJ, Thaemlitz C, et al (2019) Direct ink writing of cement structures modified with nanoscale Additive. *Adv Eng Mater* 21:1801380.

Salles LP, Gomes-Cornélio AL, Guimarães FC, et al (2012) Mineral trioxide aggregate–based endodontic sealer stimulates hydroxyapatite nucleation in human osteoblast-like cell culture. *J Endod* 38:971–976.

Sarwar MS, Huang Q, Ghaffar A, et al (2020) A smart drug delivery system based on biodegradable chitosan/poly (allylamine hydrochloride) blend films. *Pharmaceutics* 12:131.

Sato H, Ishihata H, Kameyama Y, et al (2021) Professional mechanical tooth cleaning method for dental implant surface by Agar particle blasting. *Materials* 14:6805.

Schmidt CE, Baier JM (2000) Acellular vascular tissues: Natural biomaterials for tissue repair and tissue engineering. *Biomaterials* 21:2215–2231.

Sen D, Patil V, Smriti K, et al (2022) Nanotechnology and nanomaterials in dentistry: Present and future perspectives in clinical applications. *Eng Sci* 20: 14–24. http://dx.doi.org/10.30919/es8d703

Shah M, Patel J, Dave D, Shah S (2015) Comparative evaluation of platelet-rich fibrin with demineralized freeze-dried bone allograft in periodontal infrabony defects: A randomized controlled clinical study. *J Indian Soc Periodontol* 19:56

Sheu H-S, Phyu KW, Jean Y-C, et al (2004) Lattice deformation and thermal stability of crystals in spider silk. *Int J Biol Macromol* 34:267–273.

Shimura K, Kikuchi A, Ohtomo K, et al (1976) Studies on silk fibroin of Bombyx mori. I. Fractionation of fibroin prepared from the posterior silk gland. *J Biochem* 80:693–702.

Skallevold HE, Rokaya D, Khurshid Z, Zafar MS (2019) Bioactive glass applications in dentistry. *Int J Mol Sci* 20:5960.

Soffer L, Wang X, Zhang X, et al (2008) Silk-based electrospun tubular scaffolds for tissue-engineered vascular grafts. *J Biomater Sci Polym Ed* 19:653–664.

Sreenivasalu PKP, Dora CP, Swami R, et al (2022) Nanomaterials in dentistry: Current applications and future scope. *Nanomaterials* 12:1676.

Stankovich S, Dikin DA, Piner RD, et al (2007) Synthesis of graphene-based nanosheets via chemical reduction of exfoliated graphite oxide. *Carbon N Y* 45:1558–1565.

Tanaka M, Sugimura N, Fujisawa A, Yamamoto Y (2017) Stabilizers of edaravone aqueous solution and their action mechanisms. 1. Sodium bisulfite. *J Clin Biochem Nutr* 61:159–163. https://doi.org/10.3164/jcbn.17-61

Tang Q, Huang G, Ran R, et al (2016) The application of chitosan and its derivatives as nanosized carriers for the delivery of chemical drugs and genes or proteins. *Curr Drug Targets* 17:811–816.

Turnbull G, Clarke J, Picard F, et al (2018) 3D bioactive composite scaffolds for bone tissue engineering. *Bioact Mater* 3:278–314.

Vorndran E, Wunder K, Moseke C, et al (2011) Hydraulic setting Mg3 (PO4) 2 powders for 3D printing technology. *Adv Appl Ceram* 110:476–481.

Wang P, Zhao L, Liu J, et al (2014) Bone tissue engineering via nanostructured calcium phosphate biomaterials and stem cells. *Bone Res* 2:1–13.

Wang TC, Rubner MF, Cohen RE (2002) Polyelectrolyte multilayer nanoreactors for preparing silver nanoparticle composites: Controlling metal concentration and nanoparticle size. *Langmuir* 18:3370–3375.

Wenk E, Wandrey AJ, Merkle HP, Meinel L (2008) Silk fibroin spheres as a platform for controlled drug delivery. *J Control Rel* 132 1:26–34.

Witkowska-Zimny M (2011) Dental tissue as a source of stem cells: Perspectives for teeth regeneration. *J Bioengineer Biomedical Sci* 1:1–8. http://doi.org/10.4172/2155-9538.S2-006

Wu X, Liu Y, Li X, et al (2010) Preparation of aligned porous gelatin scaffolds by unidirectional freeze-drying method. *Acta Biomater* 6:1167–1177.

Xu J, Strandman S, Zhu JXX, et al (2015) Genipin-crosslinked catechol-chitosan mucoadhesive hydrogels for buccal drug delivery. *Biomaterials* 37:395–404.

Yamaguchi Y, Matsuno T, Miyazawa A, et al (2021) Bioactivity evaluation of biphasic hydroxyapatite bone substitutes immersed and grown with supersaturated calcium phosphate solution. *Materials* 14:5143.

Yen AH-H, Sharpe PT (2008) Stem cells and tooth tissue engineering. *Cell Tissue Res* 331:359–372.

Yeo I-S, Oh J-E, Jeong L, et al (2008) Collagen-based biomimetic nanofibrous scaffolds: Preparation and characterization of collagen/silk fibroin bicomponent nanofibrous structures. *Biomacromolecules* 9:1106–1116.

Yu J, Xia H, Ni Q-Q (2018) A three-dimensional porous hydroxyapatite nanocomposite scaffold with shape memory effect for bone tissue engineering. *J Mater Sci* 53:4734–4744.

Zafar MS, Amin F, Fareed MA, et al (2020) Biomimetic aspects of restorative dentistry biomaterials. *Biomimetics* 5:34.

Zafar MS, Belton DJ, Hanby B, et al (2015a) Functional material features of Bombyx mori silk light versus heavy chain proteins. *Biomacromolecules* 16:606–614.

Zafar MS, Khurshid Z, Almas K (2015b) Oral tissue engineering progress and challenges. *Tissue Eng Regen Med* 12:387–397.

Zarkoob S, Eby RK, Reneker DH, et al (2004) Structure and morphology of electrospun silk nanofibers. *Polymer (Guildf)* 45:3973–3977.

Zhang Y, Gulati K, Li Z, et al (2021) Dental implant nano-engineering: Advances, limitations and future directions. *Nanomaterials* 11:2489.

Zhang Y-Q, Shen W-D, Xiang R-L, et al (2007) Formation of silk fibroin nanoparticles in water-miscible organic solvent and their characterization. *J Nanopart Res* 9:885–900.

Zhao Y-H, Zhang M, Liu N-X, et al (2013) The combined use of cell sheet fragments of periodontal ligament stem cells and platelet-rich fibrin granules for avulsed tooth reimplantation. *Biomaterials* 34:5506–5520.

Zheng YF, Gu XN, Witte F (2014) Biodegradable metals. *Mater Sci Eng R Rep* 77:1–34.

Zhuang H, Han Y, Feng A (2008) Preparation, mechanical properties and in vitro biodegradation of porous magnesium scaffolds. *Mater Sci Eng: C* 28:1462–1466

11 Bionanomaterials in Drug Delivery

Application of Polysaccharides

Mansi Upadhyay, Ashutosh Kumar, Ramakrishna V. Hosur, and Brahmeshwar Mishra

11.1 INTRODUCTION

Nanotechnology has emerged as an extensive field of research in the past decades (Nasra et al. 2022). Based on the nanometric scale, the term "nano" could be defined as an ultra-dispersed solid supramolecular structure composed of polymers that exhibit a size of sub-micron, preferably smaller than 500 nm (Couvreur 2013). More often, it is the area related to the synthesis, engineering and application of nanomaterials ranging in size of 100 nm (Joudeh and Linke 2022). Owing to their physical, chemical, electrical, magnetic, and biological properties, nanomaterials have extensive drug delivery applications (Jacob et al. 2018). Related to drug delivery, nanoparticles can be termed as "nanoaggregates", where the drug is physically dispersed, or "nanocapsules", the vesicular system where the drug remains either in the oily or aqueous liquid core enclosed by polymeric membrane or, thirdly, "nanosphere", a matrix system composed of spherical particles where the entrapped drug is distributed evenly (Kumari et al. 2010; Mishra et al. 2010). Application of natural polysaccharides as carrier for the fabrication of nanoparticles has numerous benefits in drug delivery and biomedical science. Biomaterials are molecular materials made up of either completely or partially biological molecules such as lipids, enzymes, proteins, cells or oligosaccharides (Honek 2013). Biomaterials extracted from plants, microbes, insects, and marine organisms are friendly to human use and the environment due to their biocompatibility and bioreactivity (Jeevanandam et al. 2022). Nanoparticles prepared using natural polysaccharides are most likely preferred due to their non-toxic nature. These nanoparticles provide better controlled and sustained drug release. Being natural in origin these nanomedicines have been found to be stable in systemic circulation, non-immunogenic, and non-inflammatory to tissues and the body, thus suitable to deliver various drugs, proteins, biologics, and nucleic acid (Panyam and Labhasetwar 2003; des Rieux et al. 2006; Fadilah et al. 2022). To date, a variety of organic- and inorganic-based natural nanoparticles have been produced from biological systems, and almost all these polymers have shown their potential theranostic use in the field of medicine. The present chapter presents an overview on polysaccharide-based biological nanoparticles and their potential application in drug delivery and related diseases. Polysaccharides are the polymeric large carbohydrates composed of simple monosaccharides units linked together through glycosidic bonds. They are generally expressed by empirical formula $(C_6H_{10}O_5)_n$ where "n" could range $40 < n < 3000$ (Zong et al. 2012). On the basis of a monosaccharides unit, polysaccharides could be linear (cellulose, chitin, and so forth) or branched. The branching in their skeletal structure could be in a form of a short saccharide unit on the linear backbone, for example starch, or it could be in a form of branch-on-branch structure, for example glycogen (Kabir et al. 2022). Polysaccharides are the most abundant molecules in nature. On the basis of structure, chemical composition, solubility and origin, they can be divided in a number of ways. Based on types of monosaccharides, it could be

DOI: 10.1201/9781003432791-14

TABLE 11.1
Techniques to Prepare Polysaccharides Nanoparticles

S.No.	Method of preparation	Polymer	Drug	Ref
1.	Ionotropic gelation method	Chitosan, Alginate	Insulin	(Avadi et al. 2010)
2.	Ionotropic gelation method	Carboxylated tamarind kernel powder	Tropicamide	(Kaur et al. 2012)
3.	Ionotropic gelation method	Chitosan	Insulin	(Pan et al. 2002)
4.	Emulsification method	Sodium alginate	Peppermint phenolic extract	(Mokhtari et al. 2017)
5.	Emulsification method	Sodium alginate	Extracted silk sericin	(Khampieng et al. 2015)
6.	Complex coacervation	Porphyra haitanensis	Resveratrol	(Xu et al. 2021)

homoglycans that contain one type of monosaccharides, for example, starch, inulin, cellulose, dextrin, cotton, and so forth (Mukherjee 2019), heteroglycans that are composed of more than one type of monosaccharides unit, for instance sodium alginate composed of peptidoglycan made up of N-acetyl- glucosamine and N-acetyl-muramic acid. Based on the glycosidic linkage they are classified as proteoglycans, glycoproteins, glycolipids, and glycoconjugates (Liu et al. 2015). On the basis of origin, polysaccharides are obtained from plants, for instance, pectin, starch, hemicellulose, and so forth (Lovegrove et al. 2017). Polysaccharides are also obtained from animals such as chitin, hyaluronic acid, chitosan, and so forth. Other bioactive sources for polysaccharides can be obtained from algae, for example agar, alginate, and so forth. Polysaccharides from bacterial origin includes dextran, gellan, xanthan, and so forth, and polysaccharides from fungal origin, for example, pullulan, glucans, and so forth (Díaz-Montes 2022). Nanoparticles made up of natural biomaterials or polysaccharides are physiologically stable, biodegradable, and have less toxicity. There are different methods by which various polysaccharides-based nanoparticles can be obtained. (Table 11.1).

11.2 POLYSACCHARIDE-BASED NATURAL NANOMATERIALS

In the past decade, polysaccharides as organic biopolymers have gained a lot of attention in the fields of nanomedicines and biomedicines due to their excellent biocompatibility, biodegradability, low toxicity, and easy-to-engineer nature (Plucinski et al. 2021a). Polysaccharides are carbohydrates that possess 10 or more monosaccharide units connected via glycosidic bond (Peptu et al. 2014). Being less or non-immunogenic in nature, polysaccharides-based bionanomaterials have prolonged the retention time of the drug, both in vitro and in vivo, thereby improving the drug encapsulation efficiency, permeability, and bioavailability (Ahmad et al. 2022). Use of polysaccharides as bionanomaterials have brought very exciting opportunities to the pharma industries. Polysaccharide-based nanoparticles have the advantages of high loading efficiencies, controlled drug release, good targeting, and stability in different physiological media, therefore, are widely used as drug carriers for the delivery of numerous synthetic and biotherapeutics (Mizrahy and Peer 2012; Heo et al. 2017; Seyedebrahimi et al. 2020; Plucinski et al. 2021b).

11.3 ORIGIN AND NATURE OF POLYSACCHARIDES

Polysaccharides, also known as biopolymeric carbohydrates, are the most abundant naturally occurring macromolecules. They can be obtained from different natural sources such as plants, animals, and microbes (Figure 11.1) (Mohammed et al. 2021). Due to their natural origin, they are strongly considered as the most vital macromolecules in nature (Ullah et al. 2019). Being an excellent biomaterial, they have imparted various biological and pharmacological activities in the field of nanosciences and biomedical engineering. Polysaccharides can be categorized in various ways, for

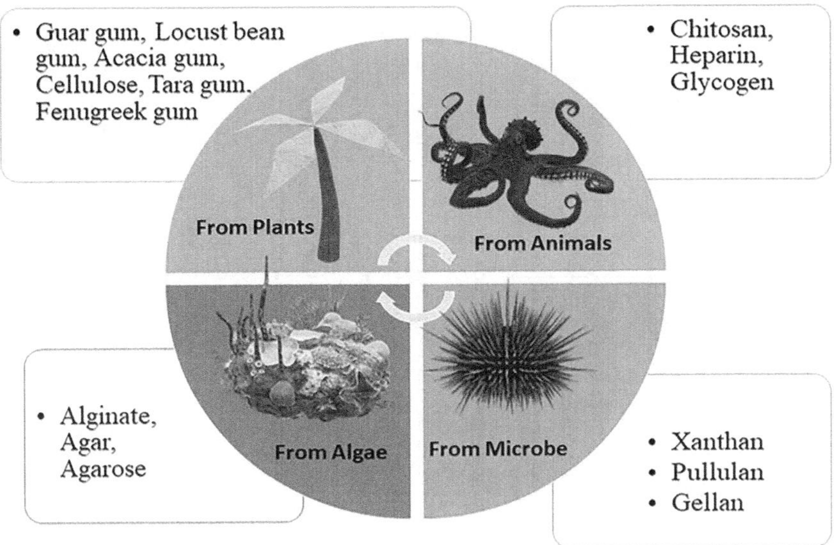

FIGURE 11.1 Natural Source of Biomaterials. (Created by the Author.)

instance, on the basis of chemical composition they are classified as homoglycans and heteroglycans. Based on origin such as polysaccharides obtained from plants for instance starch cellulose; obtained from algae for example alginate; from microbes xanthan gum, dextran, pullulan; from animals chitosan, hyaluronic acid, heparin, and so forth. Polysaccharides can also be categorized on the basis of the ionic charge that they carry, for example, cationic polysaccharides such as chitosan, and chitin; anionic polysaccharides, for instance, hyaluronic acid, alginate, chondroitin sulfate, and gellan gum; non-ionic or neutral polysaccharides such as cellulose, agarose, dextran, and so forth. Today, naturally occurring polysaccharides are in high demand in various disciplines such as heparin for blood coagulation and hyaluronic acid as a lubricator for human joints. These polysaccharides have also served in mediating various biological signals, for example, cell adhesion, and cell-to-cell communication. Their versatile nature, tuneable properties, and morphology have allowed these macromolecules to be used safely for therapeutic purposes (Li et al. 2018; Nosrati et al. 2021).

11.3.1 Bionanomaterials from Plants

Plants and their waste have always been considered safe and non-toxic sources for obtaining biopolymeric polysaccharides to form nanomaterials. Plant-based nanoparticles have the advantage over synthetic nanoparticles in terms of biocompatibility, and biodegradability, therefore most of the polysaccharides are considered by USFDA as safe to be consumed. Further, their long chain flooded with glycosidic linkages and side chains has opened an area of grafting and tuning these polysaccharides in various forms, and thus, has produced numerous bio-nanostructures. Today there is increased demand for products that are biodegradable, human-friendly, and have high safety. Polysaccharides are one of the materials of choice (Bilal et al. 2021). Recently engineering polysaccharides at the nanoscale level, especially in the pharmaceutical industry, has improved the half-life of numerous water-soluble drugs by controlling their drug release. In the food industry, polysaccharides have been found to increase the shelf life and also improve the texture taste and appearance of food (Singh et al. 2017). Plants are one of the richest sources of polysaccharides, and the development of plant-based nanomaterial is very important for human health. Natural polysaccharides and their application in the pharmaceutical industry and as biomedicines are listed in Table 11.2.

TABLE 11.2
Applications of Biomaterials Obtained from Plant Sources in Drug Delivery

S.No	Type of biomaterial	Type of delivery vehicle	Drug/ Disease	Outcome	Ref
1.	Guar gum	Polymeric NPs	Ag85A /Tuberculosis	Drug-loaded guar gum nanoparticles showed strong immune response and proved to be promising tool for needle free targeted drug delivery.	(Kaur et al. 2015)
2.	Guar gum	Polymeric NPs	Tamoxifen/ Breast cancer	Tamoxifen loaded within polymer NPs showed in vivo uptake of drug by the mammary tissues after 24 h in comparison to the free drug.	(Sarmah et al. 2011)
3.	Guar gum	Thermoresponsive magnetic nanoparticles	Doxorubicin/ solid tumor	The drug release from hydrogel magnetic nanoparticle was continuously observed for 21 days, exhibiting sustained release over long period.	(Murali et al. 2014)
4.	Locust bean gum	Polymeric nanoparticles	Allicin/ Atherosclerosis	The application of LBG NPs proved to be beneficial for targeted drug delivery.	(Soumya et al. 2018)
5.	Locust bean gum	Nanaoparticles	Ovalbumin/ immune response	BALB/c mice when administered orally with ovalbumin-loaded nanoparticle resulted into balanced immune response. The obtained NPs obtained by complexing chitosan with sulfate derivative of LBG nanoparticle acted as promising antigen mucosal delivery strategy, for oral administration.	(Braz et al. 2017)
6.	Acacia gum	Nano cargo	Hesperidin/ Rheumatoid arthritis	Hesperidin loaded gum acacia stabilized with silver nanoparticles exhibited minimal arthritic symptoms varying from mild to moderate swelling with minor degenerative changes.	(Rao et al. 2018)
7.	Acacia gum	Functionalized Colloidal Gold Nanoparticles	Letrozole/Breast Cancer	The gum acacia functionalized gold nanoparticle exhibited excellent biocompatibility and cytotoxicity against breast cancer cell line. Thus, it was concluded that acacia gum functionalized nanoparticles have potential as a promising drug delivery vehicle against breast cancer.	(Aldawsari et al. 2021)
8.	Cellulose	Nanocrystal	Curcumin	Enhancement in the drug loading and bioavailability of curcumin was observed when incorporated within nanocellulose	(Ching et al. 2019)
9.	Cellulose	Quantum dot nanocomposite hydrogel films	Doxorubicin/ cancer	The prepared quantum dot incorporated within CMC hydrogel showed no toxicity to the blood cancer cell line (K562) and proved to be efficient anticancer film.	(Javanbakht and Namazi 2018)

11.3.1.1 Guar Gum

Guar gum is a water-soluble galactomannan polysaccharide obtained from the seeds of *Cyamposis tetragonoloba* (Al-Saidan et al. 2004; Sinha et al. 2005; Soumya et al. 2010). Owing to its natural occurrence, biocompatibility, biodegradability, and easy availability guar gum is suggested to be safe to be used therapeutically as a drug delivery system. Its reasonable cost and even a mild modification in the structure can convert the polymer into the more useful form, which is making this gum more important for therapeutics. Its stability, non-toxicity, and gel-forming characteristics are responsible for the use of this gum in various forms of drug delivery vehicles such as nanoparticles, nanospheres, nanofibers, nanocomposites, and so forth (Verma and Sharma 2021). Due to its susceptibility to microbial degradation in the large intestine and excellent controlled drug-release properties, it is widely used in colon drug delivery (Soumya et al. 2010; Upadhyay et al. 2018a). Guar gum as a bionanomaterial has proved to be an outstanding candidate for pharmaceutical and biomedical applications.

11.3.1.2 Locust Bean Gum

Locust bean gum is a non-toxic, high molecular weight polysaccharide obtained from the seeds of the carob tree (Upadhyay et al. 2018a). It is a neutral non-ionic polymer composed of β-D-mannopyranosyl and α-D galactose (Upadhyay et al. 2019). Locust bean gum is widely used in pharmaceutical and biomedical applications. Many water-soluble drugs possess short elimination half-lives due to their rapidly being cleared off from the plasma, therefore, these drugs have to be administered multiple times. To delay the release of such drugs, controlled-release polymers are required, and thus, both the dose and dosing frequency can be reduced and, eventually, have minimum side effects. It is a GRAS gum (Generally Recognized as Safe) therefore extensively used as an emulsifier, thickener, and stabilizing agent in the food industry and as a controlled-release polymer for drugs and biomedical applications (Barak and Mudgil 2014).

11.3.1.3 Acacia Gum

Acacia gum or gum Arabic is an edible dried gummy exudate obtained from the trunk and branches of the trees *Acacia senegal* and *Acacia seyal* (Sanchez et al. 2018). The gum is non-ionic and slightly acidic in nature. The backbone of the gum consists of 1,3-linked β-D-galactopyranosyl units and side chains are made up of two or five 1,3-linked β-D-galactopyranosyl units that are joined to the parent chain by 1,6 linkages (Musa et al. 2019). *Gum arabic* is also used to functionalize and stabilize the nanoparticles. The gum possesses various amine and carboxyl as a charged group that physically adsorb onto the surface of the nanoparticles, which leads to the grafting of many other polymers and thus, helps in improving the drug properties. Also, its branched structures cause steric repulsion between nanoparticles thereby maintaining steric stability (Zhang et al. 2009). Due to all these important physical features, *Acacia gum* is widely used as biomaterial in biomedical engineering and pharmaceutical drug delivery.

11.3.1.4 Cellulose

Forming an efficient drug delivery system using inert material is always the first choice in pharmaceutical practice. Due to their reasonable, renewable, biodegradable, and eco-friendly nature, the application of cellulose has resulted as a promising biomaterial and bio excipient to be used for drug delivery systems. Cellulose is a natural linear water-insoluble glucose polymer. Cellulose is the major element of plant cell walls and approximately, 40 per cent of all the organic matter on earth is composed of cellulose (Raghav et al. 2021). Introducing cellulose as a nanomaterial as a drug delivery vehicle can efficiently retard the release of immediate-release drugs (Mozafari 2006). Nanocellulose as nanostructured materials and in the form of nanofibers have been widely used for oral drug delivery, transdermal, and local drug delivery.

11.3.2 Bionanomaterials from Animals

Polysaccharides have been studied for decades. Their occurrence is not restricted only to plant sources. Numerous animals and microbes have also contributed to being used as biomaterials. A few of them are enlisted in Tables 11.3 and 11.4 respectively.

11.3.2.1 Chitosan

Encapsulating the drug into naturally occurring polymer has always protected the active compound from degradation, thus improved the absorption, controlled the release, and enhanced the bioavailability and therapeutic efficacy of the drug (Ahmed and Aljaeid 2016). Chitosan is a non-toxic biodegradable polysaccharide obtained by deacetylation of chitin. Chitin is the second most abundant natural polysaccharide after cellulose. Structurally, chitosan is composed of poly [-(1,4)-2-amino-2-deoxy-D-glucopyranose] (Hejjaji et al. 2018). Chitin could be naturally obtained from the exoskeleton of crustaceans, insects, and fungi and commercially it is obtained from shrimp, lobster, and crabs (Hejazi and Amiji 2003). Chitosan and chitosan nanoparticles have been extensively used in the pharmaceutical industry for targeted drug delivery, and oral drug delivery.

11.3.2.2 Heparin

Heparin is a glycosaminoglycan mucopolysaccharide biomaterial. They are composed of repeating units of disaccharides (D-glucosamine and D-galactosamine) and monosaccharides (D-glucuronic acid and L-iduronic acid). Heparin is used as an anticoagulant and antithrombotic in medicine and is found in the human body and animals. For a long time, heparin has been used in nanoparticle synthesis. Heparin as bionanomaterials has been used for the delivery of various metals such as gold, and silver. It has also been explored for cancer diagnosis (Rodriguez-Torres et al. 2018).

11.3.3 Bionanomaterials from Algae and Microbes

11.3.3.1 Xanthan Gum

In the past decades, xanthan gum has gained attention in drug delivery and other pharmaceuticals, and in food, cosmetics, and biomedical applications due to its safe administration as declared by USFDA as GRAS. It is obtained as the fermented product of the gram-negative bacteria *Xanthomonas compestris*. Structurally, it consists of glucopyranose glucan as the backbone with mannose trisaccharide as a side chain. The gum cannot be digested other than in the colon and therefore can easily be targeted to the colonic region. When this gum initially reaches the colon, it breaks down in monosaccharides by the bacteroids and bifidobacterial species present in the colon that degrades the polymers and thus releases the encapsulated drugs directly into the colonic region (Patel et al. 2020).

11.3.3.2 Alginate

Alginate has been widely used in drug delivery systems due to its excellent physical, chemical and gel-forming properties. It is a non-toxic, low-cost, readily available, mucoadhesive, biocompatible, and non-immunogenic substance. It is a natural anionic linear biomaterial obtained from brown algae and bacteria and made up of mannuronate and guluronate sugar subunits. The polymer can be cross-linked easily, therefore has widely attracted the interest of researchers for various controlled drug delivery (Paques et al. 2014; Upadhyay et al. 2018b, 2020).

11.4 CONCLUSION

Naturally occurring polysaccharides have gained significant attention in drug delivery mainly because of their non-toxic, non-immunogenic, biocompatible, and biodegradable nature. All these characteristics assure their promising roles and future in pharmaceutical applications. Many of the

TABLE 11.3
Applications of Biomaterials Obtained from Animal Sources in Drug Delivery

S.No.	Type of biomaterial	Type of delivery vehicle	Drug/ Disease	Outcome	Ref
1.	Chitosan	Nanoparticles	Jatropha pelargonifolia (JP)	JP encapsulated within chitosan nanoparticles (JP-CHNPs) exhibited sustained release at different pH. Also, the cytotoxicity shown by JP-CHNPs exhibited high cytotoxicity on A549 human lung adenocarcinoma	(Alqahtani et al. 2021)
2.	Chitosan	Silver nanoparticles	Imatinib/ colon cancer	In this work, the co-delivery of imatinib and silver nanoparticles were delivered through folic acid conjugated chitosan nanoparticles. The apoptosis assessment performed with different techniques such as RT-PCR, flow cytometry showed the potential ability of the formed nanoparticles in inducing the apoptosis in colonic cancer cell lines. Thus, it was concluded that, the prepared NPs can be potentially used for the treatment of colon cancer.	(Azadpour et al. 2022)
3.	Chitosan/Alginate	Alginate-chitosan nanoparticles	Amygdalin/ anticancer	Amygdalin loaded within alginate chitosan nanoparticles showed a very high drug entrapment efficiency of approximately 90%. Further, they showed sustained release for 10 h. Also, they exhibited excellent anticancer effect on H1299 cell lines. Thus, it was concluded that alginate chitosan nanoparticles could be an effective drug delivery system for the treatment of cancer.	(Sohail and Abbas 2020)
4.	Heparin	Dendronized nanoparticle	Doxorubicin/ cancer	The formed nanoparticles showed pH sensitive properties i.e. fast release at acidic pH and slow and sustained release at alkaline pH. The nanoparticles effectively showed strong antitumor activity in vivo in 4T1breast tumor model mice.	(She et al. 2013)
5.	Heparin and chitosan		Ciprofloxacin/ pathogenic enteric diseases	The antibacterial activity of the prepared nanoparticles were performed against E.coli MTCC443 which proved the nanoparticles have potential antibacterial action. The *in vitro* drug release study performed at biological pH showed sustained release properties.	(Kumar et al. 2016)

TABLE 11.4
Applications of Biomaterials Obtained from Algae and Microbe Sources in Drug Delivery

S.No.	Type of biomaterial	Type of delivery vehicle	Drug/ Disease	Outcome	Ref
1.	Xanthan gum	Gold nanoparticles	Amoxicillin	The microwave-assisted gold nanoparticles embedded within natural biomaterial xanthan gum grafted with acrylic acid, showed entrapment of amoxycillin about 85% with controlled and pH-dependent drug release	(Singh et al. 2020)
2.	Xanthan gum and zein	Nanocomplex	Curcumin	In the present study, xanthan gum was used to stabilize the zein nanoparticles. The complexed zein xanthan gum was found to be stable over a wide range of temperature, pH and salt concentrations. Further, the addition of xanthan gum improved the encapsulation efficiency of curcumin and also exhibited the sustained release of curcumin in simulated intestinal fluid.	(Zhang et al. 2021)
3.	Alginate	Nanoparticles	Rifampicin	Rifampicin-loaded alginate nanoparticles were fabricated by green method. The nanoparticles showed pH-dependent drug release and swelling. The *in vitro* cytotoxicity assay performed showed the non-toxic nature of formulation. Also, the acute oral toxicity done *in vivo* in albino Wistar rats showed no systemic toxicity.	(Thomas et al. 2020)
4.	Alginate/ chitosan	Nanofibers	Capsaicin	The capsaicin-loaded alginate nanoparticles that were further embedded in polycaprolactone chitosan nanofibers extended the drug release for more than 500 h. Also, the nanoparticles effectively inhibited the proliferation of MCF-7 human breast cancer cell line.	(Ahmady et al. 2023)
5.	Alginate/ chitosan	Nanoparticles	Astaxanthin	Astaxanthin-loaded Chitosan oligosaccharide/alginate nanoparticles (ATX-COANPS) showed good storage stability. *In vitro* release study showed sustained release of the drug following the Fickian diffusion method.	(Sorasitthiyanukarn et al. 2022)

polysaccharides are recognized as GRAS by USFDA, confirming their safety for use in oral delivery applications. Moreover, polysaccharides possess numerous side chains and different types of functional groups on their backbone, which allows the polymer to modify in various ways that can further improve the properties of drug delivery vehicles. Thus, it can be concluded that the application of polysaccharides as bionanomaterial would definitely enhance the physicochemical properties of the encapsulated drugs.

REFERENCES

Ahmad A, Gulraiz Y, Ilyas S, Bashir S (2022) Polysaccharide based nano materials: Health implications. *Food Hydrocolloids Health* 2:100075. https://doi.org/10.1016/j.fhfh.2022.100075

Ahmady AR, Solouk A, Saber-Samandari S, et al (2023) Capsaicin-loaded alginate nanoparticles embedded polycaprolactone-chitosan nanofibers as a controlled drug delivery nanoplatform for anticancer activity. *J Colloid Interface Sci* 638:616–628. https://doi.org/10.1016/j.jcis.2023.01.139

Ahmed TA, Aljaeid BM (2016) Preparation, characterization, and potential application of chitosan, chitosan derivatives, and chitosan metal nanoparticles in pharmaceutical drug delivery. *Drug Des Devel Ther* 10:483–507. https://doi.org/10.2147/DDDT.S99651

Aldawsari HM, Singh S, Alhakamy NA, et al (2021) Gum acacia functionalized colloidal gold nanoparticles of letrozole as biocompatible drug delivery carrier for treatment of breast cancer. *Pharmaceutics* 13:1554. https://doi.org/10.3390/pharmaceutics13101554

Alqahtani MS, Al-Yousef HM, Alqahtani AS, et al (2021) Preparation, characterization, and in vitro-in silico biological activities of Jatropha pelargoniifolia extract loaded chitosan nanoparticles. *Int J Pharm* 606:120867. https://doi.org/10.1016/j.ijpharm.2021.120867

Al-Saidan SM, Krishnaiah YSR, Satyanarayana V, et al (2004) Pharmacokinetic evaluation of guar gum-based three-layer matrix tablets for oral controlled delivery of highly soluble metoprolol tartrate as a model drug. *Eur J Pharm Biopharm* 58:697–703. https://doi.org/10.1016/j.ejpb.2004.04.013

Avadi MR, Sadeghi AMM, Mohammadpour N, et al (2010) Preparation and characterization of insulin nanoparticles using chitosan and Arabic gum with ionic gelation method. *Nanomedicine* 6:58–63. https://doi.org/10.1016/j.nano.2009.04.007

Azadpour A, Hajrasouliha S, Khaleghi S (2022) Green synthesized-silver nanoparticles coated with targeted chitosan nanoparticles for smart drug delivery. *J Drug Deliv Sci Technol* 74:103554. https://doi.org/10.1016/j.jddst.2022.103554

Barak S, Mudgil D (2014) Locust bean gum: Processing, properties and food applications--A review. *Int J Biol Macromol* 66:74–80. https://doi.org/10.1016/j.ijbiomac.2014.02.017

Bilal M, Gul I, Basharat A, Qamar SA (2021) Polysaccharides-based bio-nanostructures and their potential food applications. *Int J Biol Macromol* 176:540–557. https://doi.org/10.1016/j.ijbiomac.2021.02.107

Braz L, Grenha A, Ferreira D, et al (2017) Chitosan/sulfated locust bean gum nanoparticles: In vitro and in vivo evaluation towards an application in oral immunization. *Int J Biol Macromol* 96:786–797. https://doi.org/10.1016/j.ijbiomac.2016.12.076

Ching YC, Gunathilake TMSU, Chuah CH, et al (2019) Curcumin/Tween 20-incorporated cellulose nanoparticles with enhanced curcumin solubility for nano-drug delivery: Characterization and in vitro evaluation. *Cellulose* 26:5467–5481. https://doi.org/10.1007/s10570-019-02445-6

Couvreur P (2013) Nanoparticles in drug delivery: Past, present and future. *Adv Drug Deliv Rev* 65:21–23. https://doi.org/10.1016/j.addr.2012.04.010

des Rieux A, Fievez V, Garinot M, et al (2006) Nanoparticles as potential oral delivery systems of proteins and vaccines: A mechanistic approach. *J Control Release* 116:1–27. https://doi.org/10.1016/j.jconrel.2006.08.013

Díaz-Montes E (2022) Polysaccharides: Sources, characteristics, properties, and their application in biodegradable films. *Polysaccharides* 3:480–501. https://doi.org/10.3390/polysaccharides3030029

Fadilah NIM, Isa ILM, Zaman WSWK, et al (2022) The effect of nanoparticle-incorporated natural-based biomaterials towards cells on activated pathways: A systematic review. *Polymers (Basel)* 14:. https://doi.org/10.3390/polym14030476

Hejazi R, Amiji M (2003) Chitosan-based gastrointestinal delivery systems. *J Control Release* 89:151–65. https://doi.org/10.1016/s0168-3659(03)00126-3

Hejjaji EMA, Smith AM, Morris GA (2018) Evaluation of the mucoadhesive properties of chitosan nanoparticles prepared using different chitosan to tripolyphosphate (CS:TPP) ratios. *Int J Biol Macromol* 120:1610–1617. https://doi.org/10.1016/j.ijbiomac.2018.09.185

Heo R, You DG, Um W, et al (2017) Dextran sulfate nanoparticles as a theranostic nanomedicine for rheumatoid arthritis. *Biomaterials* 131:15–26. https://doi.org/10.1016/j.biomaterials.2017.03.044

Honek JF (2013) Bionanotechnology and bionanomaterials: John Honek explains the good things that can come in very small packages. *BMC Biochem* 14:29. https://doi.org/10.1186/1471-2091-14-29

Jacob J, Haponiuk JT, Thomas S, Gopi S (2018) Biopolymer based nanomaterials in drug delivery systems: A review. *Mater Today Chem* 9:43–55. https://doi.org/10.1016/j.mtchem.2018.05.002

Javanbakht S, Namazi H (2018) Doxorubicin loaded carboxymethyl cellulose/graphene quantum dot nanocomposite hydrogel films as a potential anticancer drug delivery system. *Mater Sci Eng C* 87:50–59. https://doi.org/10.1016/j.msec.2018.02.010

Jeevanandam J, Ling JKU, Barhoum A, et al (2022) Bionanomaterials: Definitions, Sources, Types, Properties, Toxicity, and Regulations. In: Jeevanandam J., Danquah M. K., Barhoum A (eds) *Fundamentals of Bionanomaterials*. Elsevier, pp 1–29.

Joudeh N, Linke D (2022) Nanoparticle classification, physicochemical properties, characterization, and applications: A comprehensive review for biologists. *J Nanobiotechnology* 20:262. https://doi.org/10.1186/s12951-022-01477-8

Kabir SF, Rahman A, Yeasmin F, et al (2022) Occurrence, Distribution, and Structure of Natural Polysaccharides. In: *Radiation-Processed Polysaccharides*. Elsevier, pp 1–27.

Kaur H, Ahuja M, Kumar S, Dilbaghi N (2012) Carboxymethyl tamarind kernel polysaccharide nanoparticles for ophthalmic drug delivery. *Int J Biol Macromol* 50:833–839. https://doi.org/10.1016/j.ijbiomac.2011.11.017

Kaur M, Malik B, Garg T, et al (2015) Development and characterization of guar gum nanoparticles for oral immunization against tuberculosis. *Drug Deliv* 22:328–334. https://doi.org/10.3109/10717544.2014.894594

Khampieng T, Aramwit P, Supaphol P (2015) Silk sericin loaded alginate nanoparticles: Preparation and anti-inflammatory efficacy. *Int J Biol Macromol* 80:636–643. https://doi.org/10.1016/j.ijbiomac.2015.07.018

Kumar GV, Su C-H, Velusamy P (2016) Ciprofloxacin loaded genipin cross-linked chitosan/heparin nanoparticles for drug delivery application. *Mater Lett* 180:119–122. https://doi.org/10.1016/j.matlet.2016.05.108

Kumari A, Yadav SK, Yadav SC (2010) Biodegradable polymeric nanoparticles based drug delivery systems. *Colloids Surf B Biointerfaces* 75:1–18. https://doi.org/10.1016/j.colsurfb.2009.09.001

Li Q, Niu Y, Xing P, Wang C (2018) Bioactive polysaccharides from natural resources including Chinese medicinal herbs on tissue repair. *Chin Med* 13:7. https://doi.org/10.1186/s13020-018-0166-0

Liu J, Willför S, Xu C (2015) A review of bioactive plant polysaccharides: Biological activities, functionalization, and biomedical applications. *Bioact Carbohydr Dietary Fibre* 5:31–61. https://doi.org/10.1016/j.bcdf.2014.12.001

Lovegrove A, Edwards CH, De Noni I, et al (2017) Role of polysaccharides in food, digestion, and health. *Crit Rev Food Sci Nutr* 57:237–253. https://doi.org/10.1080/10408398.2014.939263

Mishra B, Patel BB, Tiwari S (2010) Colloidal nanocarriers: A review on formulation technology, types and applications toward targeted drug delivery. *Nanomedicine* 6:9–24. https://doi.org/10.1016/j.nano.2009.04.008

Mizrahy S, Peer D (2012) Polysaccharides as building blocks for nanotherapeutics. *Chem Soc Rev* 41:2623–2640. https://doi.org/10.1039/C1CS15239D

Mohammed ASA, Naveed M, Jost N (2021) Polysaccharides; Classification, chemical properties, and future perspective applications in fields of pharmacology and biological medicine (A review of current applications and upcoming potentialities). *J Polym Environ* 29:2359–2371. https://doi.org/10.1007/s10924-021-02052-2

Mokhtari S, Jafari SM, Assadpour E (2017) Development of a nutraceutical nano-delivery system through emulsification/internal gelation of alginate. *Food Chem* 229:286–295. https://doi.org/10.1016/j.foodchem.2017.02.071

Mozafari MR (ed) (2006) *Nanocarrier Technologies*. Springer Netherlands, Dordrecht.

Mukherjee PK (2019) Bioactive Phytocomponents and Their Analysis. In: Mukherjee PK (ed) *Quality Control and Evaluation of Herbal Drugs*. Elsevier, pp 237–328.

Murali R, Vidhya P, Thanikaivelan P (2014) Thermoresponsive magnetic nanoparticle – Aminated guar gum hydrogel system for sustained release of doxorubicin hydrochloride. *Carbohydr Polym* 110:440–445. https://doi.org/10.1016/j.carbpol.2014.04.076

Musa HH, Ahmed AA, Musa TH (2019) *Chemistry, Biological, and Pharmacological Properties of Gum Arabic.* In: Mérillon JM., Ramawat K.G. (eds) *Bioactive Molecules in Food.* Springer, pp 797–814.

Nasra S, Bhatia D, Kumar A (2022) Recent advances in nanoparticle-based drug delivery systems for rheumatoid arthritis treatment. *Nanoscale Adv* 4:3479–3494. https://doi.org/10.1039/D2NA00229A

Nosrati H, Khodaei M, Alizadeh Z, Banitalebi-Dehkordi M (2021) Cationic, anionic and neutral polysaccharides for skin tissue engineering and wound healing applications. *Int J Biol Macromol* 192:298–322. https://doi.org/10.1016/j.ijbiomac.2021.10.013

Pan Y, Li Y, Zhao H, et al (2002) Bioadhesive polysaccharide in protein delivery system: Chitosan nanoparticles improve the intestinal absorption of insulin in vivo. *Int J Pharm* 249:139–47. https://doi.org/10.1016/s0378-5173(02)00486-6

Panyam J, Labhasetwar V (2003) Biodegradable nanoparticles for drug and gene delivery to cells and tissue. *Adv Drug Deliv Rev* 55:329–47. https://doi.org/10.1016/s0169-409x(02)00228-4

Paques JP, van der Linden E, van Rijn CJM, Sagis LMC (2014) Preparation methods of alginate nanoparticles. *Adv Colloid Interface Sci* 209:163–171. https://doi.org/10.1016/j.cis.2014.03.009

Patel J, Maji B, Moorthy NSHN, Maiti S (2020) Xanthan gum derivatives: Review of synthesis, properties and diverse applications. *RSC Adv* 10:27103–27136. https://doi.org/10.1039/D0RA04366D

Peptu CA, Ochiuz L, Alupei L, et al (2014) Carbohydrate based nanoparticles for drug delivery across biological barriers. *J Biomed Nanotechnol* 10:2107–2148. https://doi.org/10.1166/jbn.2014.1950

Plucinski A, Lyu Z, Schmidt BVKJ (2021a) Polysaccharide nanoparticles: From fabrication to applications. *J Mater Chem B* 9:7030–7062. https://doi.org/10.1039/D1TB00628B

Plucinski A, Lyu Z, Schmidt BVKJ (2021b) Polysaccharide nanoparticles: From fabrication to applications. *J Mater Chem* B 9:7030–7062. https://doi.org/10.1039/D1TB00628B

Raghav N, Sharma MR, Kennedy JF (2021) Nanocellulose: A mini-review on types and use in drug delivery systems. *Carbohydr Polym Technol Appl* 2:100031. https://doi.org/10.1016/j.carpta.2020.100031

Rao K, Aziz S, Roome T, et al (2018) Gum acacia stabilized silver nanoparticles based nano-cargo for enhanced anti-arthritic potentials of hesperidin in adjuvant induced arthritic rats. *Artif Cells Nanomed Biotechnol* 46:597–607. https://doi.org/10.1080/21691401.2018.1431653

Rodriguez-Torres M del P, Acosta-Torres LS, Diaz-Torres LA (2018) Heparin-based nanoparticles: An overview of their applications. *J Nanomater* 2018:1–8. https://doi.org/10.1155/2018/9780489

Sanchez C, Nigen M, Mejia Tamayo V, et al (2018) Acacia gum: History of the future. *Food Hydrocoll* 78:140–160. https://doi.org/10.1016/j.foodhyd.2017.04.008

Sarmah JK, Mahanta R, Bhattacharjee SK, et al (2011) Controlled release of tamoxifen citrate encapsulated in cross-linked guar gum nanoparticles. *Int J Biol Macromol* 49:390–6. https://doi.org/10.1016/j.ijbiomac.2011.05.020

Seyedebrahimi R, Razavi S, Varshosaz J (2020) Controlled delivery of brain derived neurotrophic factor and gold-nanoparticles from chitosan/TPP Nanoparticles for tissue engineering applications. *J Clust Sci* 31:99–108. https://doi.org/10.1007/s10876-019-01621-9

She W, Li N, Luo K, et al (2013) Dendronized heparin–doxorubicin conjugate based nanoparticle as pH-responsive drug delivery system for cancer therapy. *Biomaterials* 34:2252–2264. https://doi.org/10.1016/j.biomaterials.2012.12.017

Singh J, Kumar S, Dhaliwal AS (2020) Controlled release of amoxicillin and antioxidant potential of gold nanoparticles-xanthan gum/poly (Acrylic acid) biodegradable nanocomposite. *J Drug Deliv Sci Technol* 55:101384. https://doi.org/10.1016/j.jddst.2019.101384

Singh T, Shukla S, Kumar P, et al (2017) Application of nanotechnology in food science: Perception and overview. *Front Microbiol* 8:. https://doi.org/10.3389/fmicb.2017.01501

Sinha VR, Mittal BR, Kumria R (2005) In vivo evaluation of time and site of disintegration of polysaccharide tablet prepared for colon-specific drug delivery. *Int J Pharm* 289:79–85. https://doi.org/10.1016/j.ijpharm.2004.10.019

Sohail R, Abbas SR (2020) Evaluation of amygdalin-loaded alginate-chitosan nanoparticles as biocompatible drug delivery carriers for anticancerous efficacy. *Int J Biol Macromol* 153:36–45. https://doi.org/10.1016/j.ijbiomac.2020.02.191

Sorasitthiyanukarn FN, Muangnoi C, Rojsitthisak P, Rojsitthisak P (2022) Chitosan oligosaccharide/alginate nanoparticles as an effective carrier for astaxanthin with improving stability, in vitro oral bioaccessibility, and bioavailability. *Food Hydrocoll* 124:107246. https://doi.org/10.1016/j.foodhyd.2021.107246

Soumya RS, Ghosh S, Abraham ET (2010) Preparation and characterization of guar gum nanoparticles. *Int J Biol Macromol* 46:267–9. https://doi.org/10.1016/j.ijbiomac.2009.11.003

Soumya RS, Sherin S, Raghu KG, Abraham A (2018) Allicin functionalized locust bean gum nanoparticles for improved therapeutic efficacy: An in silico, in vitro and in vivo approach. *Int J Biol Macromol* 109:740–747. https://doi.org/10.1016/j.ijbiomac.2017.11.065

Thomas D, KurienThomas K, Latha MS (2020) Preparation and evaluation of alginate nanoparticles prepared by green method for drug delivery applications. *Int J Biol Macromol* 154:888–895. https://doi.org/10.1016/j.ijbiomac.2020.03.167

Ullah S, Khalil AA, Shaukat F, Song Y (2019) Sources, extraction and biomedical properties of polysaccharides. *Foods* 8. https://doi.org/10.3390/foods8080304

Upadhyay M, Adena SKR, Vardhan H, et al (2018a) Development of biopolymers based interpenetrating polymeric network of capecitabine: A drug delivery vehicle to extend the release of the model drug. *Int J Biol Macromol* 115:907–919. https://doi.org/10.1016/j.ijbiomac.2018.04.123

Upadhyay M, Adena SKR, Vardhan H, et al (2019) Locust bean gum and sodium alginate based interpenetrating polymeric network microbeads encapsulating Capecitabine: Improved pharmacokinetics, cytotoxicity &in vivo antitumor activity. *Mater Sci Eng: C* 104:109958. https://doi.org/10.1016/j.msec.2019.109958

Upadhyay M, Adena SKR, Vardhan H, et al (2018b) Development and optimization of locust bean gum and sodium alginate interpenetrating polymeric network of capecitabine. *Drug Dev Ind Pharm* 44:511–521. https://doi.org/10.1080/03639045.2017.1402921

Upadhyay M, Vardhan H, Mishra B (2020) Natural polymers composed mucoadhesive interpenetrating buoyant hydrogel beads of capecitabine: Development, characterization and in vivo scintigraphy. *J Drug Deliv Sci Technol* 55:101480. https://doi.org/10.1016/j.jddst.2019.101480

Verma D, Sharma SK (2021) Recent advances in guar gum based drug delivery systems and their administrative routes. *Int J Biol Macromol* 181:653–671. https://doi.org/10.1016/j.ijbiomac.2021.03.087

Xu Y-Y, Huo Y-F, Xu L, et al (2021) Resveratrol-loaded ovalbumin/Porphyra haitanensis polysaccharide composite nanoparticles: Fabrication, characterization and antitumor activity. *J Drug Deliv Sci Technol* 66:102811. https://doi.org/10.1016/j.jddst.2021.102811

Zhang D, Jiang F, Ling J, et al (2021) Delivery of curcumin using a zein-xanthan gum nanocomplex: Fabrication, characterization, and in vitro release properties. *Colloids Surf B Biointerfaces* 204:111827. https://doi.org/10.1016/j.colsurfb.2021.111827

Zhang L, Yu F, Cole AJ, et al (2009) Gum Arabic-coated magnetic nanoparticles for potential application in simultaneous magnetic targeting and tumor imaging. *AAPS J* 11:693. https://doi.org/10.1208/s12248-009-9151-y

Zong A, Cao H, Wang F (2012) Anticancer polysaccharides from natural resources: A review of recent research. *Carbohydr Polym* 90:1395–410. https://doi.org/10.1016/j.carbpol.2012.07.026

12 Bionanomaterials in Restorative Dental Materials

S.C. Onwubu, C.S. Okonkwo, C.G. Iwuela, M.U. Makgobole, P.S. Mdluli, and T.H. Mokhothu

12.1 INTRODUCTION

Bionanomaterials have become increasingly popular in recent times due to their exceptional properties and potential applications in different areas, including restorative dental materials (Wu et al., 2020). In the field of restorative dental materials, bionanomaterials have the potential to revolutionize the way dental restorations are made and improve the longevity and biocompatibility of these restorations (Khurshid et al., 2015). Bionanomaterials are typically composed of biological molecules, such as proteins and peptides, combined with inorganic nanoparticles. This combination results in materials that have both biological and physical properties that make them suitable for use in dental restorations (Haugen et al., 2020).

Traditionally, dental restorations have been made from materials such as metal alloys, ceramics, and polymers (Sakaguchi and Powers, 2011, Cramer et al., 2011, Egbo, 2021). However, these materials have limitations in terms of biocompatibility, stability, and longevity. In recent years, there has been a growing interest in developing new types of restorative dental materials that address these limitations. Bionanomaterials have emerged as a promising new class of materials for this purpose (Padovani et al., 2015, Wu et al., 2019). Bionanomaterials are biocompatible, meaning that they are not toxic or harmful to living tissue (Jeevanandam et al., 2022). This makes them a good option for use in dental restorations that are in close contact with living tissue, such as in fillings or coatings on dental implants. Bionanomaterials are also able to bond with living tissue, which can improve the long-term stability of dental restorations. This is particularly true for bionanocomposites, which are materials composed of a mixture of biological molecules and inorganic nanoparticles (Rokaya et al., 2018). These materials can form chemical bonds with living tissue, which can help to improve the stability of dental restorations over time (Asiri and Mohammad, 2018).

Additionally, bionanomaterials are being used to develop new types of dental restorations such as self-healing composites. These materials have the ability to repair themselves in the event of a small crack or fracture, which can improve the longevity of the restoration (Padovani et al., 2015). Although the field of bionanomaterials in restorative dental materials is still in its early stages, it has a huge potential to revolutionize the field of restorative dental materials by providing new options for dental restorations that are more biocompatible, stable, and long-lasting than traditional materials (Khurshid et al., 2015, Asiri and Mohammad, 2018). This chapter gives a summary of the current trends in the use of bionanomaterials in restorative dental materials. The chapter focuses on the unique properties and potential applications of bionanomaterials in this field, as well as the challenges and future directions of research in this area. The main research questions the chapter

DOI: 10.1201/9781003432791-15

addresses is, "What are the current trends in the application of bionanomaterials in restorative dental materials, including their unique properties, potential applications, challenges, and future directions?"

12.2 OVERVIEW OF TRADITIONAL RESTORATIVE DENTAL MATERIALS AND THEIR LIMITATIONS

Traditional restorative dental materials have been used for many years to restore the function and aesthetics of teeth (Egbo, 2021). These materials include composite resin, amalgam, and glass ionomer cement (GIC). Composite resin is a type of tooth-coloured filling material that is made from a mixture of glass or quartz filler particles and a resin matrix (Voicu et al., 2018). It is widely used for the restoration of both anterior and posterior teeth, due to its ability to match the colour of natural teeth and its ability to be sculpted and polished to a high aesthetic finish (Qaiser et al., 2020). However, composite resin has some limitations, including a relatively short lifespan compared to other restorative materials and the potential for polymerization shrinkage which can lead to microleakage and marginal gap (Boaro et al., 2019).

Amalgam, also known as silver filling, is a mixture of metals, primarily silver, tin, and copper. Amalgam has been used for many years due to its durability and ability to withstand the forces of chewing (Kumar and Ajitha, 2019). However, it has some limitations, such as its appearance, which is not aesthetic, and the potential for marginal leakage (Rangreez and Mobin, 2019). Glass ionomer cement (GIC) is a tooth-coloured filling material made from a mixture of glass powder and an acidic polymer (Bhattacharya et al., 2017). GIC is known for its biocompatibility and ability to release fluoride, which can help to prevent tooth decay (Vicente et al., 2021). However, GIC has some limitations, such as its relatively low mechanical strength, which can make it less suitable for use in larger restorations (Bakhadher, 2019).

Overall, traditional restorative dental materials have been used for many years, but they have some limitations that bionanomaterials may be able to address. These limitations include aesthetic, durability, and biocompatibility issues. The use of bionanomaterials in restorative dental materials has the potential to improve the performance and longevity of these materials, as well as address some of the limitations of traditional restorative dental materials.

12.3 INTRODUCTION TO BIONANOMATERIALS AND THEIR UNIQUE PROPERTIES

Bionanomaterials are a relatively new class of materials that have been attracting a lot of attention in recent years due to their unique properties (Troy et al., 2021). Bionanomaterials are defined as materials that combine the properties of biological molecules and nanoparticles. These materials can be made from a variety of natural or synthetic materials, such as proteins, lipids, polysaccharides, and inorganic materials (Troy et al., 2021). Bionanomaterials are characterized by their small size, high surface area, and biocompatibility, which make them attractive for use in a wide range of applications, including restorative dental materials. One of the key characteristics of bionanomaterials is their small size, typically in the range of 1–100 nanometers (Jandt and Watts, 2020). This small size gives bionanomaterials a high surface area-to-volume ratio, which makes them highly reactive and able to interact with biological systems at the cellular and molecular level. The high surface area-to-volume ratio also allows bionanomaterials to have a large number of functional groups on their surface, which can be used for a variety of applications (Khan et al., 2019).

Another important characteristic of bionanomaterials is their biocompatibility. Bionanomaterials can be designed to be non-toxic and non-immunogenic, making them suitable for use in biological systems, including the human body (Singh and Singh, 2021). This makes them an attractive option for use in restorative dental materials, as they are less likely to cause adverse reactions in the body (Khan et al., 2019). Bionanomaterials also have the ability to be functionalized with various biomolecules such as proteins, DNA, and RNA, which can be used to enhance their properties and tailor them for specific applications (Abarca-Cabrera et al., 2021).

In summary, bionanomaterials are a unique class of materials that combine the properties of biological molecules and nanoparticles to create materials with unique properties such as high surface area, reactivity, and biocompatibility. These properties make bionanomaterials a promising option for use in restorative dental materials, as they have the potential to improve the performance and longevity of these materials (Khan et al., 2019).

12.3.1 Current Trends in the Application of Bionanomaterials in Restorative Dental Materials

Research in the use of bionanomaterials for restorative dental materials has been growing rapidly in recent years (Waheed et al., 2020). There are a variety of bionanomaterials that have been investigated for use in restorative dental materials, including nanoparticles, bioactive glass (Pajares-Chamorro and Chatzistavrou, 2020), and biocomposites (Chatzistavrou et al., 2018). Nanoparticles, such as hydroxyapatite (HA) and silica, have been investigated for use in restorative dental materials due to their biocompatibility and ability to mimic the structure and composition of natural tooth enamel (Elkassas and Arafa, 2017). Studies have shown that HA nanoparticles can be incorporated into composite resin to improve their mechanical properties, and silica nanoparticles have been used to improve the wear resistance of composite resin (Ai et al., 2017, Liu et al., 2021, Melo et al., 2013).

Bioactive glass is a type of bionanomaterial that has been investigated for use in restorative dental materials due to its ability to bond to tooth structure, release fluoride, and promote the formation of new tooth-like material. Studies have shown that bioactive glass can be incorporated into composite resin to improve their properties, and used as a filling material in teeth (Tiskaya et al., 2021, Pajares-Chamorro and Chatzistavrou, 2020).

Biocomposites are a type of bionanomaterial that have been investigated for use in restorative dental materials. Biocomposites are made by combining natural or synthetic polymers with nanoparticles or other bionanomaterials (Hasan et al., 2020). Studies have shown that biocomposites can be used to improve the mechanical properties and biocompatibility of restorative dental materials. In addition to these materials, there are other bionanomaterials such as self-assembling peptides, proteins, and other biomaterials, which are being researched for their potential use in restorative dental materials (Chatzistavrou et al., 2018, Rokaya et al., 2018).

Overall, the current state of research in bionanomaterials for restorative dental materials is active and promising. There are a variety of bionanomaterials that have been investigated for use in restorative dental materials, including nanoparticles, bioactive glass, and biocomposites. These materials have the potential to improve the performance and longevity of restorative dental materials, and address some of the limitations of traditional restorative dental materials. However, more research is needed to fully understand the potential of bionanomaterials and to develop clinical applications (Pajares-Chamorro and Chatzistavrou, 2020, Chatzistavrou et al., 2018).

12.4 BIONANOCOMPOSITES

12.4.1 Definition and Properties of Bionanocomposites

A bionanocomposite is a type of material that is made by combining a natural or synthetic polymer with nanoparticles or other bionanomaterials (Mhd Haniffa et al., 2016). The polymer serves as a matrix or a base material, while the nanoparticles or other bionanomaterials are dispersed within the matrix. The resulting material has properties that are different from those of the individual components. Bionanocomposites have a number of unique properties that make them useful in restorative dental materials (Rokaya et al., 2018). Some of these properties include:

- Improved mechanical properties: The inclusion of nanoparticles or other bionanomaterials within the polymer matrix can improve the mechanical properties of the resulting bionanocomposite. This includes increased strength, stiffness, and toughness.
- Enhanced biocompatibility: Bionanocomposites can be designed to mimic the structure and composition of natural tooth enamel. This can improve the biocompatibility of the material, making it more compatible with the body and less likely to cause an adverse reaction.
- Improved wear resistance: The inclusion of nanoparticles within the polymer matrix can improve the wear resistance of the resulting bionanocomposite.
- Enhanced aesthetics: Bionanocomposites can be designed to mimic the color, translucency, and surface texture of natural teeth.
- Improved chemical stability: Bionanocomposites can be designed to resist chemical degradation, such as acid erosion, which is a common problem with traditional restorative dental materials.
- Enhanced bioactivity: Some bionanocomposites are designed to interact with the body in beneficial ways, such as promoting the formation of new tooth-like material.

It is worth noting that, the properties of a bionanocomposite depend on the properties of the individual components and the way they are combined together. The properties of a bionanocomposite are usually greater than the sum of the properties of the individual components (Patra et al., 2022).

12.4.2 Types of Bionanocomposites Used in Restorative Dental Materials

There are several types of bionanocomposites that have been developed for use in restorative dental materials. These include:

1. Polymer-inorganic nanoparticle bionanocomposites: These bionanocomposites are made by combining a polymer matrix with inorganic nanoparticles such as silica, hydroxyapatite, or zirconia. These bionanocomposites have improved mechanical properties and enhanced biocompatibility (Díez-Pascual, 2022).
2. Polymer-organic nanoparticle bionanocomposites: These bionanocomposites are made by combining a polymer matrix with organic nanoparticles such as chitosan, cellulose, or proteins. These bionanocomposites have improved mechanical properties and enhanced biocompatibility (Zhao et al., 2018).
3. Polymer-bioactive nanoparticle bionanocomposites: These bionanocomposites are made by combining a polymer matrix with bioactive nanoparticles such as growth factors, enzymes or other biomolecules. These bionanocomposites have improved mechanical properties, enhanced biocompatibility, and bioactivity (Ediyilyam et al., 2022).
4. Polymer-nanocomposites: These bionanocomposites are made by combining a polymer matrix with nanoparticles such as clay, carbon nanotubes, or graphene. These bionanocomposites have improved mechanical properties and chemical stability (Tavares et al., 2017).
5. Hybrid bionanocomposites: These bionanocomposites are made by combining two or more types of bionanocomposites. These bionanocomposites have improved mechanical properties and enhanced biocompatibility (Annu et al., 2021).

It is worth noting that each of these types of bionanocomposites has its own unique properties and advantages. Researchers are still working to optimize the properties of these bionanocomposites and to develop new types of bionanocomposites for use in restorative dental materials.

12.4.3 Advantages and Limitations of Bionanocomposites in Restorative Dental Materials

The advantages of bionanocomposites in restorative dental materials:

1. Improved mechanical properties: Bionanocomposites have high strength, toughness, and modulus compared to traditional dental materials. This makes them suitable for load-bearing restorations (Taymour et al., 2022).
2. Enhanced biocompatibility: The incorporation of bioactive nanoparticles into bionanocomposites enhances their biocompatibility and reduces the risk of adverse reactions (Eivazzadeh–Keihan et al., 2020).
3. Improved bioactivity: Bionanocomposites with bioactive nanoparticles have improved bioactivity and can promote tissue regeneration (Covarrubias et al., 2018).
4. Improved chemical stability: The incorporation of nanoparticles into bionanocomposites improves their chemical stability and reduces the risk of degradation (Rokaya et al., 2018).

Limitations of bionanocomposites in restorative dental materials (Patra et al., 2022):

1. Complex processing: The preparation of bionanocomposites is a complex process that requires specialized equipment and expertise.
2. High cost: The synthesis and characterization of bionanocomposites are expensive, which increases the cost of the final product.
3. Lack of clinical data: There is a lack of clinical data on the long-term performance of bionanocomposites in restorative dental materials.
4. Lack of standardization: There is a lack of standardization in the preparation and characterization of bionanocomposites, which can affect the reproducibility of results.

Despite these limitations, bionanocomposites show great potential as restorative dental materials, and research is ongoing to improve their properties and to develop new types of bionanocomposites for use in restorative dental materials (Díez-Pascual, 2022, Zhao et al., 2018, Ediyilyam et al., 2022, Tavares et al., 2017, Annu et al., 2021).

12.5 SELF-HEALING COMPOSITES

12.5.1 Definition and Properties of Self-Healing Composites

Self-healing composites are a new class of materials that have the ability to repair themselves after damage or degradation (Diba et al., 2018, Wu et al., 2019). They are made of a host matrix and healing agents that are capable of repairing cracks or breaks that occur in the material (Wang and Urban, 2020). The self-healing ability of these materials can extend their lifetime and reduce the need for maintenance (Wu et al., 2019).

Self-healing composites are made of a host matrix and healing agents (Figure 12.1). The host matrix is typically a polymer, ceramic, or metal (Wang et al., 2015). It provides the structural support for the composite material. The healing agents, on the other hand, are responsible for repairing damage to the host matrix (Diba et al., 2018). These agents can be in the form of liquids, gels, or particles. Additionally, they can be activated by different triggers such as temperature, pH, and mechanical stress (Diba et al., 2018, Wang et al., 2015).

There are different types of self-healing composites, each with its own set of advantages and limitations. For example, some self-healing composites are activated by temperature, while others are activated by pH or mechanical stress (Wang and Urban, 2020, Pathan and Shende, 2021). Some

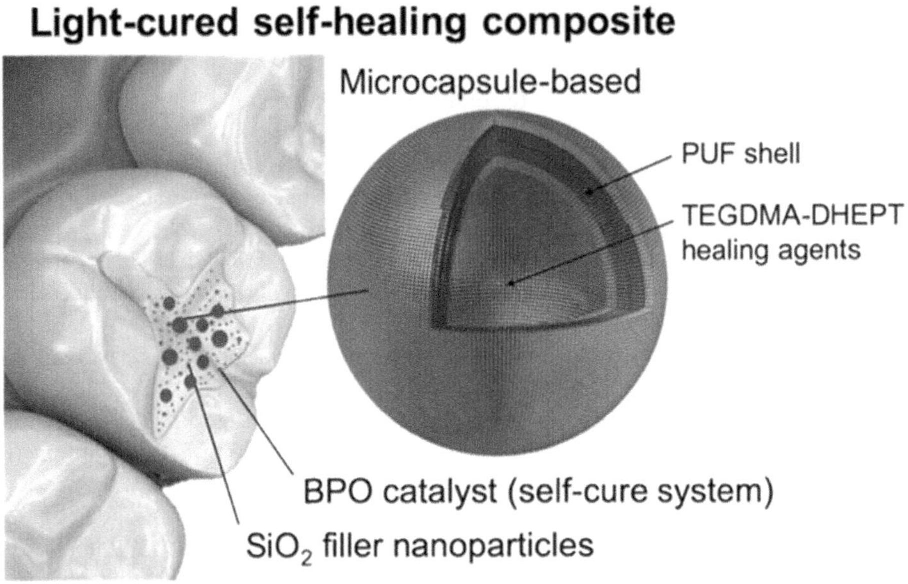

FIGURE 12.1 Self-Healing Composite (Source: Abid Althaqafi et al., 2022.)

self-healing composites are made with bioactive materials, making them biocompatible and suitable for use in medical and dental applications (Prasad, 2021).

Self-healing composites have high strength and toughness, similar to traditional composites. They also have high durability and are resistant to damage. The self-healing ability of the material makes it more durable. They are also environmentally friendly, as they can reduce the need for replacement and disposal of materials (Althaqafi et al., 2020, Wu et al., 2019). The properties include the following.

1. Self-healing ability: The most important property of self-healing composites is their ability to repair themselves after damage. This can extend the lifetime of the material and reduce the need for maintenance.
2. High strength: Self-healing composites have high strength and toughness, similar to traditional composites.
3. Durability: The self-healing ability of the material makes it more durable and resistant to damage.
4. Biocompatibility: Some self-healing composites are made with bioactive materials, making them biocompatible and suitable for use in medical and dental applications.
5. Environmentally friendly: Self-healing composites are environmentally friendly as they can reduce the need for replacement and disposal of materials.

Research on self-healing composites is ongoing, and scientists are working on developing new types of self-healing composites with improved properties and new applications. Currently, self-healing composites are being researched for use in various fields such as aerospace, civil engineering, and biomedical engineering (Chaudhary and Kandasubramanian, 2022, Wang et al., 2015). In the field of dentistry, self-healing composites are being researched for use in dental fillings, dental adhesives, and dental cements (Menikheim and Lavik, 2020). It's worth noting that self-healing composites are still in the research phase and not yet widely available for clinical use. Nonetheless, the technology is promising, and researchers are working to develop new types of self-healing composites with improved properties and new applications (Cramer et al., 2011).

12.5.2 Types of Self-Healing Composites Used in Restorative Dental Materials

There are different types of self-healing composites being researched for use in restorative dental materials. These types can be broadly classified based on the healing mechanism and the type of healing agents used.

One type of self-healing composite is the thermally activated self-healing composite. These composites are activated by an increase in temperature, usually generated by the heat of polymerization (Pathan and Shende, 2021). The healing agents in these composites are typically a liquid monomer that is encapsulated within the composite. When the composite is damaged, the monomer is released and polymerizes, filling the crack and repairing the damage (Althaqafi et al., 2020).

Another type of self-healing composite is the pH-activated self-healing composite. These composites are activated by a change in pH, usually caused by saliva or a dental adhesive (Matichescu et al., 2020). The healing agents in these composites are typically a neutral pH-sensitive polymer that is encapsulated within the composite. When the composite is damaged, the pH-sensitive polymer is activated and swells, filling the crack and repairing the damage (Zhang et al., 2022).

Bioactive self-healing composites are also being researched for use in restorative dental materials. These composites use bioactive materials such as calcium phosphates, bioactive glasses, and bioglasses as healing agents (Chatzistavrou et al., 2018). They are biocompatible and can be used for dental restorations in direct contact with living tissue (Pajares-Chamorro and Chatzistavrou, 2020).

It is worth noting that all of these types of self-healing composites are still in the research phase and not yet widely available for clinical use. Nonetheless, researchers are working to develop new types of self-healing composites with improved properties and new applications.

12.5.3 Advantages and Limitations of Self-Healing Composites in Restorative Dental Materials

Self-healing composites have several potential advantages over traditional restorative dental materials (Taymour et al., 2022). One major advantage is that they can repair small cracks and damage on their own, without the need for additional intervention from a dentist (Wu et al., 2019). This can potentially extend the lifespan of the restoration and reduce the need for frequent replacements.

Another advantage is that self-healing composites can potentially improve the mechanical properties of the restoration. For example, self-healing composites that are activated by mechanical stress may be more resistant to wear and tear caused by chewing or grinding (Wang et al., 2023). Additionally, some self-healing composites can provide additional benefits such as antimicrobial properties, bioactivity and self-sterilization (Wang et al., 2022).

Despite these advantages, self-healing composites also have several limitations. One limitation is that they may not be able to repair larger cracks or damage. Additionally, self-healing composites are still a relatively new area of research and are not yet widely available for clinical use. Further research is needed to optimize the properties and performance of self-healing composites and to understand their long-term durability (Patra et al., 2022).

Another limitation is that self-healing composites can be more complex to manufacture than traditional restorative dental materials, which can increase the cost of the restoration. Additionally, self-healing composites are generally more expensive than traditional materials (Patra et al., 2022).

In conclusion, self-healing composites have the potential to improve the performance and longevity of restorative dental materials, but more research is needed to fully understand their advantages and limitations and to develop effective and affordable self-healing composites for clinical use.

12.6 APPLICATIONS OF BIONANOMATERIALS IN RESTORATIVE DENTAL MATERIALS

12.6.1 Overview of Current and Potential Applications of Bionanomaterials in Restorative Dental Materials

Bionanomaterials have been researched extensively in recent years as potential replacements for traditional restorative dental materials. These materials are characterized by their unique properties such as high mechanical strength, biocompatibility, and bioactivity, which make them suitable for use in dental restorations (Wu et al., 2019). Bionanomaterials have a wide range of potential applications in restorative dental materials, including fillings, cements, adhesives, and coatings. Bionanomaterials have the potential to enhance the durability and lifespan of dental restorations while also providing antimicrobial properties and bioactivity. One promising area of research in bionanomaterials for restorative dental materials is tooth fillings. By using bionanocomposites, it is possible to develop fillings that are more robust and resistant to wear and tear compared to traditional fillings (Wang et al., 2022). Additionally, bionanocomposites can be designed to release antimicrobial agents to reduce the risk of secondary caries or to promote remineralization of the tooth (Komasa and Okazaki, 2022, Wang and Urban, 2020).

Bionanomaterials can also be used in dental adhesives, which are used to bond restorations such as veneers and inlays to the tooth (Vasiliu et al., 2021). Bionanocomposites have the potential to be utilized in creating adhesives that are more resilient and long-lasting compared to traditional adhesives. Bionanomaterials can also be used in dental coatings, which are used to protect the surface of restorations such as fillings, cements, and adhesives from wear and tear (Jandt and Watts, 2020). Bionanomaterials can also be used in dental cements, which are used to bond restorations such as crowns and bridges to the tooth. Bionanocomposites have the potential to be utilized in the creation of cements that exhibit greater durability and resistance to wear and tear compared to traditional cements. Furthermore, some bionanomaterials have been studied for their potential use in the regeneration of dental tissues. These materials can stimulate cell growth and differentiation, and have the potential to be used in the regeneration of dental pulps, dentin, and enamel. Additionally, research is ongoing in the field of self-healing composites being developed for restorative dental materials (Zhang et al., 2022). According to Wu et al. (2019), bionanomaterials possess the capacity to mend any cracks or harm that the restoration may encounter over time, and this capability can increase the lifespan of the restoration.

In conclusion, bionanomaterials have a wide range of potential applications in restorative dental materials, which can improve the performance and longevity of restorations as well as provide additional benefits such as antimicrobial properties and bioactivity (Bansal et al., 2021, Sreenivasalu et al., 2022). These materials are expected to be used in a wide range of dental restorations in the future, as they have the ability to solve many of the limitations of traditional dental materials.

12.6.2 Case Studies or Examples of Bionanomaterials Being Used in Restorative Dental Materials

One example of a bionanomaterial that has been used in restorative dental materials is hydroxyapatite (HAp). HAp is a biocompatible and bioactive material that is similar in composition to natural tooth structure (Irfan and Irfan, 2020). It has been used in tooth fillings, dental cements, and coatings to improve the mechanical properties and bioactivity of the restoration (Tiskaya et al., 2021).

Another example of a bionanomaterial that has been used in restorative dental materials is silica-based nanoparticles. These particles have been used to create bionanocomposites that have high mechanical strength and wear resistance (Wang et al., 2015). They have also been shown to have antimicrobial properties (Komasa and Okazaki, 2022), which can reduce the risk of secondary caries. A case study of a bionanocomposite that has been used in restorative dental materials is a composite filled with silica-based nanoparticles and a resin matrix. This material has been shown to have improved mechanical properties and wear resistance compared to traditional composite materials, and also has antimicrobial properties (Wang and Urban, 2020).

In conclusion, bionanomaterials have been used in a variety of restorative dental materials, including tooth fillings, dental cements, coatings, and self-healing composites. Examples of bionanomaterials include hydroxyapatite and silica-based nanoparticles. These materials have been shown to improve the mechanical properties, bioactivity, and antimicrobial properties of restorations, and have the potential to extend the lifespan of the restoration.

12.7 CHALLENGES AND FUTURE DIRECTIONS

12.7.1 Overview of the Current Challenges Facing the Development and Use of Bionanomaterials in Restorative Dental Materials

The use of bionanomaterials in restorative dental materials is a relatively new field of research, and as such, there are several challenges that need to be overcome in order to fully realize their potential (Jandt and Watts, 2020). One major challenge is the lack of long-term clinical data on the safety and effectiveness of bionanomaterials. While many in vitro and animal studies have shown promising results, more human clinical trials are needed to fully understand how these materials perform in a real-world setting, particularly their safety and effectiveness (AlKahtani, 2018).

Furthermore, there is a lack of standardization for bionanomaterials (Johnston et al., 2020). Different researchers and companies may use different methods to produce and test bionanomaterials, which can make it difficult to compare results and determine the best materials to use (Foulkes et al., 2020). Additionally, the biocompatibility of bionanomaterials has been a concern, as some bionanomaterials have been reported to cause inflammation and cytotoxicity, which may lead to a negative impact on the overall health of the patient (Ganguly et al., 2018). While bionanomaterials have shown great promise in the field of restorative dental materials, there are still several challenges that need to be overcome in order to fully realize their potential. These challenges include the lack of long-term clinical data, high costs, lack of standardization, and biocompatibility concerns. It is essential that these challenges are addressed in order to ensure the safe and effective use of bionanomaterials in restorative dental materials.

12.7.2 Discussion of Future Directions for Research and Development in the Field

As the field of bionanomaterials in restorative dental materials continues to evolve, several areas of research are expected to be of particular importance in the coming years. One area of focus is the development of new and improved bionanocomposites. Researchers are working on developing materials that are stronger, more durable, and more biocompatible than current materials (Zhou et al., 2019). They are also working on ways to make these materials more affordable and easier to produce. Another area of focus is the development of self-healing composites. These materials have the ability to repair themselves if they are damaged, which could greatly improve their longevity and reduce the need for frequent repairs or replacements (Wu et al., 2019).

Furthermore, there is interest in the development of bionanomaterials that can be used for more than just filling cavities. These materials could be used for orthodontic applications, such as brackets and wires, or for dental implants (Hossain et al., 2022). Equally, there is a current interest in the development of bionanomaterials that can mimic natural tooth structure more closely (Yazdanian et al., 2021). This could lead to restorations that are more comfortable for patients and have better long-term outcomes. In addition, the development of 3D printing technology has opened up a new avenue of research in bionanomaterials for restorative dental materials (Zhang et al., 2021). 3D printing allows for the production of highly precise and customized dental restorations, which can be a great advantage in terms of improving the fit and function of restorations. Moreover, the use of bioactive molecules, such as growth factors, enzymes, and proteins, can be used to improve the biocompatibility and bioactivity of bionanomaterials (Zhao et al., 2021). Additionally, researchers are also working on developing bionanomaterials that can be used for tooth regeneration. This could involve using bionanomaterials to stimulate the growth of new tooth structure, or to create scaffolds that can be used to grow new teeth in a lab (Yazdanian et al., 2021).

Overall, the field of bionanomaterials in restorative dental materials is an exciting and rapidly evolving field, with many new and innovative materials and techniques in development. As research continues and these materials and techniques are further refined, we can expect to see significant improvements in the safety, effectiveness, and affordability of restorative dental materials in the future.

12.8 CONCLUSION

In this chapter, we discussed the current trends in the application of bionanomaterials in restorative dental materials. We covered the following key points:

- The definition and properties of bionanomaterials, which are materials that combine the properties of biological materials with those of nanoscale materials.
- The limitations of traditional restorative dental materials, such as dental amalgam, composite resins, and ceramics, and how bionanomaterials can address some of these limitations.
- The types of bionanocomposites that are currently being used in restorative dental materials, including those made from synthetic polymers and natural polymers.
- The advantages and limitations of bionanocomposites, such as their improved strength, durability, and biocompatibility compared to traditional materials, as well as their potential for high production costs and difficulty of processing.
- The introduction of self-healing composites, which have the ability to repair themselves if they are damaged and can improve the longevity of restorative materials.
- The potential applications of bionanomaterials in restorative dental materials, including dental fillings, dental cements, and dental implants.
- Challenges are facing the development and use of bionanomaterials in restorative dental materials, such as issues with biocompatibility and cost.
- The future directions for research and development in the field, including the continued development of new and improved bionanocomposites, the development of self-healing composites, and the use of 3D printing technology and bioactive molecules to improve the biocompatibility and bioactivity of bionanomaterials.

Overall, bionanomaterials offer a promising new approach to restorative dental materials, with the potential to address many of the limitations of traditional materials and improve the safety, effectiveness, and affordability of restorative dental treatments.

12.8.1 Implications of the Research for the Field of Restorative Dental Materials

The research on bionanomaterials in restorative dental materials has significant implications for the field. Bionanomaterials have the potential to revolutionize restorative dental materials by offering new and improved materials that are stronger, more durable, and more biocompatible than traditional materials. One of the most significant implications of this research is the potential for bionanomaterials to improve the longevity of restorative dental treatments. Traditional restorative materials, such as dental amalgam and composite resins, often require frequent replacement due to wear and tear, which can be costly and time-consuming for patients. Bionanomaterials, on the other hand, have the potential to be more durable and resistant to wear and tear, which could lead to longer-lasting restorative treatments.

Another important implication of this research is the potential for bionanomaterials to improve the biocompatibility of restorative dental materials. Traditional materials, such as dental amalgam, have been known to cause allergic reactions and other adverse effects in some patients. Bionanomaterials, made of natural polymers, can mitigate these issues, and improve the overall safety and effectiveness of restorative dental treatments. The research on bionanocomposites and self-healing composites also has potential implications for the field. Bionanocomposites can improve the mechanical properties of restorative dental materials, making them stronger and more durable. Self-healing composites have the ability to repair themselves if they are damaged, which can improve the longevity of restorative materials.

Overall, the research on bionanomaterials in restorative dental materials has the potential to significantly improve the safety, effectiveness, and affordability of restorative dental treatments. However, further research is needed to fully understand the potential of these materials and to develop new and improved bionanomaterials for restorative dental applications.

12.8.2 Suggestions for Future Research

In the field of restorative dental materials, bionanomaterials have shown great promise as a new class of materials with unique properties that can address many of the limitations of traditional restorative materials. However, there are still challenges to be addressed in the development and use of bionanomaterials, such as developing methods for large-scale production and ensuring biocompatibility and long-term stability.

Future research in the field should focus on addressing these challenges and further exploring the potential applications of bionanomaterials in restorative dental materials. This could include developing new types of bionanocomposites with improved mechanical properties and self-healing capabilities, as well as investigating the use of bionanomaterials in specialized applications such as orthodontic materials and root canal filling materials. Additionally, more studies are needed to evaluate the long-term performance and biocompatibility of bionanomaterials in restorative dental materials. This can include clinical studies to evaluate the longevity, aesthetics, and clinical performance of bionanocomposites in restorative dental materials.

In conclusion, although bionanomaterials have demonstrated significant potential in restorative dental materials, there are still various obstacles to overcome such as insufficient long-term clinical data and expensive production costs. Nonetheless, with continued research and innovation, these challenges can be addressed, resulting in more advanced and efficient restorative dental materials that can enhance the wellbeing of patients globally and achieve their maximum potential. Future research should focus on developing new bionanocomposites and investigating their applications in specialized areas of restorative dentistry and also evaluate their long-term performance and biocompatibility.

REFERENCES

ABARCA-CABRERA, L., FRAGA-GARCÍA, P. & BERENSMEIER, S. 2021. Bio-nano interactions: Binding proteins, polysaccharides, lipids and nucleic acids onto magnetic nanoparticles. *Biomaterials Research,* 25, 1–18.

ABID ALTHAQAFI, K., ALSHABIB, A., SATTERTHWAITE, J. & SILIKAS, N. 2022. Properties of a model self-healing microcapsule-based dental composite reinforced with silica nanoparticles. *Journal of Functional Biomaterials,* 13, 19.

AI, M., DU, Z., ZHU, S., GENG, H., ZHANG, X., CAI, Q. & YANG, X. 2017. Composite resin reinforced with silver nanoparticles–Laden hydroxyapatite nanowires for dental application. *Dental Materials,* 33, 12–22.

ALKAHTANI, R. N. 2018. The implications and applications of nanotechnology in dentistry: A review. *The Saudi Dental Journal,* 30, 107–116.

ALTHAQAFI, K. A., SATTERTHWAITE, J. & SILIKAS, N. 2020. A review and current state of autonomic self-healing microcapsules-based dental resin composites. *Dental Materials,* 36, 329–342.

ANNU, AHMED, S., NIRALA, R. K., KUMAR, R. & IKRAM, S. 2021. Green synthesis of chitosan/nanosilver hybrid bionanocomposites with promising antimicrobial, antioxidant and anticervical cancer activity. *Polymers and Polymer Composites,* 29, S199–S210.

ASIRI, A. M. & MOHAMMAD, A. 2018. *Applications of nanocomposite materials in dentistry*, Woodhead Publishing.

BAKHADHER, W. 2019. Modification of glass ionomer restorative material: A review of literature. *EC Dental Science,* 18, 1001–1006.

BANSAL, V., KATIYAR, S. S. & DORA, C. P. 2021. Toxicity aspects: Crucial obstacles to clinical translation of nanomedicines. In *Direct Nose-to-Brain Drug Delivery.* Elsevier.

BHATTACHARYA, A., VAIDYA, S., TOMER, A. K. & RAINA, A. 2017. GIC at It's best–A review on ceramic reinforced GIC. *International Journal of Applied Dental Sciences,* 3, 405–408.

BOARO, L. C. C., LOPES, D. P., DE SOUZA, A. S. C., NAKANO, E. L., PEREZ, M. D. A., PFEIFER, C. S. & GONÇALVES, F. 2019. Clinical performance and chemical-physical properties of bulk fill composites resin—A systematic review and meta-analysis. *Dental Materials,* 35, e249–e264.

CHATZISTAVROU, X., LEFKELIDOU, A., PAPADOPOULOU, L., PAVLIDOU, E., PARASKEVOPOULOS, K. M., FENNO, J. C., FLANNAGAN, S., GONZÁLEZ-CABEZAS, C., KOTSANOS, N. & PAPAGERAKIS, P. 2018. Bactericidal and bioactive dental composites. *Frontiers in Physiology,* 9, 103.

CHAUDHARY, K. & KANDASUBRAMANIAN, B. 2022. Self-healing nanofibers for engineering applications. *Industrial & Engineering Chemistry Research,* 61, 3789–3816.

COVARRUBIAS, C., CÁDIZ, M., MAUREIRA, M., CELHAY, I., CUADRA, F. & VON MARTTENS, A. 2018. Bionanocomposite scaffolds based on chitosan–gelatin and nanodimensional bioactive glass particles: In vitro properties and in vivo bone regeneration. *Journal of Biomaterials Applications,* 32, 1155–1163.

CRAMER, N., STANSBURY, J. & BOWMAN, C. 2011. Recent advances and developments in composite dental restorative materials. *Journal of Dental Research,* 90, 402–416.

DIBA, M., SPAANS, S., NING, K., IPPEL, B. D., YANG, F., LOOMANS, B., DANKERS, P. Y. & LEEUWENBURGH, S. C. 2018. Self-healing biomaterials: from molecular concepts to clinical applications. *Advanced Materials Interfaces,* 5, 1800118.

DÍEZ-PASCUAL, A. M. 2022. *Inorganic-nanoparticle modified polymers.* MDPI.

EDIYILYAM, S., LALITHA, M. M., GEORGE, B., SHANKAR, S. S., WACŁAWEK, S., ČERNÍK, M. & PADIL, V. V. T. 2022. Synthesis, characterization and physicochemical properties of biogenic silver nanoparticle-encapsulated chitosan bionanocomposites. *Polymers,* 14, 463.

EGBO, M. K. 2021. A fundamental review on composite materials and some of their applications in biomedical engineering. *Journal of King Saud University-Engineering Sciences,* 33, 557–568.

EIVAZZADEH-KEIHAN, R., BAHOJB NORUZI, E., KHANMOHAMMADI CHENAB, K., JAFARI, A., RADINEKIYAN, F., HASHEMI, S. M., AHMADPOUR, F., BEHBOUDI, A., MOSAFER, J. & MOKHTARZADEH, A. 2020. Metal-based nanoparticles for bone tissue engineering. *Journal of Tissue Engineering and Regenerative Medicine,* 14, 1687–1714.

ELKASSAS, D. & ARAFA, A. 2017. The innovative applications of therapeutic nanostructures in dentistry. *Nanomedicine: Nanotechnology, Biology and Medicine,* 13, 1543–1562.

FOULKES, R., MAN, E., THIND, J., YEUNG, S., JOY, A. & HOSKINS, C. 2020. The regulation of nanomaterials and nanomedicines for clinical application: Current and future perspectives. *Biomaterials Science,* 8, 4653–4664.

GANGULY, P., BREEN, A. & PILLAI, S. C. 2018. Toxicity of nanomaterials: Exposure, pathways, assessment, and recent advances. *ACS Biomaterials Science & Engineering,* 4, 2237–2275.

HASAN, K. F., HORVÁTH, P. G. & ALPÁR, T. 2020. Potential natural fiber polymeric nanobiocomposites: A review. *Polymers,* 12, 1072.

HAUGEN, H. J., BASU, P., SUKUL, M., MANO, J. F. & RESELAND, J. E. 2020. Injectable biomaterials for dental tissue regeneration. *International Journal of Molecular Sciences,* 21, 3442.

HOSSAIN, N., ISLAM, M. A., CHOWDHURY, M. A. & ALAM, A. 2022. Advances of nanoparticles employment in dental implant applications. *Applied Surface Science Advances,* 12, 100341.

IRFAN, M. & IRFAN, M. 2020. Overview of hydroxyapatite; composition, structure, synthesis methods and its biomedical uses. *Biomedical Letters,* 6, 17–22.

JANDT, K. D. & WATTS, D. C. 2020. Nanotechnology in dentistry: Present and future perspectives on dental nanomaterials. *Dental Materials,* 36, 1365–1378.

JEEVANANDAM, J., LING, J. K. U., BARHOUM, A., SAN CHAN, Y. & DANQUAH, M. K. 2022. Bionanomaterials: Definitions, sources, types, properties, toxicity, and regulations. *Fundamentals of Bionanomaterials.* Elsevier.

JOHNSTON, L. J., GONZALEZ-ROJANO, N., WILKINSON, K. J. & XING, B. 2020. Key challenges for evaluation of the safety of engineered nanomaterials. *NanoImpact,* 18, 100219.

KHAN, I., SAEED, K. & KHAN, I. 2019. Nanoparticles: Properties, applications and toxicities. *Arabian Journal of Chemistry,* 12, 908–931.

KHURSHID, Z., ZAFAR, M., QASIM, S., SHAHAB, S., NASEEM, M. & ABUREQAIBA, A. 2015. Advances in nanotechnology for restorative dentistry. *Materials,* 8, 717–731.

KOMASA, S. & OKAZAKI, J. 2022. *Advances in dental bio-nanomaterials.* MDPI.

KUMAR, S. A. & AJITHA, P. 2019. Evaluation of compressive strength between Cention N and high copper amalgam-An in vitro study. *Drug Invention Today,* 12, 1–5.

LIU, J., ZHANG, H., SUN, H., LIU, Y., LIU, W., SU, B. & LI, S. 2021. The development of filler morphology in dental resin composites: A review. *Materials,* 14, 5612.

MATICHESCU, A., ARDELEAN, L. C., RUSU, L.-C., CRACIUN, D., BRATU, E. A., BABUCEA, M. & LERETTER, M. 2020. Advanced biomaterials and techniques for oral tissue engineering and regeneration—A review. *Materials,* 13, 5303.

MELO, M. A., GUEDES, S. F., XU, H. H. & RODRIGUES, L. K. 2013. Nanotechnology-based restorative materials for dental caries management. *Trends in Biotechnology,* 31, 459–467.

MENIKHEIM, S. D. & LAVIK, E. B. 2020. Self-healing biomaterials: The next generation is nano. *Wiley Interdisciplinary Reviews: Nanomedicine and Nanobiotechnology,* 12, e1641.

MHD HANIFFA, M. A. C., CHING, Y. C., ABDULLAH, L. C., POH, S. C. & CHUAH, C. H. 2016. Review of bionanocomposite coating films and their applications. *Polymers,* 8, 246.

PADOVANI, G. C., FEITOSA, V. P., SAURO, S., TAY, F. R., DURÁN, G., PAULA, A. J. & DURÁN, N. 2015. Advances in dental materials through nanotechnology: facts, perspectives and toxicological aspects. *Trends in Biotechnology,* 33, 621–636.

PAJARES-CHAMORRO, N. & CHATZISTAVROU, X. 2020. Bioactive glass nanoparticles for tissue regeneration. *Acs Omega,* 5, 12716–12726.

PATHAN, N. & SHENDE, P. 2021. Strategic conceptualization and potential of self-healing polymers in biomedical field. *Materials Science and Engineering: C,* 125, 112099.

PATRA, S., SINGH, M., PAREEK, D., WASNIK, K., GUPTA, P. S. & PAIK, P. 2022. Advances in the development of biodegradable polymeric materials for biomedical applications. In *Encyclopedia of Materials: Plastics and Polymers.* Elsevier: Amsterdam.

PRASAD, A. 2021. Bioabsorbable polymeric materials for biofilms and other biomedical applications: Recent and future trends. *Materials Today: Proceedings,* 44, 2447–2453.

QAISER, S., DEVADIGA, D. & HEGDE, M. N. 2020. Esthetic rehabilitation with nanohybrid composite: Case report series. *Journal of Advanced Oral Research,* 11, 251–256.

RANGREEZ, T. A. & MOBIN, R. 2019. Polymer composites for dental fillings. *Applications of Nanocomposite Materials in Dentistry.* Elsevier.

ROKAYA, D., SRIMANEEPONG, V., SAPKOTA, J., QIN, J., SIRALEARTMUKUL, K. & SIRIWONGRUNGSON, V. 2018. Polymeric materials and films in dentistry: An overview. *Journal of Advanced Research,* 14, 25–34.

SAKAGUCHI, R. L. & POWERS, J. M. 2011. *Craig's restorative dental materials-e-book,* Elsevier Health Sciences.

SINGH, R. P. & SINGH, K. R. 2021. *Bionanomaterials: Fundamentals and biomedical applications,* IOP Publishing.

SREENIVASALU, P. K. P., DORA, C. P., SWAMI, R., JASTHI, V. C., SHIROORKAR, P. N., NAGARAJA, S., ASDAQ, S. M. B. & ANWER, M. K. 2022. Nanomaterials in dentistry: Current applications and future scope. *Nanomaterials,* 12, 1676.

TAVARES, M. I. B., DA SILVA, E. O., DA SILVA, P. R. C. & DE MENEZES, L. R. 2017. Polymer nanocomposites. *Nanostructured materials-fabrication to applications.* InTech. Available at: http://dx.doi.org/10.5772/intechopen.68142.

TAYMOUR, N., FAHMY, A. E., GEPREEL, M. A. H., KANDIL, S. & EL-FATTAH, A. A. 2022. Improved mechanical properties and bioactivity of silicate based bioceramics reinforced poly (ether-ether-ketone) nanocomposites for prosthetic dental implantology. *Polymers,* 14, 1632.

TISKAYA, M., SHAHID, S., GILLAM, D. & HILL, R. 2021. The use of bioactive glass (BAG) in dental composites: A critical review. *Dental Materials,* 37, 296–310.

TROY, E., TILBURY, M. A., POWER, A. M. & WALL, J. G. 2021. Nature-based biomaterials and their application in biomedicine. *Polymers,* 13, 3321.

VASILIU, S., RACOVITA, S., GUGOASA, I. A., LUNGAN, M.-A., POPA, M. & DESBRIERES, J. 2021. The benefits of smart nanoparticles in dental applications. *International Journal of Molecular Sciences,* 22, 2585.

VICENTE, A., RODRÍGUEZ-LOZANO, F. J., MARTÍNEZ-BENEYTO, Y., JAIMEZ, M., GUERRERO-GIRONÉS, J. & ORTIZ-RUIZ, A. J. 2021. Biophysical and fluoride release properties of a resin modified glass ionomer cement enriched with bioactive glasses. *Symmetry,* 13, 494.

VOICU, G., CONSTANTIN, G. A. & SARACIN, A. 2018. Simulation of mechanical behaviour in milling and polishing of dental polymeric (resin) composites. *Material Plastics,* 54, 308–314.

WAHEED, M., YOUSAF, M., SHEHZAD, A., INAM-UR-RAHEEM, M., KHAN, M. K. I., KHAN, M. R., AHMAD, N. & AADIL, R. M. 2020. Channelling eggshell waste to valuable and utilizable products: A comprehensive review. *Trends in Food Science & Technology,* 106, 78–90.

WANG, D., HAN, S. & YANG, M. 2023. Tooth diversity underpins future biomimetic replications. *Biomimetics,* 8, 42.

WANG, J., DAI, D., XIE, H., LI, D., XIONG, G. & ZHANG, C. 2022. Biological effects, applications and design strategies of medical polyurethanes modified by nanomaterials. *International Journal of Nanomedicine*, 17, 6791–6819.

WANG, S. & URBAN, M. W. 2020. Self-healing polymers. *Nature Reviews Materials,* 5, 562–583.

WANG, Y., PHAM, D. T. & JI, C. 2015. Self-healing composites: A review. *Cogent Engineering,* 2, 1075686.

WU, J., XIE, X., ZHOU, H., TAY, F. R., WEIR, M. D., MELO, M. A. S., OATES, T. W., ZHANG, N., ZHANG, Q. & XU, H. H. 2019. Development of a new class of self-healing and therapeutic dental resins. *Polymer Degradation and Stability,* 163, 87–99.

WU, Q., MIAO, W.-S., ZHANG, Y.-D., GAO, H.-J. & HUI, D. 2020. Mechanical properties of nanomaterials: A review. *Nanotechnology Reviews,* 9, 259–273.

YAZDANIAN, M., RAHMANI, A., TAHMASEBI, E., TEBYANIAN, H., YAZDANIAN, A. & MOSADDAD, S. A. 2021. Current and advanced nanomaterials in dentistry as regeneration agents: An update. *Mini Reviews in Medicinal Chemistry,* 21, 899–918.

ZHANG, B., DE ALWIS WATUTHANTHRIGE, N., WANASINGHE, S. V., AVERICK, S. & KONKOLEWICZ, D. 2022. Complementary dynamic chemistries for multifunctional polymeric materials. *Advanced Functional Materials,* 32, 2108431.

ZHANG, Y., XU, Y., SIMON-MASSERON, A. & LALEVÉE, J. 2021. Radical photoinitiation with LEDs and applications in the 3D printing of composites. *Chemical Society Reviews,* 50, 3824–3841.

ZHAO, D., YU, S., SUN, B., GAO, S., GUO, S. & ZHAO, K. 2018. Biomedical applications of chitosan and its derivative nanoparticles. *Polymers,* 10, 462.

ZHAO, Y., ZHANG, Z., PAN, Z. & LIU, Y. 2021. *Advanced bioactive nanomaterials for biomedical applications*. ExplorationWiley Online Library, pp. 20210089.

ZHOU, X., HUANG, X., LI, M., PENG, X., WANG, S., ZHOU, X. & CHENG, L. 2019. Development and status of resin composite as dental restorative materials. *Journal of Applied Polymer Science,* 136, 48180.

13 Bionanomaterials in Tissue Engineering

Pooja Mittal, Chestha Goyal, Parteek Rana, Ramit Kapoor, and Rupesh K. Gautam

13.1 INTRODUCTION

Biomaterial is material that is used and modified for medical purposes. Biomaterials can have a benign role, such as being used for a heart valve, or they can be bioactive, such as hydroxyapatite-coated hip implants (one example is the Furlong Hip from Joint Replacement Instrumentation, Sheffield—such implants can last up to 20 years). Biomaterials are also used in dental applications, surgery, and drug delivery on a daily basis [1].

At the turn of the 1940s and 1950s, the first biomaterials as we know them were used in medicine. These materials were commonly available, originally developed for other purposes, and used in medicine by pioneering clinicians [2, 3].

Bionanomaterials are nanomaterials that are created using biomolecules or that encapsulate or immobilize a conventional nanomaterial in the presence of a biomolecule. Biomolecules originating from microorganisms, plants, agricultural wastes, insects, marine creatures and some mammals are used to create bionanomaterials. Due to improved biocompatibility, bioavailability, and bioreactivity, these bionanomaterials have shown minimal or negligible toxicity to humans, other creatures, and the environment.

Polymeric bionanomaterials are a flexible class of biomolecules used to remodel or restore the capabilities of damaged or injured tissues. Polymeric bionanomaterials have recently attracted the interest of researchers in tissue regeneration and drug delivery applications due to their biocompatibility, natural degradation rate, and chemically tunable properties. In addition, these bionanomaterials are constructed on extracellular matrix analogs that serve as an environment of regularity for cell-matrix interactions during tissue healing.

Colloidal formations made up of organic or synthetic polymers are known as polymeric nanoparticles. In comparison to other nanocarriers such liposomes, micelles, and inorganic nanosystems, they provide a number of advantages, including the ability to build at scale and in accordance with precise manufacturing procedures [4]. The creation of polymeric nanoparticles used biodegradable polymers, such as poly(methyl methacrylate) (PMMA), polyacrylamide, polystyrene, and polyacrylates as its fundamental building blocks (PNs). The presence of ongoing toxicity and inflammatory reactions was discovered, despite the fact that nanosystems made of these compounds cleaned fast and effectively. While non-degradable polymers frequently need a degradation period that is greater than their practical application duration [5], several characteristics, such as the physicochemical properties of biodegradable polymer nanoparticles, may affect the pace of breakdown (length, shape and molecular weight) and external factors such as pH and temperature.

Synthetic polymers such as poly (D, 1-lactide) (PLA), poly (D, L-glycolide) (PLG), poly(lactide-co-glycolide) (PLGA) copolymer, polyalkylcyanoacrylates and polycaprolactone are examples of

DOI: 10.1201/9781003432791-16

biodegradable polymers. They are considered safe, and several biodegradable polymer compounds have been licensed for pharmaceutical use by the US Food and Drug Administration (FDA) and the European Medicines Agency (EMA) [7]. Generally speaking, biodegradable polymer particles are more biocompatible, have lower systemic toxicity, and enable medication to release kinetics to be modified. They are often converted to oligomers and monomers, which are subsequently metabolized and eliminated from the body via normal processes [8, 9]. Polymeric nanoparticles have also been created using non-synthetic biodegradable polymers such as chitosan, alginate, gelatin, zein, and albumin [10].

Tissue engineering aims to develop tissue replacements that can be utilized to repair damaged tissues or organs. To create a tissue construct, cells are frequently planted on biomaterial scaffolds that imitate the extracellular matrix (ECM) and microenvironment. Because tissue engineering entails establishing effective replacements for injured tissues/organs, chemicals should be inert or biocompatible when in contact with living cells or tissues [11]. Recent advances in the production of nanomaterials via polymer chemistry, materials science, biomedical technology, and biomedical engineering have resulted in significant advancements in tissue engineering and regenerative medicines. 12]. Currently, several bionanomaterial-based techniques for tissue engineering have been identified, and scaffolds with beneficial properties such as the extracellular matrix are being developed for tissue synthesis.

13.2 NANOMATERIALS-BASED MATERIALS FOR TISSUE REGENERATION

Nanomaterials are typically manufactured and synthesized utilizing various agents, such as metals, polymers, and ceramics, as well as various synthesis processes, such as physical, chemical, and biological, each of which has unique properties. To achieve the desired functionality, a range of modification techniques are continually applied to specifically functionalize nanomaterials. Furthermore, as science and technology progress, those plans are regularly updated, providing consumers with more accurate information regarding nanoparticles. Furthermore, antigenicity, biocompatibility and toxicity of the product, biodegradability, drug release behaviour, physical and chemical properties of the drug, NP size, and surface charge distribution must all be considered [13]. Table 13.1. shows various types of nanomaterials and their applications.

TABLE 13.1
Types of Nanomaterials and their Application [69–70]

NANOMATERIAL	MATERIAL FOR NANOMATERIAL	APPLICATION
POLYMERIC NANOMATERIAL	Poly(glycolic acid) Polyanhydrides Gelatin/collagen	Drug delivery
	Poly(lactic acid)	Formation of nanocomposite and nanofibers
	Poly(ortho esters)	
	Chitosan	
METALLIC NANOMATERIAL	Gold	Nanorods for chemical and
	Silver	Biological sensor
	Titanium and	Nanowires in alginate scaffold
	Ti6al4v	For cardiac patches
		Wound healing
		Bone regeneration
CARBON NANOTUBES AND NANOCERAMICS	Carbon nanotubes (CNT)	Neural tissue regeneration
	Alumina and Titania	Bone regeneration
	Hydroxyapatite	Nanocrystalline HA in PLLA scaffold for bone scaffold

13.2.1 Polymeric Nanomaterial

Colloidal systems made up of synthetic or natural polymeric materials are known as polymeric nanoparticles. In comparison to other nanocarriers like liposomes, micelles, and inorganic nanosystems, they have outstanding benefits such the ability to scale up production and adherence to good manufacturing practise (GMP) [4]. In addition to the large variety of polymers available, additional significant characteristics of polymeric nanoparticles include their incredible resistance in biological fluids, the capacity to functionalize their surfaces, and the ability to tailor the breakdown of the polymer and the release of the encapsulated compound(s) in response to different stimuli [14, 15, 16].

13.2.2 Metallic Nanomaterials

Metal, metal oxide, and magnetic nanoparticles (NPs) are the three forms of metallic NPs. Metal NPs, on the other hand, have precise antibacterial capabilities in addition to catalytic activity, mechanical qualities, and electrical conductivity, which makes them particularly suitable for tissue engineering applications [17]. Due to the aforementioned natural properties, precious metal NPs are commonly used in cosmetics and medicine. To begin with, gold nanoparticles have been widely employed in cancer detection and treatment, biological probes, and drug administration, and so on due to their excellent biocompatibility and ease of production.

Nanostructured gold is an excellent option for usage in medical applications due to its low toxicity and high electric conductivity. It has been demonstrated that gold nanowires placed in alginate scaffolds may pass through the electrically inert pore walls of alginate and improve electric communication among neighbouring heart cells. Gold nanowires enhanced tissue alignment and thickness. Additionally, it was demonstrated that the cells in this tissue had the ability to synchronously contract in response to electrical stimulation [18]. Nanometals are also utilized in the regeneration of cartilage and bones. Nanophase metals made with Ti, Ti6Al4V, and CoCrMo alloys demonstrate more osteoblast adherence to the surface than traditional metals due to improved surface particle boundaries in nanophase materials. The chemistry of such nanometallic surfaces is the same as that of real tissues, with the exception of surface roughness. This enhanced cellular adhesion may also lead to a longer period of calcium-containing mineral deposition, which is crucial for bone healing [19]. Metal nanomaterials have limited utility because to non-biodegradability, which restricts high-dose uses, notwithstanding their practical diversity. Numerous non-biodegradable metals can poison adjacent tissue, which might have unfavourable effects. By applying a biocompatible material or surface treatment to the metal, the potential toxicity of nanometals can be reduced [20].

13.2.3 Nanocomposite

A nanocomposite consists of at least two components, generally polymers, fillers or inorganic materials, and should be nanoscale (up to 100 nm) for at least one component [21]. Due to the combined benefits of organic and inorganic hybrid material creation, nanocomposite substances have significantly improved mechanical, biodegradability, and dimensional balance compared to parent substances [22]. Similar to this, the composition of the nanocomposite may result in varied qualities. For instance, due to its exact features, such as flexibility, biocompatibility, antibacterial activity, wide surface area, and clean functionalization, graphene oxide (go) is very effective in tissue engineering. Gold (Au) has the ability to stimulate the synthesis of apatite, and hydroxyapatite (HAP) is appropriate for bone tissue engineering.

13.2.4 Carbon Nanotubes and Nanoceramics

The most often used natural nanomaterial in tissue engineering and regenerative medicine is carbon nanotubes (CNTs). As a result of their exceptional biocompatibility based on surface chemistry, mechanical robustness, and chemical inertness, they may be employed as a component of many tissue scaffolds. Carbon nanotubes have a very high surface-to-volume ratio because of their nanodiameter [23, 24]. Carbon nanotubes were employed to cover bone scaffolds and improve their mechanical qualities because they are one of the strongest materials known [25]. It was looked into if a thin layer of PEG-conjugated multiwalled carbon nanotubes (MWCNTs) spray-dried over warmed coverslips may affect the growth, shape, and final differentiation of human mesenchymal stem cells (hMSCs) into osteoblasts [26]. Because CNTs are extremely compatible with blood, they are used as catheters in cardiopulmonary applications. They are chemically inert and do not decompose. Due to such causes, the addition of CNTs to polyurethane greatly improved the material's tensile strength and elongation properties. Carbon nanotube-based materials may be employed in a range of applications that come into contact with blood since CNTs also significantly improved the anticoagulant properties of polyurethane [27]. After being employed in lung applications, CNTs were found to be harmful in lung tissue. It is yet unclear whether further organ toxicity or systemic toxicity exists. There is also a heated argument about whether CNTs may cause cancer. Greater in-depth research is therefore required to develop a general understanding of CNTs as a biocompatible nanomaterial [28].

Many biological materials are mechanically and chemically compatible with nanoceramics. As a result, nanoceramics are a potential choice for applications involving the direct interaction of the scaffold with biological fluids and hard tissue engineering. Because calcium phosphate (hydroxyapatite, HA) is the native mineral phase of bone and dentin, ceramics are suitable as replacements for damaged bone [29]. Ceramic nanoparticles have been researched for orthopaedic and dental applications for these reasons. Additionally, this ceramic's surface chemistry is essentially inert in biological fluids, enabling cardiovascular programmes. Alumina and titanium in the nanophase have also been found to improve osteoblast adhesion in vitro when compared to non-nanophase materials [30].

13.2.5 Nanoparticle for Biomedical Ranging

The majority of the nanoparticles being researched for application in biomedical imaging include solid lipid nanoparticles, liposomes, micelles, nanotubes, metal nanoparticles, quantum dots, dendrimers, polymeric nanoparticles, and iodinated nanoparticles [31]. Real-time assays that track cell-environment interactions at the molecular and organelle level as well as non-invasive real-time methodologies for monitoring the characteristics of engineered tissues are two examples of how biomedical imaging helps to the development of enabling technologies. For example, an implanted scaffold is designed to degrade over time and be replaced by cells that create tissue that is basically normal. For the greatest benefits, the rates of ECM creation employing cells and scaffold breakdown must be similar. Additionally, the host's reaction to the implanted construct might be impacted by scaffold degradation [32, 33].With the assistance of evaluating changes in scaffold weight [34, 35], morphological alterations, viscosity, mechanical characteristics, and molecular weight [36], all of which may depend on culture circumstances, significant work has been put into examining scaffold degradation behaviour in vitro. However, traditional in vitro approaches cannot properly evaluate in vivo scaffold degradation because the physiological circumstances cannot be replicated in vitro. In order to investigate scaffold degradation in vivo, currently animal models must be implanted. However, this requires killing many animals at different stages after implantation, which often leads to incorrect findings. Future tissue engineering applications will benefit from the usage of nanoparticles since they are very sensitive, target-specific, and secure.

13.2.6 Nanofibers

Synthetic biodegradable polymers and various polyesters are the most often utilized building blocks in tissue engineering. They might be created by a first covalent link between monomers. As a result, it is possible to precisely regulate the size and makeup of the repeating monomers. The creation of nanopolymers in nearly any shape, size, and functionality is possible thanks to the malleability of their synthesis, manufacture, and modification. On the other hand, natural polymers including collagen, elastin, chitosan, and polypeptides have been employed to build medicinal scaffolds. These substances are biocompatible and nontoxic since they are natural. These synthetic or organic materials can be employed in a number of ways to build nanofibrous scaffolds.

Electrospun nanofiber mats were discovered to improve mesenchymal stem cell adhesion and proliferation when compared to a flat, smooth surface [37]. Electrostatic fibrous scaffolds can improve biological responses including cell adhesion and preservation of cellular phenotype, according to various researchers [38, 39]. Human MSC adhesion, proliferation, motility, and differentiation were shown by Shih et al. According to these studies, nanoscale fibres may encourage early cell attachment, which might affect cellular proliferation and differentiation.

The orientation of electrospun fibres can influence cell proliferation in a manner similar to how it might influence cell orientation and tissue formation. It is well known that the musculoskeletal tissues have significant anisotropic mechanical characteristics and specifically oriented cells underneath the ECM. Designing functioning skeletal muscular tissues requires the capacity to mimic the structure of natural tissue, which includes highly oriented muscle fibres made of joined mononuclear muscle cells. It is generally accepted that muscle tissue function is governed by the form and arrangement of the fibres. For the purpose of creating aligned myotubes for skeletal muscle tissue engineering, it would be very advantageous to be able to efficiently organize muscle cells in vitro.

The technology of tissue engineering known as electrospinning has advanced significantly. The biomaterial composition, fibre diameter, fibre arrangement, form, and drug/protein incorporation into the scaffold may all be controlled utilizing this manufacturing method. A variety of cell types may adhere to and multiply on electrospun nanofiber scaffolds while still maintaining their phenotypic and functional characteristics.

13.3 ROLE OF REGENERATIVE MEDICINE IN WOUND HEALING

The biggest organ in the body, the skin naturally protects against external threats, controls temperature, and performs a number of other vital functions. Skin injuries heal by their own self healing mechanism; however, if healing is slowed down, as it is in patients with inadequate vascular supply (such as diabetics) or those who have greater regions of damaged skin (such as burns), therapy meant to help or speed up wound closure is essential to lowering morbidity [40, 41].

Hemostasis, inflammation, proliferation, and remodeling/maturation are all components of the intricate physiological process that makes up wound healing [42]. Acute wounds and chronic wounds are the two categories into which skin wounds are separated [43]. Acute wounds heal quickly and have a perforation or rupture of the epidermal layer. Chronic wounds are frequently difficult to treat in the short term, since they are frequently linked to issues like diabetes and weight. A crucial stage in the healing of a wound is angiogenesis. Vascular growth expedites wound healing and the production of granulation tissue [45] in addition to providing sufficient blood flow, nutrients, oxygen, and other essential elements [44]. While abnormal blood vessel development results in permanent wound formation and inhibits healing.

As a result, while treating skin wounds, emphasis must be given to both skin tissue and vascular regeneration. Numerous treatment approaches, such as local oxygen therapy, hyperbaric oxygen therapy, ozone therapy, negative pressure wound therapy, and others are available to speed the chronic wound healing process in a variety of skin injuries [46]. Major skin injuries are often treated with autologous transplantation. In a nutshell, the doctor removes the donor's whole thickness of skin, stretches it, and transplants it into the incision [47]. However, this strategy is constrained by the donor location and the injury area. In this chapter, an additional efficient autologous cell treatment method is suggested. After sufficient in vitro proliferation, these mixed cells are utilized for wound healing [48]. Success rate, treatment cycle, and cost are all significantly influenced by a variety of different factors. In order to improve the proliferation efficiency and biosafety of skin tissue engineering for wound healing [49, 50], nanomaterials are utilized. The ideal biomaterial should be able to do the following: (1) operate as a barrier layer for keratinocyte regeneration; (2) have strong adhesion to the lower dermis; (3) alter blood vessels in the wounded area; and (4) give the skin elastic structural support.

Skin tissue engineering in burn units continues to face difficulties with full-skin functional replacements. The surgeon may decide to utilize an autograft from the patient, which would result in non-functional transplanted tissue (missing sweat glands) depending on the degree and depth of the burn [51]. Only if the patient has an adequate number of healthy donor sites is this treatment appropriate. As a result, laboratory-produced materials that satisfy the demand for alternate approaches for wound covering and tissue regeneration in these people have been developed [40]. Additionally, lowering hospitalization and morbidity in burn patients requires controlling or preventing nosocomial infections [52].

13.3.1 Silver

Silver must be taken into account when choosing antibacterial agents. Over a century ago, silver nitrate (also known as silver ionic) was first used in eye drops for infants to prevent gonorrheal ophthalmia [53] and microsomal infection in burn patients [54]. Since then, silver has been used in treatment as an antibacterial agent. However, these techniques were abandoned because they had too many negative effects [55, 56]. The hunt for novel, more potent medications to treat bacterial infections, including silver nanoparticles (AgNPs) [58, 59], has been prompted by the worrisome rise in the frequency of antibiotic- and multidrug-resistant bacterial strains, occasionally referred to as superbugs [57]. The capacity of AgNPs to limit the infiltration of inflammatory macrophages and neutrophils, decrease the production of inflammatory cytokines, and regulate the expression of metalloproteinases was previously unknown, according to recent in vitro and in vivo investigations [60]. Therefore, it is anticipated that the inclusion of AgNPs in biomaterial dressings would not only give an antibacterial detail of the scaffold but also speed up wound healing [61, 62]. For instance, Lu et al. found that, compared to silver sulfadiazine or chitosan film alone, a chitosan-AgNP dressing significantly enhanced the rate of dermal wound healing by 10 per cent body surface area in a rat model [61]. AgNP also had a reduced proportion of silver distributed outside the wound region. Chitosan is a biodegradable, non-toxic polycationic polysaccharide that may be broken in vivo with the help of lysozyme to create secure amino sugars [63]. For instance, the commercial product HemCon dressing, which was created as a hemostatic dressing, has the potential to be effective in skin care applications [64].

13.3.2 Nanocellulose

Nanocellulose, a novel type of nanomaterial that contains nanostructures based on cellulose, is becoming more well-known. Nanomaterials based on cellulose are often employed in biomedical programmes for the treatment of skin problems because they have the capacity to absorb wound exudate and remove the dressing more quickly than conventional dressing materials (such as gauze). According to Fu et al., nanocellulose, which is created by microorganisms, has more advantages for wound care than standard dressings, such as promoting capillary formation and faster tissue regeneration. It also results in less infection and is a lower toxicity product. A hybrid composite that is antibacterial, photoluminescent, and elastomeric polypeptide-based was developed by Xi et al. to suppress multidrug-resistant bacteria (MDR) and improve wound healing.

13.3.3 Nitric Oxide

Both the immunological response and the proliferative/regenerative stage of wound healing need nitric oxide (NO). A NO-NP platform was developed by Friedman et al. [66] using glass/hydrogel NPs containing the antibacterial polysaccharide chitosan. These NPs significantly decreased methicillin-resistant S. aureus (MSRA) burden, accelerated wound closure time, and promoted wound epithelialization in a murine model of wound contamination [67]. NO-NP treatment increased cytokine production from the wound, including IL-6, TNF, and MCP-1 as well as TGF-, which aided in fibroblast migration, proliferation, and wound healing. As a result, bacterial collagen degradation within the wound decreased with time as a result of NO-NP-mediated bacterial lysis [68]. Additionally, topical NO release treatment was less effective than NO-NP therapy in encouraging wound healing in a diabetic rat model [69].

13.3.4 Gold Nanoparticles

Gold nanoparticles coated with the antimicrobial peptide surfactin provided a synergistic platform to lower the bacterial load and improve wound healing in a rat model of [MRSA-infected wounds] [70]. In rat models of cutaneous wounds, AuNPs were found to have synergistic benefits when used with hydrocolloid membrane dressings and the antioxidants lipoic acid/epigallocatechin gallate to improve wound healing by reducing oxidative stress and inflammation and promoting angiogenesis [71]. In mouse models, it has been demonstrated that the cerium oxide NP dressing alone had antioxidant and wound-healing-promoting effects, in part due to the proliferation of keratinocytes and fibroblasts and the decrease of nitrated protein end products as a result of Reactive Oxygen Species (ROS) [72].

13.3.5 Stem Cell Therapy

In order to improve revascularization in skin tissue engineering, it is possible to effectively shield molecules from nucleases and regulate gene release by designing appropriate nanocarriers for gene transfection. Stem cells with strong vascular endothelial growth factor expression and temporal alteration have been created (particularly after transplantation) to promote angiogenesis in skin regeneration [73]. Studies have shown that biodegradable nanoparticles may introduce the epidermal growth factor gene into human MSCs and cells made from human embryonic stem cells. In the meantime, the implant's effects are being felt more strongly in the target tissues since the treated cells are manufacturing epidermal growth factor with increased vigour. When VEGF-expressing stem cells were transplanted into nanostructured scaffolds as opposed to control cells, angiogenesis was 2–4 times more pronounced. It demonstrates how the regeneration of skin tissue using biodegradable nanoparticles has revolutionized the therapeutic application of stem cells [74].

13.4 ROLE OF STEM CELL THERAPY IN TISSUE ENGINEERING AND TISSUE REGENERATION

Inside the human body, stem cells create, maintain, and repair tissues. These are specialized cells that have the capacity to self-renew and give rise to a wide variety of cell types found in the human body. Stem cells have access to a variety of developmental pathways, giving them the capacity to differentiate into any form of cell in the body or a specific cell type with a known origin. Because of these characteristics, stem cells are very beneficial for biomedical applications and regenerative medicine, and they have evolved into the main molecular instrument for these goals [75, 76].

Epithelial surfaces (skin, cornea, and mucous membranes), as well as skeletal tissues, may presently be created utilizing stem cells. The pace of self-renewal and physical makeup of these systems vary, and these are important aspects in any attempt to repair tissues utilizing stem cells. The development of appropriate interventional strategies in certain fields depends on an awareness of the fundamental variations between organ systems and associated stem cells [77].

Unlike embryonic stem cells, mesenchymal cells may develop into cell lineages that are unique to particular organs and come from adult organisms. Because there are no ethical issues with using mesenchymal cells in regenerative medicine, this makes them a very good choice. As in other tissues, cartilage's capacity for self-renewal is constrained by the lack of a dense population of progenitor cells. Mesenchymal stem cells with several pluripotencies have been applied therapeutically to cartilage repair. Mesenchymal stem cells isolated from the anterior and posterior iliac crests have been employed therapeutically as connective tissue progenitor donors for osteoarthritis of the carpometacarpal joint, which is frequent in postmenopausal women. Mesenchymal stem cell therapy is a very effective therapeutic option; the patient avoids surgery while also seeing a marked improvement in joint function and a decrease in pain [78].

Induced pluripotent stem cells are adult cells with a changed genetic programme that have grown stronger by the transfection of a transcription factor. According to Mall and Wernig, 2017, cell reprogramming enables the conversion of mature skin cells into neurons, hepatocytes, or heart cells. This technique is useful for a number of scientific applications, such as examining disease progression and determining the effectiveness and safety of recently produced medications before subjecting them to animal testing in clinical trials [79].

The finest example of a stem cell-dependent self-renewing mechanism is hematopoiesis. Only a tiny compartment of long-term self-renewing stem cells is present in the hierarchy of hematopoietic progenitors equipped with diverse self-renewal capacities for permanent and complete regeneration of hematopoiesis following lethal irradiation [80]. Similar to how progenitor cells in keratinocyte culture are arranged in a hierarchy, these cells also produce three different types of colonies (holoclones, meroclones, and paraclones) under clonogenic conditions, each of which has a different level of capacity for self-renewal and differentiation of the epidermis. Only holoclones, a population of permanent multiplying cells, and paraclones, a population of aged or differentiated progenitors, are produced from real stem cells, as shown by their enormous capacity for self-replication (>140 doublings) [81]. In theory, generating epidermal grafts would only need a tiny, pure population of cells that produce holoclone. A novel tool for this technique in the epidermis might be offered by the most recent identification of a keratinocyte stem cell marker [82].

Skeletal stem cells (SSCs; also known as bone marrow stromal stem cells or mesenchymal stem cells) are a subset of adherent clonogenic marrow-derived cells that have a high capacity for in vitro proliferation. After ectopic in vivo transplantation in model systems, all primary tissues present in bone as an organ (bone, cartilage, adipocytes, and hematopoiesis-supporting stroma) are produced [83]. SSCs generated in culture should be linked to appropriate carriers prior to transplantation. They provide a three-dimensional framework for forming the vascular bed and differentiating donated progenitor cells into the bone/marrow organ. Many suitable materials are being researched and improved, including polyglycolic and polylactic acids [84, 85], synthetic hydroxyapatite/tricalcium phosphates, and polyglycolic acid. Whether administered alone or in combination with growth factors, such as bone morphogenetic proteins, the majority of them are efficient at stimulating bone repair. However, their compatibility with long-term stem cell preservation and the ultimate fate of the bone-biomaterial composite created at the transplant site have received little attention. The detailed view is mentioned in Figure 13.1.

13.5 ROLE OF HYDROGELS IN TISSUE REGENERATION

Hydrogel has been utilized for years in scientific applications based on 3D cell cultures. They mimic actual tissues because of their high water content and exceptional flexibility [86]. The hydrogels enlarge as a result of the water they have absorbed. Water is absorbed by the functional groups

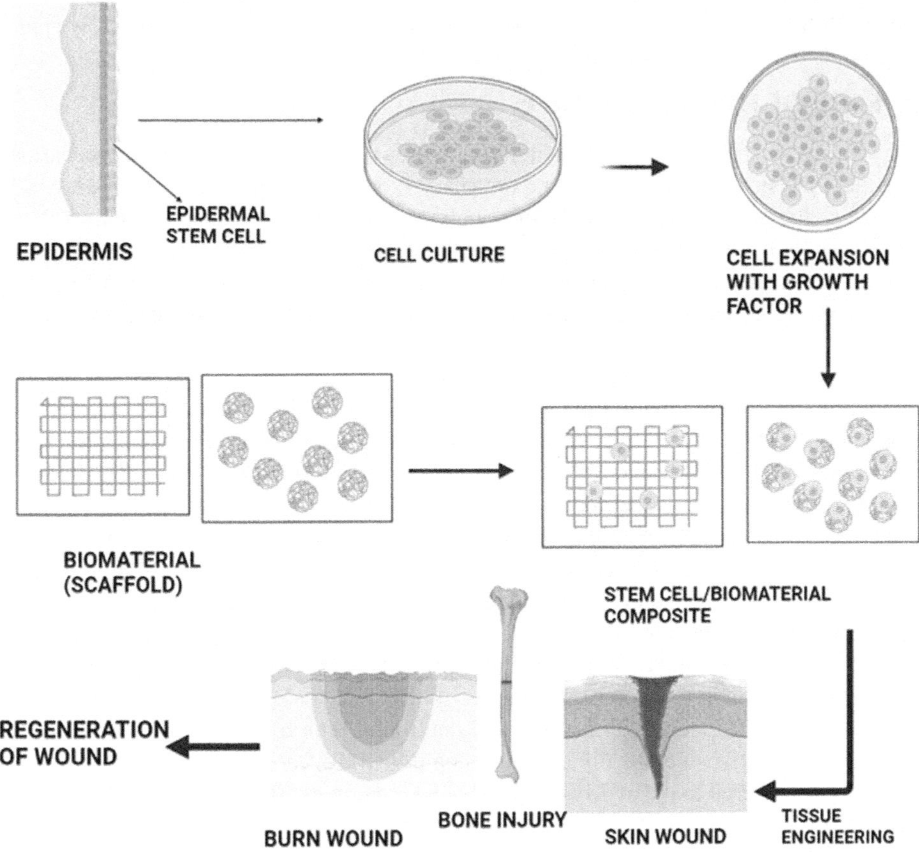

FIGURE 13.1 Stem Cell Therapy in Tissue Regeneration [26,73].

connected to the polymer backbone, but not dissolved, since intermolecular or interfibrillar cross-linked structures are generated between the network chains made by polymer molecules or fibrillar proteins [86, 87]. The mechanical characteristics of a brand-new, low-cost plant-based nanocellulose hydrogel and a revolutionary plant-based Engel-Holm Swarm matrix produced from mouse sarcoma basement membrane were comparable [88]. Alginate and nanocellulose are the two most promising candidate biomaterials for 3D bioprinting [89, 90].

Hydrogels with microengineered features or those with dimensions on the order of several microns have the potential to be very powerful engineering tools [91]. Since the earliest attempts to produce transplantable tissues, microscale hydrogels have been used in tissue engineering. For instance, hydrogel microcapsules were utilized in cell microencapsulation, a method in which transplanted cells were protected from the host's immune system by being immobilized behind a semipermeable barrier [92]. The capacity to modify the characteristics of hydrogel materials, including their adhesion, stiffness, cell signalling potential, size, and form, has allowed for more recent developments in tissue engineering. Here, we quickly review the development of microscale hydrogels for tissue engineering as well as some of its more recent applications. We first examine the procedures utilized to create hydrogels with microscale properties in order to solve the problems of vascularization, cell seeding, and tissue complexity in tissue-engineered structures.

Drug delivery and biosensing are only two biological applications where hydrogels are often employed. New avenues for addressing issues including vascularization, tissue shape, and cell seeding have emerged as a result of recent developments in tissue engineering. These difficulties include the ability to make hydrogels with sizes and structures that are physiologically appropriate.

13.6 ROLE OF HYDROCOLLOIDS

Hydrocolloid-based dressings have been developed for a number of wound healing scenarios. Many people endure physical, psychological, and monetary agony every year as a result of dressing, which is crucial for efficient burn recovery. Natural polymers like gelatin, collagen, and carboxymethyl cellulose (CMC) may be dissolved in water to create hydrocolloids. Hydrophilic synthetic polymers can also be dissolved in water to create hydrocolloids [93, 94]. Because the fluid from the burns destroys the swelling and develops a hydrocolloid form, this dressing is suitable for treating burns [95, 96].

A biocompatible hydrocolloid dressing containing Silk Fibroin (SF) nanoparticles demonstrated increased wound healing capabilities. To make the hydrocolloid dressing, we used SF, styrene-isoprene-styrene (SIS), and sodium carboxymethyl cellulose (CMC). SF particles were produced using Bombyx mori, or B mori, silk cocoons. A popular commercial product called Neoderm (Everaid, Seoul, Korea) was compared to the hydrocolloid dressing's qualities.

In one investigation, healthy human keratinocytes and fibroblasts' in vitro cell attachment and spreading were demonstrated to be dramatically impacted by SF nanofibers [97]. According to another study, SF showed fibroblast cell attachment and growth similar to collagen [98]. Similar to this, SF has been shown to have beneficial effects on the immune system in the context of wound healing. These effects include SF-induced mononuclear cell activation as indicated by lower interleukin-1 production in comparison to reference materials like poly(styrene) and poly(2-hydroxyethyl methacrylate), as well as reduced inflammation and neutrophil/lymphocyte infiltration into wounds in comparison to duo active, a clinically used dressing [99, 100].

It has been discovered that a hydrocolloid dressing containing SF NPs is acceptable and biocompatible for use as a wound dressing. It decreased burn size and increased collagen fibre density in animal models when compared to a Neoderm SFNHD dressing that is marketed for use in human patients. In comparison to Neoderm, SFNHD produced increased PCNA expression, which suggests quicker burn healing and better structural wound integrity [101].

13.7 APPLICATIONS OF NANOPARTICLES IN TISSUE ENGINEERING

Nanomaterials showing various biomedical applications are shown in Figure 13.2.

13.7.1 Cardiac Tissue

Procedures for cardiovascular tissue engineering have been investigated in an effort to regenerate or repair scar tissue and/or dormant myocardium after ischemic injury, especially myocardial infarction (MI) [102]. Heart failure following MI continues to be a common complication due to the limited capacity of cardiac muscle to regenerate [103] despite current surgical revascularization techniques. For the regeneration and repair of heart tissue, nanomaterials have been incorporated as crucial parts of experimental tissue engineering systems.

After a heart attack, injectable biomaterials composed of hydrated natural or synthetic polymer solutions have been used as a treatment for the heart [104]. These hydrogels provide a protective environment to boost transplanted cell retention and function and can lower wall strain, which causes pathological dilative remodelling of the ventricle [105]. Another technique to ECM treatment is the creation of nanofiber matrices. Nanofiber diameter may be modified to match that of native ECM fibres, offering a greater surface area for cell engraftment [106]. Contrary to traditional methods of culture, some nanofiber meshes made of synthetic polymers functionalized with native ECM

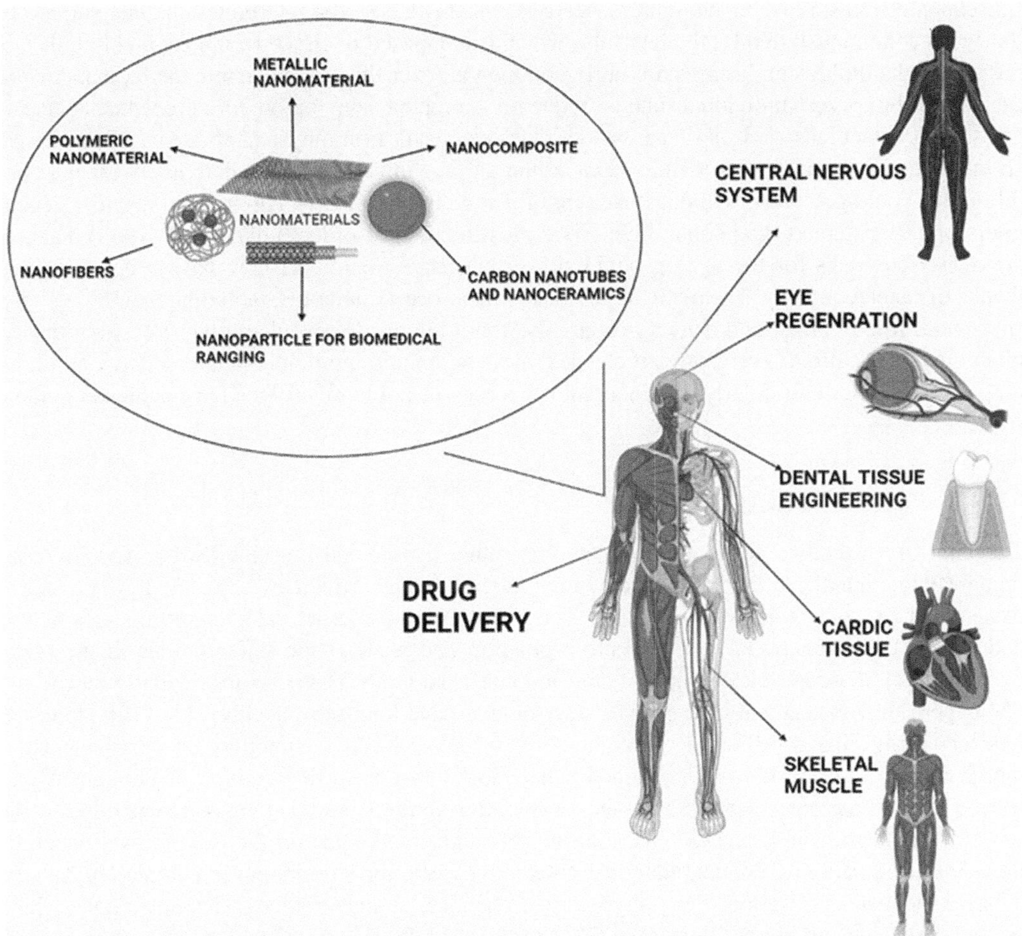

FIGURE 13.2 Nanomaterial, Its Types and Application.

proteins have shown in vitro the ability to create scaffolds with a topology resembling heart tissue, complete with aligned rabbit cardiomyocytes that express mature CM contractile protein and gap junction indicators. Similar to this, vascular grafts have been made with nanofibers to provide a bioadsorbable scaffold that promotes neo-arteriogenesis and improves long-term graft patency [107].

Future research on nanomaterials for cardiac regeneration will offer new opportunities to improve the conductivity of biomaterial scaffolds, but in vivo testing on small- and large-animal models will be necessary to rule out the possibility of arrhythmias. Nanofiber matrices have provided a bottom-up method for developing an environment that mimics the ECM to control cell development and differentiation. Nanofiber scaffolds may reduce the occurrence of restenosis and improve regeneration as biodegradable vascular grafts. For cardiac and vascular regeneration in patients with MI, heart failure, or coronary artery disease, tissue-engineered substances including nanomaterials, in particular noble metal nanoparticles (NPs), are being employed more and more.

13.7.2 Eye Regeneration

The human eye is a sophisticated and organized sensory organ. An enormous amount of light is captured, directed, and processed by the eye's anatomical and functional components, which are subsequently transmitted to the central nervous system (CNS) for interpretation. The surface of the eye is continuously fed by the tear film, which is composed of aqueous, mucin, and lipid layers [108]. Ophthalmological diseases are fairly common all over the world, despite the fact that ocular function differs across individuals and is frequently corrected using optometric procedures (such as glasses or contact lenses). Long-term vision issues can result from physical abrasions or damage in sports-related eye injuries, as well as from ocular surface infections; conjunctivitis is often a condition that could get worse without treatment or if it recurs frequently. Age-related macular degeneration [109], cataracts, glaucoma, high blood pressure, and disorders linked with type II diabetes are all serious risks for the ageing population. Alternative treatments now have a new platform thanks to nanotechnology. The distinct qualities of nano-based materials, including their bioactivity, shape and size, mobility, and delivery capability, are of interest for eye therapy [108]. Additionally, biocompatible scaffolds and nanoparticles lessen irritation and immunological reactivity, aiding in long-term healing. Particularly, the inherent properties of noble metal NPs have achieved notable success [110].

13.7.3 Skeletal Muscles

The bulk of the body's active motions are controlled by skeletal muscles (SMs). The SMs are organized into bundles of layered muscle fibres called fascicles, which are connected by connective tissues, nerves, and blood vessels [111]. The SMs play both structural (supporting soft tissues and posture) and functional (temperature regulation and skeletal movement) roles in the body. Contraction of these muscles requires a large amount of strength. These tissues regularly renew, and SM hypertrophy is necessary for growth, development, and long-term health [111, 112]. However, recovery from clinical difficulties brought on by atrophy, nerve dysfunction, or significant volumetric muscle loss (VML) is constrained and typically results in the creation of permanent scar tissue [112]. Numerous holes in SM tissue regeneration and repair are anticipated to be filled by the use of nanomaterials in tissue engineering. Important characteristics for the use of novel materials include biocompatibility, biodegradability, conductivity, and the stimulation of cell alignment, vascularization, and innervations. The physical and chemical characteristics of nanomaterials, which are regularly changed by the addition of GFs, drugs, and NPs, determine their functioning.

Regeneration of skeletal muscles is a challenging procedure. To meet this need, greater proliferation and differentiation of surrounding SM cells must be used to produce considerable tissue replacement/healing. Additionally, promoting axonal growth and innervations will improve the flow of information to damaged peripheral nerves. Nanomaterials will also need to stimulate angiogenesis and neurogenesis in addition to SM regeneration. Noble metal NPs are promising, but further study is required to fully comprehend there in vivo effects. This innovation will likely make use of hybrid/composite materials, metal NPs, and progenitor cell delivery.

13.7.4 Central Nervous System

Both inside and outside of the human body, the nervous system is crucial for the proper transmission and processing of information. The peripheral nervous system (PNS), which comprises all other neural tissues in the body, and the central nervous system (CNS), which comprises the brain and spinal cord, are its two main components [113]. The fundamental building block of the brain, a neuron, has a fast rate of impulse reception and transmission. The CNS integrates, evaluates, and coordinates sensory and motor data to enable a functional response. The PNS predominantly communicates with other cells all around the body and sends messages to the CNS via the spinal cord [113]. But much like other human tissues, the nervous system can sustain short-term or long-term harm. People of all ages can experience nerve issues brought on by illness, physical trauma, poisons, and hypoxic situations. Furthermore, neurological conditions are prevalent among the elderly. A decreased number of neurons or impairment to the connections between neurons and tissues frequently accompany nerve degeneration. Clinical therapies for brain regeneration span the spectrum from potent, 5.0 mm PNS axonal gap junctions to many, incurable neural degenerative disorders [114]. Nanomaterials are predicted to revolutionize current cellular or immunological treatments and enhance novel medications.

The burden of restoring both structural and functional brain tissue is enormous. The biochemical, cellular, and genomic/proteomic processes of brain regeneration still have a lot to be discovered. Understanding interactions between nanomaterials and tissues is significantly hampered by this. To properly understand the regenerative importance, future research should concentrate on endogenous mechanisms functioning at the material interface [114]. This information effectively preserves the originality of the ideas for nanomaterials that mimic these processes. Additionally, hybrid/composite nanomaterials could be able to use bioelectric cell characteristics to regenerate the myelin sheath. Biocompatibility, biodegradability, low toxicity, immunogenicity, and functional surfaces/interiors are crucial to transforming model instances into therapeutic potential; nanoparticles already possess these features when examined in vivo.

13.7.5 Dental Tissue

The use of nanoparticles in dental tissue engineering has gained popularity since the turn of the twenty-first century. With advancing years, periodontal or related problems (such cardiovascular disease [115], diabetes [116], and rheumatoid arthritis [117] become more prevalent. Effective treatment options are required in patients with periodontal problems to heal the damaged tissues and restore the original structure and function since the periodontal tissue is deteriorating and losing its ability to self-heal. Thankfully, the creation of several metal and polymeric nanoparticles has greatly aided in the treatment of periodontal diseases [118]. Nanogels are mostly used in dentistry to enhance or repair the mechanical characteristics and biological activities of periodontal materials. Other uses of nanoparticles in dentistry include novel coatings for implants, toothpastes, and personal care products.

Due to their distinct antibacterial characteristics and modification, metal nanoparticles have been recognized in dental tissue engineering. Antibacterial metals including Ag, gold, TiO_2, and ZnO are frequent examples, and their antibacterial capabilities can be strengthened by functionalizing the properties [119]. The materials' size and form may also contribute to their ability to kill bacteria.

13.7.6 Drug Delivery

The practise of designing tissues and organs utilizing cells, biomaterials, and bioactive substances, either individually or in combination, is known as tissue engineering. However, cell survival is low in the early days following transplantation, and attempts to increase cell survival by fusing cells with medicines or growth factors have also failed [120, 121]. It is crucial to create a potent drug delivery system that directs functional tissue regeneration in situ without endangering the rest of the body since drug delivery systems can increase the effectiveness and safety of tissue engineering [122, 123, 124]. Due to the ignorance of biological barriers, this is still exceedingly challenging and results in ineffective drug administration and low in vivo bioavailability. Enzymes, vaccines, medicines, peptides, and antibodies are all examples of therapeutic molecules that can be delivered more effectively using biomaterials through the use of loading or co-conjugation procedures. Despite great progress, there are still substantial obstacles to more effective disease treatment delivery. The growing area of nanomaterial-based medication delivery now includes dendrimers, polymer nanospheres, liposomes, lipid nanoparticles, micelles, and inorganic nanomaterials as novel materials (such as iron oxide, gold, metal, and silicon). In the interim, biomaterials that trigger drug release in response to a variety of environmental cues (such as enzymes, pH, glucose, pressure, and temperature) have been created [125, 126].

13.8 CONCLUSION AND FUTURE PROSPECTIVES

Active research is being done on the use of natural polymers and their semi-synthetic derivatives in medication delivery systems. Natural polymers continue to be the preferred option since they are affordable, easily accessible, and versatile in terms of chemical alterations, and perhaps biocompatible and degradable [124]. New techniques for tissue regeneration will be made possible by the use of nanomaterials in existing tissue engineering therapies. NPs have optical, magnetic, and photodynamic characteristics that can be used by altering their surface, biochemistry, encasing contrast compounds, and being coated by cells generated from tumours. Consequently, the area of diagnostics has undergone a revolution thanks to nanotechnology [125]. The genuine potential and effects of customized materials for the heart, skin, eyes, MS, and neurological system have been briefly discussed in this study. For the repair or regeneration of injured tissue, tissue engineering applications and the creation of nanomaterials are crucial. Regarding current nanotechnology, more and more researchers are attempting to combine various nanomaterials to produce novel biomaterials. The sensitivity of the implanted material, the ensuing immunological reaction, potential toxicity, influence on reproduction, and even impact on embryonic development must all be carefully considered when using these nanomaterials in tissue engineering to replace damaged organs. Tissue engineering has a lot of promise as a result of the development of novel nanomaterials, but it must live up to the expectations of both patients and physicians. Understanding nanoscale interactions like the dynamics of surface oxidation, the replacement of terminating agents, and the formation of supramolecular structures of nanomaterials in living organisms will pave the way for the future of tissue engineering, even though there is still much work to be done to better understand the true impact of nanotechnology. Macroscopic materials will be designed at the nanoscale.

REFERENCES

1. Tathe A, Ghodke M, Nikalje AP A brief review: Biomaterials and their application. https://mme.deu.edu.tr/wp-content/uploads/2017/10/biomaterials.pdf. Accessed 1 Jun 2023
2. Ratner BD, Hoffman AS, Schoen FJ, Lemons JE (2004) *Biomaterials Science: An Introduction to Materials in Medicine*. Elsevier
3. Ratner BD, Bryant SJ (2004) Biomaterials: Where we have been and where we are going. *Annu Rev Biomed Eng* 6:41–75. https://doi.org/10.1146/annurev.bioeng.6.040803.140027
4. Van Vlerken LE, Vyas TK, Amiji MM (2007) Poly(ethylene glycol)-modified nanocarriers for tumor-targeted and intracellular delivery. *Pharm Res* 24:1405–1414. https://doi.org/10.1007/s11095-007-9284-6
5. Anju S, Prajitha N, Sukanya VS, Mohanan PV (2020) Complicity of degradable polymers in healthcare applications. *Mater Today Chem* 16:100236. https://doi.org/10.1016/j.mtchem.2019.100236
6. Su S, Kang PM (2020) Systemic review of biodegradable nanomaterials in nanomedicine. *Nanomaterials (Basel)* 10. https://doi.org/10.3390/nano10040656
7. Palma E, Pasqua A, Gagliardi A, Britti D, Fresta M, Cosco D (2018) Antileishmanial activity of amphotericin B-loaded-PLGA nanoparticles: An overview. *Materials* 11:1167. https://doi.org/10.3390/ma11071167
8. Kwon GS (2005) Thermosensitive biodegradable hydrogels for the delivery of therapeutic agents. In: *Polymeric Drug Delivery Systems*. CRC Press, pp 275–298.
9. Vilar G, Tulla-Puche J, Albericio F (2012) Polymers and drug delivery systems. *Curr Drug Deliv* 9:367–394. https://doi.org/10.2174/156720112801323053
10. Gagliardi A, Paolino D, Iannone M, Palma E, Fresta M, Cosco D (2018) Sodium deoxycholate-decorated zein nanoparticles for a stable colloidal drug delivery system. *Int J Nanomedicine* 13:601–614. https://doi.org/10.2147/IJN.S156930

11. Kim NJ, Lee SJ, Atala A (2013) 1 – Biomedical Nanomaterials in Tissue Engineering. In: Gaharwar AK, Sant S, Hancock MJ, Hacking SA (eds) *Nanomaterials in Tissue Engineering*. Woodhead Publishing, pp 1–25e.
12. Shankar P, Jagtap J, Sharma G, Sharma GP, Singh J, Parashar M, Kumar G, Mittal S, Sharma MK, Jadhav K, Parashar D (2022) Chapter 12 – A Revolutionary Breakthrough of Bionanomaterials in Tissue Engineering and Regenerative Medicine. In: Barhoum A, Jeevanandam J, Danquah MK (eds) *Bionanotechnology: Emerging Applications of Bionanomaterials*. Elsevier, pp 399–441.
13. Zheng X, Zhang P, Fu Z, Meng S, Dai L, Yang H (2021) Applications of nanomaterials in tissue engineering. *RSC Adv* 11:19041–19058. https://doi.org/10.1039/d1ra01849c
14. Venkatraman SS, Ma LL, Natarajan JV, Chattopadhyay S (2010) Polymer- and liposome-based nanoparticles in targeted drug delivery. *Front Biosci* 2:801–814. https://doi.org/10.2741/s103
15. Goodall S, Jones ML, Mahler S (2015) Monoclonal antibody-targeted polymeric nanoparticles for cancer therapy – Future prospects. *J Chem Technol Biotechnol* 90:1169–1176. https://doi.org/10.1002/jctb.4555
16. Sarcan ET, Silindir-Gunay M, Ozer AY (2018) Theranostic polymeric nanoparticles for NIR imaging and photodynamic therapy. *Int J Pharm* 551:329–338. https://doi.org/10.1016/j.ijpharm.2018.09.019
17. Bankier C, Matharu RK, Cheong YK, Ren GG, Cloutman-Green E, Ciric L (2019) Synergistic antibacterial effects of metallic nanoparticle combinations. *Sci Rep* 9:16074. https://doi.org/10.1038/s41598-019-52473-2
18. Dvir T, Timko BP, Brigham MD, Naik SR, Karajanagi SS, Levy O, Jin H, Parker KK, Langer R, Kohane DS (2011) Nanowired three-dimensional cardiac patches. *Nat Nanotechnol* 6:720–725. https://doi.org/10.1038/nnano.2011.160
19. Webster TJ, Ejiofor JU (2004) Increased osteoblast adhesion on nanophase metals: Ti, Ti6Al4V, and CoCrMo. *Biomaterials* 25:4731–4739. https://doi.org/10.1016/j.biomaterials.2003.12.002
20. Ai J, Biazar E, Jafarpour M, Montazeri M, Majdi A, Aminifard S, Zafari M, Akbari HR, Rad HG (2011) Nanotoxicology and nanoparticle safety in biomedical designs. *Int J Nanomedicine* 6:1117–1127. https://doi.org/10.2147/IJN.S16603
21. Jordan J, Jacob KI, Tannenbaum R, Sharaf MA, Jasiuk I (2005) Experimental trends in polymer nanocomposites—A review. *Mater Sci Eng A* 393:1–11. https://doi.org/10.1016/j.msea.2004.09.044
22. Kumar SK, Krishnamoorti R (2010) Nanocomposites: Structure, phase behavior, and properties. *Annu Rev Chem Biomol Eng* 1:37–58. https://doi.org/10.1146/annurev-chembioeng-073009-100856
23. Veetil JV, Ye K (2009) Tailored carbon nanotubes for tissue engineering applications. *Biotechnol Prog* 25:709–721. https://doi.org/10.1002/btpr.165
24. Dvir T, Timko BP, Kohane DS, Langer R (2011) Nanotechnological strategies for engineering complex tissues. *Nat Nanotechnol* 6:13–22. https://doi.org/10.1038/nnano.2010.246
25. Yu MF, Files BS, Arepalli S, Ruoff RS (2000) Tensile loading of ropes of single wall carbon nanotubes and their mechanical properties. *Phys Rev Lett* 84:5552–5555. https://doi.org/10.1103/PhysRevLett.84.5552
26. Nayak TR, Jian L, Phua LC, Ho HK, Ren Y, Pastorin G (2010) Thin films of functionalized multiwalled carbon nanotubes as suitable scaffold materials for stem cells proliferation and bone formation. *ACS Nano* 4:7717–7725. https://doi.org/10.1021/nn102738c
27. Meng J, Kong H, Xu HY, Song L, Wang CY, Xie SS (2005) Improving the blood compatibility of polyurethane using carbon nanotubes as fillers and its implications to cardiovascular surgery. *J Biomed Mater Res A* 74:208–214. https://doi.org/10.1002/jbm.a.30315
28. van der Zande M, Junker R, Walboomers XF, Jansen JA (2011) Carbon nanotubes in animal models: A systematic review on toxic potential. *Tissue Eng Part B Rev* 17:57–69. https://doi.org/10.1089/ten.TEB.2010.0472
29. Park J, Lakes RS (2007) *Biomaterials: An Introduction*. Springer Science & Business Media.
30. Webster TJ, Siegel RW, Bizios R (1999) Osteoblast adhesion on nanophase ceramics. *Biomaterials* 20:1221–1227. https://doi.org/10.1016/s0142-9612(99)00020-4
31. Choi HS, Frangioni JV (2010) Nanoparticles for biomedical imaging: Fundamentals of clinical translation. *Mol Imaging* 9:291–310
32. Burg KJ, Porter S, Kellam JF (2000) Biomaterial developments for bone tissue engineering. *Biomaterials* 21:2347–2359. https://doi.org/10.1016/s0142-9612(00)00102-2

33. Hutmacher DW (2000) Scaffolds in tissue engineering bone and cartilage. *Biomaterials* 21:2529–2543. https://doi.org/10.1016/s0142-9612(00)00121-6
34. Agrawal CM, McKinney JS, Lanctot D, Athanasiou KA (2000) Effects of fluid flow on the in vitro degradation kinetics of biodegradable scaffolds for tissue engineering. *Biomaterials* 21:2443–2452. https://doi.org/10.1016/s0142-9612(00)00112-5
35. Lu L, Peter SJ, Lyman MD, Lai HL, Leite SM, Tamada JA, Vacanti JP, Langer R, Mikos AG (2000) In vitro degradation of porous poly(L-lactic acid) foams. *Biomaterials* 21:1595–1605. https://doi.org/10.1016/s0142-9612(00)00048-x
36. Oh SH, Park SC, Kim HK, Koh YJ, Lee J-H, Lee MC, Lee JH (2011) Degradation behavior of 3D porous polydioxanone-b-polycaprolactone scaffolds fabricated using the melt-molding particulate-leaching method. *J Biomater Sci Polym Ed* 22:225–237. https://doi.org/10.1163/092050609X12597621891620
37. Finne-Wistrand A, Albertsson A-C, Kwon OH, Kawazoe N, Chen G, Kang I-K, Hasuda H, Gong J, Ito Y (2008) Resorbable scaffolds from three different techniques: Electrospun fabrics, salt-leaching porous films, and smooth flat surfaces. *Macromol Biosci* 8:951–959. https://doi.org/10.1002/mabi.200700328
38. Chua K-N, Lim W-S, Zhang P, Lu H, Wen J, Ramakrishna S, Leong KW, Mao H-Q (2005) Stable immobilization of rat hepatocyte spheroids on galactosylated nanofiber scaffold. *Biomaterials* 26:2537–2547. https://doi.org/10.1016/j.biomaterials.2004.07.040
39. Badami AS, Kreke MR, Thompson MS, Riffle JS, Goldstein AS (2006) Effect of fiber diameter on spreading, proliferation, and differentiation of osteoblastic cells on electrospun poly(lactic acid) substrates. *Biomaterials* 27:596–606. https://doi.org/10.1016/j.biomaterials.2005.05.084
40. Ulrich MMW (2014) Science, Practice and education regenerative medicine in burn wound healing: Aiming for the perfect skin. *EWMA J* 14:25–27
41. Uçkay I, Aragón-Sánchez J, Lew D, Lipsky BA (2015) Diabetic foot infections: What have we learned in the last 30 years? *Int J Infect Dis* 40:81–91. https://doi.org/10.1016/j.ijid.2015.09.023
42. Landén NX, Li D, Ståhle M (2016) Transition from inflammation to proliferation: A critical step during wound healing. *Cell Mol Life Sci* 73:3861–3885. https://doi.org/10.1007/s00018-016-2268-0
43. Dreifke MB, Jayasuriya AA, Jayasuriya AC (2015) Current wound healing procedures and potential care. *Mater Sci Eng C* 48:651–662. https://doi.org/10.1016/j.msec.2014.12.068
44. Pober JS, Sessa WC (2014) Inflammation and the blood microvascular system. *Cold Spring Harb Perspect Biol* 7:a016345. https://doi.org/10.1101/cshperspect.a016345
45. Bodnar RJ (2015) Chemokine regulation of angiogenesis during wound healing. *Adv Wound Care* 4:641–650. https://doi.org/10.1089/wound.2014.0594
46. Mechanick JI (2004) Practical aspects of nutritional support for wound-healing patients. *Am J Surg* 188:52–56. https://doi.org/10.1016/S0002-9610(03)00291-5
47. Greer N, Foman NA, MacDonald R, Dorrian J, Fitzgerald P, Rutks I, Wilt TJ. Advanced wound care therapies for nonhealing diabetic, venous, and arterial ulcers: a systematic review. Annals of internal medicine. 2013 Oct 15;159(8):532-42.
48. Wood F, Martin L, Lewis D, Rawlins J, McWilliams T, Burrows S, Rea S (2012) A prospective randomized clinical pilot study to compare the effectiveness of Biobrane® synthetic wound dressing, with or without autologous cell suspension, to the local standard treatment regimen in paediatric scald injuries. *Burns* 38:830–839. https://doi.org/10.1016/j.burns.2011.12.020
49. Leonida MD, Kumar I (2016) Wound Healing and Skin Regeneration. In: Leonida MD, Kumar I (eds) *Bionanomaterials for Skin Regeneration*. Springer International Publishing, Cham, pp 17–25.
50. Chen S, Wang H, Su Y, John JV, McCarthy A, Wong SL, Xie J (2020) Mesenchymal stem cell-laden, personalized 3D scaffolds with controlled structure and fiber alignment promote diabetic wound healing. *Acta Biomater* 108:153–167. https://doi.org/10.1016/j.actbio.2020.03.035
51. Tam J, Wang Y, Farinelli WA, Jiménez-Lozano J, Franco W, Sakamoto FH, Cheung EJ, Purschke M, Doukas AG, Anderson RR (2013) Fractional skin harvesting: Autologous skin grafting without donor-site morbidity. *Plast Reconstr Surg Glob Open* 1:e47. https://doi.org/10.1097/GOX.0b013e3182a85a36
52. Norbury W, Herndon DN, Tanksley J, Jeschke MG, Finnerty CC (2016) Infection in burns. *Surg Infect* 17:250–255. https://doi.org/10.1089/sur.2013.134
53. Cse C (1881) Die Verhurtung der Augenentzundung der Neugeborenen. *Archiv fur Gynaekologie* 17:50–53

54. Klasen HJ (2000) A historical review of the use of silver in the treatment of burns. II. Renewed interest for silver. *Burns* 26:131–138. https://doi.org/10.1016/S0305-4179(99)00116-3
55. Alexander JW (2009) History of the medical use of silver. *Surg Infect* 10:289–292. https://doi.org/10.1089/sur.2008.9941
56. Lansdown ABG (2006) Silver in health care: Antimicrobial effects and safety in use. *Curr Probl Dermatol* 33:17–34. https://doi.org/10.1159/000093928
57. Tenover FC (2006) Mechanisms of antimicrobial resistance in bacteria. *Am J Infect Control* 34:S3–10; discussion S64–73. https://doi.org/10.1016/j.ajic.2006.05.219
58. Varner K, Sanford J, El-Badawy A, Feldhake D (2010) *State of the Science Literature Review: Everything Nanosilver and More*. Agency.
59. Griffith M, Udekwu KI, Gkotzis S, Mah T-F, Alarcon EI (2015) Anti-microbiological and Anti-infective Activities of Silver. In: Alarcon EI, Griffith M, Udekwu KI (eds) *Silver Nanoparticle Applications: In the Fabrication and Design of Medical and Biosensing Devices*. Springer International Publishing, Cham, pp 127–146.
60. Wright JB, Lam K, Buret AG, Olson ME, Burrell RE (2002) Early healing events in a porcine model of contaminated wounds: Effects of nanocrystalline silver on matrix metalloproteinases, cell apoptosis, and healing. *Wound Repair Regen* 10:141–151. https://doi.org/10.1046/j.1524-475x.2002.10308.x
61. Lu B, Li T, Zhao H, Li X, Gao C, Zhang S, Xie E (2012) Graphene-based composite materials beneficial to wound healing. *Nanoscale* 4:2978–2982. https://doi.org/10.1039/c2nr11958g
62. Herron M, Agarwal A, Kierski PR, Calderon DF, Teixeira LBC, Schurr MJ, Murphy CJ, Czuprynski CJ, McAnulty JF, Abbott NL (2014) Reduction in wound bioburden using a silver-loaded dissolvable microfilm construct. *Adv Healthc Mater* 3:916–928. https://doi.org/10.1002/adhm.201300537
63. Nicol S. Life after death for empty shells. New Scientist. 1991 Feb;129:46-8.
64. Kirichenko AK, Bolshakov IN, Ali-Riza AE, Vlasov AA (2013) Morphological study of burn wound healing with the use of collagen-chitosan wound dressing. *Bull Exp Biol Med* 154:692–696. https://doi.org/10.1007/s10517-013-2031-6
65. Fu L, Zhang J, Yang G (2013) Present status and applications of bacterial cellulose-based materials for skin tissue repair. *Carbohydr Polym* 92:1432–1442. https://doi.org/10.1016/j.carbpol.2012.10.071
66. Friedman AJ, Han G, Navati MS, Chacko M, Gunther L, Alfieri A, Friedman JM (2008) Sustained release nitric oxide releasing nanoparticles: Characterization of a novel delivery platform based on nitrite containing hydrogel/glass composites. *Nitric Oxide* 19:12–20. https://doi.org/10.1016/j.niox.2008.04.003
67. Martinez LR, Han G, Chacko M, Mihu MR, Jacobson M, Gialanella P, Friedman AJ, Nosanchuk JD, Friedman JM (2009) Antimicrobial and healing efficacy of sustained release nitric oxide nanoparticles against Staphylococcus aureus skin infection. *J Invest Dermatol* 129:2463–2469. https://doi.org/10.1038/jid.2009.95
68. Han G, Nguyen LN, Macherla C, Chi Y, Friedman JM, Nosanchuk JD, Martinez LR (2012) Nitric oxide–releasing nanoparticles accelerate wound healing by promoting fibroblast migration and collagen deposition. *Am J Pathol* 180:1465–1473. https://doi.org/10.1016/j.ajpath.2011.12.013
69. Blecher K, Martinez LR, Tuckman-Vernon C, Nacharaju P, Schairer D, Chouake J, Friedman JM, Alfieri A, Guha C, Nosanchuk JD, Friedman AJ (2012) Nitric oxide-releasing nanoparticles accelerate wound healing in NOD-SCID mice. *Nanomedicine* 8:1364–1371. https://doi.org/10.1016/j.nano.2012.02.014
70. Chen W-Y, Chang H-Y, Lu J-K, Huang Y-C, Harroun SG, Tseng Y-T, Li Y-J, Huang C-C, Chang H-T (2015) Self-assembly of antimicrobial peptides on gold nanodots: Against multidrug-resistant bacteria and wound-healing application. *Adv Funct Mater* 25:7189–7199. https://doi.org/10.1002/adfm.201503248
71. Kim JE, Lee J, Jang M, Kwak MH, Go J, Kho EK, Song SH, Sung JE, Lee J, Hwang DY (2015) Accelerated healing of cutaneous wounds using phytochemically stabilized gold nanoparticle deposited hydrocolloid membranes. *Biomater Sci* 3:509–519. https://doi.org/10.1039/c4bm00390j
72. Chigurupati S, Mughal MR, Okun E, Das S, Kumar A, McCaffery M, Seal S, Mattson MP (2013) Effects of cerium oxide nanoparticles on the growth of keratinocytes, fibroblasts and vascular endothelial cells in cutaneous wound healing. *Biomaterials* 34:2194–2201. https://doi.org/10.1016/j.biomaterials.2012.11.061

73. Yang F, Cho S-W, Son SM, Bogatyrev SR, Singh D, Green JJ, Mei Y, Park S, Bhang SH, Kim B-S, Langer R, Anderson DG (2010) Genetic engineering of human stem cells for enhanced angiogenesis using biodegradable polymeric nanoparticles. *Proc Natl Acad Sci U S A* 107:3317–3322. https://doi.org/10.1073/pnas.0905432106
74. Cui L, Liang J, Liu H, Zhang K, Li J (2020) Nanomaterials for angiogenesis in skin tissue engineering. *Tissue Eng Part B Rev* 26:203–216. https://doi.org/10.1089/ten.TEB.2019.0337
75. Xin T, Greco V, Myung P (2016) Hardwiring stem cell communication through tissue structure. *Cell* 164:1212–1225. https://doi.org/10.1016/j.cell.2016.02.041
76. España-Sánchez BL, Cruz-Soto ME, Elizalde-Peña EA, Sabasflores-Benítez S, Roca-Aranda A, Esquivel-Escalante K, Luna-Bárcenas G (2018) Trends in tissue regeneration: Bio-nanomaterials. *Tissue Regener* 27:218–228
77. Bianco P, Robey PG (2001) Stem cells in tissue engineering. *Nature* 414:118–121. https://doi.org/10.1038/35102181
78. Cruz-Soto ME, Fernández C, Carrillo JL (2018) Articulate cartilage enriched with autologous bone marrow aspirate as an approach of pain amelioration and functionality restoration in osteoarthritis patients older than 50 years-old. *Med Clin Res Rev* 2:1–3.
79. Mall M, Wernig M (2017) The novel tool of cell reprogramming for applications in molecular medicine. *J Mol Med* 95:695–703. https://doi.org/10.1007/s00109-017-1550-4
80. Lagasse E, Shizuru JA, Uchida N, Tsukamoto A, Weissman IL (2001) Toward regenerative medicine. *Immunity* 14:425–436. https://doi.org/10.1016/s1074-7613(01)00123-6
81. Barrandon Y, Green H (1987) Three clonal types of keratinocyte with different capacities for multiplication. *Proc Natl Acad Sci U S A* 84:2302–2306. https://doi.org/10.1073/pnas.84.8.2302
82. Pellegrini G, Dellambra E, Golisano O, Martinelli E, Fantozzi I, Bondanza S, Ponzin D, McKeon F, De Luca M (2001) p63 identifies keratinocyte stem cells. *Proc Natl Acad Sci U S A* 98:3156–3161. https://doi.org/10.1073/pnas.061032098
83. Friedenstein AJ, Piatetzky-Shapiro II, Petrakova KV (1966) Osteogenesis in transplants of bone marrow cells. *J Embryol Exp Morphol* 16:381–390
84. Langstaff S, Sayer M, Smith TJ, Pugh SM, Hesp SA, Thompson WT (1999) Resorbable bioceramics based on stabilized calcium phosphates. Part I: Rational design, sample preparation and material characterization. *Biomaterials* 20:1727–1741. https://doi.org/10.1016/s0142-9612(99)00086-1
85. Langstaff S, Sayer M, Smith TJ, Pugh SM (2001) Resorbable bioceramics based on stabilized calcium phosphates. Part II: Evaluation of biological response. *Biomaterials* 22:135–150. https://doi.org/10.1016/s0142-9612(00)00139-3
86. Ahmed EM (2015) Hydrogel: Preparation, characterization, and applications: A review. *J Advert Res* 6:105–121. https://doi.org/10.1016/j.jare.2013.07.006
87. Worthington P, Pochan DJ, Langhans SA (2015) Peptide hydrogels – Versatile matrices for 3D cell culture in cancer medicine. *Front Oncol* 5:92. https://doi.org/10.3389/fonc.2015.00092
88. Curvello R, Kerr G, Micati DJ, Chan WH, Raghuwanshi VS, Rosenbluh J, Abud HE, Garnier G (2020) Engineered plant-based nanocellulose hydrogel for small intestinal organoid growth. *Adv Sci* 8:2002135. https://doi.org/10.1002/advs.202002135
89. Mahendiran B, Muthusamy S, Sampath S, Jaisankar SN, Popat KC, Selvakumar R, Krishnakumar GS (2021) Recent trends in natural polysaccharide based bioinks for multiscale 3D printing in tissue regeneration: A review. *Int J Biol Macromol* 183:564–588. https://doi.org/10.1016/j.ijbiomac.2021.04.179
90. Mahendiran B, Muthusamy S, Sampath S, Jaisankar SN, Popat KC, Selvakumar R, Krishnakumar GS (2021) Recent trends in natural polysaccharide based bioinks for multiscale 3D printing in tissue regeneration: A review. *Int J Biol Macromol* 183:564–588. https://doi.org/10.1016/j.ijbiomac.2021.04.179
91. Chrobak KM, Potter DR, Tien J (2006) Formation of perfused, functional microvascular tubes in vitro. *Microvasc Res* 71:185–196. https://doi.org/10.1016/j.mvr.2006.02.005
92. Khademhosseini A, Langer R (2007) Microengineered hydrogels for tissue engineering. *Biomaterials* 28:5087–5092. https://doi.org/10.1016/j.biomaterials.2007.07.021
93. Abramo F, Argiolas S, Pisani G, Vannozzi I, Miragliotta V (2008) Effect of a hydrocolloid dressing on first intention healing surgical wounds in the dog: A Pilot study. *Aust Vet J* 86:95–99. https://doi.org/10.1111/j.1751-0813.2007.00243.x

94. Lim HJ, Kim HT, Oh EJ, Choi JH, Ghim HD, Pyun DG, Lee SB, Chung DJ, Chung HY (2010) Effect of newly developed pectin/CMC dressing materials on three different types of wound model. *Polym Korea* 34:363–368. https://doi.org/10.7317/pk.2010.34.4.363
95. Chakravarthy D, Rodway N, Schmidt S, Smith D, Evancho M, Sims R (1994) Evaluation of three new hydrocolloid dressings: Retention of dressing integrity and biodegradability of absorbent components attenuate inflammation. *J Biomed Mater Res* 28:1165–1173. https://doi.org/10.1002/jbm.820281007
96. Brown-Etris M, Milne C, Orsted H, Gates JL, Netsch D, Punchello M, Couture N, Albert M, Attrell E, Freyberg J (2008) A prospective, randomized, multisite clinical evaluation of a transparent absorbent acrylic dressing and a hydrocolloid dressing in the management of Stage II and shallow Stage III pressure ulcers. *Adv Skin Wound Care* 21:169–174
97. Min B-M, Lee G, Kim SH, Nam YS, Lee TS, Park WH (2004) Electrospinning of silk fibroin nanofibers and its effect on the adhesion and spreading of normal human keratinocytes and fibroblasts in vitro. *Biomaterials* 25:1289–1297. https://doi.org/10.1016/j.biomaterials.2003.08.045
98. Minoura N, Aiba S, Higuchi M, Gotoh Y, Tsukada M, Imai Y (1995) Attachment and growth of fibroblast cells on silk fibroin. *Biochem Biophys Res Commun* 208:511–516. https://doi.org/10.1006/bbrc.1995.1368
99. Santin M, Motta A, Freddi G, Cannas M (1999) In vitro evaluation of the inflammatory potential of the silk fibroin. *J Biomed Mater Res* 46:382–389. https://doi.org/10.1002/(sici)1097-4636(19990905)46:3<382::aid-jbm11>3.0.co;2-r
100. Sugihara A, Sugiura K, Morita H, Ninagawa T, Tubouchi K, Tobe R, Izumiya M, Horio T, Abraham NG, Ikehara S (2000) Promotive effects of a silk film on epidermal recovery from full-thickness skin wounds (44552). *Proc Soc Exp Biol Med* 225:58–64. https://doi.org/10.1177/153537020022500107
101. Lee OJ, Kim J-H, Moon BM, Chao JR, Yoon J, Ju HW, Lee JM, Park HJ, Kim DW, Kim SJ, Park HS, Park CH (2016) Fabrication and characterization of hydrocolloid dressing with silk fibroin nanoparticles for wound healing. *Tissue Eng Regen Med* 13:218–226. https://doi.org/10.1007/s13770-016-9058-5
102. Pfeffer MA, Braunwald E (1990) Ventricular remodeling after myocardial infarction. Experimental observations and clinical implications. *Circulation* 81:1161–1172. https://doi.org/10.1161/01.cir.81.4.1161
103. Sutton MG, Sharpe N (2000) Left ventricular remodeling after myocardial infarction: Pathophysiology and therapy. *Circulation* 101:2981–2988. https://doi.org/10.1161/01.cir.101.25.2981
104. Christman KL, Lee RJ (2006) Biomaterials for the treatment of myocardial infarction. *J Am Coll Cardiol* 48:907–913. https://doi.org/10.1016/j.jacc.2006.06.005
105. Johnson TD, Christman KL (2013) Injectable hydrogel therapies and their delivery strategies for treating myocardial infarction. *Expert Opin Drug Deliv* 10:59–72. https://doi.org/10.1517/17425247.2013.739156
106. Mukherjee S, Reddy Venugopal J, Ravichandran R, Ramakrishna S, Raghunath M (2011) Evaluation of the biocompatibility of PLACL/collagen nanostructured matrices with cardiomyocytes as a model for the regeneration of infarcted myocardium. *Adv Funct Mater* 21:2291–2300. https://doi.org/10.1002/adfm.201002434
107. Pektok E, Nottelet B, Tille J-C, Gurny R, Kalangos A, Moeller M, Walpoth BH (2008) Degradation and healing characteristics of small-diameter poly(epsilon-caprolactone) vascular grafts in the rat systemic arterial circulation. *Circulation* 118:2563–2570. https://doi.org/10.1161/CIRCULATIONAHA.108.795732
108. Rai M, Ingle AP, Gaikwad S, Padovani FH, Alves M (2016) The role of nanotechnology in control of human diseases: Perspectives in ocular surface diseases. *Crit Rev Biotechnol* 36:777–787. https://doi.org/10.3109/07388551.2015.1036002
109. Rosenthal VD, Bijie H, Maki DG, Mehta Y, Apisarnthanarak A, Medeiros EA, Leblebicioglu H, Fisher D, Álvarez-Moreno C, Khader IA, Del Rocío González Martínez M, Cuellar LE, Navoa-Ng JA, Abouqal R, Guanche Garcell H, Mitrev Z, Pirez García MC, Hamdi A, Dueñas L, Cancel E, Gurskis V, Rasslan O, Ahmed A, Kanj SS, Ugalde OC, Mapp T, Raka L, Yuet Meng C, Thu LTA, Ghazal S, Gikas A, Narváez LP, Mejía N, Hadjieva N, Gamar Elanbya MO, Guzmán Siritt ME, Jayatilleke K, INICC members (2012) International Nosocomial Infection Control Consortium (INICC) report, data summary of 36 countries, for 2004-2009. *Am J Infect Control* 40:396–407. https://doi.org/10.1016/j.ajic.2011.05.020

110. Jo DH, Kim JH, Lee TG, Kim JH (2015) Size, surface charge, and shape determine therapeutic effects of nanoparticles on brain and retinal diseases. *Nanomedicine* 11:1603–1611. https://doi.org/10.1016/j.nano.2015.04.015
111. Grasman JM, Zayas MJ, Page RL, Pins GD (2015) Biomimetic scaffolds for regeneration of volumetric muscle loss in skeletal muscle injuries. *Acta Biomater* 25:2–15. https://doi.org/10.1016/j.actbio.2015.07.038
112. Wolf MT, Dearth CL, Sonnenberg SB, Loboa EG, Badylak SF (2015) Naturally derived and synthetic scaffolds for skeletal muscle reconstruction. *Adv Drug Deliv Rev* 84:208–221. https://doi.org/10.1016/j.addr.2014.08.011
113. Goldberg JL, Barres BA (2000) The relationship between neuronal survival and regeneration. *Annu Rev Neurosci* 23:579–612. https://doi.org/10.1146/annurev.neuro.23.1.579
114. Orive G, Anitua E, Pedraz JL, Emerich DF (2009) Biomaterials for promoting brain protection, repair and regeneration. *Nat Rev Neurosci* 10:682–692. https://doi.org/10.1038/nrn2685
115. Genco RJ, Van Dyke TE (2010) Prevention: Reducing the risk of CVD in patients with periodontitis. *Nat Rev Cardiol* 7:479–480. https://doi.org/10.1038/nrcardio.2010.120
116. Lalla E, Papapanou PN (2011) Diabetes mellitus and periodontitis: A tale of two common interrelated diseases. *Nat Rev Endocrinol* 7:738–748. https://doi.org/10.1038/nrendo.2011.106
117. Bingham CO 3rd, Moni M (2013) Periodontal disease and rheumatoid arthritis: The evidence accumulates for complex pathobiologic interactions. *Curr Opin Rheumatol* 25:345–353. https://doi.org/10.1097/BOR.0b013e32835fb8ec
118. Besinis A, De Peralta T, Tredwin CJ, Handy RD (2015) Review of nanomaterials in dentistry: Interactions with the oral microenvironment, clinical applications, hazards, and benefits. *ACS Nano* 9:2255–2289. https://doi.org/10.1021/nn505015e
119. Kumar V, Choudhary AK, Kumar P, Sharma S (2019) Nanotechnology: Nanomedicine, nanotoxicity and future challenges. *Nanosci Nanotechnol Asia* 9:64–78. https://doi.org/10.2174/2210681208666180125143953
120. Vacanti J (2010) Tissue engineering and regenerative medicine: From first principles to state of the art. *J Pediatr Surg* 45:291–294. https://doi.org/10.1016/j.jpedsurg.2009.10.063
121. Siepmann J, Siegel RA, Rathbone MJ (2012) *Fundamentals and Applications of Controlled Release Drug Delivery*. Springer US.
122. Biondi M, Ungaro F, Quaglia F, Netti PA (2008) Controlled drug delivery in tissue engineering. *Adv Drug Deliv Rev* 60:229–242. https://doi.org/10.1016/j.addr.2007.08.038
123. Griset AP, Walpole J, Liu R, Gaffey A, Colson YL, Grinstaff MW (2009) Expansile nanoparticles: Synthesis, characterization, and in vivo efficacy of an acid-responsive polymeric drug delivery system. *J Am Chem Soc* 131:2469–2471. https://doi.org/10.1021/ja807416t
124. Thakur G, Singh A, Singh I (2015) Chitosan-montmorillonite polymer composites: Formulation and evaluation of sustained release tablets of aceclofenac. *Sci Pharm* 84:603–617. https://doi.org/10.3390/scipharm84040603
125. Mukhtar M, Bilal M, Rahdar A, Barani M, Arshad R, Behl T, Brisc C, Banica F, Bungau S (2020) Nanomaterials for diagnosis and treatment of brain cancer: Recent updates. *Chemosensors* 8:117. https://doi.org/10.3390/chemosensors8040117
126. Dhiman S, Singh TG, Rehni AK (2011) Transdermal patches: A recent approach to new drug delivery system. https://scholar.archive.org/work/g3vflutahfhl3jrj3htabnew4u/access/wayback/http://ijppsjournal.com/Vol3Suppl5/2695.pdf. Accessed 1 Jun 2023

14 Innovative Trends and Applications of Bionanomaterials in Skin Rejuvenation

S.C. Onwubu, M.U. Makgobole, T.H. Mokhothu, A.K. Misra, N. Mpofana, and B.N. Mkhwanazi

14.1 INTRODUCTION

The field of skin rejuvenation involves the restoration and revitalization of the skin to its youthful state. As people age, their skin loses elasticity and firmness, and wrinkles, fine lines, and age spots appear (Ahmed et al., 2020). Various cosmetic procedures and products are available to address these concerns, including chemical peels, laser treatments, and topical creams (Pathak et al., 2020, Devgan et al., 2019). In recent years, researchers have shown increasing interest in exploring the use of bionanomaterials for skin rejuvenation. This has been demonstrated in several studies conducted by Leonida and Kumar (2016), Singh et al. (2021), and Souto et al. (2022). Bionanomaterials have shown promise as a means of rejuvenating the skin. Bionanomaterials are derived from biological sources and have a size on the nanoscale (Honek, 2013, Jeevanandam et al., 2022). These materials have unique properties that make them attractive for skin rejuvenation (Leonida et al., 2016a). An instance of their capabilities is their ability to infiltrate the stratum corneum, which is the skin's topmost layer, and reach deeper layers of the skin, where they can stimulate collagen production and improve skin hydration (Leonida and Kumar, 2016).

Bionanomaterials used in skin rejuvenation include liposomes, nanoparticles, and dendrimers (Salvioni et al., 2021). Liposomes are small spheres made up of a lipid bilayer that can encapsulate active ingredients and deliver them to the skin. Nanoparticles, on the other hand, are small particles made up of materials such as gold, silver, and silica, and they can also deliver active ingredients to the skin (Leonida et al., 2016c). Dendrimers are highly branched nanoscale polymers that can encapsulate and deliver active ingredients (Leonida et al., 2016c). The use of bionanomaterials in skin rejuvenation is a rapidly evolving field, and there are many promising avenues for future research (Leonida and Kumar, 2016). For example, researchers are exploring the use of bionanomaterials in combination with other therapies, such as laser treatments and chemical peels, to enhance their effects (Lohani et al., 2014, Nath et al., 2021). Advancements in nanotechnology are also making it possible to tailor treatments to individual patients based on their skin type and specific concerns. In summary, the use of bionanomaterials in skin rejuvenation has the potential to revolutionize the field and provide patients with safer and more effective treatment options (Leonida et al., 2016c). The present chapter aims to offer a summary of the latest developments and uses of bionanomaterials in rejuvenating the skin. It aims to explore the unique properties of bionanomaterials that make them attractive for skin rejuvenation and review the current research in this rapidly evolving field.

DOI: 10.1201/9781003432791-17

The chapter is organized into six sections. The first section introduces the topic and highlights the potential of bionanomaterials to revolutionize skin rejuvenation. The second section discusses the various types of bionanomaterials, including liposomes, nanoparticles, and dendrimers and their properties. The third section explores the mechanisms of action of bionanomaterials in skin rejuvenation, including how they can penetrate the skin and stimulate collagen production. The fourth section reviews the current and potential applications of bionanomaterials in skin rejuvenation, including their use in combination with other therapies. The fifth section discusses the safety and regulation of bionanomaterials in skin rejuvenation, including potential risks and regulatory frameworks. Finally, the conclusion summarizes the key points of the chapter and provides an outlook for the future of bionanomaterials in skin rejuvenation.

14.2 OVERVIEW OF BIOMATERIALS IN SKIN REJUVENATION

The goal of bionanomaterials in skin rejuvenation is to develop new and effective ways to enhance the health and appearance of the skin (Souto et al., 2022). The research in this field is geared towards creating new materials that can transport active ingredients such as peptides, antioxidants, and other molecules deep into the skin to improve its function and overall look (Leonida et al., 2016a). Unlike traditional skin rejuvenation treatments, bionanomaterials offer several advantages. They can deliver active ingredients deep into the skin, improving its appearance and function (Leonida et al., 2016a, Milan et al., 2019). They can also be used to improve the delivery of drugs and other therapeutic agents to the skin. Additionally, bionanomaterials are biocompatible, which means they are non-toxic and non-irritating to the skin (Du and Wong, 2019, Chhabra and Bhati, 2021). These materials also have the ability to target specific cells in the skin, which can improve the effectiveness of skin rejuvenation treatments.

14.2.1 Definition and Explanation of Bionanomaterials

According to Singh and Singh (2021), bionanomaterials refer to molecular materials that incorporate biological molecules, including proteins, enzymes, nucleic acids, lipids, polysaccharides, viruses, and secondary metabolites. These materials are considered nanomaterials as they are composed of biomolecules encapsulated or immobilized around a conventional nanomaterial. Biomaterials typically consist of biological components, resulting in nano-sized molecular structures. Other studies, including Singh and Singh, 2021, Leonida et al., 2016a, Honek, 2013, have also highlighted the use of biological elements such as RNA, DNA, oligosaccharides, and cells in the fabrication of bionanomaterials. The biological molecules used to manufacture bionanomaterials are derived from a variety of sources, including microorganisms, plants, agricultural waste, insects, marine organisms, as well as some mammals (Jeevanandam et al., 2022). Because of their biocompatibility, biodegradability, non-toxicity and bioactive assets, they have been recognized as a possible alternative to hazardous conventional nanomaterials for biomedical purposes (Singh et al., 2021, Chhabra and Bhati, 2021).

One of the main advantages of bionanomaterials is their molecular size, usually between 1 and 100 nm, which allows them to easily penetrate the skin and reach the deeper layers (Honek, 2013, Leonida et al., 2016a, Du and Wong, 2019, Jeevanandam et al., 2022). Their molecular size is important in terms of having biological properties and function as it is similar in size to DNA and protein molecules (Du and Wong, 2019, Chhabra and Bhati, 2021); this allows them to easily penetrate the skin as well as target specific skin cells. Their efficacy is determined by their ability to pass the stratum corneum (SC), a rate-limiting barrier, and reach an intercellular lipid matrix (Leonida and Kumar, 2016). The characteristics and uses of bionanomaterials are the topics of continuous study and development in a variety of disciplines, including agriculture, the environment, and biomedical applications (Singh and Singh, 2021). These materials have the potential to change a wide range of industries, including cosmetics and dermatology (Leonida and Kumar, 2016). One potential application of bionanomaterials is in skin rejuvenation.

14.2.2 Types of Bionanomaterials

Bionanomaterials are increasingly being used in skincare products for their unique properties and ability to improve skin health. Various types of bionanomaterials are utilized in skin rejuvenation, including nanoparticles, and dendrimers, as stated by reference (Leonida et al., 2016c). These materials possess distinct characteristics that make them suitable for diverse applications in skincare (see Figure 14.1).

14.2.2.1 Liposomes

Liposomes are spherical structures composed of a phospholipid bilayer and are commonly used as carriers for active ingredients in skincare products (Leonida et al., 2016a). They can encapsulate both hydrophilic and hydrophobic molecules, which makes them versatile delivery systems (Van Tran et al., 2019). Liposomes can penetrate the skin's outermost layer, the stratum corneum, and release their contents in the deeper layers, where they can have a therapeutic effect, additionally, they have been used to deliver peptides, vitamins, and other active ingredients that can improve skin hydration, firmness, and texture (Costa and Santos, 2017, Khezri et al., 2018).

14.2.2.2 Nanoparticles

Nanoparticles are minute particles, measuring between 10–1000 nm, containing an active ingredient and are made from biocompatible materials. As indicated by reference (Khezri et al., 2018), nanoparticles can considerably enhance the properties of the material when compared to its bulk form. Nanoparticles can be made from various materials, such as metals, ceramics, and polymers (Khan et al., 2016). In skincare, nanoparticles have the ability to penetrate the skin more easily than

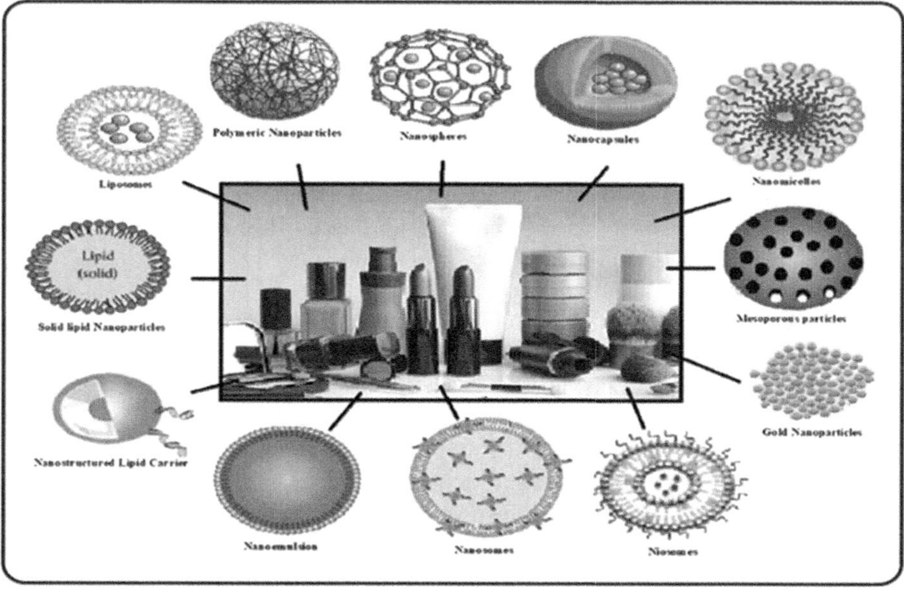

FIGURE 14.1 Illustrating the Categories of Nanocosmetic Formulations, Classified Based on Type of Nanocarrier System or Nanomaterial Employed (Source: Dhawan et al., 2020.)

larger particles and can deliver active ingredients to the deeper layers of the skin, targeting specific cellular and subcellular arears (Khezri et al., 2018, Salvioni et al., 2021). They have been used to deliver antioxidants, such as vitamin E, and to promote collagen synthesis, improving skin texture and reducing the appearance of fine lines (Leonida et al., 2016d).

14.2.2.3 Dendrimers

Dendrimers are highly branched, nanoscale polymers that can have a range of properties, depending on their composition and structure (Dianzani et al., 2014). They can be designed to have specific properties, such as high biocompatibility or the ability to penetrate cell membranes (Cao et al., 2020). Dendrimers have been used in skincare to deliver active ingredients, such as retinoids, which can improve skin texture and reduce the appearance of fine lines (Dubey et al., 2022). They can also act as carriers for nanoparticles or liposomes, enhancing their delivery to the skin (Leonida et al., 2016d, Souto et al., 2022).

14.2.2.4 Hyaluronic acid

Hyaluronic acid is a naturally occurring substance found in the human body and is a component of skin, connective tissue, and eyes (Rabah and Aslan, 2021, Walker et al., 2021). It is also used in cosmetic and medical products, such as moisturizing creams, lotions, ointments, and serums, due to its ability to retain water and hydrate the skin (Bukhari et al., 2018). Hyaluronic acid has been studied for its effects on skin rejuvenation and has been found to improve skin tone, elasticity, texture, radiance, and reduce wrinkles and scarring (Bravo et al., 2022). Hyaluronic acid has been used in different applications, such as being injectable for facial skin rejuvenation, and studies have found that it can improve skin quality (Ayatollahi et al., 2020, Magda Belmontesi et al., 2018). There are different types of hyaluronic acid products available with varying molecular weights, and these can penetrate the skin to different depths (Setthanakul, 2019, Qiu et al., 2021). Hyaluronic acid has been evaluated for safety as a topical moisturizer and has been found to be effective in protecting the skin with an antimicrobial effect that can last up to 28 days (Magda Belmontesi et al., 2018, Tort and Karakucuk, 2021, Souto et al., 2022).

14.2.2.5 Collagen

Collagen is a protein that is naturally present in the body and is an essential component of the skin. It provides structure, elasticity, and hydration to the skin (Reilly and Lozano, 2021). As one ages, the production of collagen in the body decreases, leading to wrinkles, fine lines, and a loss of skin elasticity (Van Kets, 2012, Ahmed et al., 2020). One approach to skin rejuvenation involves the use of collagen-based products to restore and enhance the skin's natural collagen (Rezvani Ghomi et al., 2021). Collagen is an effective biomaterial component because, when applied to damage skin, it is readily degradable by extracellular collagenases, which subsequently assists in tissue regeneration by collagen resorption (Lo and Fauzi, 2021).

Collagen can be administered topically, injected, or applied through other means to help improve skin texture, tone, and firmness (Rezvani Ghomi et al., 2021). Topical collagen products, like creams and serums, are widely available and can aid in enhancing the skin's appearance by offering hydration and, to some extent, replenishing collagen (Aguirre-Cruz et al., 2020). Conversely, collagen injections entail the direct injection of collagen into the skin, and they can effectively reduce wrinkles, fine lines, and scars and promote skin texture and firmness (Devgan et al., 2019). However, the effects of collagen injections are temporary, and repeat injections are necessary to maintain the results (Haneke, 2019).

According to Chhabra and Bhati (2021), collagen-based nanofilms have proven to enhance wound healing and cell regeneration by increased fibroblast migration and by enhancing cell proliferation. Additionally, collagen nanoparticles are routinely utilized. Collagen-based nanoparticles possess the ability to be absorbed by the reticuloendothelial system (RES), which facilitates enhanced absorption of the loaded medications into diverse cells. Hence, collagen-based nanoparticles exhibit great potential as a systemic drug delivery options (Razavi et al., 2017).

Overall, the use of collagen in skin rejuvenation has shown promise in improving the appearance of the skin, particularly in addressing wrinkles and fine lines. However, the efficacy of collagen-based treatments may vary depending on individual factors, such as age, skin type, and overall skin health. It is essential to consult with a dermatologist or skincare professional to determine the best approach to collagen-based skin rejuvenation for an individual's specific needs.

14.2.2.6 Nanochitin in Skin Rejuvenation

Nanochitin is a bionanomaterial that has recently gained attention in the field of skin rejuvenation due to its unique properties and potential applications (Ahmad et al., 2020). Chitin is a natural polymer found in the exoskeletons of crustaceans, insects, and some fungi (Iber et al., 2022). Nanochitin is a type of chitin that has been broken down into nanoscale particles, making it more easily absorbed by the skin. According to Souto et al. (2022) Nanochitin has been demonstrated in trials to increase collagen formation and enhance skin moisture, potentially reducing the appearance of wrinkles and fine lines. Furthermore, nanochitin has been found to possess anti-oxidant and anti-inflammatory properties, which can shield the skin from environmental stresses and alleviate inflammation that leads to skin aging (Souto et al., 2022). Nanochitin can not only function as a stand-alone treatment but can also be incorporated into other skincare products such as serums, moisturizers, and masks (Danti et al., 2019). Due to its small size and unique characteristics, nanochitin is an attractive alternative for delivering targeted therapy to specific skin layers or cells (Razavi et al., 2017).

Nevertheless, as with any new material, additional investigation is required to fully comprehend the safety and effectiveness of nanochitin in skin rejuvenation. Although studies have demonstrated its low toxicity and good skin tolerance, more research is needed to establish optimal dosages and possible long-term impacts (Razavi et al., 2017).

To revolutionize skincare, bionanomaterials offer unique properties and capabilities to deliver active ingredients to deeper skin layers. Among the promising types of bionanomaterials are liposomes, nanoparticles, and dendrimers. Liposomes can encapsulate a diverse range of active ingredients, nanoparticles can deliver antioxidants and promote collagen synthesis, while dendrimers can serve as carriers for active ingredients and enhance their delivery to the skin. Nonetheless, more research is necessary to gain a full understanding of the potential of these materials in skincare and to develop innovative applications for bionanomaterials in skin rejuvenation.

14.2.3 Mechanisms of Action of Bionanomaterials in Skin Rejuvenation

This section delves into how bionanomaterials work to rejuvenate the skin. It describes how these materials can penetrate the skin and reach the deeper layers, where they can stimulate collagen production and improve skin hydration. Bionanomaterials are a promising approach to skin rejuvenation due to their unique properties and mechanisms of action. These materials are designed to be small enough to penetrate the skin and reach the deeper layers, where they can interact with skin cells and stimulate regeneration and repair. The mechanisms of action of bionanomaterials in skin rejuvenation can be classified into several categories, including collagen synthesis, skin hydration, and antioxidant activity.

One of the main mechanisms of action of bionanomaterials is their ability to stimulate collagen production in the skin. Collagen is a key component of the skin's extracellular matrix and is responsible for its strength and elasticity (Lo and Fauzi, 2021). As people age, the production of collagen decreases, leading to the development of wrinkles and fine lines. Bionanomaterials, such as liposomes, nanoparticles, and dendrimers, can penetrate the skin and deliver peptides or other molecules that stimulate collagen synthesis (Leonida et al., 2016d). For example, liposomes can

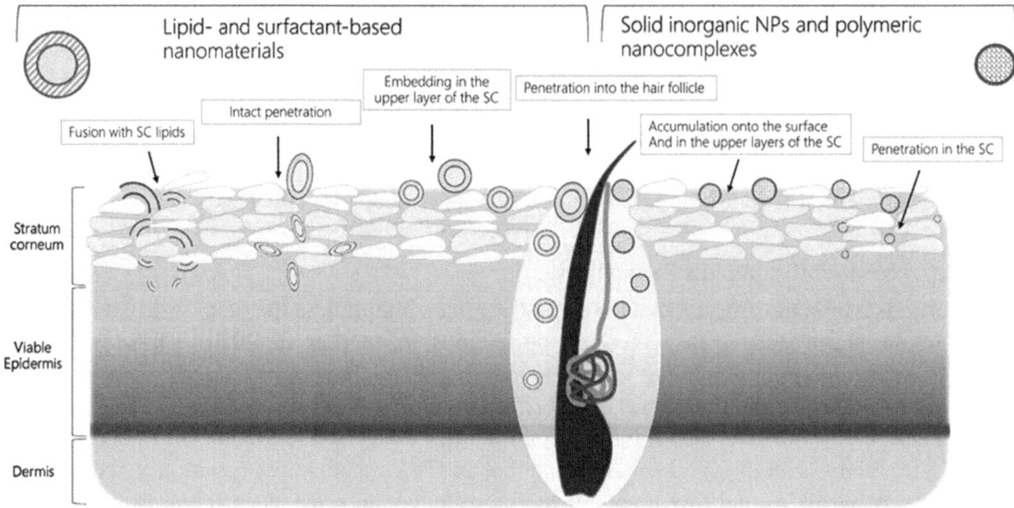

FIGURE 14.2 Potential Routes of the Penetration of Nanomaterials into the Skin (Source: Salvioni et al., 2021.)

deliver peptides that mimic the activity of collagen and stimulate its production, while nanoparticles can deliver growth factors that promote cell proliferation and collagen synthesis (Wang et al., 2021, Souto et al., 2022). Another mechanism of action of bionanomaterials is their ability to improve skin hydration. Dehydration is a common problem in aging skin and can contribute to dryness, roughness, and the development of fine lines (Kim, 2018). Bionanomaterials can enhance skin hydration by delivering moisturizing agents, such as hyaluronic acid or ceramides, to the deeper layers of the skin (Nafisi and Maibach, 2017). These materials can also form a protective barrier on the skin, reducing water loss and maintaining skin moisture (Souto et al., 2022).

Bionanomaterials have also been found to exhibit antioxidant activity, which can protect the skin from oxidative stress and prevent premature aging (Liu and Shi, 2019). Oxidative stress is a key contributor to skin aging, as it can damage cellular components and lead to the development of wrinkles, age spots, and other signs of aging (Schneider, 2021). Bionanomaterials can deliver antioxidants, such as vitamins C and E or polyphenols, to the skin and neutralize free radicals, reducing oxidative stress and preventing further damage (Leonida et al., 2016b). In summary, bionanomaterials can exert their effects on skin rejuvenation through several mechanisms of action, including collagen synthesis, skin hydration, and antioxidant activity. These materials can penetrate the skin and reach the deeper layers, where they can interact with skin cells and promote regeneration and repair (Figure 14.2). Bionanomaterials show great promise in the field of skincare and have the potential to bring a revolutionary change to skin rejuvenation.

14.2.4 Applications of Bionanomaterials in Skin Rejuvenation

The field of bionanomaterials in skin rejuvenation is a relatively new area of research that has been gaining attention in recent years. The goal of bionanomaterials in skin rejuvenation is to develop new and effective methods for improving the appearance and health of the skin. According to Leonida et al. (2016a), bionanomaterials have the potential to improve the delivery of active ingredients, such as antioxidants, peptides, and other molecules, deep into the skin to improve its overall appearance. Additionally, bionanomaterials can act as active principles, such as moisturizing agents, UV filters, or anti-wrinkle agents, or as ingredients in formulations to modify rheological or optical properties (Leonida and Kumar, 2016, Milan et al., 2019).

Various applications of bionanomaterials have been explored in the medical field including the treatment of burns, skin diseases and wound healing (Leonida and Kumar, 2016). In the cosmetic

TABLE 14.1
Benefits of Nano-Cosmetics Product on the Skin

Type of benefits	Interpretation
Protective	Offers protection against harmful external factors like pollution, dry air, and UV light
Cleansing	Eliminates dirt and microorganisms from the skin
Hydrating	Restores or maintains fluid balance by providing water to the skin
Moisturizing	Forms a barrier that effectively prevents water loss through the epidermis
Soothing	Provides a gently calming effect
Firming	Enhances skin tone and smoothness

Source: Author's Own Creation.

field, bionanomaterials are used in a variety of skincare, hair care, and nail care applications (Lohani et al., 2014, Nath et al., 2021). Bionanomaterials have been used in a variety of skin rejuvenation applications. Table 14.1 summarizes the main effect of their applications on the skin.

14.2.4.1 Anti-aging

Anti-aging is an important aspect of skin rejuvenation, as it addresses the changes that occur in the skin over time, such as wrinkles, fine lines, age spots, and loss of elasticity. These changes are caused by a combination of factors, including sun exposure, genetics, and the natural aging process. To improve the appearance of the skin and slow down the aging process, anti-aging treatments such as topical creams and serums, chemical peels, laser resurfacing, and injectable treatments (Botox and dermal fillers) are commonly used, as per Ovadia et al. (2021).These treatments have been shown to effectively enhance the skin's appearance and slow down the aging process (Ahmed et al., 2020).

14.2.4.1.1 The Challenges and Limitations of Traditional Anti-aging Products

Traditional anti-aging products, such as creams, serums, and lotions can be effective in improving the appearance of the skin and slowing down the aging process. However, they also have some challenges and limitations. One limitation is that traditional anti-aging products can be ineffective or even harmful if they contain ingredients that are not suitable for the user's skin type or if they are not used in the correct way (Ganceviciene et al., 2012). Another limitation is that traditional anti-aging products can be expensive, so not all people can afford them, and some products might be overhyped and not deliver on their promises. Another limitation is that traditional anti-aging products may not address all the signs of aging, such as deep wrinkles and loss of volume (Zouboulis et al., 2019). Another limitation is that the results of traditional anti-aging products may not be as dramatic or long-lasting as other anti-aging treatments such as laser resurfacing, chemical peels, or injectables (Ganceviciene et al., 2012).

14.2.4.1.2 Bionanomaterial-based Anti-aging Products That Have Been Developed

In terms of skin rejuvenation, bionanomaterials have become very significant due to their unique properties and capabilities. These nanoengineered materials can penetrate deep into the epidermis and efficiently distribute active ingredients as well as give physical and mechanical support, increase skin hydration and elasticity, and protect against environmental harm (Leonida and Kumar, 2016). There are several bionanomaterial-based anti-aging products that have been developed and are available in the market. Some of the most notable examples include the following:

- Nanocollagen: A bionanomaterial derived from collagen is utilized in skincare products to diminish the visibility of wrinkles and fine lines. Collagen, which naturally occurs in the body, gives the skin its form and support (Lo and Fauzi, 2021).

- Nanosilver: Anti-aging products utilize nanosilver particles due to their antimicrobial and anti-inflammatory properties, which can help in reducing skin redness and irritation. Studies have supported the use of nanosilver particles in such products (Shah et al., 2020, Rehman et al., 2023).
- Nanopeptides: Nanopeptides are tiny protein fragments that have been designed to penetrate deep into the skin and deliver active ingredients to where they are needed most. They are commonly used in anti-aging products to boost collagen production and improve skin firmness (Roberts et al., 2017).
- Nanohydroxyapatite: A mineral-based bionanomaterial, nanohydroxyapatite is often utilized in anti-aging products to enhance skin density and flexibility. It has the ability to encourage collagen production and minimize the visibility of fine lines and wrinkles.

14.2.4.2 Moisturization

Among the primary concerns of aging skin are the loss of moisture and hydration, which can lead to dryness, dullness, and fine lines. Bionanomaterials have been explored as a means of improving skin hydration and barrier function, helping to combat these signs of aging. One approach to improving skin hydration is through the use of liposomes, which are spherical vesicles composed of a lipid bilayer. These structures can encapsulate and deliver active ingredients, such as humectants and ceramides, directly to the skin. Liposomes have been shown to improve skin hydration and reduce transepidermal water loss, leading to a smoother, more youthful appearance (Costa and Santos, 2017). Another bionanomaterial that has been explored for its moisturizing properties is hyaluronic acid nanoparticles. Hyaluronic acid is a naturally occurring polysaccharide found in the skin and which plays a key role in maintaining hydration and elasticity. By encapsulating hyaluronic acid in nanoparticles, it can be delivered to the skin in a more targeted and sustained manner, providing longer-lasting hydration (Bukhari et al., 2018).

In addition to these materials, other bionanomaterials, such as solid lipid nanoparticles and dendrimers, have also been investigated for their potential to improve skin hydration and barrier function (Leonida et al., 2016c). These materials offer unique properties, such as high stability, biocompatibility, and controlled release, making them promising candidates for use in skincare products (Salvioni et al., 2021, Khezri et al., 2018). Overall, bionanomaterials show great potential for improving skin hydration and barrier function, providing a promising avenue for combatting the signs of aging and improving overall skin health.

14.2.4.3 Skin brightening

Skin brightening is a popular cosmetic treatment that involves lightening the skin tone to achieve a more even complexion. This is a common goal of skin rejuvenation, as it can help to even out skin tone, reduce the appearance of dark spots and blemishes, and give the skin a more youthful and radiant appearance (Happy et al., 2021). Bionanomaterials have emerged as a promising tool for achieving skin brightening by delivering active ingredients to the skin more effectively and safely.

One promising bionanomaterial for skin brightening is nanoliposomes. These are small vesicles made of phospholipids that can encapsulate hydrophilic and hydrophobic active ingredients. Nanocarriers such as liposomes and dendrimers have been found effective in skin brightening as these materials can encapsulate skin brightening agents and deliver them to the deeper layers of the skin, where they can have a more significant impact on skin tone and texture (Leonida et al., 2016d). For example, Xia et al. (2022) found that nanoliposomes successfully delivered Phenylethyl resorcinol (PR). PR is a tyrosinase inhibitor and used in cosmeceutical as skin lightening agent. The PR loaded nanoliposomes exhibited good chemical stability, decreased cytotoxicity, improved cell uptake as well as reduced melanin production. Other bionanomaterials with potential for skin brightening include plant-derived materials such as curcumin or resveratrol, which have been shown to have antioxidant and anti-inflammatory effects that can help to brighten the skin (Szulc-Musioł and Sarecka-Hujar, 2021). Resveratrol, a naturally occurring polyphenol found in grapes and other plants, is another active ingredient that has been shown to have skin brightening properties (Farris et al., 2016). When delivered using bionanomaterials such as liposomes or dendrimers, resveratrol can effectively inhibit tyrosinase activity and reduce melanin production (Vaishampayan and Rane, 2022).

Another approach to skin brightening with bionanomaterials involves the use of nanoparticles, which can be used to encapsulate and deliver a range of active ingredients to the skin (Abu Hajleh et al., 2021). Silver nanoparticles were discovered to inhibit melanin production by decreasing tyrosinase activity and expression in melanocytes, as well as having antimicrobial properties, making them useful in the treatment of acne and other skin infections that can result in post-inflammatory hyperpigmentation (Figueiredo et al., 2023). Because of their high skin permeability, biocompatibility, and biodegradability, as well as their function as a UV ray blocker, lipid nanoparticles were discovered to be the most extensively used nanocosmeceutical in treating hyperpigmentation (Tangau et al., 2022).

14.2.5 Mechanisms Behind How Bionanomaterials Improve Skin Rejuvenation

How bionanomaterials improve skin rejuvenation involves diverse mechanisms and depends on the specific material and active ingredient being used. The major researched mechanism is the delivery of active ingredients (Leonida and Kumar, 2016, Du and Wong, 2019, Kee et al., 2022). Through the targeted delivery of active ingredients, bionanomaterials have demonstrated potential for improving skin rejuvenation. Two popular bionanomaterials employed for this purpose are liposomes and delivery systems based on nanoparticles (Khezri et al., 2018). These materials can be made to encapsulate and deliver active ingredients, such as growth factors, antioxidants, and other skincare ingredients, directly to the skin cells (Kant et al., 2021, Salvioni et al., 2021).

When compared to oral administration, transdermal delivery has the added benefit of preventing gastrointestinal side effects, for example, diarrhea, nausea, or drug breakdown in the gut (Souto et al., 2022). Drug penetration through the skin is the limiting element in transdermal bioavailability, influenced by molecular size, lipophilicity, product pH, and skin wetness. (Souto et al., 2022, Leonida et al., 2016d, Santos et al., 2019).

The skin comprises three layers, namely the epidermis, dermis, and the subcutaneous connective tissues. The epidermis, in turn, consists of the stratum corneum (SC), keratinocytes, and other cells. The SC function as a highly effective barrier, preventing the penetration of xenobiotics hindering the delivery of therapeutic medications (Khezri et al., 2018, Leonida et al., 2016d). According to Salvioni et al. (2021) and Leonida et al. (2016a), bionanomaterial have demonstrated the effectiveness of active ingredient delivery when compared to conventional topical treatments. Lipid carriers such as Liposome, nanoemalsion, and nanoparticles are being used to deliver active ingredients to the skin.

Liposome are spheres composed of a phospholipid bilayer with diameters ranging from 20 to 100 nm (Leonida et al., 2016c). They can improve delivery by encapsulating hydrophilic active ingredients in their core, whilst hydrophobic ones can be integrated into the bilayer (Van Tran et al., 2019). The encapsulation helps to prevent the ingredients from degrading and releases them gradually over time (Costa and Santos, 2017, Khezri et al., 2018), thus, benefits of the active ingredients may have long-lasting effects due to this slow release. According to Santos et al. (2019) various studies regarding liposome and encapsulated vitamins/antioxidants have shown improvement in anti-ageing formulations.

Nanoemulsions are bilayer systems that consist of oil, water, and one or more emulsifying agents. These systems comprise nanoscale droplets of dispersed phase (20–100 nm) and are transparent in appearance (Leonida et al., 2016c). Nanoemulsions can incorporate both lipophilic and hydrophilic components since they can be synthesized as either oil-in-water (O/W) or water-in-oil (W/O) emulsions, although the majority of nanoemulsions used to date have been of the O/W type (Santos et al., 2019, Leonida et al., 2016c). Compared to conventional emulsion, nanoemulsions, have several advantages, including low light scattering, resulting in transparency or translucency, high thermodynamic stability that prevents droplet aggregation or gravitational separation, unique rheological properties like reduced viscosity, and the ability to significantly enhance the bioavailability of lipophilic compounds (Santos et al., 2019, Souto et al., 2022). Additionally, nanoemulsaions enable easy solubilization of drugs, increase therapeutic efficacy, and subsequently reduce side effects (Souto et al., 2022). Santos et al. (2019) elaborated on how several studies have demonstrated how nanoemulsions are appropriate for delivery of active ingredients while enhancing skin penetration.

14.2.6 Safety and Regulation for Bionanomaterials Use in Skin Rejuvenation

As with any new technology, the safety of bionanomaterials is an important consideration when used in skincare products. While bionanomaterials have shown promise in improving skin health, there are concerns about their potential to cause harm. These concerns relate to the potential for these materials to penetrate the skin and accumulate in organs, which could lead to toxic effects. To ensure the safety of bionanomaterials in skincare, regulatory frameworks have been put in place to assess their potential risks and ensure their safe use (Catalán and Norppa, 2017, Salvioni et al., 2021). In the United States, the Food and Drug Administration (FDA) regulates the use of bionanomaterials in cosmetics and personal care products (Rathod et al., 2021). The FDA requires that all cosmetic ingredients be safe for use and must be adequately labelled to inform consumers of their potential risks. However, the regulation of bionanomaterials is complex and challenging due to their unique properties and the lack of standardized testing methods for their safety (Dhawan et al., 2020).

To address these challenges, several organizations have developed guidelines and recommendations for the safe use of bionanomaterials in cosmetics. The International Organization for Standardization (ISO) has developed guidelines for the characterization and testing of nanoparticles in cosmetics, which includes methods for measuring their size, shape, and surface properties (Lu et al., 2018). The Cosmetic Products Regulation (CPR) established regulatory standards for nanomaterials to assure their safety, including data for physicochemical characterization and nanomaterial labelling on cosmetic items (Rasmussen et al., 2019). The Scientific Committee on Consumer Safety (SCCS) assesses the safety of nanomaterials in cosmetic items and their components. The SCCS has released two guideline publications to help with nanomaterial evaluation. These publications comprise physicochemical property lists recommended by the SCCS for nanomaterial identification and characterization (Rasmussen et al., 2019).

Overall, the safety of bionanomaterials in skincare depends on several factors, including the type of material used, its concentration, and the route of exposure. It is important to note that while bionanomaterials have the potential to improve skin health, their safety and efficacy should be carefully evaluated before being used in skincare products. Regulatory frameworks and guidelines have been put in place to ensure their safe use, but ongoing research and monitoring are needed to ensure the continued safety and effectiveness of these materials in skincare.

14.2.7 Future Directions

As research into bionanomaterials and their applications in skin rejuvenation continues to advance, there are several promising areas for future exploration. These are further illustrated in Figure 14.3 and discussed in detail below.

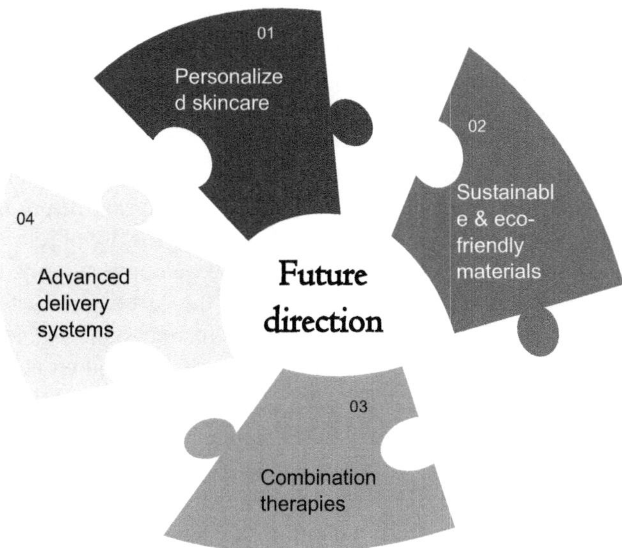

FIGURE 14.3 Trends and Future Direction of Nanomaterials Applications in Skincare (Source: Created by Authors.)

14.2.7.1 Personalized Skincare

Personalized skincare is an emerging trend in the beauty industry, and bionanomaterials offer exciting possibilities for creating customized skincare products. By incorporating nanosensors and other advanced technologies, skincare products could be designed to monitor and adjust to an individual's changing skin condition. One approach to personalized skincare is the use of smart skincare devices that incorporate nanosensors to analyze tracking outdoor UV exposure and managing thermal comfort (Patel et al., 2022). It would be an advantage to have nanosensors analyse skin properties such as hydration, oiliness, and pH levels. This data can then be used to create personalized skincare routines that target an individual's specific needs. For example, a skincare device could analyze the hydration levels of an individual's skin and recommend a customized moisturizer that contains bionanomaterials specifically designed to improve skin hydration.

In addition to smart skincare devices, personalized skincare could also be achieved through the use of skincare products containing bionanomaterials that have been designed to target specific skin concerns. For example, nanoparticles of retinol could be used to target fine lines and wrinkles, while liposomes containing antioxidants could be used to improve skin texture and reduce the appearance of age spots. The use of bionanomaterials in personalized skincare has the potential to revolutionize the way we approach skincare by allowing us to tailor our routines to our individual needs. However, it is important that these technologies are developed and used responsibly, with safety and efficacy as the top priorities. As such, regulatory bodies must establish clear guidelines and safety standards to ensure the development of safe and effective personalized skincare products.

14.2.7.2 Sustainable and Eco-friendly Materials

The development of sustainable and eco-friendly bionanomaterials for skincare is becoming increasingly important due to the growing demand for effective, environmentally friendly products (Kim et al., 2022). One way to achieve sustainability is by using natural and renewable resources in the production of bionanomaterials. For instance, using plant-based materials like cellulose and chitosan to create biodegradable nanoparticles with minimal environmental impact (Salvioni et al., 2021). Additionally, liposomes made from natural oils and waxes are non-toxic and biodegradable, making them a safe and sustainable alternative to synthetic liposomes (Leonida et al., 2016c). Another approach to sustainable bionanomaterials is the use of biodegradable and compostable packaging materials. Skincare products packaged in plastic containers contribute to the accumulation of plastic waste in the environment, but biodegradable and compostable packaging materials can help reduce this impact.

In addition to the use of sustainable and eco-friendly materials, the skincare industry is also exploring ways to reduce the environmental impact of the production and disposal of bionanomaterials (Kim et al., 2022). For example, some companies are exploring the use of renewable energy sources in the production of bionanomaterials, while others are developing systems to recycle and reuse materials to reduce waste (Li et al., 2021). In conclusion, the development of sustainable and eco-friendly bionanomaterials is an important area of focus for the skincare industry. By using natural and renewable resources, biodegradable packaging materials, and reducing the environmental impact of production and disposal, it is possible to create skincare products that improve skin health without compromising the health of the environment.

14.2.7.3 Combination Therapies

Combination therapies that incorporate bionanomaterials with other skincare technologies are an exciting area of research in skin rejuvenation (Souto et al., 2022). The idea behind combination therapies is to enhance the effects of bionanomaterials by combining them with other technologies that target different aspects of skin aging. One example of a combination therapy is the use of bionanomaterials in conjunction with light therapy. Light therapy involves the use of different wavelengths of light to penetrate the skin and stimulate collagen production, reduce inflammation, and improve skin tone and texture (Houreld, 2019). By combining bionanomaterials with light therapy, it may be possible to enhance the delivery and effectiveness of the bionanomaterials while also enhancing the effects of the light therapy (Chen et al., 2020).

Another example of a combination therapy is the use of bionanomaterials with microneedling. Microneedling involves the use of tiny needles to puncture the skin and create controlled micro-injuries that stimulate collagen production and skin rejuvenation (Sreeharsha and Mazen, 2022). By incorporating bionanomaterials into the microneedling process, it may be possible to improve the penetration of the bionanomaterials and enhance the rejuvenation effects of the microneedling (Vaishampayan and Rane, 2022). In addition to light therapy and microneedling, there are many other technologies that could potentially be combined with bionanomaterials for enhanced skin rejuvenation effects (Leonida et al., 2016d). These include ultrasound therapy, radiofrequency therapy, and even injectable fillers and botulinum toxin.

While the potential benefits of combination therapies are promising, there are still many questions that need to be addressed, including the safety and efficacy of these approaches. Nevertheless, the development of combination therapies that incorporate bionanomaterials is an exciting area of research that has the potential to revolutionize the field of skin rejuvenation.

14.2.7.4 Advanced Delivery Systems

Advanced delivery systems are a crucial area of research in the development of bionanomaterials for skin rejuvenation. The goal of these systems is to enhance the penetration and bioavailability of the bionanomaterials, allowing them to reach specific skin layers or cells for maximum efficacy (Leonida et al., 2016a). One approach to advanced delivery systems is the use of nanocarriers, such as liposomes, nanoparticles, and dendrimers. These carriers can encapsulate bionanomaterials and target them to specific skin layers or cells (Khezri et al., 2018, Salvioni et al., 2021). For example, liposomes can be designed to release their contents in response to specific stimuli, such as changes in temperature or pH (Costa and Santos, 2017). This allows for precise control over the release of the bionanomaterials and can enhance their efficacy while reducing potential side effects.

Another approach to advanced delivery systems is the use of biocompatible materials that can adhere to the skin and provide sustained release of the bionanomaterials over time (Du and Wong, 2019, Chhabra and Bhati, 2021). For example, biocompatible hydrogels can be used to deliver bionanomaterials to the skin, providing sustained release and enhancing their efficacy (Du and Wong, 2019).

In addition to these approaches, there is also ongoing research into the use of physical and chemical enhancers to improve the penetration of bionanomaterials into the skin. These enhancers include micro- and nanoneedles, iontophoresis, and sonophoresis (Patel et al., 2022). By enhancing the penetration of bionanomaterials, these systems could improve their bioavailability and reduce potential side effects.

The development of advanced delivery systems is an exciting area of research in the field of skin rejuvenation. By enhancing the penetration and bioavailability of bionanomaterials, these systems have the potential to improve their efficacy and reduce potential side effects, paving the way for more effective and safer skincare products.

14.3 Summary and Conclusion

In conclusion, bionanomaterials offer a promising avenue for skin rejuvenation, with the potential to improve skin health and reduce the signs of aging. The properties of these materials, including their small size, high surface area, and biocompatibility, make them well-suited for use in skincare products. Liposomes, nanoparticles, dendrimers, and other types of bionanomaterials have unique properties that can be harnessed to improve skin hydration, stimulate collagen production, and reduce inflammation. These materials can also be combined with other skincare technologies, such as light therapy or microneedling, for enhanced effects. However, the safety and regulation of bionanomaterials are important considerations, and regulatory frameworks need to be in place to ensure their safe use. Additionally, the development of sustainable and eco-friendly bionanomaterials is an important area of focus for the skincare industry. Looking to the future, advanced delivery systems and the integration of AI and ML technologies could accelerate the development of new and more effective bionanomaterials, while personalized skincare regimens could be tailored to an individual's unique skin needs. These trends are likely to continue to shape the skincare industry in the years to come.

In summary, bionanomaterials have enormous potential for skin rejuvenation and could lead to the development of more effective, safe, and sustainable skincare products in the future. As research in this field continues to advance, we can expect to see new breakthroughs and innovations that will further revolutionize the field of skincare—particularly, with continued research and development in those areas likely to lead to new and innovative applications and products that enhance skin health and rejuvenation. However, it is important that this research is conducted in a responsible and ethical manner to ensure the safety and efficacy of these materials for consumers.

REFERENCES

ABU HAJLEH, M. N., ABU-HUWAIJ, R., AL-SAMYDAI, A., AL-HALASEH, L. K. & AL-DUJAILI, E. A. 2021. The revolution of cosmeceuticals delivery by using nanotechnology: A narrative review of advantages and side effects. *Journal of Cosmetic Dermatology,* 20**,** 3818–3828.

AGUIRRE-CRUZ, G., LEÓN-LÓPEZ, A., CRUZ-GÓMEZ, V., JIMÉNEZ-ALVARADO, R. & AGUIRRE-ÁLVAREZ, G. 2020. Collagen hydrolysates for skin protection: Oral administration and topical formulation. *Antioxidants,* 9**,** 181.

AHMAD, S. I., AHMAD, R., KHAN, M. S., KANT, R., SHAHID, S., GAUTAM, L., HASAN, G. M. & HASSAN, M. I. 2020. Chitin and its derivatives: Structural properties and biomedical applications. *International Journal of Biological Macromolecules,* 164**,** 526–539.

AHMED, I. A., MIKAIL, M. A., ZAMAKSHSHARI, N. & ABDULLAH, A.-S. H. 2020. Natural anti-aging skincare: Role and potential. *Biogerontology,* 21, 293–310.

AYATOLLAHI, A., FIROOZ, A. & SAMADI, A. 2020. Evaluation of safety and efficacy of booster injections of hyaluronic acid in improving the facial skin quality. *Journal of Cosmetic Dermatology,* 19, 2267–2272.

BRAVO, B., CORREIA, P., GONÇALVES JUNIOR, J. E., SANT'ANNA, B. & KEROB, D. 2022. Benefits of topical hyaluronic acid for skin quality and signs of skin aging: From literature review to clinical evidence. *Dermatologic Therapy*, 35, e15903.

BUKHARI, S. N. A., ROSWANDI, N. L., WAQAS, M., HABIB, H., HUSSAIN, F., KHAN, S., SOHAIL, M., RAMLI, N. A., THU, H. E. & HUSSAIN, Z. 2018. Hyaluronic acid, a promising skin rejuvenating biomedicine: A review of recent updates and pre-clinical and clinical investigations on cosmetic and nutricosmetic effects. *International Journal of Biological Macromolecules,* 120, 1682–1695.

CAO, J., HUANG, D. & PEPPAS, N. A. 2020. Advanced engineered nanoparticulate platforms to address key biological barriers for delivering chemotherapeutic agents to target sites. *Advanced Drug Delivery Reviews,* 167, 170–188.

CATALÁN, J. & NORPPA, H. 2017. Safety aspects of bio-based nanomaterials. *Bioengineering,* 4, 94.

CHEN, J., FAN, T., XIE, Z., ZENG, Q., XUE, P., ZHENG, T., CHEN, Y., LUO, X. & ZHANG, H. 2020. Advances in nanomaterials for photodynamic therapy applications: Status and challenges. *Biomaterials,* 237, 119827.

CHHABRA, P. & BHATI, K. 2021. Bionanomaterials: Advancements in wound healing and tissue regeneration. *Recent Advances in Wound Healing*. IntechOpen.

COSTA, R. & SANTOS, L. 2017. Delivery systems for cosmetics-From manufacturing to the skin of natural antioxidants. *Powder Technology,* 322, 402–416.

DANTI, S., TROMBI, L., FUSCO, A., AZIMI, B., LAZZERI, A., MORGANTI, P., COLTELLI, M.-B. & DONNARUMMA, G. 2019. Chitin nanofibrils and nanolignin as functional agents in skin regeneration. *International Journal of Molecular Sciences,* 20, 2669.

DEVGAN, L., SINGH, P. & DURAIRAJ, K. 2019. Minimally invasive facial cosmetic procedures. *Otolaryngologic Clinics of North America,* 52, 443–459.

DHAWAN, S., SHARMA, P. & NANDA, S. 2020. Cosmetic nanoformulations and their intended use. *Nanocosmetics.* Elsevier.

DIANZANI, C., ZARA, G. P., MAINA, G., PETTAZZONI, P., PIZZIMENTI, S., ROSSI, F., GIGLIOTTI, C. L., CIAMPORCERO, E. S., DAGA, M. & BARRERA, G. 2014. Drug delivery nanoparticles in skin cancers. *BioMed Research International,* 2014, 895986

DU, J. & WONG, K. K. 2019. Nanomaterials for wound healing: Scope and advances. *Theranostic Bionanomaterials.* Elsevier.

DUBEY, S. K., DEY, A., SINGHVI, G., PANDEY, M. M., SINGH, V. & KESHARWANI, P. 2022. Emerging trends of nanotechnology in advanced cosmetics. *Colloids and Surfaces B: Biointerfaces*, 214, 112440.

FARRIS, P., ZEICHNER, J. & BERSON, D. 2016. Efficacy and tolerability of a skin brightening/anti-aging cosmeceutical containing retinol 0.5%, niacinamide, hexylresorcinol, and resveratrol. *Journal of Drugs in Dermatology: JDD,* 15, 863–868.

FIGUEIREDO, C. C. M., DA COSTA GOMES, A., ZIBORDI, L. C., GRANERO, F. O., XIMENES, V. F., PAVAN, N. M., SILVA, L. P., SONVESSO, C. D. S. M., JOB, A. E. & NICOLAU-JUNIOR, N. 2023. Biosynthesis of silver nanoparticles of Tribulus terrestris food supplement and evaluated antioxidant activity and collagenase, elastase and tyrosinase enzyme inhibition: In vitro and in silico approaches. *Food and Bioproducts Processing* 138. https://doi.org/10.1016/j.fbp.2023.01.010

GANCEVICIENE, R., LIAKOU, A. I., THEODORIDIS, A., MAKRANTONAKI, E. & ZOUBOULIS, C. C. 2012. Skin anti-aging strategies. *Dermato-endocrinology,* 4, 308–319.

HANEKE, E. 2019. Adverse effects of fillers. *Dermatologic Therapy,* 32, e12676.

HAPPY, A. A., JAHAN, F. & MOMEN, M. A. 2021. Essential oils: Magical ingredients for skin care. *Journal of Plant Sciences,* 9, 54.

HONEK, J. F. 2013. Bionanotechnology and bionanomaterials: John Honek explains the good things that can come in very small packages. *BMC Biochemistry,* 14, 29.

HOURELD, N. N. 2019. The use of lasers and light sources in skin rejuvenation. *Clinics in Dermatology,* 37, 358–364.

IBER, B. T., KASAN, N. A., TORSABO, D. & OMUWA, J. W. 2022. A review of various sources of chitin and chitosan in nature. *Journal of Renewable Materials,* 10, 1097.

JEEVANANDAM, J., LING, J. K. U., BARHOUM, A., SAN CHAN, Y. & DANQUAH, M. K. 2022. Bionanomaterials: Definitions, sources, types, properties, toxicity, and regulations. *Fundamentals of Bionanomaterials.* Elsevier.

KANT, V., KUMARI, P., JITENDRA, D. K., AHUJA, M. & KUMAR, V. 2021. Nanomaterials of natural bioactive compounds for wound healing: Novel drug delivery approach. *Current Drug Delivery,* 18, 1406–1425.

KEE, L. T., NG, C. Y., AL-MASAWA, M. E., FOO, J. B., HOW, C. W., NG, M. H. & LAW, J. X. 2022. Extracellular vesicles in facial aesthetics: A review. *International Journal of Molecular Sciences,* 23, 6742.

KHAN, W. S., HAMADNEH, N. N. & KHAN, W. A. 2016. Polymer nanocomposites–synthesis techniques, classification and properties. In: *Science and Applications of Tailored Nanostructures,* Springer, 50–67.

KHEZRI, K., SAEEDI, M. & DIZAJ, S. M. 2018. Application of nanoparticles in percutaneous delivery of active ingredients in cosmetic preparations. *Biomedicine & Pharmacotherapy,* 106, 1499–1505.

KIM, J. 2018. Clinical effects on skin texture and hydration of the face using microbotox and microhyaluronicacid. *Plastic and Reconstructive Surgery Global Open,* 6, e1935

KIM, S., OH, T., LEE, H. & NAM, J.-M. 2022. Trends and perspectives in bio-and eco-friendly sustainable nanomaterial delivery systems through biological barriers. *Materials Chemistry Frontiers,* 6, 2152–2174.

LEONIDA, M. D. & KUMAR, I. 2016. *Bionanomaterials for Skin Regeneration*, Springer.

LEONIDA, M. D., KUMAR, I., LEONIDA, M. D. & KUMAR, I. 2016a. Bionanomaterials for the skin: More than just size. In: *Bionanomaterials for Skin Regeneration*, Springer, 1–5.

LEONIDA, M. D., KUMAR, I., LEONIDA, M. D. & KUMAR, I. 2016b. Bionanomaterials with antioxidant effect for skin regeneration. In: *Bionanomaterials for Skin Regeneration*, Springer, 61–67.

LEONIDA, M. D., KUMAR, I., LEONIDA, M. D. & KUMAR, I. 2016c. Nanoparticles, nanomaterials and nanocarriers. In: *Bionanomaterials for Skin Regeneration*, Springer, 37–46.

LEONIDA, M. D., KUMAR, I.,LEONIDA, M. D. & KUMAR, I. 2016d. Transdermal and topical delivery to the skin. In *Bionanomaterials for Skin Regeneration.* Springer, 27–35.

LI, T., CHEN, C., BROZENA, A. H., ZHU, J., XU, L., DRIEMEIER, C., DAI, J., ROJAS, O. J., ISOGAI, A. & WÅGBERG, L. 2021. Developing fibrillated cellulose as a sustainable technological material. *Nature,* 590, 47–56.

LIU, Y. & SHI, J. 2019. Antioxidative nanomaterials and biomedical applications. *Nano Today,* 27, 146–177.

LO, S. & FAUZI, M. B. 2021. Current update of collagen nanomaterials—fabrication, characterisation and its applications: A review. *Pharmaceutics,* 13, 316.

LOHANI, A., VERMA, A., JOSHI, H., YADAV, N. & KARKI, N. 2014. Nanotechnology-based cosmeceuticals. *International Scholarly Research Notices,* 2014, 843687.

LU, P., FANG, S., CHENG, W., HUANG, S., HUANG, M. & CHENG, H. 2018. Characterization of titanium dioxide and zinc oxide nanoparticles in sunscreen powder by comparing different measurement methods. *Journal of Food and Drug Analysis,* 26, 1192–1200.

MAGDA BELMONTESI, M., FRANCESCA DE ANGELIS, M., CARLO DI GREGORIO, M. & IVANO IOZZO, M. 2018. Injectable non-animal stabilized hyaluronic acid as a skin quality booster: An expert panel consensus. *Journal of Drugs in Dermatology,* 17, 83–88.

MILAN, P. B., KARGOZAR, S., JOGHATAIE, M. T. & SAMADIKUCHAKSARAEI, A. 2019. Nanoengineered biomaterials for skin regeneration. *Nanoengineered Biomaterials for Regenerative Medicine.* Elsevier.

NAFISI, S. & MAIBACH, H. 2017. Nanotechnology in cosmetics. In: *Cosmetic Science and Technology: Theoretical Principles and Applications*. Elsevier, 337.

NATH, R., CHAKRABORTY, R., ROY, R., MUKHERJEE, D., NAG, S. & BHATTACHARYA, A. 2021. Nanotechnology based cosmeceuticals. *International Journal of Scientific Research in Science and Technology,* 8, 94–106.

OVADIA, S. A., EFIMENKO, I. V. & LESSARD, A. S. 2021. Dorsal hand rejuvenation: A systematic review of the literature. *Aesthetic Plastic Surgery*, 45, 1804–1825.

PATEL, V., CHESMORE, A., LEGNER, C. M. & PANDEY, S. 2022. Trends in workplace wearable technologies and connected-worker solutions for next-generation occupational safety, health, and productivity. *Advanced Intelligent Systems,* 4, 2100099.

PATHAK, A., MOHAN, R. & ROHRICH, R. J. 2020. Chemical peels: Role of chemical peels in facial rejuvenation today. *Plastic and Reconstructive Surgery,* 145, 58e–66e.

QIU, Y., MA, Y., HUANG, Y., LI, S., XU, H. & SU, E. 2021. Current advances in the biosynthesis of hyaluronic acid with variable molecular weights. *Carbohydrate Polymers,* 269, 118320.

RABAH, H. & ASLAN, S. S. 2021. Hyaluronic acid and related analysis in food supplements. In: *Research & Reviews in Health Sciences-I.* Elsevier, 857.

RASMUSSEN, K., MECH, A. & RAUSCHER, H. 2019. Characterisation of nanomaterials with focus on metrology, nanoreference materials and standardisation. In: Cornier, J., Keck, C., Van de Voorde, M. (eds) *Nanocosmetics From Ideas to Products.* Springer, 233–265.

RATHOD, S., SHINDE, K., SHINDE, N. & ALOORKAR, N. 2021. Cosmeceuticals and nanotechnology in beauty care products. *Research Journal of Topical and Cosmetic Sciences,* 12, 93–101.

RAZAVI, M., ZHU, K. & ZHANG, Y. S. 2017. Naturally based and biologically derived nanobiomaterials. *Nanobiomaterials Science, Development and Evaluation.* Elsevier.

REHMAN, H., ALI, W., KHAN, N. Z., AASIM, M., KHAN, T. & KHAN, A. A. 2023. Delphinium uncinatum mediated biosynthesis of zinc oxide nanoparticles and in-vitro evaluation of their antioxidant, cytotoxic, antimicrobial, anti-diabetic, anti-inflammatory, and anti-aging activities. *Saudi Journal of Biological Sciences,* 30, 103485.

REILLY, D. M. & LOZANO, J. 2021. Skin collagen through the lifestages: Importance for skin health and beauty. *Plastic and Aesthetic Research,* 8, 2.

REZVANI GHOMI, E., NOURBAKHSH, N., AKBARI KENARI, M., ZARE, M. & RAMAKRISHNA, S. 2021. Collagen-based biomaterials for biomedical applications. *Journal of Biomedical Materials Research Part B: Applied Biomaterials,* 109, 1986–1999.

ROBERTS, M., MOHAMMED, Y., PASTORE, M., NAMJOSHI, S., YOUSEF, S., ALINAGHI, A., HARIDASS, I., ABD, E., LEITE-SILVA, V. & BENSON, H. 2017. Topical and cutaneous delivery using nanosystems. *Journal of Controlled Release,* 247, 86–105.

SALVIONI, L., MORELLI, L., OCHOA, E., LABRA, M., FIANDRA, L., PALUGAN, L., PROSPERI, D. & COLOMBO, M. 2021. The emerging role of nanotechnology in skincare. *Advances in Colloid and Interface Science,* 293, 102437.

SANTOS, A. C., RODRIGUES, D., SEQUEIRA, J. A., PEREIRA, I., SIMÕES, A., COSTA, D., PEIXOTO, D., COSTA, G. & VEIGA, F. 2019. Nanotechnological breakthroughs in the development of topical phytocompounds-based formulations. *International Journal of Pharmaceutics,* 572, 118787.

SCHNEIDER, S. 2021. *Characterization of molecular mechanisms involved in intrinsic and extrinsic skin aging of in situ aged normal human dermal fibroblasts.* Doctoral degree Inaugural-Dissertation, Heinrich Heine University Düsseldorf.

SETTHANAKUL, M. N. 2019. *Thesis entitled "Effects of molecular weight of hyaluronic acid on biological activity and skin penetration".* Naresuan University.

SHAH, M., NAWAZ, S., JAN, H., UDDIN, N., ALI, A., ANJUM, S., GIGLIOLI-GUIVARC'H, N., HANO, C. & ABBASI, B. H. 2020. Synthesis of bio-mediated silver nanoparticles from Silybum marianum and their biological and clinical activities. *Materials Science and Engineering: C,* 112, 110889.

SINGH, K. R., NAYAK, V. & SINGH, R. P. 2021. Introduction to bionanomaterials: An overview. In: *Bionanomaterials: Fundamentals and Biomedical Applications.* IOP Publishing, 1–21.

SINGH, R. P. & SINGH, K. R. 2021. *Bionanomaterials: Fundamentals and Biomedical Applications*, IOP Publishing.

SOUTO, E. B., CANO, A., MARTINS-GOMES, C., COUTINHO, T. E., ZIELIŃSKA, A. & SILVA, A. M. 2022. Microemulsions and nanoemulsions in skin drug delivery. *Bioengineering,* 9, 158

SREEHARSHA, N. & MAZEN, A. 2022. Therapeutics of microneedling for skin repair. *Asian Journal of Research in Pharmaceutical Sciences,* 12, 199–204.

SZULC-MUSIOŁ, B. & SARECKA-HUJAR, B. 2021. The use of micro-and nanocarriers for resveratrol delivery into and across the skin in different skin diseases—A literature review. *Pharmaceutics,* 13, 451.

TANGAU, M. J., CHONG, Y. K. & YEONG, K. Y. 2022. Advances in cosmeceutical nanotechnology for hyperpigmentation treatment. *Journal of Nanoparticle Research,* 24, 155.

TORT, S. & KARAKUCUK, A. 2021. Serum type hyaluronic acid formulations: In vitro characterization and Patch Test Study. *FABAD Journal of Pharmaceutical Sciences,* 46, 271–278.

VAISHAMPAYAN, P. & RANE, M. M. 2022. Herbal nanocosmecuticals: A review on cosmeceutical innovation. *Journal of Cosmetic Dermatology,* 21, 5464–5483.

VAN KETS, V. L. 2012. *An Investigation into the Cellular Mechanisms Underlying Photodynamic Rejuvenation in Human Skin.* University of Cape Town.

VAN TRAN, V., MOON, J.-Y. & LEE, Y.-C. 2019. Liposomes for delivery of antioxidants in cosmeceuticals: Challenges and development strategies. *Journal of Controlled Release,* 300, 114–140.

WALKER, K., BASEHORE, B. M., GOYAL, A. & ZITO, P. M. 2021. Hyaluronic acid. *StatPearls [Internet].* StatPearls Publishing.

WANG, J., BAO, Y. & YAO, Y. 2021. Application of bionanomaterials in tumor immune microenvironment therapy. *Journal of Immunology Research,* 2021, 1–10.

XIA, H., TANG, Y., HUANG, R., LIANG, J., MA, S., CHEN, D., FENG, Y., LEI, Y., ZHANG, Q. & YANG, Y. 2022. Nanoliposome use to improve the stability of phenylethyl resorcinol and serve as a skin penetration enhancer for skin whitening. *Coatings,* 12, 362.

ZOUBOULIS, C. C., GANCEVICIENE, R., LIAKOU, A. I., THEODORIDIS, A., ELEWA, R. & MAKRANTONAKI, E. 2019. Aesthetic aspects of skin aging, prevention, and local treatment. *Clinics in Dermatology,* 37, 365–372.

15 Bionanomaterials in Diagnosis and Therapeutic Applications

Vidushi Ahuja, Babita Thakur, and Sukhminderjit Kaur

15.1 INTRODUCTION

Humanity is facing a serious threat from diseases like HIV, cancer, and diabetes, which are brought on by certain viruses, mutations, and extra environmental contaminants. Over the past ten years, there has been a lot of focus on the evolution of bacterial resistance to all classes of conventional antibiotics as well as the sharp rise in bacterial infections. It is highly desirable to develop unique techniques for the early detection and containment of diseases. Using nanoparticles, which are extremely tiny and have larger surface areas than their bulk equivalents, is known as nanotechnology. Nanomaterials are very good at having unique chemical, optical, and thermal properties. As a result, nanomaterials have emerged as promising candidates for several biological applications (Church, 2004). Silicates, non-oxides, and metal oxides are the materials used in the production of nanoparticles. In addition to having the necessary size and solubility, they also have higher reactivity and are more easily able to traverse biological barriers. Microbial synthesis of bionanomaterials has gained increasing attention in recent years as an environmentally friendly and cost-effective alternative to traditional methods of producing nanoparticles. Yeast, fungus, and bacteria are examples of microorganisms that may produce a variety of nanoparticles, including metallic, oxide, and sulphide particles. Microorganisms normally produce these nanoparticles by reducing metal ions in the growth media, followed by their precipitation or nucleation on the surfaces of microbial cells. The resulting bionanomaterials have unique properties and can be easily modified for a variety of applications in biomedicine, catalysis, and environmental remediation. The use of nanotechnology has sparked fresh optimism for finding solutions to the issues facing people today (Etc et al., 2019). Since nanotechnology was introduced as a factor affecting many sectors, the use of nanomaterials has rapidly risen in a range of professions. The use of nanotechnology has benefited the pharmaceutical and medical industries, resulting in the creation of new relevant products for the market. The prevention, diagnosis, and treatment of several illnesses have all benefited from the application of nanotechnology (A. Kumar et al., 2021). Researchers are still working to fully understand how nanoparticles interact with the immune system. Recent research has demonstrated that the nanoparticles' binding to blood proteins may stimulate or repress immune responses. Different immune cells are aware of the proteins' adsorption on these nanoparticles. Additionally, they influence how nanoparticles (NPs) interact with other blood constituents. In addition to improving innate immune responses, nanomaterials increase the immune system's exposure to antigens, which increases the adjuvant's efficacy. Nanomaterials' surface chemistry plays a significant role in determining how biocompatible they are with the immune system. Nanoparticles (NPs), nanotubes, and nanorods are examples of zero-dimensional nanomaterials. Nanofilms, nanolayers, and graphene are examples of one-dimensional nanomaterials. Two methods have been employed in the creation of NPs over a prolonged period of time. The first technique is the breakdown (also known as top-down)

 DOI: 10.1201/9781003432791-18

technique, in which a solid bulk is subjected to an external force that directs its fracture in the direction of smaller nanoparticles (NPs). The build-up process, also known as the bottom-up method, creates NPs starting with liquid atoms based on molecular condensations or atomic changes. (Liu et al., 2017).

Nanomaterials possess unique chemical, physical, and mechanical properties, making them useful for various technical applications, such as next-generation computer chips, high-definition TV phosphors, solar cells, flat-panel displays, cutting devices, pollutant removal, high-energy batteries, magnetic materials, biosensors, and long-lasting batteries. Recent research has focused on how nanoparticles may be used biologically, particularly in the treatment of cancer, the creation of antioxidant drugs, and the creation of nanodrug agents to fight bacteria that have developed drug resistance. In terms of diagnosis and therapy, gold and silver nanoparticles, magnetic metal oxides, carbon nanotubes, quantum dots, and naturally occurring polymeric nanoparticles have all shown promise. Nanoparticles have significant potential in bioimaging for early disease detection, as well as in biosensors and detectors. Nanoparticles have also shown potential as antimicrobial agents and antioxidants, and in cancer therapeutics (El-Abhar et al., 2008).

15.2 BIONANOPARTICLES

In the category of nanoparticles known as "bionanoparticles", substances including proteins, nucleic acids, lipids, and carbohydrates. These nanoscale particles generally have sizes between a few nanometers and a few hundred nanometers, and can be found in nature or created utilizing biological molecules. The unique physical and chemical properties of bionanoparticles arise from their composition, which can vary depending on their source. For example, some bionanoparticles are composed of self-assembling proteins, which can form complex architectures with unique properties such as catalytic activity, magnetic properties, or the ability to bind to specific molecules. Other bionanoparticles, such as liposomes, are composed of lipids, which can self-assemble into bilayer membranes that can encapsulate hydrophilic or hydrophobic molecules. Because they are biocompatible and easy to manipulate to target certain cells or tissues, liposomes have been widely exploited as drug delivery systems.

Another example of bionanoparticles is virus-like particles (VLPs), which are non-infectious particles that resemble viruses in shape and size. VLPs are advantageous as vaccines and medication delivery systems because they may be created to show certain molecules on their surface. VLPs have been investigated as possible vaccinations for a number of viruses, such as the hepatitis B and human papillomavirus, as well as for cancer immunotherapy.

Exosomes are another class of bionanoparticles that are secreted by cells and play a role in intercellular communication. Exosomes are composed of a lipid bilayer membrane that encloses proteins, nucleic acids, and other biomolecules. These particles' capacity to penetrate biological barriers and target certain cells has led to research into them as prospective medication delivery systems. Ferritin is a protein that binds and stores iron, and it can also self-assemble into nanoparticles (Crețu et al., 2021). Due to its biocompatibility and capacity to target certain organs, ferritin nanoparticles have been investigated for application in imaging and medication administration. Bacterial magnetosomes are naturally occurring nanoparticles that are produced by certain types of bacteria. Due to its magnetic characteristics, magnetosomes can be used for MRI and targeted medicine administration, among other things. Overall, because of their distinct characteristics and the possibility to manufacture them with particular functionalities, bionanoparticles offer a wide variety of potential applications in medicine, biotechnology, and materials science.

15.2.1 Bionano Medicine

Numerous areas of biology and medicine, including the identification and management of cancer, might be transformed by nanotechnologies. In recent years, there has been a significant expansion of the application of nanotechnology in medicine for both the treatment and prevention of disease. Millions of people have high hopes for better, more effective, and more reasonably priced healthcare, which potentially provides promising treatments for a wide range of diseases (Warren-Findlow et al., 2012). A branch of nanotechnology known as "nano-medicine" describes targeted medical treatment at the molecular level, these treatments aim to address illnesses or injuries by repairing damaged tissues, including those found in bone, muscle, and nerve tissue, as well as in chronic pulmonary diseases and coronary artery disease. Technologies utilizing nanoparticles have shown particularly great potential for medical applications, from illness diagnosis to the provision of innovative therapeutics. The current nanoparticle-based technologies must overcome several obstacles to be clinically applicable, including the ability to selectively administer nanoparticles, potential concerns regarding cytotoxicity, the capability to image nanoparticles, and the ability to assess their therapeutic effectiveness in real-time. Nanomaterials are also useful for in vitro and in vivo biomedical research and applications because of their size similarity to a number of biological components and structures. Applications for diagnostic tools, contrast agents, analytical tools, physical therapy methods, and drug delivery systems have all been made possible by the combination of nanomaterials with biology. A signal can be produced by modifying the size-dependent features of nanoparticles, especially those related to their optical and magnetic qualities. Attaching a label or probe to the target biomolecule is a vital step in the majority of nanoparticle-based assays since it will provide a detectable signal showing the presence of the target biomolecules (Salata, 2004). Quantum dots and semiconductor nanocrystals possess exceptional light absorption properties, enabling them to function as fluorescent markers for biomolecules. Quantum dots (QDs), carbon nanotubes, nanoshells, gold nanoparticles, and cantilevers are examples of novel nano-diagnostic tools. The most efficient diagnostic nanostructures are quantum dots (QDs), semiconductor nanocrystals with remarkable photostability, single-wavelength excitation, and size-tunable emission (Alharbi and Al-sheikh, 2014).

Based on droplets, technologies are utilized for the direct generation of particles and encapsulating several biological entities for biotechnology and medical use, with droplet sizes ranging from the nano to femtoliter range. The issues facing biomedical engineering today may be overcome by this technology, which also holds promise for innovative medicines and enhanced diagnostics. With the functionalization of particular categories for medical diagnostic applications, nanowires may enable the construction of increasingly complex biosensors with high multispecificity. By adding ligand functionalization, gold nanoparticle surfaces may be modified to bind particular biomarkers (Sohrabi et al., 2020) biomarkers. The most popular techniques include DNA thiol-linking and chemically functionalizing gold nanoparticles for particular protein/antibody binding. These bio-nanoprobes were first used to identify certain DNA sequences, but their use is now being broadened to a variety of clinical diagnostic tests using straightforward, low-cost procedures. Using a mix of polymers, fluorescent dyes, and proteins that are specific to glucose, sensitive responsiveness may be achieved by nanoengineering coated colloids and microcapsules to precisely regulate optical, mechanical, and catalytic characteristics. Several of these nanodevices are commercially available, but some of them need to be modified for the needed use (Wang and Wang, 2014).

15.3 SYNTHESIS OF NANOPARTICLES BY MICROBIAL STRAINS

Establishing a green manufacturing technique that can produce low-toxic nanoparticles is one of nanotechnology's main goals. Although biological processes for creating metal nanoparticles are quick, affordable, and environmentally benign, many researchers are interested in using them to accomplish this goal. The biological creation of nanoparticles is therefore carried out by a diverse spectrum of microorganisms, such as bacteria, fungi, algae, plants, and viruses. (Utilizing their proteins, enzymes, DNA, lipids, carbohydrates, etc.). The size and form of bionanoparticles can potentially be controlled by microorganisms, which are thought of as powerful Environment-friendly nanomanufacturing plants. Table 15.1 provides a summary of the microorganisms utilized

TABLE 15.1
Syncretism of Nanoparticles by Microorganisms

NANOPARTICLES	MICROORGANISM	THERAPEUTIC APPLICATION	REFERENCES
TiO_2	*Lactobacillus* sp.	Antibacterial activity	(Ahmad et al., 2014)
Ag	*Escherichia coli*	Antimicrobial activity	(Karimi et al., 2018)
Ag	*Exiguobacterium aurantiacumm*	Antimicrobial activity	(Cavanaugh et al., 2021)
Ag	*Brevundimonas diminuta*	Antimicrobial activity	(Ishida, 2018)
Au	*Lactobacillus kimchicus* DCY51	Antioxidant activity	(Markus et al., 2016)
Au	*Paracoccus haeundaensis* BC74171	Antioxidant activity and antiproliferative effect	(Prakash et al., 2019)
Au	*Gordonia amicalis*	Antioxidant scavenging activity	(Kumar et al. 2020)
Ag	*Streptomyces albidoflavus* (Marine source)	Acaricidal activity against *Haemaphysalis bispinosa and Rhipicephalus microplus* and *Haemaphysalis bispinosa*	(Karthik et al., 2014)
Ag	*Nocardiopsis* MBRC-11 sp.	Activity against microbes	(Manivasagan et al., 2013)
Cu	Marine endophytic actinomycetes	Antibacterial efficacy	(Palza, 2015)
Ag	*Penicillium* sp.	Antibacterial activity	(Devi et al., 2012)
Ag	*Aspergillus niger*	Antifungal activity	(Surapuram et al., 2014)
Ag	*Cladosporium perangustum*	Anticancer, antioxidant, and nano-toxicological study	(Ahmed et al., 2022)
Au	*Trichoderma harzianum*	Antibacterial activity	(Tripathi et al. 2018)
Au	*Cladosporium* sp.	Anticancer and Antitumor activity	(Munawer et al., 2020)

in the manufacture of nanoparticles and Figure 15.1 illustrates synthesis of nanoparticles by microorganisms (Khan et al., 2019).

15.4 DIAGNOSTIC APPLICATIONS OF BIONANOPARTICLES

Nanoparticles are increasingly being used in diagnostics and as biosensors, typically in conjunction with diagnostic enzymes. Recent research has looked at the usage of biogenic nanoparticles in biosensors and imaging methods like MRI. Gram negative magnetotactic bacteria (MTB) create the contrast materials utilized in MRIs in magnetosomes, endocellular organelles with a lipid bilayer encasing magnetic iron oxide crystals. Compared to artificial nanoparticles, bacterial magnetosomes have higher r2 relaxivity when utilized to target tumor cells that express the HER2 protein. The relaxivity of a contrast agent serves as a proxy for its sensitivity. In situations when two chemicals are equal, a molecule with a higher degree of relaxivity could be able to produce the same amount of contrast with less of the other chemical. A lesser dose may lessen the risk of nanoparticle toxicity. HER2-targeting bacterial magnetosomes administered intravenously improved the contrast of MR signals in orthotopic breast cancer rats (Fikriyah et al., 2020). RGD-peptide, which was depicted through genetic engineering using a different study, represents magnetosomes. By precisely targeting glioma brain tumor cells that overexpressed the v3 integrin, the *Magnetospirillum magneticum AMB-1* strain was identified using MRI imaging. Another group employed the same bacterial strain's magnetic nanoparticles to theranostically treat tumors using photothermal therapy and MRI guidance (Dhandapani et al., 2019). An intriguing study employed magneto-endosymbionts

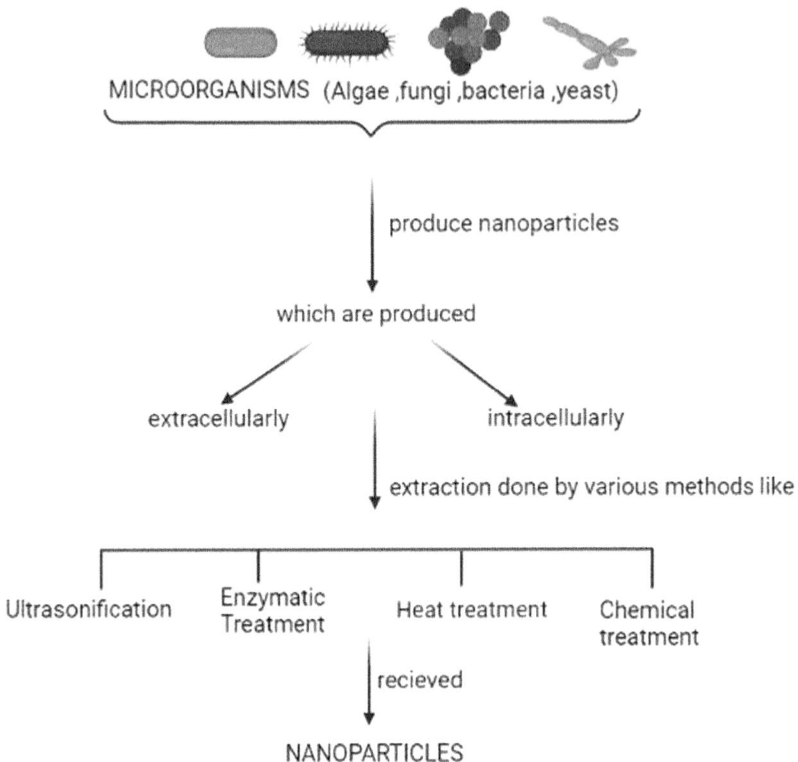

FIGURE 15.1 Mechanistic Illustration of Microorganisms Producing Nanoparticles. (Compiled by the Author.)

in iPSC-derived cardiomyocytes as a living contrast agent that could be seen by MRI and was destroyed after a week, enhancing biocompatibility. The attachment of AuNPs produced from *Candida albicans* to antibodies targeting the cell surface of liver cancer enables precise binding to the unique antigens present on malignant cells. This allows for the differentiation of cancerous cells from healthy ones when investigating liver cells. As a consequence, these nanoparticles might be used in cancer therapy or employed for diagnostic purposes.

15.4.1 Molecular Imaging

Molecular imaging (MI) is a vital area of biomedical research that deals with examining diseases or bodily functions at the level of individual molecules. With great sensitivity and specificity, the imaging methods utilized in MI allow for the straightforward observation, characterization, and quantification of bodily activity. MI includes the use of cutting-edge technologies, including magnetic resonance imaging, X-ray radiography, ultrasound, positron emission tomography, microscopy, and bioluminescence imaging, each with specific capabilities. The investigation and characterization of a variety of brain illnesses, from brain infections to tumors and neurophysiological problems, has shown that MI approaches are helpful. The addition of contrast beacons, sometimes called probes, increases the specificity of MI. Targetable probes are those that make a connection to a particular target, whereas activatable probes work with a target's unique cues to produce a visible signal (Richardson, 2021) signal. According to the previous study, oligopeptide nanoparticles (NPs) may be employed as a probe that can be activated due to their ability to emit fluorescence when driven by the low pH of the tumor microenvironment. Moreover, it has been shown that targetable probes may be carried across the BBB by PS-80 which is coated over PBCA dextran polymeric NPs, making it simpler to identify A plaques in an AD model. According to a recent study, an interaction between sulfated dextran-coated IO NPs and the class is highly expressed. Scavenger receptors can greatly improve the bioimaging of brain inflammation caused by activated microglia. Doped nanoparticles have also been shown to facilitate the emission of infrared light after binding to integrin V3 in fluorescence imaging. By showing how they may be utilized to improve MI by dispersing compounds that act as probes or bioimaging probes, the material above emphasizes the significance of NPs in the early detection of brain illnesses and deep-seated tumours (Hurley et al., 2013).

15.4.2 Radiosensitization

To increase the vulnerability of cancer cells to ionizing radiation and increase the radiotherapeutic ratio, nanoparticle-based radiosensitization is used. This approach makes use of nanoparticles with sizes between 1 and 100 nm. A radiosensitizer is a substance that "radiosensitizes" the body. Therefore, radiosensitizers are chemicals that, when subjected to ionizing radiation, accelerate DNA damage and produce free radicals, making cancer tissue more susceptible to harm or damage. They can be either active therapeutic agents (like drug-based nanoparticles, platinum-based nanoparticles, and drug-encapsulated polymeric NPs) or inactive therapeutic agents (like platinum-based nanoparticles) (like gold nanoparticles that amplify the effects of ionizing radiation) (Subiel et al., 2016)). Therapeutic NPs like cisplatin, oxaliplatin, and carboplatin, can sensitize and kill malignant cells without using ionizing radiation. To make therapeutic nanoparticles (NPs), drugs can be incorporated into high-Z nanoparticles like Gold nanoparticles (GNPs), or linked to them. In this case, the high atomic number NPs (Z NPs) enhance the dosage of ionizing radiation while the drugs act as therapeutic agents. High-Z NPs are capable of producing a sizable amount of Auger electrons, which are crucial for the dose escalation of the radiotherapeutic window due to their interactions with ionizing radiation and the absorption of photoelectrons. Energy is delivered into the tumor in a very small region as a result of the auger electrons' lower energy and shorter range (m). GNPs similar to other high-Z metal-based NPs, are chemically inert on their own but become very effective when they interact with ionizing reagents (Boateng and Ngwa, 2019).

15.4.3 Detection by Biomarkers

An identifiable chemical known as a biomarker is one that is specifically linked to a given ailment or disease. The specificity of the biomarker in identifying the illness stage and its capacity to discriminate between healthy and ill individuals are essential for disease therapy. Using many biomarkers that are identified in brain disorders and illness is challenging in the absence of adequate techniques. NPs have proven to be quite successful at spotting crucial signs of diseases and conditions of the brain. Studies have demonstrated that UCH-L1, a ubiquitin hydrolase that is located at the C-terminus, is much more prevalent in the plasma of patients with traumatic brain injury (TBI) than in healthy persons, suggesting that this protein may function as a biomarker for the illness. A novel approach that makes use of the surface plasmon resonance of Au NPs allows for the rapid and precise detection of UCH-L1 in TBI patients with 100 per cent sensitivity and specificity (Strimbu and Tavel, 2010). Additionally, there is a link between UCH-L1 levels and disorders including dementia and other congenital conditions. According to animal research, modified magnetic NPs can be used to detect Alzheimer's disease (AD)-related pathologies in a safe and reliable manner, whereas anticholesterol antibody-bound magnetic NPs may successfully detect elevated cholesterol levels, a significant reflector of AD. The identification of AD biomarkers including A, inflammatory cytokines, and

Tau proteins is also possible using fluorescent NPs, according to a recent research. Overall, these findings demonstrated that NPs are rapid and efficient methods for locating biomarkers of brain disorders and illnesses (Fernández-Cabada, 2019).

15.4.4 Therapeutic Application of Bionanomaterials

Microbial nanoparticles are used in several biological applications because of their regulated sizes, distinctive characteristics, biocompatibility, and lack of toxicity. As antimicrobials, agents against biofilm, antioxidants, agents against cancer, and agents for diagnostic or imaging, they have found widespread use in the biomedical and pharmaceutical.

15.4.4.1 Antimicrobial Agents

Many metallic nanoparticles, including those made of zinc, copper, gold, magnesium, titanium, and silver, are known to be having antibacterial capabilities. The antibacterial properties of the nanoparticles can be attributed to their capacity to modify membrane structure, create gaps on microbial cell walls, inhibit the development of biofilms, or, in the case of metal oxide nanoparticles, as seen in Figure 15.2, affect membrane reactivity. The nanoparticles' antibacterial effectiveness is greatly influenced by their size and form. The search for novel antimicrobial nanoparticles has been sparked by the growth of the multidrug resistance (MDR) trait among harmful bacteria. The benefit of biologically mediated synthesis is the pre-existence of naturally existing capping or stabilizing agents, such as proteins or polysaccharides, on the nanoparticle surface at the moment of synthesis, which considerably decreases the number of post-production procedures (Guo et al., 2018). The culture of the bacterium *Ochrobactrum rhizosphaerae sp.* was used to create glycolipoprotein-coated AuNPs with strong antibacterial activity against *Vibrio cholerae*. In general, proteinaceous substances are utilized as capping agents to cover nanoparticles made by fungi. Examples of intracellularly generated Ag-NPs from *Schizophyllum commune* and *Trichoderma viride* both

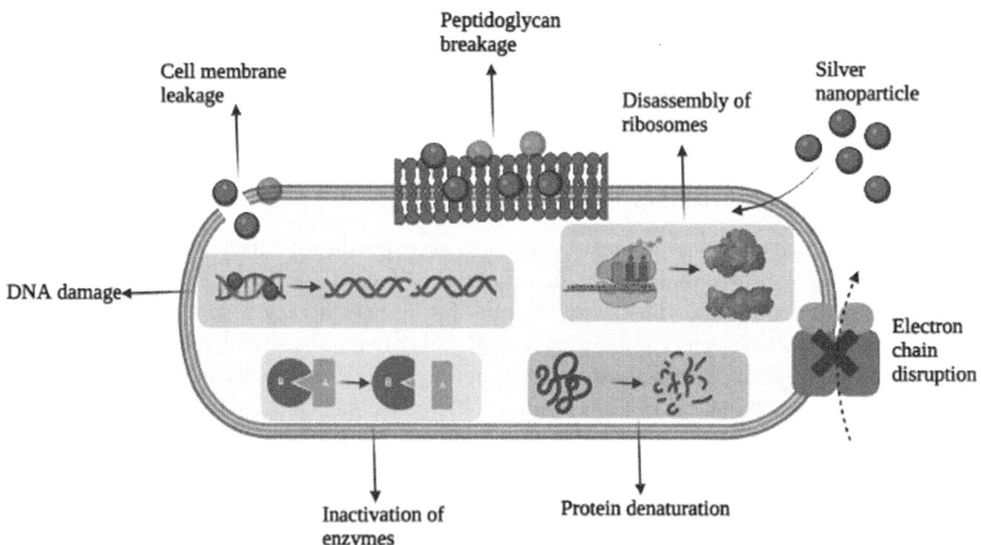

FIGURE 15.2 Antimicrobial Mechanism of Silver Nanoparticles. (Compiled by the Author.)

possess protein capping and have antibacterial activity against strains *of E. coli, Trichophyton rubrum, Pseudomonas sp., Klebsiella planticola, Bacillus subtilis, Trichophyton mentagrophytes, Trichophyton simii, and Trichophyton rubrum*, respectively (Ghosh et al., 2021).

Typically, silver nanoparticles function by releasing Ag+ ions, which can harm bacterial membranes and obstruct the production of DNA and proteins. Gold nanoparticles can be created in combination with photosensitizers for antimicrobial photodynamic treatment due to their photocatalytic capabilities. The bacterial cell wall is destroyed by the heat produced when near-infrared radiation (NIR) is exposed. It has been found that the efficacy of nanoparticles is typically enhanced by the frequent attachment of conventional antibiotic moieties. Vancomycin-linked AuNPs from the fungus *T. viride* prevented the development of vancomycin-resistant *S. aureus* and *E. coli*. It is thought that this is the case because the terminal peptides of the glycopeptidyl precursors were intended to bind to the *S. aureus* transpeptidase rather than the vancomycin-AuNPs (MBA and Nweze, 2021). Numerous medications, including ciprofloxacin, gentamycin, vancomycin, and rifampicin, are applied to NPs because they provide a greater surface area for the pharmaceuticals to attach to and inhibit the development of *S. haemolyticus* and *S. epidermidis*. Given the aforementioned instances, it is intriguing to note that nanoparticles created utilizing microbe extracts that are efficient at eliminating other microbial species and increasing the efficacy of current drugs to combat antimicrobial resistance phenotypes (Abdelghafar et al., 2022).

15.4.4.2 Antibiofilm Agent

A significant obstacle to the creation of new antibiotics and antimicrobials is the prevalence of antibiotic resistance. Bacteria's ability to form biofilms, which render the organism drug-resistant, is a critical aspect of infection and their multidrug-resistant phenotype. As shown in Figure 15.3, inhibiting the formation of biofilms is a key area of investigation in the case of biogenic nanoparticles because it is well known that opportunistic illnesses are caused by microbes like *Staphylococcus aureus, Escherichia coli*, and *Pseudomonas aeruginosa* as a result of biofilm formation. The increased risk of infection transmission associated with biomedical and dental equipment due to the development of biofilm has also been examined, and a feasible solution has been identified in

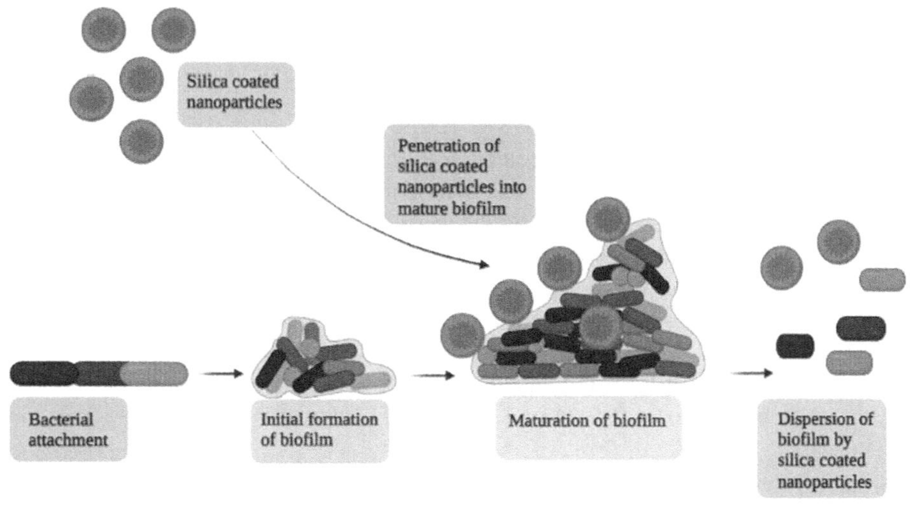

FIGURE 15.3 Antibiofilm Potential of Nanoparticles. (Compiled by the Author.)

the form of nanoparticle coating. Cellular crystal violet staining and absorbance measurements are widely used to evaluate the development of biofilms in the majority of study. Dhandapani et al. (2019) suggested the production of TiO_2 nanoparticles using biomass derived from the bacterium *Bacillus subtilis.* Then, utilizing pond water rich in microbes, nanoparticles, and polychromatic light, biofilm formation in solution or on glass slides was carried out. As a photocatalyst, the TiO_2 nanoparticles released H_2O_2, which stopped the biofilm from expanding.

Silver nanoparticles were 90 per cent effective at preventing the growth of biofilms when separated intracellularly from the biomass of *B. licheniformis.* Additionally, at a concentration of 250 M, gold-silver bimetallic nanoparticles produced biochemically with the aid of the proteobacterium *Shewanella oneidensis* MR-1 showed antimicrobial properties and were effective at preventing the formation of biofilms in *P. aeruginosa, S. aureus, E. coli, and Enterococcus faecalis* cultures. Fungi like *Phanerochaete chrysosporium* may have the ability to remove biofilms, as well. Despite the fact that *E. coli* and *C. albicans* have different cell walls from one another, the extracellular extracts of the fungus were employed to create silver nanoparticles (45 nm in diameter) that could kill the two types of bacteria (Mohanta, 2020).

15.4.4.3 Drug Delivery Agents

Traditional drug delivery methods fall short when compared to biological nanoparticles in terms of stability, biocompatibility, bioavailability, controlled drug release characteristics, targeted dispersion, and non-toxic nature. Liposomes, emulsions, micelles, or water-soluble polymers are some examples of these nano-agents' physical forms. They can encapsulate a specific medicine and release it conditionally at the location of the ailment which is required for the drug carriers as shown in Figure 15.4. To guarantee focused drug administration at the location, the delivery agents also need to be able to flow across cells and through blood tissue barriers. It is crucial to assess their efficiency in cancer cells and their safety for normal cells quite far away. Magnetosomes, bilayer membrane-bound structures that can be used to enclose and transport drugs, are known to be produced by magnetotactic bacteria from magnetic greigite Fe3S4 and/or magnetite Fe3O4. When tested on mice with the H22 tumor, doxorubicin-loaded bacterial magnetosomes showed higher tumor reduction than doxorubicin alone. Immunotherapy has been used to TC-1 animal models using anti-4-1BB agonistic antibodies and magnetosomes from *Magnetospirillum gryphiswaldense.* Applications for the delivery of anti-tumor drugs have included taxol mixed with gold nanoparticles made by the fungus *Humicola sp.* (Das et al., 2016).

15.4.4.4 Anticancer Agents

Anti-cancer drugs have also been produced using pure, drug-free biosynthesized nanoparticles. The cell line MCF-7 for breast cancer and the epidermoid carcinoma A431 may both be effectively treated with platinum nanoparticles made by the yeast *Saccharomyces boulardii.* When examined in vitro on MCF-7 breast cancer cells and HEPG-2 human liver cancer cells, gold nanoparticles made by the bacterium *Streptomyces cyaneus* both exhibited anticancer activity. DNA deterioration, longer cytokinesis delays, and mitochondrial death are just a few of the repercussions that the nanoparticles could have. It has been demonstrated that when MCF-7 cells are exposed to silver nanoparticles made from the aqueous extract of the endophytic fungus *Cladosporium perangustum*, the production of caspase-3, caspase-7, caspase-8, and caspase-9 is increased (Saravanan, 2018).

Terbium oxide nanoparticle application revealed dose-dependent cytotoxicity in MG-63 and Saos-2 cell lines, while it was not damaging to naturally occurring osteoblasts. This was accompanied by an increase in ROS generation and the confirmation of apoptosis with nanoparticle treatment. Additionally, in comparison to normal cells, HT-29 colon cancer cells experienced cell death upon exposure to ZnO nanoparticles that were biosynthesized from *Rhodococcus pyridinivorans* and loaded with anthraquinone. These particles therefore have the potential to act as anti-cancer agents.

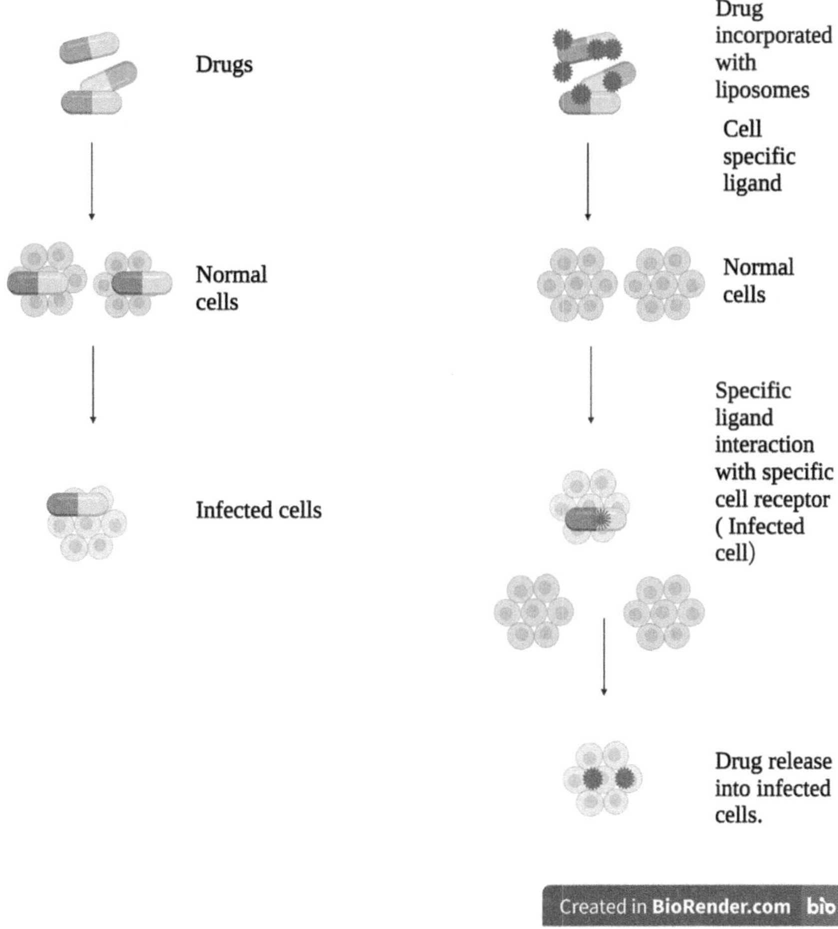

FIGURE 15.4 Bionanomaterials as Drug Delivery Agents. (Compiled by the Author.)

In HEK293 cells, *Helminthosporium solani* AuNPs combined with doxorubicin revealed a higher absorption and similar cytotoxicity in comparison to doxorubicin alone. For anti-cancer applications, taxol or doxorubicin may be conjugated to Humicola sp., which generates nanoparticles that are comparable to those made of gold and gadolinium oxide. In one exciting study, biomineralized magnetic nanoparticles guided by MRI were used to heat-ablate tumour cells without causing any known harm. The theranostic impact of the bacterial nanoparticles in this case was a photothermal effect (Tolliver et al., 2020). Numerous in-vivo studies have demonstrated the capability of bacterial magnetic nanoparticles. Bacterial magnetosomes were administered to BALB/C mice in a separate experiment to investigate.

15.4.4.5 Gene Delivery

Gene therapy is a method for treating numerous genetic illnesses that involves changing, removing, or inserting a gene at specific loci. For the gene to be delivered to the target site, vectors are needed. A virus or a non-virus may have produced the vectors. The viruses that transmit disease include adenoviruses, retroviruses, lentiviruses, and adeno-associated viruses. When using the infection's natural process, viral vectors are quite helpful. To treat and cure illnesses, it is possible to transfer genes to diseased cells or replace a faulty gene with a healthy one using the gene delivery technique. It was originally used as a tool to cure inherited disorders, but it today plays a crucial part in the treatment of diseases like cancer. Nanotechnology has the potential to address some of the constraints of gene therapy. In contrast to the viral vectors currently in use, non-viral vectors like dendrimers and liposomes have been developed through nanotechnology. The negatively charged DNA is contained by the non-viral vectors, which are typically cationic and act through electrostatic interactions (Scheller and Krebsbach, 2009). Given their safety, ease of use, and low cost of production, nanoparticles constitute effective non-viral vector systems. They are less effective in transfecting cells than viral vectors, but with a few modifications, such as the addition of ligands, they might become a potential alternative to viral vectors.

The delivery of genes can be accomplished by dendrimers. They can protect DNA from the DNAse enzyme's action. PAMAM is the dendrimer used for gene transfer most commonly due to its high transfection efficacy. Even further efficiency gains in transfection can be achieved with heat treatment in solvents like water and butanol. The greater flexibility of the dendrimers permits them to become compact when coupled with DNA. The polyamidoamines' surface binds to and absorbs the DNA, while their interior tertiary amino groups enable DNA release into the cytoplasm. There are several advantages to using liposomes as efficient gene delivery systems, which has also been studied. Their size is simply adjustable to accommodate a targeting agent. It is difficult to employ liposomes as gene delivery systems because they are poor at encasing DNA. Cationic liposomes can help to enhance this subpar effectiveness. They consist of positively charged lipid bilayers that interact with the negatively charged DNA through hydrophobic and electrostatic interactions. To improve the efficacy of transfection, these liposomes are mixed with cholesterol and then altered with functional ligands (Wang et al., Yin 2011).

15.5 THE ROLE OF BIONANOMATERIAL IN DISTINCT DISEASES

15.5.1 Brain Tumor

Brain tumors include both cancerous and benign tumor forms that impact the brain. Due to the intricacy of the brain, benign tumor development can potentially have negative effects in addition to metastatic spread. More than 298,000 new instances of brain tumors were reported globally in 2018, according to the data. Cognitive impairment is linked to the development of the illness. The disease's pathophysiology is difficult to categorize, and there is still debate about the best course of action. The therapeutic benefit of NPs in delivering potential anticancer medicines has been noted in several investigations. Yet, several studies have shown that NPs can deliver promising anticancer drugs with therapeutic efficacy. By employing polymeric NPs to carry short interfering RNA that can specifically target the genes sodium-potassium (Na-K)-chloride cotransporter, the ability of glioblastoma cells to grow and spread can be greatly reduced. Gangliclovir and modified polymeric NPs loaded with herpes simplex virus type 1 thymidine kinase dramatically reduce the survival of glioma cells and increase the lifespan of tumor-bearing animals. The aforementioned study suggests that it is advantageous to use NPs to deliver medications and genes to certain types of brain tumours (Huisman, 2009).

15.5.2 Infections

Microbe-based diseases can cause inflammation in the brain or other surrounding tissues and are usually deadly. These diseases include those caused by bacteria, viruses, fungi, and parasites. Viral and bacterial diseases are the two most typical types. These disorders are characterized by oxidative stress, neuronal damage, and acute to chronic inflammation. The ability of nerolidol-loaded nanoparticles (NPs) to restore Na-K ATPase and acetylcholine esterase activities that had been lost in mice due to *Trypanosoma evansi* infection has been demonstrated. Elvitegravir can be enclosed in PLGA nanoparticles to enhance its inhibitory effects on human monocyte-derived microglia-like cells and animal models of HIV-1 infection. Additionally, NPs can increase the bioavailability of anti-HIV medications like darunavir, indinavir, and efavirenz, enhancing their capacity to penetrate the blood-brain barrier and treat viral infections in the brain. The inhibitory effects of cobalt phosphate and hydroxide nanoparticles can diminish the severity of granulomatous amoebic encephalopathy brought on by *Acanthamoeba castellanii* (Li et al., 2000). Ag NPs coupled with anti-diabetic medications such glimepiride, repaglinide, and vildagliptin can considerably lessen encystation and cytotoxicity. Treatment with gold (Au) and zinc oxide (ZnO) NPs can successfully alleviate the oxidant/antioxidant state and brain damage in *Schistosoma mansoni*-infected mice by re-regulating many altered genes. By preventing peptides from aggregating and secreting, Au NPs have been found to lessen neurological defects brought on by herpes simplex virus-1 infection. Furthermore, the treatment of mice models of amoebic meningoencephalitis, encephalitis virus infection, and cerebral TB has showed great promise when using NP-mediated drug delivery. In general, the use of NPs to bacterial illnesses may lessen cytotoxic effects and improve the efficacy of treatments (Ngowi et al., 2021).

15.5.3 Cancer

Cancer is the aggregate name for a set of illnesses characterized by uncontrolled cell growth and invasiveness. Over the past few years, major efforts have been undertaken to uncover a variety of cancer risk factors. Radiation and pollution are two environmental (acquired) factors that have been directly related to the development of some malignancies. But unhealthy lifestyle choices like smoking, eating an imbalanced diet, being stressed, and not exercising much have a big impact on predicting cancer risk. Despite the fact that chemotherapy and radiation treatment might result in cytostasis and cytotoxicity, they are typically linked to substantial side effects and a high risk of recurrence. These drugs have the potential to result in neuropathies, bone marrow suppression, gastrointestinal and skin problems, hair loss, and exhaustion. Cardiotoxicity and pulmonary toxicity brought on by bleomycin and anthracyclines are two drug-specific side effects. Effective cancer treatment will need the creation of a drug or gene delivery system that can specifically target tumour cells while sparing healthy ones (Sudha et al., 2021). This can be achieved by strategically delivering nanoparticles to the tumor microenvironment (TME) and directing them toward cancer cells, thereby improving therapeutic effectiveness and protecting healthy cells from cytotoxicity. Despite the presence of numerous physiological and biological obstacles, nanoformulations can overcome these barriers, including endothelial, epithelial, and cellular membranes, which constitute the mechanical and physiological, chemical barriers, as well as enzymatic barriers. Mechanical, physicochemical, and enzymatic barriers limit the size, biocompatibility, and surface chemistry of NP-based drugs in order to avoid non-targeted usage. These barriers also impose constraints on the treatments' ability to be used. To design NP-based medication targeting, various research has been conducted, with more under development. These nanoparticles must have a number of characteristics in order to be targeted effectively, including stability in the bloodstream until they reach their target, avoidance of reticuloendothelial system clearance, avoidance of the mononuclear phagocyte system, accumulation in the tumour microenvironment via tumour vasculature, high-pressure penetration into the tumour fluid, and the capacity to interact only with tumour cells. Surface functionalization, physicochemical characteristics, and pathophysiological characteristics all play significant roles in regulating the process of NP drug targeting (Gavas et al., 2021).

15.5.4 Dental Therapy

Several cutting-edge and fascinating applications, including those in the medical field, have been made possible by the use of nanomaterials in nanotechnology. NPs are being researched more and more for their potential to cure and prevent oral illnesses and infections. NPs fall into two primary categories: inorganic and organic, despite the fact that there are numerous variations of each kind. In a variety of experimental dental and oral applications, each and every one of these nanomaterials has proven crucial. In order to prevent and cure oral health issues, such as dental caries, these nanoparticles may be employed in resins, metals, ceramics, and other materials used in restorative, prosthetic, endodontic, periodontal, and implantation procedures. The distinct chemical, physical, and biological properties of MNPs have generated a lot of attention (Anjum et al., 2021).

15.5.4.1 Silver Nanoparticles in Dental Therapy

The available MNPs placed a lot of emphasis on the AgNPs' biomedically significant characteristics, notably their antibacterial activity. Silver has been used for many years in dental applications, first as ionic compounds and more recently as NP formulations. The benefits of 38 per cent silver diamine fluoride (SDF) as an antibacterial agent in the fight against caries have been established in several research utilizing ionic silver and fluoride as remineralization, antibacterial, and antimicrobial agents, respectively. Sadly, it was found that the high Ag^+ concentration required to prevent dental cavities had an ugly appearance since Ag precipitated on the surface of the tooth in a dark hue. In recent years, efforts to solve this issue have been attempted; nanotechnology has shown successful in doing so by significantly increasing the surface area while decreasing the size of bulk Ag. Bioactive chemicals can be joined in this way to enhance the antibacterial actions and avoid tooth discoloration. As a result, there is increased interest in using AgNPs because they offer higher bioactivity than Ag^+ due to their unique physical and biochemical properties (Ahmed et al., 2022). Numerous applications, such as optoelectronics, water sanitation, and diagnostics, utilize AgNPs as antimicrobial agents. To cure oral malignancies and prevent tooth decay, AgNPs have been employed in several dental specialties, including periodontology, prosthetics, endodontics, and restoration. Even when using modest quantities of AgNPs, the inclusion of the particles into composite resins has a noticeable antibacterial effect while not affecting the material's intrinsic mechanical and biological qualities. It has also been demonstrated that using AgNPs as a canal irrigant can minimize microleakage in the root canal system as compared to using sodium hypochlorite (NaOCl) as a canal irrigant. AgNP-based irrigation solution was equally effective as NaOCl at eliminating *E. faecalis*. Nevertheless, irrigants containing AgNPs had the additional advantage of removing the smear layer, which the two solutions did not have, despite both being effective in terms of their antibacterial activity (Fernandez et al., 2021).

15.5.5 Autoimmune Diseases

The structural and functional compatibility of certain tissues is threatened by immune infiltration in autoimmune diseases. The nanoparticles, which have also been created to change antigen-presenting cells, downregulate the innate immunity signals that underlie adaptive autoimmune responses (APCs). Pharmacologically, glucocorticoid loaded liposomes were found to be successful in treating experimental autoimmune encephalomyelitis (EAE) at dosages lower than those used in typical glucocorticoid therapy via affecting macrophages. One of the major issues with traditional individual antigen-based approaches for the treatment of autoimmune illnesses is their reliance on specific antigens due to the antigenic complexity of autoimmune diseases and the necessity to target the many characteristics of autoreactive T cells. CD4+ regulatory T cells that are less acidic are promoted by nanoparticles coated with the Major Histocompatibility Complex (pMHC). By using autoantigen-loaded APCs in the target tissue with preference, these nanoparticles also reduce polyclonal autoimmune reactions. Future nano-based medications will be developed with the help of innovative nanoparticle compounds, such as those with various surfaces (Liu et al., 2017).

15.5.6 Nanoparticles in Immune Response

In contrast to innate immunity, which is a non-specific, naturally occurring, non-clonal, germline-encoded, non-clonal, and non-anticipatory mechanism, acquired adaptive immunity is a particular, clonal, somatic, anticipatory system. The size, hydrophobicity, surface charge, hydrophobicity, and coating agents of nanoparticles all have a role in how much they interact with the immune system.

Because of the chemicals that bind to them in specific microenvironments and trigger an inflammatory response, the NPs are recognized as foreign invaders by the innate immune system. The NPs don't interact with innate immune cells directly, save from the compounds decorated on their surface. On the other hand, leukocytes mostly absorb NPs laden with chemotherapy for anticancer treatment. Thus, a lack of innate immune response may exist. Inflammatory responses can be broadened and amplified by delayed adaptive immunity, which depends on the kind and intensity of innate immune responses. The immunogenicity, folding, and distortion of NPs caused by body molecules adhering to their surface lead to the adaptive immune response. The production of NPs modifies the peptides given to T cells by dendritic cells (DCs), which in turn influences the adaptive immunological responses. Rats were given TiO_2 NPs intravenously, which instantly triggered an immunological response that raised IFN-, IL-4, and IL-10 levels over a number of days (Ray et al., 2021).

15.6 THE ROLE OF BIONANOMATERIALS IN VACCINE DEVELOPMENT

The effectiveness of virus-based nanoparticles (VNPs) derived from the hepatitis B virus (HBV) and human papillomavirus (HPV) in humans has spurred the creation of various vaccines utilizing virus-like particles (VLPs) and virus-based nanoparticles (VNPs). The potential for toxicity of manufactured nanoparticles in the human body is one issue with their utilization. Some of the variables that contribute to the detrimental effects of many substances in the human body include the inability of biological membrane coverings, high surface area to volume ratios, low rate of biodegradability, and high reactivity. The ability of several viral capsid subunits to self-assemble in VLPs highlights advances in vaccine design. Additionally, since they may induce certain toll-like receptors (TLRs) to activate the majority of VLPs surround nucleic acids as they are being synthesized. Due to the absence of genetic material, VLPs are superior to conventional subunit vaccinations and are safer. Though the creation of synthetic particles is typically more complex and riskier than the creation of VLPs due to undesirable immunogenicity and conditions for elimination in the body, it is still easier and safer (Verdecia et al., 2021). The biocompatible and biodegradable nanoparticle injections stimulate both local and systemic immune responses when taken orally. By changing the molecular weight and composition of poly (lactic-co-glycolide) copolymers, one of the biodegradable materials, the material may be customized. PLG microspheres are often used to deliver bacterial vaccines, but only a small number of them are being studied to deliver viral vaccines. The two phospholipid layers that make up liposomes are joined with cholesterol to stabilize the synthetic membrane. In addition, liposomes commonly fail.

The micelles known as immune stimulating complexes (ISCOMs) have a range diameter of around 40 nm and are made of a mixture of a mix of phospholipids, Quil A saponins, and cholesterol.

Despite advancements in ISCOMS, challenges similar to those encountered with liposomes persist, limiting its application. In addition, vaccines based on nanoemulsions (NEs) are mucosal adjuvants that may be viable options for the vaccine platform. While there is always an opportunity for advancement in the process of developing innovative vaccines, there is little hope for their widespread application because formulations that work in lab animals may occasionally be expensive to test in human clinical trials. In order to develop a defense against diseases, human research in this area must foresee their onset. Continuous efforts are required to increase vaccines' immunity and efficacy since they are one of the most efficient methods to improve health on a global scale (Gagliardi et al., 2021).

15.7 THE ROLE OF BIONANOMATERIALS IN PHOTOACTIVATED BIOMEDICAL APPLICATIONS

Over the past several decades, a variety of biomaterials and technologies have been developed, opening up new possibilities for therapy and disease diagnosis. In particular, "smart" materials that

TABLE 15.2
Bionanomaterials Are Used in Photodynamic Biomedical Applications

BIONANOMATERIAL	SHELL	MODALITY	APPLICATION	REFERENCES
Silver/gold Gold/Silver	Silica	Photodetection Optical sensing	Labeling cancer cells labeling	(Menon et al., 2013)
Silica	Gold	Fluorescent biological-imaging bioimaging	Fluorescence detection marking of cells	(Prieto-montero et al., 2020)
Magnetic	Silica	Dual(2)- imaging	Cell Targeting, cell sorting, bioimaging, targeting bioimaging	(Hegazy et al., 2020)
CdSe/Gold/CdSe	PEG	Nanoscopic or sub-nanometer sensing	Biological imaging Bioimaging	(Haidar, 2010)
SPONs	Amphiphilic polymer	MRI	Cancer imaging	(Shavandi et al., 2021)
Nano quantum dots	-	signal to brighten	Live imaging	(Walling et al., 2009)

may be influenced by environmental factors like pH, temperature, and light are still sought after in biomedical research. The use of light-based stimulation is one of these external factors that is becoming more and more common in the development of innovative biomaterials and drug delivery systems by the biomaterial systems. As seen in Table 15.2, some of the attractive mass properties of light-responsive materials include the utilization of the near-infrared region of visible light, focus control and high spatial and focus control, the use of least intrusive processes, and the vast choice of materials that may be employed. The unique light absorption properties of the materials enable them to respond to light. Since tissue penetration is likely in the near-infrared region, the bulk of light-responsive materials are plasmonic metal complexes with significant optical characteristics (Sagar and Nair, 2018).

15.8 CONCLUSION AND FUTURE PROSPECTS

Bionanomaterials have emerged as a promising class of therapeutic and diagnostic agents in recent years. They have offered several advantages over traditional drugs and today how bio-nano materials are being used in distinct several-disease diagnoses such as HIV, cancer, and diabetes. The bionanomaterial has emerged as a liable candidate for several biological applications that application, benefitted the pharmaceuticals industries. The development of bionanomaterials holds great promise for improving the diagnosis and treatment of a wide range of illnesses, including cancer, infections, dental treatments, autoimmune diseases, and so forth. They also play a crucial role in the development of vaccines and aid in the induction of an immune response. Bionanomaterial can be used as a delivery system for vaccines. They can protect the vaccine from degradation in the body and can be designed to target specific cells or tissues, enhancing the effectiveness of the vaccine. By more effectively delivering the vaccine antigen to the immune system, nanoparticles can also be employed to boost the immunological response to a vaccination. Bionanomaterials can also be utilized in the manufacturing of vaccines themselves. Using bionanomaterials as adjuvants, which are compounds that improve the body's immunological response to the vaccination, is one approach of producing vaccines. All things considered, bionanomaterials have the potential to enhance the effectiveness, security, and accessibility of vaccinations, making them a crucial weapon in the battle against infectious diseases.

Bionanomaterials with anti-biofilm characteristics include silver nanoparticles, zinc oxide nanoparticles, and chitosan nanoparticles. These nanoparticles can penetrate the EPS matrix of the biofilm and directly target the bacterial cells, disrupting their communication and metabolism.

They can also prevent the formation of biofilms by inhibiting bacterial adhesion and colonization. Furthermore, bionanomaterials can be combined with other antibacterial agents to enhance their efficacy against biofilms. For instance, it has been demonstrated that silver nanoparticles improve the antibiofilm action of antibiotics such gentamicin and tetracycline. Overall, bionanomaterials have enormous promise for the creation of novel antibiofilm medicines, which might result in better alternatives for treating persistent infections and reducing antibiotic resistance in the pharmaceutical sector, among other things.

REFERENCES

Abdelghafar, A., Yousef, N. & Askoura, M. (2022). Zinc oxide nanoparticles reduce biofilm formation, synergize antibiotics action and attenuate *Staphylococcus aureus* virulence in host; An important message to clinicians. *BMC Microbiol* **22**, 244. https://doi.org/10.1186/s12866-022-02658-z

Ahmad, R., Khatoon, N., & Sardar, M. (2014). Antibacterial effect of green synthesized TiO2 nanoparticles. *Advanced Science Letters*, *20*(7–8), 1616–1620.https://doi.org/10.1166/asl.2014.5563

Ahmed, O., Sibuyi, N.R.S., Fadaka, A.O., Madiehe, M.A., Maboza, E., Meyer, M., & Geerts, G. (2022). Plant extract-synthesized silver nanoparticles for application in dental therapy. *Pharmaceutics*, *14*, 380. https://doi.org/10.3390/ pharmaceutics14020380

Alharbi, K. K., & Al-sheikh, Y. A. (2014). Role and implications of nanodiagnostics in the changing trends of clinical diagnosis. *Saudi Journal of Biological Sciences*, *21*(2), 109–117. https://doi.org/10.1016/j.sjbs.2013.11.001

Anjum, S., Ishaque, S., Fatima, H., Farooq, W., Hano, C., Abbasi, B. H., & Anjum, I. (2021). Emerging applications of nanotechnology in healthcare systems: Grand challenges and perspectives. *Pharmaceuticals*, *14*(8), 707. http://dx.doi.org/10.3390/ph14080707

Boateng, F., & Ngwa, W. (2019). Delivery of nanoparticle-based radiosensitizers for radiotherapy applications. *International Journal of Molecular Sciences*, *21*(1), 273. http://dx.doi.org/10.3390/ijms21010273

Cavanaugh, N. T., Parthasarathy, A., Wong, N. H., Steiner, K. K., Chu, J., Adjei, J., & Hudson, A. O. (2021). Exiguobacterium sp . is endowed with antibiotic properties against Gram positive and negative bacteria. *BMC Research Notes*, 1–7. https://doi.org/10.1186/s13104-021-05644-2

Church, D. L. (2004). Major factors affecting the emergence and re-emergence of infectious diseases. *Clinics in Laboratory Medicine*, *24*(3), 559–586. https://doi.org/10.1016/j.cll.2004.05.008

Creţu, B. E.-B., Dodi, G., Shavandi, A., Gardikiotis, I., Şerban, I. L., & Balan, V. (2021). Imaging constructs: The rise of . *Molecules*, *26*(11), 3437. http://dx.doi.org/10.3390/molecules26113437

Das, G., Patra, J. K., Singdevsachan, S. K., Gouda, S., & Shin, H. S. (2016). Diversity of traditional and fermented foods of the Seven Sister states of India and their nutritional and nutraceutical potential: A review. *Frontiers in Life Science*, *9*(4), 292–312. https://doi.org/10.1080/21553769.2016.1249032

Devi, L. S., & Joshi, S. R. (2012). Antimicrobial and synergistic effects of silver nanoparticles synthesized using soil fungi of high altitudes of eastern Himalaya. *Mycobiology*, *40*(1), 27–34. https://doi.org/10.5941/MYCO.2012.40.1.027

Dhandapani, S. S., Manju, D., Sharma, B. S., & Mahapatra, A. K. (2019). Prognostic significance of age in traumatic brain injury. Journal of Neurosciences in Rural Practice, *3*(2), 131–135. https://doi.org/10.4103/0976-3147.98208

El-Abhar, H. S., Hammad, L. N. A., & Gawad, H. S. A. (2008). Modulating effect of ginger extract on rats with ulcerative colitis. *Journal of Ethnopharmacology*, *118*(3), 367–372. https://doi.org/10.1016/j.jep.2008.04.026

Etc, M. C. S., Das C., Lucia MS, H. K., & T. J. (2019). 乳鼠心肌提取 HHS public access. *Physiology & Behavior*, *176*(3), 139–148.

Fernández-Cabada, T., & Ramos-Gómez, M. (2019). A novel contrast agent based on magnetic nanoparticles for cholesterol detection as Alzheimer's disease biomarker. *Nanoscale Research Letters*, *14*(1), 1–6.

Fernandez, C. C., Sokolonski, A. R., Fonseca, M. S., Stanisic, D., Araújo, D. B., Azevedo, V., Portela, R. D., Tasic, L. (2021). Applications of silver nanoparticles in dentistry: Advances and technological innovation. *International Journal of Molecular Sciences, 22*, 2485. https://doi.org/10.3390/ijms22052485

Fikriyah, M., Almurdani, M., Eryanti, Y., Teruna, H. Y. (2020, October). Antidiabetic Activity of Triterpenoids from Anisophyllea disticha. In *Journal of Physics: Conference Series*, *1655*(1), p. 012033. IOP Publishing.

Gagliardi, A., Giuliano, E., Venkateswararao, E., & Fresta, M. (2021). Biodegradable polymeric nanoparticles for drug delivery to solid tumors. *Frontiers in Pharmacology*, *12*, 1–24. https://doi.org/10.3389/fphar.2021.601626

Gavas, S., Quazi, S., & Karpiński, T. M. (2021). Nanoparticles for cancer therapy: Current progress and challenges. *Nanoscale Research Letters*, *16*(1), 173.

Ghosh, S., Ahmad, R., Zeyaullah, M., & Khare, S. K. (2021). Microbial nano-factories: Synthesis and biomedical applications. *Frontiers in Chemistry*, *9*, 626834. https://doi.org/10.3389/fchem.2021.626834

Guo, J., Sun, W., Kim, J. P., Lu, X., Li, Q., Lin, M., Mrowczynski, O., Rizk, E. B., Cheng, J., Qian, G., & Yang, J. (2018). Development of tannin-inspired antimicrobial bioadhesives. *Acta Biomaterialia*, *72*, 35–44. https://doi.org/10.1016/j.actbio.2018.03.008

Haidar, Z. S. (2010). *Bio-Inspired/-Functional Colloidal Core-Shell Polymeric-Based NanoSystems: Technology Promise in Tissue Engineering, Bioimaging and NanoMedicine*. 323–352. https://doi.org/10.3390/polym2030323

Hegazy, G. E., Abu-Serie, M. M., Abo-Elela, G. M., Ghozlan, H., Sabry, S. A., Soliman, N. A., & Abdel-Fattah, Y. R. (2020). In vitro dual (anticancer and antiviral) activity of the carotenoids produced by haloalkaliphilic archaeon Natrialba sp. M6. *Scientific Reports*, *10*(1), 1–14. https://doi.org/10.1038/s41598-020-62663-y

Huisman T. A. (2009). Tumor-like lesions of the brain. *Cancer Imaging: The Official Publication of the International Cancer Imaging Society*, *9 Spec No A*(Special issue A), S10–S13. https://doi.org/10.1102/1470-7330.2009.9003

Hurley, B. R. A., Ouzts, A., Fischer, J., & Gomes, T. (2013). Paper presented AT IAPRI world conference 2012 effects of private and public label packaging on consumer purchase patterns. *Packaging and Technology and Science*, *29*(January), 399–412. https://doi.org/10.1002/pts

Ishida, T. (2018). Antibacterial mechanism of Ag+ ions for bacteriolyses of bacterial cell walls via peptidoglycan autolysins, and DNA damages. MOJ *Toxicology*, *4*, 345–350.

Karimi, S., Azizi, F., Nayeb-Aghaee, M., & Mahmoodnia, L. (2018). The antimicrobial activity of probiotic bacteria Escherichia coli isolated from different natural sources against hemorrhagic E. coli O157: H7. *Electronic physician*, *10*(3), 6548.

Karthik, L., Kumar, G., Kirthi, A. V., Rahuman, A. A., & Bhaskara Rao, K. V. (2014). Streptomyces sp. LK3 mediated synthesis of silver nanoparticles and its biomedical application. *Bioprocess and Biosystems Engineering*, *37*, 261–267. https://doi.org/10.1007/s00449-013-0994-3

Khan, I., Saeed, K., & Khan, I. (2019). Nanoparticles: Properties, applications and toxicities. *Arabian Journal of Chemistry*, *12*(7), 908–931. https://doi.org/10.1016/j.arabjc.2017.05.011

Kumar, A., Choudhary, A., Kaur, H., Mehta, S., & Husen, A. (2021). Metal-based nanoparticles, sensors, and their multifaceted application in food packaging. *Journal of Nanobiotechnology*, *19*(1). https://doi.org/10.1186/s12951-021-00996-0

Kumar, H., Bhardwaj, K., Nepovimova, E., Kuča, K., Dhanjal, D. S., Bhardwaj, S., Bhatia, S. K., Verma, R., & Kumar, D. (2020). Antioxidant functionalized nanoparticles: A combat against oxidative stress. *Nanomaterials (Basel, Switzerland)*, *10*(7), 1334. https://doi.org/10.3390/nano10071334

Li, X., Kolltveit, K. M., Tronstad, L., & Olsen, I. (2000). Systemic diseases caused by oral infection. *Clinical Microbiology Reviews*, *13*(4), 547–558. https://doi.org/10.1128/CMR.13.4.547

Liu, Y., Hardie, J., Zhang, X., & Rotello, V. M. (2017). Effects of engineered nanoparticles on the innate immune system. *Seminars in Immunology*, *34*, 25–32. https://doi.org/10.1016/j.smim.2017.09.011

Manivasagan, P., Venkatesan, J., Senthilkumar, K., Sivakumar, K., & Kim, S. K. (2013). Biosynthesis, antimicrobial and cytotoxic effect of silver nanoparticles using a novel Nocardiopsis sp. MBRC-1. *BioMed Research International*, *2013*, 287638. https://doi.org/10.1155/2013/287638

Markus, J., Mathiyalagan, R., Kim, Y., Abbai, R., Singh, P., Ahn, S., Elizabeth, Z., Perez, J., Hurh, J., & Yang, D. C. (2016). Intracellular Synthesis of Gold Nanoparticles with Antioxidant Activity by Probiotic Lactobacillus kimchicus DCY51T Isolated from Korean Kimchi. In *Enzyme and Microbial Technology*. Elsevier Inc. https://doi.org/10.1016/j.enzmictec.2016.08.018

Mba, I. E., & Nweze, E. I. (2021). Nanoparticles as therapeutic options for treating multidrug - resistant bacteria: Research progress, challenges, and prospects. *World Journal of Microbiology and Biotechnology*, *37*(6), 1–30. https://doi.org/10.1007/s11274-021-03070-x

Menon, J. U., Jadeja, P., Tambe, P., Vu, K., Yuan, B., & Nguyen, K. T. (2013). Theranostics nanomaterials for photo-based diagnostic and therapeutic applications. *Theranostics 3*(3). https://doi.org/10.7150/thno.5327

Mohanta, Y. K., & Mohanta, T. K. (2020). Anti-biofilm and antibacterial activities of silver nanoparticles synthesized by the reducing activity of phytoconstituents present in the Indian medicinal plants. *Frontiers in Microbiology 11*, 1–15. https://doi.org/10.3389/fmicb.2020.01143

Munawer, U., Basavegowda, V., Ningaraju, S., Linganna, K., Raj, A., & Melappa, G. (2020). Biofabrication of gold nanoparticles mediated by the endophytic *Cladosporium species*: Photodegradation, in vitro anticancer activity and in vivo antitumor studies. *International Journal of Pharmaceutics*, *588*, 119729. https://doi.org/10.1016/j.ijpharm.2020.119729

Ngowi, E. E., Wang, Y., & Qian, L. (2021). The application of nanotechnology for the diagnosis and treatment of brain diseases and disorders. *Frontiers in Bioengineering and Biotechnology 9*, 1–19. https://doi.org/10.3389/fbioe.2021.629832

Palza, H. (2015). Antimicrobial polymers with metal nanoparticles. *International Journal of Molecular Sciences*, *16*(1), 2099–2116. https://doi.org/10.3390/ijms16012099

Prakash, M., Kang, M., Niyonizigiye, I., Singh, A., Kim, J., Bae, Y., & Kim, G. (2019). Colloids and surfaces B: Biointerfaces extracellular synthesis of gold nanoparticles using the marine bacterium Paracoccus haeundaensis BC74171 T and evaluation of their antioxidant activity and antiproliferative effect on normal and cancer cell lines. *Colloids and Surfaces B: Biointerfaces*, *183*(June), 110455. https://doi.org/10.1016/j.colsurfb.2019.110455

Prieto-Montero, R., Katsumiti, A., Cajaraville, M. P., López-Arbeloa, I., & Martínez-Martínez, V. (2020). Functionalized fluorescent silica nanoparticles for bioimaging of cancer cells. *Sensors*, *20*(19), 5590.

Ray, P., Haideri, N., Haque, I., Mohammed, O., Chakraborty, S., Banerjee, S., Quadir, M., Brinker, A. E., & Banerjee, S. K. (2021). The impact of nanoparticles on the immune system: A gray zone of nanomedicine. *Journal of Immunological Sciences*, *5*, 19–33.

Richardson, C. N. (2021). *Leveraging Induced Pluripotent Stem Cell Derived Hepatocyte Like Cells to Understand Liver Biology* (Doctoral dissertation, Weill Medical College of Cornell University).

Sagar, V., & Nair, M. (2018). Near-infrared biophotonics-based nanodrug release systems and their potential application for neuro-disorders. *Expert opinion on drug delivery*, *15*(2), 137–152. https://doi.org/10.1080/17425247.2017.1297794

Salata, O. V. (2004). Applications of nanoparticles in biology and medicine. *Journal of Nanobiotechnology*, *2*, 1–6. https://doi.org/10.1186/1477-3155-2-3

Saravanan, M. (2018). *Efficacy of Green Nanoparticles Against Cancerous and Normal Cell Lines: A Systematic Review and Meta-Analysis*. 377–391. https://doi.org/10.1049/iet-nbt.2017.0120

Scheller, E. L., & Krebsbach, P. H. (2009). Gene therapy: Design and prospects for craniofacial regeneration. *Journal of Dental Research*, *88*(7), 585–596. https://doi.org/10.1177/0022034509337480

Shavandi, A., Jafari, H., Zago, E., Hobbi, P., Nie, L., & De Laet, N. (2021). A sustainable solvent based on lactic acid and l-cysteine for the regeneration of keratin from waste wool. *Green chemistry*, *23*(3), 1171-1174.

Sohrabi, S., Kassir, N., & Keshavarz Moraveji, M. (2020). Droplet microfluidics: Fundamentals and its advanced applications. *RSC Advances*, *10*(46), 27560–27574. https://doi.org/10.1039/d0ra04566g

Strimbu, K., & Tavel, J. A. (2010). What are biomarkers?. *Current Opinion in HIV and AIDS*, *5*(6), 463–466. https://doi.org/10.1097/COH.0b013e32833ed177

Subiel, A., Ashmore, R., & Schettino, G. (2016). Theranostics standards and methodologies for characterizing radiobiological impact of high-Z nanoparticles. *Theranostics*, *6*(10). https://doi.org/10.7150/thno.15019

Sudha, P. N., Sangeetha, K., Vijayalakshmi, K., & Barhoum, A. (2021). Chapter 12 – Nanomaterials history, classification, unique properties, production and market. In *Emerging Applications of Nanoparticles and Architecture Nanostructures* (Issue December). Elsevier Inc. https://doi.org/10.1016/B978-0-323-51254-1/00012-9

Surapuram V., Setzer W. N., McFeeters R. L., McFeeters H. (2014). Antifungal activity of plant extracts against Aspergillus niger and Rhizopus stolonifer. *Natural Product Communications*. *9*(11). doi:10.1177/1934578X1400901118

Tolliver, L. M., Holl, N. J., Hou, F. Y. S., Lee, H.-J., Cambre, M. H., & Huang, Y.-W. (2020). Differential cytotoxicity induced by transition metal oxide nanoparticles is a function of cell killing and suppression of cell proliferation. *International Journal of Molecular Sciences*, *21*(5), 1731. http://dx.doi.org/10.3390/ijms21051731

Tripathi, R. M., Shrivastav, B. R., & Shrivastav, A. (2018). Antibacterial and catalytic activity of biogenic gold nanoparticles synthesised by *Trichoderma harzianum*. *IET Nanobiotechnology*, *12*(4), 509–513. https://doi.org/10.1049/iet-nbt.2017.0105

Verdecia, M., Kokai-Kun, J. F., Kibbey, M., Acharya, S., Venema, J., & Atouf, F. (2021). COVID-19 vaccine platforms: Delivering on a promise?. *Human vaccines & immunotherapeutics*, *17*(9), 2873-2893.

Walling, M., Novak, J., & Shepard, J. R. E. (2009). Quantum dots for live cell and in vivo imaging. *International Journal of Molecular Sciences*, *10*(2), 441–491. http://dx.doi.org/10.3390/ijms10020441

Wang, E. C., & Wang, A. Z. (2014). Nanoparticles and their applications in cell and molecular biology. *Integrative Biology: Quantitative Biosciences from Nano to Macro*, *6*(1), 9–26. https://doi.org/10.1039/c3ib40165k

Wang, H. U. I., Shi, H., & Yin, S. (2011). *Polyamidoamine Dendrimers as Gene Delivery Carriers in the Inner Ear: How to Improve Transfection Efficiency*. 777–781. https://doi.org/10.3892/etm.2011.296

Warren-Findlow, J., Seymour, R. B., & Brunner Huber, L. R. (2012). The association between self-efficacy and hypertension self-care activities among African American adults. *Journal of Community Health*, *37*, 15–24. https://doi.org/10.1038/nature08365.

16 Bionanomaterials in Electrochemical Biosensing Applications

Senzekile Majola and Myalowenkosi Sabela

16.1 INTRODUCTION

Bionanomaterials are all nanomaterials created from biological molecules that can be employed for medicinal applications with minimal or no adverse medical effects, excluding medications [1, 2]. Furthermore, they are an ideal substitute for typical nanomaterials in biomedical applications [2]. Biomolecules (proteins, amino acids, enzymes, and metabolites) can be extracted from sources such as microbes [3], plants [4], agricultural waste [5], or marine organisms [6] and then coated, or immobilized, with metallic ions [2]. Figure 16.1 illustrates an experimental technique for employing plant extract to synthesize metallic nanomaterials. Organic and inorganic nanomaterials are the two primary categories of bionanomaterials [1, 7]. Organic bionanomaterials contain carbonaceous components found in living organisms, while inorganic bionanomaterials have fewer carbon elements in combination, except for some fewer carbon materials, such as carbon dioxide, carbonates, carbides, cyanides, carbon monoxide, and cyanates [1, 8].

The focus on bionanomaterials to improve human health has been proven to be essential and it continues to receive attention from researchers. In this chapter, the sources of bionanomaterials and their applications in the electrochemical biosensing of cancer and diabetes biomarkers are discussed. Electrochemical methods like voltammetric approaches are frequently utilized to detect biological entities, due to their simplicity and selectivity [9]. Electrochemical biosensors are essential in today's laboratory and clinical examination of a wide range of chemical and biological targets. The recent development of functional bionanomaterials opens new avenues for improving electrochemical biosensor performance, and they have been proposed for the detection of several disease biomarkers [8, 10, 11]. Biomarkers offer a dynamic and powerful approach to understanding the disease spectrum, with applications in prognosis, epidemiology, screening, and diagnosis and randomised clinical trials [12, 13]. They make it possible to categorize diseases and risk factors uniformly, they can increase our understanding of the underlying pathophysiology of diseases, and they can map the full disease spectrum from the first signs to the end stage.

16.2 ELECTROCHEMICAL BIOSENSORS

Electrochemical sensors consist of a reference electrode and a working electrode separated by an electrolyte, and they use chemical solutions in conjunction with sensors to generate electrical signals proportional to analyte concentration. These sensors offer several benefits that make them desirable diagnostic tools in medicinal applications. They utilize low-cost electrodes that can be quickly and easily integrated with basic electronics for quantitative measurements in the environmental and health sectors where the analyte is in a complex sample [9, 14, 15]. Amperometry current voltammetry (ACV), open-circuit voltage (EOCV), adsorptive stripping voltammetry (AdsSV), square wave

DOI: 10.1201/9781003432791-19

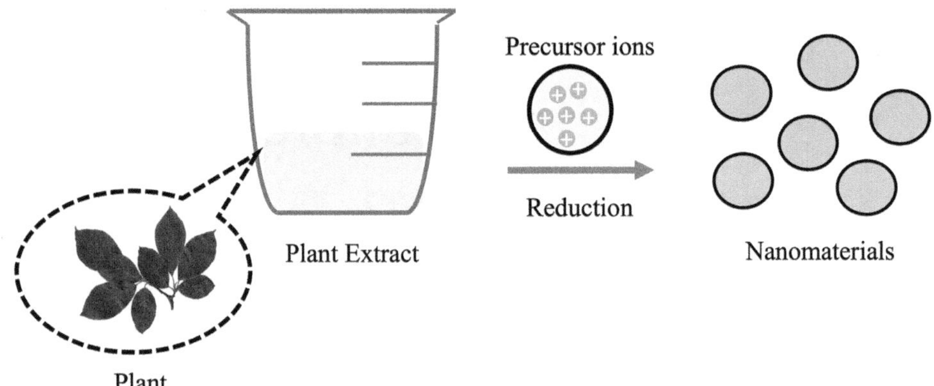

FIGURE 16.1 A Schematic Diagram Describing Green Synthesis of Nanomaterials Using Plant Extract. (Compiled by Author.)

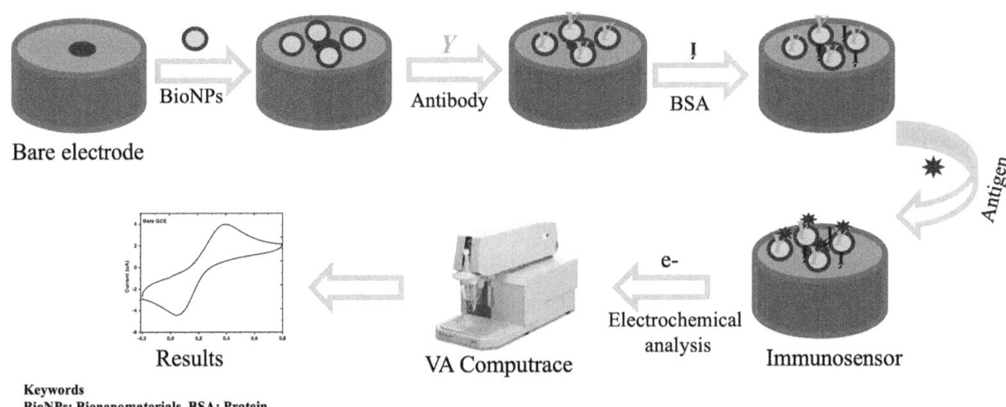

FIGURE 16.2 Schematic Diagram for the Fabrication of an Electrochemical Immunosensor. (Compiled by Author.)

voltammetry (SWV), differential pulse voltammetry (DPV), electrochemical impedance spectroscopy (EIS) and organic electrochemical transistor (OECT) are among the diagnostic techniques used in electrochemical biosensing [16]. For the detection of disease biomarkers, different types of biosensors are used, including enzyme-based biosensors, which are a good example of a basic biosensor [17]. Immunosensors based on antigen-antibody binding reactions and aptasensor based on a DNA or RNA aptamer as the biological recognition element [13, 18]. Figure 16.2. shows a typical electrochemical immunosensor production process.

16.3 BIOMARKER

Biomarkers enable the homogenous classification of diseases and risk factors and can help our fundamental understanding of disease etiology by reflecting the full disease spectrum from early symptoms to terminal stages. This sub-section focuses on biomarkers for cancer and diabetes.

16.3.1 Cancer

Early cancer detection and treatment can greatly cut mortality and save lives. In general, when a disease develops, organs or cells emit certain biomolecules biomarkers into the organism. The discovery of these biomarkers is critical for the early detection and treatment of associated diseases [19]. Onco-biomarkers are among the most effective tools for cancer detection and treatment. They are created in response to cancer or other disorders by malignant or non-cancerous cells. They are created under normal circumstances, but their concentration is substantially higher when malignancies are present. These protein-rich compounds can be detected in bodily fluids such as blood, serum, and urine [20]. Generally, cancer is the most widespread disease worldwide, responsible for millions of deaths every year, and different biomarkers are used to target different types of cancer.

Khodadoust et al. developed an electrochemical biosensor for simultaneous detection of miRNA-141 and miRNA-21 as lung cancer biomarkers [21]. The altered expression of circulating miRNA-21 and miRNA-141 in lung cancer has proved their utility as biomarkers [22]. Takita et al. used prostate-specific antigen (PSA) as a prostate cancer biomarker. Currently, the serum concentration of PSA in the blood is used to make an early diagnosis of prostate cancer [23]. Zhang et al. employed HER2, a receptor from the human epidermal growth factor receptor (EGFR) family, as a biomarker for breast cancer [24]. HER2 is released into the circulation by breast cancer cells, and a high level of it in the bloodstream is one of the indicators of breast cancer [24, 25]. Zhou et al. used circulating tumour cells (CTCs), which are promising liquid biopsy biomarkers for early cancer detection [26].

16.3.2 Diabetes

Diabetes is a chronic disease that occurs when the pancreas does not produce enough insulin or when the body does not utilize the insulin effectively. Type 1 diabetes is defined by a lack of insulin production, whereas Type 2 diabetes is characterized by incorrect insulin utilization in the body and cell resistance, an increase in the level of circulating blood glucose stimulates the release of insulin into the blood [27, 28]. The most often used biomarker for the diagnosis of prediabetes and Type 2 diabetes is hemoglobin A1c (HbA1c). Hemoglobin A1c is generated when glucose binds to the amino terminus of hemoglobin's β-subunit [29], followed by insulin antibodies, which have been shown to be a strong predictor of diabetes development in genetically susceptible individuals [30]. And glucose, an essential clinical indicator in diabetes patients, an insulin can be detected in the blood [31].

Thapa et al. created an electrochemical sensor using a screen-printed carbon electrode to measure blood glucose and HbA1c concentrations from a single sample drop (blood) on a single strip biosensor [32]. Mandali et al. measured HbA1c using electrochemical and point-of-care analyzers [33]. Singh and co-workers used a highly sensitive electrochemical sensor incorporating quantum dots to detect glucose in blood samples as a biomarker for diabetes [34]. Employing electrochemical techniques, insulin biomarker has been shown to be quantifiable in the serum of patients with Type 1 and Type 2 diabetes [35]. An electrochemical sensor constructed with a gold electrode modified with molecularly imprinted polymer (MIP) and carboxylate multiwalled carbon nanotubes (f-MWCNTs) was used to detect insulin in blood serum samples by Wardani and co-workers [35]. Wang et al. also identified an insulin antibody and a glutamate decarboxylase antibody for the early detection of Type 1 diabetes using an electrochemical technique [36]. Glutamic acid decarboxylase (GAD) catalyses the conversion of glutamic acid to gamma-aminobutyric acid in pancreatic islet cells. Autoantibodies against GAD are found in patients with Type 1 diabetes mellitus [36].

16.4 BIONANOMATERIALS, SOURCES, AND THEIR APPLICATIONS: ELECTROCHEMICAL BIOSENSING APPLICATIONS

According to their origin, nanomaterial sources are divided into three categories: (a) man-made nanomaterials with tailored properties for specific applications; (b) nanomaterials that are generated inadvertently as a by-product of an industrial process which can be identified as nanoparticles in perspiration fumes such as vehicle engine exhaust, combustion, and forest fires; (c) nanomaterials produced naturally by microorganisms [2, 4]. This section place its emphasis on category (a) nanomaterials that are specifically derived from plant extracts, proteins, and microorganisms.

16.4.1 Plant Extract-Based Nanomaterials

Plant extracts typically contain macromolecules, phytochemicals, and compounds that serve as an important source of bioactive compounds for a variety of applications, including the synthesis of nanomaterials [37, 38] and have attracted considerable interest for the synthesis of bionanomaterials due to the abundance of plant sources in nature. Recent research demonstrates that extracts derived from diverse plant tissues (leaves, foliage, seeds, fruits, roots, and tubers) can be utilized effectively to create stable nanomaterials with intriguing physicochemical and biological properties [39]. Phytochemicals present in plants, such as amino acids, tannins, flavonoids, proteins, carbonyls, alkaloids, vitamins, polysaccharides, phenols, and ketones are used as a source of reducing agents to convert the metal ions into atoms, thereby facilitating the biosynthesis of metal nanoparticles [40, 41].

Onco-biomarkers are one of the most effective methods for early cancer diagnosis and treatment, and an elevated level may indicate the presence of cancer [42]. Tran et al. developed an ultrasensitive electrochemical biosensor for the detection of breast cancer cells by depositing phytohemagglutinin-L (PHA-L) and nitrogen-doped spherical graphene quantum dots (NGQDs) onto the electrode surface [43]. *Passiflora edulis* juice extract was used as a green carbon source to create NGQDs with a particle size of 3 to 12 nm and then conjugated with PHA-L. The NGQDs enhanced the electrical conductivity of the electrochemical sensor and serve as PHA-L nanofillers. The NGQDs/PHA-L complex provided an outstanding sensor probe for early detection of breast cancer at low concentrations ($5 - 5\times10^3$ cellsmL^{-1}) of tumour cells [43]. Darvishi et al. synthesized gold (AuNPs) and silver nanoparticles (AgNPs) using Calendula officinalis L extract, which were spherical in shape with an average hydrodynamic size of 74.4 nm and 116.6 nm, respectively. From this, electrochemical immunosensor for detection of prostate cancer biomarker (PSA) was developed with high stability and reproducibility [44].

Diabetes is a chronic disease that affects the entire world and is one of the most rapidly expanding global health crises of the twenty-first century. As a result, many scientists are concentrating on developing inexpensive alternatives utilizing bionanomaterials to detect their biomarker for early diagnosis. It is another disease that has received attention in the plant-based biosensor research. Using the *Zingiber officinale* root, Dönmez et al. developed an amperometric glucose biosensor based on the green synthesis of spherical zinc oxide nanoparticles (ZnONPs) in the range of approximately 17– 40 nm. The incorporation of ZnONPs increased the electrocatalytic activity of the immobilized enzyme and enhanced its efficacy. In the presence of interfering substances such as uric acid and ascorbic acid, the biosensor exhibited adequate anti-interference properties [45]. Table 16.1 displays additional bionanomaterials derived from various plant extracts used in the development of electrochemical biosensors for disease biomarkers, including but not limited to diabetes and cancer.

16.4.2 Microorganisms-Based Nanomaterials

Microorganisms (fungi, bacteria, microalgae, and yeast) are found everywhere, including in the air, soil, and water, and are too small to be seen with the naked eye. Furthermore, some bacteria are hazardous to people and can cause illnesses, whereas others are necessary for human health [2, 20]. The synthesis of bionanomaterials using microorganisms has sparked tremendous interest, particularly for anticancer, medical diagnostic, antibacterial, and biosensor applications. The creation of bionanomaterials by microorganisms is broadly classified as intracellular or extracellular (Figure 16.3). Metal nanoparticle formation in bacteria is often triggered by the internalization of metal ions, which are deposited on the cell surface (extracellular method) or occur within the microbial cells (intracellular approach) [57, 58]. The obtained nanomaterials have specific morphology, dimensions, and chemical composition, are easy to handle and cultivate microbiological cells, can be propagated, and have great adaptability to different environmental conditions [2, 18].

TABLE 16.1
Bionanomaterials from Different Plant Extracts and Their Biomarker

Extract	Nanomaterial	Biomarker	Reference
Allium sativum	rGO-AgNPs	Cancer	[46]
E. tereticornis	AuNPs	Lung cancer	[47]
S. mukorossi	rGO/Au	Liver disease	[48]
Crocus sativus L	AgNPs	Maple syrup urine disease	[49]
Prunus persica	ZnONPs	Diabetes	[50]
Fragaria× ananassa	ZnONPs	Tuberculosis	[51]
Pyrus	AgNPs	Tuberculosis	[51]
Bambusa	AgNPs	Diabetes	[52]
Camellia Sinensis	PtNPs	Diabetes	[53]
Amaranthus spinosus	AgNPs	Diabetes	[54]
Aconitum heterophyllum	rGO	Diabetes	[55]
Citrus × aurantiifolia	rGO/TiO_2	Diabetes	[56]

Note: rGO-AgNPs:reduced graphene oxide decorated with silver nanoparticles, rGO/Au: reduced graphene oxide gold nanocomposite, AuNPs: Gold nanoparticles, AgNPs: Silver nanoparticles, ZnONPs: Zinc oxide nanoparticles. rGO/TiO_2: reduced graphene oxide titanium oxide nanocomposite.

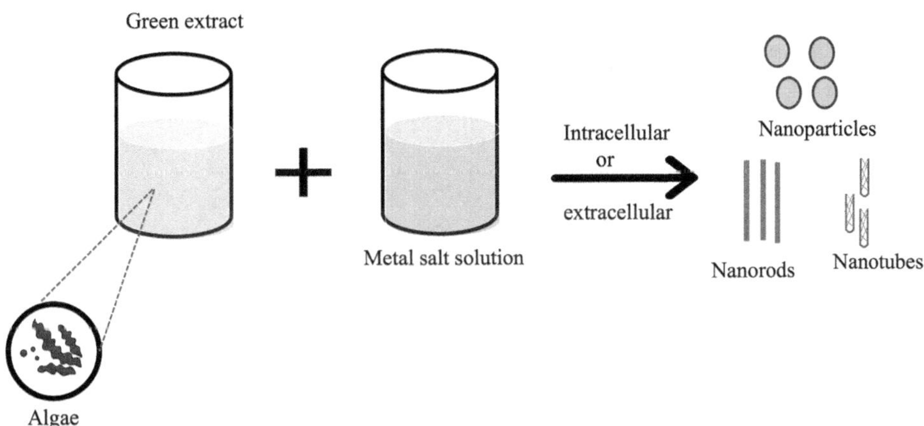

FIGURE 16.3 Schematic Representation of the Synthesis of Different Bionanomaterials from Algae. (Compiled by Author.)

Electrochemical sensing with microbial-based nanomaterials has advanced significantly. Mazloum-Ardakani et al., for example, used *saccharomyces cerevisiae* yeast to manufacture gold nanoparticles, which they then combined with graphene oxide (GO) nanosheets to create an electrochemical biosensor to detect acute lymphoblastic leukaemia, a malignancy that begins in lymphoid cells [59]. Using the identical *saccharomyces cerevisiae* yeast strains, Mazloum-Ardakani et al. synthesized Cu_2O-carbon dots and gold nanoparticles nanocomposite for detecting acute carcinoembryonic antigen, an aptasensor based on electrosynthesized conducting nanocomposite [18]. Both sensors are anticipated to have applications for the simple, rapid, and quantitative detection of cancer biomarkers. Green *algae* are organisms characterized by chlorophylls that have greater biological activity and availability and grow much more rapidly; scientists are also interested in them for bionanomaterial synthesis. Elgamouz et al. reported the fabrication of spherical with the average size of 4.5 nm AgNPs coated with extract of green *algae* (*noctiluca scintillans*) for hydrogen peroxide biosensing as a potential cancer biomarker. The green *algae* self-reduced silver ions to AgNPs without hazardous reducing chemicals. Furthermore, the decomposition of H_2O_2 on the catalytic surface of *Algae*-AgNPs demonstrated their ability to detect reactive oxygen species, such as H_2O_2, in bodily fluids like urine and blood [60].

16.4.3 Protein-Based Nanomaterial

Proteins, such as elastin (Figure 16.4a), silk, resilin, keratin, and collagen, are essential to all biological systems and can be derived from plant or animal sources. These proteins are good building blocks with well-defined binding characteristics because their side chains of amino acids are chemically distinct [2]. Nanoparticles and nanofibers are common classifications for protein-based nanomaterials, which have numerous biomedical applications, including biosensing, diagnostics and drug delivery [61, 62]. Proteins such as whey protein, milk protein, soy protein, legumin (Figure 16.4b), gelatin, albumin, gliadin (Figure 16.4c), elastin, and zein (Figure 16.4d) can be used to synthesize nanomaterials, using ultrasonication, desolvation, emulsification, complex coacervation, and electrospray [63]. Figure 16.5 shows a schematic representation of a protein-based nanoparticle produced by ultrasonication. Shamsipur et al. synthesized stable graphene nanosheets from graphite powder using sonication in a hemoglobin-capped gold nanoclusters (Hb@AuNCs) solution, and then used an electrochemical technique to form a nanocomposite for biosensing chronic myelogenous leukaemia (CML). The Hb@AuNCs were utilized as graphene stabilizers via non-covalent bonding and as dispersants for graphene multilayer nanosheets [64].

Tavakkol et al. described a nanobiocomposite made of zein nanoparticles and multiwalled carbon nanotubes that have been disseminated in water/ethanol and drop casted onto a glassy carbon electrode (GCE). The modified electrode was used as a sensor to identify HepG2 cancer cells. According to electrochemical studies, zein nanoparticles improve the catalytic activity and the conductivity of

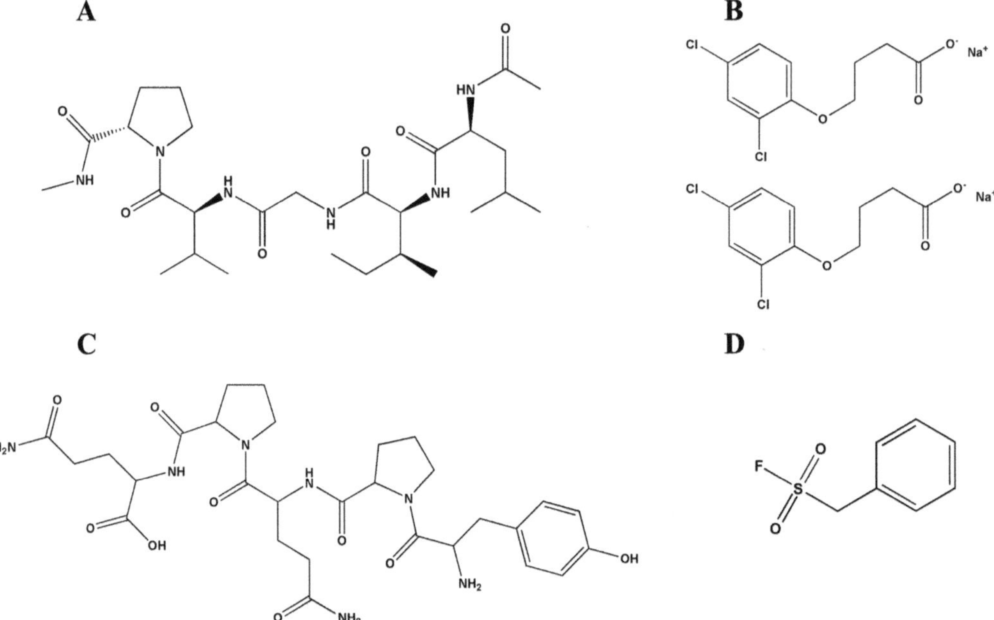

FIGURE 16.4 A Chemical Structure of (A) Elastin, (B) Legumin, (C) Gliadin, (D) Zein. (Generated from Chemdraw.)

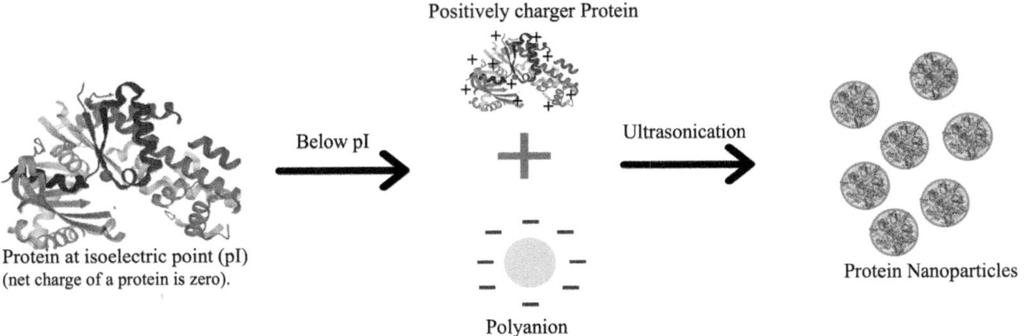

FIGURE 16.5 Schematic Representation of Biosynthesis Protein-Based Nanoparticles via Ultrasonication. (Compiled by Author.)

MWCNTs, allowing for more selectivity and sensitivity assessments of cancer cells (HepG2) [65]. Impedimetric prostate cancer aptasensor based on a glassy carbon electrode coated with titanium oxide nanoparticles (TiO_2NPs) and silk fibroin nanofibers, reported Benvidi et al., TiO_2NPs, and a silk fibroin nanofiber (SF) composite were used to modify the GCE. By using EIS, the electrochemical aptasensor reveals a reasonable dynamic range and a low detection limit for target prostate cancer under optimal conditions [66]. Rasitanon et al. synthesized AuNPs from egg white, a food-based substance rich in protein, to develop a biosensor for the electrochemical detection of glucoses. The AuNPs enhanced the transfer of electrons between the electrode and the redox centre, exhibited high sensitivity, and were able to increase the stability by more than 85 per cent [67].

16.5 CHALLENGES OF BIONANOMATERIALS

The primary challenge of electrochemical biosensors is the lack of suitable electrodes with unique recognition and high sensitivity, and the use of bionanomaterials is an appropriate solution for increasing sensitivity [68]. We have highlighted the use of green synthesized nanomaterials in electrochemical biosensing in this chapter. Green synthesis has numerous advantages over chemical and physical processes, including the fact that it is non-toxic, pollution-free, ecologically friendly, inexpensive, and more sustainable. However, there are problems with raw material extraction, reaction times, and product quality, and the raw materials are not widely available. For instance, seasonal and geographical availability affects the availability of plant sources [69]. The incorporation of bionanomaterial-based receptors into transducers without reducing or altering their activity is one of the most difficult aspects of biosensor development. Typically, bionanomaterials-based electrochemical biosensors are produced by modifying the surface of metal (Ag and Au) and carbon (graphite, graphene, and nanotubes) electrodes with bio-material. The metallic electrodes have more applications due to their stability against oxidation and lack of toxicity, but some are toxic for internal routes, such as AgNPs [70] . The main issue with bionanomaterial-based sensors is their inapplicability to biological samples due to device-to-device variation. Nanomaterials continue to pose a challenge for their integration into lab-on-a-chip systems and the elimination of matrix interference in analytical performance [71].

16.6 CONCLUSION

Recent breakthroughs in nanoscience have resulted in the invention of various new bionanomaterials, which are being used as alternatives to typical nanomaterials in a variety of biomedical applications, such as biosensors for disease biomarkers. Proteins, plant extracts, and microorganisms have successfully demonstrated their ability to synthesize various types of bionanomaterials that are stable, improve the electrical conductivity of the electrochemical sensor, and serve as the biomarker's nanocargo. Electrochemical biosensors made from these bionanomaterials, on the other hand, have an outstanding sensor probe that can detect small concentrations of biomarker cells electrochemically for early diagnosis and treatment. Additional research in this field will lead to the development of novel bionanomaterials and the discovery of more comprehensive biomarkers for use in biosensing.

REFERENCES

[1] Shojaei T R, Soltani S, Derakhshani M (2022) Fundamentals of bionanomaterials. In: Barhoum A, Jeevanandam J, Danquah M K (eds). *6 – Synthesis, properties, and biomedical applications of inorganic Bionanomaterials*: Elsevier. p. 139–174. https://doi.org/10.1016/B978-0-12-824 147-9.00006-6

[2] Jeevanandam J, Ling J K U, Barhoum A et al. (2022) Fundamentals of bionanomaterials. In: Barhoum A, Danquah M K (eds). *1 – Bionanomaterials: Definitions, sources, types, properties, toxicity, and regulations*: Elsevier. p. 1–29. https://doi.org/10.1016/B978-0-12-824147-9.00001-7

[3] Khalil A T, Ovais M, Iqbal J et al. (2022) Microbes-mediated synthesis strategies of metal nanoparticles and their potential role in cancer therapeutics *Seminars in Cancer Biology*, 86: 693–705. https://doi.org/10.1016/j.semcancer.2021.06.006

[4] Sharma S K, Sharma P R, Johnson K I et al. (2022) Separation science and technology. In: Ahuja S (ed). *6 – Plant-derived carboxycellulose: Highly efficient bionanomaterials for removal of toxic lead from contaminated water*: Academic Press. p. 87–95. https://doi.org/10.1016/B978-0-323-90763-7.00004-4

[5] Zamani A, Marjani A P, Mousavi Z (2019) Agricultural waste biomass-assisted nanostructures: Synthesis and application. *Green Processing and Synthesis*, 8(1): 421–429. https://doi.org/10.1515/gps-2019-0010

[6] Jeong G-J, Khan S, Tabassum N et al. (2022) Marine-bioinspired nanoparticles as potential drugs for multiple biological roles. *Marine Drugs*, 20(8): 527. https://doi.org/10.3390/md20080527

[7] Maduraiveeran G (2020) Bionanomaterial-based electrochemical biosensing platforms for biomedical applications. *Analytical Methods*, 12(13): 1688–1701. https://doi.org/10.1039/D0AY00171F

[8] Wang L, Xiong Q, Xiao F et al. (2017) 2D nanomaterials based electrochemical biosensors for cancer diagnosis. *Biosensors and Bioelectronics*, 89: 136–151. https://doi.org/10.1016/j.bios.2016.06.011

[9] Asadpour F, Mazloum-Ardakani M, Hoseynidokht F et al. (2021) In situ monitoring of gating approach on mesoporous silica nanoparticles thin-film generated by the EASA method for electrochemical detection of insulin. *Biosensors and Bioelectronics*, 180(113124): 1–7. https://doi.org/10.1016/j.bios.20221.113124

[10] Li S, Al-Misned F A, El-Serehy H A et al. (2021) Green synthesis of gold nanoparticles using aqueous extract of Mentha Longifolia leaf and investigation of its anti-human breast carcinoma properties in the in vitro condition *Arabian Journal of Chemistry*, 14(2): 102931. https://doi.org/10.1016/j.arabjc.2020.102931

[11] Cruz A G, Haq I, Cowen T et al. (2020) Design and fabrication of a smart sensor using in silico epitope mapping and electro-responsive imprinted polymer nanoparticles for determination of insulin levels in human plasma. *Biosensors and Bioelectronics*, 169(112536): 1–7. https://doi.org/10.1016/j.bios.2020.112536

[12] Mojsoska B, Larsen S, Olsen D A et al. (2021) Rapid SARS-CoV-2 detection using electrochemical immunosensor. *Sensors*, 21(2): 1–11. https://doi.org/10.3390/s21020390

[13] Rahmati Z, Roushani M, Hosseini H et al. (2021) Electrochemical immunosensor with Cu_2O nanocube coating for detection of SARS-CoV-2 spike protein. *Microchimica Acta*, 188(3): 1–9. https://doi.org/10.1007/s00604-021-04762-9

[14] Chupradit S, Km Nasution M, Rahman H S et al. (2022) Various types of electrochemical biosensors for leukemia detection and therapeutic approaches. *Analytical Biochemistry*, 654: 114736. https://doi.org/10.1016/j.ab.2022.114736

[15] Anusha T, Bhavani K S, Kumar J S et al. (2022) Fabrication of electrochemical immunosensor based on GCN-β-CD/Au nanocomposite for the monitoring of vitamin D deficiency. *Bioelectrochemistry*, 143(107935): 1–16. https://doi.org/10.1016/j.bioelechem.2021.107935

[16] Negahdary M, Angnes L (2022) Application of electrochemical biosensors for the detection of microRNAs (miRNAs) related to cancer Coordination. *Chemistry Reviews*, 464: 214565. https://doi.org/10.1016/j.ccr.2022.214565

[17] Geetha Bai R, Muthoosamy K, Tuvikene R et al. (2021) Highly sensitive electrochemical biosensor using folic acid-modified reduced graphene oxide for the detection of cancer biomarker. *Nanomaterials*, 11(5): 1272. https://doi.org/10.3390/nano11051272

[18] Mazloum-Ardakani M, Barazesh B, Moshtaghioun S M (2019) An aptasensor based on electrosynthesized conducting polymers, Cu_2O–carbon dots and biosynthesized gold nanoparticles, for monitoring carcinoembryonic antigen. *Journal of Nanostructures*, 9(4): 659–668. https://doi.org/10.22052/JNS.2019.04.008

[19] Chen M, Song Z, Yang X et al. (2022) Antifouling peptides combined with recognizing DNA probes for ultralow fouling electrochemical detection of cancer biomarkers in human bodily fluids. *Biosensors and Bioelectronics*, 206: 114162. https://doi.org/10.1016/j.bios.2022.114162

[20] Gomaa E Z (2022) Microbial mediated synthesis of zinc oxide nanoparticles, characterization and multifaceted applications. *Journal of Inorganic and Organometallic Polymers and Materials*, 32(11): 4114–4132. https://doi.org/10.1007/s10904-022-02406-w

[21] Khodadoust A, Nasirizadeh N, Seyfati S M et al. (2023) High-performance strategy for the construction of electrochemical biosensor for simultaneous detection of miRNA-141 and miRNA-21 as lung cancer biomarkers. *Talanta*, 252: 123863. https://doi.org/10.1016/j.talanta.2022.123863

[22] Mohammadniaei M, Koyappayil A, Sun Y et al. (2020) Gold nanoparticle/MXene for multiple and sensitive detection of oncomiRs based on synergetic signal amplification. *Biosensors and Bioelectronics*, 159: 112208. https://doi.org/10.1016/j.bios.2020.112208

[23] Takita S, Nabok A, Lishchuk A et al. (2023) Enhanced performance electrochemical biosensor for detection of prostate cancer biomarker PCA3 using specific aptamer. *Engineering*, 4(1): 367–379. https://doi.org/10.3390/eng4010022

[24] Xie S, Wang Y, Gong Z et al. (2022) Liquid biopsy and tissue biopsy comparison with digital PCR and IHC/FISH for HER2 amplification detection in breast cancer patients. *Journal of Cancer*, 13(3): 744. https://doi.org/10.7150%2Fjca.66567

[25] Zhang Y, Li N, Xu Y et al. (2023) A novel electrochemical biosensor based on AMNFs@ ZIF-67 nano composite material for ultrasensitive detection of HER2. *Bioelectrochemistry*, 150: 108362. https://doi.org/10.1016/j.bioelechem.2022.108362

[26] Zhou X, Bai D, Yu H et al. (2023) Detection of rare CTCs by electrochemical biosensor built on quaternary PdPtCuRu nanospheres with mesoporous architectures. *Talanta*, 253: 123955. https://doi.org/10.1016/j.talanta.2022.123955

[27] Shepa J, Šišoláková I, Vojtko M et al. (2021) NiO nanoparticles for electrochemical insulin detection. *Sensors*, 21(15): 1–12. https://doi.org/10.3390/s21155063

[28] Arya S, Gourley A J, Penedo J C et al. (2021) Fatty acids may influence insulin dynamics through modulation of albumin-Zn^{2+} interations. *Bio Essays*, 43(12): 1–9. https://onlinelibrary.wiley.com/doi/abs/10.1002/bies.202100172

[29] El Malahi A, Van Elsen M, Charleer S et al. (2022) Relationship between time in range, glycemic variability, HbA1c, and complications in adults with type 1 diabetes mellitus. *The Journal of Clinical Endocrinology & Metabolism*, 107(2): e570–e581. https://doi.org/10.1210/clinem/dgab688

[30] Farrokhnia M, Amoabediny G, Ebrahimi M et al. (2022) Ultrasensitive early detection of insulin antibody employing novel electrochemical nano-biosensor based on controllable electro-fabrication process. *Talanta*, 238(122947): 1–12. https://doi.org/10.1016/j.talanta.2021.122947

[31] Liu Y, Yue W, Cui Y (2023) Development of an amperometric biosensor on a toothbrush for glucose. *Sensors and Actuators Reports*, 5: 100133. https://doi.org/10.1016/j.snr.2022.100133

[32] Thapa M, Heo Y S (2023) Label-free electrochemical detection of glucose and glycated hemoglobin (HbA1c). *Biosensors and Bioelectronics*, 221: 114907. https://doi.org/10.1016/j.bios.2022.114907

[33] Mandali P K, Prabakaran A, Annadurai K et al. (2023) Trends in quantification of HbA1c using electrochemical and point-of-care analyzers. *Sensors*, 23(4): 1901. https://doi.org/10.3390/s23041901

[34] Singh A K, Jaiswal N, Tiwari I et al. (2023) Electrochemical biosensors based on in situ grown carbon nanotubes on gold microelectrode array fabricated on glass substrate for glucose determination. *Microchimica Acta*, 190(2): 55. https://doi.org/10.1007/s00604-022-05626-6

[35] Wardani N I, Kangkamano T, Wannapob R et al. (2023) Electrochemical sensor based on molecularly imprinted polymer cryogel and multiwalled carbon nanotubes for direct insulin detection. *Talanta*, 254: 124137. https://doi.org/10.1016/j.talanta.2022.124137

[36] Wang J, Zhao J, Liu Y et al. (2023) SERS-based detection of early Type 1 diabetes mellitus biomarkers: Glutamate decarboxylase antibody and insulin autoantibody. *Sensors and Actuators B: Chemical*, 133456. https://doi.org/10.1016/j.snb.2023.133456

[37] Hacke A C M, Lima D, Kuss S (2022) Green synthesis of electroactive nanomaterials by using plant-derived natural products. *Journal of Electroanalytical Chemistry*: 116786. https://doi.org/10.1016/j.jelechem.2022.116786

[38] Alzahrani E, Alkhudidy A T (2021) Synthesis, optimization, and characterization of ecofriendly production of gold nanoparticles using lemon peel extract. *International Journal of Analytical Chemistry*, 2021 (7192868): 1–13. https://doi.org/10.1155/2021/7192868

[39] Ahmad S, Munir S, Zeb N et al. (2019) Green nanotechnology: A review on green synthesis of silver nanoparticles—An ecofriendly approach. *International Journal of Nanomedicine*, 14: 5087. https://doi.org/10.2147/IJN.S200254

[40] Hosny M, Fawzy M, El-Fakharany E M et al. (2022) Biogenic synthesis, characterization, antimicrobial, antioxidant, antidiabetic, and catalytic applications of platinum nanoparticles synthesized from Polygonum salicifolium leaves. *Journal of Environmental Chemical Engineering*, 10(1): 106806. https://doi.org/10.1016/j.jece.2021.106806

[41] SinghA K(2022)A review on plant extract-based route for synthesis of cobalt nanoparticles: Photocatalytic, electrochemical sensing and antibacterial applications. *Current Research in Green and Sustainable Chemistry*: 100270. https://doi.org/10.1016/j.crgsc.2022.100270

[42] Li J, Li S, Yang C F (2012) Electrochemical biosensors for cancer biomarker detection. *Electroanalysis*, 24(12): 2213–2229. https://doi.org/10.1002/elan.201200447

[43] Tran H L, Dang V D, Dega N K et al. (2022) Ultrasensitive detection of breast cancer cells with a lectin-based electrochemical sensor using N-doped graphene quantum dots as the sensing probe. *Sensors and Actuators B: Chemical*, 368: 132233. https://doi.org/10.1016/j.snb.2022.132233

[44] Darvishi E, Ehzari H, Shahlaei M et al. (2021) The electrochemical immunosensor for detection of prostatic specific antigen using quince seed mucilage-GNPs-SNPs as a green composite. *Bioelectrochemistry*, 139: 107744. https://doi.org/10.1016/j.bioelechem.2021.107744

[45] Dönmez S (2020) Green synthesis of zinc oxide nanoparticles using zingiber officinale root extract and their applications in glucose biosensor. *El-Cezeri Journal of Science and Engineering*, 7(3): 1191–1200. https://doi.org/10.31202/ecjse.729462

[46] Moussa F B, Achi F, Meskher H et al. (2022) Green one-step reduction approach to prepare rGO@AgNPs coupled with molecularly imprinted polymer for selective electrochemical detection of lactic acid as a cancer biomarker. *Materials Chemistry and Physics*, 289: 126456. https://doi.org/10.1016/j.matchemphys.2022.126456

[47] Nazarpour S, Hajian R, Sabzvari M H (2020) A novel nanocomposite electrochemical sensor based on green synthesis of reduced graphene oxide/gold nanoparticles modified screen printed electrode for determination of tryptophan using response surface methodology approach. *Microchemical Journal*, 154: 104634. https://doi.org/10.1016/j.microc.2020.104634

[48] Kasturi S, Eom Y, Torati S R et al. (2021) Highly sensitive electrochemical biosensor based on naturally reduced rGO/Au nanocomposite for the detection of miRNA-122 biomarker. *Journal of Industrial and Engineering Chemistry*, 93: 186–195. https://doi.org/10.1016/j.jiec.2020.09.022

[49] Karastogianni S, Paraschi I, Girousi S (2022) Electrochemical sensing of the maple syrup urine disease biomarker valine, using saffron-silver nanoparticles. *Biosensors and Bioelectronics: X*, 12: 100275. https://doi.org/10.1016/j.biosx.2022.100275

[50] Muthuchamy N, Atchudan R, Edison T N J I et al. (2018) High-performance glucose biosensor based on green synthesized zinc oxide nanoparticle embedded nitrogen-doped carbon sheet. *Journal of Electroanalytical Chemistry*, 816: 195–204. https://doi.org/10.1016/j.jelechem.2018.03.059

[51] Mgwili P Y (2017) Graphenated organic nanoparticles immunosensors for the detection of TB biomarkers, mini Dissertation, University of the Western Cape.

[52] Jayarambabu N, Saraswathi K, Akshaykranth A et al. (2023) Bamboo-mediated silver nanoparticles functionalized with activated carbon and their application for non-enzymatic glucose sensing. *Inorganic Chemistry Communications*, 147: 110249. https://doi.org/10.1016/j.inoche.2022.110249
[53] Hu H, Wang F, Ding X et al. (2022) Green fabrication of Pt nanoparticles via tea-polyphenols for hydrogen peroxide detection. *Colloids and Surfaces A: Physicochemical and Engineering Aspects*, 637: 128201. https://doi.org/10.1016/j.colsurfa.2021.128201
[54] Mamuru S, Malachy M, Eseyin A et al. (2022) The use of *Amaranthus Spinosus* in the synthesis of silver nanoparticles and their application in the detection of glucose. *Journal of Chemical Society of Nigeria*, 47(2). https://doi.org/10.46602/jcsn.v47i2.727
[55] Sahu A, Chatterjee P, Chakraborty A K (2022) Non-enzymatic electrochemical sensing of glucose using phyto-extract modified reduced graphene oxide MRS. *Communications*, 12(5): 902–909. https://doi.org/10.1557/s43579-022-00271-9
[56] Gijare M, Chaudhari S, Ekar S et al. (2023) Facile green preparation of reduced graphene oxide using citrus limetta-decorated rGO/TiO2 nanostructures for enzymeless glucose sensing. *Application Electronics*, 12(2): 294. https://doi.org/10.3390/electronics12020294
[57] Mohd Yusof H, Abdul Rahman N A, Mohamad R et al. (2020) Microbial mediated synthesis of silver nanoparticles by lactobacillus plantarum TA4 and its antibacterial and antioxidant activity. *Applied Sciences*, 10(19): 6973. https://doi.org/10.3390/app10196973
[58] Gahlawat G, Choudhury A R (2019) A review on the biosynthesis of metal and metal salt nanoparticles by microbes. *RSC Advances*, 9(23): 12944–12967. https://doi.org/10.1039/C8RA10483B
[59] Mazloum-Ardakani M, Barazesh B, Khoshroo A et al. (2018) A new composite consisting of electrosynthesized conducting polymers, graphene sheets and biosynthesized gold nanoparticles for biosensing acute lymphoblastic leukemia. *Bioelectrochemistry*, 121: 38–45. https://doi.org/10.1016/j.bioelechem.2017.12.010
[60] Elgamouz A, Idriss H, Nassab C et al. (2020) Green synthesis, characterization, antimicrobial, anticancer, and optimization of colorimetric sensing of hydrogen peroxide of algae extract capped silver nanoparticles. *Nanomaterials*, 10(9): 1861. https://doi.org/10.3390/nano10091861
[61] Aljabali A A, Rezigue M, Alsharedeh R H et al. (2022) Protein-based nanomaterials: A new tool for targeted drug delivery. *Therapeutic Delivery*, 13(6): 321–338. https://doi.org/10.4155/tde-2021-0091
[62] Jain A, Singh S K, Arya S K et al. (2018) Protein nanoparticles: Promising platforms for drug delivery applications. *ACS Biomaterials Science & Engineering*, 4(12): 3939–3961. https://doi.org/10.1021/acsbiomaterials.8b01098
[63] Verma D, Gulati N, Kaul S et al. (2018) Protein based nanostructures for drug delivery. *Journal of Pharmaceutics*, 2018. https://doi.org/10.1155/2018/9285854
[64] Shamsipur M, Samandari L, Farzin L et al. (2020) Dual-modal label-free genosensor based on hemoglobin@gold nanocluster stabilized graphene nanosheets for the electrochemical detection of BCR/ABL fusion gene. *Talanta*, 217: 121093. https://doi.org/10.1016/j.talanta.2020.121093
[65] Tavakkoli H, Akhond M, Ghorbankhani G A et al. (2020) Electrochemical sensing of hydrogen peroxide using a glassy carbon electrode modified with multiwalled carbon nanotubes and zein nanoparticle composites: Application to HepG2 cancer cell detection. *Microchimica Acta*, 187: 1–12. https://doi.org/10.1007/s00604-019-4064-7
[66] Benvidi A, Banaei M, Tezerjani M D et al. (2018) Impedimetric PSA aptasensor based on the use of a glassy carbon electrode modified with titanium oxide nanoparticles and silk fibroin nanofibers. *Microchimica Acta*, 185: 1–10. https://doi.org/10.1007/s00604-017-2589-1
[67] Rasitanon N, Veenuttranon K, Thandar Lwin H et al. (2023) Redox-mediated gold nanoparticles with glucose oxidase and egg white proteins for printed biosensors and biofuel cells. *International Journal of Molecular Sciences*, 24(5): 4657. https://doi.org/10.3390/ijms24054657
[68] Sharifi M, Avadi M R, Attar F et al. (2019) Cancer diagnosis using nanomaterials based electrochemical nanobiosensors. *Biosensors and Bioelectronics*, 126: 773–784. https://doi.org/10.1016/j.bios.2018.11.026
[69] Ying S, Guan Z, Ofoegbu P C et al. (2022) Green synthesis of nanoparticles: Current developments and limitations. *Environmental Technology & Innovation*, 26: 102336. https://doi.org/10.1016/j.eti.2022.102336

[70] Kaya S I, Ozcelikay G, Mollarasouli F et al. (2022) Recent achievements and challenges on nanomaterial based electrochemical biosensors for the detection of colon and lung cancer biomarkers. *Sensors and Actuators B: Chemical*, 351: 130856. https://doi.org/10.1016/j.snb.2021.130856
[71] Kizhepat S, Rasal A S, Chang J-Y et al. (2023) Development of two-dimensional functional nanomaterials for biosensor applications: Opportunities, challenges, and future prospects. *Nanomaterials*, 13(9): 1520. https://doi.org/10.3390/nano13091520

17 Bionanomaterials in Wound Dressings

Pooja Mittal, Parteek Rana, Chestha Goyal, Ramit Kapoor, Murari Lal Soni, and Rupesh K. Gautam

17.1 INTRODUCTION

Equipoise, incendiary, augmentation, and modernization are the four biological processes that take place sequentially, but overlap, while restoring damaged tissue as part of the painstakingly planned process of wound healing [1–3]. Any interruption of any of the aforementioned stages of wound healing by internal or external stimuli may cause that stage to extend and have an unsatisfactory outcome, further resulting in the status of a chronic bruise. Complete wound healing is most usually hampered by the colonization of bacteria and fungus at the location of skin damage during the typical abrasion-healing phase, infecting bacteria [4], and putting on a bandage. The fundamental goal of a wound dressing is to shield the abrasion from exterior irritants. It also maintains moisture in the wound to aid in healing and conceal the wound's origin [5]. A wound dressing should therefore be made of materials that are biocompatible, economical, and semi-permeable to oxygen and water. Thus, technologically advanced dressing materials are needed for wound dressing rather than traditional materials like cotton and wool. The cutting-edge materials have the capacity to transfer active compounds to hasten healing while keeping the wound environment intact [6]. In this context, there are numerous wound-dressing solutions available on the market that mostly contain biodegradable components [7–12]. These goods include hydrogels, lotions, ointments, and creams containing antibacterial ingredients as well as polymers. Quinolones [13], tetracyclines [14], cephalosporins [15], neomycin [16], and polymyxin B [16], being the most often used antibiotics for abrasion and care because of their capacity to prevent bacteria growth.

17.1.1 CLASSIFICATION OF WOUND DRESSINGS

The main therapeutic use nowadays of abrasion coverings that preserve a microenvironment and safeguard the abrasion bed are considered biocompatible in abrasion healing [18]. The uses of these to actively promote wound healing has been a focus of biomaterial research. This is accomplished by immunological modulation, creation of connective tissue, and biological process [19]. Many other therapies for both new and recurring wounds have shown potential for both organic and artificial material [17].

Natural polymers like polysaccharides, due to their biomaterial, biodegradability, and resemblance to the biological process proteins (such as gelatin, trypsin, and fibronectin), hydrogels, and viscose, as well as proteoglycans, are frequently utilized in surgical sutures. According to Rho et al., human keratinocytes grown on an electrospun collagen matrix demonstrated increased adhesion and dissemination in acute wounds [20]. Due to their intrinsic qualities, naturally based biocompatibility such as alginate have shown promise in usage as a natural treatment, including hemostasis

DOI: 10.1201/9781003432791-20

management, biocompatibility, and the capacity to be modified to permit drug delivery. In an in vivo investigation, it was discovered that chitosan by itself promotes the healing of pressure ulcers in mouse models [21].

Natural materials may avoid the immunogenic consequences of the increasing use of synthetic polymers in bioactive dressings. Moreover, these materials are readily functionalized, allowing for the addition of pharmaceuticals to create bioactive dressings. Oh et al.—who created biodegradable electrospun mats that encouraged cutaneous fibroblast cell growth and had antimicrobial activities in vitro—recently proved these capabilities [22]. To do this, they mixed poly(-caprolactone) and chitosan before conjugating it with caffeic acid. Incorporating gati into electrospun nanofibres demonstrated CDDs, minimal Quick acute filled treatment improves in mice and in vitro assays [23]. By synthesizing a polycarbonate hydrogel functionalized with methyl iodide, Pascal has created biomaterials with broad-spectrum antibacterial action [24]. These substances showed effective fungal and Gram+/- bacterial killing in vitro as well as rapid bacterial breakdown [25]. In order to replicate the biological process and aid in healing, a composites chitin poly (ethylene oxide) nanofibers scaffolding was designed. These biocompatible substances displayed increased cutaneous fibroblast proliferation, controlled growth factor release, and antibacterial action in vitro. Re-epithelialization, angiogenesis, collagen deposition, and early remodelling were all significantly improved in mice models, these and structures were used to treat acute entire injuries. Gold nanoparticles and cryopreserved fibroblasts in a methylcellulose gel to speed up wound closure in a mice model of third-degree burns offered more proof of the adaptability of composite polymers [26]. That resulted in the formation of microvessels, full re-epithelialization, organization of the tendon and collagen fiber, and quick healing of the wound.

17.2 NANOMATERIAL-BASED DRESSING

Nanomaterial-based wound dressing runs the risk of causing the abrasion to become dehydrated. This dressing could mechanically cling to the surface of the abrasive, which can make some changing of the abrasive dressing less difficult and painless [27–30]. Therefore, nanomaterial-based wound-healing treatment has introduced novel methods and advantages in this subject [31]. For the sake of wound-healing treatment, there are presently two groups of NMs that may be treated. Particularly, nanomaterials with intrinsic healing properties owing to their nanoscaled nature and nanoparticles that may transport therapeutic chemicals. The physical features of nanoparticles have a significant impact on the effects they have on abrasive repair. Morphology, complexation of the surfaces, and charge density, colloidal stability, biocompatibility, and biodegradability are some essential NM features that may influence how the wound healing process is altered. Biocompatible and biodegradable nanomaterials have an edge over particles that are remaining in the body after being unable to be absorbed. The addition of a payload component that has abrasive-healing properties in nanomaterials, however, is preferable to the physicochemical characteristics. Because of their distinct characteristics, NMs offer a new area of wound-healing options. Because of their antimicrobial, NSAIDs, proliferation characteristics, nanomaterials may influence each stage of abrasive healing. Nanomaterials may also change the expression level of a number of important peptides and signalling molecules to speed up abrasive closure. Because of this, the use of nanomaterials or a mixture of elements at the submicron and nanometer may one day be efficient enough to overcome the majority of the difficulties associated with managing abrasive care. Nanoparticles, nanomaterials, coatings, and scaffolding are the four primary categories of nanomaterials that can be employed to treat wounds. The major kinds of nanomaterials may be used to heal wounds. The uses of organic and synthetic structures are support scaffolds. Among the most popular approaches in bioengineering is the use of cells. Nanotechnology offers great techniques for the biocompatibility with the ability to precisely replicate the biological process and the cellular environs by adding nanotopographic signals [33–35]. Every cell-based treatment must maintain new or host cell populations by creating a favourable microenvironment since this is crucial to wound healing. At the nanoscale, it is feasible to modify the surface and 3D structure, and nanomaterials may be linked to create a bigger scaffolding or guide framework. Thus, it is essential to create an unique abrasive dressing due to the drawbacks of present abrasive dressings. that inhibits additional damage, has excellent antibacterial activities, and promotes wound healing. Nanoparticles, nanocomposites, coatings, and sealants are the four main types of NMs that may be used to heal abrasive.

17.3 SEMIPERMEABLE FILM AS DRESSING

Semipermeable film dressings are often used to heal superficial wounds such abrasions, tears in the skin, and lacerations. They may additionally be employed to conceal full thickness incisions, transplant sites, and crochet abrasive. And, used in regions of resistance to reduce between the treatment and the supporting structure, there are snipping pressures. They may be employed to conceal intravascular catheter locations as well as incisions for ultrasound therapy because they are impermeable, allowing for cleaning [36–38].

Precautions: A complexion substance must be applied to intact skin which will be coated with a permeable sheet dressing to stop maceration. In order for bacteria to enter a wound, the edge must first be well sealed. These dressings will not adhere to greasy or wet skin. When the abrasive bed is covered, we can simply trim if the film dressing's border starts to separate from your flesh. These dressings should not be used on wounds that are infected [39–40].

17.4 ROLE OF REGENERATIVE MEDICINE IN ABRASIVE HEALING

Tissue engineering encompasses a broad spectrum of possible treatments, all of which have as their ultimate objective the encouragement of tissue repair and healing. Although it is true that there is frequently a great deal of overlap, these treatments can be generally categorized as using growth factors or changing transcription factors, autologous, nanomaterials, and biomedical engineering. In this study, we discuss the evolution and limitations of the most current research in this field, as well as potential uses of tissue engineering in abrasion healing.

17.5 THE ROLE OF STEM CELL THERAPY

The abrasion-healing drugs that are available require a longer recovery period. Pus generation and other issues are brought on by the longer wound-healing process. For instance, diabetes slows down the normal physiological healing process and raises the probability of chronic abrasion. Moreover, some medicines may have adverse effects such as diarrhea, kidney issues, and more. Hence, numerous novel wound healing therapies are currently being examined in both preclinical and clinical research. Due to its ability to encourage cell growth in the abrasion-healing process, the stem cell-based technique among these emerging treatments has drawn interest. The vaccination of isolates derived via a variety of techniques is the basis of stem cell-based therapy. Before being implanted into the patient, the adult main cells are expanded in a lab setting. Adipose Derived Stem Cells (ASC)s that can proliferate and grow into a specific kind of cell that is essential for wound healing are present in the biopsied tissues. Due to their ability to generate pro-regenerative cytokines, stem cells may have the ability to speed up the healing of wounds. Autologous Mesenchymal Stem Cells (MSC) therapy may reduce the need for amputation in those with critical limb ischemia. Because it provides a substitute to amputation, stem cell-based therapy may be regarded a successful treatment for ulcers in diabetic feet [42–50].

17.6 HYDROGELS AS WOUND HEALING DRESSING

The development of a special microenvironment at the wound sites is really the consequence of a variety of actions that occur throughout the intricate and dynamic process of abrasion healing. To enhance the general effectiveness of the healing process, multifunctional skin replacement should be developed [51]. Only two of the many advantageous properties of hydrogels are their softness and straightforward water retention [52–55]. Hydrogels may be used as bandages for burns, diabetic foot ulcers, and both acute and chronic wounds because they can reduce tissue dehydration [56–58]. Two other varieties of hydrogels—which are not soluble, yet these polymers swell with a high water content—include amorphous gels and elastic solid sheets or films [59–62]. Because they rehydrate deceased tissues, these dressings are effective for facilitating autolytic cleansing of necrosis and slough. [63]. Hydrogels only partly allow the passage of gases and liquids [64]. Hydrogel dressings are non-adherent bandages that chill the wound's surface by up to 5 °C to speed healing and lessen pain. Moreover, hydrogel dressings promote wound re-epithelialization and are completely invisible. Although the hydrogel requires a second dressing, this has no bearing on its ability to supply fluids to the wound site. Because of their flexibility, the hydrogel sheets can be trimmed to fit properly all around the wound. Hydrogel dressings are suggested for dry and barely exuding wounds, since they do not absorb much exudate despite their substantial amount of water content (between 70% and 90%). The accumulation of liquids may cause skin maceration and bacterial development. It may result in an infected wound and an unpleasant odour. Hydrogels are also difficult to handle due to their poor mechanical strength. Depending on the degree of wetness in the wound, hydrogels should normally be changed every 1–3 days [65]. To avoid maceration of the skin, care must be taken to reapply the dressing often enough.

17.7 ALGINATE AS DRESSING

Alginate, the most commonly utilized ionic polysaccharide in the development and manufacture of many wound dressing products, enhances the efficiency of wound healing [66]. In order to meet the various medical needs, alginate is used in a number of derivative materials. Wafers, foam, gauze, and fibres, among its derivative materials, are frequently utilized for wound healing. Among them, clinical applications are particularly interested in nano and microfibers. The exchange of wound gases and fluids can occur through very high porosity in nano and microfibers. Moreover, in order to absorb exudates and medications, a high surface area is needed as well as sufficient pores (1–10 m) to keep microorganisms out of the air.

17.8 MULTILAYER DRESSINGS

Most of the aforementioned dressings are combined to form multilayered dressings. For treatment of cuts, burns, abrasions, or surgical wounds, the process applying a semi- or non-adherent layer and a highly absorbent layer of fibres, such as rayon fabric, cotton, and others, is usually followed [71–75]. Moreover, pressure wounds, superficial leg ulcers, and burns are treated using a combination of hydrocolloids and alginates. The hydrogel, foam, and polyurethane layers are put together to form a multilayered dressing for the treatment of persistent wounds [76, 77]. By choosing the chemical as well as physical properties of the component biomaterials, multilayer hydrogels with more than two layers are produced to meet the specifications of wound healing. Hydrogel bandages with multiple layers combine the benefits of each component. Because of the mass transfer constraints of drug molecules across the polymeric matrix are preserved in multilayer hydrogels with a drug-loaded layer, medications can be discharged under controlled conditions over an extended period of time [78–80]. Compared to single layer dressings, multilayer wound dressings are a more preferred option because they speed up the healing process. A creatively designed layered hydrogel wound dressing was developed for antibiotic release using a study that relies on water and natural polymers for effective wound healing. Electrostatic interaction of gelatin, carboxylated polyvinyl alcohol, hyaluronic acid, and still more gelatin were spread across four polymeric layers to create layered hydrogels using Layer By Layer (LBL) self-assembly. Ampicillin release from multilayer hydrogels was visible over the course of seven days. Even though the manufacturing of wound dressings has undergone substantial improvements, there is still an interest in the development of innovative antibacterial multilayer wound dressings that are more affordable and transparent, and have a fast rate of healing. Yet multilayer foam dressings have proven quite effective in preventing pressure ulcers [81–86].

17.9 CONCLUSION

The treatment of wounds is an important and growing issue on a global scale. Two important components of a wound treatment are clinical efficiency and experience and knowledge of the different wound dressing procedures if most efficient evidence-based wound care is to be provided. There are seven types of advanced wound dressings. These include multilayer dressings, non-adherent contact layer dressings, hydrogel dressings, hydrocolloid dressings, semipermeable films, and semi permeable foam dressings. Future scope for researchers in this field may be in the development of novel biocompatible and biodegradable nanomaterials that can effectively manage all stages of wound healing as well as inclusion of the beneficial features such as antibacterial, self-healing, superior mechanical, and adhesion properties into wound dressings to their performance in clinical settings.

REFERENCES

[1] Guo S, Dipietro LA (2010) Factors affecting wound healing. *J Dent Res* 89:219–229. https://doi.org/10.1177/0022034509359125

[2] Eming SA, Martin P, Tomic-Canic M (2014) Wound repair and regeneration: Mechanisms, signaling, and translation. *Sci Transl Med* 6:265sr6. https://doi.org/10.1126/scitranslmed.3009337

[3] Gurtner GC, Werner S, Barrandon Y, Longaker MT (2008) Wound repair and regeneration. *Nature* 453:314–321. https://doi.org/10.1038/nature07039

[4] Edwards R, Harding KG (2004) Bacteria and wound healing. *Curr Opin Infect Dis* 17:91–96. https://doi.org/10.1097/00001432-200404000-00004

[5] Dreifke MB, Jayasuriya AA, Jayasuriya AC (2015) Current wound healing procedures and potential care. *Mater Sci Eng C* 48:651–662. https://doi.org/10.1016/j.msec.2014.12.068

[6] Dreifke MB, Jayasuriya AA, Jayasuriya AC (2015) Current wound healing procedures and potential care. *Mater Sci Eng C* 48:651–662. https://doi.org/10.1016/j.msec.2014.12.068

[7] Matica MA, Aachmann FL, Tøndervik A, Sletta H, Ostafe V (2019) Chitosan as a wound dressing starting material: Antimicrobial properties and mode of action. *Int J Mol Sci* 20. https://doi.org/10.3390/ijms20235889

[8] Wang X, Xu P, Yao Z, Fang Q, Feng L, Guo R, Cheng B (2019) Preparation of antimicrobial hyaluronic acid/quaternized Chitosan hydrogels for the promotion of seawater-immersion wound healing. *Front Bioeng Biotechnol* 7:360. https://doi.org/10.3389/fbioe.2019.00360

[9] Das S, Baker AB (2016) Biomaterials and nanotherapeutics for enhancing skin wound healing. *Front Bioeng Biotechnol* 4:82. https://doi.org/10.3389/fbioe.2016.00082

[10] Mir M, Ali MN, Barakullah A, Gulzar A, Arshad M, Fatima S, Asad M (2018) Synthetic polymeric biomaterials for wound healing: A review. *Prog Biomater* 7: 1—21

[11] Pinho E, Soares G (2018) Functionalization of cotton cellulose for improved wound healing. *J Mater Chem B Mater Biol Med* 6:1887–1898. https://doi.org/10.1039/c8tb00052b

[12] Zheng Y, Liang Y, Zhang D, Sun X, Liang L, Li J, Liu Y-N (2018) Gelatin-based hydrogels blended with gellan as an injectable wound dressing. *ACS Omega* 3:4766–4775. https://doi.org/10.1021/acsomega.8b00308

[13] Fukuda M, Sasaki H (2015) Effects of fluoroquinolone-based antibacterial ophthalmic solutions on corneal wound healing. *J Ocul Pharmacol Ther* 31:536–540. https://doi.org/10.1089/jop.2014.0118
[14] Pirila E, Ramamurthy N, Maisi P, McClain S, Kucine A, Wahlgren J, Golub LM, Salo T, Sorsa T (2001) Wound healing in ovariectomized rats effects of Chemically Modified Tetracycline (CMT-8) and estrogen on matrix metalloproteinases -8, -13 and Type I collagen expression. *Curr Med Chem* 8:281–294. https://doi.org/10.2174/0929867013373552
[15] Srinivasan USM, Vishnu V, Sharmila S, Kumar A (2018) Formulation and evaluation of cefixime trihydrate topical gel for wound infections. *Asian J Pharm Clin Res* 11:369. https://doi.org/10.22159/ajpcr.2018.v11i8.26150
[16] Heal CF, Banks JL, Lepper PD, Kontopantelis E, van Driel ML (2016) Topical antibiotics for preventing surgical site infection in wounds healing by primary intention. *Cochrane Database Syst Rev* 11:CD011426. https://doi.org/10.1002/14651858.CD011426.pub2
[17] Aderibigbe BA, Buyana B (2018) Alginate in wound dressings. *Pharmaceutics* 10. https://doi.org/10.3390/pharmaceutics10020042
[18] Boateng J, Catanzano O (2015) Advanced therapeutic dressings for effective wound healing—A review. *J Pharm Sci* 104:3653–3680. https://doi.org/10.1002/jps.24610
[19] Asghari F, Samiei M, Adibkia K, Akbarzadeh A, Davaran S (2017) Biodegradable and biocompatible polymers for tissue engineering application: A review. *Artif Cells Nanomed Biotechnol* 45:185–192. https://doi.org/10.3109/21691401.2016.1146731
[20] Rho KS, Jeong L, Lee G, Seo B-M, Park YJ, Hong S-D, Roh S, Cho JJ, Park WH, Min B-M (2006) Electrospinning of collagen nanofibers: Effects on the behavior of normal human keratinocytes and early-stage wound healing. *Biomaterials* 27:1452–1461. https://doi.org/10.1016/j.biomaterials.2005.08.004
[21] Park CJ, Clark SG, Lichtensteiger CA, Jamison RD, Johnson AJW (2009) Accelerated wound closure of pressure ulcers in aged mice by chitosan scaffolds with and without bFGF. *Acta Biomater* 5:1926–1936. https://doi.org/10.1016/j.actbio.2009.03.002
[22] Oh G-W, Ko S-C, Je J-Y, Kim Y-M, Oh J, Jung W-K (2016) Fabrication, characterization and determination of biological activities of poly(ε-caprolactone)/chitosan-caffeic acid composite fibrous mat for wound dressing application. *Int J Biol Macromol* 93:1549–1558. https://doi.org/10.1016/j.ijbiomac.2016.06.065
[23] Pawar MD, Rathna GVN, Agrawal S, Kuchekar BS (2015) Bioactive thermoresponsive polyblend nanofiber formulations for wound healing. *Mater Sci Eng C Mater Biol Appl* 48:126–137. https://doi.org/10.1016/j.msec.2014.11.037
[24] Pascual A, Tan JPK, Yuen A, Chan JMW, Coady DJ, Mecerreyes D, Hedrick JL, Yang YY, Sardon H (2015) Broad-spectrum antimicrobial polycarbonate hydrogels with fast degradability. *Biomacromolecules* 16:1169–1178. https://doi.org/10.1021/bm501836z
[25] Xie Z, Paras CB, Weng H, Punnakitikashem P, Su L-C, Vu K, Tang L, Yang J, Nguyen KT (2013) Dual growth factor releasing multi-functional nanofibers for wound healing. *Acta Biomater* 9:9351–9359. https://doi.org/10.1016/j.actbio.2013.07.030
[26] Volkova N, Yukhta M, Pavlovich O, Goltsev A (2016) Application of cryopreserved fibroblast culture with Au nanoparticles to treat burns. *Nanoscale Res Lett* 11:22. https://doi.org/10.1186/s11671-016-1242-y
[27] Fan Z, Liu B, Wang J, Zhang S, Lin Q, Gong P, Ma L, Yang S (2014) A novel wound dressing based on Ag/graphene polymer hydrogel: Effectively kill bacteria and accelerate wound healing. *Adv Funct Mater* 24:3933–3943. https://doi.org/10.1002/adfm.201304202
[28] Mihai MM, Dima MB, Dima B, Holban AM (2019) Nanomaterials for wound healing and infection control. *Materials* 12. https://doi.org/10.3390/ma12132176
[29] Ma Y, Lin M, Huang G, Li Y, Wang S, Bai G, Lu TJ, Xu F (2018) 3D spatiotemporal mechanical microenvironment: A hydrogel-based platform for guiding stem cell fate. *Adv Mater* 30:e1705911. https://doi.org/10.1002/adma.201705911
[30] Ho J, Walsh C, Yue D, Dardik A, Cheema U (2017) Current advancements and strategies in tissue engineering for wound healing: A comprehensive review. *Adv Wound Care* 6:191–209. https://doi.org/10.1089/wound.2016.0723
[31] Kalashnikova I, Das S, Seal S (2015) Nanomaterials for wound healing: scope and advancement. *Nanomedicine* 10:2593–2612. https://doi.org/10.2217/NNM.15.82

[32] Dumville JC, Gray TA, Walter CJ, Sharp CA, Page T, Macefield R, Blencowe N, Milne TK, Reeves BC, Blazeby J (2016) Dressings for the prevention of surgical site infection. *Cochrane Database Syst Rev* 12:CD003091. https://doi.org/10.1002/14651858.CD003091.pub4
[33] Castaño O, Pérez-Amodio S, Navarro-Requena C, Mateos-Timoneda MÁ, Engel E (2018) Instructive microenvironments in skin wound healing: Biomaterials as signal releasing platforms. *Adv Drug Deliv Rev* 129:95–117. https://doi.org/10.1016/j.addr.2018.03.012
[34] Ezzelarab MH, Nouh O, Ahmed AN, Anany MG, Rachidi NGE, Salem AS (2019) A randomized control trial comparing transparent film dressings and conventional occlusive dressings for elective surgical procedures. *Open Access Maced J Med Sci* 7:2844–2850. https://doi.org/10.3889/oamjms.2019.809
[35] Winarto N, Dosan R, Aisyah PB (2019) The benefits of occlusive dressings in wound healing. *Open Dermatol J.*
[36] Moshakis V, Fordyce MJ, Griffiths JD, McKinna JA (1984) Tegadern versus gauze dressing in breast surgery. *Br J Clin Pract* 38:149–152.
[37] Debra JB, Cheri O (1998) *Wound healing: Technological innovations and market overview*. Technology Catalysts International Corporation.
[38] Thomas S, Loveless P, Hay NP (1988) Comparative review of the properties of six semipermeable film dressings. *Pharm J.*
[39] Morgan D (2002) Wounds: what dressings should a formulary include. *Hosp Pharm.*
[40] Thomson T. (2006). Foam Composite. US Patent 7048966.
[41] Ramos-e-Silva M, Ribeiro de Castro MC (2002) New dressings, including tissue-engineered living skin. *Clin Dermatol* 20:715–723. https://doi.org/10.1016/s0738-081x(02)00298-5
[42] Rando TA (2006) Stem cells, ageing and the quest for immortality. *Nature* 441:1080–1086. https://doi.org/10.1038/nature04958
[43] Duscher D, Barrera J, Wong VW, Maan ZN, Whittam AJ, Januszyk M, Gurtner GC (2016) Stem cells in wound healing: The future of regenerative medicine? A mini-review. *Gerontology* 62:216–225. https://doi.org/10.1159/000381877
[44] Sampogna G, Guraya SY, Forgione A (2015) Regenerative medicine: Historical roots and potential strategies in modern medicine. *J Microsc Ultrastruct* 3:101–107. https://doi.org/10.1016/j.jmau.2015.05.002
[45] Garg RK, Rennert RC, Duscher D, Sorkin M, Kosaraju R, Auerbach LJ, Lennon J, Chung MT, Paik K, Nimpf J, Rajadas J, Longaker MT, Gurtner GC (2014) Capillary force seeding of hydrogels for adipose-derived stem cell delivery in wounds. *Stem Cells Transl Med* 3:1079–1089. https://doi.org/10.5966/sctm.2014-0007
[46] Procházka V, Gumulec J, Jalůvka F, Salounová D, Jonszta T, Czerný D, Krajča J, Urbanec R, Klement P, Martinek J, Klement GL (2010) Cell therapy, a new standard in management of chronic critical limb ischemia and foot ulcer. *Cell Transplant* 19:1413–1424. https://doi.org/10.3727/096368910X514170
[47] Dominici M, Le Blanc K, Mueller I, Slaper-Cortenbach I, Marini F, Krause D, Deans R, Keating A, Prockop D, Horwitz E (2006) Minimal criteria for defining multipotent mesenchymal stromal cells. The International Society for cellular therapy position statement. *Cytotherapy* 8:315–317. https://doi.org/10.1080/14653240600855905
[48] Caplan AI, Dennis JE (2006) Mesenchymal stem cells as trophic mediators. *J Cell Biochem* 98:1076–1084. https://doi.org/10.1002/jcb.20886
[49] Aggarwal S, Pittenger MF (2005) Human mesenchymal stem cells modulate allogeneic immune cell responses. *Blood* 105:1815–1822. https://doi.org/10.1182/blood-2004-04-1559
[50] Rustad KC, Wong VW, Sorkin M, Glotzbach JP, Major MR, Rajadas J, Longaker MT, Gurtner GC (2012) Enhancement of mesenchymal stem cell angiogenic capacity and stemness by a biomimetic hydrogel scaffold. *Biomaterials* 33:80–90. https://doi.org/10.1016/j.biomaterials.2011.09.041
[51] Qu J, Zhao X, Liang Y, Xu Y, Ma PX, Guo B (2019) Degradable conductive injectable hydrogels as novel antibacterial, anti-oxidant wound dressings for wound healing. *Chem Eng J* 362:548–560. https://doi.org/10.1016/j.cej.2019.01.028
[52] Martin FT, O'Sullivan JB, Regan PJ, McCann J, Kelly JL (2010) Hydrocolloid dressing in pediatric burns may decrease operative intervention rates. *J Pediatr Surg* 45:600–605. https://doi.org/10.1016/j.jpedsurg.2009.09.037

[53] Mir M, Ali MN, Barakullah A, Gulzar A, Arshad M, Fatima S, Asad M (2018) Synthetic polymeric biomaterials for wound healing: A review. *Prog Biomater* 7:1–21. https://doi.org/10.1007/s40204-018-0083-4
[54] Baljit Singh, Rajneesh, Baldev Singh, Kumar A, Aery S (2019) Polysaccharides sterculia gum/psyllium based hydrogel dressings for drug delivery applications. *Polym Sci Series A* 61:865–874. https://doi.org/10.1134/S0965545X19060105
[55] Pan H, Fan D, Duan Z, Zhu C, Fu R, Li X (2019) Non-stick hemostasis hydrogels as dressings with bacterial barrier activity for cutaneous wound healing. *Mater Sci Eng C Mater Biol Appl* 105:110118. https://doi.org/10.1016/j.msec.2019.110118
[56] Zhang L, Yin H, Lei X, Lau JNY, Yuan M, Wang X, Zhang F, Zhou F, Qi S, Shu B, Wu J (2019) A systematic review and meta-analysis of clinical effectiveness and safety of hydrogel dressings in the management of skin wounds. *Front Bioeng Biotechnol* 7:342. https://doi.org/10.3389/fbioe.2019.00342
[57] Comotto M, Saghazadeh S, Bagherifard S, Aliakbarian B, Kazemzadeh-Narbat M, Sharifi F, Mousavi Shaegh SA, Arab-Tehrany E, Annabi N, Perego P, Khademhosseini A, Tamayol A (2019) Breathable hydrogel dressings containing natural antioxidants for management of skin disorders. *J Biomater Appl* 33:1265–1276. https://doi.org/10.1177/0885328218816526
[58] Lu H, Yuan L, Yu X, Wu C, He D, Deng J (2018) Recent advances of on-demand dissolution of hydrogel dressings. *Burns Trauma* 6:35. https://doi.org/10.1186/s41038-018-0138-8
[59] Boateng J (2020) *Therapeutic dressings and wound healing applications*. John Wiley & Sons.
[60] Nešović K, Janković A, Radetić T, Vukašinović-Sekulić M, Kojić V, Živković L, Perić-Grujić A, Rhee KY, Mišković-Stanković V (2019) Chitosan-based hydrogel wound dressings with electrochemically incorporated silver nanoparticles – In vitro study. *Eur Polym J* 121:109257. https://doi.org/10.1016/j.eurpolymj.2019.109257
[61] Khampieng T, Wongkittithavorn S, Chaiarwut S, Ekabutr P, Pavasant P, Supaphol P (2018) Silver nanoparticles-based hydrogel: Characterization of material parameters for pressure ulcer dressing applications. *J Drug Deliv Sci Technol* 44:91–100. https://doi.org/10.1016/j.jddst.2017.12.005
[62] Raafat AI, El-Sawy NM, Badawy NA, Mousa EA, Mohamed AM (2018) Radiation fabrication of Xanthan-based wound dressing hydrogels embedded ZnO nanoparticles: In vitro evaluation. *Int J Biol Macromol* 118:1892–1902. https://doi.org/10.1016/j.ijbiomac.2018.07.031
[63] Khorasani MT, Joorabloo A, Moghaddam A, Shamsi H, MansooriMoghadam Z (2018) Incorporation of ZnO nanoparticles into heparinised polyvinyl alcohol/chitosan hydrogels for wound dressing application. *Int J Biol Macromol* 114:1203–1215. https://doi.org/10.1016/j.ijbiomac.2018.04.010
[64] Boonkaew B, Barber PM, Rengpipat S, Supaphol P, Kempf M, He J, John VT, Cuttle L (2014) Development and characterization of a novel, antimicrobial, sterile hydrogel dressing for burn wounds: Single-step production with Gamma irradiation creates silver nanoparticles and radical polymerization. *J Pharm Sci* 103:3244–3253. https://doi.org/10.1002/jps.24095
[65] Thirumaleshwar S, K. Kulkarni P, V. Gowda D (2012) Liposomal hydrogels: A novel drug delivery system for wound dressing. *Curr Drug ther* 7:212–218. https://doi.org/10.2174/157488512803988021
[66] Hansson C (1997) Interactive wound dressings. A practical guide to their use in older patients. *Drugs Aging* 11:271–284. https://doi.org/10.2165/00002512-199711040-00003
[67] Thomas R (2002) ProGuide Dressings Datacard. http://www.worldwidewounds.com/WMPRC/DataCards/HTML/proguide.html. Accessed 9 Jun 2023
[68] Williams C (1994) Kaltostat. *Br J Nurs* 3:965–967. https://doi.org/10.12968/bjon.1994.3.18.965
[69] Williams C (1994) Sorbsan. *Br J Nurs* 3:677–680. https://doi.org/10.12968/bjon.1994.3.13.677
[70] O'Meara S, Martyn-St James M, Adderley UJ (2015) Alginate dressings for venous leg ulcers. *Cochrane Database Syst Rev* 2015:CD010182. https://doi.org/10.1002/14651858.CD010182.pub3
[71] Choi J, Konno T, Takai M, Ishihara K (2009) Smart controlled preparation of multilayered hydrogel for releasing bioactive molecules. *Curr Appl Phys* 9:e259–e262. https://doi.org/10.1016/j.cap.2009.06.054
[72] Zhou B, Li Y, Deng H, Hu Y, Li B (2014) Antibacterial multilayer films fabricated by layer-by-layer immobilizing lysozyme and gold nanoparticles on nanofibers. *Colloids Surf B Biointerfaces* 116:432–438. https://doi.org/10.1016/j.colsurfb.2014.01.016
[73] Aruan NM, Sriyanti I, Edikresnha D, Suciati T, Munir MM, Khairurrijal (2017) Polyvinyl alcohol/soursop leaves extract composite nanofibers synthesized using electrospinning technique and their potential as antibacterial wound dressing. *Procedia Engineering* 170:31–35. https://doi.org/10.1016/j.proeng.2017.03.006

[74] Kumar SKS, Prakash C, Vaidheeswaran S, Kumar BK, Subramanian S (2018) Design and characterization of secondary and tertiary layers of a multilayer wound dressing system. *J Test Eval* 48:2683–2698
[75] Shokrollahi M, Bahrami SH, Nazarpak MH, Solouk A (2020) Multilayer nanofibrous patch comprising chamomile loaded carboxyethyl chitosan/poly(vinyl alcohol) and polycaprolactone as a potential wound dressing. *Int J Biol Macromol* 147:547–559. https://doi.org/10.1016/j.ijbiomac.2020.01.067
[76] Groves AR, Lawrence JC (1986) Alginate dressing as a donor site haemostat. *Ann R Coll Surg Engl* 68:27–28
[77] Berry DP, Bale S, Harding KG (1996) Dressings for treating cavity wounds. *J Wound Care* 5:10–17. https://doi.org/10.12968/jowc.1996.5.1.10
[78] Matthew IR, Browne RM, Frame JW, Millar BG (1995) Subperiosteal behaviour of alginate and cellulose wound dressing materials. *Biomaterials* 16:275–278. https://doi.org/10.1016/0142-9612(95)93254-b
[79] Chiu C-T, Lee J-S, Chu C-S, Chang Y-P, Wang Y-J (2008) Development of two alginate-based wound dressings. *J Mater Sci Mater Med* 19:2503–2513. https://doi.org/10.1007/s10856-008-3389-2
[80] Tamahkar E, Özkahraman B, Süloğlu AK, İdil N, Perçin I (2020) A novel multilayer hydrogel wound dressing for antibiotic release. *J Drug Deliv Sci Technol* 58:101536. https://doi.org/10.1016/j.jddst.2020.101536
[81] Chopra H, Kaur A, Singh I, Sharma RK, Emran TB (2022) Nano-based targeting strategies for cancer treatment. *Int J Surg* 105:106864. https://doi.org/10.1016/j.ijsu.2022.106864
[82] Khan H, Tiwari P, Kaur A, Singh TG (2021) Sirtuin acetylation and deacetylation: A complex paradigm in neurodegenerative disease. *Mol Neurobio*l 58:3903–3917. https://doi.org/10.1007/s12035-021-02387-w
[83] Grewal AK, Singh TG, Sharma D, Sharma V, Singh M, Rahman MH, Najda A, Walasek-Janusz M, Kamel M, Albadrani GM, Akhtar MF, Saleem A, Abdel-Daim MM (2021) Mechanistic insights and perspectives involved in neuroprotective action of quercetin. *Biomed Pharmacother* 140:111729. https://doi.org/10.1016/j.biopha.2021.111729
[84] Grewal AK, Singh N, Singh TG (2019) Neuroprotective effect of pharmacological postconditioning on cerebral ischaemia-reperfusion-induced injury in mice. *J Pharm Pharmacol* 71:956–970. https://doi.org/10.1111/jphp.13073
[85] Sharma A, Khanna S, Kaur G, Singh I (2021) Medicinal plants and their components for wound healing applications. *Future J Pharm Sci* 7:53. https://doi.org/10.1186/s43094-021-00202-w
[86] Chopra H, Bibi S, Kumar S, Khan MS, Kumar P, Singh I (2022) Preparation and evaluation of chitosan/PVA based hydrogel films loaded with honey for wound healing application. *Gels* 8. https://doi.org/10.3390/gels8020111

18 Bionanoparticles Impact on Human Health, an In Vitro and In Vivo Status

Hemavathi Brijesh, Medini Bheemappa, Ramya Manjunath, B.K. Nithin Gowda, H.V. Shambhavi, N.G. Manjula, and Shilpa Borehalli Mayegowda

18.1 INTRODUCTION

Nano-sized particles less than 100 nm, also known as nanoparticles (NPs), could enter the lungs, which then joins the bloodstream via inhalation, ingestion, or skin contact. However, the chance of eventually being exposed to nanoparticles during production by workers, consumers, and the environment is high. Several research findings have revealed the harmful effect of NPs on biological organisms is triggered by bigger particles rather than their chemically comparable particles (Catalán et al., 2016). The entry routes include lungs, skin, or the gastrointestinal tract (Gnach et al. 2015). The toxicity of bionanomaterials should be carefully studied, much as with other medications. Nanomaterials can easily penetrate cells, tissues, and organs compared to traditional medications. The majority of in vitro research demonstrates that cytotoxicity will not be caused by bionanomaterials at low doses. Significant cytotoxicity will be caused by bionanomaterials at higher concentrations and for longer periods of time. Second, because NPs are particulate, they differ from conventional soluble compounds in their kinetics in biological environments and in terms of the potential harm they may cause. This presents an exceptional challenge in detecting the hazards of NPs to the animals used in experiments, including humans and other species in the environment (Mayegowda et al., 2023). Given the enormous potential of NPs, the toxic effects of human exposure and the associated ecotoxicological effects must be summarized. In this review, the biological applications of metallic and non-metallic nanoparticles related to toxicity at cellular level (in vitro) and organ toxicity (in vivo) are discussed.

18.2 TOXICITY MECHANISMS OF NANOPARTICLES

NPs gain entry through inhalation, ingestion, and cutaneous exposure and are found to be the most common ways that humans come in contact with NPs (blood circulation). The mechanism of nanoparticle toxicity is described in Figure 18.1. The alteration of gene expression, chromosomal breakage, fragmentation of DNA strands, point mutations, and DNA adducts due to oxidation have all been linked to several nanomaterials, including oxides of metal NPs, quantum dots (QDs), and fibre type NPs. In some cases, these effects have even been observed to occur through cellular barriers. The safety profile becomes very important in these situations. NPs when entering the human body can pass through profuse cellular barriers and could reach the delicate organs (kidney, lung and liver) leading to damage of mitochondria, inducing mutations in deoxyribonucleic acid (DNA), resulting in cellular apoptosis. The reactive oxygen species (ROS) produced during NPs exposure leads to toxicity inducing oxidative stress, DNA strand breakage, proteins, and cell membrane disruption.

 DOI: 10.1201/9781003432791-21

The amount of ROS generation brought on by NPs is include various factors, such as particle uptake, surface, presence of mutagens, aggregation/agglomeration, size, composition, shape, associated transition metals with the particles, and solubility. Another crucial factor to take into account when determining a nanomaterial's toxicity is its degradability. Nondegradable nanomaterials can collect in organs and cells and cause harm if there is no system for removing them from the body. Rheumatoid arthritis has been treated using injectable gold compounds, and patients may have toxic consequences when gold compounds build up in their bodies over time. Yet, if the degraded components of the substance are hazardous, biodegradable compounds may likewise have harmful consequences (Fischer and Chan, 2007). Current use of several therapeutic approaches was seen also on the master organ, the brain, and the major concern of nanomaterial application is damage to the nervous system. The blood brain barrier (BBB), which is incredibly effective, prevents most chemicals gaining entry into the brain through blood. Although direct entry of NPs through inhalation to the brain is impossible, it is nevertheless possible for them to do so via the bloodstream, lymphatic system, olfactory mucosa, and cerebrospinal fluid. Although being well-established, certain routes are less effective. As the development of nanomedicines used in the treatment of fatal disorders progresses, nanotoxicity must be taken into account (Teleanu et al. 2018) (Figure 18.1).

Due to the fact that only silver NPs smaller than 10 nm were found to be cytotoxic to cells derived from human lungs (20 g/mL dosage), Gliga et al. (2014) showed that the cytotoxic effect posed by these silver NPs is size-dependent. However, there was no difference in cytotoxicity between polyvinylpyrrolidone and citrate encapsulated silver NPs with a size of 10 nm. According to Rizk et al. (2017), genetic imbalance began when the experimental animals were exposed to a dosage of 500 mg/kg body weight for an extended period of time and titanium oxide NPs of 21 nm size displayed remarkable effect as analyzed by liver histological pattern (45 days), functional enzymes of liver and

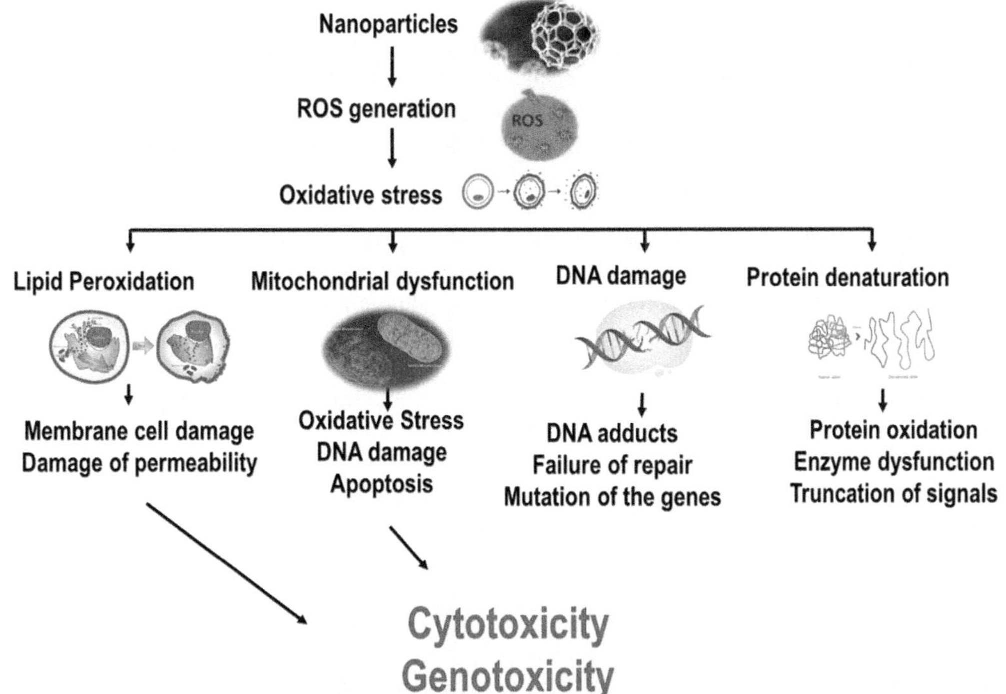

FIGURE 18.1 Nanoparticles Mediated Toxicity Mechanism through ROS-Induced Cytoplasmic and Nuclear Alteration Leading to Death of the Cell. (Compiled by the Authors.)

markers related to oxidative stress (Figure 18.1). Further, it was found that the kidney was the organ primarily targeted by NPs (6 nm). When CdS nanoparticles are delivered via gavage, nephrotoxicity is seen (Zhang et al. 2014). The renal system usually eliminates the tiniest nanoparticles. The zwitterionic layer on QDs prevented serum proteins from adhering to them, preserved their small size, and allowed them to pass through the kidney's distinct, multilayered structure of glomeruli with ease. The nephrotoxicity is low when the interaction between the renal tissues and nanoparticles is low. There are boundaries to the relationship between renal clearance and body size. On the other hand, reproductive toxicity could result in different ways, such as through interference at fertility level, which affects both the sexes or with the outcome of pregnancy, where the mother and the newborn are affected. In the first type, where fertility is affected, the majority of the evidence is indirect and consists of gonads that have undergone organic and functional change. Several studies have shown that nearly all nanomaterials can result negatively in pregnancy outcomes (Campagnolo et al. 2017). Additionally, a study on mice revealed that nanomaterials exposure at the prenatal stage may lead to the development of postnatal diseases, some of which may not become clinically apparent until adulthood (Stapleton et al. 2013). Furthermore, the form of gold nanoparticles also has an impact on their cytotoxicity, as demonstrated by Steckiewicz et al. (2019). Gold NPs with cytotoxic effect were known as gold nano stars with high anticancer ability, whereas the least toxic gold NPs were known as gold nanospheres having low anticancer activity (De Jong et al., 2008). In the section that follows, we will go over the toxicity effects of both metallic and non-metallic NPs.

18.3 TOXICITY OF METALLIC NPS IN VITRO

18.3.1 Aluminum Oxide

Twenty per cent of all nanosized chemicals are made up of NPs based on aluminum. Aluminum oxide NPs, according to Chen et al. (2008), have negative impacts on viability of cells, oxidative stress, function of mitochondria, and the expression of BBB's tight junction protein. Some studies have noted that aluminum oxide nanoparticles (NPs) did not exhibit observable harmful effects on the mammalian cells progression at various concentrations (10, 50, 100, 200, and 400 g/mL), according to Radziun et al. (2011). Cell viability analysis was carried out using EZ4U assay instead of 3-[4,5-dimethylthiazol-2-yl]-2,5 diphenyl tetrazolium bromide (MTT) assay. Similar results were reported from another study, which described toxicity of aluminum oxide NPs (160 nm) on human-derived mesenchymal stem cells were dosage dependent when exposed at different concentrations ranging from 25 to 40 μg/mL. Studies on cellular toxicity were performed using MTT assay (Alshatwi et al. 2012). Using rat blood cells, they performed a micronucleus and comet assay to evaluate the genotoxicity. In another study, the results indicated that aluminum oxide NPs with a size of 50 nm exhibited only genotoxicity on DNA and did not induce mutagenicity in mouse-derived lymphoma cells (Kim et al. 2009).

Previously, many studies demonstrated NPs interaction with immune cells conducted in cell lines and animal models and revealed that exposing the immune cells to NPs causes inflammation that is characterized by the activation of the inflammasome thus producing proinflammatory cytokines (Murray et al. 2013; Dobrovolskaia et al. 2016). While nanomaterials do not naturally cause inflammation, they can do so if they aggregate to create larger particles that are easily detected by phagocytes or by the adsorption of bacterial products like LPS that are potent inflammatory inducers on their surface (Geiser, 2016). Nanomaterials' immunotoxicity can also result in immunosuppression in addition to immune response activation. The bionanomaterial implanted into the body may be identified as foreign matter and trigger the immune response. The agglomeration of bionanomaterials is one of the primary factors responsible for cytotoxicity and tissues under normal physiological conditions. In in vitro investigations, bionanomaterials had an effect on cell morphology and cellular functions that include proliferation, differentiation, and mineralization (Roodbar et al. 2022).

The negative effect of bionanomaterials in vivo includes oxidative stress, inflammation, granulomas and fibrosis (Khanna et al. 2015). Immune cells that have been loaded with nanoparticles are less effective at scavenging and defending the body from pathogens. Also, nanoparticles loaded cells undergo apoptosis sooner than normal cells (Porta et al. 2013). The immunological responses to nanomaterials applied in therapeutic and diagnostic fields are currently poorly understood, and further research is needed to determine which properties allow for recurrent systemic delivery without side effects (Bachmann and Gary, 2010).

Only few reported in vivo studies have examined this feature of NPs. The literature that is currently available indicates that aluminum oxide NPs have been tested primarily for their effects on cytotoxicity and genotoxicity. It is very important to detect aluminum-derived NPs toxicity through screening that cause toxic health hazards in humans according to standard protocols as they are used extensively in numerous areas subsequently resulting in human exposure. By observing variation in enzymes of lysosome and mitochondrial dehydrogenase activity over the course of 24 hours, Di Virgilio et al. (2010) found cytotoxicity in CHO cells was dosage dependent when the cells were exposed to TiO_2 and AlO_2 NPs at lower concentrations (5g/mL). Additionally, they reported that TiO_2 and AlO_2 NPs concentrations of 0.5–10 g/mL significantly increased the genotoxic effects on micronucleus frequencies. Additionally, AlO_2 NPs cause toxicity to cells and damaged membrane in Chinook salmon cells, resulting in morphological abnormalities, peroxidation of lipids, and oxidative stress, according to Srikanth et al. (2015).

18.3.2 Gold

Gold nanoparticles have different physicochemical properties. They possess the capacity to readily functionalize in addition to attaching to thiol and amine groups. Gold NPs are being studied as contrast substances, medication carriers in the cancer and thermal therapies because of these properties. Gold nanoparticles are thought to be usually harmless because they are inert and non-toxic. One study looked at the cytotoxicity of different gold NPs with sizes ranged from 4–18 nm when fused with varied caping agents (Jain et al. 2012). The study demonstrated that spherical gold NPs enter cells and do not impair cellular function. The cytotoxicity of gold nanoparticles is influenced by several factors, including cell type, physical and chemical characteristics, and a kind of toxicity assay. The toxicity in normal as well as cancer cells from human lung and liver are found to vary (Patra et al. 2007). Senut et al. (2016) discovered that when human-derived embryonic stem cells exposure to gold NPs (1.5 nm) capped with mercaptosuccinic acid displayed cell death due to loss of cohesiveness, rounding up, and detachment at a high concentration (0.1 µg/mL). However, gold NPs with 4 and 14 nm size exhibited virtually zero toxicity on human embryonic stem cells at a higher (10 µg/mL) concentration. In addition, Steckiewicz et al. (2019) conducted a study that demonstrated gold NPs stars (215 nm) with a strong anticancer activity also exhibited high cytotoxicity among all studied nanoparticle forms, while gold nanoparticle spheres (6.3 nm) showed poor potential for anticancer effect.

In another study, natural killer (NK) cells derived from human blood was exposed to various doses of AgNPs and Ag+ to study the AgNPs cytotoxicity on NK cells. NK cells' exposure to AgNPs resulted in decreased survival and reduction in cytotoxic capacity, with an increased expression of inhibitory receptor (CD159a) as compared to Ag+ controls (Müller et al. 2018). Furthermore, while gold NPs may not cause inhibition of cell growth in the short term, long-term exposure could, however, hamper cell metabolic activity and also energy homeostasis. Gold NPs taken up by cells could be retained in endosomes/lysosomes or enter nuclei via surface functionalization. The functionalization and cell type affect how well gold NPs are absorbed. While gold NPs absorbed by cancer cells result in cell death, they will eventually be removed if taken up by healthy cells (Mayegowda et al., 2022a).

18.3.3 Copper Oxide

In many different types of technologies—including semiconductors, antimicrobials, intrauterine contraceptives, and heat transferring fluids—copper oxide NPs are commonly utilized. Experimental evidence suggests that copper NPs promote toxicity effects on kidney and liver. Experimental animal liver, kidney, and spleen have been severely impaired by nano-copper. When supplied orally, the highly reactive copper ion interacts with gastric juice and builds up in kidneys of experimental animals used in copper NPs exposure (Manjula et al. 2022: Kumar et al., 2022; Mayegowda et al., 2024a: Mayegowda et al., 2024b Meng et al. 2007). In another in vitro investigation, copper oxide NPs induced oxidative stress, impaired cell membrane integrity, and demonstrated cytotoxic and genotoxic effects (Ananda et al. 2022; Ahamed et al. 2010).

18.3.4 Silver

In the past, silver was widely recognized as an antibacterial agent (Adarsha et al. 2022; Shilpa et al. 2022; Rotti et al. 2023). Several different commercial products use its NPs. Silver nanoparticles are used as a coating for surgical instruments, wound dressings, and prosthetics. They gain entry to the human body in a variety of ways, build up in various organs, and enter the brain by crossing the blood brain barrier (BBB). In one study, rats who either inhaled or were given subcutaneous injections of silver NPs had silver nanoparticles in their liver, lungs, spleen, brain, and kidneys. Foldbjerg et al. (2011) observed cytotoxicity following the administration of silver NPs to lung cancer cell line of human origin at various dosages, which led to the formation of cellular DNA adducts when measured by the in vitro MTT assay. The production of spherical silver NPs having sizes ranged from 20–25 nm was demonstrated by Khan et al. (2019), and its toxicity towards embryos of zebrafish and endothelial cells derived from human umbilical vein through an apoptosis mechanism was evaluated. The NPs toxicity study in human cell lines and a zebra fish model demonstrated that 10 g mL concentration of synthesized NPs induced oxidative stress, cell senescence, and apoptosis. The prior study has some backing from a different study. These scientists claimed that silver NPs with peptide coatings (20 nm) are significantly more harmful than silver NPs with citrate coatings with similar size. In this study, human-derived leukemia cell lines were used to perform WST-1 assay to assess the cytotoxicity of the cells (Haase et al. 2011). Even though silver NPs are being employed more and more in biomedical applications, harmful consequences have been observed. Silver NPs exhibit greater toxicity than other nanosized metals in terms of increased ROS synthesis and leakage of lactate dehydrogenase enzyme. Although only minimal concentration of 6.2 g/mL silver ions under oxic and 0.2 g/mL in anoxic conditions are released from silver NPs, Mulenos et al. (2020) discovered that these concentrations are significantly less harmful than those observed in naturally occurring environments.

18.3.5 Zinc Oxide

Sunscreens, paints, UV detectors, wave filters, gas sensors, and many personal care items are all created using zinc oxide NPs. Several studies were conducted to understand the toxicity of zinc NPs, which showed cytotoxicity, disruption of cell membranes, and severe oxidative stress in different mammalian cell types. Higher levels of zinc oxide NPs with a concentration of 49 mg/mL were employed to treat human-derived mesothelioma cells and mouse-derived fibroblast cells, which resulted in 100 per cent cell death when seen in cell culture (Brunner et al. 2006). After being identified using the MTT and comet assays, respectively. Zinc oxide based NPs exhibited few cell structural abnormalities like shape irregularity, DNA breakage, and alteration in the activity of mitochondria in human-derived embryonic kidney and hepatic cells. In another study Meyer et al. (2011) demonstrated the exposure of fibroblasts from human-derived dermal cells to zinc oxide NPs (20 nm) and analyzed the DNA damage by MTT assay technique. Zinc oxide NPs caused DNA damage in HEp-2 cell line in an in vitro study when assessed using conventional methods like comet assay conducted by Osman et al. (2010). Experimental mice exposed to zinc oxide NPs (300 mg/kg) for prolonged period caused DNA breakage and altered several enzymes of liver as revealed by the comet assay adopted for determining the breakage of DNA. Further, Jain et al. (2019) demonstrated that zinc oxide NPs caused a Chinese hamster's lung fibroblast cells to experience oxidative stress, which resulted in genotoxicity, mutation of hypoxanthine-guanine phosphoribosyl transferase gene, and ultimately cell death/apoptosis. According to Xiao et al. (2016), zinc oxide NPs caused stress at oxidative level, which resulted in elevated levels of ROS and malondialdehyde leading to kidney toxicity.

18.3.6 Iron Oxide

NPs synthesized using iron oxide are employed extensively in various biomedical applications and therapeutic applications. The reticulo-endothelial system is one of the organs where these NPs have been identified to bioaccumulate. Reduced cell viability was shown to be the most frequent harmful impact generated by iron oxide NPs when they were coated with diverse compounds, according to a number of in vitro investigations. Super magnetic iron oxide NPs (30 nm) coated with Tween have shown toxic effects in murine-based macrophage cells (Dulińska-Litewka et al. 2019). They attribute this effect to low concentrations of iron oxide NPs exposure for 2 h (25–200 g/mL) causing more cellular toxicity than high concentrations exposure of 6 h (300–500 g/mL). Nevertheless, human macrophage cell viability was 20 per cent lower after seven days when incubated with dextran-encapsulated iron oxide NPs (100–150 nm) with a dosage of 0.1 mg/mL (Pawelczyk et al. 2008). Similarly, chitosan-fused iron oxide NPs having a size of 13.8 nm only demonstrated 10 per cent of cell progression in human-derived hepatocellular carcinoma cell line when exposed for 12 hours at a concentration of 123.52 g/mL (Jeng et al. 2006). Previously, a study suggested that toxicity effects of iron oxide NPs are caused by high ROS production which further lead to DNA breakage and peroxidation of lipid (Liu et al. 2013). Magdolenova et al. (2015) tested human lymphoblastoid cells and primary peripheral lymphocytes for cytotoxicity and genotoxicity in vitro using oleate fused with or without iron oxide NPs. Cytotoxic or genotoxic effect was not present when cells were exposed to oleate not fused with iron oxide NPs. However, iron oxide NPs coated with oleate showed cytotoxic effect and also caused DNA damage, suggesting its genotoxicity potential. Sadeghi et al. (2015) also investigated the iron oxide NPs toxicity in human hepatic cells. The results revealed that increased oxidative damage causes cells to undergo apoptosis thus reducing the cell survival post exposure of 12-h and 24-h to iron oxide nanoparticles at various concentrations 925, 50, 75, and 100 g/ml). Biological dispersion along with iron oxide NPs toxicity were examined by Feng et al. (2018) in relation to surface coating and nanoparticle size in cell lines and also animal models. They discovered that the dispersion, cytotoxicity, and absorption of iron oxide NPs were influenced by their size and coating. In addition, iron oxide NPs coated with polyethyleneimine demonstrated increased cellular uptake by macrophages and cancer cells, cytotoxicity, and low tumor spreading when compared to iron oxide NPs coated with polyethylene glycol. Furthermore, smaller (10 nm) iron oxide NPS coated with polyethylene glycol demonstrated greater uptake by cells and tumor aggregation as compared to larger ones (30 nm).

18.3.7 Titanium Oxide

Several reports have demonstrated TiO_2 NPs toxicity on health upon exposure. Formation of adducts in DNA together with induction of oxidative stress resulted when exposed TiO_2 NPs. TiO_2 NPs can cause the following types of toxicity in organisms: (1) generation of ROS and light-dependent electron-hole pairs formation; (2) cell membrane binding of TiO_2 NPs through the electrostatic attachment, which results in lipid peroxidation and cell wall damage; (3) TiO_2 NPs attachment with cell organs and macromolecules (Hou et al. 2019).

Genotoxicity refers to the TiO_2 NPs, potential to alter the genetic information which causes gene mutations. TiO_2 NPs genotoxicity in cells is mostly studied through the respiratory or circulatory systems. Despite TiO_2 NPs crystallinity, their genotoxicity is mainly determined by particle size. Smaller TiO_2 NPs are found to be more genotoxic compared to larger ones because they can easily enter cell nucleus and cytoplasm. Agglomerations of TiO_2 NPs larger in size induce DNA damage. Several investigations have shown that human amniotic epithelial cells, lymphocytes, lung fibroblasts, and hepatoma (HepG2) cells are genotoxic and cytotoxic to TiO_2 NPs, with genotoxicity leading to DNA breakage and cellular mutation under in vitro conditions. TiO_2-induced immunotoxicity occurred via receptor activation, which then activated specific signaling pathways thereby reducing antioxidant activity via ROS production as demonstrated with evidence. When exposed to more ROS, the mitochondrial membrane potential ($\Delta\Psi m$) decreases, eventually causing cell death/apoptosis, and inducing toxicity in immune cells by causing imbalance in immune redox process. In neuronal cells TiO_2 NPs can damage glial cells and neurons over time (U373), and this damage may eventually result in neurotoxicity even at low doses (Baranowska-Wójcik et al. 2020). TiO_2 NPs could affect functions of brain by increasing oxidative stress, lowering the antioxidant enzymes activity and increasing the release of nitric oxide (NO) and ROS (Figures 18.1 and 18.2) (Grissa et al. 2020). Thus, in the cerebral cortex, TiO_2 NPs caused a neurotoxic injury that was accompanied by an increase in degenerating and apoptotic neurons. Toxicity on cardiovascular system occurred through reduction in the activity of mitochondrial dehydrogenase significantly when treated with TiO_2 NPs of 100 nm in human-derived lymphocytes (Ghosh et al. 2013). This in turn caused DNA damage, leading to mitochondria-mediated apoptosis-based cell death by TiO_2 NPs (Figure 18.2). Haemoglobin when interacted with TiO_2 NPs, can impair red blood cell oxygen transportation. After being incubated for 24 hours, TiO_2 NPs drastically reduced cellular viability and elevated the generation of toxic inducers, demonstrating that they were hazardous to human derived PBMCs (peripheral blood mononuclear cells) (Kongseng et al. 2016). With elevated TiO_2 NP concentrations (25 g m-1), PBMCs' "proinflammatory cytokine production" and cell death elevated because of ROS induced oxidative stress. TiO_2 NPs, effect on the production of neopterin and the degradation of tryptophan in PBMCs indicated that tryptophan breakdown was suppressed while production of neopterin increased in unstimulated as well as stimulated PBMCs, indicating that TiO_2 NPs overall effect was strongly pro-inflammatory (Becker et al. 2014).

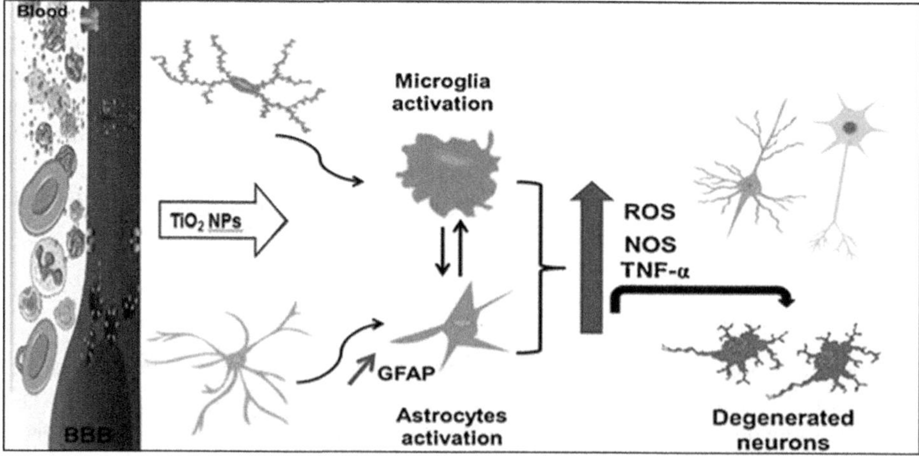

FIGURE 18.2 TiO_2 NPs Induced Neurotoxicity Mechanism in Cells. (Adapted with Permission from Grissa et al. 2020.)

Further, toxicity on respiratory systems was analyzed by numerous studies that revealed damage to human-derived lung epithelial cells (A549) causing inflammation of the alveoli due to TiO_2 exposure (Kose et al. 2021). These events cause fibrosis and the growth of tumors in addition to irreversible cell changes that result in lung dysfunction. The characteristics of TiO_2 NPs viz, shape, size, nature of crystallinity, aggregation, and mode of surface binding are related to the toxicity of lungs and inflammatory effects. Human-derived A549 and BEAS-2B cell lines were used to test the toxic effect at the cellular level and inflammatory reactions of TiO_2 NPs anatase and rutile forms. They discovered that anatase TiO_2 NPs had high cytotoxic effect on bronchial cells as compared to rutile TiO_2 NPs. Moreover, pre-irradiated TiO_2 NPs had increased lethal effect than non-irradiated TiO_2 NPs did in lung cells (Figure 18.2).

18.3.8 Cobalt

Cobalt-derived NPs are applied in pigments and other devices such as energy storage and sensor devices. It has been demonstrated that cobalt-coated NPs causes oxidative stress, morphological abnormalities, DNA damage, and inflammatory effects in varied cell lines when exposed. Possible toxic effects of cobalt NPs were demonstrated previously using a variety of models cell lines and animals models. Chattopadhyay et al. (2015), for instance, analyzed intracellular signaling transduction mechanisms underlying oxidative stress caused by cobalt oxide nanoparticles. It was observed that cobalt oxide NPs significantly increased ROS induced tumor necrosis factor alpha leading to cell death. Cell death is greatly influenced because of caspase-8, caspase-3, p38 MAPK activation, and tumor necrosis factor alpha. It was discovered that, toxicity of cobalt oxide NPs also affects human immune cells. Moreover, Cappellini et al. (2018) demonstrated that lung cells ingested three different forms of cobalt nanoparticles (NPs) and that oxidative damage and DNA strand breaks were brought on by NPs prepared using cobalt and cobalt (II) oxide.

18.4 TOXICITY OF NON-METALLIC NANOPARTICLES

Silica nanoparticles (SiNPs), composed of silicon dioxide, has been widely used in the food industry, medical diagnosis, synthetic industry, and drug delivery. This inorganic material is unique as its surface area is large, and particle size and biocompatibility make it a preferred nanoparticle for biomedical applications (Archana et al. 2021; Kim et al. 2015; Yadav et al. 2021). The stability of SiNPs renders it to be the second largest nanomaterial produced globally. The fluorescent SiNPs are applied for imaging, magnetic resonance imaging, radio-labelled imaging, and ultrasound imaging in medical diagnosis. As nanomaterials have higher penetrance and large surface to volume ratios, it can enter the living systems readily through breathing and ingestion. This poses a toxic impact on the living tissues. Many toxicity studies have been reported about SiNPs on humans. Several characteristic features of SiNPs such as particle type, size, dose of NPs, and administration time are just a few of the variables that can have an impact on how SINPs cause cytotoxicity and organ damage. Additional important considerations include individual variations in cells and in vivo, as well as the type of the cell, state of the cell, distribution of organ, state of animal, and delivery period. Oxidative stress, inflammation, and apoptosis are some of the toxicity mechanisms induced by SiNPs; however, studies on the relevant signal pathways are still in their infancy. There is not yet a set criteria for toxicity research because there are so many different variables involved. The extraordinary class of recently created quasi-spherical carbonaceous nanomaterials known as carbon nanoparticles (CNPs) has diameters below 10 nm, which are synthesized by carbonization, heating, activation, and grinding. They are typically below the size of 10–45 nm. Like other nanoparticles, large surface area and adsorbing characteristics attributes to them a toxic potential.

CNPs have the potential to be cytotoxic and genotoxic to a variety of tissues and systems, according to experimental and epidemiological research. Moreover, it has been demonstrated that they have detrimental consequences on the kidneys, immunological system, skin, and eyes, and they also demonstrate developmental nanotoxicity (Di Ianni et al. 2022; Borehalli Mayegowda et al., 2022). Damage is usually generated when CNPs interact with the cell membrane, nucleus, or mitochondria. Reactive oxygen species signal pathways that trigger inflammation, and increased production of proinflammatory cytokines are all associated with this damage (TNF, IL-6). Inflammation, reactive oxygen species (ROS), and direct interactions with carbon nanoparticles (CNPs) can affect DNA repair mechanisms. (Mayegowda et al. 2022b, c). This could result in the accumulation of mutations and malignant transformation established for the first time the progression of CNPs toxicity upon exposure in zebrafish embryos (Mutalik et al., 2024; Lavanya et al., 2024; Shi et al., 2022). CNPs, exposure mostly altered the expression of genes involved in phototransduction, metabolic pathway of amino acids, and the biosynthetic pathway, according to RNASeq study and RT-qPCR results. Several metal NPs made of biopolymers are frequently used in many different scientific disciplines. The majority of research related to biopolymer-based metal NPs is focused on their prospective applications, like other emerging technologies (Ramakrishnappa et al. 2022). The toxicity of biopolymer-based metal NPs regarding environment, animals, and cells is poorly understood. Despite the fact that numerous reports have demonstrated that NP deposition causes toxicity in a number of organs, the potential dangers of exposure to biopolymer-based metal NPs, their methods of gaining entry into the body, and their mechanisms of metabolism have not yet been thoroughly investigated.

18.5 IN VIVO TOXICITY

Several studies on the behavior of NPs have been extensively done using in vivo models. Although NPs are considered to be highly promising agents and, applied in the medical field, they exhibit potential side effects as well. The most widely employed NPs, notably in analyses of environmental protection, are metal oxides like TiO_2. Consequently, it is crucial to assess TiO_2 toxicity after injection in experimental animals. The laboratory rats used for experiment were administered with a dose of 15 µg/cm2 TiO_2 NPs suspension and the results revealed that the animals encountered the inflammation induced by TiO_2 NPs within 24 h, suggesting the relative hazardous effect of TiO_2 NPs. (Kiss et al. 2008). AgNPs are widely utilized in medical applications due to their potential antibacterial activity (Jadimurthy et al. 2022). The Guinea pigs were used to study the toxicity and bio distribution by setting an experiment exposing them with different sizes of silver NPs (100, 1000, 10,000 ppm). The findings showed a strong link between cutaneous exposure and tissue AgNP levels which accumulated in various organs. When experimental animals were injected with given optimal and high doses of AgNPs, histological investigations revealed extensive deterioration of kidneys proximal convoluted and distal convoluted tubules (Korani et al. 2013).

A recent study indicated that silica nanoparticles (SiNPs) were consumed through intratracheal exposure in mice to assess SiNP toxicity. Exposure to SiNPs induced impairment in the lysosome subsequently causing blocked autophagic flux, and production of ROS. This further resulted in autophagy dysfunction and triggered the production of cytokines leading to lung inflammation in mice (Figure 18.3) (Wang et al. 2020).

The toxic effects of gold NPs on mice caused weight loss, low hematocrit, and decreased count of red blood cells. Because the advantages of using gold nanoparticles in medication delivery must outweigh the disadvantages, it is also crucial to understand their hazardous qualities. According to the findings of one investigation, modified gelatin nanoparticles (NPs) containing polyethylene glycol and intended for delivering ibuprofen sodium salt were safe at the concentration (1 mg/Kg) necessary for efficient drug administration. These results were confirmed by estimating the cytokine levels produced during inflammation process using an in vivo model and also studying their organs using histological analysis (Narayanan et al. 2013). Similar observations were reported when drilling holes into the skulls of male Fischer 344 rats and injecting magnetic nanoparticles (MPNs) coated

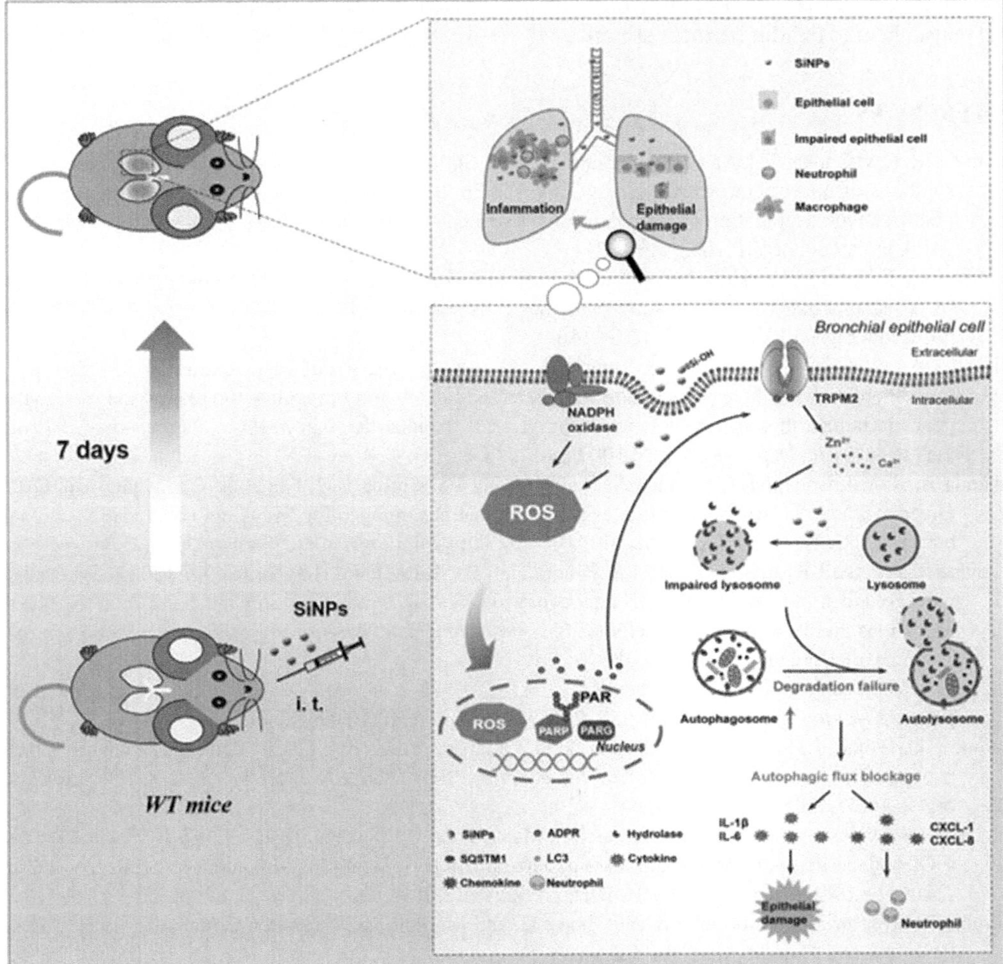

FIGURE 18.3 Diagrammatic Illustration of Pulmonary Inflammation in Mice Induced by SiNPs Exposure. (Reproduced with permission from Wang et al., 2020.)

with polyethylene glycol (PEG) into intracerebral brain tumors. The findings showed that spleen bio-distribution and PEG-MNP accumulation raised questions about toxicity, and the research team suggested that a lengthy amount of time be required to evaluate the hazards associated with PEG-MPNs toxicity (Cole et al. 2011).

18.6 CONCLUSION

Toxicity of NPs arises from various means such as the precursors used in the preparation of NPs, the disturbances of the site-specific cellular mechanisms, and body cycle functioning of an organism. From previous studies, it has been observed that the same NPs show different cytotoxic and genotoxic effects when different models were used for experimental studies. Proteomics, transcriptomics, genomes, and metabolomics research are needed to quantify the NPs toxicity at exact levels and to better understand the underlying mechanisms. Knowing the safety parameters and appropriate application of NPs is crucial for comprehending their toxicity, cellular function, and internal fate. Similar to this, established and precise processes should be created to determine the particular particle size while interacting with the surface surrounds and the characteristics of NPs that cause them to exhibit harmful effects.

REFERENCES

Adarsha JR, Ravishankar TN, Ananda A, Manjunatha CR, Shilpa BM, Ramakrishnappa T (2022) Hydrothermal synthesis of novel heterostructured Ag/TiO_2/$CuFe_2O_4$ nanocomposite: Characterization, enhanced photocatalytic degradation of methylene blue dye, and efficient antibacterial studies. *Water Environ Res* 94(6). https://doi.org/10.1002/wer.10744

Ahamed M, Siddiqui MA, Akhtar MJ, Ahmad I, Pant AB, Alhadlaq HA (2010) Genotoxic potential of copper oxide nanoparticles in human lung epithelial cells. *Biochem Biophys Res Commun* 396(2):578–583. https://doi.org/10.1016/j.bbrc.2010.04.156

Alshatwi AA, Vaiyapuri Subbarayan P, Ramesh E, Al-Hazzani AA, Alsaif MA, Alwarthan AA (2012) Al_2O_3 nanoparticles induce mitochondria-mediated cell death and upregulate the expression of signaling genes in human mesenchymal stem Cells: AL_2O_3 nanoparticles in hMSCs. *J Biochem Mol Toxicol* 26(11):469–476. https://doi.org/10.1002/jbt.21448

Ananda A, Ramakrishnappa T, Archana S, Reddy Yadav LS, Shilpa BM, Nagaraju G, Jayanna BK (2022) Green synthesis of MgO nanoparticles using Phyllanthus emblica for Evans blue degradation and antibacterial activity. *Mater Today Proc* 49:801–810. https://doi.org/10.1016/j.matpr.2021.05.340

Archana S, Jayanna BK, Ananda A, B.M S, Pandiarajan D, Muralidhara HB, Kumar KY (2021) Synthesis of nickel oxide grafted graphene oxide nanocomposites – A systematic research on chemisorption of heavy metal ions and its antibacterial activity. *Environ Nanotechnol Monitoring Manage* 16:100486. https://doi.org/10.1016/j.enmm.2021.100486

Bachmann MF, Jennings GT (2010) Vaccine delivery: A matter of size, geometry, kinetics and molecular patterns. *Nat Rev Immunol* 10(11):787–796. https://doi.org/10.1038/nri2868

Baranowska-Wójcik E, Szwajgier D, Oleszczuk P, Winiarska-Mieczan A (2020) Effects of titanium dioxide nanoparticles exposure on human health—A review. *Biol Trace Elem Res* 193(1):118–129. https://doi.org/10.1007/s12011-019-01706-6

Becker K, Schroecksnadel S, Geisler S, Carriere M, Gostner JM, Schennach H, Herlin N, Fuchs D (2014) TiO2 nanoparticles and bulk material stimulate human peripheral blood mononuclear cells. *Food Chem Toxicol* 65:63–69. https://doi.org/10.1016/j.fct.2013.12.018

Borehalli Mayegowda S, Bhoomika S, Nagshetty K, and Manjula NG. "Environmental Adequacy of Green Polymers and Biomaterials." In *Polymeric Biomaterials*, by Pooja Agarwal, Divya Bajpai Tripathy, Anjali Gupta, and Bijoy Kumar Kuanr, 193–214, 1st ed. New York: CRC Press, 2022. https://doi.org/10.1201/9781003240884-10

Brunner TJ, Wick P, Manser P, Spohn P, Grass RN, Limbach LK, Bruinink A, Stark WJ (2006) In vitro cytotoxicity of oxide nanoparticles: Comparison to asbestos, silica, and the effect of particle solubility. *Environ Sci Technol* 40(14):4374–4381. https://doi.org/10.1021/es052069i

Campagnolo L, Massimiani M, Vecchione L, Piccirilli D, Toschi N, Magrini A, Bonanno E, Scimeca M, Castagnozzi L, Buonanno G, Stabile L, Cubadda F, Aureli F, Fokkens PH, Kreyling WG, Cassee FR, Pietroiusti A (2017) Silver nanoparticles inhaled during pregnancy reach and affect the placenta and the foetus. *Nanotoxicology* 11(5):687–698. https://doi.org/10.1080/17435390.2017.1343875

Cappellini F, Hedberg Y, McCarrick S, Hedberg J, Derr R, Hendriks G, Odnevall Wallinder I, Karlsson HL (2018) Mechanistic insight into reactivity and (geno)toxicity of well-characterized nanoparticles of cobalt metal and oxides. *Nanotoxicology* 12(6):602–620. https://doi.org/10.1080/17435390.2018.1470694

Catalán J, Siivola KM, Nymark P, Lindberg H, Suhonen S, Järventaus H, Koivisto AJ, Moreno C, Vanhala E, Wolff H, Kling KI, Jensen KA, Savolainen K, Norppa H (2016) *In vitro* and *in vivo* genotoxic effects of straight versus tangled multi-walled carbon nanotubes. *Nanotoxicology* 10(6):794–806. https://doi.org/10.3109/17435390.2015.1132345

Chattopadhyay S, Dash SK, Tripathy S, Das B, Mandal D, Pramanik P, Roy S (2015) Toxicity of cobalt oxide nanoparticles to normal cells; An in vitro and in vivo study. *Chem Biol Interactions* 226:58–71. https://doi.org/10.1016/j.cbi.2014.11.016

Chen L, Yokel RA, Hennig B, Toborek M (2008) Manufactured aluminum oxide nanoparticles decrease expression of tight junction proteins in brain vasculature. *J Neuroimmune Pharmacol* 3(4):286–295. https://doi.org/10.1007/s11481-008-9131-5

Cole AJ, David AE, Wang J, Galbán CJ, Yang VC (2011) Magnetic brain tumor targeting and biodistribution of long-circulating PEG-modified, cross-linked starch-coated iron oxide nanoparticles. *Biomaterials* 32(26):6291–6301. https://doi.org/10.1016/j.biomaterials.2011.05.024

De Jong WH, Hagens WI, Krystek P, Burger MC, Sips AJAM, Geertsma RE (2008) Particle size-dependent organ distribution of gold nanoparticles after intravenous administration. *Biomaterials* 29(12):1912–1919. https://doi.org/10.1016/j.biomaterials.2007.12.037

Di Ianni E, Jacobsen NR, Vogel UB, Møller P (2022) Systematic review on primary and secondary genotoxicity of carbon black nanoparticles in mammalian cells and animals. *Mutation Res /Rev Mutation Res* 790:108441. https://doi.org/10.1016/j.mrrev.2022.108441

Di Virgilio AL, Reigosa M, Arnal PM, Fernández Lorenzo De Mele M (2010) Comparative study of the cytotoxic and genotoxic effects of titanium oxide and aluminium oxide nanoparticles in Chinese hamster ovary (CHO-K1) cells. *J Hazard Mater* 177(1–3):711–718. https://doi.org/10.1016/j.jhazmat.2009.12.089

Dobrovolskaia MA, Shurin M, Shvedova AA (2016) Current understanding of interactions between nanoparticles and the immune system. *Toxicol Appl Pharmacol* 299:78–89. https://doi.org/10.1016/j.taap.2015.12.022

Dulińska-Litewka J, Łazarczyk A, Hałubiec P, Szafrański O, Karnas K, Karewicz A (2019) Superparamagnetic iron oxide nanoparticles—Current and prospective medical applications. *Materials* 12(4):617. https://doi.org/10.3390/ma12040617

Feng Q, Liu Y, Huang J, Chen K, Huang J, Xiao K (2018) Uptake, distribution, clearance, and toxicity of iron oxide nanoparticles with different sizes and coatings. *Sci Rep* 8(1):2082. https://doi.org/10.1038/s41598-018-19628-z

Fischer HC, Chan WC (2007) Nanotoxicity: The growing need for in vivo study. *Curr Opini Biotechnol* 18(6):565–571. https://doi.org/10.1016/j.copbio.2007.11.008

Foldbjerg R, Dang DA, Autrup H (2011) Cytotoxicity and genotoxicity of silver nanoparticles in the human lung cancer cell line, A549. *Arch Toxicol* 85(7):743–750. https://doi.org/10.1007/s00204-010-0545-5

Geiser M (2010) Update on macrophage clearance of inhaled micro- and nanoparticles. *J Aerosol Med Pulmon Drug Del* 23(4):207–217. https://doi.org/10.1089/jamp.2009.0797

Ghosh M, Chakraborty A, Mukherjee A (2013) Cytotoxic, genotoxic and the hemolytic effect of titanium dioxide (TiO_2) nanoparticles on human erythrocyte and lymphocyte cells *in vitro*: Hematotoxicity of TiO_2 nanoparticles. *J Appl Toxicol* 33(10):1097–1110. https://doi.org/10.1002/jat.2863

Gliga AR, Skoglund S, Odnevall Wallinder I, Fadeel B, Karlsson HL (2014) Size-dependent cytotoxicity of silver nanoparticles in human lung cells: The role of cellular uptake, agglomeration and Ag release. *Particle Fibre Toxicol* 11(1):11. https://doi.org/10.1186/1743-8977-11-11

Gnach A, Lipinski T, Bednarkiewicz A, Rybka J, Capobianco JA (2015) Upconverting nanoparticles: Assessing the toxicity. *Chem Soc Rev* 44(6):1561–1584. https://doi.org/10.1039/C4CS00177J

Grissa I, ElGhoul J, Mrimi R, Mir LE, Cheikh HB, Horcajada P (2020) In deep evaluation of the neurotoxicity of orally administered TiO2 nanoparticles. *Brain Res Bull* 155:119–128. https://doi.org/10.1016/j.brainresbull.2019.10.005

Haase A, Tentschert J, Jungnickel H, Graf P, Mantion A, Draude F, Plendl J, Goetz ME, Galla S, Mašić A, Thuenemann AF, Taubert A, Arlinghaus HF, Luch A (2011) Toxicity of silver nanoparticles in human macrophages: Uptake, intracellular distribution and cellular responses. *J Phys: Conf Ser* 304:012030. https://doi.org/10.1088/1742-6596/304/1/012030

Hou J, Wang L, Wang C, Zhang S, Liu H, Li S, Wang X (2019) Toxicity and mechanisms of action of titanium dioxide nanoparticles in living organisms. *J Environ Sci* 75:40–53. https://doi.org/10.1016/j.jes.2018.06.010

Jadimurthy R, Mayegowda SB, Nayak SC, Mohan CD, Rangappa KS (2022) Escaping mechanisms of ESKAPE pathogens from antibiotics and their targeting by natural compounds. *Biotechnol Rep* 34:e00728. https://doi.org/10.1016/j.btre.2022.e00728

Jain AK, Singh D, Dubey K, Maurya R, Pandey AK (2019) Zinc oxide nanoparticles induced gene mutation at the HGPRT locus and cell cycle arrest associated with apoptosis in V-79 cells. *J Appl Toxicol* 39(5):735–750. https://doi.org/10.1002/jat.3763

Jain S, Hirst DG, O'Sullivan JM (2012) Gold nanoparticles as novel agents for cancer therapy. *BJR* 85(1010):101–113. https://doi.org/10.1259/bjr/59448833

Jeng HA, Swanson J (2006) Toxicity of metal oxide nanoparticles in mammalian cells. *J Environ Sci Health Part A* 41(12):2699–2711. https://doi.org/10.1080/10934520600966177

Khan I, Bahuguna A, Krishnan M, Shukla S, Lee H, Min S-H, Choi DK, Cho Y, Bajpai VK, Huh YS, Kang SC (2019) The effect of biogenic manufactured silver nanoparticles on human endothelial cells and zebrafish model. *Sci Total Environ* 679:365–377. https://doi.org/10.1016/j.scitotenv.2019.05.045

Khanna P, Ong C, Bay B, Baeg G (2015) Nanotoxicity: An interplay of oxidative stress, inflammation and cell death. *Nanomaterials* 5(3):1163–1180. https://doi.org/10.3390/nano5031163

Kim Y-J, Choi H-S, Song M-K, Youk D-Y, Kim J-H, Ryu JC (2009) Genotoxicity of aluminum oxide (Al2O3) nanoparticle in mammalian cell lines. *Mol Cell Toxicol* 5(2):172–178.

Kiss B, Br T, Czifra G, Tth BI, Kertsz Z, Szikszai Z, Kiss RZ, Juhsz I, Zouboulis CC, Hunyadi J (2008) Investigation of micronized titanium dioxide penetration in human skin xenografts and its effect on cellular functions of human skin-derived cells. *Exp Dermatol* 17(8):659–667. https://doi.org/10.1111/j.1600-0625.2007.00683.x

Kongseng S, Yoovathaworn K, Wongprasert K, Chunhabundit R, Sukwong P, Pissuwan D (2016) Cytotoxic and inflammatory responses of TiO_2 nanoparticles on human peripheral blood mononuclear cells: High concentrations of TiO_2 –NPs could induce cytotoxicity in PBMCs. *J Appl Toxicol* 36(10):1364–1373. https://doi.org/10.1002/jat.3342

Korani M, Rezayat SM, Arbabi Bidgoli S (2013) Sub-chronic dermal toxicity of silver nanoparticles in Guinea Pig: Special emphasis to heart, bone and kidney toxicities. *Iran J Pharm Res* 12(3):511–519

Kose O, Tomatis M, Turci F, Belblidia N-B, Hochepied J-F, Pourchez J, Forest V (2021) Short preirradiation of TiO_2 nanoparticles increases cytotoxicity on human lung coculture system. *Chem Res Toxicol* 34(3):733–742. https://doi.org/10.1021/acs.chemrestox.0c00354

Kumar KM, Ram S, Manjula NG, and Borehalli Mayegowda S. (2022) Biopolymers and Their Applications in Biomedicine. In Agarwal P, Tripathy DB, Gupta A, Kuanr Bijoy Kumar (eds) *Polymeric Biomaterials*, 1st ed. New York: CRC Press, pp. 35–62. https://doi.org/10.1201/9781003240884-3

Liu G, Gao J, Ai H, Chen X (2013) Applications and potential toxicity of magnetic iron oxide nanoparticles. *Small* 9(9–10):1533–1545. https://doi.org/10.1002/smll.201201531

Magdolenova Z, Drlickova M, Henjum K, Runden-Pran E, Tulinska J, Bilanicova D, Pojana G, Kazimirova A, Barancokova M, Kuricova M, Liskova A, Staruchova M, Ciampor F, Vavra I, Lorenzo Y, Collins A, Rinna A, Fjellsbø L, Volkovova K, Marcomini A, Amiry-Moghaddam M, Dusinska M (2015) Coating-dependent induction of cytotoxicity and genotoxicity of iron oxide nanoparticles. *Nanotoxicology* 9(sup1):44–56. https://doi.org/10.3109/17435390.2013.847505

Manjula NG, Sarma G, Shilpa BM, Suresh Kumar K (2022) Environmental Applications of Green Engineered Copper Nanoparticles. In: Shah MP, Roy A (eds) *Phytonanotechnology*. Springer Nature, Singapore, pp 255–276.

Mayegowda KMK Shristi Ram, NG Manjula, Shilpa Borehalli (2022a) Biopolymers and Their Applications in Biomedicine. In: *Polymeric Biomaterials*. CRC Press, New York, 35–62.

Mayegowda SB, Ng M, Alghamdi S, Atwah B, Alhindi Z, Islam F (2022b) Role of antimicrobial drug in the development of potential therapeutics. *Evid Based Compl Alter Med* 2022:1–17. https://doi.org/10.1155/2022/2500613

Mayegowda SB, Sarma G, Gadilingappa MN, Alghamdi S, Aslam A, Refaat B, Almehmadi M, Allahyani M, Alsaiari AA, Aljuaid A, Al-Moraya IS (2023) Green-synthesized nanoparticles and their therapeutic applications: A review. *Green Process Synthesis* 12(1). https://doi.org/10.1515/gps-2023-0001

Mayegowda SB, Sureshkumar K, Yashaswini R, Ramakrishnappa T (2022c) Phytonanotechnology for the Removal of Pollutants from the Contaminated Soil Environment. In: Shah MP, Roy A (eds) *Phytonanotechnology*. Springer Nature, Singapore, pp 319–336.

Meng H, Chen Z, Xing G, Yuan H, Chen C, Zhao F, Zhang C, Zhao Y (2007) Ultrahigh reactivity provokes nanotoxicity: Explanation of oral toxicity of nano-copper particles. *Toxicol Lett* 175(1–3):102–110. https://doi.org/10.1016/j.toxlet.2007.09.015

Meyer K, Rajanahalli P, Ahamed M, Rowe JJ, Hong Y (2011) ZnO nanoparticles induce apoptosis in human dermal fibroblasts via p53 and p38 pathways. *Toxicol In Vitro* 25(8):1721–1726. https://doi.org/10.1016/j.tiv.2011.08.011

Mulenos MR, Liu J, Lujan H, Guo B, Lichtfouse E, Sharma VK, Sayes CM (2020) Copper, silver, and titania nanoparticles do not release ions under anoxic conditions and release only minute ion levels under oxic conditions in water: Evidence for the low toxicity of nanoparticles. *Environ Chem Lett* 18(4):1319–1328. https://doi.org/10.1007/s10311-020-00985-z

Müller L, Steiner SK, Rodriguez-Lorenzo L, Petri-Fink A, Rothen-Rutishauser B, Latzin P (2018) Exposure to silver nanoparticles affects viability and function of natural killer cells, mostly via the release of ions. *Cell Biol Toxicol* 34(3):167–176. https://doi.org/10.1007/s10565-017-9403-z

Murray AR, Kisin E, Inman A, Young S-H, Muhammed M, Burks T, Uheida A, Tkach A, Waltz M, Castranova V, Fadeel B, Kagan VE, Riviere JE, Monteiro-Riviere N, Shvedova AA (2013) Oxidative stress and dermal toxicity of iron oxide nanoparticles in vitro. *Cell Biochem Biophys* 67(2):461–476. https://doi.org/10.1007/s12013-012-9367-9

Narayanan D, M.G. G, H. L, Koyakutty M, Nair S, Menon D (2013) Poly-(ethylene glycol) modified gelatin nanoparticles for sustained delivery of the anti-inflammatory drug Ibuprofen-Sodium: An in vitro and in vivo analysis. *Nanomedicine* 9(6):818–828. https://doi.org/10.1016/j.nano.2013.02.001

Osman IF, Baumgartner A, Cemeli E, Fletcher JN, Anderson D (2010) Genotoxicity and cytotoxicity of zinc oxide and titanium dioxide in HEp-2 cells. *Nanomedicine* 5(8):1193–1203. https://doi.org/10.2217/nnm.10.52

Patra HK, Banerjee S, Chaudhuri U, Lahiri P, Dasgupta AKr (2007) Cell selective response to gold nanoparticles. *Nanomedicine* 3(2):111–119. https://doi.org/10.1016/j.nano.2007.03.005

Pawelczyk E, Arbab AS, Chaudhry A, Balakumaran A, Robey PG, Frank JA (2008) In vitro model of bromodeoxyuridine or iron oxide nanoparticle uptake by activated macrophages from labeled stem cells: Implications for cellular therapy. *Stem Cells* 26(5):1366–1375. https://doi.org/10.1634/stemcells.2007-0707

Porta F, Lamers GEM, Morrhayim J, Chatzopoulou A, Schaaf M, Den Dulk H, Backendorf C, Zink JI, Kros A (2013) Folic acid-modified mesoporous silica nanoparticles for cellular and nuclear targeted drug delivery. Advanced *Healthc Mater* 2(2):281–286. https://doi.org/10.1002/adhm.201200176

Radziun E, Dudkiewicz Wilczyńska J, Książek I, Nowak K, Anuszewska EL, Kunicki A, Olszyna A, Ząbkowski T (2011) Assessment of the cytotoxicity of aluminium oxide nanoparticles on selected mammalian cells. *Toxicol in Vitro* 25(8):1694–1700. https://doi.org/10.1016/j.tiv.2011.07.010

Ramakrishnappa SBM Kempahanumakkagari Sureshkumar, M Sahana, Thippeswamy (2022) pH and Thermo-Responsive Systems. In: *Polymeric Biomaterials*. CRC Press.

Rizk MZ, Ali SA, Hamed MA, El-Rigal NS, Aly HF, Salah HH (2017) Toxicity of titanium dioxide nanoparticles: Effect of dose and time on biochemical disturbance, oxidative stress and genotoxicity in mice. *Biomed Pharmacother* 90:466–472. https://doi.org/10.1016/j.biopha.2017.03.089

Roodbar ST, Soltani S, Derakhshani M (2022) Synthesis, properties, and biomedical applications of inorganic bionanomaterials. In: Barhoum A, Jeevanandam J, Danquah MK. (eds) *Fundamentals of Bionanomaterials*. Elsevier, pp 139–174. https://doi.org/10.1016/B978-0-12-824147-9.00006-6.

Rotti RB, Sunitha DV, Manjunath R, Roy A, Mayegowda SB, Gnanaprakash AP, Alghamdi S, Almehmadi M, Abdulaziz O, Allahyani M, Aljuaid A, Alsaiari AA, Ashgar SS, Babalghith AO, Abd El-Lateef AE, Khidir EB (2023) Green synthesis of MgO nanoparticles and its antibacterial properties. *Front Chem* 11:1143614. https://doi.org/10.3389/fchem.2023.1143614

Sadeghi L, Tanwir F, Yousefi Babadi V (2015) In vitro toxicity of iron oxide nanoparticle: Oxidative damages on Hep G2 cells. *Exp Toxicol Pathol* 67(2):197–203. https://doi.org/10.1016/j.etp.2014.11.010

Senut M-C, Zhang Y, Liu F, Sen A, Ruden DM, Mao G (2016) Size-dependent toxicity of gold nanoparticles on human embryonic stem cells and their neural derivatives. *Small* 12(5):631–646. https://doi.org/10.1002/smll.201502346

Shi L, Qian Y, Shen Q, He Y, Jia Y, Wang F (2022) The developmental toxicity and transcriptome analyses of zebrafish (Danio rerio) embryos exposed to carbon nanoparticles. *Ecotoxicol Environ Safety* 234:113417. https://doi.org/10.1016/j.ecoenv.2022.113417

Shilpa BM, Rashmi R, Manjula NG, Sreekantha A (2022) Bioremediation of Heavy Metal Contaminated Sites Using Phytogenic Nanoparticles. In: Shah MP, Roy A (eds) *Phytonanotechnology*. Springer Nature, Singapore, pp 227–253.

Srikanth K, Mahajan A, Pereira E, Duarte AC, Venkateswara Rao J (2015) Aluminium oxide nanoparticles induced morphological changes, cytotoxicity and oxidative stress in Chinook salmon (CHSE-214) cells: Capping agent affects uptake and toxicity. *J Appl Toxicol* 35(10):1133–1140. https://doi.org/10.1002/jat.3142

Stapleton SR, Osborne C, Illuzzi J (2013) Outcomes of care in birth centers: Demonstration of a durable model. *J Midwifery Women's Health* 58(1):3–14. https://doi.org/10.1111/jmwh.12003

Steckiewicz KP, Barcinska E, Malankowska A, Zauszkiewicz–Pawlak A, Nowaczyk G, Zaleska-Medynska A, Inkielewicz-Stepniak I (2019) Impact of gold nanoparticles shape on their cytotoxicity against human osteoblast and osteosarcoma in in vitro model. Evaluation of the safety of use and anti-cancer potential. *J Mater Sci: Mater Med* 30(2):22. https://doi.org/10.1007/s10856-019-6221-2

Teleanu D, Chircov C, Grumezescu A, Volceanov A, Teleanu R (2018) Impact of nanoparticles on brain health: An up to date overview. *JCM* 7(12):490. https://doi.org/10.3390/jcm7120490

Wang M, Li J, Dong S, Cai X, Simaiti A, Yang X, Zhu X, Luo J, Jiang L-H, Du B, Yu P, Yang W (2020) Silica nanoparticles induce lung inflammation in mice via ROS/PARP/TRPM2 signaling-mediated lysosome impairment and autophagy dysfunction. *Part Fibre Toxicol* 17(1):23. https://doi.org/10.1186/s12989-020-00353-3

Xiao L, Liu C, Chen X, Yang Z (2016) Zinc oxide nanoparticles induce renal toxicity through reactive oxygen species. *Food Chem Toxicol* 90:76–83. https://doi.org/10.1016/j.fct.2016.02.002

Yadav LSR, Shilpa BM, Suma BP, Venkatesh R, Nagaraju G (2021) Synergistic effect of photocatalytic, antibacterial and electrochemical activities on biosynthesized zirconium oxide nanoparticles. *Eur Phys J Plus* 136(7):764. https://doi.org/10.1140/epjp/s13360-021-01606-6

Zhang Y, Bai Y, Jia J, Gao N, Li Y, Zhang R, Jiang G, Yan B (2014) Perturbation of physiological systems by nanoparticles. *Chem Soc Rev* 43(10):3762–3809. https://doi.org/10.1039/C3CS60338E

19 Cytotoxicity and Biocompatibility of Bionanomaterials

Mehmethan Yıldırım, Durmus Burak Demirkaya, Cansu Aydın, and Serap Yalcın

19.1 INTRODUCTION

Today, as a result of the development of the field of nanotechnology, the use of nanometer-sized devices and the materials that make them up has become widespread. In order to produce biological materials with new functional nanostructures, it is necessary to take advantage of the complexes formed by the structures we call biomolecules, the use, usage processes, and functions of nanosystems that develop accordingly. Thus, rapidly growing sub-fields of nanotechnology and biotechnology were created (Salata, 2004; Feynman, 1991).

The created systems host many functional structures that are used differently, from biochips, bio-nanocomposites, and bio-barcodes. Biomarkers used to detect macromolecules such as DNA and proteins, or to make clinical diagnoses, are some examples of this process. When the use of bionanomaterials comes into play in this regard, nanoparticles are produced with nano-scale hybrid materials. In this way, we are talking about bionanomaterials in bioimaging, biosensing, and therapeutic and diagnostic applications, especially in health (Salata, 2004; Sinani et al., 2003).

Bionanomaterials are materials that are selected by the biological structure and used in living organisms so that their toxicity is minimized. It is produced and applied for living organisms to perform their tissue functions and to restore the functions of lost or needed tissues to the organism in a healthy way. These materials, which are produced naturally or synthetically, must meet the expected requirements structurally and functionally and must be biocompatibility. This is because the living organism will come into contact with the body flow continuously or at specific intervals (Salata, 2004; Feynman, 1991; Sinani et al., 2003).

The development of nano-biotechnology has led to the formation of various designed structures. For example, nanoparticles, nanotubes or nanofibers and lipids, polysaccharides, oligonucleotides, biological cofactors, and so forth. Biological molecules have interacted in many applications. Bionanomaterials offer different functions in different fields with their features, such as multifunctionality, quantum structure, high surface-to-volume ratio, improved resolution and macroscopic quantum tunneling (Bruchez et al., 1998).

Bionanomaterials, which are based on molecular interactions between nucleotides, proteins, lidips and even small ligands, and where the highest level of biocompatibility is desired, are now designed, produced, and used from the food industry to surgical operations. Bionanomaterials such as bioorthogal chemical conjugations, protein chemical ligations, technologies including DNA/RNA nanotechnology, gene mutagenesis, and DNA sequence amplification are examined in the literature, and studies are questioned about their cytotoxic structures. The formation, cytotoxicity, and cell interactions of these materials, which have a great impact on both molecular biology and genetic engineering and biotechnology, will be covered in the following sections (Salata, 2004).

DOI: 10.1201/9781003432791-22

19.2 RAW MATERIALS

Bionanomaterials consist of different nano conformations such as nanoparticles, nanotubes, nanolayers, nanofibers, and nanocomposites. They can be designed in different morphological structures in different nano sizes such as rings, hollow or filled spheres, micelles, nanospheres, and dendrimers (Mah et al., 2000).

Successful in vivo maturation of these designs requires the realization of biomedical applications in tissue engineering and the precise control of the delivery of so-called bioactive agents within a scaffold (Edelstein et al., 2000). From beginning to end, drug loading capacities should be improved, not only biologically but also chemically, and a shift to multifunctional systems should be made instead of simple delivery systems. Bionanomaterial products designed at the nanoscale, like implants used at the macroscale, are of great importance for living organisms (Ma et al., 2003).

19.2.1 Metallic Bionanomaterials

Crystal structures, strong metallic and mechanical properties, metal, and metal alloys reveal many different uses when entering the field of biomaterials. For instance, as a joint prosthesis, and bone regeneration material; it is used in field applications as artificial heart parts, catheters, valves, and heart valves (Melchor et al., 2021; Eliaz, 2019).

The metallic parts of the devices produced for diagnostic and therapeutic purposes are made of biomaterials (Eliaz, 2019). The metal Vanadium Steel can be said to be the first example used in the human body. This substance, which is used as a screw and plate in bone fractures, was found to cause corrosion in the body after the 1960s and was excluded from the application due to its toxicity (Eliaz et al., 2019; Eliaz, 2012).

The most important criterion is that the biomaterial to be used is compatible with the body. Metals are known in the literature as they corrode in the body. Corrosion is the event that metals lose their properties by forming compounds such as oxygen and hydroxide as a result of unwanted chemical reactions with their environment. Since ions such as fluid, water, dissolved oxygen, chloride, and hydroxide in the human body have a highly corrosive environment, biomaterials in this environment may weaken as a result of corrosion or may damage cells (Melchor et al., 2021; Eliaz, 2019).

Apart from corrosion, galvanic battery formation is also very important (Eliaz, 2012). When different metals come into contact with each other, a galvanic cell is formed in the body fluid and galvanic corrosion occurs (Virtanen, 2008; Anderson et al., 2008).

The points to be considered in the selection of materials are:

- Biocompatibility,
- Manufacturability,
- Formability,
- Toxic effect,
- Corrosive effects of body fluids.

Biomaterials should be selected according to the biological structure of the place to be used. To achieve this, the compatibility test of the biomaterial must be done with samples taken from body fluids and solutions prepared very close to these samples (Melchor et al., 2021; Virtanen, 2008).

A biocompatible material should not impair normal bodily functions as much as possible and should certainly minimize toxicity. Although there are multiple factors affecting biocompatibility, especially in implants; surface heat ability, implant size, surface smoothness, surface charge and material composition are the primary factors.

The use of metal as a raw material in the use of bionanomaterials (especially in implants) may cause corrosion of the used implants. This corrosion can affect the surrounding tissues in three ways.

One of them is that the corrosion process can change the chemical environment, and metal ions can change cellular metabolism as electric current affects the behavior of cells. The problems caused by the release of corrosion products into the body actually lead to the emergence of systematic and distant side effects. In many cases, even mild corrosion can also cause symptoms ranging from tenderness in the worn area to redness, pain, and swelling in the entire area around the device. The use of metallic bionanomaterial causes deposition due to the flux of these metal ions released by corrosion. It can lead to simultaneous accumulation of organ-specific ions in the body and very high concentrations of special alloying elements of implant structures (Anderson et al., 2008; Gilbert et al., 2012).

For this reason, the effect of physiological tolerance to toxicity occurs and can disrupt the general balance of the body. Bionanomaterials with cobalt-based or stainless steel raw materials cause longer-term changes in the blood composition of patients and animals with orthopedic total joint replacement components. When histological examination of metallic content in tissues is investigated, discoloration and FBRs that begin to form in surrounding tissues indicate implant corrosion, with no clear evidence of slow release of metal ions (Malchor et al., 2021; Gilbert et al., 2012; Blau, 1992).

The concept of metallosis in the literature includes the local destruction of soft or hard tissues due to the use and placement of metals, known as raw materials for bionanomaterials, in implants. As it is usually associated with osteolysis, it can sometimes constitute a wear condition of joint arthroplasty. There is a certain effect, manifested by the infiltration of the resulting metallic residues into the periprosthetic soft tissues and even into the bone. Metal ions by themselves do not carry the structural complexity required to challenge the immune system. However, they interact with biological molecules found in the skin, fat, and adipose tissues, and even in the blood. Immune responses are thus induced, considered harmful. The concept of type hypersensitivity is given to the most typical reaction of a metal sensitive person in such situations (Gilbert et al., 2012; Blau, 1992).

Bionanomaterial raw materials used for the human body do not cause permanent damage to the body (Eliaz, 2019). It is indispensable in the combinations designed in this regard. The mass and number of the molecule, the active or inactive ion of the ion, and its interaction with water molecules determine whether it can react with anions, which we call inorganic (Eliaz et al., 2019). For example, titanium ion is a very active ion and can react effectively with hydroxyl radicals and anions to form salts and oxides in body fluids. Therefore, it reveals a low probability of combination with biomolecules. On the other hand, to give another example, inactive ions in the structure of copper or nickel do not immediately combine with water molecules and inorganic anions and remain ionic for a long time. This may further increase the likelihood of these ions combining with biomolecules, increasing toxicity (Eliaz, 2012; Virtanen, 2008).

So while corrosion itself is not a major concern, there are factors that can lead to significant implant failure, such as mechanical action, inflammation, limited slit-like geometries, and any combination of these. For this reason, the metal raw material used as bionanomaterial is not preferred today. However, there are many studies on this substance in the literature (Eliaz, 2019; Virtanen, 2008; Blau, 1992).

19.2.2 Polymeric Bionanomaterials

Polymeric-based bionanomaterials are frequently used materials in recent years due to their small-size properties. Polymeric bionanomaterials, which we call drug carriers, have certain advantages. For example, by interacting with other molecules with biological activity, the drug's ability to protect against the environment is enhanced. This reveals their potential uses for controlled release, namely their bioavailability and therapeutic indices. Therefore, both nanospheres and nanocapsules, which contain the term "nano" and differ according to their morphology, are included in this group of polymer-based bionanomaterials (Zielińska et al., 2020; Jawahar et al., 2012).

The nanocapsules are encapsulated with the aid of a polymeric acceptor, in which the drug is usually dissolved upon controlling its profile by the release of the drug from the core, and this consists of an oily core. The structures we call nanospheres are not based solely on the polymeric network that the drug can hold or somehow adsorb to their surface. As a result, it is placed in a separate category from other species known as nanospheres besides the reservoir system (Jawahar et al., 2012; Krishnan et al., 2020; Singh et al., 2009).

Polymer-based bionanomaterials differ in their surface properties, structures, sizes, and even crystallinity and distribution, along with different physical properties in their concentration or composition. These properties are generally evaluated by various methods attempting to achieve full characterization of nanoparticles. Methods such as dynamic light, electron microscopy, various electrophoresis methods, and photon correlation spectroscopy are included in this characterization (Krishnan et al., 2020; Singh et al., 2009; Mourdikoudis et al., 2018).

Characterization of polymer-based bionanomaterials is very important. Being well characterized increases their applicability, but also requires attention to its nanotoxicology.

Transmission and scanning electron microscopy is a device used in research to reveal the size, structure and shape of polymer-derived bionanomaterials. These can generally be used to perform morphological analyzes of bionanomaterials and are combined with cryofraction techniques. The device, which we will briefly talk about as TEM, has a widespread use. In addition to determining the wall thickness of the nanocapsule, it can distinguish between nanospheres and nanocapsules. The structures we call nanospheres can be distinguished by their solid polymeric structure and their spherical presence. Nanocapsules, on the other hand, are formed from a thin polymeric envelope of about 5 nanometers around the oily core. Characterizing the surface morphology of polymer-based bionanomaterials is also performed by atomic force microscopy. This device, which shows surface details with a high resolution in three dimensions and nanometric scale, can also solve surface details in an atomic way. With the application of this technique, a topography observation is also provided on the surfaces of the nanoparticles. Sections of the samples are analyzed and the presence of pores is also revealed (Jawahar et al., 2012; Krishnan et al., 2020).

In fact, in general, polymer-based bionanomaterials obtained by different methods can have average ages between 100–300 nm. Multi-dispersion has as low a dispersion capacity as possible and a low percentage of size is also unimodal. Particles with diameters between about 60–70 nm and even smaller than 50 nm are currently being designed and produced (Jawahar et al., 2012).

The dimensions of polymer-based bionanomaterials can be measured using various techniques and can be designed according to the differences in application areas. The most commonly used methods are techniques called AFM, dynamic, and static light scattering, SEM, and TEM. Dimensional measurements of designs vary according to the method used. To give an example, polymer-based bionanomaterials used in electron microscopy provide an image of a particle isolated from the environment; dynamic light scattering, abbreviated as DLS, allows the hydrodynamic radius to be determined (Bundschuh et al., 2018).

In addition, DLS and TEM are also capable of measuring larger sizes and provide information on the aggregation state of a nanoparticle by detecting changes in particle size distribution. Many factors such as quantitative and qualitative composition can affect the size of polymer-based bionanomaterials. Nanocapsules can be given as an example, and a factor that affects the particle diameter during their manufacture is the oil used as the nucleus. Nanoparticles also have different factors that can affect their lifetime. A wider size distribution may cause variations in the amount of drug for large particles (Zielińska et al., 2020; Mourdikoudis et al., 2018).

When the chemical compositions of polymer-based bionanomaterials are examined, it is seen that they express natural or formed functional groups as well as atomic elements. Atomic absorption-based spectroscopy is one of the most widely used ensemble techniques. With this technique, the ground state electrons of atoms are given a certain amount of energy, which can be measured from light of a certain wavelength, and are excited. A signal associated with the amount of light absorbed and the number and type of atoms in the light path is estimated by determining the sample mass concentration and compared to signals from calibration standards at known concentrations (Zielińska et al., 2020; Krishnan et al., 2020; Singh et al., 2009).

Determination of the chemical composition of a particle is a method known as time-of-flight mass spectrometry, which generally serves to provide minimal separation and gas phase ionization of small to large organic analytes. Determination of the chemical composition takes place by detecting their separation using a certain period of time. To summarize, TOFMS determines the arrangement of element atoms in a nanoparticle by electron diffraction with a transmission electron microscope, which is used as a crystal structure. With X-ray diffraction involved, about one gram of material is examined for analysis. However, electron diffraction can be done on a single particle, and this method cannot provide more (Mourdikoudis et al., 2018; Bundschuh et al., 2018).

Quantification of drugs associated with polymer-based bionanomaterials is only due to the small-size substances associated with the free fraction of the drug. Therefore, it would not be wrong to say that this process has a complex side. Usually in a separation technique procedure, after centrifugation, the suspension in the supernatant undergoes ultracentrifugation, where free drug is determined. The main concentration of the total drug is completed by dissolving a fraction of the nanoparticles with the help of a suitable solvent. The drug concentration associated with these bionanomaterials is calculated based on the difference in the main doses between free and total drug concentrations. There is also ultrafiltration centrifugation, in which a membrane is used to separate some of the dispersed aqueous phases from the colloidal suspension, to mention another method. According to studies in the literature, several factors can affect the amount of drug associated with nanostructured systems. To give an example, many factors, such as pH in the environment, physicochemical properties of the drug, and surface-dependent properties of bionanomaterials, the nature of the polymer, the order of addition, and dissolution of the drug to the formulation, and the surfactant type factors adsorbed to the polymeric surface can be counted. Although the parameters are quite large, these methods lead to the purpose of determining the drug isotherm on the surface of bionanomaterials, since they provide information about how the drug is dispersed on the particle surface or its binding capacity. To give information about nanospheres, different drug conjugation forms have taken place in the literature. One of them is that the drug can be dissolved and dispersed in a chemical called polymeric matrix or adsorbed on the polymer. Speaking of nanocapsules, they are designed and manufactured to load lipophilic drugs that must be retained by the soluble polymeric membrane in the fatty nucleus (Zielińska et al., 2020; Mourdikoudis et al., 2018; Bundschuh et al., 2018).

19.2.3 Composite-Based Bionanomaterials

Composite materials are prepared by adding various reinforcing materials into the material called matrix (Jain et al., 2019). Generally, various polymers are used as matrix, glass, carbon, or polymer fibers as reinforcement, sometimes mica and various powder ceramics are used (Chenthamara et al., 2019; Velu et al., 2019).

19.2.3.1 Bionanocomposites Containing Metallic-based Nanofillers

Metallic nano-fillers are used to improve electrical conductivity and antimicrobial properties in tissues. Especially in biomedical applications, these substances have an interesting structure. To give as an example the incorporation of gold nanoparticles into extracellular matrices, namely the ECM, musculoskeletal and other tissue engineering applications have been brought to the literature for their work, where it improves cellular function by enhancing the ECM while reducing inflammation. Functionalization with magnetic particles to obtain a material with both magnetic properties and antibacterial properties, which we call bio-based polyurethane backbones, provides more benefits for its intended use (Jain et al., 2019; Chenthamara et al., 2019; Velu et al., 2019).

19.2.3.2 Ceramic Nanofilled Bionanocomposites

To give an example of strengthening the polymer with ceramic nanofillers, reinforcement with chitosan pill nanopowders can be given. Pure chitosan is known to have poor mechanical properties. For this reason, its use in load-bearing applications is excluded and the use of bionanomaterial with a crystal conformation that can be said to be close to the natural bone; demonstrated the synthesis of scaffolds with high strength, controlled pore structures. Similar to chitosan, its cellulose structure is functionalized to produce artificial bone tissue scaffolds (Velu et al., 2019; Blum et al., 2015; Kumar et al., 2016).

19.2.3.3 Bionanocomposites Containing Carbon-based Nanofillers

Carbon nanotubes, or CNTs for short, can be designed from polymers such as polyurethanes and collagen. It has been found in studies that the incorporation of CNTs into biocompatible polymers increases the growth and differentiation of different cells such as neurons or bone cells. For example, the integration of single-walled carbon nanotubes into hydroxyapatite-doped chitosan improves osteoblast adhesion to scaffolds and their cytotoxicity in CNTs is fully addressed before administration to the body (Velu et al., 2019; Bhowmick et al., 2016).

19.2.3.4 Bionanocomposites Containing Cellulose-based Nanofillers

Cellulose-based bionanomaterials are gaining more and more attention due to their large area, mechanical properties, and ability to self-assemble into networks to form a complementary structure, as well as their low cytotoxicity and biocompatibility. However, they have a hydrophilic nature that allows them to challenge their distribution in polymer-based matrices. Cellulose can be divided into three different categories containing fillers:

- Cellulose Nanocrystals (CN),
- Bacterial Cellulose (BC),
- Cellulose Nanofibrils (CNF).

It is known that carbon nanotubes can be used as additives in cement-based materials and have the potential to improve mechanical strength. Besides the biocompatibility of this material, its biodegradability, tissue repair and even healing, and its use in biomedical applications such as medical implants are very interesting. For bacterial-based cellulose materials, a pure cellulose nanofiber network structure woven by bacteria, which we usually call vascular implant, is used. This material, which offers extraordinary durability with its ability to be designed at nano, micro, and macro scales, is evaluated as an advantage as a physical barrier that reduces bacterial infection by including its biocompatibility at an optimal three-dimensional level and microfibrillar structure (Velu et al., 2019; Blum et al., 2015).

19.3 THE ROLE OF BIONANOMATERIALS IN CANCER IMMUNOTHERAPY

Cancer immunotherapy has changed cancer treatments from top to bottom since its emergence, and the most important difference of this treatment method from other treatment methods is to minimize the death rate of non-target normal cells as well as cancer cells directly, and to improve patients by stimulating antitumor immunity (Kumar et al., 2016; Bhowmick et al., 2016). In cancer immunotherapy, the activation of the immune system cells by using agents of the cancer cells that escape from the immune system cells is to reveal them again against these cells. For this reason, immunotherapy is considered to be a very promising treatment method in the treatment of certain types of cancer (Thomas et al., 1986).

To give an example of some of the studies used in immunotherapy, recombinant interleukin-2, or IL-2, as we can abbreviate, was studied as an immunotherapy agent in cancer-based studies and was later approved by the FDA primarily for kidney cancer and metastatic melanoma (Bhowmick et al., 2016). The use of IL-2 created a lot of excitement in a short time because patients started to get permanent responses. However, the troublesome part of this treatment was that the doses used were very high, and this high dose caused very serious side effects in some patients (Rosenberg et al., 2014). Looking at the recent studies in the literature, there are checkpoint inhibitors called programmed cell death receptor 1, which we will briefly refer to as PD-1, and which target its ligand, PD-L1. In addition, the chimeric antigen receptor, that is, CAR, has been clinically developed and validated with the advancement of technological methods in T cell therapies (Riley et al., 2019; Lee et al., 2011). There are many immunotherapy systems currently in use, and they can be divided into multiple classes: checkpoint inhibitors, lymphocyte-activating cytokines, cancer vaccines, and antibodies against co-stimulatory receptors.

New approaches are constantly being developed to make cancer immunotherapy more effective and less toxic. Studies in the delivery systems of agents used in these treatments may reduce the accumulation of less agents and reduce toxicity, or they may affect or stimulate target cells more effectively, reducing the risk of occurrence of undesirable conditions. In this section, we will focus on the advantages, disadvantages, and future perspectives of the systems used in cancer immunotherapy.

19.3.1 Checkpoint Inhibitors

When we look at checkpoint inhibitors, inhibition of the PD-1/PD-L1 complex and CTLA4 structure are among the most studied and available immunotherapy treatment strategies in the literature (Riley et al., 2019; Alwan et al., 2014).

Under normal conditions, the purpose of immune checkpoint inhibitors is to protect healthy tissues from the attacks of immune system cells caused by all kinds of pathogenic effects that enter the cell (Ribas et al., 2018). For example, they show PD-1 activation to protect normal tissues from T cell activation, especially CDT8+ activity, which occurs when the cell is inflamed (Couzin, 2013; Alsaab et al., 2017).

Another immune checkpoint, CTLA4, can be defined as a co-inhibitor molecule that regulates the active or inactive T cell. The reactions between CTLA4 and its ligands are involved in the inhibition of T-cell activity and inhibit the growth of tumor cells (Alsaab et al., 2017).

19.3.2 Cytokines

Cytokines has been a treatment method at the peak of immunotherapy since the day IFN-α was approved (Munn et al., 2016). Once cytokines enter the cell, or rather, are injected, they benefit immune system cells and allow them to grow and become active. They can be directly involved in this regard (Rosenberg et al., 2014). There are three main types of cytokines followed as cytokine therapy in immunotherapy: Interferons, Interleukins, and Granulocyte-macrophage colony stimulating factor (GM-CSF) (Rosenber et al., 2014). As a new example, interferon can be administered for use in these treatments. Interferons are chemical secretions that respond to a microbial pathogen under normal conditions and are stimulated by immune system cells (Ahmed et al., 2003). In addition, molecules of many structures, including macrophages, natural killer cells, as well as lymphocytes and dendric cells naturally found in cells, are involved in the activation of a large number of immune system cells (Lee et al., 2011; Sun et al., 2014; Müller et al., 2017).

19.3.3 Engineered T Cells

In recent studies, the success of clinical trials of engineered T cells is quite remarkable. In this treatment method, first, blood samples are taken from patients and T cells are collected from patients, and then these T cells are genetically redesigned to target a cancer-specific antigen, and then these designed T cells are reinjected into the same patient (Rosenberg et al., 2014). These T cells, which begin to form after the injection process, recognize the antigens on the designed structures. Thus, it ensures the death of tumor cells (Katze et al., 2002). In addition to other treatment methods, the difference of this treatment is that it is injected only once, and these cells can maintain their activities for many years (Lim et al., 2017).

19.3.4 Stimulant Receptor Agonists

Antibodies, which we call agonists, are highly specific to receptors on the surfaces of T cells, which play an important role in the immune system. This reveals that there are surface-specific receptors here. Their aim is to bind to the receptors on this T surface and increase the growth and survival time of these cells and enable them to perform a more effective activation (Scholler et al., 2012). To give examples of these receptors, CD28, TNFR, 4-1BB, OX40 can be given as examples. Ligand binding to these receptors aims to shorten the growth and life span of cancer cells by increasing anticancer activation (Peggs et al., 2009; Chester et al., 2018).

Agonist antibodies currently have no candidate approved by the FDA. This treatment method, in addition to other treatment methods, is still under investigation and development (Riley et al., 2019).

19.3.5 Cancer Vaccines

Cancer vaccines used in immunotherapy include tumor cell lysate, dendric cells, nucleic acids and neo-antigens (Yang et al., 2019). Dendric cell vaccines are one of the most widely used vaccine methods in this treatment method. Functionally, it provides the expression of cancer-related antigens and is designed to attack cancer cells and activate T cells directly, and this is done with dendric cells collected from patients (Riley et al., 2019). Although dendritic cell vaccines are highly safe and highly effective, they have unfortunately failed in clinical trials (Guo et al., 2013).

19.3.6 Bionanomaterials in Cancer Immunotherapy

19.3.6.1 Lipid-based Biomaterials

Dendric cells, which are among the most potent antigen presenting cells, have very strong potentials in both innate immunity and in providing additive immunity (Rosenberg et al., 2004; Palucka et al., 2012). In the induction of the CD8+ immune system, which is an antigen-specific cytotoxic T cell, antigen-presenting cells must be handled heavily by dendric cells (Schlitzer et al., 2015; Steinman et al., 2012). The method, which is based on injecting dendric cells into patients, is very time consuming and costly, so it delivers antigens designed with lipid-based nanomaterials such as liposome technology directly into dendric cells in vivo (Lee et al., 2015; Karaki et al., 2016; Sabado et al., 2017; Le Moignic et al., 2018).

Liposomes are nano-sized bubble-shaped bionanomaterials made of phospholipid bilayer as its main component (Yang et al., 2019; Torchilin, 2005). They are immunotherapy vectors that have been used continuously in the market in vaccines such as hepatitis A, on which studies have been conducted for many years (Xing et al., 2016). To give an example of some of the studies on this system, our first example is Lai et al. In a study, synthesized liposomes designed with dendric cell targeting and immune adjuvant CpG-ODN loaded with the Melanoma-specific TRP2180-188 peptide, these liposomes allowed to further increase CD8+ cells specific for tumor antigens and to proliferate more than they should in the diseased area and, as a result, it provides the vascular formation of tumor cells and inhibition of proliferation of tumor cells (Pattni et al., 2015).

In another study, it has been shown that TGFβ uses small molecules to prevent TGFβ receptor from interacting with each other, and that liposomes pre-inserted into T cells modified with inhibitors will enable T cells to access B16F10 melanoma tumors more efficiently (Lai et al., 2018).

In their studies, Nikpoor et al. placed antibodies containing CTLA-4 inhibition in a liposomal bionanomolecule and examined its antitumor effects.

19.3.6.2 Polymers-based Bionanomaterials

When micelles reach critical micelle concentrations, they tend to coalesce as amphiphilic nanosized polymer fragments, namely micelles (Munn et al., 2016; Ahmed et al., 2003). Thanks to this formation, these particles, which are smaller in size and become neutral, tend to reach the lymph nodes more easily (Zheng et al., 2017; Nikpoor et al., 2017; Letchford et al., 2007).

If we take a look at the studies, Zeng et al., by combining and modifying polyethylene (PEI)-2K with stearic acid, produced an amphiphilic molecule (PSA) and a PSA micelles that cause selective specificity accumulation in drainable lymph nodes instead of the systemic site, and this system exerts antitumor effects by inducing their expression in lymph nodes and certain proteins provided (Liu et al., 2018; Zhan et al., 2012; Reddy et al., 2007; Yang et al., 2020).

In another study, researchers investigated an arousal in the immune system. For this, they measured and investigated the capacities of cationic liposomes and PLGA nanoparticles, which we can call long-based synthetically (Jiang et al., 2018; Min et al., 2017). In this study, the researchers discovered that liposomes were more successful in stimulating T cells than PLGA particles (Schmid et al., 2017; Li et al., 2012).

Antigen-capturing nanoparticles, aka AC-NPs, have demonstrated that PLGA polymers designed for reproducible use support an antitumor immune stimulus. It has also been proven to provide high-dose activation of aPD-1 immunotherapy (Peng et al., 2018; Ye et al., 2016). PLGA has been modified by different kinds of chemical groups to bind the surfaces of nanoparticles and tumor antigens (Su et al., 2020; Choi et al., 2015). Protein is abundant in all AC-NPs except mPEG AC-NPs. This allows PLGA and Mal AC-NPs to improve immunotherapeutic efficacy, demonstrating a higher ability in another respect (Jung et al., 2015; Zhou et al., 2015).

19.3.7 Inorganic Bionanomaterials

Silica nanoparticles with mesopores can be prepared for use in condensation and hydrolysis reactions by using a substance called organosilane to participate in certain reactions (Wang et al., 2017; Kempe et al., 2017). Moreover, by modifying the surfaces of these particles, various reactive groups can be studied in different applications (Doronin et al., 2009; Wagner et al., 2020). It has been suggested that silica nanoparticles modified using amino acids are effective in promoting cytokine production, and silica nanospheres supplemented with Zn, Ca, and Mg (MS-Zn, MS-Ca and MS-Mg) can stimulate the Th1 anticancer immune response (Parky et al., 2010; Zeng et al., 2015). Silica with mesopores and particles in hollow particulate form target an intact lymph node (Moffett et al., 2017). It was designed to load adjuvants and antigens to determine how immune cell stimulation would affect their volume (Kim et al., 2015; Levy et al., 2011; Li et al., 2018; Kyriakides et al., 2021). Yang et al. succeeded in the synthesis of hollow spheres in dendritic mesoporous organosilica structure for the first time, and conducted work to trigger an antitumor immune response (Yu et al., 2020; Sunshine et al., 2014; Zeng et al., 2015).

In the research of Mooney et al., GM-CSF and CpGs were studied to elicit cellular immunity, humoral and specific responses against tumors (De Koker et al., 2016; Wang et al., 2016). In the presence of these two complexes, dendric cells were replaced by a macroporous structure and self-assembled high aspect ratio mesoporous silica rods were found (Gause et al., 2015; Levy et al., 2012).

Iron oxide nanoparticles are not approved for use as MRI agents for the treatment of human diseases. Iron oxide nanoparticles are the more preferred nanoparticles in cancer immunotherapy and imaging, as well as concurrent use, as different effects on the human body's iron storage have been observed (Gause et al., 2015). There is a high probability of integrating treatment as well as imaging.

In this section, we analyzed the molecular mechanisms of different treatment strategies used in immunotherapy and the treatment methods applied using nanomaterials. Although methods based on cancer immunotherapy are advancing at a fast pace, the use of biomaterials to produce optimal systems for many tumor structures is still in its infancy. The development of these systems will become more effective with future studies. With this chapter, it is hoped that biomaterials can be designed broadly and innovatively in cancer immunotherapy, thus increasing their effectiveness, resulting in less toxic but more potent effects, and reducing immune-related side effects (Jeevanandam et al., 2018).

19.4 BIOCOMPATIBILITY AND TOXICITY OF BIONANOMATERIALS

Today, bionanomaterials designed for biomedical treatments and applications should be considered in two different categories in terms of their toxicity and general biocompatibility. In particular, it refers to its effects on cytotoxicity, viability and cell functions, and it is of great importance in every biological study. Second, it should include a comparative evaluation of biocompatibility by performing comprehensive in vivo and in vitro studies of bionanomaterials. It is important that they are tested according to standard protocols to ensure their biosafety. For example, in accordance with the guideline of ISO, ISO/TR 10993-22:2017 Chapter 22, bionanomaterials:

- Cytotoxicity,
- Carcinogenic structure,
- Reproductive toxicity,
- Genotoxicity,
- Pyrogenicity,
- Immunotoxicity,
- İrritation and sensitization,
- Implantation,
- Certain results should be obtained by testing hemocompatibility parameters. Precautions should be taken to prevent adverse effects. Although we talk about the importance of toxicology tests, understanding the basic physical and chemical properties of bionanomaterials, as well as some specific bioeffects and the relationships between these effects and biomolecules, is of particular importance to expand their applications. Studies and research on the applications of bionanomaterials in therapeutic treatments have demonstrated their ability to reduce fibrosis, control the scarring process, and inhibit cancer growth (Padmanabhan et al., 2015; Banik et al., 2016).

In addition, various types of materials, analytical measurements, and biological systems that measure their interactions can complicate such assessments. When looking at some situations in the literature research, a number of difficulties arise in understanding the bio-interaction states of bionanomaterials. In order to understand the combined state of the physical and chemical properties of nanomaterials through their interactions with biological systems, it is necessary to look at the relationship structure;

- There is a wide variety of bionanomaterials that vary in size, chemical composition, shape, and surface;
- There are quite a few different model mechanisms available for use in experiments. Thus, different nanomaterials are used that interact in different ways.
- Availability of standardized procedures such as incubation times.
- Varies and branches according to the increasing complexity associated with in vivo testing. In fact, this keeps the amount of material to be tested optimally and may not provide a full

understanding of the effects of nanomaterials (Peres et al., 2017; Reis et al., 2006; Cha et al., 2013; Mochalin et al., 2011).

Considering that the main theme discussed in this chapter is well reinforced, we should consider the most basic test methods and use them according to the working capacity needed as special applications. We will present the effects and interactions of bionanomaterials on immunomodulation, fibrosis, cytotoxicity and genotoxicity of occupational exposures in the production phase mentioned in the literature, in sub-sections (Torres et al., 2019; Yadav, 2016; Lin et al., 2014; Ngobili et al., 2016).

19.4.1 Cytotoxicity and Genotoxicity in Bionanomaterials

The engineered bionanomaterial can influence cytotoxicity in the organism and alter its effects through several characteristics, including size, conformation, and kinetics. When we look at the induced apoptosis in the cell cycle, it can directly counteract necrosis and different problems caused by antibodies triggered immune clearance. Cytotoxicity tests are used in studies, and studies are correlated based on the correct assessment of the following criteria associated with injured/dead cells;

- A marker for cell death,
- Impaired metabolism,
- Damaged cell membrane.

Now to look at examples: Lactate dehydrogenase (LDH) is an assay that detects the release of the LDH structure, which is usually found in intact cells. The test, called the caspase 3/7 test, evaluates the production of caspase expression, which are mediators of apoptosis, but other categories such as the MTT (3-(4,5-Dimethylthiazol-2-yl)-2,5-Diphenyltetrazolium Bromide) test measure changes in anabolic or catabolic activities to assess cell viability. In the MTT assay, redox indicators associated with cell circulation need to be examined. Thus, by measuring the reduction in tetrazolium salt, a similarly resazurin reduction-based test may be required to assess cell viability if a different alternative is desired. In addition, in vitro assays in different studies use lipid models that mimic the cell membrane very well to study certain specific damage caused by reaction with bionanomaterials. In this type of study and research, the reaction mechanism behind these interactions includes membrane disruption that leads to cell death, that is, apoptosis (Jeevanandam et al., Lin et al., 2014; Cha et al., 2013).

A closer look at bionanomaterials can reduce cell viability, depending on their composition or conformation, by somehow damaging DNA and otherwise disrupting the flow of their metabolic activities. To give an example, it has been reported that CuO, which is a semiconductor and also used in chemical reactions, is cytotoxic although it is a catalyst. When this chemical formula is compared with Fe2O3, Fe3O4 and TiO_2, it has been found in the literature that CuO causes damage to DNA and mitochondria and causes apoptosis in the use of bionanomaterials (Jeevanandam et al., 2018; Mochalin et al., 2011). To say something additional, SiO_2 has proven to be more cytotoxic than TiO_2 for the epithelial cells of bionanomaterials and CNTs. Bionanomaterial geometry has also been shown to influence cytotoxicity at reactive oxygen levels. Again, if we make an example, it has been revealed in studies in the literature that cell viability increases with a decrease in silver nanoparticle size. When compared to larger scales, it has been shown that nanometer-sized particles of CuO have more toxicity than micrometer particles (Jeevanandam et al., 2018).

When a substance used in treatments has no effect on cell death, it may still have different effects on the genome. Since the human body includes a wide variety of mechanisms, certain effects may occur at low doses. Studies on the cytotoxicity and genotoxicity of bionanomaterials have previously been brought to the literature. Standard methods exist to comparatively test the effects on such cytotoxicity and genotoxicity. In addition to tests such as Comet, Ames, DNA ladder, micronuclei test, there is also chromosome aberration test. Today, next-generation sequencing techniques have also been used to measure chemical mutagenicity. It has been understood that bionanomaterials can directly enter the cell by endocytosis or passive diffusion pathways and cause some changes in the genomic structure, as well as catalyze intracellular expression. -OH radicals can cause increased reactive oxygen level and destruction of DNA conformation. Also as an example, bionanomaterials that show a certain increase in reactive oxygen level include carbonanotubes, quantum dots, Ag, TiO_2, and silica. Aside from substance formation and design, studies also show that surface potential charges affect the genotoxicity of liposomes. This means that positively charged bio nanocarriers showed lower genotoxicity than negatively charged bio nanocarriers. In fact, chromosomal aberrations caused by Ag nanoparticles are given as an example (Jeevanandam et al., 2018).

19.4.2 Bionanomaterial Immunomodulation

Bionanomaterials are designed according to their usage areas. By design, they can induce indirect or direct immunomodulation, which includes immunosuppression and immunostimulation. If we look at the role of bionanomaterials in immunomodulation, they are used in drug delivery applications by infusing nanoparticles and enabling them to act as vehicles for drugs. Acquired immune responses, together with innate responses in the human body, produce an immune response by in vivo injection of bionanomaterials. Innate immunity recognizes interactions between macrophage-structured and non-specific immune cells such as bionanomaterials by natural killer cells. From a design point of view, it is therefore very important to recognize bionanomaterials by the body's natural protection mechanisms. The adaptive immune response elicits antigen-specific effects, thereby activating T cells and B cells. Zinc oxide and silver nanoparticles significantly increased IL-6 and IL-8 production in kidney cells, while increasing both innate and adaptive responses in parallel (Zolnik et al., 2010; Feng et al., 2019).

Bionanomaterials have many parameters such as chemical composition, surface chemistry, molecular conformation and size. Detailed descriptions of bionanomaterial categories and compositions are available in the literature for studies focusing on some immune constructs and their therapeutic responses. In addition, the role of physical and chemical properties at the bio-nano interface, such as protein adsorption and corona formation, which is sought to enhance immunity, of the cell in immunomodulation in therapy should also be addressed (Jeevanandam et al., 2018; Zolink et al., 2010).

To give an example of the size of a bionanomaterial, it can be a nanoparticle with a size of approximately 193 nanometers and a nanoparticle with a micro size of 1530 nanometers. For the larger one, a greater immune response will occur. Studies have proven that a larger size can elicit a stronger serum immunoglobulin response. Therefore, material with needle nanostructures has been found to modulate the immune response (Feng et al., 2019; Porras et al., 2016).

In vitro studies have shown an increased CD40 expression range in DCs after 12 hours of incubation in needle-tipped particles in in vitro experiments. Based on this, expression of CD40 is useful in comparatively measuring a correspondingly higher immunity level of APCs.

Nanostructures in a bulky material such as needle-tipped nanostructures create mechanical stress on cells. This enables potassium influx in DCs, eliciting inflammatory activation, offering a variety of immunomodulatory responses elicited by various bionanomaterials. Because of the somewhat inconsistent methods used to evaluate it, it is difficult to find specific characteristics for results. In addition, it is known that more than one parameter has synergistic effects in this process. It can be said that a standard interpretation cannot be accepted on this subject (Jeevanandam et al., 2018; Porras et al., 2016).

19.4.3 Bionanomaterial Induced Fibrosis

The structure we call fibrosis is formed with the excessive accumulation of dense ECM. Therefore, fibrosis may occur as a response to bionanomaterials from the moment they enter the body. In fact, judging by the circulation, the system operator in this regard is not fully understood, because this can occur in the absence of an acute response or in the absence of standard predictive tests performed. In vitro tests for fibrotic responses can be performed. Evaluation of the expression of fibrosis-related proteins such as a-SMA and TGF-β in cells will provide convenience in studies. However, cells are usually cultured in two dimensions on a sewage substrate. With these investigations, which may require in vivo studies, altered gene expression may occur due to pore opening from the nucleus caused by mechanical forces (Jeevanandam et al., 2018; Oberdürster et al., 2000).

Bionanomaterials have small dimensions. As a result, they can easily reach the lung alveoli. As soon as bionanomaterials come into contact with lung cells, they can affect cells in various ways, causing fibrosis. To give an example in this regard, it can be said that bionanomaterials induce the production of reactive oxygen levels and increase TGF-β production. However, the emergence of such effects can cause more than a short-term discomfort in the person. This therefore entails associating bionanomaterials with these adverse events. The substance studied should be suitable for the location where it is studied and thus contribute to the treatment processes. It has been determined that carbon nanotubes activate type 2 immune mechanisms and play a role in triggering fibrosis, IL-1, IL-7, TGF-β and cancer. However, elevated reactive oxygen levels of metal oxides such as titanium, copper, cerium and similar can induce fibrosis through information and increased TGF-β expression (Oberdürster et al., 2000; Jeyarani et al., 2020).

19.4.4 Therapeutic Use of Bionanomaterials

Bionanomaterials have different abilities to cross biological membranes. Photodynamic therapy or therapies involving drug delivery, which are now widespread, have been developed. This makes the cytotoxic properties of bionanomaterials particularly desirable for use in the treatment of diseases associated with excessive and rapid cell growth. Looking at research, there are bionanomaterials used to treat hypertrophic scarring. Another example is some bionanomaterials developed to treat solid cancer. A gold nanoparticle by biosynthesis was applied to breast cancer cell for the evaluation of its cytotoxicity. Bionanomaterials, including gold and quantum dots, are employed to elevate reactive oxygen species and mitigate genomic damage. In short, precise targeting in these therapies is known to be challenging. However, topical and systematic evaluation of these therapies with proven techniques still continues (Harmatys et al., 2019; Bhattacharya et al., 2016).

19.4.5 Systematic and Topical Interactions of Bionanomaterials

When we need to explain the concept of system, it refers to an organism or a structure that encompasses multiple organ systems. When it comes to topical, there must be a state of interaction with bionanomaterials in tissue types formed by the formation of a single organ or cells. Topical affects a more specific part of the body, namely its location. Bionanomaterials can have different types of improved delivery routes, including implantable, injectable, or oral administration. These are formed systemically or topically, according to the treatment needed by the patient. A tissue that evades innate immune responses and oxidative inflammatory mechanisms, it is exposed to certain effects in surface modifications. To illustrate, while carbon nanotubes and graphene oxide are designed for effective drug delivery, they face enzymatic degradation or other immunomodulatory problems. Bionanomaterials used for topical reparative treatments can induce systemic immune responses compared to regenerative treatments. Materials with pores up to carbon-based bionanomaterials, one of the biopolymeric fillers, may not be able to fully repair damaged musculoskeletal tissues due to the emerging innate immune response, because the body's immune system can accept them as antigens. That is, it is important to pay attention to the interaction between bionanomaterials and immunity and to examine it in the appropriate context for intended and unintended systemic and topical effects (Jeevanandam *et al.,* 2018; Egli *et al.*, 2012; Lanone *et al.,* 2006; Foroozandeh *et al.,* 2015; Rossignoli et al., 2019).

19.4.6 Bionanomaterials and Their Molecular Interactions

The cytotoxicity and biocompatibility of bionanomaterials vary according to the complex processes taking place in the organism. Physicochemical properties, such as conformation and chemical characterization of bionanomaterials, are subject to varying degrees of toxicity and require extensive literature knowledge and procedure before they can be applied in vivo. With the development of technology, bionanomaterials used in biomedical sense should provide identity characterization in biological environments and the physiological response triggered in the body should be followed based on different processes of the bionanomaterial such as design and use (Egli et al., 2012; Lanone et al., 2006).

19.4.6.1 Biological Identity Including Molecular Interactions and the Concept of "Nano"

When bionanomaterials enter a biological environment, biomolecules adhere to their surfaces. In recent studies, the protein affinity of three core cross-linked polymeric nanoparticles with long circulation times has been investigated. It has been discovered that none of the PEG, polysarcosine and methacrylamide are related to the corona virus, which has brought new studies to the medical community in recent years (Jeevanandam et al., 2018; Foroozandeh et al., 2015; Cai et al., 2013).

In another study, we wanted to show that internalization in macrophages and monocytes affects corona proteins differently and was carried out using disulfide-characterized methacrylic acid polynanoporous bionanomaterial (Lanone et al., 2006).

In the last 20 years, carbon-based bionanomaterials have gained great importance both academically and commercially. Therefore, the physical and chemical properties of the designed material provide the variety of application areas (Simon et al., 2018; Qie et al., 2016). In current sources, it has been proven that corona formation of human serum albumin in CODs reduces their cytotoxicity and changes energy metabolism in mouse cells (Gorshkov et al., 2019). However, the effect of the size of carbon-based bionanomaterials on protein binding has been proven. Larger CNT2s are consistent with previous studies showing that binding with bionanomaterials is more efficient (Palchetti et al., 2016; Behzadi et al., 2017).

A specific and serious problem arises with regard to drug release via bionanomaterials. During interaction with immune cells, the proteins bound to the bionanomaterial are denatured and cannot fulfill their functions. As a potential way of solving this problem, silica-based bionanomaterials have been used. It has been recently determined that corona formation has a latent effect on macrophage recognition. By examining the properties of the coronavirus protein, researchers turned to a bionanomatting design similar to these properties, and studied polystyrene nanoparticles characterized by immunoglobulin-depleted plasma. They evaluated the in vitro cytotoxicity and biocompatibility level of this bionanomaterial by designing the PEG-based surfactant AT50, which showed a precoating and marked reduction in cellular uptake by immune cells (Cai et al., 2013; Simon et al., 2018; Qie et al., 2016).

To give another example, liposome technology is a widely used technology in the health sector and studies have been carried out in peripheral blood mononuclear cells until the sequestration of these cells. As a result, it was discovered that it plays an important role in the coronavirus protein. Antigen lagging of bionanomaterials by immune cells remains one of the major barriers to in vivo antitumor drug delivery, but to overcome this, different circulating leukocyte populations have been studied and designs that reduce the capture of bionanomaterials are being developed (Jeevanandam et al., 2018; Gorshkov et al., 2019).

19.4.7 Cellular Interactions in Bionanomaterials

Before engaging in specific cellular interactions with bionanomaterials, it is necessary to know the general mechanism of bionanomaterial uptake by cells. The interactions between the cell and the bionanomaterial can generally be classified into two different categories: receptor recognition-based and endocytic pathway. The concept we call endocytic uptake: clathrin-dependent endocytosis, caveola-dependent endocytosis, phagocytosis and macropinocytosis are divided into different options with many subheadings. For larger sized bionanomaterials (larger than the 200nm size mentioned here), phagocytosis and micropinocytosis are the generally used methods of uptake. Phagocytosis is a process performed by structures such as neutrophils and macrophages. With the help of receptors, actin cytoskeleton activity takes place, surrounding the bionanomaterial around the plasma membrane with a cup-shaped protrusion. Then it compresses inward, turning it into a phagosome (Paillard et al., 2010; Liu et al., 2017).

However, macropinocytosis is different. It does not have a system based on receptor-mediated bionanomaterial recognition. It attracts the outside structure into the cell in a non-specific manner. Smaller bionanomaterials (we have talked about those larger than 200 nm earlier), clathrin- or caveolae-dependent endocytosis are the structures that enter the cell right here. Both clathrin and caveolin are protein complexes that facilitate the process of invigination of bionanomaterials into the cell at the plasma membrane. These vesicle structures are then transported to lysosomes for degradation. Thus, they can perform their functions by being broken down by the acidic environment of lysosomes.

The uptake of bionanomaterials into the cell takes place in a different way by internalization. Bionanomaterials can interact with cells with specific membrane receptor mediators that can lead to therapeutic effects. Bionanomaterials activate certain signaling cascades with the help of certain receptors and they can only perform their functions through interaction without being taken into the cell. Some bionanomaterials have specific receptors to stimulate other bionanomaterials that can naturally activate certain signaling pathways. Each bionanomaterial can be designed not to act directly on the cell, but to find the way of the other bionanomaterial. It can also be designed based on antibodies and other ligands. Specific receptors in the context of the immune response include, but are not limited to toll-like receptors, TLRs, scavenger receptors, integrins, and Fc receptors. We will see the place of the raw materials we have seen before in the concepts of cytotoxicity and biocompatibility in the sub-headings of the chapters (Jeevanandam et al., 2018; Liu et al., 2017; Wang et al., 2017).

19.4.7.1 Cellular Interactions of Polymer-based Bionanomaterials

Polymer-based bionanomaterials can be adjusted for their loading capacity, size, and conformation. This allows for sensitive bionanomaterial optimization specific to individual cell types and has gained a place in the literature as an immune modulation method due to its biodegradability. Polymer bionanomaterials are generally designed to stimulate the immune response. They are loaded with cytokines, coated with specific antibodies specific to immune response receptors, and are mostly polystyrene based on a bionanomaterial. Especially PLGA nanoparticles are preferred in this category. PBAE nanoparticles were tested and concluded on immune cells, which are mainly used for test mouse cancer models. To give an example for this, PBAE bionanoparticles loaded with IL-12 were injected into a subject mouse model with melanoma and caused repolarization of M2 macrophages to M1 macrophages, limiting subsequent tumor progression. This proved that the biocompatibility continued without any disturbance in the test mouse (Jeevanandam et al., 2018).

In another study, PLGA bionanoparticle coated with DC-specific receptor DEC205 increased the levels of IL-10 produced by T cells in an in vitro study. In recent studies, PLGA bionanoparticles allow the development of nucleic acid-based approaches specifically to induce changes in the immune system. PLGA bionanoparticles carrying siRNA against PD-1 and PDL1 targeted tumor lymphocytes in a test mouse model of colon cancer. Thus, it suppressed the growth of tumor cells and prevented their development.

19.4.7.2 Cellular Interactions of Carbon-based Bionanomaterials

Carbon-based bionanomaterials consist of various structures such as fullerene, carbon nanoscales, nanographite and carbon nanotubes, and are being investigated for their prevalence in the communication between each of these materials and cells and their environment (Palomäki et al., 2011). It is stimulated through the uptake of macrojas into the cellular structure of carbon nanotubes, detecting it by molecular recognition receptors such as complement receptors, integrin and lectin morphologically similar receptors (Jeevanandam et al., 2018; Pondman et al., 2017).

A problem with carbonnanotubes is that, in therapeutic use, they can be compromised by the immune system and thus unstable. Being sensitive to this is unfortunately a disadvantage. It is thought that the activation of nicotinamide adenine dinucleotide phosphate (NADPH) oxidase, which stimulates reactive oxygen species that contribute to macrophages' uptake and degradation of carbonanotubes, positively affects the activation. However, certain changes can be seen in different organ systems. In vivo studies in test mice have discovered that multi-walled carbon nanotubes injected into the cerebral cortex degrade as early as the second day after entry, and those injected into the lungs do not degrade at the location for up to six months (Lee et al., 2018).

19.4.7.3 Cellular Interactions of Metal–Metal Oxide-based Bionanomaterials

Including metal oxide bionanoparticles, integrins, carbohydrate receptors, TLRs and even the nucleotide oligomerization domain can directly bind to similar receptors via endocytosis and induce macrophage activation. Before cell death, pro-inflammatory or anti-inflammatory cytokine production takes place, producing the toxic compound, thereby initiating a downward movement including membrane component translocation. Movement processes vary depending on the intrinsic properties of bionanomaterials, their design, conformation, and activation states of macrophages. Looking at gold-loaded bionanomaterials, M2 polarization of porous silica is induced. It promotes the secretion of anti-inflammatory factors by macrophages. While silver and gold bionanomaterials are involved in the stimulation of neutrophils, they also interact with their extracellular traps (Lee et al., 2018).

19.5 CONCLUSIONS AND FUTURE PERSPECTIVES

The proliferation of multidisciplinary fields with the developing technology for years has increased the need for the use of bionanomaterials. Bionanomaterials are distinguished from each other by physical factors such as size, molecular weight, surface area, and biological factors related to cell and tissue interactions. The suitability of the designed bionanomaterial for its intended use should first be determined and in vitro studies should be performed before being subjected to an in vivo treatment sequence. It is one of the most important parameters for a bionanomaterial that the material to be found in living organisms causes minimal damage to tissues and cells, and does not cause local or systemic toxicity. The nano-sized coronavirus, which entered our lives with the pandemic, has played a major role in the development of the literature narrowness about nanoparticles. Bionanomaterials have started to gain new technological designs and usage areas such as the corona virus, which is taken as a role model in terms of cellular relationships. Clarification of the molecular mechanisms of these interactions has facilitated their use in the relevant fields, both clinically and therapeutically.

New bionanomaterials are being designed and produced with current studies and discovered biocompatible materials. Just as it is understood that metal raw materials cause corrosion over time, research in the branches of medicine and other biomedical fields will have the least cytotoxicity and will least affect the cellular mechanisms in the living body. The process of researching technology and natural resources shows that bionanomaterials will take place more in our lives in the future perspective.

REFERENCES

Ahmed, S., & Rai, K. R. (2003). Interferon in the treatment of hairy-cell leukemia. *Best Practice & Research Clinical Haematology*, 16, 69–81.

Alsaab, H. O. et al. (2017). PD-1 and PD-L1 checkpoint signaling inhibition for cancer immunotherapy: Mechanism, combinations, and clinical outcome. *Frontiers in Pharmacology*, 8, 1–15.

Alwan, L. et al. (2014). Comparison of acute toxicity and mortality after two different dosing regimens of high-dose interleukin-2 for patients with metastatic melanoma. *Targeted Oncology*, 9, 63–71..

Anderson, J. M., Rodriguez, A., & Chang, D. T. (2008, April). Foreign body reaction to biomaterials. In *Seminars in immunology* (Vol. 20, No. 2, pp. 86–100). Academic Press.

Banik, B. L., Fattahi, P., & Brown, J. L. (2016). Polymeric nanoparticles: The future of nanomedicine. *Wiley Interdisciplinary Reviews: Nanomedicine and Nanobiotechnology*, 8(2), 271–299.

Behzadi, S., Serpooshan, V., Tao, W., Hamaly, M. A., Alkawareek, M. Y., Dreaden, E. C., Brown, D., Alkilany, A. M., Farokhzad, O. C., & Mahmoudi, M. (2017). Cellular uptake of nanoparticles: Journey inside the cell. *Chemical Society Reviews*, 46(14), 4218–4244.

Bhattacharya, K., Mukherjee, S. P., Gallud, A., Burkert, S. C., Bistarelli, S., Bellucci, S., Bottini, M., Star, A., & Fadeel, B. (2016). Biological interactions of carbon-based nanomaterials: From coronation to degradation. *Nanomedicine: Nanotechnology, Biology, and Medicine*, 12(2), 333–351.

Bhowmick, S., & Koul, V. (2016). Assessment of PVA/silver nanocomposite hydrogel patch as antimicrobial dressing scaffold: Synthesis, characterization and biological evaluation. *Materials Science and Engineering: C*, 59, 109–119.

Blau, P. J. (1992). Friction, lubrication, and wear technology. ASM International, 175–235.

Blum, A. P., Kammeyer, J. K., Rush, A. M., Callmann, C. E., Hahn, M. E., & Gianneschi, N. C. (2015). Stimuli-responsive nanomaterials for biomedical applications. *Journal of the American Chemical Society*, 137(6), 2140–2154.

Bruchez Jr, M., Moronne, M., Gin, P., Weiss, S., & Alivisatos, A. P. (1998). Semiconductor nanocrystals as fluorescent biological labels. *science*, 281(5385), 2013–2016.

Bundschuh, M., Filser, J., Lüderwald, S., McKee, M. S., Metreveli, G., Schaumann, G. E., ... & Wagner, S. (2018). Nanoparticles in the environment: Where do we come from, where do we go to?. *Environmental Sciences Europe*, 30(1), 1–17.

Cai, X., Ramalingam, R., Wong, H. S., Cheng, J., Ajuh, P., Cheng, S. H., & Lam, Y. W. (2013). Characterization of carbon nanotube protein corona by using quantitative proteomics. *Nanomedicine: Nanotechnology, Biology, and Medicine*, 9(5), 583–593.

Cha, C., Shin, S. R., Annabi, N., Dokmeci, M. R., & Khademhosseini, A. (2013). Carbon-based nanomaterials: multifunctional materials for biomedical engineering. *ACS Nano*, 7(4), 2891–2897.

Chenthamara, D., Subramaniam, S., Ramakrishnan, S. G., Krishnaswamy, S., Essa, M. M., Lin, F. H., & Qoronfleh, M. W. (2019). Therapeutic efficacy of nanoparticles and routes of administration. *Biomaterials Research*, 23(1), 1–29.

Chester, C., Sanmamed, M. F., Wang, J., & Melero I. (2018). Immunotherapy targeting 4-1BB: Mechanistic rationale, clinical results, and future strategies. *Blood*, 131, 49–57.

Choi, J. W., Nam, J. P., Nam, K., Lee, Y. S., Yun, C. O., & Kim, S. W. (2015). Oncolytic Adenovirus coated with Multidegradable Bioreducible core-cross-linked Polyethylenimine for cancer gene therapy. *Biomacromolecules*, 16(7), 2132–43.

Couzin-Frankel, J. (2013). Cancer immunotherapy. *Science*, 342, 1432–1433.

De Koker, S., Cui, J., Vanparijs, N., Albertazzi, L., Grooten, J., & Caruso, F., et al. (2016). Engineering polymer hydrogel nanoparticles for lymph node-targeted delivery. *Angewandte Chemie International Edition English*, 55, 1334–9.

Doronin, K., Shashkova, E. V., May, S. M., Hofherr, S. E., & Barry, M. A. (2009). Chemical modification with high molecular weight polyethylene glycol reduces transduction of Hepatocytes and increases efficacy of intravenously delivered Oncolytic Adenovirus. *Human Gene Therapy,* 20(9), 975–88.

Edelstein, R. L., Tamanaha, C. R., Sheehan, P. E., Miller, M. M., Baselt, D. R., Whitman, L., & Colton, R. J. (2000). The BARC biosensor applied to the detection of biological warfare agents. *Biosensors and Bioelectronics*, 14(10–11), 805–813.

Egli, R. J., & Luginbuehl, R. (2012). Tissue engineering – nanomaterials in the musculoskeletal system. *Swiss Medical Weekly*, 142, w13647.

Eliaz N. (2019). Corrosion of metallic biomaterials: A review. *Materials (Basel, Switzerland)*, 12(3), 407.

Eliaz, N. (Ed.). (2012). *Degradation of implant materials*. Springer Science & Business Media.

Eliaz, N., & Gileadi, E. (2019). *Physical electrochemistry: Fundamentals, techniques, and applications*. John Wiley & Sons.

Feng, X., Xu, W., Li, Z., Song, W., Ding, J., & Chen, X. (2019). Immunomodulatory nanosystems. *Advanced Science*, 6(17), 1900101.

Feynman, R. (1991). There's plenty of room at the bottom. *Science*, 254, 1300–1301.

Foroozandeh, P., & Aziz, A. A. (2015). Merging worlds of nanomaterials and biological environment: Factors governing protein corona formation on nanoparticles and its biological consequences. *Nanoscale Research Letters*, 10, 221.

Gause, K. T., Yan, Y., Cui, J., O'Brien-Simpson, N. M., Lenzo, J. C., & Reynolds, E. C., et al. (2015). Physicochemical and immunological assessment of engineered pure protein particles with different redox states. *ACS Nano*, 9, 2433–44.

Gilbert, J. L., & Mali, S. A. (2012). Degradation of implant materials. *Degradation of implant materials*, 1st ed., Springer US, New York, 195–252.

Gorshkov, V., Bubis, J. A., Solovyeva, E. M., Gorshkov, M. V., & Kjeldsen, F. (2019). Protein corona formed on silver nanoparticles in blood plasma is highly selective and resistant to physicochemical changes of the solution. *Environmental Science. Nano*, 6(4), 1089–1098.

Guo, C et al. (2013). Therapeutic cancer vaccines; past, present and future. *Advances in Cancer Research*, 119, 421–475.

Harmatys, K. M., Overchuk, M., & Zheng, G. (2019). Rational design of photosynthesis-inspired nanomedicines. *Accounts of Chemical Research*, 52(5), 1265–1274.

Jain, A. K., & Thareja, S. (2019). In vitro and in vivo characterization of pharmaceutical nanocarriers used for drug delivery. *Artificial Cells, Nanomedicine, and Biotechnology*, 47(1), 524–539.

Jawahar, N., & Meyyanathan, S. N. (2012). Polymeric nanoparticles for drug delivery and targeting: A comprehensive review. *International Journal of Health & Allied Sciences*, 1(4), 217.

Jeevanandam, J., Barhoum, A., Chan, Y. S., Dufresne, A., & Danquah, M. K. (2018). Review on nanoparticles and nanostructured materials: History, sources, toxicity and regulations. *Beilstein Journal of Nanotechnology*, 9(1), 1050–1074.

Jeyarani, S., Vinita, N. M., Puja, P., Senthamilselvi, S., Devan, U., Velangani, A. J., Biruntha, M., Pugazhendhi, A., & Kumar, P. (2020). Biomimetic gold nanoparticles for its cytotoxicity and biocompatibility evidenced by fluorescence-based assays in cancer (MDA-MB-231) and non-cancerous (HEK-293) cells. *Journal of photochemistry and photobiology. B, Biology*, 202, 111715.

Jiang, L., Liang, Y., Huo, Q., Pu, Y., Lu, W., & Han, S., et al. (2018). Viral capsids mimicking based on pH-sensitive biodegradable polymeric micelles for efficient anticancer drug delivery. *Journal of Biomedical Nanotechnology*, 14, 1409–19.

Jung, S. J., Kasala, D., Choi, J. W., Lee, S. H., Hwang, J. K., Kim, S. W, et al. (2015). Safety profiles and antitumor efficacy of Oncolytic Adenovirus coated with Bioreducible polymer in the treatment of a CAR negative tumor model. *Biomacromolecules*,16(1), 87–96.

Karaki, S., Anson, M., Tran, T., Giusti, D., Blanc, C., & Oudard, S, et al. (2016). Is there still room for cancer vaccines at the era of checkpoint inhibitors. *Vaccines*, 4, 37.

Katze, M. G., He, Y., & Gale, M. (2002). Viruses and interferon: A fight for supremacy. *National Reviews Immunology*, 2, 675–687.

Kempe, K., Xiang, S. D., Wilson, P., Rahim, M. A., Ju, Y., & Whittaker, M. R., et al. (2017). Engineered hydrogen-bonded glycopolymer capsules and their interactions with antigen presenting cells. *ACS Applied Materials Interfaces*, 9, 6444–52.

Kim, J., Li, W. A., Choi, Y., Lewin, S. A., Verbeke, C. S., & Dranoff, G., et al. (2015). Injectable, spontaneously assembling, inorganic scaffolds modulate immune cells in vivo and increase vaccine efficacy. *Nature Biotechnology*, 33, 64–72.

Krishnan, A., & Chuturgoon, A. (Eds.). (2020). *Integrative nanomedicine for new therapies*. Springer International Publishing.

Kumar, S., Raj, S., Jain, S., & Chatterjee, K. (2016). Multifunctional biodegradable polymer nanocomposite incorporating graphene-silver hybrid for biomedical applications. *Materials & Design*, 108, 319–332.

Kyriakides, T. R., Raj, A., Tseng, T. H., Xiao, H., Nguyen, R., Mohammed, F. S., Halder, S., Xu, M., Wu, M. J., Bao, S., & Sheu, W. C. (2021). Biocompatibility of nanomaterials and their immunological properties. *Biomedical Materials (Bristol, England)*, 16(4), 10.1088/1748-605X/abe5fa

Lai, C., Duan, S., Ye, F., Hou, X., Li, X., & Zhao, J., et al. (2018). The enhanced antitumor-specific immune response with mannose- and CpG-ODN-coated liposomes delivering TRP2 peptide. *Theranostics*, 8, 1723–39.

Lanone, S., & Boczkowski, J. (2006). Biomedical applications and potential health risks of nanomaterials: Molecular mechanisms. *Current Molecular Medicine*, 6(6), 651–663.

Le Moignic, A., Malard, V., Benvegnu, T., Lemiegre, L., Berchel, M., & Jaffres, P. A., et al. (2018). Preclinical evaluation of mRNA trimannosylated lipopolyplexes as therapeutic cancer vaccines targeting dendritic cells. *Journal of Controlled Release*, 278, 110–21.

Lee, J., Breton, G., Oliveira, T. Y., Zhou, Y. J., Aljoufi, A., & Puhr, S., et al. (2015). Restricted dendritic cell and monocyte progenitors in human cord blood and bone marrow. *Journal of Experimental Medicine*, 212, 385–99.

Lee, S. B., Lee, Y. J., Cho, S. J., Kim, S. K., Lee, S. W., Lee, J., Lim, D. K., & Jeon, Y. H. (2018). Antigen-Free Radionuclide-Embedded Gold Nanoparticles for Dendritic Cell Maturation, Tracking, and Strong Antitumor Immunity. *Advanced Healthcare Materials*, 7(9), e1701369.

Lee, S., & Margolin, K. (2011). Cytokines in cancer immunotherapy. *Cancers (Basel)*, 3, 3856–3893.

Lee, S., & Margolin, K. (2011). Cytokines in cancer immunotherapy. *Cancers (Basel*), 3, 3856–3893.

Letchford, K., & Burt, H. (2007). A review of the formation and classification of amphiphilic block copolymer nanoparticulate structures: Micelles, nanospheres, nanocapsules and polymersomes. *European Journal of Pharmaceutics and Biopharmceuatics*, 65, 259–69.

Levy, M., Luciani, N., Alloyeau, D., Elgrabli, D., Deveaux, V., Pechoux, C., et al. (20112). Long term in vivo biotransformation of iron oxide nanoparticles. *Biomaterials*, 32, 3988–99.

Li, A. W., Sobral, M. C., Badrinath, S., Choi, Y., Graveline, A., & Stafford, A. G., et al. (2018). A facile approach to enhance antigen response for personalized cancer vaccination. *Nature Materials*, 17, 528–34.

Li, A. W., Sobral, M. C., Badrinath, S., Choi, Y., Graveline, A., Stafford, A. G., et al. (2018). A facile approach to enhance antigen response for personalized cancer vaccination. *Nature Materials*, 17, 528–34.

Li, Z., Barnes, J. C., Bosoy, A., Stoddart, J. F, & Zink, J. I. (2012). Mesoporous silica nanoparticles in biomedical applications. *Chemical Society Reviews*, 41, 2590–605.

Lim, W. A., & June, C. H. (2017). The principles of engineering immune cells to treat cancer. *Cell*, 168, 724–740.

Lin, D., Feng, S., Huang, H., Chen, W., Shi, H., Liu, N., … & Chen, R. (2014). Label-free detection of blood plasma using silver nanoparticle based surface-enhanced Raman spectroscopy for esophageal cancer screening. *Journal of Biomedical Nanotechnology*, 10(3), 478–484.

Liu, T., Liu, Z., Chen, J., Jin, R., Bai, Y., & Zhou, Y., et al. (2018). Redox-responsive supramolecular micelles for targeted imaging and drug delivery to tumor. *Journal of Biomedical Nanotechnology*, 14, 1107–16.

Liu, Y., Hardie, J., Zhang, X., & Rotello, V. M. (2017). Effects of engineered nanoparticles on the innate immune system. *Seminars in Immunology*, 34, 25–32.

Ma, J., Wong, H., Kong, L. B., & Peng, K. W. (2003). Biomimetic processing of nanocrystallite bioactive apatite coating on titanium. *Nanotechnology*, 14(6), 619.

Mah, C., Zolotukhin, I., Fraites, T. J., Dobson, J., Batich, C., & Byrne, B. J. (2000). Microsphere-mediated delivery of recombinant AAV vectors in vitro and in vivo. *Molecular Therapy*, 1(5), S293.

Melchor-Martínez, E. M., Torres Castillo, N. E., Macias-Garbett, R., Lucero-Saucedo, S. L., Parra-Saldívar, R., & Sosa-Hernández, J. E. (2021). Modern world applications for nano-bio materials: Tissue engineering and COVID-19. *Frontiers in Bioengineering and Biotechnology*, 9, 597958.

Min, Y., Roche, K. C., Tian, S., Eblan, M. J., McKinnon, K. P., Caster, J. M. et al. (2017). Antigen-capturing nanoparticles improve the abscopal effect and cancer immunotherapy. *Nature Nanotechnology*, 12, 877–82.

Mochalin, V. N., Shenderova, O., Ho, D., & Gogotsi, Y. (2011). The properties and applications of nanodiamonds. *Nature Nanotechnology*, 7(1), 11–23.

Moffett, H. F., Coon, M. E., Radtke, S., Stephan, S. B., McKnight, L., & Lambert, A., et al. (2017). Hit-And-Run Programming of therapeutic Cytoreagents using mRNA Nanocarriers. *Nature Communications*, 8(1), 389.

Mourdikoudis, S., Pallares, R. M., & Thanh, N. T. (2018). Characterization techniques for nanoparticles: comparison and complementarity upon studying nanoparticle properties. *Nanoscale*, 10(27), 12871–12934.

Müller, L., Aigner, P., & Stoiber, D. (2017). Type I interferons and natural killer cell regulation in cancer. *Frontiers in Immunology*, 8, 1–11.

Munn, D. H., & Bronte, V. (2016). Immune suppressive mechanisms in the tumor microenvironment. *Current Opinion in Immunology*, 39, 1–6.

Ngobili, T. A., & Daniele, M. A. (2016). Nanoparticles and direct immunosuppression. *Experimental Biology and Medicine*, 241(10), 1064–1073.

Nikpoor, A. R., Tavakkol-Afshari, J., Sadri, K., Jalali, S. A., Jaafari, M. R. (2017). Improved tumor accumulation and therapeutic efficacy of CTLA-4-blocking antibody using liposome-encapsulated antibody: In vitro and in vivo studies. *Nanomedicine*, 13, 2671–82.

Oberdürster, G. (2000). Toxicology of ultrafine particles: In vivo studies. *Philosophical Transactions of the Royal Society of London. Series A: Mathematical, Physical and Engineering Sciences*, 358(1775), 2719–2740.

Padmanabhan, J., & Kyriakides, T. R. (2015). Nanomaterials, inflammation, and tissue engineering. *Wiley Interdisciplinary Reviews: Nanomedicine and Nanobiotechnology*, 7(3), 355–370.

Paillard, A., Hindré, F., Vignes-Colombeix, C., Benoit, J. P., & Garcion, E. (2010). The importance of endo-lysosomal escape with lipid nanocapsules for drug subcellular bioavailability. *Biomaterials*, 31(29), 7542–7554.

Palchetti, S., Digiacomo, L., Pozzi, D., Peruzzi, G., Micarelli, E., Mahmoudi, M., & Caracciolo, G. (2016). Nanoparticles-cell association predicted by protein corona fingerprints. *Nanoscale*, 8(25), 12755–12763.

Palomäki, J., Välimäki, E., Sund, J., Vippola, M., Clausen, P. A., Jensen, K. A., Savolainen, K., Matikainen, S., & Alenius, H. (2011). Long, needle-like carbon nanotubes and asbestos activate the NLRP3 inflammasome through a similar mechanism. *ACS Nano*, 5(9), 6861–6870.

Palucka, K., & Banchereau, J. (2012). Cancer immunotherapy via dendritic cells. *Nature Reviews Cancer*, 12, 265–77.

Park, Y., Kang, E., Kwon, O. J., Hwang, T., Park, H., & Lee, J. M., et al. (2010). Ionically Crosslinked Ad/chitosan Nanocomplexes Processed by Electrospinning for targeted cancer gene therapy. *Journal of Controlled Release*, 148(1), 75–82.

Pattni, B. S., Chupin, V. V., Torchilin, V. P. (2015). New developments in liposomal drug delivery. *Chemical Reviews*, 115, 10938–66.

Peggs, K. S., Quezada, S. A., & Allison, J. P. (2009). Cancer immunotherapy: Co-stimulatory agonists and co-inhibitory antagonists. *Clinical and Experimental Immunology*, 157, 9–19.

Peng, J., Xiao, Y., Li, W., Yang, Q., Tan, L., & Jia, Y., et al. (2018). Photosensitizer Micelles together with IDO inhibitor enhance cancer photothermal therapy and immunotherapy. *Advanced Science*, 5(5), 2198–3844 (Print)).

Peres, C., Matos, A. I., Conniot, J., Sainz, V., Zupančič, E., Silva, J. M., ... & Florindo, H. F. (2017). Poly (lactic acid)-based particulate systems are promising tools for immune modulation. *Acta Biomaterialia*, 48, 41–57.

Pondman, K. M., Salvador-Morales, C., Paudyal, B., Sim, R. B., & Kishore, U. (2017). Interactions of the innate immune system with carbon nanotubes. *Nanoscale Horizons*, 2(4), 174–186.

Porras, A. M., Hutson, H. N., Berger, A. J., & Masters, K. S. (2016). Engineering approaches to study fibrosis in 3-D in vitro systems. *Current Opinion in Biotechnology*, 40, 24–30.

Qie, Y., Yuan, H., von Roemeling, C. A., Chen, Y., Liu, X., Shih, K. D., Knight, J. A., Tun, H. W., Wharen, R. E., Jiang, W., & Kim, B. Y. (2016). Surface modification of nanoparticles enables selective evasion of phagocytic clearance by distinct macrophage phenotypes. *Scientific Reports*, 6, 26269.

Reddy, S. T., van der Vlies, A. J., Simeoni, E., Angeli, V., Randolph, G. J., O'Neil, C. P., et al. (2007). Exploiting lymphatic transport and complement activation in nanoparticle vaccines. *Nature Biotechnology*, 25, 1159–64.

Reis, C. P., Neufeld, R. J., Ribeiro, A. J., & Veiga, F. (2006). Nanoencapsulation I. Methods for preparation of drug-loaded polymeric nanoparticles. *Nanomedicine: Nanotechnology, Biology and Medicine*, 2(1), 8–21.

Ribas, A., & Wolchok, J. D. (2018). Cancer immunotherapy using checkpoint blockade. *Science*, 359, 1350–1355.

Riley, R. S., June, C. H., Langer, R., & Mitchell, M. J. (2019 Mar). Delivery technologies for cancer immunotherapy. *Nature Reviews Drug Discovery*, 18(3), 175–196.

Rosenberg, S. A. (2014). IL-2: The first effective immunotherapy for human cancer. *Journal of Immunology*, 192, 5451–5458.

Rosenberg, S. A., Yang, J. C., & Restifo, N. P. (2004). Cancer immunotherapy: Moving beyond current vaccines. *Nature Medcine*, 10, 909–915.

Rossignoli, F. et al. (2019). MSC-delivered soluble TRAIL and paclitaxel as novel combinatory treatment for pancreatic adenocarcinoma *Theranostics*, 9, 436–48.

Sabado, R. L., Balan, S., & Bhardwaj, N. (2017). Dendritic cell-based immunotherapy. *Cell Research*, 27, 74–95.

Salata, O. (2004). Applications of nanoparticles in biology and medicine. *Journal of Nanobiotechnology*, 2(1), 3.

Schlitzer, A., Sivakamasundari, V., Chen, J., Sumatoh, H. R., Schreuder, J., & Lum, J., et al. (2015). Identification of cDC1- and cDC2-committed DC progenitors reveals early lineage priming at the common DC progenitor stage in the bone marrow. *Nature Immunology*, 16, 718–28.

Schmid, D., Park, C. G., Hartl, C. A., Subedi, N., Cartwright, A. N., & Puerto, R. B., et al. (2017). T Cell-targeting nanoparticles focus delivery of immunotherapy to improve antitumor immunity. *Nature Communications*, 8(1), 1747.

Scholler, J. et al. (2012). Decade-long safety and function of retroviral-modified chimeric antigen receptor T cells. *Science Translational Medicine*, 4, 132ra53.

Simon, J., Müller, L. K., Kokkinopoulou, M., Lieberwirth, I., Morsbach, S., Landfester, K., & Mailänder, V. (2018). Exploiting the biomolecular corona: Pre-coating of nanoparticles enables controlled cellular interactions. *Nanoscale*, 10(22), 10731–10739.

Sinani, V. A., Koktysh, D. S., Yun, B. G., Matts, R. L., Pappas, T. C., Motamedi, M., ... & Kotov, N. A. (2003). Collagen coating promotes biocompatibility of semiconductor nanoparticles in stratified LBL films. *Nano Letters*, 3(9), 1177–1182.

Singh, R., & Lillard Jr, J. W. (2009). Nanoparticle-based targeted drug delivery. *Experimental and Molecular Pathology*, 86(3), 215–223.

Steinman, R. M. (2012). Decisions about dendritic cells: Past, present, and future. *Annual Review of Immunology*, 30, 1–22.

Su, Z., Xiao, Z., Wang, Y., Huang, J., An, Y., & Wang, X., et al. (2020). Codelivery of Anti-PD-1 antibody and Paclitaxel with matrix Metalloproteinase and pH dual-sensitive Micelles for enhanced tumor chemoimmunotherapy. *Small*, 16(7), 1906832.

Sun, T. et al. (2014). Inhibition of tumor angiogenesis by interferon-γ by suppression of tumor-associated macrophage differentiation. *Oncology Research*, 21, 227–235.

Sunshine, J. C., Perica, K., Schneck, J. P., Green, J. J. (2014). Particle Shape Dependence of CD8+T cell activation by artificial antigen presenting cells. *Biomaterials*, 35(1), 269–77.

Thomas, B., Coates, D., Tzeng, V., Baehner, L., & Boxer, A. Treatment of hairy cell leukemia with recombinant alpha-interferon. *Blood*, 68, 493–497.

Torchilin, V. P. (2005). Recent advances with liposomes as pharmaceutical carriers. *Nature Reviews Drug Discovery*, 4, 145–60.

Torres, F. G., Troncoso, O. P., Pisani, A., Gatto, F., & Bardi, G. (2019). Natural polysaccharide nanomaterials: an overview of their immunological properties. *International Journal of Molecular Sciences*, 20(20), 5092.

Velu, R., Calais, T., Jayakumar, A., & Raspall, F. (2019). A comprehensive review on bio-nanomaterials for medical implants and feasibility studies on fabrication of such implants by additive manufacturing technique. *Materials (Basel, Switzerland)*, 13(1), 92.

Virtanen, S. (2008). *Corrosion of biomedical implant materials.*

Wagner, J., Li, L. J., Simon, J., Krutzke, L., Landfester, K., Mailander, V., et al. (2020). Amphiphilic Polyphenylene Dendron conjugates for surface remodeling of Adenovirus 5. *Angewandte Chemie International Edition,* 59(14), 5712–20.

Wang, X., Li, X., Ito, A., Sogo, Y., Watanabe, Y., & Tsuji, N. M., et al. (2017). Biodegradable metal ion-doped mesoporous silica nanospheres stimulate anticancer Th1 immune response in vivo. *ACS Applied Materials & Interfaces*, 9, 43538–44.

Wang, X., Li, X., Ito, A., Yoshiyuki, K., Sogo, Y., & Watanabe, Y., et al. (2016). Hollow structure improved anti-cancer immunity of mesoporous silica nanospheres in vivo. *Small*, 12, 3510–5.

Wang, Y., Lin, Y. X., Qiao, S. L., An, H. W., Ma, Y., Qiao, Z. Y., Rajapaksha, R. P., & Wang, H. (2017). Polymeric nanoparticles promote macrophage reversal from M2 to M1 phenotypes in the tumor microenvironment. *Biomaterials*, 112, 153–163.

Xing, H., Hwang, K., & Lu, Y. (2016). Recent developments of liposomes as nanocarriers for theranostic applications. *Theranostics*, 6, 1336–52.

Yadav, S. K. (Ed.). (2016). *Nanoscale materials in targeted drug delivery, theragnosis and tissue regeneration*. Berlin, Germany: Springer.

Yang, F., Shi, K., Jia, Y. P., Hao, Y., Peng, J. R., & Qian, Z. Y. (2020 Jul). Advanced biomaterials for cancer immunotherapy. *Acta Pharmacologica Sinica*, 41(7), 911–927.

Yang, W., Zhu, G., Wang, S., Yu, G., Yang, Z., & Lin, L., et al. (2019). In situ dendritic cell vaccine for effective cancer immunotherapy. *ACS Nano*, 13, 3083–94.

Ye, Y. Q., Wang, J. Q., Hu, Q. Y., Hochu, G. M., Xin, H. L., & Wang, C., et al. (2016). Synergistic transcutaneous immunotherapy enhances antitumor immune responses through delivery of checkpoint inhibitors. *ACS Nano*, 10(9), 8956–63.

Yu, Q. R., Zhang, M. X., Chen, Y. T., Chen, X. L., Shi, S. Y., Sun, K., et al. (2020). Self-Assembled Nanoparticles Prepared From Low-Molecular-Weight PEI and Low-Generation PAMAM for EGFRvIII-Chimeric Antigen Receptor Gene Loading and T-Cell Transient Modification. *International Journal of Nanomedicine*, 15, 483–95.

Zeng, Q., Jiang, H., Wang, T., Zhang, Z., Gong, T., & Sun, X. (2015). Cationic micelle delivery of Trp2 peptide for efficient lymphatic draining and enhanced cytotoxic T-lymphocyte responses. *Journal of Controlled Release*, 200, 1–12.

Zhan, X., Tran, K. K., Shen, H. (2012). Effect of the poly(ethylene glycol) (PEG) density on the access and uptake of particles by antigen-presenting cells (APCs) after subcutaneous administration. *Molecular Pharmacology*, 9, 3442–51.

Zheng, Y., Tang, L., Mabardi, L., Kumari, S., & Irvine, D. J. (2017). Enhancing adoptive cell therapy of cancer through targeted delivery of small-molecule immunomodulators to internalizing or noninternalizing receptors. *ACS Nano*, 11, 3089–100.

Zhou, L., Huang, J., Yu, B., Liu, Y., & You, T. (2015). A novel electrochemiluminescence immunosensor for the analysis of HIV-1 p24 antigen based on P-RGO@Au@Ru-SiO2 composite. *ACS Applied Materials & Interfaces*, 7, 24438–45.

Zielińska, A., Carreiró, F., Oliveira, A. M., Neves, A., Pires, B., Venkatesh, D. N., Durazzo, A., Lucarini, M., Eder, P., Silva, A. M., Santini, A., & Souto, E. B. (2020). Polymeric nanoparticles: production, characterization, toxicology and ecotoxicology. *Molecules (Basel, Switzerland)*, 25(16), 3731.

Zolnik, B. S., González-Fernández, Á., Sadrieh, N., & Dobrovolskaia, M. A. (2010). Minireview: Nanoparticles and the immune system. *Endocrinology*, 151(2), 458–465.

Part 4

Bionanomaterials in Environmental Applications

20 Bionanomaterials in Environmental Protection

Shilpa Borehalli Mayegowda, Abhishek Gowda, Tajunnisa, and N.G. Manjula

20.1 INTRODUCTION

The emergent industrialization and modernization in the developing nations has also contributed to the environmental contamination which is a major concern today. With intensive food, textile, widespread industries, and agricultural activities remaining the major cause of pollution, releasing varied organic contaminants like phenols, greases, dyes, carbohydrates, oils, detergents, biphenyls, and so forth are a grave threat to living beings (Wang and Zhuang, 2017). Even though a variety of traditional approaches are used for the treatment of diverse environmental pollutants they have certain drawbacks, with incomplete eradication. As a result, a better and newer approach or technology that has greater efficiency and is ecofriendly, is needed to clean-up the environmental pollutants. This led to the development of nanotechnology that helps to generate enormous possibilities to overcome the disadvantages caused by the traditional methods and also has created immense interest among researchers in nanoparticles (NPs). NPs outperform these traditional cleaning techniques since they have a higher reactivity along with the option of adding new functionalities that are a high number of reactive sites due to a high surface area with volume ratio enabling greater contact with pollutants causing a rapid decrease in contaminant concentration. The qualities of metal NPs that have been used for remediating pollutants, including copper, gold, silver, zinc oxide, and iron oxide, play an essential role in domestic and industrial waste water, polluted soil, and natural waters treatment (Khin et al., 2012).

Green synthesized NPs are the context for the principle of green chemistry featuring the novel, eco-friendly, cost-effective NPs, which also have health benefits. Basically, this has been an advantage of nanotechnology over the conventional methods and has been used in varied applications. This field has been helpful in bioremediating greenhouse gases, volatile organic compounds, and antimicrobial agents (Jadimurthy et al., 2022; Mayegowda et al., 2022a, b; Rotti et al., 2023), photocatalytic degradation (Adarsha et al., 2022), phytoremediation, filtration and so forth. In agriculture use, NPs have been applied to the soil as modifying agents that help as phytoremediator, as stabilizers to boost the efficiency of waste water treatment in removal of organic pollutants by various processes, like membrane filtration, adsorption, photocatalysis, and disinfection (Yadav et al., 2021; Rani and Sridevi, 2017) (Figure 20.1). Finally, these bionanomaterials can be used to eliminate radioactive contaminants from the environment, like uranium and plutonium, which can pose potential risks if not properly handled.

DOI: 10.1201/9781003432791-24

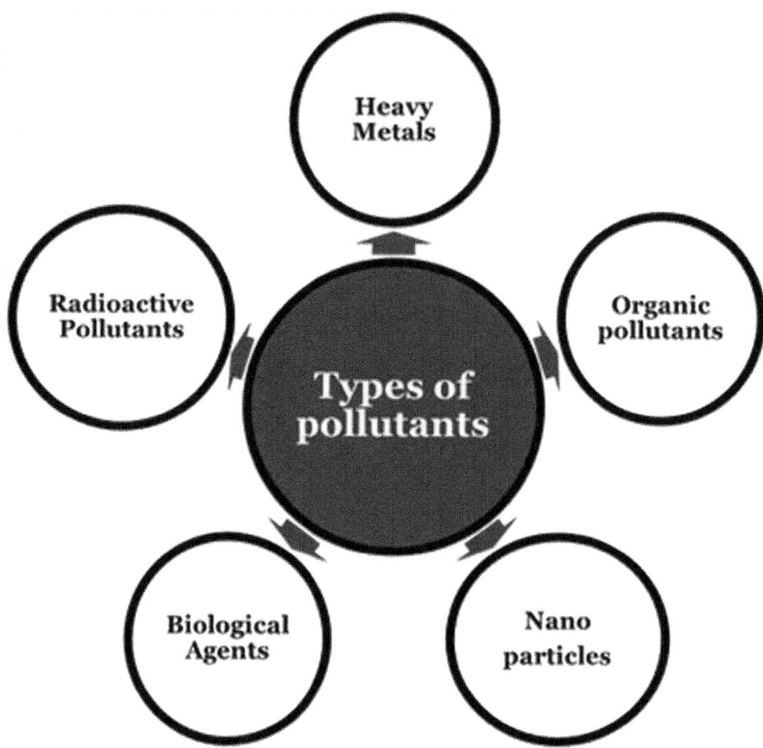

FIGURE 20.1 Various Agents Responsible for Contaminating the Environment. (Created by the Authors.)

20.2 ENVIRONMENTAL AND HEALTH CONCERNS DUE TO POLLUTANTS

The most widely employed methods for food and agricultural production to satisfy global food demand has increased the various contaminants in nature. The most common contaminant is pesticides with their negative impact on the environment's water quality, biodiversity, and public health. An estimated global topography of ecological contamination risk is induced by 92 different active components in many nations that has been recorded due to its usage inducing synergistic toxicity to target both acute and chronic exposure. Nevertheless, even bionanomaterials might potentially affect ecosystems if they are not properly discharged into the environment as they are so tiny that they can readily enter into the food chain and accumulate. In addition, they can pollute water and soil, endangering both social well-being and the environment. Therefore, they need to be properly handled before use and disposal in the environment (Boros et al., 2020).

The greatest menace, air pollutants, consume a larger influence on the climatic change and human health with increasing rates of morbidity. These pollutants are basically smaller diameter particulate matters (PM) that penetrate the respiratory tract causing respiratory, cardiovascular, and central nervous disorders and reproductive dysfunction. The main pollutants that act on the ozone layer include the volatile organic compounds (VOCs), sulphur dioxide, nitrogen oxide and polyaromatic hydrocarbons (PAHs), which cause human health issues. Heavy metals such as lead (Pb) can cause toxification in the human the body, which can be chronic, further causing obstructive pulmonary disease (COPD), bronchitis, lung cancer, and asthma along with being the leading cause of infectious diseases (Manisalidis et al., 2020) (Figure 20.1).

Pollutants of water and soil are basically man-made, such as domestic and agricultural wastes, fertilizers, oil spills and radioactive substances that have damaging impact on the environment and human health (Figure 20.2) (Ainsworth et al., 2018). Polluted water endangers aquatic plants and

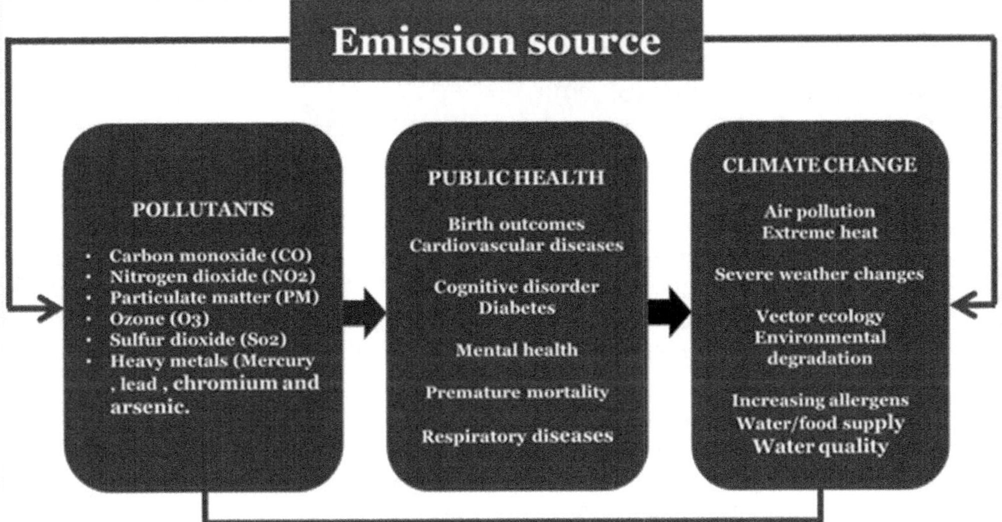

FIGURE 20.2 Air Pollutants Affecting Human Health and Environmental Damage. (Created by the Authors.)

animals, while soil and water pollutants cause toxicity, carcinogenesis, mutagenesis, and teratogenesis in living beings, including humans. The only way to understand and tackle such issues lies in public awareness along with alternative approaches or techniques to remediate these pollutants using sustainable approaches without introducing emerging threats to the environment and (Rodgers et al., 2018).

20.3 CLASSIFICATION OF BIONANOMATERIALS

20.3.1 Based on Organic Bionanoparticles

(a) **Metal Nanomaterials (NMs):** This includes gold (Au), iron (Fe), silver (Ag) and so forth, with remarkable properties as chemical, optical, and electrical. AuNPs is synthesized using sonoelectrochemical along with ultrasonic vibration techniques (Shiau et al., 2018). However, light absorbance depends on the metallic NPs size through the interband transition as in the case of Pt (platinum), Ni (nickel), Pd (palladium) while intraband transition occurs in Cu (copper), Al (aluminum), Ag (silver) and Au (gold) due to their NPs catalytic properties, further enhanced by irradiation with a light source (Kim and Lee, 2018).

(b) **Metal oxide NMs:** These are made up of titanium oxide (TiO_2), iron oxide (Fe_2O_3), aluminum oxide (Al_2O_3), zinc oxide (ZnO) and SiO_2 synthesized using hydrothermal or sol-gel methods with surface properties having an impact on the band gap energy that is used in various applications as semiconductors, chemical sensors, and catalytic converters (Adarsha et al., 2022; Saleh and Fadillah, 2019). The most important properties are active area with high-level surface biocompatibility modified by various chemical reactions with the addition of polymer chains, coupling factors and so forth (Das et al., 2020). Various organic compounds used to modify NPs for surface properties include epoxides, amines, thiols, and anionic compounds (Zeng et al., 2019).

(c) **Composite NMs:** These include the solid materials consisting of several phases, dimension smaller than 100 nm or nanoscale, spaced with repeating structures (Lozhkomoev et al., 2019). These composites of multiple material combination create various characteristics such as optical property and water absorption (Abozaid et al., 2019). Combining materials helps to create nano composites with increased surface area help to improve the adsorption capacity of the NPs with various interactions like electrostatic, dipole-dipole hydrogen bonds and so forth (Parker et al., 2012).

(d) **Carbon-based NMs:** This is an allotrope found in diverse forms, such as diamond, amorphous carbon, and graphite used to synthesize carbon based-NPs with different sizes (Li et al., 2019; Nehra et al., 2019; Xie et al., 2019). It exhibits various characteristics of conductivity, physical and chemical, mechanical and thermal stability. An allotrope of carbon, fullerene with unique characteristics neutralizes reactive species like oxygen and nitrogen (Sumi and Chitra, 2019). Carbon nanotube (CNT) (1D), provides modified structure to be used for direct synthesis by chemical vapour deposition (CVD) with specific size and homogeneity (Battaglia et al., 2020). A thin film graphene sheet (2D) is obtained by mechanical and chemical exfoliation for their unique properties as high electrical conductivity, stability, chemical reactivity and high surface area (Silva et al., 2018). Graphite consists of sp2 arranged in hexagonal shape while, nanodiamonds have layered spherical shape with optical and magnetic properties that has been used in semiconductors, coatings and abrasive materials (Sudha et al., 2018; Yan et al., 2016).

(e) **Silica-based NMs:** Zeolite-NPs, mesoporous silica-NPs are used widely for their surface morphology, structure and pore size produced in acidic medium by changing pH, concentration of matrix and use of hydrophobic compounds (Kumar et al., 2018). The nanoclays are used widely due to their weak Vander Waal bonds that can break easily and intercalate polymer chain (Figure 20.3). Though most polymers are nanoclay incompatible with surface energy difference, nanoclay-based Pd (Hal-Pd) complex is an active heterogenous catalyst. Others modified NMs used include the CNT) with unique properties (Saleh and Fadillah, 2019).

(f) **Polymeric NMs:** A nanometer-sized solid particle made of either natural or synthetic polymers used in medicine and pharmaceuticals as drug release regulators for body sensing (Yang et al., 2020). These include polymeric micelles in specific solvent of self-assembly amphiphilic block copolymers like chitosan used in drug delivery that are nanometer size, stable, biocompatible, low toxic properties. The synthesis of polymeric NPs has impact on certain characteristics and are divided into three categories: dispersion of polymers, ionic gel formation of hydrophilic polymers, and synthesis of monomers (Kumar et al., 2023; Mayegowda et al., 2023a, b). Dendrimers, a novel class of macromolecules with 3D shapes are smaller than 15 nm, widely used in healthcare and pharmaceutical industries because of their adaptability while, nanofillers are combined to form polymer nanocomposites (Mayegowda et al., 2023a, b) (Figure 20. 2).

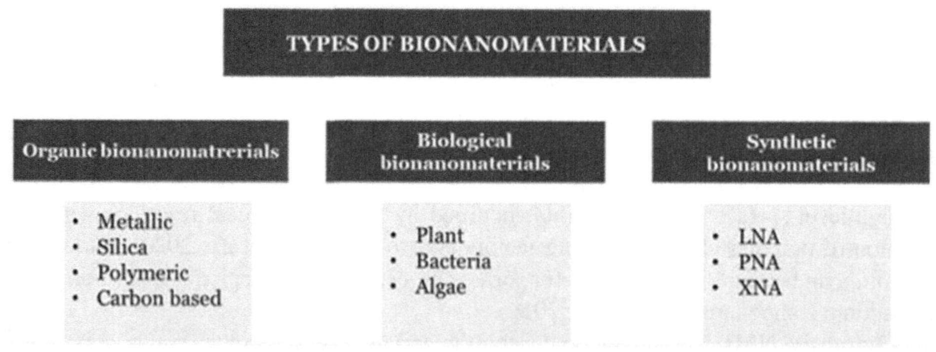

FIGURE 20.3 Classification of Nanomaterials from Different Sources of Biological Origin. (Created by the Authors.)

20.3.2 Biological Nanomaterials

Numerous reports suggest that creating NPs chemically and physically is expensive and environmentally harmful. Therefore, an increasing demand for creating techniques that are both ecofriendly and economically responsible has led to the use of green synthesized NPs using prokaryotes to multicellular fungi and plants. Bacterial NPs can be derived from gold, silver, cadmium, zinc, iron, and magnetite. While algal NPs are generated by silver and gold, fungal NPs uses gold, silver, cadmium and plants with silver, gold, palladium zinc oxide, magnetite, and platinum (Li et al., 2022; Senapati et al., 2012). Bio-based protocols used to create highly stable and thoroughly characterized NPs with significant factors, such as the types of organisms, inheritable and genetic characteristics of organisms, optimal requirements for cellular metabolism and enzymatic activities, suitable reaction conditions, and the choice of the biocatalyst state, have been taken into account (Mayegowda et al., 2022a; Ananda et al., 2022; Rotti et al., 2023). The concentration of the substrate, pH, light, temperature, strength of buffer, amount of an electron donor (such as glucose or fructose), accumulation of biomass and material, charge, and exposure time are some critical factors that can be changed to alter NPs size and morphologies (Figure 20.3 and Table 20.1) (Inchara et al., 2022; Iravani et al., 2014).

(a) By Bacteria

An effective and affordable method for removing the complex industrial tannery discharge contains heavy metals. Metal toxicity is a major environmental issue because of the inability to degrade and bioaccumulation of metals in environment, which have proven to be a series threating the ecosystem. By generating metallic NPs and biomedicating heavy metals, bacteria can serve as green biofactories and provide environmentally friendly substitutes (Igiri et al., 2018). Different protective mechanisms contribute to bacterial cells' ability to withstand heavy metals like extracellular capture and obstruction, this mechanism is monitored by bioreduction of metals in their ionic state through the effective transport of metallic ions (efflux) and intracellular captur (Jin et al., 2018; Choudhury and Srivastava, 2001). The arrangement of the prokaryotic cell wall, the sorption sites, and ionization based on chemical components of cellular wall support the integrity of a bacteria with metals combination (Kapahi and Sachdeva, 2019). In the presence of toxic metals, bacteria perform important tasks like biotransformation, enzymatic reactions, and production of exopolysaccharide and metallothionein that associates and helps the bacterial cell for its survival (Wu et al., 2010).

TABLE 20.1
Various Pollutants and Their Effects on Human Health

Pollutants	Effects
Sulfur dioxide (SO_2)	Nose and throat irritation, respiratory disease with prolonged exposure leads to chronic bronchitis.
Nitrogen dioxide (NO_2), Nitric oxide (NO)	Irritation of eyes, throat and nose. Reduces the supply of oxygen to tissues.
Carbon monoxide (CO)	Binds to hemoglobin, reduced supply of oxygen to tissues leading to pulmonary and cardiovascular diseases.
Heavy metals	Damage to kidney, liver, brain. Infertility issues, causes anemia and neurological disorders.
Aliphatic hydrocarbons and polycyclic organic compounds	Cancer.
Radioactive substances	Anemia, cancer, genetic disorders.
Suspended particulate matter	Asthma, bronchitis and pulmonary disorders.

Source: Created by the Authors.

TABLE 20.2
Microbial Synthesis of NPs with Applications

Metals	Microorganisms	Size	References
Ag	*B. subtilis*	5–50 nm	Saifuddin et al., 2009
CdS	*K. aerogenes*	20–200 nm	Holmes et al., 1995
Ti	*Lactobacillus strains*	40–60 nm	Prasad et al., 2007
Au	*P. aeruginosa*	15–30 nm	Husseiny et al., 2007
CdS, Ag	*Escherichia coli*	2–5nm, 8–9 nm	Sweeney et al., 2004; Mahanty et al., 2013
Pt	*Fusarium oxysporum*	100 nm	Dhillon et al. 2012
Pd	*Penicillium Chrysosporium*	10–14 nm	Tarver et al., 2019
TiO_2	*Saccharomyces Cerevisiae*	6–7 nm	Peiris et al., 2018
Ti	*Trichoderma harzianum*	2–16 nm	Li et al., 2022
CuO	*Bifurcaria bifurcate*	5–45 nm	Senapati et al., 2012
AgCl	*Sargassum Plagiophyllum*	18–42 nm	Desai et al., 2022
Ag	*Caulerpa racemose*	5–25 nm	Jena et al., 2014
Au	*Galaxaaura elongata*	3.85–77.13 nm	Abboud et al., 2014

Microbes have efficient mechanisms for cleansing of metals and resilience in response to their environmental harmfulness. This involves a variety of techniques, such as ion exchange, precipitation, interface complexation, redox process, and electrostatic contact (Yang et al., 2015). By converting their valence, vaporizing compounds, and precipitating them extracellularly, bacteria can remediate metals from environment (Ramasamy et al., 2006). *P. fluorescens* and *B. safensis* were found to be able to breakdown chromium by greater than 84 and 72 per cent, respectively (Table 20.2 and Figure 20.2) (Kalaimurugan et al., 2020).

(b) By Fungi

Through a number of interconnected biomechanics and biochemical processes, fungi exhibit different capacities to affect the formation, toxicity, mobility, and mineral speciation, disintegration, or degradation. Nanomaterials can be in elemental, mineral or complex form, a byproduct of many metal-mineral interactions. Some myogenic NPs have ability to act as nanoenzymes that imitate other catalysts like peroxidase, benefitting defense against environmental stress and elimination of toxic metals in microbes. The majority of fungi are amenable to controlled culture and are well known for secreting compounds and catalysts involved in NP or nano mineral genesis with some benefits. Both intracellular and extracellular processes produce NPs, with extracellular processes being more easily harvested and cell wall providing a huge number of nucleation area for NPs creation (Shilpa et al., 2022a; Mayegowda et al., 2022b). Nanoscale zerovalent (nZVI) iron has been used as an effective substance for contaminant removal due to its large surface area, strong reactivity, and mobility. Although actual applications showed that *Neurospora crassa* is used in synthesizing nZVI with a distinctive branching substance derived from the hyphae of a fungus, conventional nanoscale zerovalent tends to aggregate, which restricts reaction efficiency. The biosynthesized nZVI displayed good firmness and high breaking down of carbon tetrachloride (Manjula et al., 2022).

(c) By Algae

Different groups of algae, including diatoms and euglenoids as well as Chlorophyceae, Phaeophycean, Cyanophycean and Rhodophyceae, are being involved in the processes to produce metallic NPs. Algae are the great option to biosynthesize NPs because of its capacity to collect metals and decreases ions of the metals. It also has a number of other benefits, including the ability to produce at minimum temperatures with improved energy efficiency and less risk to the ecosystem and toxicity. Different widely available surfactants are utilized in the synthesis of NPs with various composition using both chemical and physical techniques as templates, and capping agents, but the elimination of residual elements becomes a significant problem (Table 20.2) (Sharma et al., 2016).

20.4 DIFFERENT METHODS OF REMEDIATION

The most significant physical methods are evaporation-condensation and laser ablation with several benefits over chemical processes. Atmosphere pressure tube furnace-based physical synthesis of silver NPs has some drawbacks, such as requiring a great deal of power to heat the area around the main substance, requiring considerable time to reach heat balance and take up a lot of space. Additionally, a typical tube furnace needs to reach a stable temperature range, preheating demands a couple of minutes and moreover a few KW of electricity (Kruis et al., 2000; Magnusson et al., 1999). In chemical actions, dodecane thiol-capped silver NPs were created using the Burst method involving transferring a gold cationic compound in a two-phased fluid system, moving from the liquid to the organic phase. Next, sodium borohydride is reduced with the help of dodecanethiol enacting as a stabilizing mediator by adhering to the exterior of the NPS, preventing from aggregating and soluble in particular fluids. It has been stated, even minor adjustments to the synthetic variables cause significant changes in the assembly patterns, stability, and size dispersion of NPs (Oliveira et al., 2005; Brust and Kiely, 2002). Uniform and size-controllable NPs can be produced using microencapsulation procedures, but one of their major drawbacks is highly hazardous organic solvents used that are carcinogenic, exhibit reproductive hazards, and release neurotoxins (Figure 20.1) (Shilpa et al., 2022a; Krutyakov et al., 2008).

Organic pollutants like herbicides, biphenyls, aromatic hydrocarbons, solvents, pharmaceutical and hygiene products, and so forth permeate the soil because of improper disposal, unintentional spills and chemical leaks (Paria, 2008). Several of these are toxic and lipophilic, only marginally either soluble or enduring in sediments that are taken up by food products, and accumulate throughout the food chain, endangering the wellbeing of people (Megharaj et al., 2011). The majority of naturally occurring NPs, such as hematite, silicate flakes, clay garins and bismuth in air cohabit tranquilly with the surrounding environment at levels that are safe for the ecosystem. Artificial materials with a rapidly rising global production rate, manufactured NPs are utilized extensively in industry and have attracted significant research interest. Recently, synthesized NPs have been investigated as innovative and effective decontamination techniques for biological toxins in soil and water (Mayegowda et al., 2022b). NPs have efficient high soil stability and dispersibility in their distribution to polluted regions for cleanup purposes (Karn et al., 2009). Due to their exceptional strong reactivity, they have received much attention in the clean-up of organic contaminants in soil. However, certain NPs are quickly corroded by water in the pore area or quickly oxidized by dissolved oxygen (DO) when discharged into the soil (Karn et al., 2009).

Nano zero-valent iron: Because of its affordability and minimal harm towards the ecosystem and human health, minimal amounts of iron, particularly nZVI, is the most extensively used in decontaminating pollutants in water and ground ecosystems. nZVI has a substantially higher reactivity to break down organic contaminants than millimeter (mm) size zero-valent iron. It has a comparatively bigger area surface of about 140 m2/g, whereas commercial mZVI only has a specified surface area of 1.8 m2/g (Ming-Chin et al., 2005, Fang et al., 2012) examined deterioration of 2,4-dichloro phenoxy acetic acid by iron oxide with the help of native soil microorganisms showing that when compared to their singular treatments, the combined effect produced a greater decontamination efficacy and a shorter 2,4-D half-life. The breakdown of organic contaminants by NPs is favored by high temperatures as shown in CMC-modified Pd/Fe NPs that destroyed 94 per cent of lindane in 3 hours at 45°C, compared to 88 per cent and 70 per cent at 35 and 25°C, respectively (Singh et al., 2012). Numerous reports suggest that creating NPs chemically and physically is both expensive and environmentally harmful. There is an increasing necessity for creation of economically and environmentally responsible biological methods that do not use hazardous substances common to product manufacturing procedures. A broad range of organisms, from prokaryotic to eukaryotic microbes to plants, are capable of producing NPs.

20.4.1 Aerobic Remediation

Activated sludge is produced by a procedure where microbes break down pollutant, organic materials, or wastewater sludge under the presence of air for creating a solids and fluids separation. The leftover organic matter is oxidized into sulphates, carbon dioxide, and phosphates and transformed into metabolic energy to supply energy for synthesis and cellular processes. This results in an increase in biomass because only a small part of the organic material is used for the production of new microbes. To keep the biomass concentration high, the biosolids that have settled then are recycled into aeration receptacles, where the supernatant is then dumped. The organisms start natural respiration to oxidize cellular material once the organic waste material runs out (Inchara et al., 2022; Whiteley et al., 2006).

20.4.2 Anaerobic Remediation

In anaerobic digestion methane is produced by the breakdown of organic matter, and compounds other than oxygen act as the process' final electron acceptor. In contrast, some bacteria such as nitrate reducers transfer the electrons to nitrate, reducing it to nitrites, nitrous oxide and nitrogen gas. In particular, sulphate-reducing microbes reduce sulphate to H2S by transferring electrons. FeIII and MnIV are additional electron acceptors. Anaerobic treatment has emerged as the most advantageous stabilization method among the many treatment options because it maximizes cost effectiveness, is ecologically friendly, minimizes the amount of final sludge disposal, and has the potential to produce methane gas, which has a positive net energy gain (De Baere, 2000).

20.5 BIONANOMATERIALS IN ENVIRONMENTAL APPLICATIONS

20.5.1 Heavy Metals and Other Pollutants Remediation

Heavy metals in sediment and ground water can be immobilized by nanoscale particles which has drawn significant attention. When utilizing NPs as amendment substances, the following conditions are used: a static vessel for collecting liquid metals, NPs must meet requirements of being deliverable to the contaminated zones, and they must stay inside the restricted domain once the external injection pressure is gone (under normal groundwater circumstances). NPs quickly start to combine into micron to millimeter scale, aggregates loosely in distinctive qualities, such as high specific surface area and soil delivery in the process. Organic polymers like starch and carboxymethyl cellulose (CMC) are frequently used as a stabilizer on NPs to improve physical stability and movement of soil, avoid agglomeration via steric or electrostatic anchoring mechanisms, and create a greater surface specific area to address these issues (Liang and Zhao, 2014). The efficiency of starch stabilized magnetite NPs for in-situ improved adsorption as well as immobilization of arsenate (As) were studied and the finding showed water-leachable arsenate and toxicity characteristics leaching process (TCLP) were both significantly reduced (Liang and Zhao, 2014). As shown by the immobilization of lead (Pb), phosphate substances may be utilized as efficient agents for the in-situ immobilization of heavy metals in polluted soils. Phosphate was commonly treated to soil either in its soluble forms, like phosphoric acid, or its solid structures, like artificial apatite and naturally occurring mineral rocks (Yang et al., 2001). Many researchers have used nanochitosan to eliminate polyaromatic

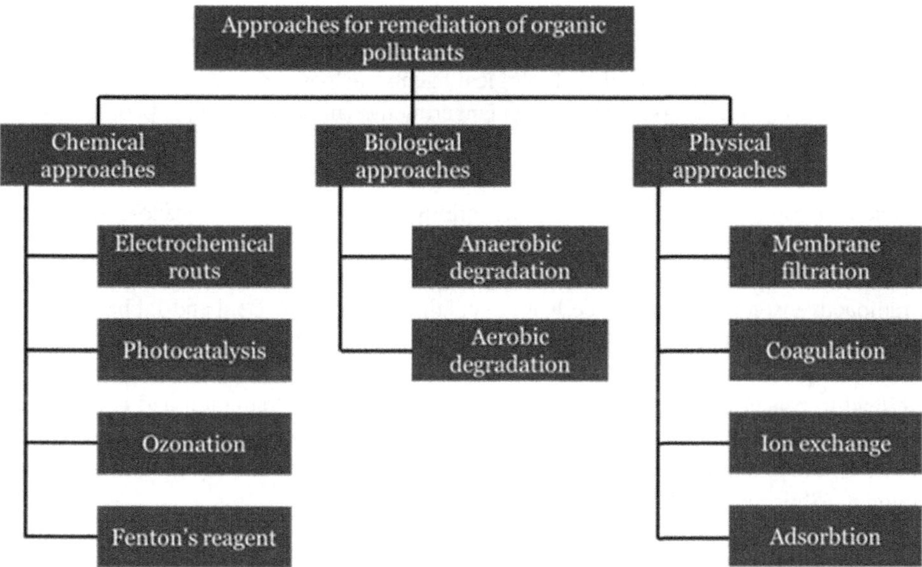

FIGURE 20.4 Bionanomaterials in Environmental Applications. (Created by the Authors.)

hydrocarbons, and the chitosan-infused nano materials have produced excellent outcomes for the eradication of contaminants at pH5 and 30°C (Tahvildari and Mojahedi, 2016).

Cleaning up contaminated water and dirt from heavy metals can be accomplished using bionanomaterials. Bacteria NPs can attach to heavy metals and be removed from the environment and used for cleanup (Shilpa et al., 2022a). Small amounts of heavy metals such as lead (Pb), cadmium (Cd), and arsenic (As) are harmful to living creatures. Heavy metals can impair people's health causing cancer, neurological problems, and abnormal developmental stages. Furthermore, they have the ability to accumulate in the atmosphere, endure for a very long time, and disrupt the ecosystem. To reduce the environmental and health dangers associated with contamination by heavy metals in bionanomaterials, it is critical to guarantee that these materials are created and used in a contaminant-free manner. Strong quality assurance measures, adequate handling processes, and proper disposal can help achieve this (Figure 20.4)(Mayegowda et al., 2023a, b).

Organic pollutants are substances that come from living things or artificial sources comprising industrial chemicals, medications, and insecticides. Even in traces these contaminants can be toxic with wide range of negative effects on human and wildlife health (Hassaan et al., 2020). Organic pollutants such as pesticides, polychlorinated biphenyls (PCBs), and polycyclic aromatic hydrocarbons (PAHs) can harm human health by causing cancer, reproductive issues and build up in the environment harming the ecosystem. Organic contaminants may be eliminated from water and soil using bionanomaterials, some bacterial enzymes can break down organic contaminants, making them less harmful.

20.5.2 Biological Agents

Bionanomaterials can be used to identify and destroy biological threats like bacteria and viruses. For instance, certain biological agents can be selectively targeted by NPs to attach to them, making it possible to detect and eliminate them from the environment (Singh et al., 2012). They can be used to cleanse biologically contaminated surfaces or surroundings, and utilized to neutralize toxins generated by bacteria or viruses by breaking down their cell walls (Jadimurthy et al., 2022). Microbes degrade biological substances in the environment and can thrive on bionanomaterials and be employed to assist their growth for use as bacterial feed that can decompose organic waste, for instance, bionanomaterials. Bionanomaterials used in medical or biological applications can potentially be dangerous if contaminated by biological agents such as viruses, bacteria, or prions. These substances may trigger infectious illnesses, endangering the safety and health of people (Figure 20.4).

20.5.3 Radioactive Pollutants

Environmental radioactive contaminants are eliminated using algal bionanomaterials that can attach to radioactive isotopes and remove them from the environment. Radioactive substances can be delivered to particular areas for medicinal purposes using bionanomaterials, for instance to carry radioactive isotopes to cancer cells for specialized treatment (Beni and Jabbari, 2022). These substances can be used to promote the development of plants by degrading radioactive contaminants in soil and help utilizing feed nutrients to plants (Sohni et al., 2018). Uranium, plutonium, and cesium lead to cancer and radiation illness. By being exposed to radiation and contaminating the environment, bionanomaterials that are polluted with radioactive contaminants might be dangerous for human health and have the potential to sustain in ecosystem for a very long time and harm the environment (Shilpa et al., 2022b).

(a) Bioremediation of Water Pollution

Wastewater containing organic pollutants is dangerous, causing cancer, a fatal disease that affects both humans and animals due to everyday consumption of contaminated water (Sarkar et al., 2017). Water recirculation and wastewater remediation are the main foci for development of advanced and clean water by the present advancements in nanotechnology, particularly green nanotechnology with a specific dimension, molecular makeup, solubility/dissolution, and surface charge (Karimi-Maleh et al., 2020; Das et al., 2015). Novel nanoscale compounds, such as metal oxide, graphene nano sheets, chitin, nanoscale zeolite, and carbon nanotubes have been used for bioremediation over the years. Several organic nanomaterials, including calcium alginate, modified dendrimers, and CNT with multiple walls are employed for hydrogenation of toluene, breakdown of crude oil, and elimination of metals. The cleanup of uranium, nitrates, biphenyls, and arsenic in addition to organic pollutants use silver and iron oxide, chitosan and polysulfone-zero-valent-iron (pZVI) NPs (Das et al., 2018). Magnetic NPs owe to their capacity to manage a high volume of wastewater when combined with the proper magnetic separation (Hu et al., 2010). The tea extract derived from biologically synthesized iron oxide NP is used to recover as well as release metal ions from wastewater effluents; iron oxide NPs demonstrated the highest sorption dimensions for arsenic cation As (III) and As (V), respectively greater than the traditional technique (Lunge et al., 2014). Numerous ions of heavy metals, including Cr^{6+}, Pd^{2+}, Cu^{2+}, and Cd^{2+}, have been adsorbed from water solutions using the produced carbon foam with equilibrium period of 5 minutes, had the highest sorption capacities of Cr^{6+}, Pd^{2+}, Cu^{2+}, and Cd^{2+} were 25.2, 45.6, 49.5, and 30.7 mg/g with a spontaneous and endothermic process (Wang et al., 2022). Green vegetable waste was utilized to create a modified form of carbon adsorption serving an environmentally beneficial sorbent for the elimination of Cu (II) ions (Sabela et al., 2019). Green nanomaterials are categorized as nano-photocatalyst, adsorbent, and structure membrane has a major impact on discharge of pollutants from effluent.

(b) Removal of Pathogens

Bionanomaterials are derived from biological sources and have nanoscale dimensions, typically between 1–100 nm. These materials possess special qualities that enable them to be used in a wide range of ways, including natural applications such as removal of pathogens, environmental remediation through contaminated water sources (Mayegowda et al., 2022; Shilpa et al., 2022a, b). The different sources of water, such as rivers, groundwater and lakes, need to create an effective technique for removing pathogens from water sources (Jiang et al., 2018). Several types of bionanomaterials is used to remove pathogens from water sources by adsorbing, thereby reducing their concentration in the water due to their capacity (Zhan et al., 2014). Silver NPs are most frequently used, have unique antibacterial and antiviral properties that disrupt the cell membranes of pathogens, thereby killing them or inhibiting their growth. They are effective in removing bacteria and viruses from water sources, including *E. coli, Salmonella,* and *Norovirus*. However, long-term results of AgNPs on humans and the ecosystem are still under investigation (Zhang et al., 2021).

Chitosan NPs are potentially used to remove pathogens from water sources. A biopolymer made from chitin, found in shrimp, crab and other invertebrates' exoskeletons. Chitosan NPs have a positive charge that allows them to bind to negatively charged microbes such as viruses and bacteria with effective removal of bacteria and viruses like *S. aureus, Escherichia coli* and *Norovirus* (Heinemann et al., 2021). Graphene oxide NPs are a potential pathogen remover, a single-layered sheet of carbon atoms with unique properties, high surface area and strong antibacterial properties is successful in eliminating *E. coli* and *Salmonella* from contaminated water (Huang et al., 2018). Bionanomaterials are potential environmental remediators employed for removal of heavy metals in contaminated soil and water, including pesticides and pharmaceuticals (Guerra et al., 2018). However, it is necessary to note that the long-term symptoms of NPs on the nature and human health are still not well understood. Further research is needed to determine the potential risks in environmental remediation applications. Additionally, it is important to ensure that the production and disposal are done in an environmentally responsible manner to minimize any potential negative impacts on the environment.

(c) Solar Cells

Solar/photovoltaic cells transform photons into electric charges used as energy sources, water pumps, water heaters, satellites, distant structures, distant structures, domestic solar systems, and so forth. The solar cells are classified into three generations: the majority are first-generation in market with high production costs and efficiency levels of 15–20 per cent. Thin films of amorphous SiO_2, and CdTe are used in second generation. While, third generation are still in the development stage and are made on the basis of nanoporous and nanocrystalline materials (Xin et al., 2019). Microbial synthesis is used to create the composite TiO_2 and $TiCl_4$ structured anatase and rutile-colored solar cells, rod-shaped titanium dioxide as a photoanode with a good conversion efficiency. The color produced by pomegranate was used to create a nature friendly and cost-efficient dye-sensitized solar cell exhibiting highest conversion power efficiency of 2 per cent. Outstanding photovoltaic performance was produced by the anthocyanin concentration found in the pomegranate color (Ghann et al., 2017).

(d) Decolorization of Dyes

The quality of surface and ground water has deteriorated owing to impurities, including dyes, and it was discovered that effluent was released into the ecosystem which contains about 15 per cent of dyes, those of which were determined being hazardous to present species (Bhuiyan et al. 2020). NMs from chitosan is employed for the elimination of dyes as they have comparatively bigger surface area and a high sorption capacity make an efficient adsorbent material for the breakdown of different colours or decolorization of dyes in textile wastewaters. The ability of chitosan NPs to adsorb and select is boosted further when the surface is changed with the functional groups or a suitable grafting chemical by utilizing magnetic chitosan to ease segregation using a magnet and improve recyclability (Tanhaei et al., 2015). Glutaraldehyde crosslinked chitosan nanocomposite with magnetic energy generated by precipitation reducing technique is frequently used for Acid Red 2 elimination. The adsorption process is based on the electron–interaction of anionic Acid Red 2, which possesses a positive charge for chitosan nanocomposite and happens as a result of addition of protons in amine group (Kadam and Lee, 2015).

In textile industries, metallic NPs have also grown in popularity due to their enzymatic characteristics for degradation of dyes that refract and receive sunlight and are believed to be interfering with aquatic organism's photosynthetic activities. In humans and animals, these chemicals may also lead to blood problems, irritation on skin, CNS disruption, and damage of hepatic and nephrotic organs. Many investigations on the enzymatic activity of NPs on dye degradation by cadmium sulphide (CdS), titanium dioxide gold (Au), Iron (Fe) doped TiO_2, along with silica gel beads coated with titanium dioxide are extensively employed for dye degradation of methyl orange and methylene blue (Archana et al., 2021).

NPs synthesized biologically were employed for decolorizing and degrading the dye (Archana et al., 2021; Adarsha et al. 2022). Fe-NPs have excellent enzymatic effects and are considered to eliminate contaminants from sewage in contrast to other NPs. Fe- NPs, such as iron oxide, utilize different biological systems such as microbes and extracts of plants like *Andean blackberry* and *Sorghum bran* used to decolorize different dyes such as malachite green, remazol yellow, saffranin, and methyl orange (Bhuiyan et al. 2020; Hassan et al. 2020). NPs being synthesized by copper, selenium, gold and silver are produced by bio-based sources for dye decolorization like 4-(dimethylamino) benzophenone, a less harmful by- product formed by methyl orange dye which when treated with *Deinococcus radiodurans* protein (Haque et al. 2020; Kim et al. 2021; Weng et al. 2020). *Chlorophytum comosum* leaf extract utilized to create Fe-NPs was used for decomposing methyl orange, with hydrogen peroxide (H_2O_2) is present approximately 77 per cent of deterioration occurred in 360 minutes (Ardakani et al. 2021). *Ruellia tuberosa*, leaf extract creates nanoscale zerovalent iron supported by kaolin produced NPs showed increased efficiency in decolorizing reactive dye azo black 5 (Khunjan and Kasikamphaiboon, 2021). However, bio-produced Pd-NPs palladium based on *Konjac glucomannan* showed remarkable efficiency with sodium borohydride for decolorization of azo dyes at 45°C (Chen et al. 2021). Leaf extract of *Phoenix dactylifera* – silver oxide NPs indicate 84.50 per cent congo red decolorization was discovered in produced NPs after an hour treatment (Laouini et al. 2021). To decolorize dyes, not only NPs alone were utilized, but also a combination of NPs and enzyme, laccase enzyme and Cu NPs were used to degrade azo dyes, and the findings show that the integrated strategy at 60 minutes resulted in the maximum (62.3%) degradation of azo dyes. Since, green produced NPs are less harmful than chemically synthesized NPs, they are utilized in the remediation of various contaminants present in wastewater.

(e) Environmental Nanobiosensors

Increased process control, ecosystem monitoring, and environmental decision-making occur when pollutant detection technology is more available and cheaper (Brahmkhatri et al., 2021). Quick and precise sensors capable of detecting contaminants at the molecular level improve human capabilities to support long-term human health and environmental sustainability (Rasheed et al., 2020). A sensor is essentially a form of energy converter that can detect physical, chemical, mechanical, and other qualities or events in its surroundings and show them as an output signal (often an electrical or optical signal). As a result, numerous sensors in various domains have been developed and have many uses. One of the most used sensors are the nanosensors with chemical, physical, or biological features that can monitor changes in the nanoscale with extremely high sensitivity and precision, either qualitatively or quantitatively. The most essential characteristics that have led to widespread acceptance in data derived from sensors and nanosensors are high-sensitivity, high-detection power, and the capacity to assess many species concurrently (Beni and Jabbari, 2022).

Continuous monitoring of air pollution is one of the most fundamental and basic necessities in terms of environmental pollution control. Controlling air pollution has made significant progress mainly due to the application of nanosensors (Brahmkhatri et al., 2021). The primary goal of creating smart dust is to create a collection of sensing devices in the form of very light nanocomputers (Niccolai et al., 2019). These nanosensors can readily float in the air for hours (Balaga et al., 2021). These tiny silicon particles may send acquired data to a central base through their own wireless network. The prototypes have a data transfer rate of roughly one kilobyte per second (Chaulya et al., 2021).

Toxic gases emission is one of the hazards of modern industrial life that disseminates lethal and poisonous gases. CNT sensors are composed of 1 nm thick single-layer nanotubes that may absorb harmful gas molecules (Meng, 2018). They can also identify a tiny number of dangerous gas molecules in the atmosphere. These sensors can be used to detect military biological agents, air pollution, and even organic molecules in space (Sebek et al., 2021; Truchot et al., 2018). Other sensors that have been widely used due to their tiny size and great accuracy are the new three-dimensional nanostructures. Ultra-thin SnO_2 films in the 3D structure nanosensor can detect poisonous gases (SO2 and H_2S) with great sensitivity (Griessler et al., 2011). The inclusion of Pt NPs in the structure of carbon nanotubes improves a sensor's sensitivity to NO_2, the use of different metals in the structure of multi-walled carbon nanotubes provides sensors with high capacities for selective detection of harmful gases like AgNPs and CuNPs detect ammonia and H_2S gases, respectively (Sharafeldin et al., 2021).

The presence of heavy metal ions in water, soil, and air emphasize need of developing sensors that can detect before the concentrations reach dangerous levels (Kharwar and Singh, 2021). The most accurate and advanced sensors for detecting heavy metals are nanomaterials based on quantum dots because of their unique physiochemical features, high specific surface area, and high reactivity (Chowdhury, 2021). By combining quantum dots nanomaterials with optical or chemical sensor converters, strong detection systems capable of detecting many metals in complicated conditions may be constructed (Wang et al., 2022). The zero-dimensional graphene dots are used in optical detection for heavy metal ions due to their optical properties, adjustable surface groups for adsorption, good stability, and processing. Biomass is also employed in quantum dot nanomaterials because of their superior environmental and biocompatibility as in fluorescent nanosensors derived from green algal waste used to detect Fe (III) in effluents (Chandrashekar et al., 2022; Liu et al., 2020).

Bacteria is used to synthesize intriguing inorganic nanomaterials (mostly Se, Au, Ag) for the development of voltametric sensoristic devices, third-generation biosensors, diagnostic applications, like bio-labelling and cell imaging, antibacterial effectiveness in vitro against pathogens (Suresh et al., 2010; Roychoudhury et al., 2016; Al-Dhabi et al., 2018). A novel self-assembled and spherical nano sized non-glucan exopolysaccharide for *Lactobacillus plantarum*-605 for quick biosynthesis of superb mono-dispersed Au and AgNPs. In the extracellular electron transport systems of anaerobic dissimilatory metal-reducing *Geobacter* and *Shewanella*, bacterial nanowires are conductive proteinaceous pills-like nano materials. *G. sulfurreducens* bacterial nanowires, redox-based conductivity mediated by cytochrome Omcs present on fiber surface and metal-like conductivity owing to aromatic amino acid richness in Pill protein fibers have been suggested (Li et al. 2017; Liu et al., 2019).

"Myconanotechnology" is an emerging field for biosynthesis of nano materials by fungi-like yeasts and molds with the ability of yeast to produce silicon NPs (Hulkoti and Taranath, 2014). *Saccharomyces cerevisiae* produces safe and highly water-soluble cadmium telluride quantum dots with interesting characteristics such as variable size, culture time and temperature, emission and photoluminescence quantum yield, making them an intriguing option for bio-imaging and bio-labelling applications. Additionally, biosynthesis with the Au-Ag alloy, NPs is used in electrochemical sensing (Luo et al., 2014; Wei et al., 2017). The species of *Penicillium*, *Aspergillus* and *Fusarium* have many benefits over bacteria like higher metal binding, higher absorption capacities, easy culture and faster development, and higher extracellular nano-synthesis (Dhillon et al., 2012). Microalgae such as *Tetraselmis kochinensis*, *Scenedesmus*, and *Desmodesmus* for the synthesis of noble metal NPs with strong antibacterial activity used in biomedical tool design as well as medication delivery, catalysis, and electronics (Senapati et al., 2012; Jena et al., 2014; Öztürk- et al., 2019).

The key benefits of nanobiosensors over traditional analytical techniques for environmental applications are mobility, compactness, and workability. They are employed as environmental quality monitoring tools for biological/ecological quality evaluation as well as chemical monitoring of both inorganic and organic priority pollutants (Sharpe et al., 2003). The key benefits are the potential of continuous monitoring, work on-site, and the capacity to assess contaminants in complicated matrices with minimum samples. They successfully detect a wide range of fertilizers, herbicides, pesticides, insecticides, pathogens, moisture, soil, and pH levels, and regulate sustainable agriculture by increasing crop productivity (Sekhon, 2014). Additionally, crop yield is threatened on a regular basis by pests, weeds, and viruses that affect the relative farm economics so plants must be safeguard by appropriate action. Nanostructured biosensors might help intelligent farming not just by tracking soil conditions and plant development over wide fields, but also by detecting infectious diseases in crops before symptoms appear (Antonacci et al. 2018).

20.6 FUTURE PERSPECTIVE

The use of bionanomaterials for environmental cleanup is due to their affordability and nontoxicity. An appropriate cleanup strategy that can readily safeguard the environment is needed for the low-cost remediation resolutions that have helped to boost the capacity, selectivity, and affinity for the pollutants, by using target-specific method and creating molecular design through the synthesis of bionanomaterials. The goal of many environmental agencies is to lessen the amount and release of pollutants into the environment that has increased due to human activities and been a global threat. However, NPs can accumulate in the body, causing long-term health issues like kidney nervous system damage and developmental abnormalities (Briffa et al., 2020). Bionanomaterials can build up in plants, animals, and soil microbes leading to DNA damage, oxidative stress, immune system malfunction, cancer, and respiratory issues through food, drink, and air (Malakar et al., 2021). The inadequacy of underlying mechanism of bionanomaterials, synthesis is challenging and needs further research. These biogenic NPs are largely restricted to lab phase due to lack of experimental effectiveness approach, which needs commercialization for biosynthesis along with the thorough understanding of environmental safety and protection being the major goal that needs to be emphasized to lower the toxicity risk, storage, and disposal.

20.7 CONCLUSION

The environmental friendly NMs is a rapidly growing area with potential for extensive research into their varied roles in biological fields as they offer clean, non-hazardous, relatively cost effective and environmental friendly aspects. Biogenic NPs have abundant compatibility for their potential efficiency to be used in biomedicine, agriculture, environmental bioremediation, food industries and so forth. These NPs have been successfully used in medical imaging, cancer therapy, drug delivery, antimicrobial activities. While, on the other hand it has greater scope in waste water treatment, soil bioremediation and so forth. Metallic NPs propagate innovative initiatives and help analyze a novel approach for their applications. However, there is still a dearth of knowledge involved in the parameter optimization, mechanism of synthesis and unfamiliar potentials that needs to be addressed for the biologically synthesized NPs in order to scale up the process. Future research needs to be directed towards the importance of bionanomaterials for the commercial applications with the better management and planning of NPs as they are crucial in mitigating the adverse impacts on the human health and environment.

REFERENCES

Abboud, Y., Saffaj, T., Chagraoui, A., El Bouari, A., Brouzi, K., Tanane, O., & Ihssane, B. (2014). "Biosynthesis, characterization and antimicrobial activity of copper oxide nanoparticles (CONPs) produced using brown alga extract (Bifurcaria bifurcata)." *Applied Nanoscience*, 4, 571–576.

Abozaid, R. M., Lazarević, Z. Ž., Radović, I., Gilić, M., Šević, D., Rabasović, M. S., Radojević, V. (2019). Optical properties and fluorescence of quantum dots CdSe/ZnS-PMMA composite films with interface modifications. *Optical Materials*, 92, 405–410.

Adarsha, J. R., Ravishankar, T. N., Ananda, A., Manjunatha, C. R., Shilpa, B. M., & Ramakrishnappa, T. (2022). "Hydrothermal synthesis of novel heterostructured Ag/TiO2/CuFe2O4 nanocomposite: Characterization, enhanced photocatalytic degradation of methylene blue dye, and efficient antibacterial studies." *Water Environment Research*, 94(6), e10744.

Ainsworth, C. H., Paris, C. B., Perlin, N., Dornberger, L. N., Patterson III, W. F., Chancellor, E., … & Perryman, H. (2018). Impacts of the Deepwater Horizon oil spill evaluated using an end-to-end ecosystem model. *PloS one*, 13(1), e0190840.

Al-Dhabi, N. A., Mohammed Ghilan, A. K., & Arasu, M. V. (2018). Characterization of silver nanomaterials derived from marine Streptomyces sp. al-dhabi-87 and its in vitro application against multidrug resistant and extended-spectrum beta-lactamase clinical pathogens. *Nanomaterials*, 8(5), 279.

Ananda, A., Ramakrishnappa, T., Archana, S., Reddy Yadav, LS, Shilpa, B. M., Nagaraju, G., & Jayanna, B. K. (2022). "Green synthesis of MgO nanoparticles using Phyllanthus emblica for Evans blue degradation and antibacterial activity." *Materials Today: Proceedings*, 49, 801–810.

Antonacci, A., Arduini, F., Moscone, D., Palleschi, G., & Scognamiglio, V. (2018). Nanostructured (Bio) sensors for smart agriculture. *TrAC Trends in Analytical Chemistry*, 98, 95–103.

Archana, S., Jayanna, B. K., Ananda, A., Shilpa, B. M., Pandiarajan, D., Muralidhara, H. B., & Yogesh Kumar, K. (2021). "Synthesis of nickel oxide grafted graphene oxide nanocomposites-A systematic research on chemisorption of heavy metal ions and its antibacterial activity." *Environmental Nanotechnology, Monitoring & Management*, 16, 100486.

Ardakani, L. S., Alimardani, V., Tamaddon, A. M., Amani, A. M., & Taghizadeh, S. (2021). Green synthesis of iron-based nanoparticles using Chlorophytum comosum leaf extract: Methyl orange dye degradation and antimicrobial properties. *Heliyon*, 7(2), e06159.

Bałaga, D., Siegmund, M., Kalita, M., Williamson, B. J., Walentek, A., & Małachowski, M. (2021). Selection of operational parameters for a smart spraying system to control airborne PM10 and PM2. 5 dusts in underground coal mines. *Process Safety and Environmental Protection*, 148, 482–494.

Battaglia, S., Evangelisti, S., Leininger, T., Pirani, F., & Faginas-Lago, N. (2020). A novel intermolecular potential to describe the interaction between the azide anion and carbon nanotubes. *Diamond and Related Materials*, 101, 107533.

Beni, A. A., & Jabbari, H. (2022). "Nanomaterials for environmental applications." *Results Engineering*, 15, 100467.

Bhuiyan, M. S. H., Miah, M. Y., Paul, S. C., Aka, T. D., Saha, O., Rahaman, M. M., … & Ashaduzzaman, M. (2020). Green synthesis of iron oxide nanoparticle using Carica papaya leaf extract: Application for photocatalytic degradation of remazol yellow RR dye and antibacterial activity. *Heliyon*, 6(8), e04603.

Boros, B. V., & Ostafe, V. (2020). Evaluation of ecotoxicology assessment methods of nanomaterials and their effects. *Nanomaterials*, 10(4), 610.

Brahmkhatri, V., Pandit, P., Rananaware, P., D'Souza, A., & Kurkuri, M. D. (2021). Recent progress in detection of chemical and biological toxins in Water using plasmonic nanosensors. *Trends in Environmental Analytical Chemistry*, 30, e00117.

Briffa, J., Sinagra, E., & Blundell, R. (2020). Heavy metal pollution in the environment and their toxicological effects on humans. *Heliyon*, 6(9), e04691. https://doi.org/10.1016/j.heliyon.2020.e04691

Brust, M., & Kiely, C. J. (2002). Some recent advances in nanostructure preparation from gold and silver particles: a short topical review. *Colloids and Surfaces A: Physicochemical and Engineering Aspects*, 202(2–3), 175–186.

Chandrashekhar Prerana, H. G. K., Ramanjinappa, R., Manjula, K. R., Mayegowda, S. B. "Implementation of Biological Fuel Cells in Treating Pharmaceutical Effluents" *Microbial Fuel Cell: Electricity Generation and Environmental Remediation. Renewable Energy: Research, Development and Policies.* New York: Nova Science Publishers, 2022.

Chaulya, S. K., Chowdhury, A., Kumar, S., Singh, R. S., Singh, S. K., Singh, R. K., … & Banerjee, G. (2021). Fugitive dust emission control study for a developed smart dry fog system. *Journal of Environmental Management*, 285, 112116.

Chen, J., Wei, D., Liu, L., Nai, J., Liu, Y., Xiong, Y., … & Liu, H. (2021). Green synthesis of Konjac glucomannan templated palladium nanoparticles for catalytic reduction of azo compounds and hexavalent chromium. *Materials Chemistry and Physics*, 267, 124651.

Choudhury, R., & Srivastava, S. (2001). Zinc resistance mechanisms in bacteria. *Current Science*, 81, 768–775.

Chowdhury, P. (2021). Functionalized CdTe fluorescence nanosensor for the sensitive detection of water borne environmentally hazardous metal ions. *Optical Materials,* 111, 110584.

Das, L., Das, P., Bhowal, A., & Bhattachariee, C. (2020). Synthesis of hybrid hydrogel nano-polymer composite using Graphene oxide, Chitosan and PVA and its application in waste water treatment. *Environmental Technology & Innovation*, 18, 100664.

Das, S., Chakraborty, J., Chatterjee, S., & Kumar, H. (2018). Prospects of biosynthesized nanomaterials for the remediation of organic and inorganic environmental contaminants. *Environmental Science: Nano*, 5 (12), 2784–2808.

Das, S., Sen, B., & Debnath, N. (2015). Recent trends in nanomaterials applications in environmental monitoring and remediation. *Environmental Science and Pollution Research*, 22 (23), 18333–18344

De Baere, L. (2000). Anaerobic digestion of solid waste: State of the art. *Water Science and Technology*, 41, 283–90.

Desai, N., Pawar, U., Aparadh, V., Dethe, U., & Gaikwad, D. "Seaweeds: A Potential Source in Progressing Nanotechnology." In *Bioprospecting Algae for Nanosized Materials*, pp. 139–152. Cham: Springer International Publishing, 2022.

Dhillon, G. S., Brar, S. K., Kaur, S., & Verma, M. (2012). Green approach for nanoparticle biosynthesis by fungi: Current trends and applications. *Critical Reviews in Biotechnology*, 32(1), 49–73.

Fang, G., Si, Y., Tian, C., Zhang, G., & Zhou, D. (2012). Degradation of 2, 4-D in soils by Fe 3 O 4 nanoparticles combined with stimulating indigenous microbes. *Environmental Science and Pollution Research*, 19, 784–793.

Ghann, W., Kang, H., Sheikh, T., Yadav, S., Chavez-Gil, T., Nesbitt, F., & Uddin, J. (2017). Fabrication, optimization and characterization of natural dye sensitized solar cell. *Scientific reports*, 7(1), 41470.

Griessler, C., Brunet, E., Maier, T., Steinhauer, S., Köck, A., Jordi, T., … & Schrems, M. (2011). Tin oxide nanosensors for highly sensitive toxic gas detection and their 3D system integration. *Microelectronic Engineering*, 88(8), 1779–1781.

Guerra, F. D., Attia, M. F., Whitehead, D. C., & Alexis, F. (2018). Nanotechnology for environmental remediation: materials and applications. *Molecules*, 23(7), 1760.

Haque, A., Kiran, S., Nosheen, S., Afzal, G., Gulzar, T., Ahmad, S., … & Tariq, M. H. (2020). Degradation of reactive blue 19 dye using copper nanoparticles synthesized from Labeo rohita fish scales: a greener approach. *Polish Journal of Environmental Studies*, 29(1), 609–616.

Hassan, A. K., Al-Kindi, G. Y., & Ghanim, D. (2020). Green synthesis of bentonite-supported iron nanoparticles as a heterogeneous Fenton-like catalyst: Kinetics of decolorization of reactive blue 238 dye. *Water Science and Engineering*, 13(4), 286–298.

Heinemann, M. G., Rosa, C. H., Rosa, G. R., & Dias, D. (2021). Biogenic synthesis of gold and silver nanoparticles used in environmental applications: A review. *Trends in Environmental Analytical Chemistry*, 30, e00129.

Holmes, J. D., Smith, P. R., Evans-Gowing, R., Richardson, D. J., Russell, D. A., & Sodeau, J. R. (1995). Energy-dispersive X-ray analysis of the extracellular cadmium sulfide crystallites of Klebsiella aerogenes. *Archives of Microbiology*, 163, 143–147.

Hu, H., Wang, Z., & Pan, L. (2010). Synthesis of monodisperse Fe3O4@ silica core–shell microspheres and their application for removal of heavy metal ions from water. *Journal of Alloys and Compounds*, 492(1–2), 656–661.

Huang, D., Li, B., Wu, M., Kuga, S., & Huang, Y. (2018). Graphene oxide-based Fe–Mg (Hydr) oxide nanocomposite as heavy metals adsorbent. *Journal of Chemical & Engineering Data*, 63(6), 2097–2105.

Hulkoti, N. I., & Taranath, T. C. (2014). Biosynthesis of nanoparticles using microbes—a review. *Colloids and surfaces B: Biointerfaces*, 121, 474–483.

Husseiny, M. I., Abd El-Aziz, M., Badr, Y., & Mahmoud, M. A. (2007). Biosynthesis of gold nanoparticles using Pseudomonas aeruginosa. *Spectrochimica Acta Part A: Molecular and Biomolecular Spectroscopy*, 67(3–4), 1003–1006.

Igiri, B. E., Okoduwa, S. I., Idoko, G. O., Akabuogu, E. P., Adeyi, A. O., & Ejiogu, I. K. (2018). Toxicity and bioremediation of heavy metals contaminated ecosystem from tannery wastewater: A review. *Journal of Toxicology*, 2018, 1–16.

Inchara M. N., Mayegowda, S. B., & Manjula, N. G. "Microbial Fuel Cells for Electricity Generation and Environmental Bioremediation" *Microbial Fuel Cell: Electricity Generation and Environmental Remediation. Renewable Energy: Research, Development and Policies*. New York: Nova Science Publishers, 2022.

Iravani, S., Korbekandi, H., Mirmohammadi, S. V., & Zolfaghari, B. (2014). Synthesis of silver nanoparticles: Chemical, physical and biological methods. *Research in Pharmaceutical Sciences*, 9(6), 385.

Jadimurthy, R., Mayegowda S. B., Chandra Nayak, S., Mohan, C. D., & Rangappa, K. S. (2022). "Escaping mechanisms of ESKAPE pathogens from antibiotics and their targeting by natural compounds." *Biotechnology Reports* 34, e00728.

Jena, J., Pradhan, N., Nayak, R. R., Dash, B. P., Sukla, L. B., Panda, P. K., & Mishra, B. K. (2014). Microalga Scenedesmus sp.: A potential low-cost green machine for silver nanoparticle synthesis. *Journal of Microbiology and Biotechnology*, 24(4), 522–533.

Jiang, B., Lian, L., Xing, Y., Zhang, N., Chen, Y., Lu, P., & Zhang, D. (2018). Advances of magnetic nanoparticles in environmental application: environmental remediation and (bio) sensors as case studies. *Environmental Science and Pollution Research*, 25, 30863–30879.

Jin, Y., Luan, Y., Ning, Y., & Wang, L. (2018). Effects and mechanisms of microbial remediation of heavy metals in soil: A critical review. *Applied Sciences*, 8(8), 1336.

Kadam, A. A., & Lee, D. S. (2015). Glutaraldehyde cross-linked magnetic chitosan nanocomposites: Reduction precipitation synthesis, characterization, and application for removal of hazardous textile dyes. *Bioresource Technology*, 193, 563–567.

Kalaimurugan, D., Balamuralikrishnan, B., Durairaj, K., Vasudhevan, P., Shivakumar, M. S., Kaul, T., … & Venkatesan, S. (2020). Isolation and characterization of heavy-metal-resistant bacteria and their applications in environmental bioremediation. *International Journal of Environmental Science and Technology*, 17, 1455–1462.

Kapahi, M., & Sachdeva, S. (2019). Bioremediation options for heavy metal pollution. *Journal of Health and Pollution*, 9(24).

Karimi-Maleh, H., Cellat, K., Arıkan, K., Savk, A., Karimi, F., Şen, F. (2020). Palladium-nickel nanoparticles decorated on functionalized- MWCNT for high precision non-enzymatic glucose sensing. *Materials Chemistry and Physics*, 250, 123042.

Karn, B., Kuiken, T., & Otto, M. (2009). Nanotechnology and in situ remediation: A review of the benefits and potential risks. *Environmental Health Perspectives*, 117(12), 1813–1831.

Kharwar, S., & Singh, S. (2021). First-principles investigation of zigzag graphene nanoribbons based nanosensor for heavy metal detector. *Materials Today: Proceedings*, 47, 2227–2231.

Khin, M. M., Nair, A. S., Babu, V. J., Murugan, R., & Ramakrishna, S. (2012). A review on nanomaterials for environmental remediation. *Energy & Environmental Science*, 5(8), 8075–8109.

Khunjan, U., & Kasikamphaiboon, P. (2021). Green synthesis of kaolin-supported nanoscale zero-valent iron using ruellia tuberosa leaf extract for effective decolorization of azo dye reactive black 5. *Arabian Journal for Science and Engineering*, 46, 383–394.

Kim, B., Song, W. C., Park, S. Y., & Park, G. (2021). Green synthesis of silver and gold nanoparticles via Sargassum serratifolium extract for catalytic reduction of organic dyes. *Catalysts*, 11(3), 347.

Kim, C., & Lee, H. (2018). Light-assisted surface reactions on metal nanoparticles. *Catalysis Science & Technology*, 8(15), 3718–3727.

Kruis, F. E., Fissan, H., & Rellinghaus, B. (2000). Sintering and evaporation characteristics of gas-phase synthesis of size-selected PbS nanoparticles. *Materials Science and Engineering: B*, 69, 329–334.

Krutyakov, Y. A., Olenin, A. Y., Kudrinskii, A. A., Dzhurik, P. S., & Lisichkin, G. V. (2008). Aggregative stability and polydispersity of silver nanoparticles prepared using two-phase aqueous organic systems. *Nanotechnologies in Russia*, 3, 303–310

Kumar, K. M., Shristi Ram, Manjula, N. G., & Borehalli, S. (2023). "3 Biopolymers and their applications in." *Polymeric Biomaterials: Fabrication, Properties and Applications*. CRC Press, pp. 35–62.

Kumar, S., Malik, M. M., & Purohit, R. (2018). Synthesis of high surface area mesoporous silica materials using soft templating approach. *Materials Today: Proceedings*, 5(2), 4128–4133.

Laouini, S. E., Bouafia, A., Soldatov, A. V., Algarni, H., Tedjani, M. L., Ali, G. A., & Barhoum, A. (2021). Green synthesized of Ag/Ag2O nanoparticles using aqueous leaves extracts of Phoenix dactylifera L. and their azo dye photodegradation. *Membranes*, 11(7), 468.

Li, C., Zhou, L., Yang, H., Lv, R., Tian, P., Li, X., ... & Lin, F. (2017). Self-assembled exopolysaccharide nanoparticles for bioremediation and green synthesis of noble metal nanoparticles. *ACS Applied Materials & Interfaces*, 9(27), 22808–22818.

Li, Q., Liu, F., Li, M., Chen, C., & Gadd, G. M. (2022). Nanoparticle and nanomineral production by fungi. *Fungal Biology Reviews*, 41, 31–44.

Li, Z., Wang, L., Li, Y., Feng, Y., & Feng, W. (2019). Carbon-based functional nanomaterials: Preparation, properties and applications. *Composites Science and Technology*, 179, 10–40.

Liang Q, & Zhao, D. (2014). Immobilization of arsenate in a sandy loam soil using starch-stabilized magnetite nanoparticles. *Journal of Hazardous Materials*, 271, 16–23.

Liu, F., Zhu, S., Li, D., Chen, G., & Ho, S. H. (2020). Detecting ferric iron by microalgal residue-derived fluorescent nanosensor with an advanced kinetic model. *Iscience*, 23(6), 101174.

Liu, X., Wang, S., Xu, A., Zhang, L., Liu, H., & Ma, L. Z. (2019). Biological synthesis of high-conductive pili in aerobic bacterium Pseudomonas aeruginosa. *Applied Microbiology and Biotechnology*, 103, 1535–1544.

Lozhkomoev, A. S., Pervikov, A. V., Chumaevsky, A. V., Dvilis, E. S., Paygin, V. D., Khasanov, O. L., & Lerner, M. I. (2019). Fabrication of Fe-Cu composites from electroexplosive bimetallic nanoparticles by spark plasma sintering. *Vacuum*, 170, 108980.

Lunge, S., Singh, S., & Sinha, A. (2014). Magnetic iron oxide (Fe3O4) nanoparticles from tea waste for arsenic removal. *Journal of Magnetism and Magnetic Materials*, 356, 21–31.

Luo, Q. Y., Lin, Y., Li, Y., Xiong, L. H., Cui, R., Xie, Z. X., & Pang, D. W. (2014). Nanomechanical analysis of yeast cells in CdSe quantum dot biosynthesis. *Small*, 10(4), 699–704.

Magnusson, M. H., Deppert, K., Malm, J. O., Bovin, J. O., & Samuelson, L. (1999). Gold nanoparticles: Production, reshaping, and thermal charging. *Journal of Nanoparticle Research*, 1, 243–251.

Mahanty, A., Bosu, R. A. N. A. D. H. I. R., Panda, P., Netam, S. P., & Sarkar, B. (2013). Microwave assisted rapid combinatorial synthesis of silver nanoparticles using E. coli culture supernatant. *Internatrional Journal of Pharma and Bio Sciences*, 4(2), 1030–1035.

Malakar, A., Kanel, S. R., Ray, C., Snow, D. D., & Nadagouda, M. N. (2021). Nanomaterials in the environment, human exposure pathway, and health effects: A review. *Science of the Total Environment*, 759, 143470.

Manisalidis, I., Stavropoulou, E., Stavropoulos, A., & Bezirtzoglou, E. (2020). Environmental and health impacts of air pollution: A review. *Frontiers in Public Health*, 14.

Manjula, N. G., Sarma, G., Mayegowda Shilpa, B., & Suresh Kumar, K. "Environmental Applications of Green Engineered Copper Nanoparticles." In *Phytonanotechnology*, pp. 255–276. Singapore: Springer Nature Singapore, 2022.

Mayegowda, S. B., Kempahanumakkagari Sureshkumar, M. S., & Ramakrishnappa, T. (2023a). "4 pH and Thermo-responsive Systems." In *Polymeric Biomaterials: Fabrication, Properties and Applications*, pp. 63–84. Abingdon, UK: Routledge.

Mayegowda, S. B., Sureshkumar, K., Yashaswini, R., & Ramakrishnappa, T. "Phytonanotechnology for the Removal of Pollutants from the Contaminated Soil Environment." In *Phytonanotechnology*, pp. 319–336. Singapore: Springer Nature Singapore, 2022b.

Mayegowda, S. B., Ng, M., Alghamdi, S., Atwah, B., Alhindi, Z., & Islam, F. (2022a). "Role of antimicrobial drug in the development of potential therapeutics." *Evidence-Based Complementary and Alternative Medicine* 2022.

Mayegowda, S. B., Bhoomika, S., Kavita, N., & Manjula, N. G. (2023b). " 10 Environmental adequacy of green polymers and biomaterials." *Polymeric Biomaterials: Fabrication, Properties and Applications.*

Megharaj, M., Ramakrishnan, B., Venkateswarlu, K., Sethunathan, N., & Naidu, R. (2011). Bioremediation approaches for organic pollutants: A critical perspective. *Environment International*, 37(8), 1362–1375.

Meng, Q. (2018). Rethink potential risks of toxic emissions from natural gas and oil mining. *Environmental Pollution*, 240, 848–857.

Ming-Chin, C., Hung-Yee, S., Hsieh, W. P., & Min-Chao, W. (2005). Using Nanoscale Zero-Valent Iron for the remediation of polycyclic aromatic hydrocarbons contaminated soil. *Journal of the Air & Waste Management Association*, 55(8), 1200.

Nehra, M., Dilbaghi, N., Hassan, A. A., & Kumar, S. Carbon-based Nanomaterials for the Development of Sensitive Nanosensor Platforms. In *Advances in Nanosensors for Biological and Environmental Analysis*, pp. 1–25. Elsevier, 2019.

Niccolai, L., Bassetto, M., Quarta, A. A., & Mengali, G. (2019). A review of Smart Dust architecture, dynamics, and mission applications. *Progress in Aerospace Sciences*, 106, 1–14.

Oliveira, M. M., Ugarte, D., Zanchet, D., & Zarbin, A. J. (2005). Influence of synthetic parameters on the size, structure, and stability of dodecanethiol-stabilized silver nanoparticles. *Journal of Colloid and Interface Science*, 292(2), 429–435.

Öztürk, B. Y. (2019). Intracellular and extracellular green synthesis of silver nanoparticles using Desmodesmus sp.: Their antibacterial and antifungal effects. *Caryologia*, 72(1), 29–43.

Öztürk-Atar, K., Eroğlu, H., Gürsoy, R. N., & Çaliş, S. (2019). Current advances in nanopharmaceuticals. *Journal of Nanoscience and Nanotechnology*, 19(7), 3686–3705. https://doi.org/10.1166/jnn.2019.16764

Paria, S. (2008). Surfactant-enhanced remediation of organic contaminated soil and water. *Advances in Colloid and Interface Science*, 138(1), 24–58.

Parker, H. L., Hunt, A. J., Budarin, V. L., Shuttleworth, P. S., Miller, K. L., Clark, J. H. (2012). The importance of being porous: Polysaccharide-derived mesoporous materials for use in dye adsorption. *RSC Advances*, 2, 8992–8997.

Peiris, M. M. K., Gunasekara, T. D. C. P., Jayaweera, P. M., & Fernando, S. S. N. (2018) TiO2 nanoparticles from Baker's yeast: A potent antimicrobial. *Journal of Microbiology and Biotechnology*, 28(10), 1664–1670. https://doi.org/10.4014/jmb.1807.07005

Prasad, K., Jha, A. K., & Kulkarni, A. R. (2007). Lactobacillus assisted synthesis of titanium nanoparticles. *Nanoscale Research Letters*, 2(5), 248–250.

Ramasamy, R. P., Feger, C., Strange, T., & Popov, B. N. (2006). Discharge characteristics of silver vanadium oxide cathodes. *Journal of Applied Electrochemistry*, *36*(4), 487–497. https://doi.org/10.1007/s10800-005-9103-x

Rani, V. Usha, and J. Sridevi. "Loss minimization and voltage profile improvement with network reconfiguration and distributed generation." *International Journal of Computer Engineering In Research Trends*. 4, no. 10 (2017): 449–455.

Rasheed, T., Hassan, A. A., Kausar, F., Sher, F., Bilal, M., & Iqbal, H. M. (2020). Carbon nanotubes assisted analytical detection–Sensing/delivery cues for environmental and biomedical monitoring. *TrAC Trends in Analytical Chemistry*, 132, 116066.

Rodgers, K. M., Udesky, J. O., Rudel, R. A., & Brody, J. G. (2018). Environmental chemicals and breast cancer: An updated review of epidemiological literature informed by biological mechanisms. *Environmental Research*, 160, 152–182.

Rotti, R. B., Roy, A., Manjunath, R., Shilpa, B. M., Gnanaprakash, A P, Sunitha, D. V, & Alghamdi, S. et al. (2023) "Green synthesis of MgO nanoparticles and its antibacterial properties." *Frontiers in Chemistry*, 11, 145.

Roychoudhury, P., Gopal, P. K., Paul, S., & Pal, R. (2016). Cyanobacteria assisted biosynthesis of silver nanoparticles—a potential antileukemic agent. *Journal of Applied Phycology*, 28, 3387–3394.

Sabela, M. I., Kunene, K., Kanchi, S., Xhakaza, N. M., Bathinapatla, A., Mdluli, P., Sharma, D., & Bisetty, K. (2019). Removal of copper (II) from wastewater using green vegetable waste derived activated carbon: An approach to equilibrium and kinetic study. *Arabian Journal of Chemistry*, 12(8), 4331–4339.

Saifuddin, N., Wong, C. W., & Yasumira, A. A. (2009). Rapid biosynthesis of silver nanoparticles using culture supernatant of bacteria with microwave irradiation. *E-journal of Chemistry*, 6(1), 61–70.

Saleh, T. A., & Fadillah, G. (2019). Recent trends in the design of chemical sensors based on graphene–metal oxide nanocomposites for the analysis of toxic species and biomolecules. *TrAC Trends in Analytical Chemistry*, 120, 115660.

Sarkar, S., Banerjee, A., Halder, U., Biswas, R., & Bandopadhyay, R., 2017. Degradation of synthetic azo dyes of textile industry: A sustainable approach using microbial enzymes. *Water Conservation Science and Engineering*, 2(4), 121–131.

Sebek, M., Peppel, T., Lund, H., Medic, I., Springer, A., Mazierski, P., … & Steinfeldt, N. (2021). Thermal annealing of ordered TiO2 nanotube arrays with water vapor-assisted crystallization under a continuous gas flow for superior photocatalytic performance. *Chemical Engineering Journal*, 425, 130619.

Sekhon, B. S. (2014). "Nanotechnology in agri-food production: An overview." *Nanotechnology, Science and Applications*, 31–53.

Senapati, S., Syed, A., Moeez, S., Kumar, A., & Ahmad, A. (2012). Intracellular synthesis of gold nanoparticles using alga Tetraselmis kochinensis. *Materials Letters*, 79, 116–118.

Sharafeldin, I., Garcia-Rios, S., Ahmed, N., Alvarado, M., Vilanova, X., & Allam, N. K. (2021). "Metal-decorated carbon nanotubes-based sensor array for simultaneous detection of toxic gases." *Journal of Environmental Chemical Engineering*, 9(1), 104534.

Sharma, A., Sharma, S., Sharma, K., Chetri, S. P., Vashishtha, A., Singh, P., ... & Agrawal, V. (2016). Algae as crucial organisms in advancing nanotechnology: A systematic review. *Journal of Applied Phycology*, 28, 1759–1774.

Sharpe, M. (2003). "It's a bug's life: Biosensors for environmental monitoring." *Journal of Environmental Monitoring: JEM*, 5(6), 109N–113N.

Shiau, B. W., Lin, C. H., Liao, Y. Y., Lee, Y. R., Liu, S. H., Ding, W. C., & Lee, J. R. (2018). The characteristics and mechanisms of Au nanoparticles processed by functional centrifugal procedures. *Journal of Physics and Chemistry of Solids*, 116, 161–167.

Mayegowda, S. B., Priyanka More, R., Bhavan, K. S., Suresh Kumar, K., Thippeswamy, R. "Environmental Applications of Microbial Fuel Cells". *Microbial Fuel Cell: Electricity Generation and Environmental Remediation. Renewable Energy: Research, Development and Policies*. New York: Nova Science Publishers, 2022b.

Shilpa, Borehalli Mayegowda, R. Rashmi, N. G. Manjula, and Athreya Sreekantha. "Bioremediation of Heavy Metal Contaminated Sites Using Phytogenic Nanoparticles." In *Phytonanotechnology*, pp. 227–253. Singapore: Springer Nature Singapore, 2022a.

Silva, A. A., Pinheiro, R. A., Rodrigues, A. C., Baldan, M. R., Trava-Airoldi, V. J., & Corat, E. J. (2018). Graphene sheets produced by carbon nanotubes unzipping and their performance as supercapacitor. *Applied Surface Science*, 446, 201–208.

Singh, R., Misra, V., Mudiam, M. K. R., Chauhan, L. K. S., & Singh, R. P. (2012). Degradation of γ-HCH spiked soil using stabilized Pd/Fe0 bimetallic nanoparticles: Pathways, kinetics and effect of reaction conditions. *Journal of Hazardous Materials*, 237, 355–364.

Sohni, S., Norulaini, N. A. N., Hashim, R, Khan, S. B., Fadhullah, W., and Mohd Omar, A. K. (2018). "Physicochemical characterization of Malaysian crop and agro-industrial biomass residues as renewable energy resources." *Industrial Crops and Products*, 111, 642–650.

Sudha, P. N., Sangeetha, K., Vijayalakshmi, K., & Barhoum, A. Nanomaterials history, classification, unique properties, production and market. In *Emerging Applications of Nanoparticles and Architecture Nanostructures*, pp. 341–384. Elsevier, 2018.

Sumi, N., & Chitra, K. C. (2019). Fullerene C60 nanomaterial induced oxidative imbalance in gonads of the freshwater fish, Anabas testudineus (Bloch, 1792). *Aquatic Toxicology*, 210, 196–206.

Suresh, A. K., Pelletier, D. A., Wang, W., Moon, J. W., Gu, B., Mortensen, N. P., ... & Doktycz, M. J. (2010). Silver nanocrystallites: Biofabrication using Shewanella oneidensis, and an evaluation of their comparative toxicity on gram-negative and gram-positive bacteria. *Environmental Science & Technology*, 44(13), 5210–5215.

Sweeney, R. Y., Mao, C., Gao, X., Burt, J. L., Belcher, A. M., Georgiou, G., & Iverson, B. L. (2004). Bacterial biosynthesis of cadmium sulfide nanocrystals. *Chemistry & Biology*, 11(11), 1553–1559.

Tahvildari, K., & Mojahedi, H. (2016). Investigation of poly aromatic hydrocarbons adsorption using chitosan and synthetic derivatives. *Trends Life Science*, 5, 2319.

Tanhaei, B., Ayati, A., Lahtinen, M., & Sillanpää, M. (2015). Preparation and characterization of a novel chitosan/Al2O3/magnetite nanoparticles composite adsorbent for kinetic, thermodynamic and isotherm studies of Methyl Orange adsorption. *Chemical Engineering Journal*, 259, 1–10.

Tarver, S., Gray, D., Loponov, K., Das, D. B., Sun, T., & Sotenko, M. (2019). Biomineralization of Pd nanoparticles using Phanerochaete chrysosporium as a sustainable approach to turn platinum group metals (PGMs) wastes into catalysts. *International Biodeterioration & Biodegradation*, 143, 104724.

Truchot, B., Fouillen, F., & Collet, S. (2018). An experimental evaluation of toxic gas emissions from vehicle fires. *Fire Safety Journal*, 97, 111–118.

Wang, J., & Zhuang, S. (2017). Removal of various pollutants from water and wastewater by modified chitosan adsorbents. *Critical Reviews in Environmental Science and Technology*, 47(23), 2331–2386.

Wang, X., Kong, L., Zhou, S., Ma, C., Lin, W., Sun, X., ... & Wang, P. (2022). Development of QDs-based nanosensors for heavy metal detection: A review on transducer principles and in-situ detection. *Talanta*, 239, 122903.

Wei, R. (2017). Biosynthesis of Au–Ag alloy nanoparticles for sensitive electrochemical determination of paracetamol. *International Journal of Electrochemical Science,* 12, 9131–9140.
Weng, Y., Li, J., Ding, X., Wang, B., Dai, S., Zhou, Y., ... & Hua, Y. (2020). Functionalized gold and silver bimetallic nanoparticles using Deinococcus radiodurans protein extract mediate degradation of toxic dye malachite green. *International Journal of Nanomedicine*, 1823–1835.
Whiteley, C. G., & Lee, D. J. (2006). Enzyme technology and biological remediation. *Enzyme and Microbial Technology*, 38(3–4), 291–316.
Wu, G., Kang, H., Zhang, X., Shao, H., Chu, L., & Ruan, C. (2010). A critical review on the bio-removal of hazardous heavy metals from contaminated soils: Issues, progress, eco-environmental concerns and opportunities. *Journal of Hazardous Materials*, 174(1–3), 1–8.
Xie, C., Niu, Z., Kim, D., Li, M., & Yang, P. (2019). Surface and interface control in nanoparticle catalysis. *Chemical reviews*, 120(2), 1184–1249.
Xin, D., Wang, Z., Zhang, M., Zheng, X., Qin, Y., Zhu, J., & Zhang, W. H. (2019). Green anti-solvent processed efficient flexible perovskite solar cells. *ACS Sustainable Chemistry & Engineering*, 7(4), 4343–4350.
Yadav, L. S., Reddy, B. M., Shilpa, B. P., Suma, R., Venkatesh, and Nagaraju, G. (2021). "Synergistic effect of photocatalytic, antibacterial and electrochemical activities on biosynthesized zirconium oxide nanoparticles." *The European Physical Journal Plus*, 136: 1–17.
Yan, Q.-L., Gozin, M., Zhao, F.-Q., Cohen, A., Pang, S.-P. (2016). Highly energetic compositions based on functionalized carbon nanomaterials. *Nanoscale*, 8, 4799–4851.
Yang, J., Mosby, D. E., Casteel, S. W., Blanchar, R. W. (2001). Lead immobiliza- tion using phosphoric acid in a smelter-contaminated urban soil. *Environmental Science and Technology*, 35, 3553–3559.
Yang, T., Chen, M. L., & Wang, J. H. (2015). Genetic and chemical modification of cells for selective separation and analysis of heavy metals of biological or environmental significance. *TrAC Trends in Analytical Chemistry*, 66, 90–102.
Yang, X., Lian, K., Tan, Y., Zhu, Y., Liu, X., Zeng, Y., ... & Hu, F. (2020). Selective uptake of chitosan polymeric micelles by circulating monocytes for enhanced tumor targeting. *Carbohydrate Polymers*, 229, 115435.
Zeng, T., Zhang, P., Li, X., Yin, Y., Chen, K., Wang, C. (2019). Facile fabrication of durable superhydrophobic and oleophobic surface on cellulose substrate via thiol-ene click modification. *Applied Surface Science*, 493, 1004–1012.
Zhan, S., Yang, Y., Shen, Z., Shan, J., Li, Y., Yang, S., & Zhu, D. (2014). Efficient removal of pathogenic bacteria and viruses by multifunctional amine-modified magnetic nanoparticles. *Journal of Hazardous Materials*, 274, 115–123.
Zhang, H., Li, Z., Dai, C., Wang, P., Fan, S., Yu, B., & Qu, Y. (2021). Antibacterial properties and mechanism of selenium nanoparticles synthesized by Providencia sp. *DCX. Environmental Research*, 194, 110630.

21 Bionanomaterials in Removing Organic Dyes from Water

Sabarish Radoor, Amritha Bemplassery, Aswathy Jayakumar, Jasila Karayil, Jun Tae Kim, and Suchart Siengchin

21.1 INTRODUCTION

Water pollution is one of the most serious issues faced by modern life. This is mainly because of the rapid increase in industrialization and population growth. Polluted water contains a high amount of pharmaceutical waste, organic dyes, heavy metals, fertilizers, pesticides, petroleum by-products, plastic wastes, microbes, and so forth. The accumulation of these effluents in water causes a serious problem for human life as well as aquatic organisms. Hence, it is not surprising that the present global community suffers from a lack of access to safe drinking water (Peramune et al. 2022; Simelane et al. 2022).

Dyes have been widely used in many industries such as leather tanning, textile, cosmetics and so forth (Wastewater Control and Management in a Leather Tanning District 1999; Ahmad et al. 2015; Sabarish and Unnikrishnan 2018b). The presence of these dyes in water causes serious danger to aquatic life owing to their visibility at very low concentrations (Radoor et al. 2020c). The adverse consequences would be reduced sunlight, penetration, and resistance of photochemical reactions (Yagub et al. 2014; Sabarish and Unnikrishnan 2018c). Several mutagenic, carcinogenic, and toxic waste products can significantly affect the COD and BOD levels of aquatic sources, which create a threat to human health. Moreover, they are biologically and chemically stable, non-biodegradable organic pollutants (Radoor et al. 2021b). Dye effluents from the textile industry would be in the form of a mixture and removal of such admixed dyes is very essential (Yaseen and Scholz 2017; Radoor et al. 2020a).

Extensive research has been carried out for the development of new materials for purifying wastewater. Over the years, a petroleum-based polymer has been commonly employed for wastewater treatment. However, the poor biodegradability and high cost make it the least interesting in the field of wastewater treatment. So, special attention has been given to replacing petroleum-based polymers with natural polymers such as chitosan, agar, starch, cellulose, carrageenan, alginate, and so forth (Mandal et al. 2023; Sirajudheen et al. 2021; Akinterinwa et al. 2022; Kausar et al. 2023; Quesada et al. 2020).

Bionanocomposite is a high-performance composite material made from biopolymers such as cellulose, chitosan, starch, and so forth. The properties of the bionanocompsites could be further improved by chemical modification and the inclusion of nanoparticles such as nanofillers. The interest in bionanocomposite has extensively increased worldwide due to its superior mechanical property, thermal resistance, barrier properties, good transparency, surface functionality, high surface area, and hydrophilicity. In recent years, there is a tremendous increase in the usage of bionanocomposites for efficient water treatment applications (Zhao et al. 2023; Saya et al. 2021; Zare et al. 2018; Joshi and Gururani 2022). Schematic representation of pollutants discharged from

 DOI: 10.1201/9781003432791-25

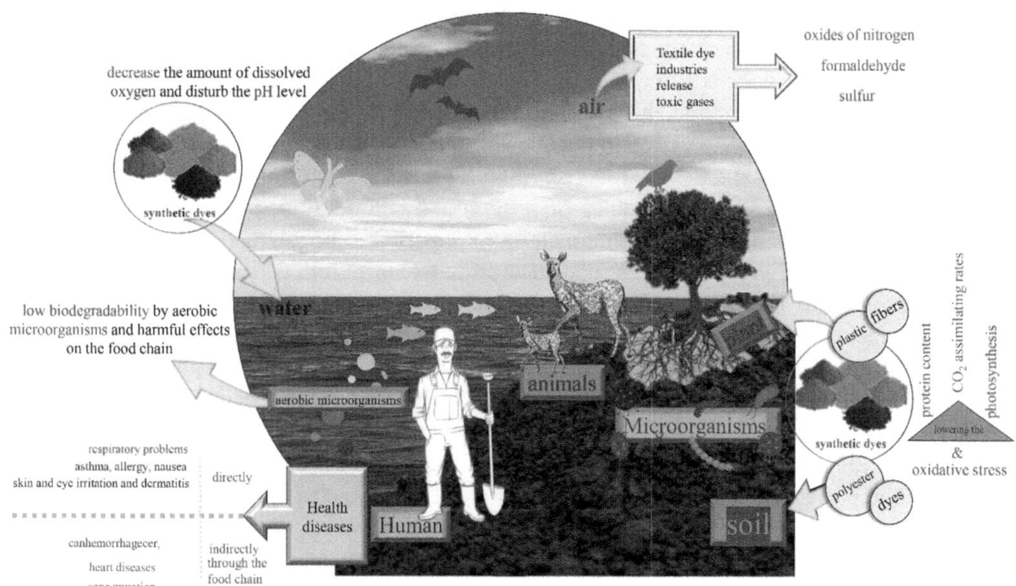

FIGURE 21.1 Direct and Indirect Impacts of Textile Dyes on Several Substrates. (Slama et al. 2021.)

various industries and their ecotoxicological and toxic effect on living organisms are illustrated in in Figure 21.1.

21.2 CHITOSAN-BASED MANO MATERIALS FOR DYE REMOVAL

Polysaccharide derivatives like agar, alginate, chitin/chitosan, cellulose, starch, pectin and so forth are some common examples for environmentally friendly and sustainable biopolymers (Sabarish and Unnikrishnan 2019). They showed high removal efficiency of pollutants preferably via the adsorption method since that can avoid toxic intermediate formation. Chitin is one of the most explored biopolymers for various applications such as wound healing, flocculants, ion exchanger and so forth. They are the second most available natural polymer in the earth's crust and are typically obtained from the shells of crabs, lobsters, shrimp, and other sea crustaceans. Chitosan is derived by enzymatic or deacetylation of chitin. It is a linear polysaccharide consisting of β-(1,4)-linked d-glucosamine (diacetyl unit) and acetyl N-acetyl-d-glucosamine (acetyl unit) (Sajid 2022). The salient features of chitosan, which attract its usage in wastewater treatment are non-toxicity, biodegradability, eco-friendly nature, antibacterial activity, and low cost (Sirajudheen et al. 2021; Dragan and Dinu 2020; Jayakumar et al. 2021). Also, the amine and hydroxyl group, functional groups of chitosan, serve as active sites for removing the toxic pollutants (of synthetic dyes and heavy metal ions) from water through electrostatic or ionic interactions. The hydrophilic nature of chitosan is also an added benefit in water purification as it improves the water absorption and swelling nature of the bio composites. In spite of its superior features, chitosan biocomposites suffer from limitations such as low mechanical properties, low surface area, low solubility in water, low pore size, and so forth. This limitation could be overcome by including metal nanoparticles (Ag, TI), graphene, carbon nanotubes, and so forth (Alsamman and Sánchez 2021; Aramesh et al. 2021; Radoor et al. 2021c).

In recent decades, there has been a huge escalation in the production of biocomposites for wastewater treatment. Babacan and co-workers (Babacan et al. 2022) studied the chitosan-based magnetic nanoparticles for the removal of Reactive Red 120 and Indigo Carmine dyes from an aqueous solution. The effect of contact time, pH, temperature, adsorbent dosage, ionic strength, and initial dye concentration on the adsorption capacity of the adsorbent was studied. In the initial contact time, there is sudden adsorption of dye molecules on the PGMA/mCHT adsorbent surface. However, with the passage of time, the adsorption occurs very slowly due to the saturation of adsorption sites. The maximum dye adsorption occurs at 180 and 60 min contact time for RR120 and IC dyes respectively. On the other hand, ionic strength (NaCl) has a detrimental effect on the adsorption rate. This is ascribed to the presence of sodium and chloride ions on the surface adsorbent. The maximum adsorption capacity of the material is reported to be 241 and 185 mg/g for RR120 and IC dyes respectively (Figure 21.2). The adsorption isotherm follows pseudo-second order and Freundlich isotherms. The recyclability property for PGMA/mCHT adsorption was excellent while maintaining the same adsorption capacity. In another study, anionic (Amaranth Red, (AR)) and cationic (Methylene Blue (MB)) dyes were successfully removed from the aqueous stream using a chitosan/bentonite hybrid composite. For industrial application, the mechanical stability of the composite was an important factor. The tensile strength and elongation at break for the chitosan/bentonite hybrid composite is studied. The excellent mechanical properties of the composites are ascribed to the presence of the bentonite nanoparticle. The pH of the solution medium is found to play a crucial role in the dye adsorption process. It is observed that the anionic dye (AR) has high removal efficiency at a low pH value due to the protonation of the amino group of chitosan and the Si-O group of bentonite. This facilitated the interaction between the negatively charged anionic dye and positively charged adsorbent. Meanwhile, high pH is suitable for the adsorption of cationic dye (MB). The basic medium decreases the $-NH_3^+$ moiety of the chitosan and also induces a negative charge on the Si-O group of bentonite. As a result, strong interaction between the adsorbent and the cationic dye could be achieved (Figure 21.3) (Dotto et al. 2016).

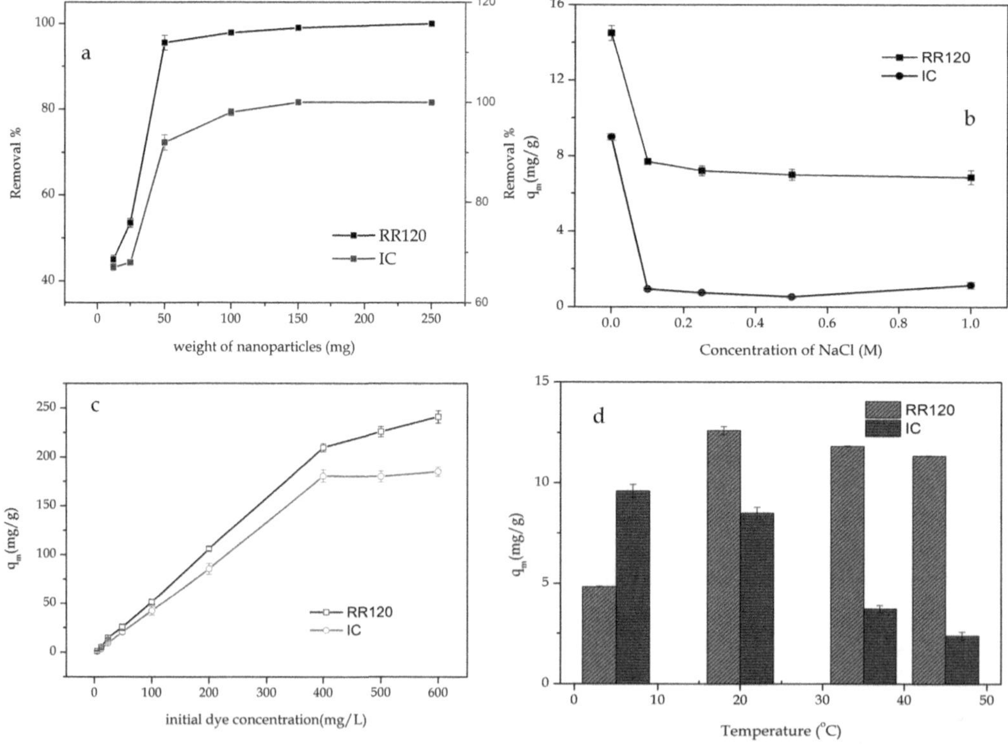

FIGURE 21.2 (a) Effect of dosage; (b) NaCl Concentration; (c) MB Dye Concentration; (d) Temperature on the PGMA/mCHT Nanoparticles. (Reproduced with permission from Elsevier, License no.: 5503931213608.)

In another study, a novel foam adsorbent from polyvinyl alcohol/chitosan composite was explored for the removal of malachite green and Cu^{2+} pollutants from an aqueous solution (Li et al. 2012). The excellent removal efficiency of the composite is ascribed to the synergistic effect of PVA and chitosan polymer. The active functional groups (hydroxyl and amino groups) on the surface of the adsorbent enhance the adsorption capacities. Furthermore, the good thermal, mechanical, and chemical stability of the adsorbent enables it to act as a better adsorbent. The maximum adsorption capacity of MG and Cu^{2+} on PVA/CS foam was 380.65 and 193.39 mg/g. Graphene oxide is an excellent material with extraordinary properties like high surface area, and superior electronic, mechanical, thermal, and optical properties. It has been widely used in water purification applications. For instance, the chitosan/acrylamide/itaconic acid semi-interpenetrating network hydrogel is reported to possess good thermo-mechanical properties (Tamer et al. 2022). High surface area and large pores enhance its ability to increase the adsorption of large molecular MB dyes. A high adsorption capacity (247.47 mg/g) was obtained using a 25 mg adsorbent dosage at pH 7 within a short contact time (60 min). The CS/Aam/IA/GO follows the pseudo-first order kinetic and Langmuir isotherm model. The hydrogel shows excellent regeneration properties (90%) even after seven consecutive adsorption-desorption cycles. Tahari and co-workers (Tahari et al. 2022) utilized chitosan/tannin/montmorillonite (Cs/Tn/MMT) for methyl orange dye removal. The removal efficiency of the composite could be tuned by the MMT dosage. A high removal efficiency of 95.62% was obtained for composite with high MMT dosage (1.0 wt%). The dominant chemical interaction between the dye and the hydrogel is hydrogen and electrostatic attraction. Apart from this, it is noted that the highly porous nature and swelling property of the hydrogel make a high tendency to adsorb MB dye molecules. The chemical interaction between the MMT and the polymer matrix was confirmed by FTIR analysis. The adsorption of MO followed pseudo-second order and Langmuir isotherm with good correlation. The maximum adsorption capacity of the film was 57.37 mg/g at pH 7. The presence of GO significantly enhanced the surface area.

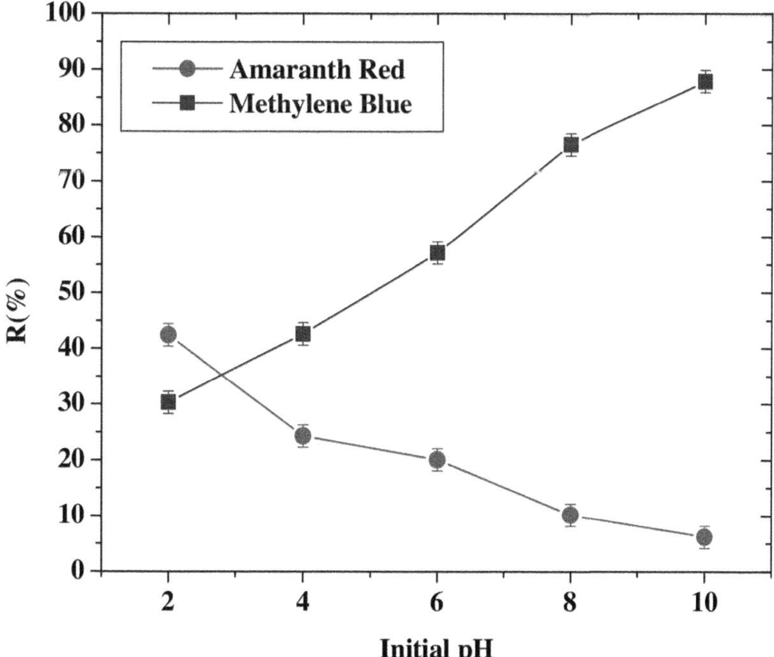

FIGURE 21.3 Effect of pH on the Chitosan/Bentonite Composite of AR and MB Dyes. (Reproduced with permission from Elsevier, License no.: 5503950661786.)

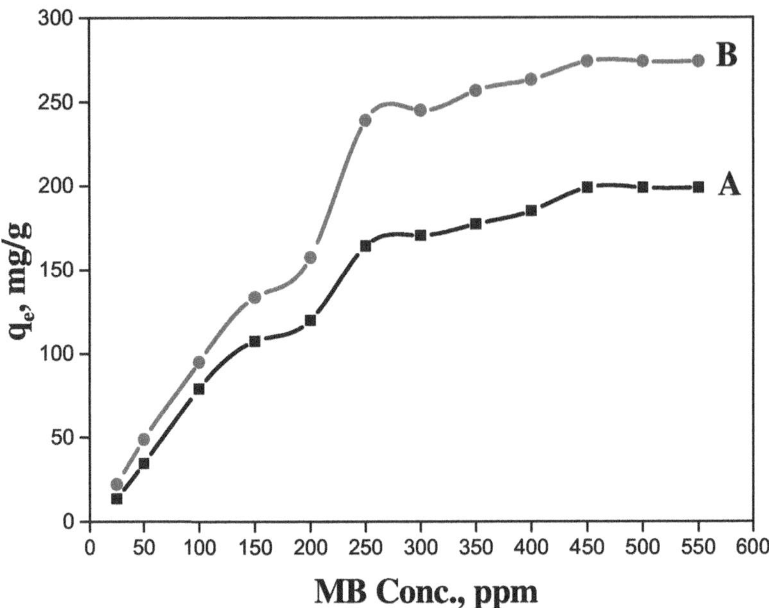

FIGURE 21.4 Effect of the MB Concentration and Contact Time on Chitosan/Silica and ZnO Incorporated Chitosan/Silica Nanocomposite. (Reproduced with permission from Elsevier, License no.: 5503960266627.)

The inclusion of metal nanoparticles in bionanocomposite improves its physical properties and improves its dye adsorption capacity. There are ample reports on metal nanoparticle-based bionanocomposites for water treatment applications. The effect of ZnO on the adsorption performance of chitosan/silica biocomposites was studied by Hassan et al. The ZnO particles were uniformly distributed on the polymer matrix without any agglomeration which is beneficial for the dye adsorption process. Here, maximum dye adsorption (293.3 mg/g) was achieved at a basic medium. (Hassan et al. 2019). The ZnO was successfully incorporated into chitosan/silica nanocomposite which was confirmed by XRD and FTIR techniques. The ZnO particles were uniformly distributed on the matrix without any agglomeration could highly influence the adsorption efficiency of dyes. The presence of a multifunctional group (amino and hydroxyl) which favorable for the penetration of cationic MB dye molecule. Similar behaviours were reported by Zango and co-workers for MB dye removal using ZnO chitosan nanocomposite. The maximum removal adsorption capacity 97.93 mg/g was reported (Figure 21.4) (Zango et al. 2022).

21.3 AGAR-BASED NANO MATERIALS FOR DYE REMOVAL

Agar is a polysaccharide comprising of a mixture of agarose and agaropectin. It possess the peculiar features as chemical inertness, non-toxicity, solubility at high temperature (>50 ^{0}C) in water (Kanmani et al. 2017). The cross-linking properties of agar make them easily form rigid and thermoreversible hydrogel and also, they are cheap and environmentally friendly (Radoor et al. 2020b; Radoor et al. 2021a). These remarkable properties make agar a perfectly suitable material for nanocomposite matrix development which is highly stable and easily stored for long periods. Agar modified nanomaterials are observed to be efficient in removing dye from waste waters. Small size, extremely large surface area, more surface active sites, unique electron conduction property and so forth extremely benefit these nanomaterials as adsorbent, catalyst, sensor and so forth (Ghosh et al. 2010; Sabarish and Unnikrishnan 2018a). Some of the studies proved that nanomaterials could promote the biodegradability of dyes in wastewater. There are several studies regarding the application of agar based bionanomaterials in dye removal (Radoor et al. 2022b; Sabarish et al. 2020; Radoor et al. 2022a). S. Patra et al. synthesized mono (Fe, P, Cu) and bimetallic (Fe/Cu and Fe/Pd nanoparticles) modified with agar by the reduction of their corresponding salts (Patra et al. 2016). They use UV-vis spectroscopy, EDAX, TEM, XRD and so on to characterize the morphology and composition of these nanoparticles. They applied agar modified bimetallic nanoparticles (agar@ Fe/Cu an agar@ Fe/Pd) for adsorption, degradation, and removal of two dyes of industrial importance, that is, methylene blue (MB) and Rhodamine B (Rh B). The adsorption and dye degradation studies pointed out that bimetallic nanoparticles are more efficient when compared to monometallic systems. However, the adsorption and removal capacity for both the dyes were excellent for agar@ F/Pd nanoparticles compared to agar@ Fe/Cu nanoparticles. The maximum adsorption capacity showed by Fe/Pd nanoparticles were 875.0 and 705.0 mgg^{-1} for MB an Rh B respectively with high dye degradation rate ($k = 4.8 \times 10\text{-}2\ s^{-1}$ for MB and $3.16 \times 10\text{-}2\ s^{-1}$ for Rh B) in 60 s of time span (Figure 21.5). Real wastewater samples from several industries were also analyzed to inspect the real sample applicability of synthesized material. Bimetallic nanoparticles were found to have almost 100 per cent efficiency in real sample analysis. Agar-based bimetallic nanoparticles launched a way to prepare a promising, efficient, low-cost, and fast adsorbent and degradation from industrial wastewaters.

O. Duman et al. (Duman et al. 2020) prepared an agar/κ-carrageenan composite hydrogel adsorbent for effective removal of methylene blue, taking advantage of fine properties of hydrogel adsorbents. Great swelling capacity, excellent adsorption performance and ease of removal after adsorption make hydrogel adsorbents unique candidates for adsorption applications (Patra et al. 2016; Narula and Rao 2019). Here, they used agar and κ-carrageenan as two candidate polysaccharides owing to their remarkable properties. κ-carrageenan is biodegradable, biocompatible, renewable non-toxic polysaccharide that can remove cationic candidate dye due to their anionic structure (Upadhyay et al. 2018; Rhein-Knudsen et al. 2017; Yu et al. 2019). Also, anti-viral and anti-coagulation properties are also there. They prepared agar/κ-carrageenan composite hydrogel adsorbent through free radical cross-linking reaction with tri (ethylene glycol) divinyl ether as a cross-linker. The characterizations of the hydrogel were performed using SEM, FTIR, and TGA experiments. The thermal degradation temperature of 217 °C was observed for the hydrogel. The FTIR studies supported that electrostatic interactions and hydrogen bonding were the main interactions in the methylene blue removal from water by the hydrogel. Increase in certain factors such as contact time, solution pH, initial dye concentration and temperature caused an enhancement in adsorption capacity of hydrogel (Figure 21.6). Most appropriate kinetic and isotherm models were found to be pseudo-second order and Langmuir confirming the monolayer coverage of dye molecules. Thermodynamic parameters were evaluated to get insights into the nature of adsorption process. The results suggested the spontaneity and endothermic nature of adsorption process. Of the several solvents tested for the regeneration of dye adsorbed-agar/κ-carrageenan hydrogel, ethyl alcohol was found to be the most suitable one. All the results obtained illustrated that agar/κ-carrageenan hydrogel is an effective, ecofriendly, and promising candidate for removal of cationic dyes from water.

The catalytic degradation of dyes by using nanostructures in wastewater treatment proved to be superior to conventional methods with absence of polycyclic products and quick oxidation. H.E. Emam et al. (Emam and Ahmed 2019) prepared homo-metallic (Ag) and hetero-metallic (Ag-Au an Ag-Au-Pd) nanostructures with agar as nano-synthesizer. Slight enlargement of small size homo metallic nanostructures (9.1–12.8 nm) to 13.6–23.7 nm for hetero-metallic nanostructures was done. Catalytic activity of these nanostructures was investigated using reductive degradation of methylene blue and methyl orange dyes. Here, the composition of nanostructures was observed to have an important role in catalytic activity. In the case of trimetallic nanostructures Ag-Au-Pd, the values of $t_{1/2}$ for the degradation of dyes were significantly diminished from 47.15 and 45.90 min to 0.69 an

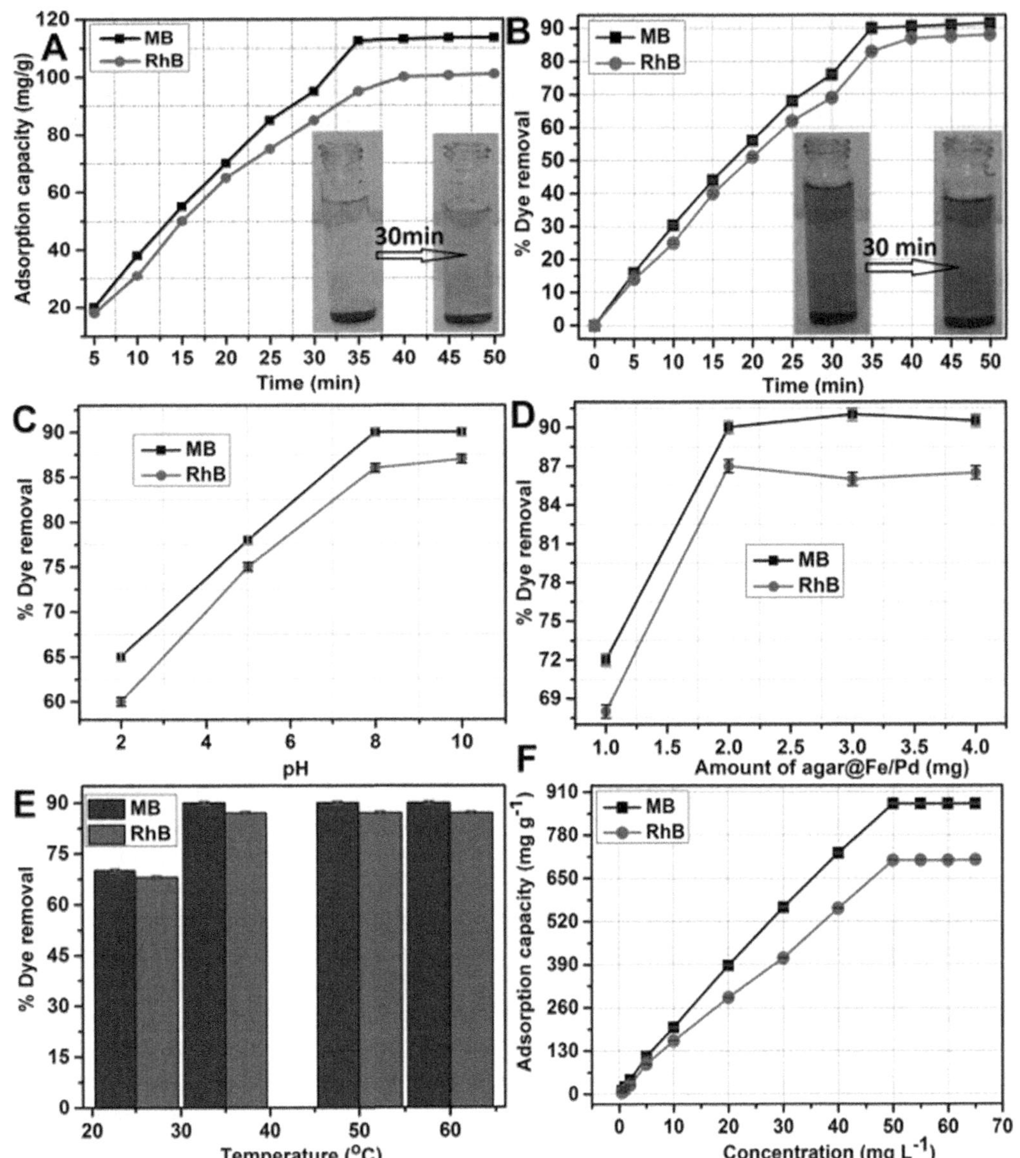

FIGURE 21.5 Effect of Adsorption Capacity (a); Removal Efficiency (b); pH (c); Dosage (d); Temperature (e); Dye Concentration (f) of MB and Rh B on the agar@Fe/Pd Bimetallic Nanoparticles. (Reproduced with permission from Elsevier, License no.: 5503960453917.)

0.97 min, for MB and MO respectively. These findings definitely demonstrate the excellent catalytic activity of trimetallic nanostructures in wastewater treatment for reductive degradation of synthetic dyes. Catalytic action of metal nanoparticles was affected by several factors with passage of time. In order to overcome agglomeration and other problems such as ease of handling, stability and so forth, different stabilizers are employed as supporting agents. Among the supporting agents, hydrogels are of great prominence owing to their ease of synthesis, tunable particle size, handling, and so forth (Kamal et al. 2019). In addition to that, hydrogels are examples for quasi-homogeneous catalysts with high catalytic activity and ease of recovery from reaction systems (Welsch et al. 2010). A. Khalil et al. (Khalil et al. 2020) prepared agar-biopolymer-cobalt oxide nanocomposite hydrogel AG-Co_3O_4

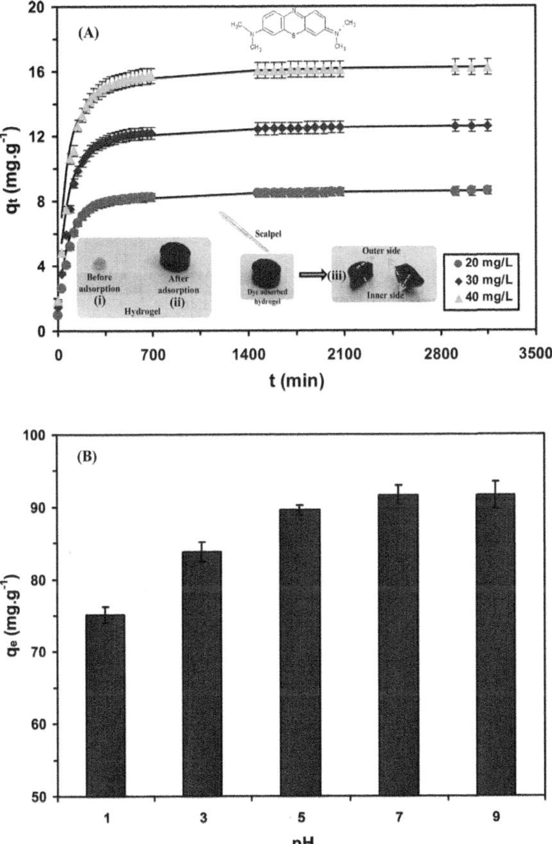

FIGURE 21.6 (a) Effect of Initial Dye Concentration and Contact Time of MB on Agar/κ-Carrageenan Hydrogel. Inset: Photographs of (i) Agar/κ-carrageenan hydrogel, (ii) Methylene Blue-adsorbed Agar/κ-carrageenan Hydrogel; (b) Effect of solution pH of MB on Agar/κ-carrageenan Hydrogel. (Reproduced with permission from Elsevier, License no.: 5503960587025.)

by heating the mixture of ex-situ synthesized Co_3O_4 nanoparticles and agar powder in an aqueous medium followed by cooling. The hydrogel was used as a catalyst for catalytic reduction of toxic dyes, Congo red (CR), methylene blue (MB), and 4-nitrophenol (4-NP). Reaction rate constant (k_{app}) of 0.3623, 0.2114 and 0.2893 or MB, 4-NP and CR, respectively, showed the excellent performance of AG-Co_3O_4. Catalytic reduction studies were performed with different dye concentrations and catalyst doses. Reactions were conducted with 0.2, 0.4 and 0.7 g of AG-Co_3O_4 and it was concluded that k_{app} value increase with increase of catalyst amount. Experimental inferences established AG-Co_3O_4 to be an efficient catalyst for the reduction of CR, MB and 4-NP.

21.4 ALGINATE-BASED NANOMATERIALS FOR DYE REMOVAL

Alginate is one of the low-cost biomaterials which are linear block copolymers consisting of two uronic acid residues, for example, β-D-mannuronic acid and α-L-guluronic acid linked by β-1,4-gycosidic bonds (Ahmed et al. 2014; An et al. 2015; Bhat and Aminabhavi 2006). Unique features such as stability, biodegradability, high water permeability and nontoxic nature make this biopolymer an excellent candidate for wastewater remediation for the adsorption of contaminants, mainly heavy metal cations (Xu et al. 2019; Kuang et al. 2015; Li et al. 2013). Alginate-based adsorbents are remarkable as they form stable bio hydrogel beads with a lot of benefits such as network surface, high surface area, and so forth, which provide the ability to be used as catalytic support material (An et al. 2015).

K. Zhao et al. (Zhao et al. 2014) synthesized TiO_2/calcium alginate (T/CA) membranes by cross-linking the mixture of sodium alginate and TiO_2 in aqueous solution. In the preparation of molecularly imprinted T/CA composite membranes, T/CA was used as supporting matrix and ethylenetri (β-methoxy)ethoxysilane (KH-570) and γ-amidopropyltri-ethoxysilane (KH-550) was used as functional monomers with methyl orange (MO) as template. The adsorption and photocatalytic degradation of methyl orange imprinted membranes were investigated. It was found that adsorption capacity and rate were higher for imprinted membranes compared to non-imprinted membranes (Figure 21.7). MIP possessed a good selectivity feature also for photodegradation of MO choosing MR as competitive molecules. In conclusion, the combination of membranes, molecular imprinting, and photocatalytic degradation can bring out fast and easy operating wastewater treatment with good selectivity. S. Thakur and so forth (Thakur and Arotiba 2017) reported the fabrication of hydrogel nanocomposites

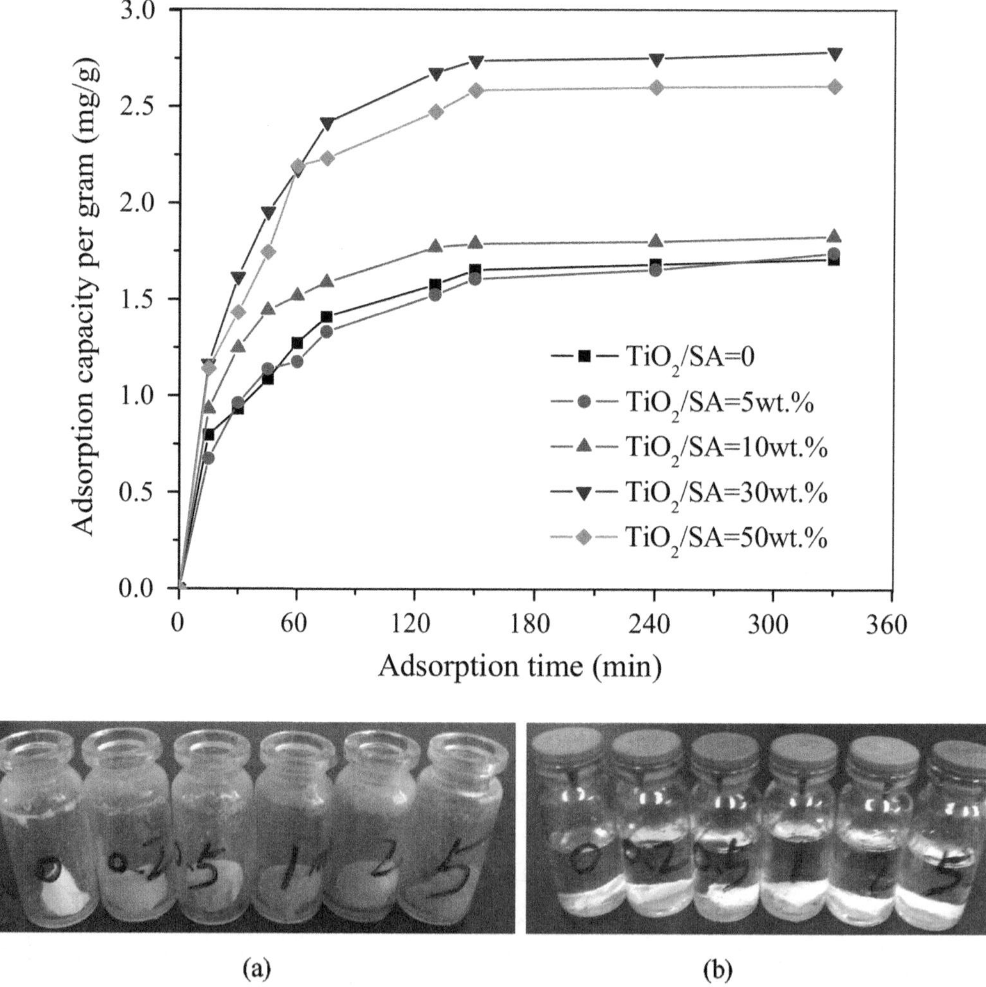

FIGURE 21.7 Effect of Contact Time Against Adsorption Capacity on Different wt% TiO_2/SA on MO Solution. Digital Photograph of NIP and MIP on with Different Concentration of MO (a) before Removal (b) after Removal of MO. (Reproduced with permission from Elsevier, License no.: 5503951328290.)

by solution polymerization of acrylic acid in the presence of sodium alginate biopolymer and TiO_2 nanoparticle. They used N, N-methylene-bis-acrylamide, and TiO_2 nanoparticle as organic and inorganic crosslinker respectively. Block copolymerization method was employed, where TiO_2 nanoparticle was incorporated to the hydrogel (SA-g-PAA) matrix and used this hydrogel nanocomposite, SA-g-PAA/TiO_2 hydrogel nanocomposite for removal of dye methyl violet from the water sources. Different characterization methods like X-ray Diffraction, Scanning Electron Microscopy, Fourier Transform IR spectroscopy and so forth, prove the presence of TiO_2 in the hydrogel matrix. The absorption capacity of SA-g-PAA/TiO_2 hydrogel was remarkable, and it showed excellent capacities for removal of methyl violet dye from water. The effect of factors like sodium alginate content, TiO_2 nanoparticle, and grafting on adsorption were also investigated. It was inferred that pseudo-second order kinetics was prominent in adsorption of methyl violet on to the nanocomposite. About 99.6 per cent efficiency in adsorption of methyl violet was calculated for hydrogel nanocomposite which surpassed the conventional adsorbents used for methyl violet removal. Several advantages like high surface area, unique structure, high pore volume, and so forth contributed to the high adsorption capacity of hydrogel nanocomposite. Therefore, SA-g-PAA/TiO_2 hydrogel nanocomposite adsorbent was proved to be highly effective in water treatment for selective removal of cationic dyes.

M. Kumar et al. (Kumar et al. 2018) studied the adsorptive removal of rhodamine-B (RMN-B) from aqueous solution. For this purpose, they synthesized novel alginate/Zn-Al-Fe_3O_4/Ca nano beads from novel Zn-Al- Fe_3O_4 nanoparticle. Several modelling equations like Freundlich, Langmuir, pseudo-first order, pseudo-second order, intraparticle diffusion, Gibbs free energy, Enthalpy and Entropy were used to analyze the adsorption efficiency of RMN-B. The experimental results showed that the process follows Langmuir and pseudo second-order equations. High temperature withstanding capacity of AZNB beads were proved by TGA studies. The diffusion between RMN-B and nano beads was significant throughout the process as established from intraparticle diffusion. Strong adsorption of RMN-B on entire surface of the nano material was well proven by surface morphology studies and FTIR spectral analyses. The values of thermodynamic parameters also provide evidence regarding the strong adsorption of RMN-B by nano beads. The negative value of ΔG^0, positive values of ΔH^0 and ΔS^0 confirmed the spontaneity of the process and also the endothermic nature of adsorption process. All the studies pointed out that there occurred a strong adsorption between RMN-B an AZNB nano beads. H. Helmiyati et al. (Ramadhani and Helmiyati 2020) successfully synthesized a biopolymer-based nanocomposite, Alginate/CMC/ZnO nanocomposite using carboxymethyl cellulose (CMC), sodium alginate and ZnO. Barium ion was employed as the crosslinking agent. The nanocomposite was characterized using FTIR, which exhibited a peak of wavenumber, 400–600 cm^{-1} which was typical of ZnO. XRD studies also proved the presence of ZnO. The zinc oxide was dispersed as granules attached to the alginate fibers as revealed by SEM analysis. The nanocomposite was used to inspect the photocatalytic activity of dye, Congo red. Different optimum condition were set for the photocatalytic reaction and those were weight of nanocomposite, pH, alginate/CMC:ZnO ratio and reaction time which were 60 mg, pH 3, ratio 1: 0.8 and 110-min reaction time respectively. The alginate/CMC/ZnO nanocomposite was subjected to reusability test which showed that there is a decrease in degradation percentage from first, second and third cycle from 90.12 per cent to 72.82 per cent and 69.63 per cent, respectively. The best degradation results of 90.12 per cent obtained was by using sunlight. The alginate/CMC/ZnO nanocomposite was found to reduce the dye Congo red very efficiently, and the process was environmentally friendly and economically viable (see Table 21.1).

21.5 CONCLUSION

Bio nanocomposites are low-cost, non-toxicity, and biodegradable with exceptional physical and chemical properties. Polysaccharide-based bio nanocomposite has a remarkable ability to remove toxic dyes and heavy metals from contaminated solutions. The diverse functional groups such as

TABLE 21.1
The Literature Data of Bionanocomposite Adsorbent for Dye Removal

Adsorbents	Pollutants	Qmax (mg/g)	Reference
Chitosan-based ZnO nanocomposite	Eriochrome Black-T	40.9	(Raval et al. 2022)
Poly(glycerol sebacate)/chitosan/graphene oxide	Methylene blue	179	(Rostamian et al. 2022)
Chitosan grafted benzaldehyde/montmorillonite/ algae	Brilliant green	899.5	(Jawad et al. 2023)
	Reactive blue 19	213.6	
Magnetic crosslinked chitosan-TPP /MgO/Fe_3O_4	Reactive blue 19	120.3	(Jawad et al. 2022)
Chitosan/silica/zinc oxide nanocomposite	Methylene blue	293.3	(Hassan et al. 2019)
Chitosan-lignin-titania nanocomposites	Brilliant Black	15.8	(Masilompane et al. 2018)
Graphene oxide/chitosan/acrylamide/itaconic acid	Methylene blue	247.47	(Tamer et al. 2022)
PAAm-Agar/Clay@r-GO	Methylene blue	189	(Sarkar et al. 2020)
Agar/zeolite/CMC/PVA	Congo red	15.30	(Radoor et al. 2022a)
PVA/agar/ ZSM-5 zeolite	Congo red	7.2	(Sabarish et al. 2020)
Agar@Fe/Pd BMNPs	Methylene blue	875.0	(Patra et al. 2016)
	Rhodamine B	780.0	
PVA/Agar/D-glucose	Methylene blue	29	(Nguyen et al. 2021b)
Agar/maltodextrin/poly(vinyl alcohol) walled montmorillonite composites	Methylene blue	71.51	(Nguyen et al. 2021a)
Agar/κ-carrageenan composite	Methylene blue	242.3	(Duman et al. 2020)
MZ/CS/AL bio composite	Methylene blue	6.14	(Kazemi and Javanbakht 2020)
bentonite – alginate composite	Crystal violet	601.93	(Fabryanty et al. 2017)
Acrylamide/graphene oxide bonded sodium alginate nanocomposite	Crystal violet	100	(Pashaei-Fakhri et al. 2021)
DAA/Ge/Ag nanocomposite	Methylene blue	625	(Abou-Zeid et al. 2019)
Zeolite/nickel ferrite/sodium alginate bio nanocomposite	Methylene blue	54.05	(Bayat et al. 2018)
SA-poly(AA)/ZnO	Methylene blue	1529.6	(Makhado et al. 2020)

hydroxyl, amino, carboxyl, sulfonic, and epoxy groups of the composite indicate strong adsorption nature to it. This chapter discussed the adsorption behavior of bio composites derived from chitosan, alginate, and agar-based biopolymers for water treatment. The effect of nanoparticles on the properties of the bio nanocomposites is also discussed. The nanoparticles embedded with composites displayed superior mechanical, thermal, barrier, antibacterial, and adsorption capacity, thereby making them promising candidates for wastewater treatment.

REFERENCES

Abou-Zeid RE, Awwad NS, Nabil S, Salama A, Youssef MA (2019) Oxidized alginate/gelatin decorated silver nanoparticles as new nanocomposite for dye adsorption. *International Journal of Biological Macromolecules* 141:1280–1286. doi:10.1016/j.ijbiomac.2019.09.076

Ahmad A, Mohd-Setapar SH, Chuong CS, Khatoon A, Wani WA, Kumar R, Rafatullah M (2015) Recent advances in new generation dye removal technologies: novel search for approaches to reprocess wastewater. *RSC Advances* 5 (39):30801–30818. doi:10.1039/c4ra16959j

Ahmed AE-SI, Moustafa HY, El-Masry AM, Hassan SA (2014) Natural and synthetic polymers for water treatment against dissolved pharmaceuticals. *Journal of Applied Polymer Science* 131(13):n/a–n/a. doi:10.1002/app.40458

Akinterinwa A, Reuben U, Atiku JU, Adamu M (2022) Focus on the removal of lead and cadmium ions from aqueous solutions using starch derivatives: A review. *Carbohydrate Polymers* 290. doi:10.1016/j.carbpol.2022.119463

Alsamman MT, Sánchez J (2021) Recent advances on hydrogels based on chitosan and alginate for the adsorption of dyes and metal ions from water. *Arabian Journal of Chemistry* 14 (12). doi:10.1016/j.arabjc.2021.103455

An B, Lee H, Lee S, Lee S-H, Choi J-W (2015) Determining the selectivity of divalent metal cations for the carboxyl group of alginate hydrogel beads during competitive sorption. *Journal of Hazardous Materials* 298:11–18. doi:10.1016/j.jhazmat.2015.05.005

Aramesh N, Bagheri AR, Bilal M (2021) Chitosan-based hybrid materials for adsorptive removal of dyes and underlying interaction mechanisms. *International Journal of Biological Macromolecules* 183:399–422. doi:10.1016/j.ijbiomac.2021.04.158

Babacan T, Doğan D, Erdem Ü, Metin AÜ (2022) Magnetically responsive chitosan-based nanoparticles for remediation of anionic dyes: Adsorption and magnetically triggered desorption. *Materials Chemistry and Physics* 284. doi:10.1016/j.matchemphys.2022.126032

Bayat M, Javanbakht V, Esmaili J (2018) Synthesis of zeolite/nickel ferrite/sodium alginate bionanocomposite via a co-precipitation technique for efficient removal of water-soluble methylene blue dye. *International Journal of Biological Macromolecules* 116:607–619. doi:10.1016/j.ijbiomac.2018.05.012

Bhat SD, Aminabhavi TM (2006) Novel sodium alginate–Na+MMT hybrid composite membranes for pervaporation dehydration of isopropanol, 1,4-dioxane and tetrahydrofuran. *Separation and Purification Technology* 51 (1):85–94. doi:10.1016/j.seppur.2005.12.025

Dotto GL, Rodrigues FK, Tanabe EH, Fröhlich R, Bertuol DA, Martins TR, Foletto EL (2016) Development of chitosan/bentonite hybrid composite to remove hazardous anionic and cationic dyes from colored effluents. *Journal of Environmental Chemical Engineering* 4(3):3230–3239. doi:10.1016/j.jece.2016.07.004

Dragan ES, Dinu MV (2020) Advances in porous chitosan-based composite hydrogels: Synthesis and applications. *Reactive and Functional Polymers* 146. doi:10.1016/j.reactfunctpolym.2019.104372

Duman O, Polat TG, Diker CÖ, Tunç S (2020) Agar/κ-carrageenan composite hydrogel adsorbent for the removal of Methylene Blue from water. *International Journal of Biological Macromolecules* 160:823–835. doi:10.1016/j.ijbiomac.2020.05.191

Emam HE, Ahmed HB (2019) Comparative study between homo-metallic & hetero-metallic nanostructures based agar in catalytic degradation of dyes. *International Journal of Biological Macromolecules* 138:450–461. doi:10.1016/j.ijbiomac.2019.07.098

Fabryanty R, Valencia C, Soetaredjo FE, Putro JN, Santoso SP, Kurniawan A, Ju Y-H, Ismadji S (2017) Removal of crystal violet dye by adsorption using bentonite – alginate composite. *Journal of Environmental Chemical Engineering* 5(6):5677–5687. doi:10.1016/j.jece.2017.10.057

Ghosh S, Kaushik R, Nagalakshmi K, Hoti SL, Menezes GA, Harish BN, Vasan HN (2010) Antimicrobial activity of highly stable silver nanoparticles embedded in agar–agar matrix as a thin film. *Carbohydrate Research* 345 (15):2220–2227. doi:10.1016/j.carres.2010.08.001

Hassan H, Salama A, El-ziaty AK, El-Sakhawy M (2019) New chitosan/silica/zinc oxide nanocomposite as adsorbent for dye removal. *International Journal of Biological Macromolecules* 131:520–526. doi:10.1016/j.ijbiomac.2019.03.087

Jawad AH, Abdulhameed AS, Surip SN, Alothman ZA (2023) Hybrid multifunctional biocomposite of chitosan grafted benzaldehyde/montmorillonite/algae for effective removal of brilliant green and reactive blue 19 dyes: Optimization and adsorption mechanism. *Journal of Cleaner Production* 393. doi:10.1016/j.jclepro.2023.136334

Jawad AH, Sahu UK, Jani NA, Alothman ZA, Wilson LD (2022) Magnetic crosslinked chitosan-tripolyphosphate/MgO/Fe3O4 nanocomposite for reactive blue 19 dye removal: Optimization using desirability function approach. *Surfaces and Interfaces* 28. doi:10.1016/j.surfin.2021.101698

Jayakumar A, Kumar VP, Radoor S, Nair IC, Siengchin S, Parameswaranpillai J, Radhakrishnan EK (2021) Chitosan-based bionanocomposites for cancer therapy. In: *Bionanocomposites in Tissue Engineering and Regenerative Medicine*. pp. 277–292. doi:10.1016/b978-0-12-821280-6.00012-x

Joshi NC, Gururani P (2022) Advances of graphene oxide based nanocomposite materials in the treatment of wastewater containing heavy metal ions and dyes. *Current Research in Green and Sustainable Chemistry* 5. doi:10.1016/j.crgsc.2022.100306

Kabdaşli, O. Tünay, D. Orhon, Wastewater control and management in a leather tanning district. *Water Science and Technology* 40(1), 1999, 261–267.

Kamal T, Khan MSJ, Khan SB, Asiri AM, Chani MTS, Ullah MW (2019) Silver nanoparticles embedded in gelatin biopolymer hydrogel as catalyst for reductive degradation of pollutants. *Journal of Polymers and the Environment* 28 (2):399–410. doi:10.1007/s10924-019-01615-8

Kanmani P, Aravind J, Kamaraj M, Sureshbabu P, Karthikeyan S (2017) Environmental applications of chitosan and cellulosic biopolymers: A comprehensive outlook. *Bioresource Technology* 242:295–303. doi:10.1016/j.biortech.2017.03.119

Kausar A, Zohra ST, Ijaz S, Iqbal M, Iqbal J, Bibi I, Nouren S, El Messaoudi N, Nazir A (2023) Cellulose-based materials and their adsorptive removal efficiency for dyes: A review. *International Journal of Biological Macromolecules* 224:1337–1355. doi:10.1016/j.ijbiomac.2022.10.220

Kazemi J, Javanbakht V (2020) Alginate beads impregnated with magnetic Chitosan@Zeolite nanocomposite for cationic methylene blue dye removal from aqueous solution. *International Journal of Biological Macromolecules* 154:1426–1437. doi:10.1016/j.ijbiomac.2019.11.024

Khalil A, Ali N, Khan A, Asiri AM, Kamal T (2020) Catalytic potential of cobalt oxide and agar nanocomposite hydrogel for the chemical reduction of organic pollutants. *International Journal of Biological Macromolecules* 164:2922–2930. doi:10.1016/j.ijbiomac.2020.08.140

Kuang Y, Du J, Zhou R, Chen Z, Megharaj M, Naidu R (2015) Calcium alginate encapsulated Ni/Fe nanoparticles beads for simultaneous removal of Cu (II) and monochlorobenzene. *Journal of Colloid and Interface Science* 447:85–91. doi:10.1016/j.jcis.2015.01.080

Kumar M, Vijayakumar G, Tamilarasan R (2018) Synthesis, characterization and experimental studies of Nano Zn–Al–Fe3O4 Blended Alginate/Ca Beads for the Adsorption of Rhodamin B. *Journal of Polymers and the Environment* 27 (1):106–117. doi:10.1007/s10924-018-1318-0

Li X, Li Y, Zhang S, Ye Z (2012) Preparation and characterization of new foam adsorbents of poly(vinyl alcohol)/chitosan composites and their removal for dye and heavy metal from aqueous solution. *Chemical Engineering Journal* 183:88–97. doi:10.1016/j.cej.2011.12.025

Li X, Qi Y, Li Y, Zhang Y, He X, Wang Y (2013) Novel magnetic beads based on sodium alginate gel crosslinked by zirconium(IV) and their effective removal for Pb2+ in aqueous solutions by using a batch and continuous systems. *Bioresource Technology* 142:611–619. doi:10.1016/j.biortech.2013.05.081

Makhado E, Pandey S, Modibane KD, Kang M, Hato MJ (2020) Sequestration of methylene blue dye using sodium alginate poly(acrylic acid)@ZnO hydrogel nanocomposite: Kinetic, Isotherm, and Thermodynamic Investigations. *International Journal of Biological Macromolecules* 162:60–73. doi:10.1016/j.ijbiomac.2020.06.143

Mandal S, Hwang S, Shi SQ (2023) Guar gum, a low-cost sustainable biopolymer, for wastewater treatment: A review. *International Journal of Biological Macromolecules* 226:368–382. doi:10.1016/j.ijbiomac.2022.12.039

Masilompane TM, Chaukura N, Mishra SB, Mishra AK (2018) Chitosan-lignin-titania nanocomposites for the removal of brilliant black dye from aqueous solution. *International Journal of Biological Macromolecules* 120:1659–1666. doi:10.1016/j.ijbiomac.2018.09.129

Narula A, Rao CP (2019) Hydrogel of the supramolecular complex of Graphene Oxide and Sulfonatocalix[4]arene as reusable material for the degradation of organic dyes: Demonstration of adsorption and degradation by spectroscopy and microscopy. *ACS Omega* 4 (3):5731–5740. doi:10.1021/acsomega.9b00545

Nguyen TT, Hoang BN, Tran TV, Nguyen DV, Nguyen TD, Vo D-VN (2021a) Agar/maltodextrin/poly(vinyl alcohol) walled montmorillonite composites for removal of methylene blue from aqueous solutions. *Surfaces and Interfaces* 26. doi:10.1016/j.surfin.2021.101410

Nguyen TT, Phung TK, Bui X-T, Doan V-D, Tran TV, Nguyen DV, Lim KT, Nguyen TD (2021b) Removal of cationic dye using polyvinyl alcohol membrane functionalized by D-glucose and agar. *Journal of Water Process Engineering* 40. doi:10.1016/j.jwpe.2021.101982

Pashaei-Fakhri S, Peighambardoust SJ, Foroutan R, Arsalani N, Ramavandi B (2021) Crystal violet dye sorption over acrylamide/graphene oxide bonded sodium alginate nanocomposite hydrogel. *Chemosphere* 270. doi:10.1016/j.chemosphere.2020.129419

Patra S, Roy E, Madhuri R, Sharma PK (2016) Agar based bimetallic nanoparticles as high-performance renewable adsorbent for removal and degradation of cationic organic dyes. *Journal of Industrial and Engineering Chemistry* 33:226–238. doi:10.1016/j.jiec.2015.10.008

Peramune D, Manatunga DC, Dassanayake RS, Premalal V, Liyanage RN, Gunathilake C, Abidi N (2022) Recent advances in biopolymer-based advanced oxidation processes for dye removal applications: A review. *Environmental Research* 215. doi:10.1016/j.envres.2022.114242

Quesada HB, de Araújo TP, Vareschini DT, de Barros MASD, Gomes RG, Bergamasco R (2020) Chitosan, alginate and other macromolecules as activated carbon immobilizing agents: A review on composite adsorbents for the removal of water contaminants. *International Journal of Biological Macromolecules* 164:2535–2549. doi:10.1016/j.ijbiomac.2020.08.118

Radoor S, Karayil J, Jayakumar A, Lee J, Nandi D, Parameswaranpillai J, Pant B, Siengchin S (2022a) Efficient removal of organic dye from aqueous solution using hierarchical Zeolite-based biomembrane: Isotherm, kinetics, thermodynamics and recycling studies. *Catalysts* 12 (8). doi:10.3390/catal12080886

Radoor S, Karayil J, Jayakumar A, Nandi D, Parameswaranpillai J, Lee J, Shivanna JM, Nithya R, Siengchin S (2022b) Adsorption of Cationic Dye onto ZSM-5 Zeolite-Based Bio Membrane: Characterizations, kinetics and adsorption isotherm. *Journal of Polymers and the Environment* 30 (8):3279–3292. doi:10.1007/s10924-022-02421-5

Radoor S, Karayil J, Jayakumar A, Parameswaranpillai J, Siengchin S (2021a) An efficient removal of malachite green dye from aqueous environment using ZSM-5 zeolite/polyvinyl alcohol/carboxymethyl cellulose/sodium alginate bio composite. *Journal of Polymers and the Environment* 29 (7):2126–2139. doi:10.1007/s10924-020-02024-y

Radoor S, Karayil J, Jayakumar A, Parameswaranpillai J, Siengchin S (2021b) Efficient removal of methyl orange from aqueous solution using mesoporous ZSM-5 zeolite: Synthesis, kinetics and isotherm studies. *Colloids and Surfaces A: Physicochemical and Engineering Aspects* 611. doi:10.1016/j.colsurfa.2020.125852

Radoor S, Karayil J, Jayakumar A, Siengchin S, Parameswaranpillai J (2021c) A low cost and eco-friendly membrane from polyvinyl alcohol, chitosan and honey: Synthesis, characterization and antibacterial property. *Journal of Polymer Research* 28 (3). doi:10.1007/s10965-021-02415-2

Radoor S, Karayil J, Parameswaranpillai J, Siengchin S (2020a) Adsorption of methylene blue dye from aqueous solution by a novel PVA/CMC/halloysite nanoclay bio composite: Characterization, kinetics, isotherm and antibacterial properties. *Journal of Environmental Health Science and Engineering* 18 (2):1311–1327. doi:10.1007/s40201-020-00549-x

Radoor S, Karayil J, Parameswaranpillai J, Siengchin S (2020b) Adsorption Study of Anionic Dye, Eriochrome Black T from Aqueous Medium Using Polyvinyl Alcohol/Starch/ZSM-5 Zeolite Membrane. *Journal of Polymers and the Environment* 28 (10):2631–2643. doi:10.1007/s10924-020-01812-w

Radoor S, Karayil J, Parameswaranpillai J, Siengchin S (2020c) Removal of anionic dye Congo red from aqueous environment using polyvinyl alcohol/sodium alginate/ZSM-5 zeolite membrane. *Scientific Reports* 10 (1). doi:10.1038/s41598-020-72398-5

Ramadhani S, Helmiyati H (2020) *Alginate/CMC/ZnO Nanocomposite for Photocatalytic Degradation of Congo Red Dye*. Paper presented at the Proceedings of the 5th International Symposium on Current Progress in Mathematics and Sciences (2019), Depok, Indonesia.

Raval NP, Priyadarshi GV, Mukherjee S, Zala H, Fatma D, Bonilla-Petriciolet A, Abdelmottaleb BL, Duclaux L, Trivedi MH (2022) Statistical physics modeling and evaluation of adsorption properties of chitosan-zinc oxide nanocomposites for the removal of an anionic dye. *Journal of Environmental Chemical Engineering* 10 (6). doi:10.1016/j.jece.2022.108873

Rhein-Knudsen N, Ale MT, Ajalloueian F, Yu L, Meyer AS (2017) Rheological properties of agar and carrageenan from Ghanaian red seaweeds. *Food Hydrocolloids* 63:50–58. doi:10.1016/j.foodhyd.2016.08.023

Rostamian M, Hosseini H, Fakhri V, Talouki PY, Farahani M, Gharehtzpeh AJ, Goodarzi V, Su C-H (2022) Introducing a bio sorbent for removal of methylene blue dye based on flexible poly(glycerol sebacate)/chitosan/graphene oxide ecofriendly nanocomposites. *Chemosphere* 289. doi:10.1016/j.chemosphere.2021.133219

Sabarish R, Jasila K, Aswathy J, Jyotishkumar P, Suchart S (2020) Fabrication of PVA/agar/modified ZSM-5 zeolite membrane for removal of anionic dye from aqueous solution. *International Journal of Environmental Science and Technology* 18 (9):2571–2586. doi:10.1007/s13762-020-02998-1

Sabarish R, Unnikrishnan G (2018a) Novel biopolymer templated hierarchical silicalite-1 as an adsorbent for the removal of rhodamine B. *Journal of Molecular Liquids* 272:919–929. doi:10.1016/j.molliq.2018.10.093

Sabarish R, Unnikrishnan G (2018b) Polyvinyl alcohol/carboxymethyl cellulose/ZSM-5 zeolite biocomposite membranes for dye adsorption applications. *Carbohydrate Polymers* 199:129–140. doi:10.1016/j.carbpol.2018.06.123

Sabarish R, Unnikrishnan G (2018c) PVA/PDADMAC/ZSM-5 zeolite hybrid matrix membranes for dye adsorption: Fabrication, characterization, adsorption, kinetics and antimicrobial properties. *Journal of Environmental Chemical Engineering* 6 (4):3860–3873. doi:10.1016/j.jece.2018.05.026

Sabarish R, Unnikrishnan G (2019) Synthesis, characterization and evaluations of micro/mesoporous ZSM-5 zeolite using starch as bio template. *SN Applied Sciences* 1 (9). doi:10.1007/s42452-019-1036-9

Sajid M (2022) Chitosan-based adsorbents for analytical sample preparation and removal of pollutants from aqueous media: Progress, challenges and outlook. *Trends in Environmental Analytical Chemistry* 36. doi:10.1016/j.teac.2022.e00185

Sarkar N, Sahoo G, Swain SK (2020) Nanoclay sandwiched reduced graphene oxide filled macroporous polyacrylamide-agar hybrid hydrogel as an adsorbent for dye decontamination. *Nano-Structures & Nano-Objects* 23. doi:10.1016/j.nanoso.2020.100507

Saya L, Malik V, Singh A, Singh S, Gambhir G, Singh WR, Chandra R, Hooda S (2021) Guar gum based nanocomposites: Role in water purification through efficient removal of dyes and metal ions. *Carbohydrate Polymers* 261. doi:10.1016/j.carbpol.2021.117851

Simelane NP, Asante JKO, Ndibewu PP, Mramba AS, Sibali LL (2022) Biopolymer composites for removal of toxic organic compounds in pharmaceutical effluents – a review. *Carbohydrate Polymer Technologies and Applications* 4. doi:10.1016/j.carpta.2022.100239

Sirajudheen P, Poovathumkuzhi NC, Vigneshwaran S, Chelaveettil BM, Meenakshi S (2021) Applications of chitin and chitosan based biomaterials for the adsorptive removal of textile dyes from water — A comprehensive review. *Carbohydrate Polymers* 273. doi:10.1016/j.carbpol.2021.118604

Slama HB, Chenari Bouket A, Pourhassan Z, Alenezi FN, Silini A, Cherif-Silini H, Oszako T, Luptakova L, Golińska P, Belbahri L (2021) Diversity of synthetic dyes from textile industries, discharge impacts and treatment methods. *Applied Sciences* 11 (14). doi:10.3390/app11146255

Tahari N, de Hoyos-Martinez PL, Izaguirre N, Houwaida N, Abderrabba M, Ayadi S, Labidi J (2022) Preparation of chitosan/tannin and montmorillonite films as adsorbents for Methyl Orange dye removal. *International Journal of Biological Macromolecules* 210:94–106. doi:10.1016/j.ijbiomac.2022.04.231

Tamer Y, Koşucu A, Berber H (2022) Graphene oxide incorporated chitosan/acrylamide/itaconic acid semi-interpenetrating network hydrogel bio-adsorbents for highly efficient and selective removal of cationic dyes. *International Journal of Biological Macromolecules* 219:273–289. doi:10.1016/j.ijbiomac.2022.07.238

Thakur S, Arotiba O (2017) Synthesis, characterization and adsorption studies of an acrylic acid-grafted sodium alginate-based TiO2hydrogel nanocomposite. *Adsorption Science & Technology* 36(1–2):458–477. doi:10.1177/0263617417700636

Upadhyay A, Kandi R, Rao CP (2018) Injectable, self-healing, and stress sustainable hydrogel of BSA as a functional biocompatible material for controlled drug delivery in cancer cells. *ACS Sustainable Chemistry & Engineering* 6 (3):3321–3330. doi:10.1021/acssuschemeng.7b03485

Wastewater control and management in a leather tanning district (1999). *Water Science and Technology* 40 (1). doi:10.1016/s0273-1223(99)00393-5

Welsch N, Ballauff M, Lu Y (2010) Microgels as nanoreactors: Applications in catalysis. In: *Chemical Design of Responsive Microgels. Advances in Polymer Science*. pp. 129–163. doi:10.1007/12_2010_71

Xu C, Nasrollahzadeh M, Sajjadi M, Maham M, Luque R, Puente-Santiago AR (2019) Benign-by-design nature-inspired nanosystems in biofuels production and catalytic applications. *Renewable and Sustainable Energy Reviews* 112:195–252. doi:10.1016/j.rser.2019.03.062

Yagub MT, Sen TK, Afroze S, Ang HM (2014) Dye and its removal from aqueous solution by adsorption: A review. *Advances in Colloid and Interface Science* 209:172–184. doi:10.1016/j.cis.2014.04.002

Yaseen DA, Scholz M (2017) Treatment of synthetic textile wastewater containing dye mixtures with microcosms. *Environmental Science and Pollution Research* 25 (2):1980–1997. doi:10.1007/s11356-017-0633-7

Yu F, Cui T, Yang C, Dai X, Ma J (2019) κ-Carrageenan/Sodium alginate double-network hydrogel with enhanced mechanical properties, anti-swelling, and adsorption capacity. *Chemosphere* 237. doi:10.1016/j.chemosphere.2019.124417

Zango ZU, Dennis JO, Aljameel AI, Usman F, Ali MKM, Abdulkadir BA, Algessair S, Aldaghri OA, Ibnaouf KH (2022) Effective removal of Methylene Blue from simulated wastewater using ZnO-Chitosan Nanocomposites: Optimization, Kinetics, and Isotherm Studies. *Molecules* 27(15). doi:10.3390/molecules27154746

Zare EN, Motahari A, Sillanpää M (2018) Nanoadsorbents based on conducting polymer nanocomposites with main focus on polyaniline and its derivatives for removal of heavy metal ions/dyes: A review. *Environmental Research* 162:173–195. doi:10.1016/j.envres.2017.12.025

Zhao C, Liu G, Tan Q, Gao M, Chen G, Huang X, Xu X, Li L, Wang J, Zhang Y, Xu D (2023) Polysaccharide-based biopolymer hydrogels for heavy metal detection and adsorption. *Journal of Advanced Research* 44:53–70. doi:10.1016/j.jare.2022.04.005

Zhao K, Feng L, Lin H, Fu Y, Lin B, Cui W, Li S, Wei J (2014) Adsorption and photocatalytic degradation of methyl orange imprinted composite membranes using TiO2/calcium alginate hydrogel as matrix. *Catalysis Today* 236:127–134. doi:10.1016/j.cattod.2014.03.041

22 Bionanomaterials in Wastewater Treatment

Kirti Garg, Divya Sharma, Rutika Sehgal, and Reena Gupta

22.1 INTRODUCTION

Wastewater treatment is posing an urgent threat that needs to be addressed for maintaining the modern-day requirements of quality water. Municipal wastewater treatment facilities that are currently in operation do not allow for the balanced development of the major microorganism groups that break down wastewater components and municipal facilities are more often the source of organic pollutants, particularly nutrients (nitrogen and phosphorus) that further lead to the eutrophication of water bodies (Sirotkin et al., 2019). Potable water being the immediate necessity of every habitation needs to be carefully inspected for chemical pollutants such as heavy metals, poisons and organic or inorganic pollutants. The current methods of water treatment, distribution and disposal, along with the massive administrative schemes and structures, are no longer viable (Trivedi and Jay, 2021).

The removal of biologically resistant contaminants from industrial effluent is, nevertheless, the most pressing issue. The stable sedimentation of microbiocenosis from purified water and the subsequent disposal of surplus biomass of activated sludge are crucial part of wastewater treatment (Jeevanandam et al., 2022). The effectiveness of biological treatment facilities that have already been established and must be improved. Hence, for such purpose a combination of nano-methods and bio-nanotechnologies, that is, bionanomaterials, can prove to be highly beneficial (Rathore et al., 2022). Bionanomaterials are the molecular substances made up of the entire or some part of biological molecules, such as proteins, antibodies, enzymes, nucleic acids, lipids, oligo- and polysaccharides, viruses, secondary metabolites, and so forth. Originally bionanomaterials gained attention due to their applicability in the biomedical field, as they exhibited features like biocompatibility, large surface area, great mechanical properties, significant chemical reactivity, low toxicity, lower cost and energy usage, efficient regeneration and reusability (Singh et al., 2021).

In the last few years, many compounds, including activated carbon, biochar and inexpensive adsorbents made from various resources, have been investigated for the purification of water (Dwivedi and Dwivedi, 2022). The high antimicrobial activity of nanoparticles against multidrug-resistant bacterial strains and other waterborne pathogenic organisms make them potential water disinfectants. Nanoparticles can efficiently adsorb and degrade metal ions, dyes, surfactants and other harmful organic pollutants to produce high quality treated water (Gautam et al., 2019). Due to their incredibly small size, they display distinctive structural, chemical, physical, optical, biological, mechanical and electrical characteristics that distinguish them from the bulk matter. These characteristics are identified using a variety of complex characterization techniques such as spectroscopy, microscopy and so forth and further enable them in to prove their utility in a variety of fields including wastewater treatment (Singh et al., 2021).

DOI: 10.1201/9781003432791-26

The development of nanoscale adsorbents and catalysts with bio-inspired design are beneficial for the removal and degradation of a variety of contaminants in water. The small size of nanoparticles enables their properties to enhance their utility and preferability in different domains. But nanoparticles must be useful to remove toxins from wastewater while being harmless to ecosystems (Khan et al., 2022). Therefore, in order to enable their employment in a wider range of fields, a "green", non-toxic method of producing metallic nanoparticles is required (Trivedi and Jay, 2021). Recent studies have devised some biogenic methods for the synthesis of bio nanoparticles, providing with ecological and economic advantages. The naturally occurring substances, such as plants, bacteria, algae, fungi, yeast, actinomycetes and so forth, are the source of these natural products. It has become quite common to create metallic nanoparticles using polyphenols, vitamins, amino acids, carbohydrates, biopolymers, and natural surfactants. Additionally, biological synthesis may produce nanomaterials on a large scale much faster than traditional approaches (Gautam et al., 2019).

This chapter deals with the treatment of pollutants present in wastewater including heavy metals, dyes, organic, inorganic and radioactive pollutants, pharmaceutical pollutants, aromatics compounds and toxins using different types of bionanomaterials and related techniques.

22.2 PROPERTIES OF BIOGENIC NANOMATERIALS

Nanotechnology is also referred to as "engineering at a molecular level" since it involves synthesis, design, characterization and potentiality at the nanoscale. The use of biological techniques to synthesize nanomaterials is favoured because they are quick, simple, environmentally friendly, require fewer harmful chemicals and are more economical (Barhoum et al., 2022). In addition to the involvement of diverse living organisms like bacteria, fungi, yeast, plants and animals, biological techniques also make use of natural substances like honey, pectin, glucose, starch, and so forth. The green synthesis of bionanomaterials refers to the use of plant extracts. Synthesis of such nanomaterials, that is, from the biological sources has been shown in Figure 22.1.

Numerous organisms, including some plants, bacteria, and fungus, have the ability to create nanoparticles coated in metal ions. These synthesized particles have both advantages and disadvantages (Annamalai et al., 2022). However, there are many factors that influence the production of bionanoparticles, such as extracellular or intracellular production of bionanoparticles, temperature, and time needed for growth, the method of extraction, and the percentage of contaminants removed, the ratio of removal and time, selection of the ideal microbial strain and metal ions (Singh et al., 2021).

The biologically synthesized nanomaterials exhibit a wide range of properties that could be proven beneficial for the treatment of wastewater. In contrast to other conventionally used homogeneous materials, bionanomaterials exhibit special thermal, optical, physical, chemical, magnetic, and electrical properties, which further benefit next-generation biosensors, catalysts and antimicrobial use (Trivedi and Jay, 2021).

Due to their distinctive size-dependent characteristics like large surface area, short intra-particle diffusion distance, compressibility with negligible surface area reduction, excellent stability, remarkable reusability and recyclability, the use of nanostructured materials for scavenging and degradation of toxic water contaminants is gaining enormous importance (Gautam et al., 2019). Nano-metered materials have a large surface area, which provides active locations for the chemical reactions and physical interactions necessary for the elimination of pollutant molecules (Khan et al., 2022).

Bionanomaterials exploit the potential of microorganisms to break down components of wastewater. They are used to design sophisticated tools and techniques through self-assembly in environment friendly conditions (Singh et al., 2021). The nanoparticles with size between 1 and 100 nm are used for creation and application of operational and diagnostic procedures for eliminating contaminants from wastewater. Implementation of bionanotechnological techniques in the

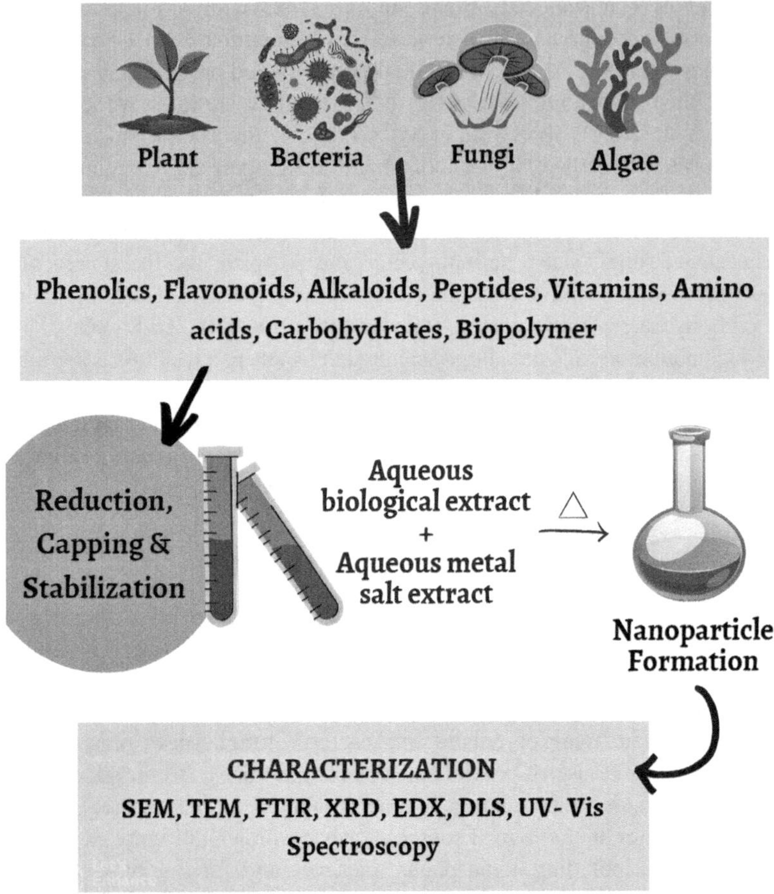

FIGURE 22.1 Representation of Green Synthesis of Bionanoparticles Using Living Organisms. (Created by the Authors.)

treatment of wastewater involves a set of operations that are technological, physicochemical, microbiological and molar biological (Sirotkin et al., 2019).

22.3 ANALYSIS OF MICROBIAL AGGREGATES IN BIOFILM

Various types of bionanoparticles comprise different categories of wastewater treatment, targeting particular group of contaminants. There are mainly three classes of techniques used for the treatment of wastewater which are technological treatments, physicochemical treatment, microbiological and molecular biological treatment techniques (Sirotkin et al., 2019). A brief description of these techniques has been given in Figure 22.2, followed by their detailed explanation.

22.3.1 Technological Treatment Techniques

Several advancements in the conventional methods of wastewater treatment are required. These include effective removal of phosphorus compounds and other wastewater constituents using reagent wastewater treatment, enhancement of biomass sedimentation and adequate use of any surplus

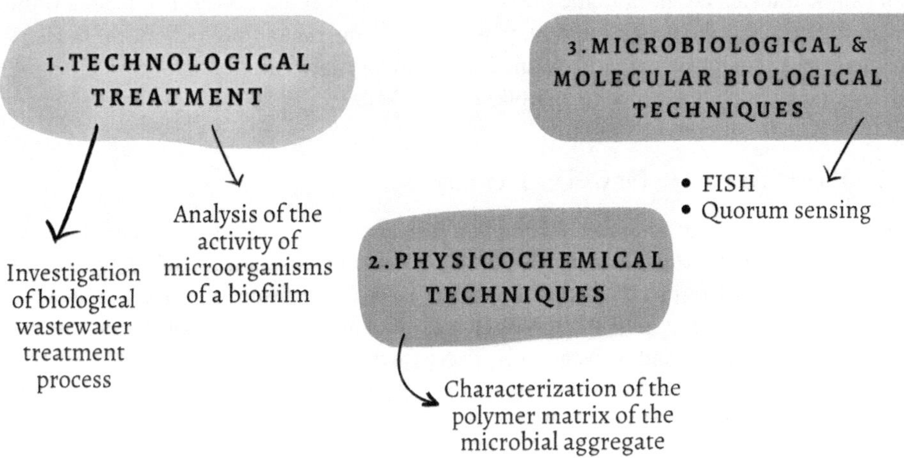

FIGURE 22.2 Various Techniques for Analysis and Characterization of Microbial Aggregates in Biofilm. (Created by the Authors.)

biomass in association with biological treatment and provision of appropriate amount of oxygen to the microorganisms (Sirotkin et al., 2019). Hence, to achieve these goals, bionanomaterials that can be used are listed as follows below.

22.3.1.1 Nanostructured Iron

The bionanomaterials enable the elimination of phosphorus compounds using nanoscale iron, for which the magnetic and catalytic properties of iron are exploited. Chloromethane, ethane, ethene, benzene, organochlorine pesticides, chlorinated phenols, polychlorinated biphenyls, dyes, numerous inorganic chemicals, and metal ions can be converted by nanosized iron particles into nontoxic substances. Iron particles at a nanoscale have high specific surface area, and high redox potential, which makes them very active for the conversion of pollutants (Sun et al., 2006).

It was proposed by researchers that the structure of sludge flakes created in contact with reagent solutions may depend on the particle size of the reagent in the working solution, which may fall within a limited range between 30 and 40 nm. Use of naturally occurring nanostructured materials such as zeolite, clinoptilolite and so forth to purify water of a variety of pollutants and enhance the capabilities of microbial communities engaged in water purification (Li et al., 2006). The use of such reagents is done in a technique commonly called as biosorption (combined physico-chemical and biological parameters).

22.3.1.2 Nanostructured Minerals

Zeolites and other minerals, activated carbon and ash from thermal power plants are all used as adsorbents in biological wastewater treatment system. The adsorption of chemicals to the surface and micro(nano) pores of the adsorbent and the immobilization of microorganisms on its surface determine the outcome of adsorption in wastewater treatment (Kobeleva et al., 2017). This method improved the stability of bio-oxidation of organic materials, nitrification and activated sludge biomass sedimentation process.

22.3.1.3 Oxygen Transfer Membranes

Nanostructured membranes with selective oxygen permeability can be used to efficiently transfer oxygen for aerobic microbes, from gas-air mixtures. Without the creation of gas bubbles, molecular oxygen can be transferred through the membrane (Nurullina et al., 2002). The benefit of biological treatment facilities for the wastewater treatment from chemical and petrochemical industries using bubble-free oxygen supply, is the avoidance of secondary air pollution and adversities of volatile organic compounds from wastewater (Sirotkin et al., 2019).

22.3.2 Physicochemical Treatment Techniques

Problems relating to the conventional treatment processes: For the deep purification of wastewater through adsorption methods and removal of dissolved organic substances (pollutants from inorganic synthesis facilities, oil refineries, pulp, and paper mills, textile facilities and many other industries), new methods are required that enhance the efficiency of conventional adsorption methods for wastewater treatment and substantially decrease the quantities of activated carbon (Singh et al., 2021).

Since no substitutes for activated carbon have been found yet, and the cost of adsorbent regeneration, which essentially determines the price of purification is generally high, so the current processes need advancements (Sirotkin et al., 2002). Methods of thermal regeneration require high reactivation temperatures, that is 650–1000°C and hence it is not a cost-effective process.

The products of thermal regeneration are typically dibenzodioxins and furans, which require additional neutralization of desorption products and replenishment of carbon losses from burning and abrasion in each regeneration cycle is up to 10 per cent. These losses all together make it more urgent to devise new methods for cost-effective wastewater purification systems (Sirotkin et al., 2019).

Solutions to the aforementioned issues using physicochemical processes: Striking as a solution to these setbacks are the intra- and the extra-cellular enzymes that are biochemical (proteinaceous) in nature and are also native nanoobjects. In such processes the strong enzyme activity of microorganisms is exploited to degrade the pollutants by making them participate in the biocatalytic oxidation-reduction reactions for the degradation of harmful components of wastewater (Khan et al., 2022). By doing so, an enhancement in the biological functionality of activated carbon filters was noted. This might be linked to the increased amount of substrate on the surface of carbon.

The biodegradation of a substance adsorbed in nanopores under the presence of extracellular enzymes is one of the reasons for the biological regeneration of a sorbent (Rathore et al., 2022). Since bacteria are too large to enter true nanopores of activated carbon, biodegradation takes place in the pores using extracellular enzymes, which can easily fit through even the smallest pores and interact with the adsorbed substrate (Stewart, 2003). These extracellular enzymes aid in the hydrolytic breakdown of the substance. Adsorption affinity of the products (after decay) decreases, causing them to desorb and become available for cellular biodegradation on the outer surface of activated carbon.

It is well known that the biofilm matrix limits the diffusion rate of both high and low-molecular weight compounds. This limited diffusion in the biofilm matrix volume generates a concentration gradient of substrates from the bacteria and their metabolic products, as well as oxygen, which causes the growth of biofilm microorganisms to be spatially heterogeneous (Rathore et al., 2022).

22.3.3 Microbiological and Molecular Biological Techniques

Tasks that need to be accomplished: Diagnosis of microbial communities of treatment facilities is conducted as per the major microbial groupings, genera, and species involved in constitution of microbial communities (for example nitrifying agents, phosphate accumulating microbes, flocculating and filamentous bacteria, and so forth) (Sirotkin et al., 2019). It is necessary to develop quick, precise, and non-destructive techniques for real-time monitoring of microbial aggregates.

Molecular biological process for the diagnosis of microbial aggregates: FISH and CLSM: Giving easy access and reliable results, Fluorescent In-Situ Hybridization (FISH) technique allows the incubation of a biofilm with a solution of gene probes to identify a particular bacterial cell. A gene probe comprises a fluorescent marker and a segment of RNA which is between nearly 10–18 bases long (Schaule, 2000). In the solution where incubation of bacterial cells and the oligonucleotides (i.e., probe) takes place, hybridization leads to the formation of DNA-RNA duplex by virtue of complementarity between nucleotides and formation of hydrogen bonds. These cells containing hybridized nucleotides are vizualized under the microscope at a wavelength that is particular to individual florescent marker. Hence, this method is highly efficient for diagnosing microbes in the given microbial aggregate (Kumar et al., 2023).

In present-day studies, such gene probes are being made that can differentiate microorganisms based on their phylogenetic levels. Several such hybridization techniques can be used to find the microbes and their location in aggregate using Confocal Laser Scanning Microscopy (CLSM). Quantitative analysis of the microbes and their imaging on a screen can also be done using this technique (Sirotkin et al., 2019). A flow chart of both techniques, used together, is given in Figure 22.3.

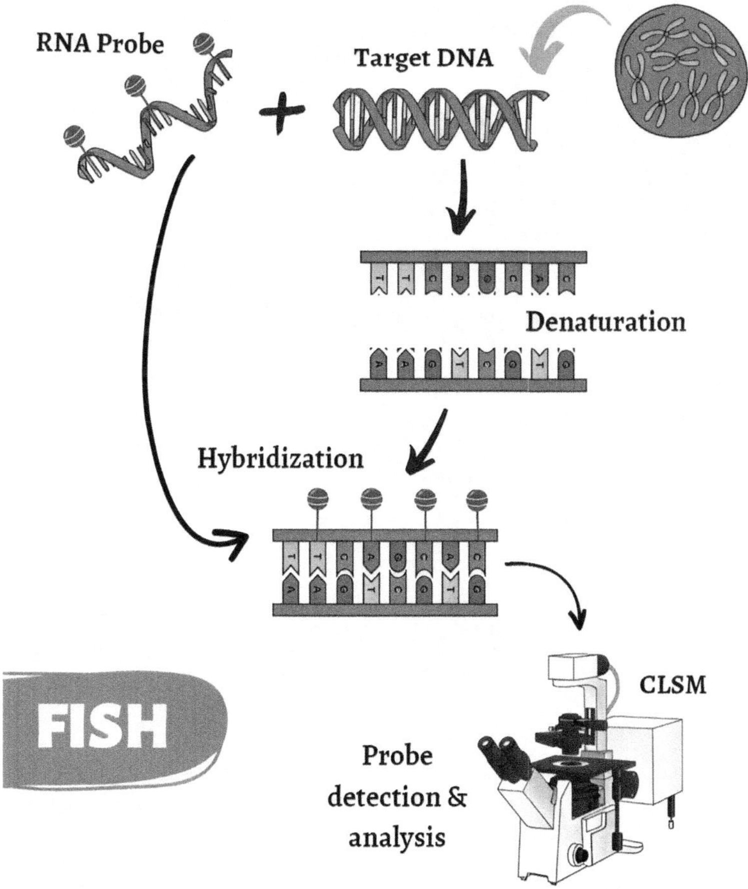

FIGURE 22.3 Diagrammatic Flow Chart Representation of Fluorescent in-situ Hybridization (FISH). (Created by the Authors.)

22.4 BIOMATERIALS IN WASTEWATER TREATMENT

In order to bioremediate heavy metals, poisonous dyes, pharmaceutical wastes and other developing contaminants from various industrial, agricultural, and municipal wastes, nanoparticles were produced with the help of bacteria, fungi and microalgae (Jeevanandam et al., 2022). Some heavy metals are well recognized for having negative effects on both aquatic and terrestrial organisms, including increased DNA damage, membrane damage, enzyme inactivation, genotoxicity and oxidative stress and disruption of numerous signalling pathways. Most frequently used nanomaterials are carbon nano-tubes, different oxide-based nano-adsorbents (i.e., manganese oxide, zinc oxide, magnesium oxide, titanium oxide, palladium oxide and oxides of iron) and graphene-based nanoparticles (Khan et al., 2022). Moreover, different types of nanomaterials that have been so far used in wastewater treatment are given in the Figure 22.4.

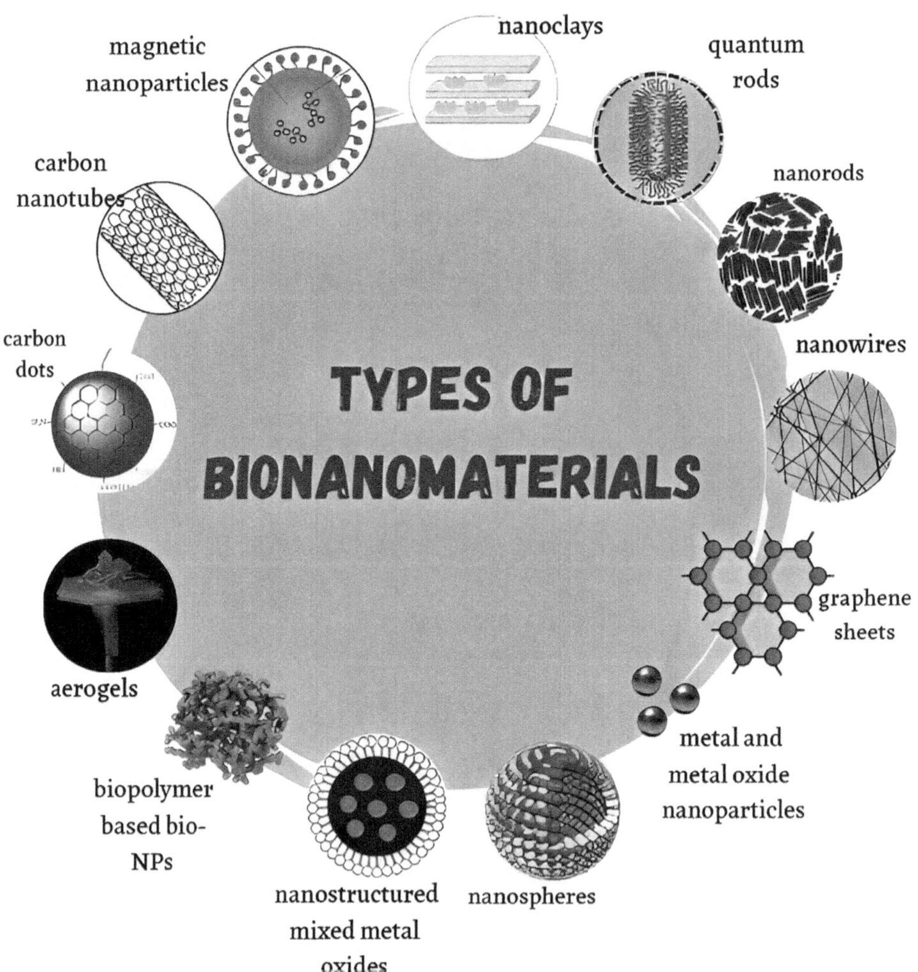

FIGURE 22.4 This Figure Shows Types of Bionanomaterials Used in Wastewater Treatment.

22.4.1 Nanoadsorbents

Over many years, the potential of nanoparticles as adsorbents has been explored. As earlier known the smaller the size of the nanoparticles, the larger will be their surface area, which in turn increases their reactivity and adsorption capacity (Annamalai et al., 2022). The adsorption coefficient and recitation partitioning of heavy metals (or other pollutant) determines the efficacy of adsorption process. It has been found by many researchers that changing the redox condition also changes toxicity of the contaminant (Falconer and Humpage, 2005).

Factors that affect the process of adsorption are huge surface area of nanoparticles, their great adsorption capacity, chemical reactivity, and surface chemistry of nanoparticle, localization of atoms on the nanoparticle surface, absence of internal resistance by diffusion, high binding energy, agglomeration state of adsorbent, crystal structure of adsorbent, shape and fractal dimension of the adsorbent and nanoparticle's chemical composition and its solubility (Ozaki, 2004).

However, an ideal nanoadsorbent, with a high efficacy for heavy metal elimination must possess properties like nontoxicity, high adsorption capacity, easy removability of pollutants from the surface of adsorbent, high reusability, and recyclability of the nanoadsorbents and should be able to adsorb even very low concentrations of pollutants (Annamalai et al., 2022).

22.4.1.1 Iron-based Nanoadsorbents

Ferric oxide is a low-cost substance for adsorbing the toxic metals because it is made of iron and is easy to synthesize. It is an environment-friendly material that has a lower risk of subsequent contamination when applied to a contaminated area (Barhoum et al., 2022). The pH, temperature, adsorbent dose, and duration of incubation are the variables that determine adsorption of various heavy metals on Fe_2O_3 as nanoadsorbent.

Various scientists have modified the surface of Fe_2O_3 to boost its capability for adsorption (Servos et al., 2005). Researchers reported surface modification of ferric oxide using 3-aminopropyltrimethoxysilane and they also observed that these modified nanoadsorbents have excellent affinity for removing many contaminants from wastewater, including Cr^{3+}, Co^{2+}, Ni^{2+}, Cu^{2+}, Cd^{2+} and As^{3+} (Singh et al., 2021).

Aspergillus tubingensis was used to bio-manufacture iron oxide nanoparticles (IONPs), which were extraordinarily stable due to the accumulated extracellular fungal components. For Ni^{2+}, Pb^{2+}, Cu^{2+}, and Zn^{2+}, these nanoparticles demonstrated the efficacy of metal removal of >90 per cent, with the reusability of up to five treatment cycles (Mahanty et al., 2020). For researchers, lowering the Cr concentration in wastewater has presented as a significant problem. Supermagnetic iron oxide nanoparticles are utilized to treat wastewater that has been contaminated with Cr (VI). The remarkable Cr (VI) removal effectiveness of the myco-fabricated IONPs was >99 per cent (Singh et al., 2021).

22.4.1.2 Manganese Oxide Nanoadsorbents

As a result of their large surface area and polymorphism structure, manganese oxide (MnOs) nanoparticles exhibit excellent capacity for adsorption. They have been frequently utilized to remove several heavy metals (such as arsenic) from wastewater (Khan et al., 2022). The most often used modified MnOs are hydrous MnO (HMO) and nano-porous/nanotunnel MnO. Researchers have prepared hydrous manganese oxide by adding $MnSO_4H_2O$ into NaClO. The inner-sphere formation mechanism (characterized by the ion-exchange process) is typically responsible for the adsorption of numerous heavy metals on HMOs, including Pb^{2+}, Cd^{2+} and Zn^{2+}. The adsorption process on the surface of HMO for any divalent metal takes place in two basic steps. The divalent metal ions adsorb on the outer surface of the adsorbent and intra-particular diffusion takes place (Urase and Kikuta, 2005).

22.4.1.3 Zinc Oxide Nanoadsorbents

The porous nanostructure of zinc oxide (ZnO) nanoparticles has a large BET surface area for the pollutant (particularly, heavy metals) adsorption. For the removal of pollutants from wastewater, nano-plates, nano-assemblies, microspheres containing nanosheets and Zinc oxide nanorods are frequently utilized as nanoadsorbents (Khan et al., 2022). The heavy metal removal efficiency of modified zinc oxide nanoadsorbents is very high as compared with the conventional ZnO (commercially available form). To eliminate Cu^{2+} from the effluent, several researchers use porous nanosheets and ZnO nanoplates. Compared to conventionally used zinc oxide, the modified forms of ZnO exhibit a high Cu^{2+} removal efficiency because of their distinctive nanostructure (Malik et al., 2022).

Additionally, the removal of several heavy metals such as Co^{2+}, Ni^{2+}, Cu^{2+}, Cd^{2+}, Pb^{2+}, Hg^{2+}, As^{3+} and so forth, was accomplished using nano-assemblies (Westerhof et al., 2005). Due to their electropositive character, microporous nano-assemblies show a huge adsorption affinity for Pb^{2+}, Hg^{2+} and As^{3+}. The use of mesoporous ZnO nanorods has also been found to have a good removal capacity for Pb^{2+} and Cd^{2+} (Malik et al., 2022).

22.4.1.4 Magnesium Oxide Nanoadsorbents

For removing several types of heavy metals from polluted water, magnesium oxide (MgO) is utilized. New structures called magnesium oxide microspheres can increase the affinity of adsorption for removal of heavy metals. Morphological alterations of various nanoparticles were made in order to boost the adsorption capability of magnesium oxide (Foley et al., 2005). Modified MgO nanomaterials include nanorods, nanobelts and fishbone fractal nanostructures produced by nanowires, nanotubes and nano-cubes. On flower-like mesoporous magnesium oxide, scientists demonstrated the efficacious adsorption of Pb^{2+} and Cd^{2+} (Malik et al., 2022).

22.4.1.5 Graphene Oxide-based Nanoadsorbents

An allotrope of carbon with unique properties, graphene is particularly advantageous in having a wide range of applications relating to the environment. By chemically oxidizing the graphite layer, 2-D structure of graphene oxide (a carbon-based nanoparticle) is created. The Hummer's method is one of the most frequently used to synthesize graphene oxide (GO) (Sadiq and Rodriguez, 2004). The hydroxyl group, carboxyl group, and other functional groups present in GO affect the adsorption of pollutants on its surface.

The features of graphene oxide, due to which it has been gaining huge attention as an effective adsorbent, are its large surface area, high mechanical strength, light weight, excellent flexibility, chemical stability, and simple process of synthesis. Some researchers have prepared a graphene oxide-based sand filter and used it in a column bioreactor for the elimination of heavy metals. Other researchers have prepared a hybrid of GO and TiO_2 and further used it for the adsorption of metallic ions like Pb^{2+}, Cd^{2+} and Zn^{2+} (Walha et al., 2007).

Graphene alone, and along with its composites, have so far shown promising results when it comes to adsorption and elimination of heavy metals from the polluted water, due to the presence of two-dimensional basal planes in every single layer. However, reduction of graphene oxide to the pristine form of graphene is still to be perfected because it might reduce the mechanical and electrical properties of graphene (Khan et al., 2022).

22.4.2 Carbon Nanotubes

Due to their high adsorption capacities carbon-based nanomaterials have been immensely used for the wastewater treatment. Since the discovery of carbon structures, the potential of carbon nanotubes (CNTs) as great adsorbent for heavy metals removal has increased. They have been commonly manufactured by chemical vapour deposition method (Jawed et al., 2020).

Some demerits of CNTs to be used as adsorbents are its poor ability for dispersion, problem with the separation, and very small size of the particles (Westerhof et al., 2005). To overcome these issues, modified magnetic CNTs called multiwalled carbon nanotubes (MWCNTs) are used that exhibit high dispersion ability and can be very easily removed from the treated water (Singh et al., 2021). They have been used to remove heavy metals like Pb^{2+}, Mn^{2+}, Cu^{2+}, and so forth. One such study revealed that the alumina coated CNT showed better ability for the removal of Pb than the uncoated nanotubes (adsorption capacity enhanced overall) (Malik et al., 2022).

Some surface modification techniques that have been used are:

- ***Acid treatment***
  By using strong acids (like HNO_3, $KMnO_4$, H_2O_2, H_2SO_4, and HCl), it aims to remove the impurities present and add more functional groups on the surface of the nanotubes.
- ***Metal impregnation***
  Using metal or metal oxides (such as MnO_2, Al_2O_3, and iron oxide).
- ***Functional molecules grafting***
  Addition of the functional groups on the surface improve its characteristics which is done by plasma method (requires less energy and is eco-friendly), chemically modifying and thermal synthesis techniques.

Surface modifications alter the BET surface area, surface charge, dispersion ability, hydrophobicity, and similar characteristics of CNT (Foley et al., 2005).

22.4.3 Nanocatalysts

22.4.3.1 Palladium Nanomaterials

Heavy metal removal: Another hazardous heavy metal found in several industrial wastewaters is palladium. In a study, *Spirulina plantensis* alga extract was used to biosynthesize palladium nanoparticles (PdNPs) (Malik et al., 2022). Various characterization techniques (such as UV-vis spectroscopy, FTIR spectroscopy, XRD, and TEM) were used to analyze the biologically synthesized PdNPs with a size of 10–20 nm. Elimination efficiency for Pb was stated as maximum 90 per cent. Additionally, the removal effectiveness varied significantly depending on a number of variables, including pH, exposure time, initial Pb content, and adsorbent concentration (Sayadi et al., 2018).

For degradation of dyes: A study examined the ability of chemically and microbially produced nanoparticles to remove dye from waste. In comparison to PdNPs, biogenic Pd based nanoparticles produced with the help of *Caldicellulosiruptor saccharolyticus* showed superior methyl orange and diatrizoate degradation capability (Srivastava et al., 2022). Bio-PdNPs were able to totally eliminate 100 mg per litre of methyl orange in 15 minutes, whereas chemically manufactured PdNPs needed 30 minutes for the same (Shen et al., 2015).

22.4.3.2 Silver Nanomaterials

Persistent organic pollutants, being highly harmful, enter waterways through home sewage, industrial emissions and agricultural runoff water. If these POPs penetrate the food chain, they could have adverse impacts such as cancer, mutagenesis, and allergies, and so forth (Malik et al., 2022). Researchers used silver nanoparticles produced from *C. japonicum* to catalytically reduce bromophenol blue dye. These Bio-Ag NPs had diameter of 8–10 nm and showed great catalytic ability in successfully eliminating 98 per cent dye in 12 minutes. Composite of carbon material with Ag nanoparticles (i.e., CM-Ag-NPs) eliminated 91 per cent of methylene blue dye in the contact time of 9 hrs (Khan et al., 2022).

In other research, noxious Victoria blue dye was removed from the wastewater sample using silver nanoparticle produced using *A. agallocha* leaf juice (Devi et al., 2016). As analyzed using transmission electron microscope (TEM), these Ag-NPs were spherically shaped, exhibited 99.46 per cent efficiency with contact time of 2 hrs, and could be reused up to four cycles.

Production of silver nanoparticles from various plant extracts has been explored over the years. One such research involved creating Ag nano-catalysts from fenugreek seeds in order to eliminate dangerous pollutants like methyl orange, methylene blue and eosin-Y. However, another study used *S. acuminata* fruit extract to eliminate dyes like 4-nitrophenol, methylene blue, methyl orange, phenol red and direct blue 4 (Bogireddy et al. 2016).

22.4.3.3 Iron Nanomaterials

In some studies, extract from the tea plant has also been employed for the production of bio-NPs. Huang et al. used oolong tea to synthesize Bio-Fe-NPs, which could eliminate Malachite green dye with 75.5 per cent efficiency (Malik et al., 2022). Similarly, green tea was also used for production of bio-Fe-NPs, which are useful in absolute and rapid removal of malachite green, methylene blue, and methyl orange from the polluted sewage (Huang et al., 2014). Scientists have been exploiting the polyphenols present in tea for the production of nanocatalysts, which are highly beneficial against organic pollutants as well as azo dyes.

22.4.3.4 Other Metallic Nanomaterials

Bacteria *E. harbicola* was used as a synthesizer of Se NPs which used the mechanism of photocatalytic reduction to degrade and eliminate dyes like methylene blue, methyl orange, and erichrome black T from the sewage water (Srivastava and Mukhopadhyay, 2014). Similarly, naphthalene was removed photocatalytically by Zn NPs and Fe coated Zn NPs (spherical and less agglomerated), produced by using *A. dubius* leaves extract (Malik et al., 2022).

22.4.4 Surfactants and Nanomaterials

We very well know about the potency of nanoparticles for the efficient removal of heavy metals from wastewater, by acting as nanoadsorbents. The mechanism of adsorption of metal ions on the surface of nanomaterials is successfully done by the chelate formation or electrostatic attraction forces (Malik et al., 2022). Nanoadsorbents also show four continuous adsorption desorption cycles. However, nanoparticles with the surfactins successfully eliminated phenolic compounds from wastewater and also showed antimicrobial potential against bacteria with very low cytotoxicity (Kundu et al., 2016).

The combination of bionanoparticles and biosurfactants is being explored for the removal of oil from water. For example, biosurfactants have been combined with nanoparticles to increase their activity rather than using them to synthesize NPs. Nanoparticles due to their small size seep into the pores in oil stocks and biosurfactants facilitate the recovery and elimination process (Rathore et al., 2022). Researchers presented an environment-related application of biosurfactants in one of their studies. They illustrated the production of poly (methyl methacrylate) surfactin nanoparticle by using the method of emulsion polymerization. The formation of a biosurfactant (BS) monolayer at the particle surface as a result of BS adsorption into SiO_2 nanoparticle surface increases the hydrophobicity of the particles and improves oil recovery (Amani, 2017).

Tetradecane or crude oil emulsions could be formed and stabilized using silica nanoparticles and rhamnolipids in freshwater and saltwater. In contrast to emulsions stabilized by either nanoparticle or biosurfactant alone, those produced by the nanoparticles and biosurfactant working together produced emulsions with smaller oil droplets and greater stability to coalescence. This combination of silica NPs and rhamnolipids have proven to be eco-friendly since it greatly helps in clearing up the oil spills (Pi et al., 2015). In many studies so far, scientists have explored the combination of

NPs and BS for the treatment of oil spills and their synergistic effect has only favoured the elimination of trapped oil. It has been further known that the decrease in interfacial tension caused by biosurfactants has in turn decreased the capillary forces and facilitated the passage of nanoparticles and biosurfactants in between the pores (Kundu et al., 2016).

22.4.5 Microorganisms in Nanomaterials

Magneto-tactic bacteria: In 1975, first study reporting the presence of magneto-tactic bacteria was published (Moe and Rheingans, 2006), after which different types of bacteria with different morphologies have been studied—such as coccus, spirilla, vibrio, ovoid, rod-shaped; both unicellular and multi-cellular—inhabiting different aquatic environments (Weber, 2002; Savage and Diallo, 2005).

A substantial number of magneto-tactic bacteria have been found to date, using different growth media and magnetic isolation procedures. Most cultured magneto-tactic bacteria grow above 30°C and are mesophilic. Not many studies have reported about thermophilic magneto-tactic bacteria (Singh et al., 2021). Other than oxides with the magnetic properties, many other oxides and sulphites have been produced by the microorganisms. Sulphite nanoparticles are famous for fundamental research as well as for their technical usages, such as quantum-dot fluorescent markers and labelling agents, and for their excellent optical and electrical properties (Kim et al., 2008). The following Table 22.1 contains a list of studies that have reported about the production of some oxides and sulphites-based nanoparticles.

Ahmad et al. discovered that eukaryotic organisms like fungus are a viable option for extracellularly manufacturing metal sulfide nanoparticles (Dotzauer et al., 2006). When fungus *F. oxysporum* was

TABLE 22.1
Oxide- and Sulphide-based Nanoparticles from Microorganisms

Nanoparticle	Source organism	Features	Reference
CdS	*Clostridiumthermoaceticum*	Precipitated from $CdCl_2$, both on the cell surface and in the medium	(Chae et al., 2009)
CdS	*Klebsiella pneumoniae*	Size range- 20–200 nm; produced when Cd^{2+} ions were present in the growth medium	(Barhate and Ramakrishna, 2007)
ZnS	*Rhodobacter sphaeroides* and *Desulfobacteraceae*	Diameters of 8 nm and 2–5 nm; intracellularly produced	(Chatterjee et al., 2005)
PbS	*Rhodobacter sphaeroides*	Diameter influenced by its growth time	(Xu et al., 2005)
Fe_3S_4	Uncultured magnetotactic bacteria	Magnetic FeS NPs produced by sulfate-reducing bacteria	(Obare and Meyer, 2004)
CdS	*Schizosaccharomycspombe* and *Candida glabrata* (yeasts)	Produced intracellularly, with cadmium salt crystals	(Stanton et al., 2003)
Zn3(PO_4)$_2$	Microstructures that resemble a butterfly	Length ranges from 80 to 200 nm and width ranges from 10 to 80 nm	(Bottino et al., 2002)
SiO_2 and TiO_2	*Fusarium oxysporum* (fungus)	From the aqueous anionic complexes of SiF_6 and TiF_6 respectively.	(Gaponenko, 1998)
$BaTiO_3$	*F. oxysporum*	Tetragonal; size range of 4–5 nm	(Gaponenko, 1998)
ZrO_2	*F. oxysporum*	Spherical; size range of 3–11 nm	(Gaponenko, 1998)
Sb_2O_3	*Saccharomyces cerevisiae*	Sustainable and affordable	(Alivisatos, 1996a)
CdS (wurtzite crystal)	*Esherichia coli*	$CdCl_2$ and Na_2SO_4 in the growth medium	

added to aqueous solution of metal sulphate, it extracellularly created several stable metal sulphides, such CdS, ZnS, PbS and MoS_2 (Singh et al., 2021).

In biological systems, a wide range of organisms use biological polymers like proteins and microbial cells to create orderly structured organic/inorganic blends. In addition to the nanoparticles that have already been described, microorganisms have also been used to produce carbonates of lead, cadmium, strontium, PHB, $Zn_3(PO_4)_2$ and CdSe nanoparticles (Singh et al., 2021). When fungi (that were being tested) were treated with aqueous Sr_{2+} ions, $SrCO_3$ crystals were produced. Yeasts were used as bio-templates in the synthesis of nano-powders of zinc phosphates.

22.4.5.1 Immobilization of Microbes on Bionanomaterials

The method of degrading resistant compounds through the inclusion of microorganisms is known as bioaugmentation. Pollutant removal using this method has been done in groundwater, surface water, and in soil—for example, removing chlorinated ethenes from groundwater using the anaerobic bacterial group, Dehalococcoides (Stroo et al., 2013). Although the bioaugmentation method has generally been found to be highly effective, certain failures have also been observed. It has been discovered that a variety of abiotic (temperature variations, less substrate, scarcity of nutrients, pH changes) and biotic stressors (shock of contaminant load, infectious phages, protozoan grazing, quorum sensing) have a significant negative impact on newly introduced microbes and might further lead to the failure of bioaugmentation (Barhoum et al., 2022). The immobilization of microorganisms on the surface of nanoparticles can be a successful strategy to get beyond the drawback of bioaugmentation for the aim of treating wastewater. From a technological standpoint, immobilization techniques are regarded as feasible for addressing the operational difficulties of freely moving microbial cells by providing ideal size, mechanical strength, and porosity (Zeng et al., 2007).

Some techniques for immobilization of microbial cells that have been so far explored are flocculation, cross-linking, encapsulation in polymer gel, adsorption on nanomaterials, covalent binding to nanocarriers, and entrapment in nanomaterial matrix. Large surface areas of nanomaterials facilitate in absorbing contaminants and rising temperatures would cause the pollutants to slowly leak into the area around the attached microbe, increasing biodegradation process (Barhoum et al., 2022). The use of nanoparticles proving as effective matrices for the immobilization of microbes has been the subject of numerous investigations. For the biosorption of Pb from wastewater, *Phanerochaete chrysosporium* cells were immobilized on the calcium-alginate-IONPs matrix. Efficiency of removal of Pb was 96.03 per cent at pH 5 and 35 °C temperature as per the study. In other set of studies, nanomaterials like carbon nanotubes (CNTs), graphene nanosheets, silica nanoparticles and chitosan nanofibers have been used as a matrix for immobilizing microbes to effectively remove contaminants from sewage water (Kishore et al., 2022).

22.5 BIOPOLYMERS AS BIONANOMATERIALS

22.5.1 Elastin Like Polypeptides (ELP)

ELPs are biopolymers that bind to metals and are composed of a pentapeptide (Val-Pro-Gly-Val-Gly) which has a structure like that of a human protein (elastin). The capacity of ELP to undergo reversible phase transition under a variety of temperature, pH and ionic strength conditions is one of its key characteristics which in turn increases the potential to recover it. By changing the chain and peptide sequence, it is possible to influence its phase transition (Kostal et al., 2001).

For the removal of cadmium from polluted soil, ELP with polyhistidine chain (Prabhukumar et al., 2004) or phytochelatin metal-binding domain have been utilized. Other than cadmium, ELP was also able to remove mercury (Lao et al., 2007). ELP has higher extraction efficiency, hence, can also function well at very lesser amount, unlike other biosurfactants. To produce a biopolymer that is temperature-responsive, a metalloregulatory protein (MerR) from bacteria that binds mercury with

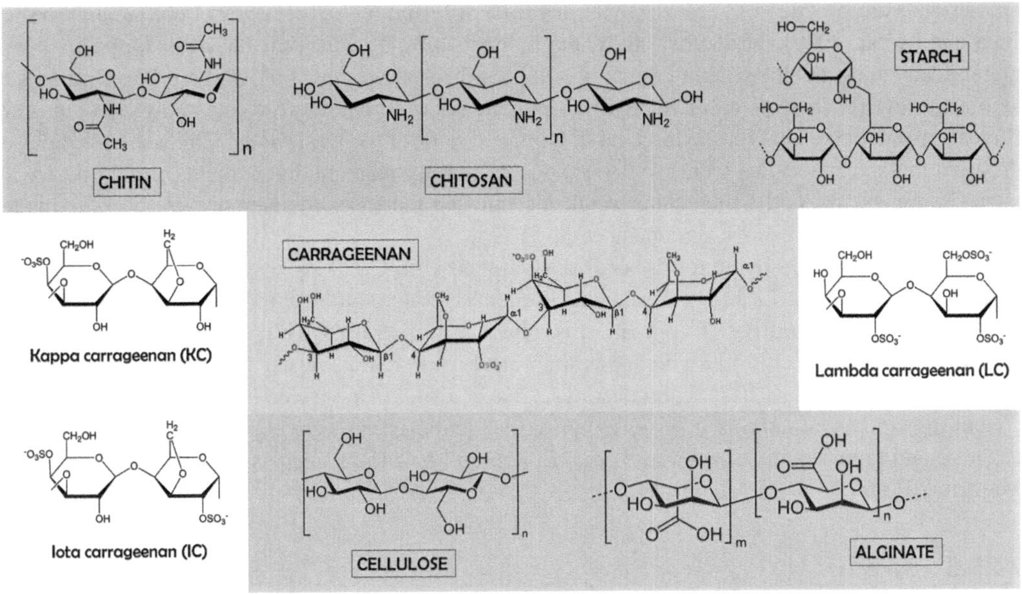

FIGURE 22.5 Structures of Some Biopolymers Used as Nanoparticles in Wastewater Treatment. (Created by the Authors.)

high affinity and selectivity, was coupled with ELP. This biopolymer showed selective binding with mercury while minimal binding for heavy metals like Ni, Cd, and Zn (Fulden et al., 2021).

Chitin and Chitosan: Chitin is a polysaccharide composed of 1–4 linked 2-acetamido-2-deoxy-β-d-glucopyranose. Chitosan is a derivative of chitin, which is composed of β-(1 4) linked D-glucosamine, which are randomly distributed (Fulden et al., 2021). Structures of both the polymers are given in Figure 22.5. Together with cellulose, both of them make up the most prevalent biopolymers found in nature, having applications in the agricultural field and other many industries mostly due to their high sorption ability and flexibility for various modifications mainly chemical and functional. However, efficiency of adsorption depends on many other factors like chemistry of the contaminant, degree of cross-linkages, crystallinity, and the toughness of chitosan-linkages and so forth (Annamalai et al., 2022).

Methods such as hydration and/or enzyme regulated hydrolysis (using chitin de-acetylase) can be used to produce chitosan from chitin. When it comes to living organisms, it can be further generated using exoskeleton of insects, arthropod shells, cephalopod breaks, and cell wall of fungi (Fulden et al., 2021). An ideal approach would be to transform this waste (like animal feeds, calcium carbonate or chitin) into protein, given that crab-shell and lobster production amounted to 6–8 million tons waste annually (Yan and Chen, 2015). Groups like -OH and -NH_2 on the molecular chain of chitosan could be beneficial in removing heavy metals (like Al^{3+}, Zn^{2+}, Cr^{3+}, Hg^{+}, Ag^{+}, Pb^{2+}, Ca^{2+}, Cu^{2+} and Cd^{2+}) (Malik et al., 2022). Additionally, chitosan has been observed to flocculate small particles into larger flocs, removing organic contaminants or solids that were floating on water (Bolto et al., 1995). The only disadvantages with chitosan is that it is extremely sensitive to pH changes and its high cost in comparison to other traditional flocculants (Annamalai et al., 2022).

Chemical modifications of chitin and chitosan: There are many studies, which have been conducted on chitin and chitosan to effectively purify wastewater (Russo et al., 2021). One such study included usage of cellulose nanofibers based-biohybrid aerogels having chitin nanocrystals (CNC). These bio-aerogels are produced by freeze drying method, differing the percentages of chitin nanocrystals (Lee and Moon, 2020). In another study, the potential of magnetic chitosan nanocomposites with glutaraldehyde as the crosslinking agent was explored for the removal of Acid red 2 dye present in textile wastewater (Kadam and Lee, 2015). Magnetic nano-, including carboxymethylated-chitosan, showed promising results for the removal of anionic dye from the wastewater. Chitosan molecules of different molecular weights were used to produce different magnetic bio-sorbents, crosslinked with kappa carrageenan. As the molecular weight of chitosan increases the nanoparticle become bigger and shows more magnetism (Annamalai et al., 2022).

Photocatalysts made by chitosan beads cross linked with Cu along with TiO_2 in nanoform, was proposed by Pincus et al. (2019). Researchers produced water soluble chitosan flocculants prepared by the technique of grafting. To prepare this, chitosan was grafted with (2-methacryloyloxyethyl) trimethyl ammonium chloride which was then used to treat wastewater from pulp mill (Lee and Moon, 2020). Chitosan based photocatalysts have also been viewed as antimicrobial agents, containing zinc oxide nanoparticles as the primary agents. These were explored against both gram-negative and gram-positive bacteria, that is, *E. coli* and *S. aureus*, respectively (Muraleedaran and Mujeeb, 2015).

22.5.3 Starch

Starch has two main components which are amylopectin (branched (1→6) α-D-glucan) and amylose (linear (1→4)-linked α-D-glucan) (Figure 22.5). The main advantages of starch as a biopolymer includes its natural origin, biocompatibility, biodegradability, high availability, huge surface area, nontoxicity, low-cost, and renewability (Fulden et al., 2021). A porous structure was prepared using corn starch crosslinking with epichlorohydrin through hydrolyzation technique with α-amylase, which enabled the removal of Methylene blue dye (Guo et al., 2013). Another crosslinked starch structure containing tertiary amine groups was presented with high sorption capabilities toward dyes present in textile industries' wastewater (Fulden et al., 2021). As compared to carbon nanotubes and graphene nanoparticles, starch provides more stability and sustainability. Various aromatic derivatives (such as chloro-phenols, nitro-phenols, benzoic acid derivatives and dyes, and so forth) have been shown to have strong sorption abilities for removal of pollutants (Annamalai et al., 2022).

One of the well-known derivatives of starch is cyclodextrin, which is has been highly useful in the preparation of membranes for wastewater treatment. Cyclodextrin-based materials let water pass through the cavities present in them, which causes the improvement of the hydrophobic property and permeability of the membranes (Annamalai et al., 2022). They have also enhanced the porosity, flux, efficiency and, hence, overall performance of the membrane. Keeping their properties in mind, cyclodextrins were added in polyamide Thin film composite (TFC) membranes to enhance its hydrophobicity, swelling, flux, and rejection capabilities (Mansoori et al., 2020).

22.5.4 Carrageenan

Also called "little rocks", carrageenans (or carrageenins) are linear sulphur containing polysaccharides derived from seaweeds, having their applications in the food industry (gelling, thickening and stabilizing agents) (Figure 22.5). There are three types of carrageenans, that is, kappa-carrageenan, iota-carrageenan, and lambda-carrageenan, differing on the basis of amount of sulfation (Russo et al., 2021). Supermagnetic FeO having the coating of kappa-carrageenan were reported for their high removal efficiency against methylene blue (Fulden et al., 2021). In a recent study, a composite of zeolite hydrogel (ZHC) based on kappa carrageenan was prepared through the method of graft copolymerization for the adsorption of dye methylene blue. It showed six back-to-back adsorption desorption cycles, proving their excellent potential for adsorption and can hence be used against cationic dye in polluted water (Russo et al., 2021).

Kappa-carrageenan and nanoclay laponite have been used to produce underwater membranes, produced by chemically modifying hydrolyzed polyacrylonitrile, which have been proved highly useful in the elimination of soluble and insoluble organic pollutants (Annamalai et al., 2022). For preparing self-assembled membrane using nanopore sized nanoclay laponie and modified carrageenan, in a layer by layer deposition method, greatly helps in the removal of the water-soluble dyes. Environmentally friendly carbon aerogels made of sodium lignin sulfonate along with kappa carrageenan have been investigated for effective adsorption of dye, underlining the need for lignin sulfonate in the elimination of methylene blue dye (Russo et al., 2021).

22.5.5 Alginate

Alginates are linear polysaccharides constituting two uronic acid residues (i.e., β-D-mannuronic and α-L-guluronic acid) connected by β-1,4-glycosidic linkages (Figure 22.5). It is extensively used for its nontoxic nature, low cost, stable structure, biodegradability, permeability to water, excellent network of structure, and hydrophilic surface moieties (making then potential catalytic support material) and high surface area (Sehgal et al., 2019; Fulden et al., 2021).

Alginate-based biosorbents constitute of multivalent ions with selective cationic interactions along with the macromolecular hydrogels arranged in a 3D network (Russo et al., 2021). Using a range of multivalent cations, a bio-polyanionic network was created by crosslinking alginate and polyacrylamide hydrogels (Yang et al., 2013). In order to effectively remove methylene blue, malachite green and methyl violet from diluted aqueous solutions, sol-gel chemistry was used and perlite beads coated with alginate were created. It was shown how adsorption efficiency could be improved by adjusting the time for mixing, initial dye concentration, dose of adsorbent, pH and temperature (Mansoori et al., 2020). Hence many studies have been devoted to find the biocompatible, cheap and sustainable composites. In one such study, researchers created Fe_3O_4 nanoparticles enclosed in alginate beads for adsorption of methylene blue from synthetic wastewater, assessing the impact of concentration of dye, dose of Fe_3O_4 in beads, time of adsorption, and magnetic field intensity, as well as utilizing the Taguchi method for optimizing operating conditions in dye adsorption onto the magnetic alginate beads (Russo et al., 2021).

22.5.6 Cellulose

Cellulose is a linear polymer constituting of repeating units of D-glucopyranose which are linked by β-1,4-glycosidic bonds (the repeating units are also called cellobiose) whose natural sources are trees, annual plants, animals, fungi, algae, and bacteria (Figure 22.5) (Russo et al., 2021). Due to their high surface-to-volume ratio, low price, large availability in nature, and intrinsic environmental inertness, cellulose NPs are regarded as a promising prospective replacement adsorbent. They have efficient surface functionalization, increasing the binding affinity of pollutants (Fulden et al., 2021). Additionally, favourable characteristics including a large aspect ratio, low density, and strong hydrophilicity make it easier to improve the anti-fouling abilities of membranes coated with cellulose based nanoparticles. Cellulose nanocrystals (CNC) can be used as secure and eco-friendly NPs in the thin film polyamide active layer (Salama et al., 2021).

To avoid safety and health problems brought on by toxic NPs entering the food chain, it is recommended to incorporate ecologically friendly cellulose (Salama et al., 2021). Furthermore, extraordinary qualities like great mechanical strength, large specific surface area, high hydrophilicity, and availability of functional groups have made cellulose NPs an exceptional filler to alter membrane structure (Annamalai et al., 2022). Pollutants can be attracted to anionic cellulose nanoparticles via electrostatic attractions, which have the potential to improve the membrane's anti-fouling resistance. The main obstacle preventing the use of cellulose-based nanoparticles is similar to that of other nanoparticles: their agglomeration into the membrane matrix might impair the membrane's permeability, water flux, mechanical stability, and separation performance (Putro et al., 2017).

22.5.7 Polyhydroxyalkanoate (PHA)

Polyhydroxyalkanoate (PHA) is a biodegradable polyester produced by bacteria as their carbon and energy source under nutrient limiting environment. PHA is a polymer consisting of hydroxyalkanoic acid as a repeating unit. An ester bond is formed between the carboxyl group of one monomer and the hydroxyl group of the next monomer (Sehgal and Gupta, 2022). Besides their environment friendly, biodegradable nature, they are non-toxic and biocompatible (Sehgal and Gupta, 2020; Sehgal et al., 2023). PHAs have been proposed as solid substrate for denitrification of water and wastewater. PHAs serve not only as constant sources of reducing power for denitrification but also as solid matrices favorable for development of microbial films (Hiraishi and Khan, 2003; Basset et al., 2016). In addition, the use of PHAs has no potential risk of release of dissolved organic carbon with the resultant deterioration of effluent water quality. In a study, a film combining olive mill wastewater (OMW) with PHB and its varying properties (chemical, thermal, mechanical, and so forth) was studied, along with its degradation. To investigate the degradation of the OMW/PHB film; soil burial technique was used with different types of soil. Due to the presence of hydrophilic groups in PHB based films, the fillers (OMW) showed noticeable results with accelerated rate of degradation (Carofiglio et al., 2017; Kumar et al., 2021).

22.6 BIOSENSORS FOR DETECTION OF POLLUTANTS

A biosensor is an integrated device that consists of a biological element in close proximity to a transducer that transforms the signal produced by the biological event into an understandable output (Singh et al., 2022). Biosensors, which are categorized based on the transducer, are electrochemical, piezoelectric, optical, and thermal. Moreover, biosensors which are categorized on the basis of biorecognition principle are immunochemical, whole cell, enzymatic, non-enzymatic receptor, and DNA biosensors. For the production of biosensors biological element must be immobilized on the surface of the transducer (Sharma et al., 2018).

22.6.1 Detection of Heavy Metals

The development of sensitive and focused heavy metal detection techniques are essential. Chromatography, spectroscopy, and electrochemistry are the established conventional techniques for the identification of heavy metals with great benefits like excellent sensitivity and precision (Singh et al., 2021). But they are not able to fulfil the requirements of portability and user-friendly detection because they require high-end, expensive equipment, intricate sample processing, and a skilled staff for operation (Singh et al., 2022). Many efforts for the construction of a fast, effective, and selective biosensor are being made by researchers these days. Some such reports are listed below as to which highlights different types of biosensors used.

Mercury is highly toxic and can cause many life-threatening problems if it enters into the food chain. Among all the heavy metals, it has been highly ranked for causing toxicities (be it neurological, renal, environmental or developmental) even when present in very lesser amounts (Malik et al., 2022). For the determination of presence of mercury, thymine is connected with Hg(II) to produce Thymine-Hg-Thymine complex with very high binding constant. To modify this, recently T-rich and specific to mercury oligonucleotide probes have been used to enhance selectivity (Lang et al., 2016).

A DNA-based electrochemical biosensor to detect the presence of Hg^{2+} ions was made, using electrodes containing poly-thymine along with methylene blue (as a redox probe). This type of biosensor has provided a highly efficient, quick, and cost-effective way for the detection of Hg^{2+}

ions (Sharma et al., 2018). Thymine-Mercury-Thymine complex leads to "hairpin-like" folding of the oligonucleotide, causing an improved electronic exchange between methylene blue dye and the electrode surface.

Platinum NPs with a diameter of 5 nm have been deposited on the surface of oxidized silicon substrates using the sputtering method. Using DNAzymes (self-cleave in the presence of Pb^{2+} ions), the nanoparticle layer was functionalized (Lang et al., 2016). With good reproducibility and stability, the resulting device has demonstrated its ability to function as an environmental monitoring sensor for the detection of Pb ions even at concentrations of 10 nm. When exposed to Pb^{2+}, the DNAzymes' conductive bridging, which promotes charge transport across distinct NPs, "collapse" as a result of their own self-cleavage, causing a change in the sensor's measured resistance (Singh et al., 2022).

Bi-enzymatic biosensors have also been studied for their reactions with heavy metals and pesticides. Although, three-enzyme biosensors, such as the one involving invertase, mutarotase, and glucose oxidase, immobilized on the surface of the transducer have also been introduced. Such biosensors showed great sensitivity for Hg^{2+} and Ag^{+} ions (Singh et al., 2022).

Another study reported the development of disposable, yet low-cost electrochemical sensor based on paper, for determining presence of both NO^{3-} and Hg^{2+} in contaminated water (such as polluted lake water and agricultural runoff). The carbon paper electrodes constituted Se particles and AuNPs (catalyst for reducing nitrate ions and site for nucleation of mercury ions) (Lang et al., 2016). Reduced graphene oxide has also been used to prepare electrochemical biosensors since it can provide large surface area for gold nanoparticles which is used for determining presence of trace Fe ions in coastal waters. Researchers used a fluorescence quenching technique and reported a green plan for developing copper nanoparticles capped with D-penicillamine. These nanoparticles show red fluorescence and disperse by Hg^{2+} ions, causing a large fluorescence quench (Barhoum et al., 2022).

22.6.2 Detection of Pesticides

Due to their high acute toxicity and ability to harm the environment and people even at trace levels, pesticides have garnered a lot of attention in research, since the traditional analytical techniques (such as gas chromatography, high-performance liquid chromatography, capillary electrophoresis, and mass spectrometry and so forth) have some major drawbacks (Saha, 2021). In this context, an amperometric biosensor based on acetylcholinesterase (AChE) including gold nanorods (AuNRs), was made for determining organophosphate pesticide. It shows a major electro-catalytic characteristic, such as oxidation of thiocholine and so forth. Hence, these amperometric biosensors exhibit low applied potential, better sensitivity, and stability (Lang et al., 2016). Using multilayer films with multiwall carbon nanotubes (MWCNTs), chitosan and an AChE liposome bioreactor, a novel AChE biosensor (ALB), was created. In order to assemble different layers of multilayer films [(MWCNTs/ALB)n/GCE], the glassy carbon electrode was alternatively submerged in MWCNTs, CS, and ALB solution (Singh et al., 2022). The best biosensor among those produced was the one based on six bilayers of multilayer sheets. In order to immobilize AChE through covalent interactions with an oxidized exfoliated graphite nanoplatelet-chitosan crosslinked composite, a sensitive biosensor for the organophosphorus insecticide chloropyrifos was created (Saha, 2021). A new, highly sensitive and selective sensing device based on the direct deposition of electrons of electrochemically reduced graphene oxide (GO), has been reported. For effectively fixing AChE to create an OPs biosensor, (ERGO)-Au nanoparticles (AuNPs)-cyclodextrin (CD) and Prussian bluechitosan (PB-CS) were used on a glass carbon electrode (Lang et al., 2016).

22.6.3 Detection of Gases

Due to its low price, low power requirements, portable design, and straightforward conduction, biosensors for the detection of gases may serve as a miniature replacement for standard ozone-based monitoring systems (Singh et al., 2022). Indium oxide nanoparticle-based photon-stimulated ozone sensor was developed by a group of scientists. A wide range of ozone concentrations in synthetic air may be monitored by this small, energy-efficient photo stimulated ozone sensor. So, in a high humidity gas environment, scientists employed a photo stimulated ozone sensor to find ozone (Lang et al., 2016). It is advised to seal the sensor chamber with a porous hydrophobic membrane because this prevents liquid water from escaping, since repeatability is crucial, and this is the best option. To detect, distinguish and categorize harmful chemicals (dimethyl formamide, isopropanol, xylene and toluene), magnonic sensor arrays based on metallic nanoparticles (specifically of $CuFe_2O_4$, $MnFe_2O_4$, $ZnFe_2O_4$ and $CoFe_2O_4$) were designed and evaluated (Matatagui et al., 2017). Spinels (AB_2O_4) metal oxides have also received a lot of attention recently for uses in hydrogen gas sensing. The use of Sm-doped cobalt ferrite ($Smx\text{-}CoFe_2\text{-}xO_4$) for hydrogen detection has recently been disclosed (Lang et al., 2016).

Core shell amino terminated hyper branched chitosan NPs (HBC-NH2 NPs), which were used as a platform for an easy and controlled synthesis of silver nanoparticles, were developed into an optical sensor for effective detection of the ammonia concentration in solutions based on the change in the surface plasmon resonance (AgNPs). The HBC-NH2 NP suspension's noticeable transformation from colorless to yellow and the emergence of a surface plasmon resonance peak at 400 nm provided conclusive evidence that AgNPs had formed (Zhuang et al., 2009). With ZnO NPs produced by alginate (A) in a quick and simple method in an aqueous solution at room temperature, ammonia gas can also been identified (Singh et al., 2022).

22.7 TREATMENT OF WASTEWATER FROM THE TEXTILE INDUSTRY

One of the major contributors of industrial wastewater is the textile industry. Wastewater from the textile industry has a complicated composition because there are variances due to different raw materials (such as cotton waste, proteins, oils, fiber debris, acids, alkalis and dyes) (Barhoum et al., 2022). For instance, preparing cotton and wool uses a significant amount of water during pre-treatment. On the other hand, synthetic fibers, require a limited amount of water (Dwivedi and Dwivedi, 2022). Various processes demand different amounts of water. Additionally, numerous chemicals are added during textile processing so as to improve the quality of the textiles, which may result in excessive amount of organic pollutants and heavy metal ions in the wastewater. The textile sector is therefore thought to have the largest water consumption (Srivastava et al., 2022). As there is rapid development in the field of bio nanotechnology, applications of nanoparticles for elimination of dyes from industrial wastewater are being explored by the researchers and the studies are being conducted to analyze different features of bio nanoparticles such as their large surface areas, high adsorption abilities and small diffusion resistance (Li et al., 2020).

Simultaneously, research based on advanced functional nanomaterials is also taking place, such as metal organic frameworks which bind to both anionic and cationic dyes using electrostatic bonding. Many reports have emphasized the elimination of dyes using metal organic frame works, showing excellent results (Dwivedi and Dwivedi, 2022). Latest research has reported use of Ag NPs, produced from *Solanum nigrum* and *Cannabis sativawere*, in degradation of yellow and red dyes present in textile wastewater. However, the time of incubation, concentration, pH, and temperature are different. Another similar study included the used of biochar, the palm waste for the synthesis of carbon dots and nano zerovalent composites of iron enabling the elimination of methylthionine chloride dye (Srivastava et al., 2022).

22.8 TREATMENT OF PHARMACEUTICAL WASTEWATER

Wastewater from pharmaceutical industries and hospitals majorly contain micro-pollutants, such as steroids, antibiotics and hormones, which pose threat to the life forms (Sinha and Dahiya,

2022). The commercial wastewater treatment techniques are mostly concerned about the removal of biological and chemical oxygen demand of water and other contaminants (present in higher amounts). But their inability to remove the low concentration of pollutants (at ppb level), make them inefficient (Kumari et al., 2019). Furgal along with his team (2015) made biogenic manganese oxide NPs (produced by *Pseudomonas putida*) beneficial in removing a wide range of microcontaminants. These biologically produced MnONPs showed complete elimination of steroids (namely, estrone and 7-α-ethinylestradiol) (Furgal et al., 2015). The procedures used for the elimination of such pharmaceutical chemicals (such as, ozonation) showed high efficiencies but also produced harmful and mutagenic by-products. Another study made nano-scaled biological manganese oxides (BioMnOx) and biogenic-palladium (Bio-Pd), used in a small sized membrane bioreactors for the elimination of various recalcitrant pollutants from pharmaceutical industry wastewater (Patel et al., 2022). Among 29 kinds of sewage effluent pollutants, 14 were removed by biogenic MnOx-MBR by oxidative treatment. By using catalytic reduction, bio-Pd was able to adsorb and eliminate some of the contaminants like iomeprol, opromide, iohexol with a 97 per cent removal rate, and obstinate diatrizoate with a 90 per cent removal rate (Sinha and Dahiya, 2022). In order to eliminate the three major pharmaceutical pollutants ciprofloxacin, sulfamethoxazole, and 17b-estradiol, biogenic-Pt and biogenic-Pd nano-catalysts were created utilizing *Desulfovibrio vulgaris* (Martins et al., 2017).

Diatrizoate is regarded as a stubborn micropollutant due to its strong chemical stability and resistance to biodegradation, making its isolation from effluent absolutely crucial. *Shewanella oneidensis* was used to synthesis Pd-NPs (De Gusseme et al., 2011). The quantity of diatrizoate was greatly reduced by improving electrochemical reduction due to bio-Pd-NPs in the cathodic side of a microbial electrolysis cell (high cell voltages speed up the dehalogenation reaction) (Patel et al., 2022). While some studies reported the failure of biogenic PdNPs and AuNPs as nanocatalysts for pollutants' removal, some studies showed their high pollutant removal efficiency (78% in contact time of 24 h) when combined together. A recent report showed the increased electrochemical degradation reactions when Bio-Pd NPs were combined with the chemicals MnO_2 and Fe_3O_4 (Sinha and Dahiya, 2022).

22.9 RECENT ADVANCEMENTS IN WASTEWATER TREATMENT

22.9.1 Algal Membrane Bioreactor

Algae farming and wastewater treatment can be combined to create a sustainable environment. Algae require nutrients from wastewater, including macronutrients such as salts of phosphates, nitrates, ammonia, sodium, calcium, potassium, and micronutrients such as vitamins (thiamine, cyanocobalamin, and so forth), for their growth. Hence, as algae grows in wastewater it removes these nutrients from it (Zhang et al., 2020). In addition to chemical flocculation, there are other methods for accumulating algal biomass—centrifugation, air floatation, and sedimentation—but these techniques are generally expensive. In order to produce algae and treat wastewater, advanced technology has been created (Jeevanandam et al., 2022).

An improved technique for treating wastewater using algae is the algal membrane bioreactor (AMBR), which allows the growing of algae at high density. Harvesting of algae uses less energy than conventional methods and allows a larger recovery of the algal biomass without damaging the cells (Hu et al., 2015). Membranes are made from substances like polyvinylidene fluoride (PVDF), polysulfone (PSF) and polyethersulfone (PES), which have excellent membrane-forming properties and are also chemically and physically stable. Although, membrane fouling or the loss in membrane performance caused by the deposition of particles or solution in the membrane pores, is a significant downside of AMBR (Zhuang et al., 2020). The reasons for membrane fouling could be a hydrophobic binding of microbial cells or solutes with the membrane or physicochemical properties of membrane—surface charge, roughness, and so forth (Jeevanandam et al., 2022). Making the membrane more hydrophilic by enhancing surface polarity through techniques like plasma treatment, nanomaterial incorporation (using nano-silver, nano-silica and zeolites), and surface coating, can reduce hydrophobic interactions in the membrane and further decrease the chances of membrane fouling (Madaeni and Ghaemi, 2007). A group of researchers prepared hollow membrane fibers by combining them with multi-walled carbon nanotubes, which increased the hydrophilicity of the membrane surface, its antifouling properties and water flux (Hu et al., 2015).

TiO_2 has been studied as a photocatalyst for a number of applications, such as the production of self-cleaning surfaces and pollution (primarily air and water) management systems. Because of its hydrophilic characteristics and large surface area, TiO_2 nanoparticles increase the hydrophilicity of membranes, reducing fouling of membrane and increasing permeate flux (Khan et al., 2022). PVDF hollow fibre membranes containing TiO_2 were produced using a specially designed single head spinning machine and these membranes were tested in an AMBR for the treatment of sewage. After utilizing AMBRs, 75 per cent of the phosphorus and nitrogen were removed from the wastewater, and there was less membrane fouling and an improvement in the hydrophilic PVDF characteristics (Hu et al., 2015). Nanoparticle-based AMBR has been highly beneficial in wastewater-treatment practices such as eliminating nutrients and polishing of wastewater. These membrane bioreactors not only require fewer chemicals for the filtration, they also allow easy separation of algae from the treated water and the reuse of the filtered water (Khan et al., 2022).

22.9.2 Microbial Fuel Cells Based on Bionanomaterials

As the name suggests, microbial fuel cells (MFCs) exploit the catalysing properties of microbes and use them as biocatalysts. With the help of microbes, they also produce energy using wastewater containing inorganic and organic compounds. Hence, use of microbial cells reduces the dependence on other conventional techniques for energy production (Jeevanandam et al., 2022). Microbes have the potential to break various chemical compounds (such as acetate, propionate, butyrate and so forth) and produce water and carbon dioxide. The cost of MFCs is quite high and it is less efficient as compared to other fuel cells, hence, it is not in much use these days (Khan et al., 2022). To make the technique economically feasible, the cost of maintenance of membrane must be decreased or a cost-effective membrane and catalyst for cathode must be used. Selecting the right material is also essential for maximizing the performance for microbial fuel cell. The oxygen reduction reaction requires a surface of cathode with great catalytic activity (Zhuang et al., 2009). Nanoparticles have large surface area and are biologically active, hence, they provide excellent alternative: to be used in microbial fuel cell. Nanocomposites (nano-structured carbon) can be used as less expensive electrodes which in turn enhance the performance of the cells (Khan et al., 2022). A study conducted by scientists did a comparison of platinum and carbon nanotubes (CNT) as the catalyst cathode in MFC. The results showed that CNT/Pt had the best outcome as compared to both the cathodes individually. CNT was found to increase catalytic activity of platinum and gave excellent outcomes because of its unique structure and powerful electrical conductivity (Ghasemi et al., 2013). To enhance microbial adhesion and decrease toxicity, use of polymers like polypyrrole and polyaniline to coat carbon nanotubes, hence, enabled the building of an extremely efficient nanocomposite. Such nanocomposites are polycationic anodes electrostatically bound to negatively charged CNTs in MFCs. These electrodes offer a large surface area, which enables bacteria to adhere to PPy-CNT composites with ease, creating a greater surface area for the attachment of microbes and for conduction of electrochemical processes. The production of a biofilm is necessary for the transmission of electrons to the anode surface. MFCs offer incredible opportunities for the production of electricity and wastewater treatment (Malik et al., 2022).

22.9.3 Electrospun Nanofibrous Webs

Electrospinning is developing as the latest technology for creating nanofibers or nanowebs, due to its affordability and special qualities. Their large surface area and porosity at nanoscale make electrospun nanofibrous webs (ENFW) ideal for application as membranes and materials for filtration (Jeevanandam et al., 2022). By combining these nanofibers and microorganisms, purification and filtration processes can be upgraded. Significant environmental impacts were demonstrated as a result of the inclusion of microbial nanofibers with bacteria or algae (San et al., 2014). Algal cells were immobilized on electrospun chitosan nanofiber mats in the hybrid approach, which removed nitrates from wastewater. Therefore, it was concluded that immobilization of microbial cells is preferable since it takes up less room, is simpler to handle, and requires a less volume of growth medium (Jeevanandam et al., 2022). The researchers found that electrospun chitosan nanofiber mats successfully trap microbial cells, which are insoluble in water and non-toxic to algal development. They determined that 87 per cent of nitrates in polluted water were removed by chitosan's physiochemical adsorption mechanism and nutrient uptake by algae (Jeevanandam et al., 2022). *Acinetobacter calcoaceticus* STB1 cells were immobilized using electrospun cellulose acetate nanofibrous webs to eliminate ammonium ions by converting them into nitrogen (accumulates as microbial biomass with great potential for reusability) (Khan et al., 2022). Another similar study carried out by a group of scientists, reported decolorization of the dye methylene blue by using ENFW involving immobilization of various bacterial species (such as *Clavibacter michiganensis*, *Pseudomonas aeruginosa*, and *Aeromonas eucrenophila* and so forth). Results showed 95 per cent elimination of methylene blue dye within 24 h (Jeevanandam et al., 2022). When the fourth cycle of the nanofibrous biocomposite was complete, it was found that bacteria entrapped inside the nanofibrous web could decolorize the dye. These biocomposites could be used to treat industrial wastewater because of its easy use, durability and porous qualities (Khan et al., 2022). Immobilized bacteria are more cost-effective than free bacteria because they require less space, and the resulting biofilm is resistant to salt, toxicity caused by metals, and harsh environmental conditions (Anjum et al., 2019).

22.10 CONCLUSION

This chapter explains the techniques based on bionanotechnology for the efficient removal of pollutants from industrial effluent and wastewater from other sources. The key lies in manipulation of compounds at a nanoscale. Usage of nanoparticles in different fields has been practiced from a very long time, but keeping their downsides in mind, researchers have found the greener and eco-friendly alternatives. Major attention has always been given to the utilization of biomolecules from different living organism instead of toxic synthetic chemicals for the production of biocompatible and biodegradable nanomaterials. Moreover, turning bio-waste into useful bionanomaterials is also a great strategy to reduce its toxicity. Microbial nanoparticles have shown great results in the elimination of most toxic and recalcitrant pollutants because of their huge surface area, unique morphology, wide range of diffusion and reusability. Nanoparticles synthesized by microbes also showed greater catalytic and absorptive efficiencies as compared to the chemically produced nanoparticles. In the recent times, the approach of immobilization of microbes on the surface of nanomaterials and using various ways for integration of bionanotechnology and other green technologies, have been studied immensely. Adopting different biopolymers for the treatment of wastewater is greatly beneficial due to their availability and abundance, along with their tunable structural, physicochemical, biomechanical and biological properties. Progress made in this field has also led to development of biosensors for the detection of the major environmental contaminants, which include chemical toxins, heavy metals, dyes, and so forth. Such nano-scaled biosensors provide selective and sensitive ways for detection of pollutants, eliminating the cumbersomeness associated with conventional techniques.

Research is currently being done on the use of novel living organisms for the quick generation of NPs. To increase the yield, specificity and their efficacy for removal of a particular class of water contaminants, more research must be done on optimizing the synthesis of nanomaterials from biological sources and their practical application in wastewater treatment. Reusability of the nanomaterials is another important issue that requires major attention. All novel biomaterials produced for wastewater treatment must be viewed in comparison to the linear "take-make-waste" model. Additionally, more research is needed to optimize production processes in terms of cost, microbe culture, and handling.

REFERENCES

Alivisatos AP. (1996). Perspectives on the physical chemistry of semiconductor nanocrystals. *The Journal of Physical Chemistry* 100: 13226–13239.

Amani H. (2017). Synergistic effect of biosurfactant and nanoparticle mixture on microbial enhanced oil recovery. *Journal of Surfactants and Detergents* 20: 589–597.

Anjum M, Miandad R, Waqas M, Gehany F and Barakat MA. (2019). Remediation of wastewater using various nano-materials. *Arabian Journal of Chemistry* 12: 4897–4919.

Annamalai K, Rameshbabu H, Mahendhran K and Ramanathan M. (2022). Smart bionanomaterials for the removal of contaminants from wastewater. In *Bionanotechnology: Emerging Applications of Bionanomaterials*, pp. 45–74. Elsevier.

Barhate RS and Ramakrishna S. (2007). Nanofibrous filtering media: Filtration problems and solutions from tiny materials. *Journal of Membrane Science* 296: 1–8.

Barhoum A, Jeevanandam J and Danquah MK, eds. (2022). *Bionanotechnology: Emerging Applications of Bionanomaterials*. Elsevier.

Basset N, Katsou E, Frison N, Malamis S, Dosta J and Fatone F. (2016). Integrating the selction of PHA storing biomass and nitrogen removal via nitrite in the main wastewater treatment line. *Bioresource Technology,* 200: 820–829.

Bogireddy, Reddy NK, Kumar HAK and Mandal BK. (2016). Biofabricated silver nanoparticles as green catalyst in the degradation of different textile dyes. *Journal of Environmental Chemical Engineering* 4: 56–64.

Bolto BA. (1995) Soluble polymers in water purification. *Progress in Polymer Science* 20: 987–1041.

Bottino A, Capannelli G and Comite A. (2002). Preparation and characterization of novel porous PVDF-ZrO2 composite membranes. *Desalination* 146: 35–40.

Carofiglio VE, Stufano P, Cancelli N, De Benedictis VM, Centrone D, Benedetto ED, Cataldo A, Sannino A and Demitri C. (2017). Novel PHB/Olive mill wastewater residue composite based film: Thermal, mechanical and degradation properties. *Journal of Environmental, Chem. Eng.* 5: 6001–6007.

Chae SR, Wang S, Hendren ZD, Wiesner MR, Watanabe Y and Gunsch CK. (2009). Effects of fullerene nanoparticles on *Escherichia coli* K12 respiratory activity in aqueous suspension and potential use for membrane biofouling control. *Journal of Membrane Science* 329: 68–74.

Chatterjee AN, Cannon DM, Gatimu EN, Sweedler JV, Aluru NR and Bohn PW. (2005). Modeling and simulation of ionic currents in three-dimensional microfluidic devices with nanofluidic interconnects. *Journal of Nanoparticle Research* 7: 507–516.

De Gusseme B, Hennebel T, Vanhaecke L, Soetaert M, Desloover J, Wille K, Verbeken K, Verstraete W and Boon N. (2011). Biogenic palladium enhances diatrizoate removal from hospital wastewater in a microbial electrolysis cell. *Environmental Science & Technology* 45: 5737–5745.

Devi TB, Begum S and Ahmaruzzaman M. (2016). Photo-catalytic activity of Plasmonic Ag@ AgCl nanoparticles (synthesized via a green route) for the effective degradation of Victoria Blue B from aqueous phase. *Journal of Photochemistry and Photobiology B: Biology* 160: 260–270.

Dotzauer DM, Dai J, Sun L and Bruening ML. (2006). Catalytic membranes prepared using layer-by-layer adsorption of polyelectrolyte/metal nanoparticle films in porous supports. *Nano Letters* 6: 2268–2272.

Dwivedi N and Dwivedi S. (2022). *Bionanotechnology Towards Sustainable Management of Environmental Pollution*. CRC Press.

Falconer IR and Humpage AR. (2005) Health risk assessment of cyanobacterial (blue-green algal) toxins in drinking water. *International Journal of Environmental Research and Public Health* 2: 43–50.

Foley JA, DeFries R, Asner GP, Barford C, Bonan G, Carpenter SR and Chapin FS. (2005). Global consequences of land use. *Science* 309: 570–574.

Fulden UK and İlke KC. (2021) Polymeric bionanomaterials for agricultural and environmental applications. *In Bionanomaterials for Environmental and Agricultural Applications*. IOP Publishing.

Furgal KM, Meyer RL and Bester K. (2015). Removing selected steroid hormones, biocides and pharmaceuticals from water by means of biogenic manganese oxide nanoparticles in situ at ppb levels. *Chemosphere* 136: 321–326.

Gaponenko SV. (1998). *Optical Properties of Semiconductor Nanocrystals*. No. 23. Cambridge university press.

Gautam PK, Singh A, Misra K, Sahoo AK and Samanta SK. (2019). Synthesis and applications of biogenic nanomaterials in drinking and wastewater treatment. *Journal of Environmental Management* 231: 734–748.

Ghasemi M, Ismail M, Kamarudin SK, Saeedfar K, Wan Daud WR, Hassan SHA, Heng LY, Alam J and Oh SE. (2013). Carbon nanotube as an alternative cathode support and catalyst for microbial fuel cells. *Applied Energy* 102: 1050–1056.

Guo L, Li G, Liu J, Meng Y and Tang Y. (2013). Adsorptive decolorization of methylene blue by crosslinked porous starch. *Carbohydrate Polymers* 93: 374–379.

Hiraishi A and Khan S.T. (2003). Application of polyhydroxyalkanoates for denitrification in water and wastewater treatment. *Applied Microbiology and Biotechnology* 61: 103–109.

Hu W, Yin J, Deng B and Hu Z. (2015). Application of nano TiO2 modified hollow fiber membranes in algal membrane bioreactors for high-density algae cultivation and wastewater polishing. *Bioresource Technology* 193: 135–141.

Huang L, Weng X, Chen Z, Megharaj M and Naidu R. (2014). Synthesis of iron-based nanoparticles using oolong tea extract for the degradation of malachite green. *Spectrochimica Acta Part A: Molecular and Biomolecular Spectroscopy* 117: 801–804.

Jawed A, Saxena V and Pandey LM. (2020). Engineered nanomaterials and their surface functionalization for the removal of heavy metals: A review. *Journal of Water Process Engineering* 33: 101009.

Jeevanandam J, Vadanasundari V, Pan S, Barhoum A and Danquah MK. (2022). Bionanotechnology and bionanomaterials: Emerging applications, market, and commercialization. *Bionanotechnology: Emerging Applications of Bionanomaterials*, pp. 3–44. Elsevier.

Kadam AA and Lee DS. (2015). Glutaraldehyde cross-linked magnetic chitosan nanocomposites: Reduction precipitation synthesis, characterization, and application for removal of hazardous textile dyes. *Bioresource Technology* 193: 563–567.

Khan SH, Alaie SA and Shah MP. (2022). Removal of emerging contaminants from wastewater through bionanotechnology. In *Development in Wastewater Treatment Research and Processes*, pp. 669–688. Elsevier.

Kim J, Davies SHR, Baumann MJ, Tarabara VV and Masten SJ. (2008). Effect of ozone dosage and hydrodynamic conditions on the permeate flux in a hybrid ozonation–ceramic ultrafiltration system treating natural waters. *Journal of Membrane Science* 311: 165–172.

Kishore S, Malik S, Shah MP, Bora J, Chaudhary V, Kumar L, Sayyed RZ and Ranjan A. (2022) A comprehensive review on removal of pollutants from wastewater through microbial nanobiotechnology-based solutions. *Biotechnology and Genetic Engineering Reviews*: 1–26. doi: 10.1080/02648725.2022.2106014

Kobeleva IV, Sirotkin AS, Vdovina TV, Petrova EV, Voznesensky EF and Miftakhov IS. (2017). Morphological analysis of activated sludge in combined biological and reagent wastewater treatment. KOBELEVA IV, SIROTKIN AS, VDOVINA TV and others. *YA Ovchinnikov Biotechnology and Physics-Chemical Biology Bulletin* 2: 22–28.

Kostal J, Mulchandani A and Chen W. (2001). Tunable biopolymers for heavy metal removal. *Macromolecules* 34: 2257–2261.

Kumar V, Sehgal R and Gupta R. (2023). Microbes and wastewater treatment. *In: Development in Wastewater Treatment Research and Processes*, pp. 239–255. Elsevier.

Kumar V, Sehgal R and Gupta R. (2021). Blends and composites of polyhydroxyalkanoates (PHAs) and their applications. *European Polymer Journal* 161: 110824.

Kumari S, Tyagi M and Jagadevan S. (2019). Mechanistic removal of environmental contaminants using biogenic nano-materials. *International Journal of Environmental Science and Technology* 16: 7591–7606.

Kundu D, Hazra C, Chatterjee A, Chaudhari A, Mishra S, Kharat A and Kharat K. (2016). Surfactin-functionalized poly (methyl methacrylate) as an eco-friendly nano-adsorbent: From size-controlled scalable fabrication to adsorptive removal of inorganic and organic pollutants. *RSC Advances* 6: 80438–80454.

Lang Q, Han L, Hou C, Wang F and Liu A. (2016). A sensitive acetylcholinesterase biosensor based on gold nanorods modified electrode for detection of organophosphate pesticide. *Talanta* 156: 34–41.

Lao UL, Chen A, Matsumoto MR, Mulchandani A and Chen W. (2007). Cadmium removal from contaminated soil by thermally responsive elastin (ELPEC20) biopolymers." *Biotechnology and Bioengineering* 98: 349–355.

Lee YC and Moon JY. (2020). *Introduction to Bionanotechnology*. Springer.

Li H, Wu S, Du C, Zhong Y and Yang C. (2020). Preparation, performances, and mechanisms of microbial flocculants for wastewater treatment. *International Journal of Environmental Research and Public Health* 17: 1360.

Li XQ and Zhang WX. (2006) Iron nanoparticles: The core– shell structure and unique properties for Ni (II) sequestration. *Langmuir* 22: 4638–4642.

Madaeni SS and Ghaemi N. (2007). Characterization of self-cleaning RO membranes coated with TiO2 particles under UV irradiation. *Journal of Membrane Science* 303: 221–233

Mahanty S, Chatterjee S, Ghosh S, Tudu P, Gaine T, Bakshi M, Das S, Das P, Bhattacharyya S, Bandyopadhyay S and Chaudhuri P. (2020). Synergistic approach towards the sustainable management of heavy metals in wastewater using mycosynthesized iron oxide nanoparticles: Biofabrication, adsorptive dynamics and chemometric modeling study. *Journal of Water Process Engineering* 37: 101426.

Malik S, Kishore S, Prasad S and Shah MP. (2022). A comprehensive review on emerging trends in industrial wastewater research. *Journal of Basic Microbiology* 62: 296–309.

Mansoori S, Davarnejad R, Matsuura T and Ismail AF. (2020). Membranes based on non-synthetic (natural) polymers for wastewater treatment. *Polymer Testing* 84: 106381.

Martins M, Mourato C, Sanches S, Noronha JP, Crespo MTB and Pereira IAC. (2017). Biogenic platinum and palladium nanoparticles as new catalysts for the removal of pharmaceutical compounds. *Water Research* 108: 160–168.

Matatagui D, Kolokoltsev OV, Qureshi N, Mejía-Uriarte EV, Ordoñez-Romero CL, Vázquez-Olmos A and Saniger JM. (2017). Magnonic sensor array based on magnetic nanoparticles to detect, discriminate and classify toxic gases. *Sensors and Actuators B: Chemical* 240: 497–502.

Moe CL and Rheingans RD. (2006). Global challenges in water, sanitation and health. *Journal of Water and Health* 4: 41–57.

Nurullina EN, Sirotkin AS, Ponkratova SV, Shaginurova GI and Emelyanov VM. (2002). Intensification of biological oxidation of contaminants in the biosorption systems for wastewater purification. *Biotechnology in Russia* 1: 45–51.

Obare SO and Meyer GJ. (2004). Nanostructured materials for environmental remediation of organic contaminants in water. *Journal of Environmental Science and Health, Part A* 39: 2549–2582.

Ozaki H. (2004). Rejection of micropollutants by membrane filtration. In *Proceedings of the Regional Symposium on Membrane Science and Technology*. Malaysia: Johor.

Patel HK, Kalaria RK, Jokhakar PH, Mehta AA and Patel HV. (2022). An application of bionanotechnology in removal of emerging contaminants from pharmaceutical waste. In *Development in Wastewater Treatment Research and Processes*, pp. 371–384. Elsevier.

Pi G, Mao L, Bao M, Li Y, Gong H and Zhang J. (2015). Preparation of oil-in-seawater emulsions based on environmentally benign nanoparticles and biosurfactant for oil spill remediation. *ACS Sustainable Chemistry & Engineering* 3: 2686–2693.

Prabhukumar G, Matsumoto M, Mulchandani A and Chen W. (2004). Cadmium removal from contaminated soil by tunable biopolymers. *Environmental Science & Technology* 38: 3148–3152.

Putro JN, Kurniawan A, Ismadji S and Ju YH. (2017). Nanocellulose based biosorbents for wastewater treatment: Study of isotherm, kinetic, thermodynamic and reusability. *Environmental Nanotechnology, Monitoring & Management* 8: 134–149.

Rathore D, Singh A, Prasad S, Malaviya P and Sevda S. (2022). Emerging nano-bio material for pollutant removal from wastewater. In *Nano-biotechnology for Waste Water Treatment: Theory and Practices*, pp. 77–87. Cham: Springer International Publishing.

Russo T, Fucile P, Giacometti R and Sannino F. (2021). Sustainable removal of contaminants by biopolymers: A novel approach for wastewater treatment. *Current State and Future Perspectives. Processes* 9: 719.

Sadiq R and Rodriguez MJ. (2004). Disinfection by-products (DBPs) in drinking water and predictive models for their occurrence: A review. *Science of the Total Environment* 321: 21–46.

Saha A. (2021). Bionanotechnology-based nanopesticide application in crop protection systems. *Functionalized Nanomaterials for Catalytic Application*, pp. 89–107 .Wiley.

Salama A, Abouzeid R, Leong WS, Jeevanandam J, Samyn P, Dufresne A, Bechelany M and Barhoum A. (2021) Nanocellulose-based materials for water treatment: Adsorption, photocatalytic degradation, disinfection, antifouling, and nanofiltration. *Nanomaterials* 11: 3008.

San NO, Celebioglu A, Tümtaş Y, Uyar T and Tekinay T. (2014). Reusable bacteria immobilized electrospun nanofibrous webs for decolorization of methylene blue dye in wastewater treatment. *RSC Advances* 4: 32249–32255.

Savage N and Diallo MS. (2005). Nanomaterials and water purification: Opportunities and challenges. *Journal of Nanoparticle Research* 7: 331–342.

Sayadi MH, Salmani N, Heidari A and Rezaei MR. (2018). Bio-synthesis of palladium nanoparticle using Spirulina platensis alga extract and its application as adsorbent. *Surfaces and Interfaces* 10: 136–143.

Schaule G, Griebe T and Flemming HC. (2000). Steps in biofilm sampling and characterization in biofouling cases. *Biofilms. Technomic, Lancaster*, pp. 1–21. Taylor and Francis.

Sehgal R, Mehta A and Gupta R. (2019). Alginates: General introduction and properties. *In: Alginates: Applications in the Biomedical and Food Industries*. Ahmed, S . (Ed.) John Wiley & Sons. 3, pp. 1–20.

Sehgal, R and Gupta, R. (2020). Polyhydroxyalkanoate and its efficient production: An eco-friendly approach towards development. *3 Biotech* 10: 549–563. https://doi.org/10.1007/s13205-020-02550-5

Sehgal R and Gupta R. (2022). Characterization of medium chain length polyhydroxyalkanoates from *Priestia megaterium* POD1. *Research Journal of Chemistry and Environment* 26, no. 10: 46–54.

Sehgal R, Kumar A and Gupta R. (2023). Bioconversion of rice husk as a potential feedstock for fermentation by Priestia megaterium POD1 for the production of polyhydroxyalkanoate. *Waste and Biomass Valorization*.14: 3657–3670.

Servos MR, Bennie DT, Burnison BK, Jurkovic A, McInnis R, Neheli T, Schnell A, Seto P, Smyth SA and Ternes TA. (2005). Distribution of estrogens, 17β-estradiol and estrone, in Canadian municipal wastewater treatment plants. *Science of the Total Environment* 336: 155–170.

Sharma D, Kanchi S and Sabela M. (2018). Bionanomaterials as emerging sensors in environmental management. *Nanotechnology in Environmental Science*: 515–542.

Shen N, Xia XY, Chen Y, Zheng H, Zhong YC and Zeng RJ. (2015). Palladium nanoparticles produced and dispersed by Caldicellulosiruptor saccharolyticus enhance the degradation of contaminants in water. *RSC Advances* 5: 15559–15565.

Singh KRB, Nayak V and Singh RP. (2021). Introduction to bionanomaterials: an overview. Bionanomaterials: fundamentals and biomedical applications. *IOP Science*, pp. 1–22. IOP Science

Singh M, Yadav A, Kushik S, Gupta N, Singh G and Gupta P. (2022). Air and water pollution monitoring and control through bionanomaterial-based sensors. In *Bionanotechnology Towards Sustainable Management of Environmental Pollution*, pp. 183–208. CRC Press.

Sinha N and Dahiya P. (2022). Removal of emerging contaminants from pharmaceutical wastewater through application of bionanotechnology. *Development in Wastewater Treatment Research and Processes*, 247–264. Elsevier.

Sirotkin AS, Koshkina LY, Ippolitov KG and Emelyanov VM. (2002). Biological regeneration of activated charcoal upon wastewater purification from nonionogenic surfactants. *Biotechnology in Russia* 1: 40–44.

Sirotkin AS, Kobeleva YV, Vdovina TV, and Zakirov RK. (2019). Bionanotechnologies for the waste water treatment. In *Proceedings of the Seventh International Environmental Congress (Ninth International Scientific-Technical Conference) Ecology and Life Protection of Industrial-Transport Complexes ELPIT 2019*, pp. 134–141.

Srivastava A, Rani RM, Patle DS and Kumar S. (2022). Emerging bioremediation technologies for the treatment of textile wastewater containing synthetic dyes: A comprehensive review. *Journal of Chemical Technology & Biotechnology* 97: 26–41.

Srivastava N and Mukhopadhyay M. (2014). Biosynthesis of SnO2 nanoparticles using bacterium *Erwinia herbicola* and their photocatalytic activity for degradation of dyes. *Industrial & Engineering Chemistry Research* 53: 13971–13979.

Stanton BW, Harris JJ, Miller MD and Bruening ML. (2003). Ultrathin, multilayered polyelectrolyte films as nanofiltration membranes. *Langmuir* 19: 7038–7042.

Stewart PS. (2003). Diffusion in biofilms. *Journal of Bacteriology* 185: 1485–1491.

Stroo HF, Major DW, Steffan RJ, Koenigsberg SS and Ward CH. (2013). Bioaugmentation with Dehalococcoides: A decision guide. In *Bioaugmentation for Groundwater Remediation*, pp. 117–140. Springer, New York, NY.

Sun YP, Li XQ, Cao J, Zhang WX and Wang HP. (2006). Characterization of zero-valent iron nanoparticles. *Advances in Colloid and Interface Science* 120: 47–56.

Trivedi R and Bergi J. (2021). Application of bionanoparticles in wastewater treatment. In *Advanced Oxidation Processes for Effluent Treatment Plants*, pp. 177–197. Elsevier.

Urase T and Kikuta T. (2005). Separate estimation of adsorption and degradation of pharmaceutical substances and estrogens in the activated sludge process. *Water Research* 39: 1289–1300.

Walha K, Amar RB, Firdaous L, Quéméneur F and Jaouen P. (2007). Brackish groundwater treatment by nanofiltration, reverse osmosis and electrodialysis in Tunisia: Performance and cost comparison. *Desalination* 207: 95–106.

Weber Jr WJ. (2002). Distributed optimal technology networks: A concept and strategy for potable water sustainability. *Water Science and Technology* 46: 241–246.

Westerhoff P, Yoon Y, Snyder S and Wert E. (2005). Fate of endocrine-disruptor, pharmaceutical, and personal care product chemicals during simulated drinking water treatment processes. *Environmental Science & Technology* 39: 6649–6663.

Xu J, Dozier A and Bhattacharyya D. (2005). Synthesis of nanoscale bimetallic particles in polyelectrolyte membrane matrix for reductive transformation of halogenated organic compounds. *Journal of Nanoparticle Research* 7: 449–467.

Yan N and Chen X. (2015). Sustainability: Don't waste seafood waste. *Nature* 524: 155–157.

Yang J, Chen J, Pan D, Wan Y and Wang Z. (2013). pH-sensitive interpenetrating network hydrogels based on chitosan derivatives and alginate for oral drug delivery. *Carbohydrate Polymers* 92: 719–725.

Zeng GM, Huang DL, Huang GH, Hu TJ, Jiang XY, Feng CL, Chen YN, Tang L and Liu HL. (2007). Composting of lead-contaminated solid waste with inocula of white-rot fungus. *Bioresource Technology* 98: 320–326.

Zhang C, Wang X, Ma Z, Luan Z, Wang Y, Wang Z and Wang L. (2020). Removal of phenolic substances from wastewater by algae. A review. *Environmental Chemistry Letters* 18: 377–392.

Zhuang L, Zhou S, Wang Y, Liu C and Geng S. (2009). Membrane-less cloth cathode assembly (CCA) for scalable microbial fuel cells. *Biosensors and Bioelectronics* 24: 3652–3656.

Muraleedaran K and Mujeeb VA. (2015). Applications of chitosan powder with in situ synthesized nano ZnO particles as an antimicrobial agent. International *Journal of Biological Macromolecules* 77: 266–272.

Pincus LN, Lounsbury AW and Zimmerman JB. (2019). Toward realizing multifunctionality: Photoactive and selective adsorbents for the removal of inorganics in water treatment. *Accounts of Chemical Research* 52: 1206–1214.

Index

Note: Page numbers indicating figures are in *italics*. Page numbers indicating tables are in **bold** type.

E

X

Z

For Product Safety Concerns and Information please contact our EU representative GPSR@taylorandfrancis.com Taylor & Francis Verlag GmbH, Kaufingerstraße 24, 80331 München, Germany

Batch number: 10392095

Printed by Printforce, the Netherlands